ISBN 978-1-5283-1058-1
PIBN 10910765

1 MONTH OF
FREE
READING

at

www.ForgottenBooks.com

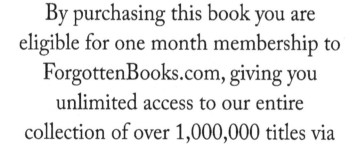

By purchasing this book you are eligible for one month membership to ForgottenBooks.com, giving you unlimited access to our entire collection of over 1,000,000 titles via our web site and mobile apps.

To claim your free month visit:

www.forgottenbooks.com/free910765

English
Français
Deutsche
Italiano
Español
Português

www.forgottenbooks.com

Mythology Photography **Fiction**
Fishing Christianity **Art** Cooking
Essays Buddhism Freemasonry
Medicine **Biology** Music **Ancient
Egypt** Evolution Carpentry Physics
Dance Geology **Mathematics** Fitness
Shakespeare **Folklore** Yoga Marketing
Confidence Immortality Biographies
Poetry **Psychology** Witchcraft
Electronics Chemistry History **Law**
Accounting **Philosophy** Anthropology
Alchemy Drama Quantum Mechanics
Atheism Sexual Health **Ancient History**
Entrepreneurship Languages Sport
Paleontology Needlework Islam
Metaphysics Investment Archaeology
Parenting Statistics Criminology
Motivational

STATE OF OHIO

THE MIAMI CONSERVANCY DISTRICT

Storm Rainfall of Eastern United States

BY

THE ENGINEERING STAFF OF THE DISTRICT

TECHNICAL REPORTS

Part V

DAYTON, OHIO

1917

THE MIAMI CONSERVANCY DISTRICT

DAYTON, OHIO

PREFATORY NOTE

This volume is the fifth of a series of Technical Reports issued in connection with the planning and execution of a notable system of flood protection works in the Miami Valley.

This valley which forms a part of the large interior plain of the central United States and comprises about 4,000 square miles of gently rolling topography in southwestern Ohio, is one of the leading industrial centers of the country. The immense damage which it sustained during the flood of March, 1913, amounting to 100 million dollars of property and over 360 lives, led to an energetic movement to prevent future disasters of this kind. This movement developed gradually into a great cooperative enterprise for the protection of the valley by one comprehensive project. The Miami Conservancy District, established in June, 1915, under the newly enacted Conservancy Law of Ohio, became the agency for securing this protection. On account of the size and character of the undertaking, the plans of the district have been developed with more than usual care.

A Report of the Chief Engineer, submitting a plan for the protection of the district from flood damage, was printed, March, 1916, in 3 volumes of about 200 pages each. Volume I contains a synopsis of the data on which the plan is based, a description of its development, and a statement of the plan in detail. Volume II contains a legal description of all lands affected by the plan. Volume III contains the contract forms, specifications, and estimates of quantities and cost.

After various slight modifications this report of the Chief Engineer was adopted by the Board of Directors as the Official Plan of the District, and was republished in May, 1916.

In order to plan the project intelligently, many thorough investigations and researches had to be carried out, the results of which have proved of great value to the District, and which will also be of widespread value to the whole engineering profession. The object of publishing this series of Technical Reports is to make available to the residents of the state and to the technical world at large, all data of interest relating to the history, investigations, design and construction of the project.

The following reports, prepared by the engineering staff of the District, have been completed:

5

PREFATORY NOTE

This volume is the fifth of a series of Technical Reports issued in connection with the planning and execution of a notable system of flood protection works in the Miami Valley.

This valley which forms a part of the large interior plain of the central United States and comprises about 4,000 square miles of gently rolling topography in southwestern Ohio, is one of the leading industrial centers of the country. The immense damage which it sustained during the flood of March, 1913, amounting to 100 million dollars of property and over 360 lives, led to an energetic movement to prevent future disasters of this kind. This movement developed gradually into a great cooperative enterprise for the protection of the valley by one comprehensive project. The Miami Conservancy District, established in June, 1915, under the newly enacted Conservancy Law of Ohio, became the agency for securing this protection. On account of the size and character of the undertaking, the plans of the district have been developed with more than usual care.

A Report of the Chief Engineer, submitting a plan for the protection of the district from flood damage, was printed, March, 1916, in 3 volumes of about 200 pages each. Volume I contains a synopsis of the data on which the plan is based, a description of its development, and a statement of the plan in detail. Volume II contains a legal description of all lands affected by the plan. Volume III contains the contract forms, specifications, and estimates of quantities and cost.

After various slight modifications this report of the Chief Engineer was adopted by the Board of Directors as the Official Plan of the District, and was republished in May, 1916.

In order to plan the project intelligently, many thorough investigations and researches had to be carried out, the results of which have proved of great value to the District, and which will also be of widespread value to the whole engineering profession. The object of publishing this series of Technical Reports is to make available to the residents of the state and to the technical world at large, all data of interest relating to the history, investigations, design and construction of the project.

The following reports, prepared by the engineering staff of the District, have been completed:

Part I.   The Miami Valley and the 1913 flood.
Part II.   History of the Miami flood control project.
Part III.   Theory of the hydraulic jump and backwater curves.
Experimental investigation of the hydraulic jump as a means of dissipating energy.
Part IV.   Calculation of flow in open channels.
Part V.   Storm rainfall of eastern United States.   .

The following are in the course of preparation:

Rainfall and runoff in the Miami Valley.

Laws relating to flood prevention work.

Flood prevention works in other localities.

Earth dams.

Selection of general type of improvement and design of retarding basin system.

Construction of protection system.

Contracts and specifications.

ARTHUR E. MORGAN,
Chief Engineer

DAYTON, OHIO,
November 1, 1917.

CONTENTS

	PAGE
Officers of the Miami Conservancy District	4
Prefatory Note	5
Contents	7
List of Tables	10
List of Illustrations	11

CHAPTER I. INTRODUCTION

Origin of this investigation	15
Scope of these studies	17
Acknowledgments	19

CHAPTER II. SUMMARY

Meteorology of storms	20
Sources of data and methods of compilation	23
Frequency of excessive precipitation	24
Selecting and sizing 160 great storms	25
Geographical location, seasonal distribution, and frequency of the 160 great storms	26
Descriptions and time-area-depth relations of the 33 most important storms	27
Mapping and preparing time-area-depth curves for the 33 most important storms	28
Variation in mean annual rainfall over eastern United States	28
European storm rainfall	29
Application to Miami Valley of study of great storms	31

CHAPTER III. METEOROLOGY OF STORMS

Source of energy of wind circulation	32
Physical properties of the atmosphere	33
Motion upon the earth's surface	40
Centripetal force and radius of curvature	41
Radial barometric gradient in a rotating atmosphere	42
Effect of the earth's rotation upon frictionless motion	43
Influence of earth's rotation on barometric gradient	47
General wind circulation for eastern United States	48
Storms	50
Cyclones	52
Rates of travel of cyclones and precipitation areas	54
West Indian hurricanes	54
Thunderstorms	56
List of books on meteorology	57

CHAPTER IV. SOURCES OF DATA AND METHODS OF COMPILATION

Meteorological observations in the United States........................... 60
Existing Weather Bureau records....................................... 62
Compilation of summary of excessive precipitation for entire United States... 64
 Definition of excessive precipitation................................. 65
Computing maximum accumulated precipitation.......................... 66
Indexing the data... 69
Variation in number and duration of rainfall records for different quadrangles.. 70
Card index of excessive precipitation periods............................ 71

CHAPTER V. FREQUENCY OF EXCESSIVE PRECIPITATION

Relation between storms and excessive precipitation........................ 80
Determining the frequency of excessive precipitation...................... 81
The pluvial index... 82
Isopluvial charts.. 85
Interpretation of isopluvial charts....................................... 100
Limitations of isopluvial charts... 101
Records of most intense rainfall... 102
Practical application of excessive precipitation data...................... 106

CHAPTER VI. SELECTING AND SIZING THE 160 GREAT STORMS

Period covered.. 110
Selecting the storms... 111
Determining the relative sizes of the storms............................. 112

CHAPTER VII. GEOGRAPHICAL LOCATION, SEASONAL DISTRIBUTION, AND FREQUENCY OF THE 160 GREAT STORMS

Northern storms... 118
Southern storms... 119
Frequency.. 129
Winter storms... 129
Summer storms.. 130

CHAPTER VIII. DESCRIPTIONS AND TIME–AREA–DEPTH RELATIONS OF THE 33 MOST IMPORTANT STORMS

Outline of procedure followed... 132
Descriptions of individual storms.. 133
Discussion of time-area-depth relations.................................. 181
Use of the time-area-depth curves....................................... 202
Selecting the 33 most important storms.................................. 203
Storms which occurred from 1843 to 1872................................ 208
Storms which occurred from 1873 to 1891................................ 210
Storms which occurred from 1892 to 1916................................ 211

CHAPTER IX. MAPPING AND PREPARING TIME–AREA–DEPTH CURVES FOR THE 33 MOST IMPORTANT STORMS

Assembling and computing data for mapping storms....................... 213
Mapping the storms... 214

Sources of error in original data.. 217
Experimental maps... 219
Conclusions from experimental maps...................................... 221
Criticism of using maximum storm periods............................... 222
Errors due to personal equation... 223
Assembling, computing, and platting data for time-area-depth curves......... 224
Degree of accuracy attained... 228

CHAPTER X. VARIATION IN MEAN ANNUAL RAINFALL OVER EASTERN UNITED STATES

Data used... 230
Results of study... 231
Effect of storms and excessive precipitation............................ 235

CHAPTER XI. EUROPEAN STORM RAINFALL

Precipitation in river basins of north Germany........................... 241
Relation between the maximum 24-hour rainfall and the monthly and annual
    rainfall in Germany... 244
Maximum 24-hour rainfalls of 3.94 inches or more....................... 246
Heavy local rains of short duration.................................... 248
Storms and floods in European river basins............................. 249

CHAPTER XII. APPLICATION TO MIAMI VALLEY OF STUDY OF GREAT STORMS

Size of flood selected as basis of design............................... 268
Geographic location of Miami Valley in its relation to storms............. 268
Great storms of the past as applied to Miami Valley conditions.............. 269
Relation of great storms to maximum possible........................... 272
Reasons for choosing, as basis of design, a flood 40 per cent greater than that
    of March, 1913... 273

APPENDIX. TIME-AREA-DEPTH DATA........................... 275

LIST OF TABLES

TABLE		PAGE
1	Humidity, vapor pressure, weight, and heat increments for saturated space and dry air	35
2	Number of rainfalls causing excessive precipitation, 1871–1914	77
3	Number of storms of various classes east of the 103d meridian, 1875–1894 and 1895–1914	78
4	Chronological list of 49 great northern storms	113
5	Chronological list of 114 great southern storms	114
6	Greatest average rainfall of 17 important northern storms	204
7	Greatest average rainfall of 17 important southern storms	206
8	Form of time-area-depth computations	226
9	Average monthly rainfall on four different areas, 1888–1916	231
10	Number of great storms occurring each year	238
11	Intensities of 12 storms applicable to the Miami Valley	271
	Time-area-depth data. APPENDIX	277

LIST OF ILLUSTRATIONS

FIGURE   PAGE

1   Pressure and temperature at various altitudes...................... 37
2   Radius of curvature.. 41
3   Motion on the earth.. 44
4   Motion in a plane tangent to the earth........................... 45
5   Normal annual rainfall, eastern United States.................... 51
6   Excessive precipitation data, sample form........................ 67
7   Number of stations used in studying excessive precipitation........... 69
8   Periods of record of stations in quadrangles 3–D and 4–C.............. 72
9   Periods of record of stations in quadrangles 9–E, 9–F, and 10–E........ 73
10   Periods of record of stations in quadrangles 12–B, 12–C, 12–D, and 12–E. 74
11   Periods of record of stations in quadrangles 12–F, 12–G, 12–H, and 12–I. 75
12   Periods of record of stations in quadrangles 12–J, 15–E, 15–G, and 15–J. 76
13   Isopluvial chart for 15-year period and 1-day rainfall.................. 86
14   Isopluvial chart for 15-year period and 2-day rainfall................. 86
15   Isopluvial chart for 15-year period and 3-day rainfall................. 87
16   Isopluvial chart for 15-year period and 4-day rainfall................. 87
17   Isopluvial chart for 15-year period and 5-day rainfall................. 88
18   Isopluvial chart for 15-year period and 6-day rainfall................. 88
19   Isopluvial chart for 25-year period and 1-day rainfall................. 89
20   Isopluvial chart for 25-year period and 2-day rainfall................. 89
21   Isopluvial chart for 25-year period and 3-day rainfall................. 90
22   Isopluvial chart for 25-year period and 4-day rainfall................. 90
23   Isopluvial chart for 25-year period and 5-day rainfall................. 91
24   Isopluvial chart for 25-year period and 6-day rainfall................. 91
25   Isopluvial chart for 50-year period and 1-day rainfall................. 92
26   Isopluvial chart for 50-year period and 2-day rainfall................. 92
27   Isopluvial chart for 50-year period and 3-day rainfall................. 93
28   Isopluvial chart for 50-year period and 4-day rainfall................. 93
29   Isopluvial chart for 50-year period and 5-day rainfall................. 94
30   Isopluvial chart for 50-year period and 6-day rainfall................. 94
31   Isopluvial chart for 100-year period and 1-day rainfall................ 95
32   Isopluvial chart for 100-year period and 2-day rainfall................ 95
33   Isopluvial chart for 100-year period and 3-day rainfall................ 96
34   Isopluvial chart for 100-year period and 4-day rainfall................ 96
35   Isopluvial chart for 100-year period and 5-day rainfall................ 97
36   Isopluvial chart for 100-year period and 6-day rainfall................ 97
37   Aggregate years of record in each quadrangle........................ 98
38   Maximum 1-day rainfall recorded in each quadrangle.................. 103
39   Maximum 2-day rainfall recorded in each quadrangle.................. 103
40   Maximum 3-day rainfall recorded in each quadrangle.................. 104
41   Maximum 4-day rainfall recorded in each quadrangle.................. 104
42   Maximum 5-day rainfall recorded in each quadrangle.................. 105
43   Maximum 6-day rainfall recorded in each quadrangle.................. 105
44   Frequency of excessive precipitation in quadrangle 3–D.............. 107
45   Frequency of excessive precipitation in quadrangle 9–E.............. 107

46 Frequency of excessive precipitation in quadrangle 12–J............... 108
47 Frequency of excessive precipitation in quadrangle 15–E............... 108
48 The 160 great storms arranged in order of size..................Facing 116
49 The 160 great storms arranged by season of occurrence..........Facing 118
50 Great storms over northern states during November, December, and
      January... 120
51 Great storms over northern states during February, March, and April... 121
52 Great storms over northern states during May, June, and July......... 122
53 Great storms over northern states during August, September, and October 123
54 Great storms over southern states during November, December, and
      January... 124
55 Great storms over southern states during February, March, and April... 125
56 Great storms over southern states during May, June, and July......... 126
57 Great storms over southern states during August, September, and October 127
58 Rainfall map for maximum day, storms of December 1895, March 1897,
      and April 1900.. 134
59 Rainfall map for maximum 2 days, storms of October 1869, December
      1895, March 1897, and April 1900............................. 135
60 Rainfall map for maximum 3 days, storms of December 1895 and April
      1900.. 136
61 Rainfall map for maximum 4 days, storm of December 1895.......... 137
62 Rainfall map for maximum day, storms of July 1887, May 1894, June 1899,
      and September 1905.. 139
63 Rainfall map for maximum 2 days, storms of July 1887, May 1894, June
      1899, and September 1905..................................... 140
64 Rainfall map for maximum 3 days, storms of July 1887, May 1894, June
      1899, and September 1905..................................... 141
65 Rainfall map for maximum 4 days, storms of July 1887, May 1894, June
      1899, and September 1905..................................... 142
66 Rainfall map for maximum 5 days, storms of July 1887, May 1894, June
      1899, and September 1905..................................... 143
67 Rainfall map for maximum day, storms of September 1894, March 1902,
      August 1903, and October 1903................................ 146
68 Rainfall map for maximum 2 days, storms of September 1894, March 1902,
      August 1903, and October 1903................................ 147
69 Rainfall map for maximum 3 days, storms of September 1894, March 1902,
      and August 1903... 148
70 Rainfall map for maximum 4 days, storms of March 1902 and August 1903 149
71 Rainfall map for maximum day, storms of May 1889, July 1897, July 1900,
      and November 1906.. 152
72 Rainfall map for maximum 2 days, storms of May 1889, July 1897, July
      1900, and November 1906..................................... 153
73 Rainfall map for maximum 3 days, storms of July 1897, July 1900, and
      November 1906.. 154
74 Rainfall map for maximum 4 days, storm of November 1906.......... 155
75 Rainfall map for maximum 5 days, storm of November 1906.......... 156
76 Rainfall map for maximum day, storms of June 1905, July 1908, August
      1908, October 1908, June 1909, and October 1910............... 158
77 Rainfall map for maximum 2 days, storms of July 1908, August 1908,
      October 1908, June 1909, and October 1910.................... 159
78 Rainfall map for maximum 3 days, storms of July 1908, August 1908,
      October 1908, June 1909, and October 1910.................... 160

79 Rainfall map for maximum 4 days, storms of July 1908, October 1908, and June 1909... 161
80 Rainfall map for maximum 5 days, storms of July 1908, October 1908, and June 1909... 162
81 Rainfall map for maximum day, storms of July 5–7, 1909, July 20–22, 1909, January 1913, October 1913, and October 1914............... 165
82 Rainfall map for maximum 2 days, storms of July 5–7, 1909, July 20–22, 1909, January 1913, October 1913, and October 1914............... 166
83 Rainfall map for maximum 3 days, storms of July 5–7, 1909, July 20–22, 1909, and January 1913..................................... 167
84 Rainfall map for maximum day, storms of March 1913, December 1913, and July 14–16, 1916... 170
85 Rainfall map for maximum 2 days, storms of March 1913, December 1913, and July 14–16, 1916... 171
86 Rainfall map for maximum 3 days, storms of March 1913, December 1913, and July 14–16, 1916... 172
87 Rainfall map for maximum 4 days, storms of March 1913 and December 1913, and for 1 day (July 16), storm of July 14–16, 1916.............. 173
88 Rainfall map for maximum 5 days, storm of March 1913.............. 174
89 Rainfall map for maximum day, storms of August 1915 and July 6–10, 1916.. 175
90 Rainfall map for maximum 2 days, storms of August 1915 and July 6–10, 1916.. 176
91 Rainfall map for maximum 3 days, storms of August 1915 and July 6–10, 1916.. 177
92 Rainfall map for maximum 4 days, storms of August 1915 and July 6–10, 1916.. 179
93 Rainfall map for maximum 5 days, storms of July 1912 and July 6–10, 1916 180
94 Time-area-depth curves for storms over northern states, 1-day period.... 182
95 Time-area-depth curves for storms over northern states, 2-day period.... 184
96 Time-area-depth curves for storms over northern states, 3-day period.... 186
97 Time-area-depth curves for storms over northern states, 4-day period.... 188
98 Time-area-depth curves for storms over northern states, 5-day period.... 190
99 Time-area-depth curves for storms over southern states, 1-day period.... 192
100 Time-area-depth curves for storms over southern states, 2-day period.... 194
101 Time-area-depth curves for storms over southern states, 3-day period.... 196
102 Time-area-depth curves for storms over southern states, 4-day period.... 198
103 Time-area-depth curves for storms over southern states, 5-day period.... 200
104 Form showing method of compiling storm data....................... 215
105 Map illustrating method of platting storms......................... 216
106 Time-area-depth curves for storm of July 14–17, 1916................. 227
107 Monthly and annual rainfall over eastern United States, 1888–1916.... 232
108 Monthly and annual rainfall over 5 interior states, 1888–1916.......... 233
109 Maximum departures of monthly rainfall, eastern United States........ 236
110 Maximum departures of annual rainfall, eastern United States.......... 236
111 Maximum departure of monthly rainfall, 5 interior states.............. 237
112 Maximum departure of annual rainfall, 5 interior states............... 237
113 Chronological distribution of 160 great storms..................Facing 238
114 Relation between average greatest annual 24-hour rainfall and mean annual rainfall, Europe.. 245

CHAPTER I.—INTRODUCTION

This report sets forth the results of what is probably the most extended study of storm rainfall that has ever been undertaken.

When an engineering examination of the Miami Valley was begun, immediately after the subsidence of the great flood of March, 1913, for the purpose of determining the best plan for preventing damage by future floods, an investigation of rainfall and runoff conditions was one of the first lines of attack. Although the 1913 flood seems to have been by far the largest that has occurred in this valley since its first occupation by white settlers a little over 100 years ago, it was soon apparent that the records of daily discharge of the Miami River going back only to 1892, and the local rainfall records of somewhat longer duration, afforded but a poor basis for the determination of the most probable size, frequency, and distribution of future floods.

The magnitude of the interests jeopardized by floods justified the use of an elaborate system of protection works if simpler ones could not be devised; and the valley is so densely populated, the areas subject to flood damage so separated and diverse, and the conditions affecting the development of plans so numerous and complicated, that only an extensive and costly system could adequately secure effective regulation. To meet this intricate situation it was necessary to determine not only the largest flood that could ever possibly occur, but also, so far as possible, the frequency of all smaller floods which would cause damage.

In default of more extensive hydrological records in the Miami Valley, it seemed necessary to study the records of the whole region of the United States, which is at all similar in its situation with respect to rainfall and runoff conditions. Since runoff records are so scant in comparison with rainfall records in this country, it became evident that chief reliance had to be placed upon the latter alone in determining from past experience what is the most probable future distribution of extreme flood conditions.

In beginning this investigation it was uncertain what geographic limits should be set to obtain the largest amount of useful and pertinent data. These limits at first were assumed to include the entire United States, but during the progress of the investigation it became apparent that rainfall conditions in the western part of the country differ so radically from those in the eastern part, that by limiting the area to that east of the 103d meridian, running through Texas, Colorado, Nebraska, and North and South Dakota, no applicable data would be excluded. So much information was compiled that will be useful to the engineering profession in general, that it was considered best to put it into systematic order for publication in permanent form. The matter here presented has far outgrown in quantity and in degree of analysis what was contemplated when the investigation was begun. Although no precise record has been kept of the time consumed in this study, it is estimated that it represents the equivalent of one man's entire time for more than ten years.

As now presented the investigation covers storm rainfall over the whole of the eastern United States as far west as the 103d meridian. Every storm of consequence ever recorded within that area by the Weather Bureau, by the Signal Service, or by any organization or individual, the record of which is on file at the Washington office of the Weather Bureau, has been utilized.

The work of compiling rainstorm data was begun in the spring of 1913, soon after the Morgan Engineering Company undertook the study of the problem of flood protection for the Miami Valley. As there were nowhere in the State of Ohio complete rainfall records for the whole country, after a season's work at Dayton and Columbus, a party of three computers was sent to the Weather Bureau at Washington, D. C., to collect such data for the entire period covered by existing records in the United States. The data abstracted included everything recorded up to December 31, 1914. The work at Washington was begun on July 28, 1914, and occupied the time of three men until March 15, 1915.

The major part of the work in Washington consisted in abstracting the excessive precipitation data, from which were prepared later the tables, curves, maps, and diagrams shown in this volume. A search for additional storm rainfall data was made in the library of the Weather Bureau, the Library of Congress, the Washington Public Library, and the Smithsonian Institution, and later a similar search of the leading engineering periodicals and engineering society publications was made by the Morgan Engineering Company at Memphis, Tennessee.

After collecting so large an array of data, the best methods of compiling and analyzing it had to be made the subject of careful study. There was no precedent or other helpful guide in this matter. While the direction of movement of storms, the tracks which they most frequently follow, their origin, size, and other characteristics have all been the subject of much study by meteorologists, the same amount of attention has not been given to the duration, intensity, and distribution of precipitation which form a part of storm phenomena. It is the latter elements which are, perhaps, of most interest to engineers in general, and are of vital importance in investigations pertaining to flood control. Previous studies for the determination of the probable maximum storm, as a basis for engineering design, have as a rule been limited to a consideration of the rainfall records at a few local stations, often without respect to the size of area covered, as for instance, in the design of sewer systems. Arbitrary empirical formulas are used extensively to determine the runoff from restricted areas, as in the design of culverts or bridges.

SCOPE OF THESE STUDIES

Throughout this investigation special stress has been laid on the three primary factors which affect storms, namely: First, *Time*, or the duration of the rain at different points within the storm area; Second, *Area* covered by given depths of rainfall; Third, *Depth*, or amount of precipitation and manner of its distribution. The manner in which these three factors are related to each other, the way this interrelation varies with geographic location and in different seasons, we believe has never heretofore been comprehensively investigated. One of the principal objects of this investigation has been to determine the relation of these storm factors of time, area, and depth, for the eastern United States, and to present the results in such form as to be readily available to the engineering profession. This has been done by means of many maps and diagrams in chapters VI to IX.

For the benefit of readers who have not made a special study of meteorological matters, it has been deemed desirable to summarize in brief form those well established meteorological facts most important to an understanding and interpretation of the investigations which are described in detail in this report. Accordingly, a chapter has been devoted to meteorology in its relation to storms, from which everything has been omitted which does not bear directly upon the studies described. This leaves only a very limited and partial view of the science of meteorology as a whole, but any one interested in the general subject can pursue it further by means of general treatises, a list of which is given at the end of chapter III.

The frequency of excessive precipitations at individual stations has been heretofore studied in a limited way at numerous places where the results were needed for use in drainage improvements or in the design of city storm sewers, but it is believed that the method explained in chapter V for dealing with this subject is more compre-

2

hensive and generally useful than any that has been proposed hitherto. The results are shown on a series of 30 charts, and should be interesting and useful to hydraulic engineers over the whole country.

The amount of runoff resulting from a given rainfall depends largely upon the season of the year in which the storm occurs. Furthermore, the damage caused by high water also varies greatly with the season. For these reasons, after a list of 160 great storms in 25 years had been compiled, as explained in chapter VI, their geographical distribution, seasonal distribution, and frequency were investigated with very interesting results, as explained in chapter VII.

The most important, fundamental, and laborious part of the whole investigation was the detailed analysis of the time-area-depth relations of 33 important storms, the aggregate extent of which reached to nearly every part of the eastern United States. This study in chapter VIII greatly surpasses in extent any similar work that has ever been published, and the conclusions are applicable to every part of that region. Maps showing the location and peculiar characteristics of each storm, and a series of diagrams enable ready comparison of all the storms as to duration in time, extent of territory covered, and intensity of precipitation. The complicated nature of this study developed so many difficulties in procedure that chapter IX is devoted to an explanation in detail of the methods tried and those adopted. This will prove of great assistance to any one contemplating further similar work.

The much mooted subjects of cyclic variation in rainfall, and of possible permanent climatic changes, were carefully looked into. It was felt that such considerations had an important bearing upon the proper interpretation of the results from the various studies made, as well as upon the flood control policy as a whole. The danger of being misled by the conclusions reached by other investigators, together with the wide divergence of opinion prevailing among the latter, prompted the Morgan Engineering Company to undertake a certain amount of independent investigation. This was aimed largely at ascertaining whether any cyclic or other variations might exist which would be likely to vitiate the conclusions indicated by the data. The methods adopted and the results obtained are set forth in chapter X. They effectually serve to clear away all apprehensions on this score.

A thorough search of foreign literature was made to find whether similar studies had been published by foreign writers. This search demonstrated that but little investigation of precisely this character had ever been previously carried out. What pertinent data could be found is discussed in chapter XI.

In pursuing the several studies above enumerated, it was kept clearly in mind that the primary object of the entire investigation was to reach safe and logical conclusions as to probable size and frequency of floods in the Miami River.   Extensive as was the scope of the investigation, it did not include the use of data which did not directly or indirectly bear on the problem.   Thus, the absence of any apparent relation between storm rainfall conditions in the western and eastern halves of the United States caused the western rainfall data to be left out from consideration.   Snowfall was not taken up because its contributing effect on floods in the Miami Valley has been found to be negligible.   The special application of the results of the studies to the situation in the Miami Valley is set forth in the last chapter.

The detailed records of excessive precipitation which were copied at Dayton, Columbus, and Washington, about 4,300 sheets in all, are not printed in this report, but blue-print copies can be obtained from the Chief Engineer of The Miami Conservancy District, at the actual cost of reproduction.

ACKNOWLEDGMENTS

Acknowledgments are due to the officials and staff of the U. S. Weather Bureau for their cordial attitude throughout this investigation; to the staff of the Dayton Public Library for securing many books from other libraries; to the staffs of numerous other libraries whose facilities were used; and to numerous engineers for advice and information.

The collection of data was begun at Dayton by G. C. Cummin and was continued in Washington, D. C., under the local charge of R. H. Merkel.   Subsequent studies were begun by N. H. Sayford and were completed by W. P. Watson with the assistance of K. B. Bragg, E. W. Lane, and F. R. Roche; the illustrations were prepared by C. H. Shea and O. Froseth; and the examination of foreign literature was made by K. C. Grant.   The report was finally prepared for publication under the personal direction of S. M. Woodward and G. H. Matthes.   The project was initiated and directed throughout by A. E. Morgan.

CHAPTER II.—SUMMARY

METEOROLOGY OF STORMS

Rainfall is controlled by the winds whose circulation is due to radiant energy received from the sun. This energy is received in greatest amount in the tropics and in least amount at the poles. The winds and ocean currents help to distribute the energy over the earth's surface.

Air is a mixture of permanent gases and a small amount of water vapor. The water vapor necessary to saturate a given space depends upon its temperature, and increases rapidly for the higher temperatures. Table 1 shows for all temperatures from 0 to 100 degrees fahrenheit, the absolute humidity of saturated air in grains per cubic foot and in inches per mile, the vapor pressure in pounds per square inch, the weight of dry air in pounds per cubic foot and the weight of saturated air having the same total pressure, and finally the heat increment of saturated vapor, for 5 degree increases of temperature, expressed in British thermal units.

In figure 1, Curve A shows the variation of atmospheric pressure with altitude, Curve B shows the variation of observed average temperature with altitude, Curve C shows the theoretical variation of temperature of dry air rising under a condition of adiabatic expansion, Curves D and E show the same changes for saturated air.

These curves show the disturbing forces introduced into the atmosphere by the water vapor. When air at the surface becomes well warmed and well saturated with moisture a state of unstable equilibrium is reached, whose disturbing effect upon the winds and weather often results in thundershowers and other forms of precipitation. The convection currents caused by such unstable equilibrium exert an influence in all storms. However, other factors also exert a large influence, and no one yet has been able satisfactorily to show just what weight is due to convection.

If air is moving in a whirl the barometric pressure must be lowered at the center. In such a case the barometric gradient along a radial line, expressed in inches of mercury per 100 miles, is

$$\frac{.25V^2}{g\rho} = \frac{V^2}{128\rho} \tag{6}$$

This shows that for ordinary conditions at the surface of the earth, the effect of circular motion upon the barometric pressure is slight. However, for the extreme conditions reached in a tropical hurricane the effect becomes very marked. Since the wind velocity is greater above than at the earth's surface, the barometric gradient which just balances centrifugal force at one altitude cannot produce equilibrium at another altitude. Hence, in the case of a whirl, there is always a creeping in of air towards the center near the earth's surface, and an outward motion at higher altitudes.

If the earth were a frictionless sphere not in rotation, a body moving upon its surface would pass entirely around it, traversing a great circle. The relative velocity of the body would be constant in amount but would, in general, be continually changing in geographical direction. If this spherical earth be rotating, the path in space traversed by the moving body would be the same as before, but the path relative to the earth would become a complicated curve. The velocity relative to the earth would vary both in amount and direction. The ellipticity of the earth introduces a component of the force of gravity which acts towards the north in the northern hemisphere and towards the south in the southern hemisphere. This force changes the motion of a frictionless body so that its velocity relative to the earth again becomes constant in amount, although it still varies in geographical direction. In the northern hemisphere the deflection is towards the right and in the southern hemisphere, towards the left. The radius of curvature is given by the equation

$$\rho = \frac{6V}{\tau \sin \phi} \text{ miles} \tag{7}$$

Interference between different portions of the atmosphere prevents motion of all parts in the paths just described. Such interference requires the path of any portion to approximate a straight line. The transverse pressure necessary to change the motion from its natural curved path to a straight line relative to the rotating earth, is just equal to the pressure that would change the motion on a still earth from a straight line to the curved path. The barometric gradient at right angles to the direction of motion necessary when the atmosphere moves in a straight line is given by the equation

$$\frac{\tau V \sin \phi}{24g} = .004V \sin \phi \tag{8}$$

in inches of mercury per 100 miles. In the northern hemisphere the higher pressure is always to the right of the line of motion. With a strong wind this necessary pressure difference is considerable.

The immediate cause of the general wind circulation is the variation in air temperature. The relatively high temperature prevailing in the equatorial regions maintains a continual circulation with the temperate zones, as indicated by the trade winds. The same cause produces a belt of fairly permanent high surface pressure in latitudes 30 to 35 degrees each side of the equator. In these belts westerly winds prevail, but disturbing factors lead to local horizontal wind movements thus producing fluctuations in weather conditions.

A storm is a widespread movement of the air accompanied by precipitation. The normal annual rainfall over eastern United States varies between wide limits. It tends to decrease with the latitude, with the distance from the Gulf of Mexico and Atlantic Ocean, but to increase with the altitude. Storms are classified as cyclones, West Indian hurricanes, and thunderstorms. A cyclone is an atmospheric disturbance several hundred miles in diameter which passes across the country from west to east and is accompanied by a characteristic distribution of air pressures and precipitation. Around the accompanying low pressure area the air circulation is counterclockwise. This circulation produces precipitation and definite characteristic changes of temperature. The precipitation occurs chiefly to the southeast of the low area. Cyclones are much more frequent, definite, and energetic in winter than in summer, with the result that the temperature changes are more violent and frequent in the winter. However, the precipitation is much greater in summer on account of the greater capacity of the air for transporting water vapor at the higher temperatures.

The average hourly movement of cyclonic centers varies from less than 20 miles in summer to over 40 miles in winter. Although the precipitation areas move at about the same rate, their movement seems much more erratic.

The West Indian hurricane originates in the tropics and moves northward. It is accompanied by extremely violent winds and torrential rains. Upon reaching a land area it soon takes on the general characteristics of an extratropical cyclone. These hurricanes are most likely to occur in late summer. In the United States they affect only the Gulf and south Atlantic states. They are relatively infrequent,—on the average only 10 a year as compared with 120 cyclonic disturbances.

Thunderstorms are brief storms of relatively small area in which the precipitation is produced by vertical convection currents. These storms usually travel from west to east at an average rate of 30 or 40 miles an hour.

Means for pursuing the subject of meteorology farther is provided in a selected list of twenty books on this subject.

SOURCES OF DATA AND METHODS OF COMPILATION

Early fragmentary weather records were kept by private individuals, United States Army surgeons, officers of the General Land Office, and educational institutions; but the first comprehensive system using uniform methods to cover the country as a whole was organized under the Smithsonian Institution about 1848. Many of the instruments and records of this service were destroyed by fire in 1865.

In 1870 Congress placed the work under the direction of the Chief Signal Officer of the United States Army. In 1890 the work was reorganized as the Weather Bureau under the U. S. Department of Agriculture.

At present there are about 200 regular Weather Bureau stations, with paid observers, and a complete equipment of standard instruments. This service is supplemented by about 4500 cooperative stations operated by voluntary observers using instruments supplied by the Weather Bureau. There is also a relatively small number of river stations, cotton stations, corn and wheat stations where observations are made for the Weather Bureau. A list of all the various kinds of rainfall records now in the possession of the Weather Bureau is given in chapter IV.

All existing records down to Dec. 31, 1914, were copied for all storms exceeding a certain intensity. A storm period is defined as that length of time in which the average precipitation does not fall below 1 inch per 24 hours.

For each station where the normal annual precipitation is more than 20 inches, the records were abstracted of all storm periods having a 1-day rainfall amounting to 10 per cent or more of the normal annual precipitation, or having a total rainfall amounting to 15 per cent of the normal annual precipitation. Where the normal annual precipitation is less than 20 inches, the records were abstracted of all storm periods in which there was a total rainfall of 4 inches. Only stations having a complete record for 5 consecutive years were included.

Specially prepared blanks were used for abstracting the storm data, one sheet for each station. On these sheets the maximum accumulated precipitation for 2-day and longer intervals was also entered. Sheets were made out for 4316 stations, of which 1262 are west of and 3054 east of the 103d meridian. The subsequent study relates solely to the stations east of the 103d meridian.

For indexing the data the United States is divided into 2-degree quadrangles. Rows running north and south are numbered consecutively from east to west, and rows running east and west are lettered consecutively from north to south. Hence each quadrangle is identified by a coordinate number and letter. After plotting all the rainfall stations on a large scale map, by quadrangles, each station was assigned a lower case letter, in order, beginning in the northeast corner and proceeding towards the southwest corner of the quadrangle. The records are assembled, first by quadrangles, and second with index letters in alphabetical order so as to bring records together of stations near each other.

Neglecting fractional quadrangles, the number of stations whose records have been used varies from 75 for each of two quadrangles in the northeastern part of the country, to 5 and 6 stations per quadrangle along the 103d meridian. The average number of stations per quadrangle is 25. In general, the number of station records available is proportional to the density of population. The duration of records for different stations and quadrangles also shows great variations as indicated on figures 8 to 12.

The records for each separate storm were next brought together on one card. These cards show records of 2641 storms east of the 103d meridian, occurring in the years 1870 to 1914. Of these 1236, nearly 50 per cent, were recorded at but one station; 996 storms, over 35 per cent, were recorded at more than 1 but less than 6 stations; while 409, about 15 per cent, were recorded at more than 6 stations. The number recorded in each of these three classes in each year is shown in table 2.

FREQUENCY OF EXCESSIVE PRECIPITATION

The frequency of a phenomenon, as the term is used in this discussion, signifies the average length of the period, in years, during which the phenomenon has happened once. In determining the frequency of excessive precipitation, each quadrangle is treated as a unit, by averaging the records at all the stations within the quadrangle, giving to each station record a weight equal to the number of years of its record. Thus the records at all the stations are combined and treated in some respects as equivalent to a record at a single station extending over a period as long as the aggregate of all the years of record at the separate stations.

The pluvial index for any quadrangle is defined as the depth of rainfall which probably will be equaled or exceeded at any point in the quadrangle once in a given number of years. By the process of averaging described above, the pluvial indices for each quadrangle for precipitation lasting 1 day, 2 days, 3 days, 4 days, 5 days, and 6 days has been computed for periods of 100 years, 50 years, 25 years, and 15 years. These values are shown on a series of 24 maps, figures 13 to 36. On these maps also are drawn isopluvial lines for successive 1-inch depths of rainfall. These lines show that the isopluvial index decreases very markedly with increase of latitude and with distance from the ocean, and that it usually decreases with altitude.

Since the rainfall records available for study are still scanty in some parts of the country, it is to be expected that when longer records become available, the values of the pluvial index shown on the map will be subject to some modification. Figure 37 is a map showing the aggregate years of record in each of the 133 quadrangles as used in computing the various pluvial indices.

The isopluvial charts furnish a convenient means of determining the pluvial index for any locality within the eastern United States, by properly interpolating between the isopluvial lines. For any particular area it is also convenient to construct frequency curves, such as are illustrated in the 4 figures, 44 to 47. These are based on the isopluvial maps and show a curve for each period of precipitation from 1 day to 6, for all frequencies. These curves become nearly horizontal at a frequency of 100 years, indicating that the maximum depths of rainfall already recorded will probably not be greatly exceeded in the future.

SELECTING AND SIZING 160 GREAT STORMS

Preceding the year 1892, the number of rainfall observing stations was too few to furnish sufficient records to determine with precision the areas covered by all important storms. Since that date the rainfall stations have been relatively more numerous, constant, and uniformly distributed. Hence for the purpose of studying seasonal and geographical distribution, frequency, and cyclic variation of large storms, the 25-year period, 1892–1916, was adopted.

In this study all storms were included that had not less than five 3-day precipitation records equaling or exceeding 6 inches. Records of 160 such storms were found. They were divided into two groups, 47 northern storms and 113 southern storms, and are listed chronologically in tables 4 and 5, together with 3 other important previously occurring storms.

After comparing the storms graphically in various ways, the most satisfactory simple basis of comparison was found to be the amount of the fifth highest 3-day precipitation record. Figure 48 shows the storms arranged in order of size according to this basis, and also shows the value of the highest, tenth highest, and twentieth highest record for each storm.

GEOGRAPHICAL LOCATION, SEASONAL DISTRIBUTION, AND FREQUENCY OF 160 GREAT STORMS

Figure 49 has a horizontal time axis to represent the length of a calendar year. The diagram is divided into 12 strips to represent the calendar months. In the appropriate place to correspond with its date, each of the 160 storms is platted as a vertical line whose ends show the highest rainfall record and the twentieth highest record. Thus figure 49 shows at a glance the season of occurrence of each of the great storms, and indicates that the best division of the year for the purpose of studying storm types is into quarters beginning with November.

To show geographical distribution the 6-inch isohyetal for the maximum 3-day period of each storm was mapped. For this purpose all the storms are shown on 8 maps, figures 50 to 57, 4 maps being used for the northern storms and 4 for the southern. On one map are gathered all the storms occurring in a single quarter of the year. These maps show marked variations in storm types with reference to location and season.

During the first quarter, November to January, all of the 21 storms shown occurred in the lower Missouri and Mississippi Valleys except 2 unimportant storms in Florida.

During the second quarter, February to April, only 2 storms occurred in the northern group. One of these was the great storm of March, 1913; the other was relatively unimportant. During this quarter 25 storms occurred in the south, all of them interior storms.

During the third quarter, May to July, the northern group contains numerous thunderstorms lying chiefly in the Mississippi and Missouri River valleys. This is the season of numerous and the most intense southern storms, 32 being shown for this quarter.

During the fourth quarter the greatest number of both northern and southern storms occur, some of which are among the largest recorded. For this quarter 24 northern and 41 southern storms are shown.

Both in location and in other characteristics storms of the first two quarters show close similarity; likewise those of the third and fourth quarters.

Of the winter storms, on the average, one occurs in the upper Mississippi Valley every three years, and one or two every year in the southern states. Of the summer storms, on the average, one occurs every two years on the north Atlantic coast, one every year west of the Mississippi north of Arkansas, and three every year in the lower Mississippi Valley and along the Gulf and south Atlantic coasts.

The summer storms are normally short and violent, of the thunderstorm type, although occasionally a southern storm is both violent and long continued. The winter storms are normally of long duration with correspondingly reduced rates of precipitation.

DESCRIPTIONS AND TIME-AREA-DEPTH RELATIONS OF THE 33 MOST IMPORTANT STORMS

The relation between area covered and average depth of rainfall in respective periods of 1, 2, 3, 4, and 5 days was studied in detail for 33 of the largest and most important storms recorded in the United States east of the 103d meridian.

The following procedure was followed for each storm:

First, the rainfall data was assembled.

Second, the dates of greatest 1-day rainfall, greatest 2-day rainfall, and so on were determined.

Third, the rainfall figures were platted on a large scale map.

Fourth, the isohyetals were drawn.

Fifth, the areas within the isohyetals were measured.

Sixth, the average depth within each isohyetal was computed.

Seventh, on coordinate paper the time-area-depth curves were drawn, using as coordinates the area in square miles and the average depth of rainfall over the corresponding area.

The 33 storms included 15 from the northern group and 15 from the southern group, all of which occurred in the 25-year interval, 1891–1916, and also the 3 greatest storms during the preceding 49 years.

The 3 early storms are the only ones of that period for which satisfactorily complete data now exists, but this data indicates that these storms were of very unusual magnitude, and that they were probably among the largest during the 49-year period.

The storms selected from the 25-year period, 1892–1916, include all the greatest storms in both northern and southern groups, and such additional great storms as were required to give a satisfactory geographical distribution over the whole country.

Each storm is represented by its even numbered isohyetals on a map for each of the periods considered, figures 58 to 93. These maps are accompanied by brief notes and descriptions calling attention to the most striking features of each storm.

The time-area-depth curves for the northern storms mapped are shown for 1- to 5-day periods on figures 94 to 98, those for the southern storms on figures 99 to 103. These curves show that almost invariably more than half of the total storm rainfall occurs on the maximum day. Among the northern group there is but relatively little difference between the largest storms. In the southern group, with much larger absolute values of the greatest precipitation, there also appear more erratic divergences between different storms.

From these curves the average depth of precipitation over any given size of area, at the storm center, may be read for any of the storms for comparison with a similar area under consideration elsewhere. Tables 6 and 7 give the values read from the curves for areas of 1, 500, 1000, 2000, 4000, and 6000 square miles for each of the storms during the maximum consecutive periods of 1 to 5 days.

MAPPING AND PREPARING TIME-AREA-DEPTH CURVES FOR THE 33 MOST IMPORTANT STORMS

In the process of mapping the storms and obtaining the data for the time-area-depth curves many difficulties were encountered which could not have been foreseen before undertaking the work. To determine the best procedure numerous experiments were tried. The various experimental methods are described in detail, and the reasons for adopting the methods finally chosen are fully discussed for the benefit of other investigators.

All the rainfall records for each storm were compiled on special sheets, from which the maximum 1-day period, the maximum 2-day period, and so on for the period of the storm, were determined. The rainfall was plotted on a large scale outline map and the 1-inch isohyetals drawn. To minimize the errors due to arbitrary location of observing stations, inaccuracy of printed records, arbitrary division of time between days, and variations in time of observations, four different ways of plotting the data were tried and the simplest method was adopted for use.

From each map, after considerable labor in measuring and computation, a time-area-depth curve was prepared. All the data from which these curves are plotted is given in the appendix. The important operations were all checked to insure accuracy. In some cases these curves contain angles and breaks due to distribution of separate storm peaks and occurrence of different peaks on different days.

VARIATION IN MEAN ANNUAL RAINFALL OVER EASTERN UNITED STATES

For each month and year of the 29-year period, 1888–1916, the average rainfall over the eastern United States was computed and is shown in figure 107. The exact area used included North and South Dakota, Nebraska, Kansas, Oklahoma, and Texas, and all states east of these. The mean annual rainfall for the whole 29-year period was 36.85 inches. The amount of variation from this value for any year may be read from the figure.

A similar study was made for the smaller area included within the states of Illinois, Indiana, Ohio, Kentucky, and West Virginia. In this case the extreme variations from the average are considerably greater than those for the larger area; and in about one fourth of the years the annual variation is in the opposite direction from what it is for the greater area. This shows that any deductions as to the cyclic nature of rainfall, based upon records at but few stations, must be very cautiously drawn.

A comparison of the mean annual rainfall over the eastern United States with table 2, page 77, giving the number of records of excessive precipitation in each year, and with table 10, page 238, giving the number of great storms in each year, shows that the variations in total annual rainfall depend chiefly upon the occurrence of great storms.

EUROPEAN STORM RAINFALL

Search of foreign literature revealed no comprehensive studies of storm rainfall in Europe, although records of rainfall and floods are available in profusion. In Germany, omitting from consideration the extremely mountainous portion, the annual rainfall seems to be less than over most of the eastern United States. Furthermore, at any single station the rainfall is subject to less variation than in this country, and heavy daily rainfalls are not so extreme. The absolute maximum 24-hour rainfall ever recorded in a given month varies between 65 and 80 per cent of the mean rainfall for that month. Similarly, the greatest single day's rainfall in a year is on the average 1.3 inches for a station whose mean annual rainfall is 20 inches; 1.5 inches where the mean annual rainfall is 30 inches; and 1.7 inches where the mean annual is 40 inches.

Extreme values of 24-hour rainfall have reached nearly 10 inches at mountain stations, but in the level country have rarely exceeded 6 inches. In the drier regions the highest maxima amount to about 20 or 30 per cent of the mean annual; but in the moister regions are between 15 and 20 per cent.

The upper branches of the Elbe rise in the mountains of northern Austria. Maximum rates of runoff recorded have been 14.7 second feet per square mile from 10,850 square miles, and 10.0 second feet per square mile from 19,730 square miles.

The upper tributaries of the Oder River in southeastern Germany are subject to destructive floods. In the lowlands below the mountain tributaries the rainfall is small, and out of a dozen stations with records covering 30 to 40 years, all having mean annual rainfalls less than 30 inches, there is but one record of a 24-hour rainfall exceeding 4 inches.

On the mountain tributaries, however, destructive summer floods occurred in August, 1888, June, 1894, and July, 1897, during which the maximum 24-hour rainfall records obtained were, respectively, 5.64, 4.37, and 6.61 inches. The maximum measured rate of runoff was 109 second feet per square mile from an area of 389 square miles, and 85 second feet per square mile from an area of 793 square miles.

Flood records on the Danube at Vienna extend back to the year 1000. The greatest flood was in 1501, with a discharge of 12.6 second feet per square mile from the drainage area of 39,400 square miles above Vienna. The second greatest was in 1787 with a discharge of 10.6 second feet per square mile. The third greatest in 1899 had a maximum discharge of 9.5 second feet per square mile.

Flood records on the Seine at Paris have been kept for several centuries. All the largest floods have occurred during the winter months and have followed periods of long continued rains sometimes lasting a month. The maximum definitely recorded flood occurred in 1658. The next greatest was in 1910, with a maximum rate of runoff of 5.26 second feet per square mile for a drainage area of nearly 17,000 square miles. Preceding the latter flood the maximum 24-hour rainfall at three stations was 2.6, 3.0, and 3.1 inches, respectively. The corresponding maximum 4-day rainfalls were 6.1, 7.5, and 7.8 inches, but the average over the whole drainage area was much less.

The upper Loire in France is subject to considerable floods following heavy storms. On October 8–9, 1878, the rainfall over a considerable area averaged 7.88 inches. On September 24–25, 1866, it likewise averaged 5.5 inches.

The Garonne in southern France receives the drainage from considerable areas of mountainous country, and is subject to heavy floods. Some of the largest rainfall records in September, 1875, at single stations are: 31.2 inches in 1 day; 15.7 inches in 1 day; 22.8 inches in 5 days; 12.6 inches in 5 days.

The Durance in southeastern France drains a portion of the Alps and is the most torrential of the great French rivers. During the greatest flood, in 1886, the maximum rate of runoff was 57 second feet per square mile from an area of 4150 square miles.

Floods in the Tiber at Rome have been recorded antedating the Christian era. The greatest on record occurred in 1598, with a maximum rate of discharge of 16.4 second feet per square mile from an area of 6455 square miles. In 1870 occurred a flood but little lower, which inundated about half the inhabitated area of the city.

For additional data regarding maximum rates of flood runoff see Vol. I, page 76, of the Official Plan of The Miami Conservancy District.

The records of storm rainfall in Europe which were consulted indicate that the range of intensities there is not materially different from that found in the eastern United States. The scope of available European data is not comprehensive enough, however, to warrant drawing final conclusions on this subject.

APPLICATION TO MIAMI VALLEY OF STUDY OF GREAT STORMS

The determination of the maximum possible flood in the Miami Valley depended upon topographic conditions, geographic location, and past records of great storms in the same section of the country. The 1913 flood was the greatest during the 100-year record for the Miami River. From a thorough consideration of the above factors the official plan for flood protection was based on a hypothetical storm great enough to cause a maximum flood runoff nearly 40 per cent greater than that of the storm of March 23–27, 1913.

Although heavy storms frequently move from west to east up the Ohio Valley, the latitude of the Miami Valley and its distance from the Gulf of Mexico and the Atlantic Ocean preclude the possibility of a maximum rainfall nearly as great as occurs in the southern states.

Of the 33 most important storms given detailed study in chapters V to VIII, 12 occurred in the northern Mississippi Valley. Table 11 shows the date of each, the location, and the amount of the most intense rainfall as compared with the storm of March 23–27, 1913.

The few cases in which any of these storms exceeded materially the storm of March, 1913, occurred during the summer months when the runoff is a much smaller fraction of the rainfall than in March. Furthermore, most of these heaviest storms occurred farther to the south. In the storm evidence of the last 75 years, there is no indication that the rainfall of March, 1913, will ever be greatly exceeded in the region of the Miami Valley.

From the records of floods in the Danube at Vienna covering 900 years, in the Seine at Paris covering 300 years, and in the Tiber at Rome covering 1500 years, it appears that the greatest flood of 1000 years is not much in excess of the greatest in 100 years.

After making the extensive investigation of storms in the eastern United States, it is believed that the March, 1913, flood is one of the great floods of centuries in the Miami Valley. In the course of three or four hundred years, however, a flood 15 or 20 per cent greater may occur. To cover our ignorance and uncertainty on this point, and to insure that the engineering works shall be absolutely safe in every respect, in planning the flood protection works provision is made for a maximum flood nearly 40 per cent greater than that of March, 1913. This is 15 or 20 per cent in excess of what is believed to be the greatest possible flood that will ever occur.

CHAPTER III.—METEOROLOGY OF STORMS

SOURCE OF ENERGY OF WIND CIRCULATION

The amount and distribution of the rainfall over the surface of the earth are controlled almost entirely by the circulation of the earth's atmosphere, commonly called the winds. The source of energy which maintains this circulation is the radiant energy received from the sun. This energy strikes that portion of the earth turned toward the sun in a nearly constant stream. To compensate for this reception of energy there is a continual loss by radiation into space from all portions of the earth's surface. The amount received and the amount lost seem in the long run to be practically equal, so that as a whole the earth is neither a gainer nor a loser in the transaction.

The amount of insolation, that is, energy of the sun's rays, received on a given day by any part of the earth depends upon its position. The amount is greatest near the equator and least near the poles. At any instant the intensity of the insolation depends upon the obliquity of the sun's rays striking the portion of the earth under consideration. The amount of insolation striking the earth at every instant is equal to the amount that would fall upon a plane within a circle whose circumference is equal to a great circle of the earth. Since the surface of the whole earth is equivalent to the area of four great circles, it follows that at a point where the sun is directly overhead, so that the rays are striking the earth perpendicularly, the radiant energy is being received at a rate just four times the average rate for all points on the earth's surface.

The daily rotation of the earth and its annual revolution around the sun produce a continual change in the amount of insolation received at any given point, but the total during a year is much greater in equatorial regions than near the poles. The amount of energy lost from the earth by radiation, although also a variable quantity for different locations, is more nearly constant than the total amount received. This becomes possible only through the transfer of energy from equatorial to polar regions by means of the winds, assisted by the ocean currents.

The numerous factors affecting in detail the variation in the reception and emission of radiant energy afford an extensive field for research, but are of little interest here. Suffice it to say that the winds act as an agency which automatically tends to even up over the whole earth the supply of radiant energy.

The general mechanism of the winds for the whole earth has been much studied and has been fairly well determined. When it comes, however, to a consideration of the innumerable details in wind circulation, the problems encountered become exceedingly complex, and in many respects have not yet been solved. Voluminous treatises have been written on this subject. During recent years several reliable, brief, elementary treatises have been published dealing with the meteorology of the whole earth in general and with conditions in the United States in greater detail. Many special studies also have been issued by the Weather Bureau and by private authors. A list at the end of this chapter contains such of these publications as seem to be most useful to any one desiring to pursue further the matters treated in this report.

To discuss intelligently the movements of the atmosphere which determine the occurrence and distribution of rainfall, it is necessary first to understand something of the physical properties of the atmosphere and the laws which govern its movements and changes.

PHYSICAL PROPERTIES OF THE ATMOSPHERE

The atmosphere is a mechanical mixture of nitrogen, oxygen, and small amounts of a number of other gases including water vapor. All of these except water vapor belong to the class called permanent gases, and hence the mixture of these permanent gases, so far as its effect upon rainfall is concerned, may be considered as having the qualities of a single perfect gas, homogeneous in composition.

The relatively small proportion of water vapor present in the atmosphere, on the other hand, obeys laws in some respects entirely different from those controlling the properties of the so-called permanent gases; and a comprehension in a general way of the laws relating to vapor pressure and temperature is absolutely necessary to an adequate understanding of the changes in the atmosphere which accompany and control evaporation and rainfall.

A vapor is by definition a substance in the gaseous state, which passes freely at ordinary temperatures and pressures into either the solid or liquid state, or in the reverse direction. The distinction between vapors and gases while very convenient for ordinary use, in reality is only relative, since all the so-called permanent gases become first vapors and then liquids if the conditions are made sufficiently extreme.

Although in the language used above a vapor is said to pass *freely* from the gaseous to the liquid state, this change is in reality strictly

3

limited by certain definite relations of temperature and pressure. These limitations can be most easily stated in terms of conditions of saturation. Let there be imagined an air-tight vessel of one cubic foot capacity from which all air has been removed. Into this vessel let a small quantity of a volatile liquid, such as water, be introduced. Some of the water will quickly change into the gaseous state and fill the whole space. Soon a definite constant condition of equilibrium will be reached, the vapor and the unvaporized water will have the same temperature, and the space in the vessel is said to be saturated with the water vapor. Measurements show that at a given temperature the amount of water vapor in one cubic foot of saturated space is exactly constant, and that adding more water in the liquid state cannot increase the quantity of water vapor so long as the temperature remains unchanged. Of course, the amount of water vapor present *may* be less than that of saturation, and it *will* be less if the amount of water originally supplied in the imaginary experiment described above is insufficient, when all evaporated, to produce the saturated state; but if water is present in the liquid state, evaporation will continue until the definite state known as saturation is produced.

Now imagine the containing vessel and its contents to be warmed by external means to a definite higher temperature. Measurements show that then the quantity of water vapor present per cubic foot, when a state of saturation is reached, will be greater than before. Thus for every temperature there is a definite quantity of water vapor necessary to saturate one cubic foot of space, this quantity increases as the temperature rises, and by experiment it is found that if the temperature is raised by constant increments, the corresponding increment in required water vapor continually increases.

The amount of water vapor present in a cubic foot of space is called its absolute humidity. Table 1 shows in column 3 the absolute humidity at saturation, in grains per cubic foot of space, at various temperatures from zero to 100 degrees fahrenheit. When a given space contains less water vapor than the amount required to produce saturation at the existing temperature, the absolute humidity is correspondingly less than the value shown for the given temperature in the above table.

Column 4 shows for each temperature the depth in inches of a layer of water which, when converted into vapor, would suffice to saturate the space above for a height of one mile. Or, it shows the amount of rain in inches that would result if all the moisture could be removed by condensation from a layer of the saturated atmosphere at uniform temperature one mile thick.

For each condition of saturated space the water vapor has a definite vapor pressure, which would be evident as an internal pressure against the walls of the containing vessel. For conditions in which space is not saturated, the vapor pressure follows practically the laws of

Table 1.—Humidity, Vapor Pressure, Weight, and Heat Increments for Saturated Space and Dry Air at Various Temperatures.

1	2	3	4	5	6	7	8
Temperature		Absolute Humidity		Vapor Pressure	Weight in Pounds per Cubic Foot		Heat Increment of Saturated Vapor per Cu. Ft.
Fahrenheit	Absolute	Grains per Cu. Ft.	Inches per Mile	Pounds per Sq. In.	Dry Air	Saturated Air	British Thermal Units
0	460	0.5	0.07	.02	.086	.086	
5	465	0.6	0.09	.02	.085	.085	
10	470	0.8	0.11	.03	.084	.084	
15	475	1.0	0.14	.04	.084	.084	
20	480	1.2	0.18	.05	.083	.083	
25	485	1.6	0.22	.06	.082	.082	
30	490	1.9	0.28	.08	.081	.081	
32	492	2.1	0.31	.09	.081	.080	
35	495	2.4	0.34	.10	.080	.080	
40	500	2.8	0.41	.12	.079	.079	.07
45	505	3.4	0.50	.15	.079	.078	.08
50	510	4.1	0.59	.18	.078	.078	.10
55	515	4.8	0.70	.21	.077	.077	.11
60	520	5.7	0.83	.25	.076	.076	.13
65	525	6.8	0.98	.30	.076	.075	.15
70	530	8.0	1.16	.36	.075	.074	.17
75	535	9.4	1.36	.42	.074	.073	.20
80	540	10.9	1.59	.50	.074	.072	.22
85	545	12.7	1.85	.59	.073	.072	.26
90	550	14.8	2.15	.69	.072	.071	.29
95	555	17.1	2.48	.81	.072	.070	.33
100	560	19.8	2.87	.94	.071	.069	.37

permanent gases; that is, if the temperature remains constant the vapor pressure is proportional to the absolute humidity; and if the temperature varies but the absolute humidity is constant, the vapor pressure is proportional to the absolute temperature. The absolute temperature is the fahrenheit temperature plus 460 degrees. Table 1 above gives in column 5 the vapor pressure in pounds per square inch, for saturation for each temperature.

Column 6 gives the weight in pounds of a cubic foot of dry air, under a pressure of 14.7 pounds per square inch, for the various temperatures. Column 7 gives, similarly, the weight of a saturated mixture of air and water vapor, when part of the dry air has been replaced by the vapor, so that the total pressure remains as before at 14.7 pounds per square inch.

Recurring again to the imaginary experiment described above, let us consider the heat changes involved in passing from one saturated state to another at a higher temperature. The heat that would of necessity be supplied to the containing vessel would be required for three purposes: first, a certain amount to raise the temperature of the liquid water present; second, a certain amount to raise the temperature of the water vapor present; and third, a certain amount to supply the latent heat of evaporation of the additional amount of water converted to vapor to produce the saturated state at the higher temperature. In these calculations the second portion is negligible, so that assuming the changes to take place by 5-degree steps, table 1 contains in column 8 the amount of heat for the third portion mentioned above, the amount required to bring the water vapor again to a condition of saturation from a saturated condition five degrees colder. It is important to note that if saturated vapor is cooled it is necessary to remove exactly corresponding amounts of heat to produce a given change of temperature. During such cooling the excess vapor will be converted into the liquid state in the form of cloud, mist, or fog.

It will be noticed that nothing has so far been said about the effect of the presence of air in our imaginary experimental vessel. The presence of such air produces only incidental and relatively insignificant effects. It retards somewhat the rate of evaporation and the diffusion of the resulting water vapor into a uniform distribution throughout the enclosed space. The pressure of the air would be added to the vapor pressure to give the pressure reading on such an instrument as a barometer. When the temperature is raised the pressure of the confined air would increase in proportion to the absolute temperature, and a certain amount of heat would be required to raise the temperature of the confined air. For example, to raise the temperature of one cubic foot of dry air at constant volume from 75 to 80 degrees fahrenheit requires .06 British thermal units of heat, while table 1 shows that to raise one cubic foot of saturated vapor through the same change of temperature requires .22 British thermal units.

It is desirable here to state the definitions of two terms much used in meteorology, namely, dew point and relative humidity. Given a certain state of unsaturated space with a definite absolute humidity, the relative humidity is the ratio of the given absolute humidity to the absolute humidity necessary to produce saturation without change of temperature, and the dew point is the temperature to which the space would have to be reduced to become saturated without any change in the absolute humidity.

If the atmosphere were a quiet uniform layer, its pressure at the surface of the earth would exactly equal its weight. Disturbances and movements of the atmosphere cause its pressure at the earth's surface to vary, but the pressure averages about 14.7 pounds per square inch at sea level. The pressure is ordinarily measured by a barometer in inches of mercury column and is indicated in this unit on the daily weather maps issued by the Weather Bureau. The value corresponding to 14.7 pounds per square inch is 29.9 inches of mercury.

Since one cubic foot of air at sea level weighs on the average 0.078 pounds, if the atmosphere were of uniform density at all elevations a thickness of about five miles would suffice to produce the pressure of 14.7 pounds per square inch. But as one rises above the earth's surface the pressure decreases by exactly the weight of the

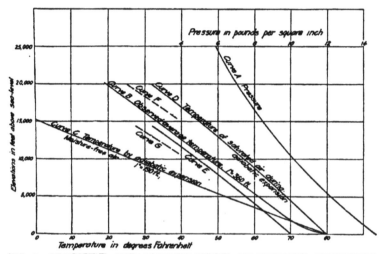

FIG. 1.—PRESSURE AND TEMPERATURE AT VARIOUS ALTITUDES.

These curves show observed pressure and temperature at elevations above the earth's surface, and temperature changes resulting from the adiabatic expansion of dry and moist air.

layers of air passed through. As the pressure decreases, the density or weight of one cubic foot decreases correspondingly. Hence, the successive layers of air weigh less and therefore the change of pressure for each 1000 feet rise constantly decreases. If the composition and temperature of the air were constant it would be possible to calculate the pressure at any altitude. But the temperature decreases with the altitude at an average rate of about 15 degrees fahrenheit per mile. This makes the upper layers more dense than they would be if at a uniform temperature and so tends to maintain a more uniform change of pressure with altitude. In figure 1, Curve *A* shows the average pressure and Curve *B* the average temperature of the air at various elevations above sea level as it might exist on a clear hot summer afternoon. The lower half of the atmosphere is contained within a layer 3.6 miles thick, but the thin outer layers extend to a height of many miles.

The sun's heat keeps the lower atmosphere in continual motion, with the result that the layers at different elevations constantly mix with each other more or less. It may be profitable to begin the discussion of the effects of this mixing by considering first the simplest possible theoretical case, that is, that of an atmosphere which is moisture free, and hence composed of permanent gases. If a cubic foot of such a gas is carried from sea level to higher altitudes, the pressure upon it gradually decreases and it undergoes a corresponding expansion in volume. As the gas thus expands it performs work upon its enclosing envelope which is measured as a product of force and distance; the distance being the increase in volume, and the force being the average pressure of the gas during the expansion. The source of energy for performing this work is the heat initially present in the gas. If no heat is received by the gas from an external source, then its temperature will fall just such an amount that its loss of heat energy will be exactly equivalent to the external work performed during expansion. Such expansion is called adiabatic expansion. Since the specific heat of air is well known, it is possible to calculate closely the temperature of such a cubic foot of air as it rises to various altitudes. Figure 1, Curve *C*, shows the temperature reached at various altitudes on the assumption that the air started from sea level at a temperature of 80 degrees. This process is reversible, so if the cubic foot were returned to its original position it would be again compressed and warmed up to its original temperature.

This curve in figure 1 may be considered a curve of neutral equilibrium. If the temperature of a dry atmosphere could be imagined to agree throughout with this curve, then if any portion of that atmosphere either rose or fell its temperature would change automatically and would be in agreement with the temperature of the surrounding layer in its new position. In contrast with this, consider a cubic foot at some altitude which should have artificially a higher temperature than that corresponding to the surrounding layer at this altitude. It would then be lighter than the surrounding air and would rise under the buoyant force. However, at each new altitude it would be likewise warmer and lighter than the surrounding layer and hence it would keep on rising indefinitely. On the other hand, a cubic foot heavier than its surrounding layer would sink to the earth's surface. By an extension of this reasoning it is obvious that if above some elevation the remaining atmosphere is warmer than the temperature shown in figure 1, then it tends to remain in its superior position, and is in stable equilibrium; but if above the given elevation the remaining atmosphere is colder than the temperature shown by the curve of equilibrium, then the whole atmosphere is in unstable equilibrium and would be likely to be overturned.

The final condition to be considered and the most complex is an atmosphere containing water vapor. There is, of course, an endless variety of cases depending upon the initial temperature condition and degree of saturation. One case only will be considered in detail; perhaps it is the simplest, but it may serve in a general way as an illustration of all related cases. Assume as an initial condition a large mass of saturated air at sea level at a temperature of 80 degrees, a rather extreme condition, but one that may, perhaps, be reached in summer. Assume that some impulse starts this mass into upward motion into layers of air whose temperature and density vary according to the average observed conditions as shown in figure 1. As the saturated atmosphere rises into regions of reduced pressure, it will expand, and thus perform work. Its temperature will be reduced by the abstraction of sufficient heat to perform the external work. Since the air is originally saturated, lowering the temperature implies the condensation of some of the water vapor into the liquid state. Now, to lower the temperature of one cubic foot of saturated air one degree requires the removal of much more heat than would be necessary if the air were moisture-free; or, what is the same thing, to supply a certain amount of heat energy, the cubic foot of saturated air will have its temperature reduced only a fraction of the amount by which a cubic foot of moisture-free air would be reduced. As a consequence, the temperature of the rising moist air will be decidedly higher than that of the surrounding layers of stationary air.

The condensed moisture in this case will first appear as fog. If the particles are small enough they will fall at an exceedingly slow rate. As their size increases their vertical velocity will likewise increase. For the sake of simplicity let it be assumed that the condensed moisture is removed by falling as soon as it forms. On this assumption the temperature can be calculated for each pressure and thus for each altitude. This temperature is shown on figure 1, Curve D. This is then an irreversible process, due to the removal of the condensed moisture, and if through some agency the saturated air should be caused to fall, it would return to its original elevation along a curve parallel to Curve C. It is supposed that fog or cloud particles are usually between 0.001 and 0.0002 inch in diameter, and that a particle of the larger size falls through the air at a rate of 2 inches per second or 1 mile in 9 hours.

The distance between Curves B and D, showing the difference in temperature at any elevation, is in a way a measure of the difference in weight or density of air under the two conditions and so indicates the buoyant force. In hot summer weather warm air from the ground does rise every clear afternoon under such forces producing the cumulus clouds so constantly visible. Such vertical currents are called convection currents. When the supply of warm moist air is sufficient, a thundershower often results. By means of measurements on the clouds it has been determined that these ascending currents often rise to an altitude of several miles.

For purposes of comparison, Curve E has been added on figure 1, similar to Curve D except starting with a temperature of 70 degrees at sea level.

One point remains for discussion. In drawing Curve D, it was assumed that all condensed moisture was removed from the air as soon as formed. Actually, condensed moisture in the form of cloud or fog remains largely with the air in which it is produced. This cloud or fog increases the weight of the air containing it and so reduces the buoyant force. To indicate something as to the amount of this effect Curve F has been added showing by its position between Curves B and D the relative decrease in the buoyancy if none of the condensed moisture had been removed from Curve D. Curve G has been similarly drawn with respect to Curve E.

The data in table 1 and the curves in figure 1 show the nature and importance of the disturbing forces introduced into the atmosphere by the presence of water vapor. When a mass of air at the surface becomes well warmed and well saturated with moisture, a state of unstable equilibrium is reached whose disturbing effects upon the winds and weather often results in thundershowers and other forms of precipitation. The convection currents which result from such a state of unstable equilibrium seem to play an important part in all storms, but other factors also have a large influence, and no one yet has been able satisfactorily to show just what weight in producing precipitation must be ascribed to the effect of convection.

MOTION UPON THE EARTH'S SURFACE

In order to comprehend the most fundamental relations existing between the horizontal motions of the atmosphere upon the earth's surface and the forces which control these motions, it is necessary to have in mind some of the underlying laws of mechanics. The more elementary of these laws will be stated and illustrated briefly using graphical methods so far as possible to illustrate the relations existing between the differential quantities. For a complete mathematical treatment by analytical methods the reader is referred to the treatises enumerated at the end of this chapter.

CENTRIPETAL FORCE AND RADIUS OF CURVATURE

In figure 2 let P represent at the beginning of a certain instant of time the position of a body moving in the direction of PA with a velocity V. If no force acts upon the moving body it will move in a straight line with unchanging velocity. This law of motion is commonly said to result from the body's possessing the property of inertia. In a brief time, dt, the body moves a distance equal to $V\,dt$.

Let

$$PA = V\,dt \tag{1}$$

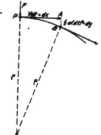

Now suppose that during this same time, dt, the body is acted upon by a constant force, F, acting always at right angles to the direction of the velocity. The effect of the force, F, is to turn the body from the straight line PA, so that at the end

FIG. 2. — RADIUS OF CURVATURE.

of the time, dt, it is at a position, B. The effect of the force, F, is to move the body sideways a distance AB in the time, dt. If a is the acceleration produced by the force, F, in the direction AB, then

$$AB = \tfrac{1}{2}a(dt)^2 \tag{2}$$

If F is measured in pounds, W is the weight of the body in pounds, and g is the acceleration of gravitation, the relation between F and a is given by

$$a = \frac{F}{W}g \tag{3}$$

When the body arrives at B it is traveling in a direction slightly different from what it had at the point, P, but its speed is unchanged, because the deflecting force always acts normal to the direction of motion of the body and therefore has no tangential component. The path from P to B is curved. For such a short length as we are here considering the curve would be sensibly a circle and it is often convenient to know under such circumstances the size of this circle of curvature, or the radius of curvature. For any such case, where the circle is known to be tangent to the line, PA, at the point, P, and passes through the point, B, whose position is known with respect to P, the following rule, derived from plane geometry, and accurate only for differential quantities, is very convenient.

Let

$$PA = dx$$

$$AB = dy$$

$$\rho = \text{radius of curvature}$$

Then

$$\rho = \frac{(dx)^2}{2(dy)} \tag{4}$$

From this and equations 1 and 2 the ordinary expression for the acceleration of centripetal force is obtained:

$$a = \frac{V^2}{\rho} \tag{5}$$

RADIAL BAROMETRIC GRADIENT IN A ROTATING ATMOSPHERE

In order to study each element in its simplest separate form let us next consider the effect of circular motion in the atmosphere upon barometer pressures as they would exist upon an earth without rotation. It will later be shown that the diurnal rotation of the earth has a decidedly important influence on the distribution of barometer pressures in a moving atmosphere, and it is simpler, and hence advantageous, to consider the earth at first without rotation.

Assume on some part of the earth's surface that over a large area the air is moving in an eddy or whirl in a counter-clockwise direction about some fixed center. This eddy might be 500 miles or more in diameter, with the wind having everywhere a velocity of 20 miles per hour. Such a circulation is called in meteorological parlance, a cyclone. If the air is at each point moving in a true circle around the center, this shows that at the center the barometric pressure must be lower than at the outskirts by the amount required by the formula for centripetal force. With a constant velocity, V, the barometric pressure along a radial line will change at a varying rate dependent upon the value of ρ. This rate or barometric gradient at the surface of the earth, expressed in inches of mercury difference in barometer reading per 100 miles of horizontal radial distance is

$$\text{Barometric gradient} = .25 \frac{V^2}{g\rho} = \frac{V^2}{128\rho} \tag{6}$$

in inches per 100 miles, in which V is the velocity of the wind in miles per hour, g is the acceleration of gravity in feet per second, and ρ is the radius in miles. For a velocity of 20 miles per hour, and a radius of 200 miles this gives a barometric gradient of .015 inch per 100 miles. This result shows that for ordinary conditions centripetal force has but small effect upon the distribution of barometric pressure, but for such extreme conditions as are reached in a West Indian hurricane the effect of rotation becomes very marked.   In the above example with the circulation in a counter-clockwise direction the higher pressure would be on the right hand side of the moving air current.

An important relation in connection with a revolving cyclone such as that described above should be noted at this point. If such a whirling mass or disk of air should be, say, two miles thick and several hundred miles in diameter, the friction of the surface of the earth would be a disturbing element which would constantly tend to reduce the velocity of the moving air at the surface to a value less than that existing at higher altitudes above the earth's surface.   This would upset any previously existing equilibrium between centrifugal force and barometric gradient.   If at a height of 1000 feet, the atmosphere were revolving in circular paths with the horizontal barometric gradient just balancing the horizontal centrifugal force, then at lower elevations where the linear velocity of the air was less, the barometric gradient would more than suffice to overcome the centrifugal force, with the result that the lower air would be pushed in toward the center of the cyclone.   Similarly at higher levels the barometric gradient would not suffice to overcome centrifugal force and the air would tend to move away from the center of rotation.   The radial barometric gradient is so slight a quantity that these variations are perhaps not of prime importance, but observations in the air at different levels indicate a relative motion inward at the ground and an outward motion at high altitudes.

EFFECT OF THE EARTH'S ROTATION UPON FRICTIONLESS MOTION

First, let us consider the earth as a perfect sphere not in rotation. If a body could move on the earth's surface without any friction, then such a body once put into motion would move around the earth, under the force of gravity, pulling it constantly toward the earth's center, in a path which would be a great circle. For example, in figure 3, representing the earth, if a body were started at *P* on the equator in the direction *PA* it would move along the great circle represented by the line *PABC* until it passed entirely around the earth and came back to the point of beginning.

With such an earth not in rotation, the moving point, *P*, would actually traverse a great circle on the earth's surface and its velocity would be constant. Such a great circle might be said to correspond to a straight line on the earth.   A peculiarity of the motion would be, however, that its geographical direction would be constantly changing.

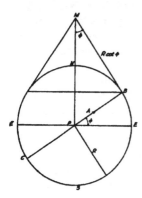

FIG. 3.—MOTION ON THE EARTH.

At P it might be going northeast, at B due east, later southeast, etc., although it would all the time have what corresponds to a straight line motion.   That is, it would remain in the plane of the same great circle of the earth.

If now the earth is supposed to have a rotation from west to east, in the absence of friction the moving body would follow exactly the same path in space as before, and its absolute velocity would continue unchanged.   Its velocity relative to the earth, however, would be much different from what it was before, and the path which it would follow or trace upon the moving surface of the earth would become a complicated curve.   The velocity of the body relative to the earth would be variable in both magnitude and direction.   For example, if the angular velocity of the earth were such that the linear velocity of a point at the equator was somewhat greater than the easterly component of the moving body at the point P, then the moving body would fall behind the earth and would seem to move relatively to the earth in a northwesterly direction from the point P.   As the body reaches more northerly latitudes, the linear velocity of points on the earth's surface becomes less, while the easterly component of the velocity of the body becomes greater.   Thus a place may be reached where the body is keeping up with easterly motion of the earth, and beyond which the moving body actually has an easterly component of motion relative to the earth.   In such a case the path traced on the earth's surface would be a series of loops.   The consideration of this case serves to introduce in a simple manner the idea of relative motion on the earth's surface, but need not be followed further, for the reason that the actual surface of the earth is not a sphere but an ellipsoid with the polar diameter appreciably less than the equatorial. This introduces a new element which complicates the situation, but which in the end simplifies the result expressing motion relative to the moving surface of the rotating earth.

On account of the ellipticity of the earth, the attraction of gravity directed towards the center of the earth acting on a body moving with the earth such as B, figure 3, is not perpendicular to a level surface at B, but has in the northern hemisphere a component directed toward the north pole. The direction of the attraction of gravity at the point B should not be confused with the direction which a plumb line at B indicates. The latter, acting normal to the tangent plane at B, is only one component of gravity and is balanced by the reaction of the earth at this point. The other component, acting at right angles to the axis of rotation of the earth, is a centripetal force, also called in technical mechanics an unbalanced, accelerating, or deviating force, although it is sometimes said to be balanced by the so-called centrifugal force. This centripetal force is what causes a body, at rest on the surface of the earth, to deviate from a straight path as the earth revolves and thus to remain on the earth instead of flying off into space. The so-called centrifugal force is not a real force at all as this word is defined in mechanics. Hence, if the elliptical earth were not rotating, a frictionless body moving due east on the earth at the point B would not continue in the great circle, as it would for a spherical earth, but would be deflected towards the north by gravity. In such case the deflection would be towards the left.

To examine in detail the motion of a body on the earth's surface it is necessary to represent the motion in a plane. This may be conveniently done by means of the cone tangent to the earth on the parallel of latitude through the point under investigation. Thus to study the relative motion at the point B, figure 3, draw the tangent cone whose vertex is at M. Any motion on the earth's surface at B may be considered as equally a motion in the conical surface.

Let figure 4 represent the development in a plane of the portion of the conical surface near B.

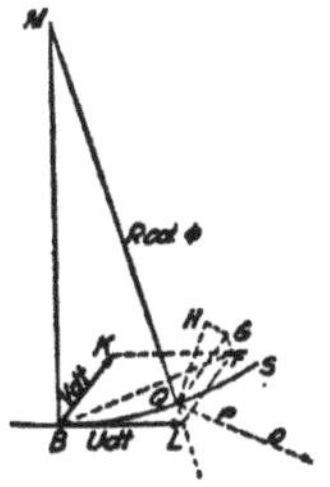

Let R = radius of the earth in miles
   ϕ = latitude of the point B
   U = linear velocity in miles per hour of a point of the earth's surface at the latitude ϕ
   V = linear velocity relative to the earth's surface in miles per hour of a frictionless moving body at the point B.

FIG. 4. — MOTION IN A PLANE TANGENT TO THE EARTH.

In figure 4, the development of the line of tangency of the cone and sphere, which is the parallel of latitude through B, will be the arc of a circle BQS, whose center is M and whose radius, BM, equals $R \cot \phi$. At B, a point of the earth's surface is moving due east in the direction of BL with a velocity, U. If no force were acting to deflect this point from the straightest path, it would move along a great circle which would show in the development as the straight line

BL. But actually on account of the eccentricity of the earth there is an unbalanced component of gravity urging it toward the north. This force is just sufficient to deflect the motion of the point from the great circle to the circle *BQS* corresponding to the parallel of latitude.

In the short time *dt*, the velocity *U* if unaffected would carry the point a distance *U dt* represented in figure 4 by *BL*. In the same space of time the ellipticity of the earth produces a deflection towards the north equal to *LQ*, so that at the end of the time *dt*, the point is not at *L* but at *Q*.

Now, let us consider the case of a frictionless body at the point *B* moving in any direction on the earth's surface, with an initial relative velocity *V*. If the earth were stationary and spherical the body would in the time, *dt*, move a distance *V dt* to a position *K*. Since the earth is in rotation the absolute velocity of the moving body is the resultant of *U* and *V*. In the time, *dt*, therefore, if no deflecting force were acting the motion of the body would be found by combining *BL* and *BK*, which would take the body to the point *F*. But the ellipticity of the earth produces during this same interval of time, *dt*, a movement *FG* equal to *LQ*. Hence, at the end of the time, *dt*, the body actually is at the point, *G*. In figure 4, *BK*, *LF*, and *QG* are all equal and parallel.

It is next desired to find the path described on the earth's surface by the moving body during the motion described above. Since at the point *B* the body has a relative velocity *V* in the direction *BK*, the path, if a curve, must be tangent at *B* to the line *BK*, and the center of curvature must be on a line drawn through *B* perpendicular to *BK*. After the time, *dt*, the point *B* has moved to the position, *Q*. During this time the line *BK* has turned through an angle so that it is no longer parallel to its original position like *QG*, but occupies a position *QH* such that the angle *GQH* equals the angle *BMQ*. Hence, drawing *QO* perpendicular to *QH*, the center of curvature, *O*, lies on this line at such a distance from *Q* that a circle drawn through *Q* shall pass through the point, *G*. Hence, referring to the equations on page 42,

$$QH = dx$$

$$HG = dy$$

But *QH* = *V dt*, and from the similar triangles *QHG* and *MQB*,

$$HG = \frac{VU(dt)^2}{R \cot \phi}$$

Substituting these values in equation 4 there is obtained

$$\rho = \frac{VR \cot \phi}{2U}$$

Since $U = \dfrac{2\pi R \cos \phi}{24}$ miles per hour

$$\rho = \frac{6}{\pi} \frac{V}{\sin \phi} \text{ miles} \tag{7}$$

This is based upon the assumption that the earth rotates once upon its axis in 24 hours. Since the sidereal day is about four minutes shorter than the mean solar day, this assumption introduces a small error.

Equation 7 shows that a slowly moving body on the earth's surface, when not near the equator, tends naturally to move around a closed curve, approximately a circle, of relatively small size. For example, with a velocity of 10 miles per hour in latitude 30 degrees, the radius of the circle would be about 40 miles. In the northern hemisphere the deflection from a straight line is toward the right, or in other words the motion is in the clockwise direction. In the southern hemisphere the deflection is toward the left, or the circular motion is counter-clockwise.

INFLUENCE OF EARTH'S ROTATION ON BAROMETRIC GRADIENT

When the movements of the atmosphere follow the curved paths described in the preceding paragraph the rotation of the earth has no influence on the pressure distribution. However, when they depart from these paths, as they generally do, there must be a barometric gradient at right angles to the line of motion of suffiicient magnitude to cause the departure.

The force necessary to change the motion of a mass of air from the curved path derived in the preceding pages to a straight line motion relative to the rotating earth, is just equal to the force that would change the motion on a still earth from a straight line to the curved path. Therefore, by substituting the value of ρ given by equation 7 in equation 6, there is at once obtained an expression for the barometric gradient at right angles to the direction of motion which is necessary, on account of the rotation of the earth, when the atmosphere moves in an approximately straight line. The equation thus obtained is:

Barometric gradient

$$= \frac{\pi V \sin \phi}{24g} = .004V \sin \phi \text{ inches per 100 miles} \quad (8)$$

For many conditions this will be a larger gradient than that in simple circular motion given above by equation 6. For example, with a velocity of 20 miles per hour in a latitude of 30 degrees the barometric gradient at right angles to straight line motion would be .04 inch of mercury per 100 miles. If the wind is moving in a curved path in a counter-clockwise direction, the total radial barometric gradient will be given by the sum of equations 6 and 8. If the motion is in the clockwise direction the gradient is found by taking the difference of the two equations.

GENERAL WIND CIRCULATION FOR EASTERN UNITED STATES

The immediate cause of the general wind circulation is the variation in air temperature. As a hypothetical illustration consider the relative pressure distribution above two points where the air temperatures average 60 and 80 degrees, respectively. At the colder point the barometric pressure would decrease upward at a rate at first of about 0.1 inch of mercury for each 100 feet of vertical rise and would continue to decrease at a rate which would be not quite a constant but a slowly decreasing quantity. At the warmer point the air would be about 4 per cent lighter for the same pressure, and at all higher altitudes would continue to be lighter than the colder air in about the same ratio. The consequence of this is that the decrease of pressure in rising a stated vertical distance would not be so much at the warmer point as at the colder. In rising 1 mile the pressure at the colder point would decrease 0.2 inch of mercury more than the decrease at the warmer point. If the pressure at the surface of the earth were the same at the two points then the pressure 3 miles above the earth's surface would be 0.6 inch of mercury less at the colder point. Or, if the pressure at the earth's surface were 0.3 inch greater at the colder point, at an altitude of 3 miles it would be 0.3 inch less than for the same altitude at the warmer point. In this latter case, then, there would be a surface wind from the colder to the warmer region and at high altitudes a wind from the warmer towards the colder region. Similarly, any variation in temperature between two points on the earth prevents the existence of a state of equilibrium and starts an air circulation.

The latter case typifies the condition existing over a large part of the earth's surface during much of the time. The high temperatures prevailing in the equatorial region lead to an unequal pressure distribution with the result that around the earth there is a fairly permanent belt of high pressure at the surface in latitude 30 to 35 degrees north, accompanied by corresponding low pressures at great altitudes above the earth. The resulting surface winds blowing constantly towards the equator over the oceans are called the trade winds. Where deepest they attain an altitude of about 2½ miles. The belt of high surface pressure is affected by the seasons as the sun moves north and south of the equator, and by the relations of the continents to the oceans, since land areas are warmer than water areas during the summer, and correspondingly colder during the winter.

The region of the trade winds only barely touches the most southern part of the United States, but our winds are affected by their influence. As previously explained on page 47, any wind in the northern hemisphere tends to be deflected toward the right. Hence the trade winds in blowing toward the equator turn toward the southwest. Similarly, the air currents at high altitudes moving north from the equator blow toward the northeast, and since they are probably much less retarded by friction on account of their separation from the earth's surface, their easterly component of velocity persists sufficiently to give a decided drift towards the east to all the atmosphere throughout the range of latitude occupied by the United States. The latter is said to lie in the region of the prevailing westerlies and this constant steady drift of our atmosphere toward the east is the basis of all our weather predictions.

Throughout the region of the prevailing westerlies the air temperatures, and their resulting winds, are much affected by the presence of the water vapor and clouds in the air, and by the vertical movements in the air masses. The mechanism controlling changes in these elements has not yet been completely determined, and for this reason long-range weather forecasts are still in only a partially developed state.

The general circulation of the atmosphere between tropical and extra tropical regions by means of high-altitude currents in one direction and low-altitude currents in the reverse direction, might be spoken of as a continuous circulation in vertical planes. In the latitudes of the United States this simple circulation, during much of the time, becomes converted into what might be called for contrast a circulation in horizontal planes. Over a large area the air will be moving toward the north, at a particular instant, while over a different but correspondingly large area at the same time the air will be moving toward the south. The whole system has a continual drift towards

4

the east in correspondence with the prevailing westerlies of these latitudes, with the result that at a given geographical station the winds are intermittently southerly and northerly.

Neither the origin of these changes nor the mechanism which controls their movements is well enough worked out yet so that accurate predictions of their movements can be made for any extended advance period. But the principle of continuity and the data which is being accumulated each year by the Weather Bureau are continually extending the range of possible prediction. The disturbance of temperature conditions by continental areas and by water vapor and clouds in the air doubtless have a large effect in the intermittent wind circulation. Also, the region of prevailing westerlies, by producing a constant circulation around the north pole, sometimes called the circumpolar whirl, must produce a constant tendency for a low altitude current to creep along the earth's surface in a northerly direction as explained on page 43.

STORMS

A storm may be defined as a movement of the air accompanied by precipitation. Precipitation in any storm is very unevenly distributed, even within limited areas. "A difference of 50 per cent within five miles or 100 per cent within ten miles can easily occur. From a study of many gages located within ten miles of Berlin, Hellman found that even in monthly totals, gages 1500 feet apart were as likely as not to differ 5 per cent, while of course in special rains they would differ 100 per cent."[*]

The distribution of normal annual rainfall over the eastern United States is shown on figure 5, a map of the country east of the 103d meridian, on which are drawn isohyetal lines (lines of equal rainfall) for each 10 inches. The map is the latest issued by the Weather Bureau, and is based on records of about 1600 stations for the 20-year period 1895–1914, and 2000 additional records from 5 to 19 years in length, uniformly adjusted to the same period.

Rainfall is dependent originally upon the evaporation of water into the air. In the form of water vapor it may then be transported long distances by the winds. Before precipitation can occur condensation must be produced by cooling. The cooling may be caused by contact with cold surfaces or with cold layers of air, by adiabatic expansion while rising to greater altitudes, by radiation, and by mixing with colder air. The oceanic areas of the torrid zone are the most extensive source for evaporation, although it is to be remembered that even on land surfaces a large fraction of the annual rainfall is re-

[*] Willis L. Moore, Descriptive Meteorology, New York, 1910, page 209.

evaporated again into the air.   The water precipitated as rain comes
therefore from these two distinct sources, each of which may pre-
ponderate according to the nature of the storm.   Thus, thunderstorms
derive their supply mainly from land surface evaporation, while
cyclonic storms which are caused primarily by warm moisture bearing
winds blowing from the tropics, are fed largely by evaporation from
the ocean.

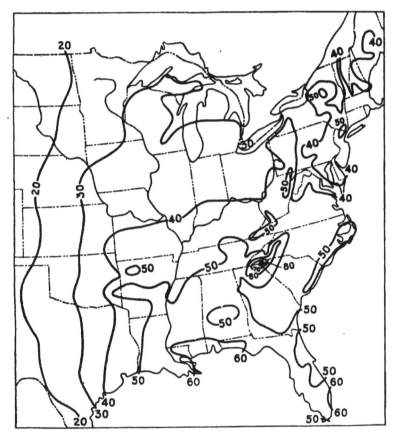

FIG. 5.—NORMAL ANNUAL RAINFALL OVER THE EASTERN
UNITED STATES.

The normal annual precipitation, as will be seen from an inspection
of the isohyetals, exhibits a decided variation in amount for different
parts of the eastern United States.   This variation is a climatic feature
traceable to factors, both general and local in character, the most
important of which are latitude, proximity to oceans, and altitude.

The annual amount of rainfall tends to diminish as the latitude increases. Broadly stated, in the eastern United States, the normal annual precipitation ranges from about 60 inches along the gulf coast to about 30 inches along the Canadian line.

Rains caused by moisture-laden air from the Gulf of Mexico and the Atlantic Ocean exert a marked influence in swelling the annual precipitation in the southeastern parts of the United States. Conversely, as the distance from these bodies of water increases this influence becomes less. This is particularly noticeable in the great plains region immediately west of the Mississippi River, where the lack of moisture from the gulf region causes the annual rainfall to diminish rapidly in a westward direction. The isohyetals here run nearly north and south with a marked easterly inclination in the northerly latitudes caused by the influence of these latitudes tending to decrease rainfall.

Increase of altitude usually is accompanied by increased rainfall, but this is largely local in effect. Except for these local variations, which may be very appreciable in amount, the effect of topography on rainfall is popularly much overestimated. Mountain ranges, by forcing air currents to higher altitudes, induce precipitation and so cause the annual precipitation to run high on the sides exposed to prevailing winds. This very action is, however, responsible for greatly decreased rainfall on the leeward sides of such mountain ranges, although the elevation there may also be considerable. The great plains region west of the Mississippi River rises 5000 feet in about 750 miles to the foot of the Rocky Mountains. Yet its normal annual precipitation shows a decrease in going westward of over 25 inches. This shows that latitude and proximity to the ocean may be more potent factors than altitude.

It is convenient to consider storms in the eastern United States as being of three classes: cyclones, West Indian hurricanes, and thundershowers.

Cyclones

Cyclones, frequently called extratropical cyclones, to distinguish them from the tropical cyclones or West Indian hurricanes, are disturbances of varying intensity which pass over the United States with more or less regularity, in a general easterly direction, and are usually accompanied by rainfall over extensive areas. Their action will be understood from the following considerations.

When a mass of air covering a large area moves from south to north there must be, as explained on page 47, a tendency towards the accumulation of pressure on the right hand or east side of the moving mass; or, what is the same thing, the production of an area of low pressure on the western side. Conversely, if air moves from north to south the pressure will be low to the east of the moving mass and high to the west of it.

On the daily weather map, isobars or lines of equal atmospheric pressure at the earth's surface are drawn in accordance with the daily simultaneous observation of the pressure at stations scattered over the whole country. The high and low pressure areas are usually inclosed by oval shaped isobars, and are commonly spoken of as the *highs* and *lows* of the weather maps. The lows are generally more definite than the highs. The lows and highs are usually from 500 to 1000 miles apart and drift across the country from west to east in a more or less regular and constant succession.

A low pressure area, as shown on a weather map, furnishes a most convenient and useful epitome or short-hand indication of the weather conditions existing and impending over an area extending for several hundred miles in every direction from its center. Around a low the wind circulation is in general counter-clockwise, that is, in a northerly direction to the east of, and in a southerly direction to the west of the low. This gyratory or cyclonic movement has given rise to the use of the term cyclone as applied to weather disturbances accompanying low pressure areas. The result of this general wind circulation is that east of a low the southerly winds bring a slowly rising temperature, while west of a low the northerly winds produce cold. As a low in its easterly motion drifts over a given station, there is often a sudden change in wind and temperature, producing the marked cold waves particularly noticeable in winter.

In the area to the east of a low the southerly winds bring moisture from the Gulf of Mexico and Atlantic Ocean into the cooler northern latitudes, leading to the formation of clouds and precipitation. Hence, in this area there is always a more or less general rainfall. The intensity of rainfall is dependent on a variety of causes, sometimes it seems somewhat proportional to the strength of the winds. The condensation of water vapor apparently takes place chiefly in the lower mile of the atmosphere, and is probably assisted by local vertical displacements of masses of air in a manner similar to the action in tropical hurricanes and thundershowers to be described later. The prediction of weather changes is based chiefly upon the geographical movement of low pressure areas as determined by observation of previous similar cases. The lows are much more frequent, definite, and energetic in winter than in summer, with the result that our alternations of warm threatening weather and cold clear weather are correspondingly more frequent and marked in the winter; but notwithstanding this fact, our precipitation is, on the average, greater in summer because of the greater capacity of the air for carrying water vapor on account of the higher temperatures prevailing during that season.

About 120 cyclonic disturbances cross the country per year. In passing across the continent, cyclonic centers show generally a tendency to avoid crossing areas of high elevation, such as mountain ranges and high plateaus, except in the cold season.

Rates of Travel of Cyclones and Precipitation Areas

Cyclones have a fairly definite rate of motion in passing across the continent. An examination of the charts published by the Weather Bureau, in Supplement No. 1 of the Monthly Weather Review, 1914, which illustrates types of storms and their movements, shows that the average hourly movement of a cyclone center varies from less than 20 miles in summer to over 40 miles in winter.

It might be supposed that the center of precipitation would have about the same velocity of translation as the center of the low pressure area, since it generally occurs in the south or southeast quarter of the latter in the eastern part of the United States. A study was made of 33 of the most important rainstorms in the eastern United States, for which data are available. Further reference to these, together with maps, will be found in a subsequent chapter. This study showed that the rate of travel of the area of precipitation was so erratic as to be practically indeterminable, although definitely greater in winter than in summer. About all that can be said is that the direction of movement is generally the same as that of the low. Several reasons suggest themselves for the erratic nature of rainfall travel: (a) the lows themselves are far from uniform in velocity of movement; (b) in the Upper Mississippi and Ohio River valleys the direction from which moisture bearing winds can approach a low is restricted; (c) rainfall gaging stations are too far apart to permit drawing accurate conclusions; (d) the rate of precipitation is far from uniform; (e) the distance from the center of the low to the center of precipitation varies greatly in the same and in different storms.

West Indian Hurricanes

West Indian hurricanes, or tropical cyclones, have many features in common with the extratropical cyclones just described, but differ from them in some important respects. The differences, however, are of consequence only when the tropical cyclone appears in its typical form. When a tropical cyclone reaches a continental land area its prime characteristics of enormous wind velocities, calm central eye, and uniformly distributed cloud area, are rapidly transformed by the changed conditions and it takes on the characteristics of an extratropical cyclone. Thus it is only the Gulf and South Atlantic Coast regions in the United States which receive the tropical cyclone in its full force. The reasons for this will appear more clearly after this type of storm is described.

As its name implies the tropical cyclone originates in the tropics. Those which reach the United States are always formed north of the Equator, usually in from 8 to 12 degrees, north latitude, a region of calm or very light variable winds. The capacity of the air for water vapor is great and it is almost completely saturated, producing an ideal condition for vigorous convection currents. This is what actually takes place: The superheated moist air rises, some of the water vapor condenses at a relatively low altitude, releasing latent heat, thus giving new impetus to the rising current and consequently to the spirally inflowing air at the base of the convection flue. This of course results in a greater rate of condensation, release of latent heat, and increased velocity of convection current. Thus the embryo tropical cyclone grows and builds itself up, feeding on the energy stored in water vapor as latent heat. The winds soon attain great velocities, frequently more than 100 miles an hour, and as the velocity increases a very low pressure area is formed at the center of the whirl. Nimbus clouds, evenly distributed over the entire area of disturbance, except for the calm central eye, are formed by the rapid condensation, and torrential rains fall. The temperature and moisture of the atmosphere in all the quadrants is the same, the former being greater and the latter less in the calm central eye than in other parts of the area.

The necessary conditions for the development of a tropical cyclone, that is, great absolute humidity, a surface of little frictional resistance, and undisturbed terrestrial winds, are to be found only on a water surface near the equator, and, since in the northern hemisphere the trade winds come from the northeast, the most favorable conditions exist on the west side of an ocean. Furthermore, since tropical conditions extend farthest north in late summer, this is the most favorable time for the development of tropical cyclones in this hemisphere, and it is then that most of them occur. The direction of the spirally inflowing winds at the bottom of the convection flue is in a counterclockwise direction, due to the influence of the terrestrial wind system. In general, the cyclone moves in a northwesterly direction, slowly at first, and then with somewhat increased velocity. About latitude 30 the direction of movement of the tropical cyclone gradually changes to northeast to conform to the changed direction of terrestrial winds. Thus the cyclone track is a parabola, convex westward, with its vertex at about 30 degrees north latitude.

When the cyclone moves on to the North American continent, its characteristics are rapidly changed. The high wind velocity is reduced by friction, the moisture is now supplied from only one instead of from all four quarters, and its amount from this quarter rapidly decreases as the cyclone moves inland, thus cutting off the source of its violent energy. As stated before, the tropical cyclone quickly assumes the characteristics of the extratropical cyclone. About .10 West Indian hurricanes touch the United States each year, most of them occurring in late summer. On the average less than one per annum is violent or destructive.

Thunderstorms

Thunderstorms are so familiar to everyone that they hardly need description. They are comparatively local in extent, the cloud area rarely being larger than 150 or 200 miles long by 40 or 50 miles wide and rarely causing precipitation over an area more than 300 miles long. They are not confined to any particular region or time of year, but occur most frequently in the hottest season of the year and warmest part of the day.

The conditions requisite for the origin and growth of a thunderstorm are a pocket of warm moist air, with some source of energy to cause this to rise rapidly. These conditions are most frequent and pronounced in the southern part of low barometric areas during the afternoon of hot summer days, and it is under such conditions that thunderstorms are most frequent and violent.

There are three fairly well defined causes for warm moist air to rise and cool with sufficient rapidity to result in a thunderstorm. In the order of their importance, these are: (1) Convection; (2) Overrunning or underrunning cool dry currents of air; (3) Local topography, such as mountains.

Convection is so well understood that it requires no further explanation here. When convection alone is active in forcing upward a moist quantity of air it is simply a question as to whether its action is vigorous enough to cause precipitation.

In the southern quadrant of lows there is frequently developed a wind shift line. The normal direction of wind in that quadrant is of course from west to east. When the wind shift line is most completely developed it is a northeast-southwest line, along the southeast side of which the wind direction is parallel to the line, and on the northwest side of which the wind direction is approximately perpendicular to the line. As a natural result of this condition, the cool dry northwest wind frequently underruns the warm moist southern wind, forcing the latter upward, or overruns it and the greater weight of this cool dry air above the moist warm air sets up vigorous convection currents. In either case rapid rising and condensation of the water vapor in the moist warm air is the result. Thunderstorms from this source may occur at any time of year, though of course they are most violent in summer, due to the fact that the absolute humidity of the atmosphere is greatest at this time of year.

The third cause of thunderstorms, that of local conditions, is generally confined to mountain or coastal regions, and is principally due to warm moist air blowing against mountains and thus being forced to rise, cool, and lose its moisture.

The clouds and storm characteristics of all three processes are the same. As condensation proceeds a dense cumulus cloud area is formed having a characteristic anvil shape when seen from the side. This cloud mass sometimes becomes two or three miles thick, or even more. Condensation continues, the cloud area grows, becomes denser, is transformed into the cumulo-nimbus type, and rain begins to fall. This frequently becomes very intense over considerable areas. The storm usually travels from west to east, that being the direction of the winds in the southern part of a low, at an average rate of 30 or 40 miles an hour. Lightning and thunder are characteristic incidental phenomena. The cooling of the air under the cloud, due to rain passing through it and the exclusion of the sun's rays, sets up vigorous eddy currents and produces squall winds in front of the cloud, which may attain destructive velocities. Such storms usually last only a few hours.

LIST OF BOOKS ON METEOROLOGY

From the large number of works dealing with this subject, the following have been selected as likely to prove most useful to those wanting some technical information but not desiring to make an exhaustive study. Besides the books here listed, the Weather Bureau publishes the daily weather maps, the Monthly Weather Review, and numerous pamphlet instructions and bulletins.

In the following list the more general and elementary works are put first, the specialized and technical treatises last, arranged somewhat in order of difficulty. Many of those in the first part of the list are much alike, but the large number is included because they are the ones most likely to be found in the smaller public libraries, which might have some but not all of them. Where no other distinction is drawn the books follow date of publication beginning with the earlier.

GREELY, A. W., American Weather, 8°, XII + 286 pp., 32 figures and 24 plates. Dodd, Mead & Co., New York, 1888. (Out of print.)

DAVIS, W. M., Elementary Meteorology, 8°, XII + 355 pp., 106 figures and 6 charts. Ginn & Co., Boston, 1894, $2.50.

WALDO, FRANK, Modern Meteorology, 12°, XXIII + 460 pp., 112 figures, Charles Scribner's Sons, New York, 1893, $1.50. The text contains a fuller technical treatment of circulation of the atmosphere than do most of the general works.

WALDO, FRANK, Elementary Meteorology, 12°, 374 pp., American Book Co., New York, 1896, $1.50.

MOORE, W. L., Descriptive Meteorology, 8°, XVIII + 344 pp., 81 figures and 45 plates. D. Appleton & Co., New York, 1910, $3.00 net.

MILHAM, W. I., Meteorology, 8°, XVIII + 600 pp., 157 figures and 50 plates. The Macmillan Co., New York, 1912, $4.50 net.

WARD, R. DE C., Practical Exercises in Elementary Meteorology, 8°, XIII + 199 pp., Ginn & Co., Boston, 1899, $1.12. An elementary work prepared for use as a high school text.

WARD, R. DE C., Climate, 8°, XVI + 372 pp., 34 figures. Putnam's, New York, 1908, $2.00 net.

ABBE, CLEVELAND, "Meteorology" in the Encyclopedia Britannica, 11th edition. A very satisfactory, condensed, technical article.

ABBE, CLEVELAND, The Mechanics of the Earth's Atmosphere, 8°, 324 pp. (Smithsonian miscellaneous collections—843). The Smithsonian Institution, Washington, 1891. This is a second collection of translations of important separate articles from European sources dealing with meteorology. Most of the articles are mathematical in treatment.

ABBE, CLEVELAND, The Mechanics of the Earth's Atmosphere, 8°, IV + 618 pp. (Smithsonian miscellaneous collections, volume 51, number 4, publication 1869.) The Smithsonian Institution, Washington, 1910. This is the third collection of translations from foreign sources by Mr. Abbe and is by far the most useful and important of the various collections. Most of the articles contain much mathematics.

SALISBURY, R. D., Physiography, 8°, XX + 770 pp., 707 figures and 26 plates. Henry Holt & Co., New York, 1907, $3.50. About 200 pages of this work deal with meteorology.

SMITHSONIAN METEOROLOGICAL TABLES, 8°, LX + 280 pp. (Smithsonian miscellaneous collections—1032). The Smithsonian Institution, Washington, third revised edition, 1907. A very useful collection of tables. (Out of print.)

WEATHER FORECASTING IN THE UNITED STATES, by a board composed of A. J. Henry, chairman, E. H. Bowie, H. J. Cox, and H. C. Frankenfield, 8°, 370 pp., 199 figures and plates. An official publication of the U. S. Weather Bureau, Government Printing Office, Washington, 1916. This work is intended as an official textbook covering all the forecasting methods in use by the Weather Bureau officials. Parts of the work are contributed by others than those named above. A very complete treatise on the subject of forecasting.

FERREL, WILLIAM, Recent Advances in Meteorology, 8°, 440 pp., numerous illustrations. Published as part 2 or appendix 71 of Annual Report of the Chief Signal Officer of the Army for the year 1885. Government Printing Office, Washington, 1886. A technical treatise systematically arranged in the form of a textbook for use in the Signal Service school of instruction and also as a handbook in the office of the chief signal officer.

FERREL, WILLIAM, A Popular Treatise on the Winds, second edition, 8°, VIII + 505 pp., numerous illustrations, John Wiley & Sons, New York, 1889, $4.00. This was written for readers familiar with only elementary mathematics, but it is doubtful whether the effort to simplify the treatment has made the subject intelligible to those who are not specialists in the field.

BIGELOW, F. H., Report on the International Cloud Observations, May 1, 1896, to July 1, 1897, 4°, 787 pp., many illustrations and plates. Published as volume II of the Report of the Chief of the Weather Bureau for 1898–99. Government Printing Office, Washington, 1900. This report is very technical and mathematical in parts and doubtless is the most complete treatment in English of considerable portions of the whole science of meteorology.

HANN, JULIUS, Lehrbuch der Meterologie, 4°, XI + 642 pp.; first edition, Leipzig, 1901; second edition, Leipzig, 1906.

HANN, JULIUS, Handbook of Climatology, Part I, General Climatology, translated from the second German edition by R. De C. Ward, 8°, XIV + 437 pp., 11 figures. The Macmillan Company, New York, 1903. This work covers volume 1 of the German edition, is complete in itself, and is the standard work on the subject. (Out of print.)

HANN, JULIUS, Handbuch der Klimatologie, second edition, 3 vols., 404, 384, and 576 pp., Stuttgart, 1897; third edition, Vol. 1, XIV + 394 pp., Stuttgart, 1908; third edition, Vol. 2, XII + 426 pp., Stuttgart, 1910.

CHAPTER IV.—SOURCES OF DATA AND METHODS OF COMPILATION

METEOROLOGICAL OBSERVATIONS IN THE UNITED STATES

The earliest record of systematic daily weather observations made in America, of which we have any knowledge, is that for the years 1644–1645, made by John Companius, a Swedish clergyman, living near Wilmington, Delaware. From that time until the early part of the nineteenth century, a few men possessing scientific curiosity were attracted to the study of weather conditions and made daily observations for varying lengths of time, but rarely or never contemporaneously. These observations were too meager in scope and of too short duration to be of general utility.

In 1814 the Surgeon General of the United States Army made it the duty of each hospital surgeon and director of a department to keep a diary of the weather. No attempt was made to systematize the observations or to standardize the instruments with which they were made, and the results were necessarily vague. This service was continued until the outbreak of the Civil War. A part of the results were published in three volumes entitled Meteorological Registers; the last, published in 1851, covers the period from 1831 to 1842.

Contemporaneous with the foregoing observations were those begun in 1817 by Josiah Meigs, then Commissioner of the General Land Office. He prepared blank forms for taking meteorological data, and issued them to the local land offices scattered throughout the states. This service was later transferred to the U. S. Patent Office, and continued until 1859.

Early in the nineteenth century the State of New York established a literature fund, a part of which was devoted to collecting meteorological data at several educational institutions in the state. Albany Academy began collecting such data under this fund in 1825, and is still continuing this work. It has cooperated, first, with the Smithsonian Institution; second, with the Signal Service of the United States Army, and last, with the U. S. Weather Bureau. At the time this station was established, Joseph Henry was connected with the Academy and became familiar with the meteorological work being done there. When he was chosen as the first Secretary of the Smithsonian Institution, upon the founding of the latter in 1846, he made plans for organizing "a system of extended meteorological observations for solving the problem of American storms."* He called to his aid James P. Espy and Elias Loomis, two scientists already well known for their studies in meteorology. On December 15, 1847, the Board of Regents of the Smithsonian Institution appropriated $1000 for inaugurating the work, but Mr. Henry realized the impossibility of establishing stations with so small an amount and used it for the purchase of observing instruments. In August of the following year, Espy was appointed Meteorologist of the Navy Department, was given an appropriation by Congress, and was directed to cooperate with the Smithsonian Institution. Henry and Espy then formulated plans for establishing a large number of observing stations scattered over the entire country. These stations were of three classes:

1. Stations without instruments, where observations were made of frost, cloudiness, the direction of the wind, and the time of beginning and ending of rain.

2. Stations supplied with thermometers, where in addition to the above the daily maximum and minimum temperatures were recorded.

3. Stations equipped with full sets of instruments, consisting of barometer, thermometers, hydrometer, wind vane, and snow and rain gage.

These instruments were all standardized. In February, 1849, a call for voluntary observers was sent out and 412 responded, 143 of whom had formerly worked under Espy's direction. This is the real origin of the United States Weather Bureau. The work was almost totally discontinued during the Civil War, and in 1865 many valuable records and instruments were destroyed by fire, but the importance of the service had been too clearly shown to allow the work to be discontinued.

In 1870 Congress passed a joint resolution placing the work under the direction of the Chief Signal Officer of the United States Army. Later by act of Congress, approved October 1, 1890, the work was reorganized as the Weather Bureau and placed under the U. S. Department of Agriculture. All meteorological records previously taken were transferred to the new bureau.

There are a number of observing stations in the United States which were established and conducted by private individuals prior to 1846, when the Smithsonian Institution undertook the work. The long time records of these isolated stations when supplemented by the later records are of great value in determining the frequency of storms of different degrees of intensity.

* Report of the Smithsonian Institution, 1846.

As stated above, the Smithsonian Institution records are incomplete prior to 1865, but after that year there are no serious interruptions. A large number of new stations, made possible by the more generous appropriations, were established between 1870 and 1875 by the Army Signal Service. There are now about 200 so-called Regular Weather Bureau Stations. These are the most important and completely equipped stations of the service, each one representing a territory of about 15,000 square miles. Observations are taken daily at 8 a. m. and 8 p. m., 75th meridian time, by trained observers employed by the Weather Bureau. There is a continuous recording attachment to the rain gage, by which it is possible to determine the rate of precipitation and the total for the storm at any time. By this means it is possible to compute the precipitation from midnight to midnight, and it is this figure which appears on the reports. In addition to the regular Weather Bureau stations, the service includes about 4500 cooperative stations throughout the United States, operated by voluntary observers. These stations are equipped with fewer instruments than the regular stations, but those used are tested and are of the same high grade. The rain gages do not give a continuous record of the precipitation. Observations are made once each day at about 8 a. m. or 8 p. m., 75th meridian time.

The Weather Bureau also maintains so-called River Stations, where river gage readings are taken; these generally perform the functions of the cooperative stations also. The records for many of the stations are incomplete for one or more years, due to various causes, but these interruptions of the service at individual stations are of slight importance when the records for all the stations are considered.

In 1883 cooperative stations, known as Cotton Stations, operated by voluntary observers, began to be established in the cotton growing sections; observations are made daily at 8 a. m., 75th meridian time, during the crop growing season. In 1896 similar service was extended to the corn and wheat sections. Observations are not made at these stations during the winter months from November to April, and for this reason the data taken at such stations has not been of much value in these investigations.

EXISTING WEATHER BUREAU RECORDS

The records of the Weather Bureau available at Washington, D. C., and consulted for this investigation are as follows:

I. Smithsonian Institution Records: About 400 volumes. All records prior to 1873 except military records, taken principally during the period from 1848 to 1873; bound by months, each volume containing the data for all the stations in the United States for one month. Voluntary observers.

II. Military Station Records: (1) Original monthly reports from 1843 to 1859 for all the military posts in the United States; about 32 volumes. The data for each station for one year is together, 2 volumes containing the data for all the stations for each year. (2) Copies of the original reports from 1860 to 1892; about 600 volumes. The originals were kept at the posts and copies sent to Washington from time to time. Each volume contains the complete record for one station from 1860, or from date of establishment if after 1860, to 1892.

III. Compiled Data from 1860 to Date: About 36 volumes, bound by states, contain the data for each station bound together for the period from 1860, or from the date of establishment if after 1860, to 1891. Another set of volumes contains compiled data for regular Weather Bureau, cooperative, and river stations, generally for the period subsequent to 1891. The records for one month for all the stations in a given state appear together.

IV. Regular Weather Bureau Station Records: Original monthly reports (daily observations) of the regular stations from 1872 to date, bound as follows: From 1872 to 1887, for each month there is one volume containing the daily observations for all the stations in the United States; from 1888 to date, for each year there are 10 to 12 volumes containing the daily observations for all the stations in the United States, the stations being arranged alphabetically, and the records at each station for the year appearing together.

V. Data Compiled from the Records of the More Important Regular Weather Bureau Stations: Daily precipitation figures are for periods from midnight to midnight. (1) From 1871 to 1902 the data for stations A to K is arranged alphabetically and bound in 10 volumes; data for stations L to Z was never compiled. The data for each station for the whole period is bound together. (2) From 1871 to 1891 the data is arranged alphabetically by stations and bound in 6 volumes. The data for each station for the whole period is bound together, each page containing the summarized data of one station for one year.

VI. Voluntary Observers' Records: Original monthly report sheets, showing daily precipitation from 1874 to date, bound as follows: (1) From 1874 to 1888, about 285 volumes; the data for the whole period for any given station is together, and the data for each state is arranged alphabetically according to station name. (2) From 1889 to 1892, about 60 volumes; the data for one year for any given station is together and the stations in each state are arranged alphabetically. (3) Subsequent to 1892, about 350 volumes; the data for a period of 3 years for any given station is together and the stations in each state are arranged alphabetically. There are about 50 volumes for each 3-year period.

VII. River Station Records: Original monthly report sheets, showing daily precipitation from 1880 to date. Data from 1880 to 1910, about 80 volumes, bound as follows: The data for the whole period at any given station is together, arranged alphabetically according to station name.

VIII. Cotton Station Records: Original monthly report sheets, showing daily precipitation from 1883 to date; about 24 volumes; bound by states.

IX. Corn and Wheat Station Records: Original monthly report sheets, showing daily precipitation from 1896 to date; bound by states.

X. The Monthly Weather Review and Climatological Data, two monthly publications of the Weather Bureau, were also referred to.

COMPILATION OF SUMMARY OF EXCESSIVE PRECIPITA-TOIN FOR ENTIRE UNITED STATES

From all the data in existence concerning storms in the United States, down to December 31, 1914, the precipitation records were copied, as part of the Miami Valley flood prevention studies, at the Weather Bureau in Washington, D. C., for all storms exceeding a certain intensity.

At the outset it was manifestly inadvisable to undertake the work of making a complete copy of all the precipitation records available at the Weather Bureau at Washington. Fortunately, the data is of such a nature and is in such shape that it is not an especially difficult operation to abstract only that part which is pertinent to the investigation in which it is to be used.

Before the abstraction of the data was begun it was necessary to determine as far as possible the uses to which it might later be put. The entire investigation is concerned primarily with rainfalls of such intensity, duration, or extent as to endanger life or cause damage to property, and consequently only the records of relatively unusual precipitation were of interest. One of the principal objects of the investigation is to determine the frequency with which a damaging storm may be expected to occur in a given location.

From the foregoing considerations it was determined to consult the record of each station throughout the entire United States and to abstract all the precipitation records exceeding an arbitrarily fixed minimum. In determining this minimum it was necessary to have in mind that a sufficient number of greater values should be available to make frequency determinations possible for these excessive precipitations, and yet the minimum should be great enough to avoid having to copy a large amount of data, of which no later use would be made. In each instance the record for the entire period of excessive precipitation was copied. This often extended over a number of days. It will be observed that this method of collecting excessive precipitation does not take into account the area over which it may have occurred.

Definition of Excessive Precipitation

Excessive precipitation is defined, for the purpose of this investigation, as follows: 1. Where the normal annual precipitation is 20 inches or more, (a) a 1-day rainfall at one station amounting to 10 per cent or more of the normal annual precipitation; or (b) a total rainfall at one station amounting to 15 per cent or more of the normal annual precipitation, regardless of the number of days in the period of excessive precipitation, so long as the average rainfall for the period is not less than 1 inch in 24 hours. 2. Where the normal annual rainfall is less than 20 inches, a total rainfall of 4 inches or more, regardless of the length of the period, so long as the average rainfall for the period is not less than 1 inch in 24 hours.

As there is little or no precedent for an investigation of this nature, it was difficult to determine beforehand just what limits to choose and the methods to be pursued in abstracting the data. A percentage of the normal annual rainfall at each station, as the limit above which all rainfall records were to be copied, was selected because it was thought a fairly definite and constant relation might exist between unusual records of precipitation and the normal annual rainfall at any station. If this were true, the number of values recorded in the different quadrangles above the limits chosen would vary approximately as the aggregate years of record of all the stations in the several quadrangles, regardless of the normal annual precipitation in the latter. This relation has proved to be much less marked than was anticipated, and the result is that for some quadrangles a superabundance of records were copied, while for others too few were abstracted. In general, the greatest number of records above the limits assumed occur in those sections of the country where the normal annual rainfall is greatest, that is along the southern Atlantic and Gulf of Mexico. The fewest number of records above the limits chosen occur in the sections where the normal annual rainfall is least, that is in the Green Mountains and Adirondack region, along the northern border of the United States, and in the three tiers of quadrangles between the 97th and 103d meridians.

Another undesirable element was introduced by choosing a percentage of the normal annual rainfall as the lower limit of values to be abstracted. Such a procedure means that, since the normal annual

5

rainfall is different at every station, the constant percentages of 10 and 15 are also different. This varying lower limit proved to be very inconvenient in using the data for frequency studies, as will be explained later. A much more simple criterion for choosing the data to be abstracted, and one which would have facilitated the later use of it, would have been to select arbitrary values, above which all data should be copied at all the stations in a region over which rainfall conditions are relatively constant. Thus that part of the United States, east of the 103d meridian, could have been divided into four or five districts over each of which rainfall conditions are sufficiently constant to warrant treating in this manner.

By the criteria chosen, that is all values for 1-day rainfall above 10 per cent of the normal annual, and for 2 or more days' rainfall above 15 per cent, approximately the same number of values were found for 1-day and for 5-day and 6-day precipitation, but there was a decided scarcity of values in many quadrangles for the 2-day, 3-day, and 4-day periods. This was especially noticeable in the extreme northern and western quadrangles, where the rainfall is frequently violent but generally limited in duration to one or two days. In the western and northwestern quadrangles there are relatively few stations, and these are of short duration, hence it was especially desirable to obtain as complete data as its character would make possible. These objects could have been accomplished by fixing arbitrary values for the several days, above which records were to be copied from all the stations throughout the region in question.

Records were abstracted for only such stations as had a complete record for 5 consecutive years. An exception to this rule was made of stations which were established in 1910, and had complete records to date of search.

COMPUTING MAXIMUM ACCUMULATED PRECIPITATION

Specially prepared blanks, as illustrated in figure 6, were used for abstracting this data. A separate sheet was made out for each station which has a complete record for 5 consecutive years (with the additional stations noted above), regardless of whether or not a period of excessive precipitation had been recorded. This was done so that the record would show positive evidence that no excessive precipitation data had been overlooked.

From the summarized data of the Weather Bureau the blanks at the tops of these excessive precipitation sheets were filled in for all the stations whose periods of observation came up to the requirements. To insure the accuracy of this operation it was checked by reading back to the summarized data. Later the daily precipitation records for these stations were carefully searched and the record abstracted for all periods of excessive precipitation whose intensity attained the adopted minimum severity.

The figures in the upper left-hand corner of the sheet following *Estab.* indicate the year and month in which the station was first established. For a large number of the stations continuous records are not available for the entire period since the station was first established. In many cases this is due to interruptions in the service of making observations; in other cases the records were lost or destroyed before or after reaching Washington. The years for which complete

MORGAN ENGINEERING COMPANY
MEMPHIS
SUMMARY OF EXCESSIVE PRECIPITATION
9-B-0

36 July
43-47
50-55
Estab 74-91 __ Mean Annual Precip. (35-40) Elev. 600 State Michigan
Decea. 92 Feb. Mean Annual Temp. _____ District No. _____ Station Ft. Brady

(37-47)
(50-55)
(74-91)

STORM PERIOD			DAILY PRECIPITATION IN INCHES								MAX PRECIPITATION			
Year	Month	Day										1 Day	2 Days	3 Days	Storm
1850	7	13-17	3.00	-	-	2.40	.03				5.43 5.43 5.40	3.00	3.00	3.00	5.43
1852	7	5-12	*	1.12	2.24	4.50	-	-	1.05	8.35 7.86 7.86 .05	4.50	6.74	7.30	8.96
1852	7	20-29	1.12	-	-	-	-	5.00		6.12 5.00 5.00	5.00	5.00	5.00	6.12
1876	6	16-20	3.60	1.70	.24	.10	.02			5.66 5.66 5.64	3.60	5.30	5.54	5.66
1883	7	11-14	*	†4.20	-	.20				4.40 4.40 4.40	4.20	4.20	4.20	4.40

* Precipitation included in record for following day.
† In 14 hours.

FIG. 6.—FORM USED IN COMPILING EXCESSIVE PRECIPITATION DATA.

continuous rainfall records are available are written directly beneath the year and month of first establishment. The years during which the station was in existence, but during a part of which the rainfall records are either fragmentary or entirely wanting, are shown in parenthesis to the right.

In abstracting the excessive precipitation data it was found that the years of complete precipitation record, as shown by the summarized data and copied at the top of the sheets, did not always agree with the actual records on file, and it was necessary to correct the period of complete record to correspond with the data on file. This discrepancy occurred chiefly in the early records, that is, records prior to 1893. All the records prior to that time were carefully checked to eliminate the errors that would otherwise have entered as to the length of complete record. After similarly checking the periods of complete record after 1893 for stations whose names begin with the letters from N to Z, it was discovered that very few errors existed, and consequently this check was not performed for the period 1893 to 1914 for stations whose names begin with the letters from A to M.

The mean annual precipitation was taken directly from the Weather Bureau records in cases for which it had been computed; where this had not been done, it was taken from the map, figure 5, showing lines of normal annual rainfall. When taken from the map and not computed from the record it is shown in parenthesis, thus: (35–40) or (35±).

Under the heading *Daily Precipitation in Inches* are 10 columns. Beginning with the first of these columns on the left, the figures indicate the amount of precipitation on the first day of each storm, followed in the succeeding columns by the figures for the successive days in order. A dash (—) shows that no precipitation occurred at that station on the day in question. An asterisk (*) indicates that the precipitation for that day is included in the figure next to the right of it, the latter being the precipitation for its day and the day or days marked by the asterisk. It was necessary to use this, since the total precipitation for two or more days was frequently recorded without giving the amounts for the separate days. The work of filling in these columns was not in general completely checked by reading back to the original data. Such a complete check was begun, but so few errors were discovered after spending several days checking, it was decided to discontinue it. A further reason for not verifying each figure is the obvious one that an error is not cumulative, but can affect only the period of excessive precipitation in which it occurs.

The maximum accumulated precipitation provided for on the blank form was computed as follows: The maximum 2-day precipitation is the maximum for 1 day plus that for the day preceding or following (depending on which amount is the larger), and the maximum 3-day precipitation is the 2-day maximum plus that for the day preceding or following. In the last column to the right is recorded the total rainfall for the entire period of excessive precipitation. For some purposes the maximum accumulated precipitation was desired for periods of 4, 5, and 6 days. As the prepared form provides only for periods of 1, 2, and 3 days, the figures for the longer periods were placed consecutively from right to left in the final columns under the heading of *Daily Precipitation in Inches* and just above any figures which might already be in these columns. The maximum accumulated precipitation for periods of 1 to 6 days, and the total rainfall for the entire period of excessive precipitation were completely checked to insure their accuracy. In this way excessive precipitation records were copied and maximum accumulated precipitation computed for 4316 stations, of which 1262 stations are west of, and 3054 east of the 103d meridian. The record contains on the average about ten excessive precipitation periods for each station. No use has yet been made of the data for the stations west of the 103d meridian. All of the subsequent discussion relates solely to the territory east of that line.

INDEXING THE DATA

To make this data ready for use, the following method of indexing the sheets was adopted. The United States is divided into 2-degree quadrangles, bounded by the odd degree meridians and parallels.

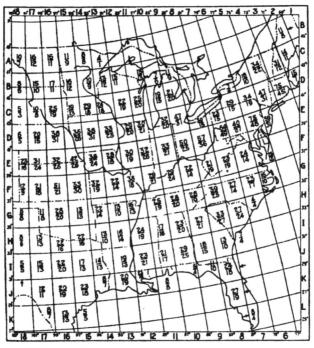

FIG. 7.—NUMBER OF STATIONS USED IN STUDYING EXCESSIVE PRECIPITATION.

The upper figure, in each quadrangle, shows the number of stations whose storm records are abstracted on the summary of excessive precipitation sheets. The lower figure shows the number of such stations still in existence at the end of 1914.

The north and south rows of quadrangles are numbered consecutively from east to west, and the east and west rows of quadrangles are lettered consecutively with capital letters from north to south. Each quadrangle is thus identified by a coordinate number and letter. The rows of quadrangles, properly numbered and lettered, are shown on the map, figure 7, and on numerous subsequent maps in this report.

On a large scale outline map of the United States, divided into quadrangles as above, each rainfall gaging station was located by a dot. In order to identify the various stations within a quadrangle an index letter was placed beside each station, beginning in the northeast corner and progressing toward the southwest corner of the quadrangle, lower case letters of the alphabet being used. When there were more than 26 observing stations in one quadrangle, double index letters were used, as *aa*, *ab*, etc., and if necessary *ba*, *bb*, *bc*, etc. These outline maps were later used repeatedly in the construction of storm maps, and consequently it was deemed advisable to check carefully the location of all the gaging stations. These maps are not reproduced in this report on account of the large scale necessary to show the individual stations. In the upper right hand corner of the excessive precipitation sheet for each station was placed the quadrangle number and its index letter, for example 13–*G*–*m*. This enabled the sheets to be assembled, first, by quadrangles, and second, with index letters in alphabetical order. Thus the records for all the stations in any quadrangle are together, arranged in the order of the station index letters from the northeast to the southwest corner of the quadrangle.

VARIATION IN NUMBER AND DURATION OF RAINFALL RECORDS FOR DIFFERENT QUADRANGLES

Figure 7 is a map of the United States east of the 103d meridian. The number in the upper part of each quadrangle is the number of stations in that quadrangle the excessive precipitation records of which were abstracted on the form shown in figure 6. The lower figure in each quadrangle shows the number of these stations that were still being continued at the close of the year 1914. Stations established subsequent to 1910 are not included in either case. The lower number is obtained from the upper number by deducting all the stations which had been discontinued.

Along the international boundaries and coast lines a number of fractional quadrangles contain so few rainfall stations that it has seemed advantageous in the compilation and subsequent study to join their records to those of adjacent quadrangles. Where such has been done, an arrow drawn on figure 7 points to the quadrangle to which the records have been joined.

Neglecting fractional quadrangles, figure 7 shows that the number of stations whose records have been used varies from 75 for quadrangle 3–D containing most of Massachusetts, Rhode Island, and Connecticut, and for quadrangle 5–E containing Philadelphia and Baltimore, to 5 and 6 stations for quadrangles along the 103d meridian. In general the number of station records available is somewhat proportional to the density of the population. Leaving out the fractional quadrangles, the average number of records per quadrangle is 25.

Not only do the number of rainfall records vary greatly in different parts of the country but the periods during which records were taken at a given station vary through as wide a range. Figures 8 to 12 are diagrams showing graphically the periods of time covered by complete records for all the stations in certain selected quadrangles. The records for 45 quadrangles were plotted in this way, from which the 17 shown have been selected to show the range and nature of the variation. The record for each station appearing on each of these charts was taken from the excessive precipitation sheet for that station.

Figure 8 shows the available records for quadrangle 3–D including most of Massachusetts, Rhode Island, and Connecticut, with the maximum number of stations, 75, and in contrast with it, those for quadrangle 4–C, covering the Adirondack region and containing relatively few stations. Figure 9 shows the records in three quadrangles, that including Dayton in southwestern Ohio, the one immediately to the west of it in Indiana, and one in the mountainous region of Kentucky to the south. Figures 10, 11, and the upper portion of figure 12 cover the row of quadrangles, 12–B to 12–J beginning in northern Wisconsin and running south down the Mississippi valley to the Gulf. The great variation of records with latitude and density of population is graphically shown. Quadrangles 15–E, 15–G, and 15–J in figure 12 typify the nature of the records in the western part of the region studied.

CARD INDEX OF EXCESSIVE PRECIPITATION PERIODS

To bring together the data taken at all stations during a single period of excessive precipitation, a separate card 4 by 6 inches, was prepared for each such period which has occurred since 1870. The data for periods of excessive precipitation prior to 1870 was not compiled in this manner because of the paucity of recording stations and their irregular distribution. The data on these cards taken from the excessive precipitation sheets shows, for each period of excessive

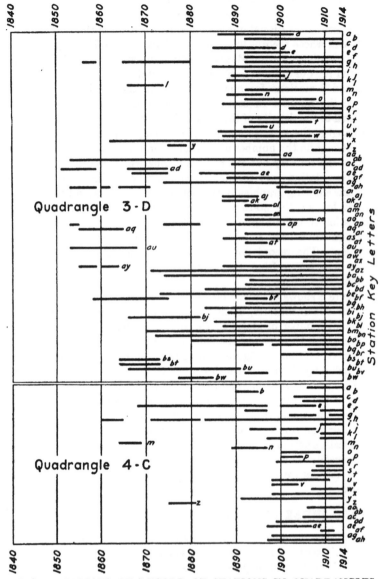

FIG. 8.—PERIODS OF RECORD OF STATIONS IN QUADRANGLES
3–D AND 4–C.

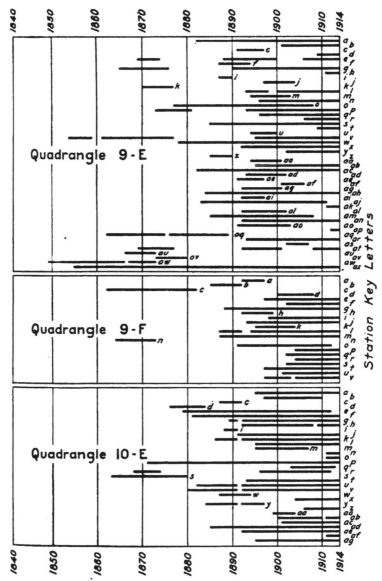

FIG. 9.—PERIODS OF RECORD OF STATIONS IN QUADRANGLES
9–E, 9–F, AND 10–E.

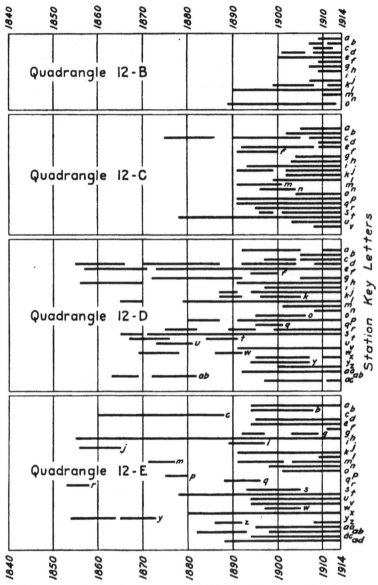

FIG. 10.—PERIODS OF RECORD OF STATIONS IN QUADRANGLES
12–B, 12–C, 12–D, AND 12–E.

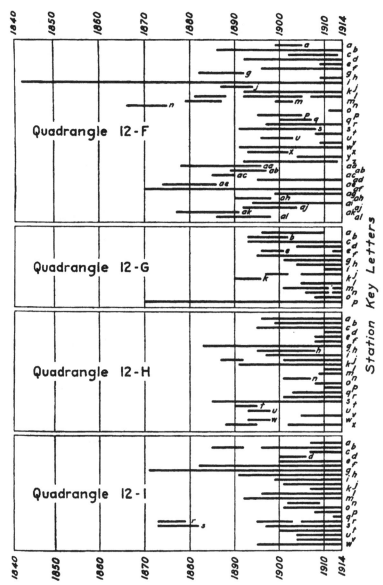

FIG. 11.—PERIODS OF RECORD OF STATIONS IN QUADRANGLES
12–F, 12–G, 12–H, AND 12–I.

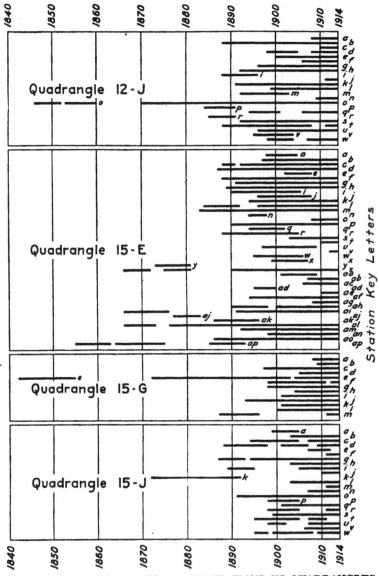

FIG. 12.—PERIODS OF RECORD OF STATIONS IN QUADRANGLES
12–J, 15–E, 15–G, AND 15–J.

precipitation, the quadrangles, dates of observations, and total precipitation at each station during such period. The card index so compiled includes a total of 2641 periods of excessive precipitation, which occurred during the years 1870 to 1914 inclusive.

The cards were next separated into three classes, and each class was arranged in chronological order as follows: (a) Cards which contain only 1 station record, indicating storms with excessive precipitation areas so small that only 1 station was included. These cards number 1236 or nearly 50 per cent of the total. On account of the restricted storm areas which they represent no further use has yet been made of them. (b) Cards on which records from more than 1 but less than 6 stations appeared. These number 996 or over 35 per cent of the total, and represent comparatively local storms. (c) The remaining 409 cards, about 15 per cent of the total number, represent storms for which 6 or more stations reported excessive precipitation. They approach general storms in their characteristics, and to the heaviest of them extensive subsequent study was given, as described in the next chapter. Table 2 shows the number of storms in each class recorded for each year from 1871 to 1914.

Table 2.—Total number of rainfalls causing excessive precipitation, recorded in eastern United States at 1 station, at 2 to 5 stations, and at 6 or more stations, from 1871 to 1914 inclusive.

1	2	3	4	5	1	2	3	4	5
Year	Storms Recorded at			Total Number of Storms	Year	Storms Recorded at			Total Number of Storms
	1 Station	2 to 5 Stations	6 or More Stations			1 Station	2 to 5 Stations	6 or More Stations	
1871	20	3	0	23	1893	23	24	2	49
1872	16	13	0	29	1894	30	19	7	56
1873	24	8	0	32	1895	24	31	8	63
1874	29	13	1	43	1896	19	32	16	67
1875	25	20	1	46	1897	32	33	12	77
1876	35	17	1	53	1898	38	31	16	85
1877	30	19	1	50	1899	17	30	17	64
1878	27	16	3	46	1900	37	31	24	92
1879	32	13	2	47	1901	28	23	13	64
1880	29	23	2	54	1902	26	33	25	84
1881	29	11	3	43	1903	29	22	22	73
1882	39	11	3	53	1904	38	29	14	81
1883	33	11	1	45	1905	35	39	25	99
1884	32	16	1	49	1906	33	30	20	83
1885	22	13	2	37	1907	40	30·	16	86
1886	28	12	1	41	1908	30	41	18	89
1887	19	7	3	29	1909	19	29	21	69
1888	31	14	4	49	1910	24	31	10	65
1889	26	21	3	50	1911	22	29	21	72
1890	26	20	4	50	1912	32	41	16	89
1891	22	18	8	48	1913	36	32	15	83
1892	27	28	6	61	1914	23	29	21	73
					Totals..	1236	996	409	2641

Table 2 shows that the total number of periods of excessive precipitation recorded yearly between 1874 and 1891 is fairly uniform. The years 1892, 1893, and 1894 show a tendency toward an increase and appear to be years of transition, and again from 1895 to 1914 the yearly totals are fairly uniform but appreciably greater than during the first named period. The effect of the number of observing stations on the number of excessive precipitation records is reflected in columns 4 and 5, but less so in columns 2 and 3. The explanation is that prior to 1892 the sparsity of stations was such that it was rare for even a large storm area to cause excessive precipitation at as many as 6 stations. This accounts not only for the small number of storms listed in column 4 prior to that year, and for the total absence of records of such storms during the years 1871, 1872, and 1873, but coincidently operated in placing many such storms in the 1-station and 2 or more station classes, thus swelling the figures in columns 2 and 3 out of normal proportion. It is therefore largely accidental that the number of storms in column 2 varies but little. It is fair to presume that the later figures in column 2 represent almost solely small area storms, such as thunderstorms, while the earlier figures include both this type and the storm covering larger areas.

It appears to be satisfactory for purposes of comparison to separate the records into two 20-year periods; the first period embracing years from 1875 to 1894; the second period from 1895 to 1914. Table 3 shows the results of comparing the two groups.

Table 3.—Comparative number of storms recorded in various classes east of the 103d meridian during the two 20-year periods 1875-1894 and 1895-1914.

Storms Causing Excessive Precipitation at	Years 1875 to 1894 Average Number per Year	Years 1895 to 1914				
		Average Number per Year	For Minimum Year		For Maximum Year	
			Number	Year	Number	Year
1 station only..........	28	29	17	1899	40	1907
2 to 5 stations.........	17	31	22	1903	41	1908 & 1912
6 or more stations......	3	18	8	1895	25	1902 & 1905
Total..............	50	78	63	1895	99	1905

The conclusions which it is possible to draw from the card index of periods of excessive precipitation are somewhat meager. The anticipation was that the card index would furnish conclusive evidence of the relative sizes of different storms, and that it would place the excessive precipitation data in much more convenient shape for making frequency and seasonal studies. None of these objects was realized. The only record shown at any station for a period of excessive precipitation is the total rainfall for the entire period. The length of this period, as well as the dates over which it extends, varies at each station. As a consequence so many variables are introduced that any comparison or use of the data except in the most general way is likely to be very misleading.

The foregoing study as to the number of excessive precipitation records in different storms, and the manner in which these have varied since 1871, is the most important use made of the card index. It is probable that the computations for this study could have been compiled in a less laborious manner. The compilation of this card index is an example of the difficulty in foreseeing the obstacles which may arise to prevent the use of compiled data in the manner originally intended.

CHAPTER V.—FREQUENCY OF EXCESSIVE PRECIPITATION

RELATION BETWEEN STORMS AND EXCESSIVE PRECIPITATION

From the definition given in the preceding chapter it is apparent that the records of excessive precipitation include intense rainfalls occurring during local cloudbursts as well as those occurring in general cyclonic storms. Table 3 shows that about 50 per cent of the excessive precipitation data was recorded at single stations, indicating heavy rains or cloudbursts over probably small areas. About 35 per cent of the data covers storms which caused excessive precipitation at from 2 to 5 stations; and about 15 per cent represents storms causing excessive precipitation at 6 or more stations.

It is also shown in chapter IV that prior to 1895 the sparsity of stations was such that only very large storms were recorded at as many as 6 stations. It may be argued from this that many of the so-called local storms were, in fact, of wide extent, and that the great distance between observing stations failed to bring out the full extent of area covered by their high rates of precipitation. Whatever doubt there may be as to the applicability of such reasoning to the 1-day and 2-day excessive precipitation records, there is small doubt that it does apply to 3-day records, and with increasing force to 4, 5, and 6-day records. Local thunderstorms are of short duration, rarely lasting over a consecutive period covering portions of two calendar days. On the other hand a 3-day or longer period of rainfall of great intensity is most often an indication of heavy rains over extensive areas. Exceptions are local thundershowers falling on successive days in the same locality.

The various aspects of 160 great storms in the eastern United States during the 25-year period 1892–1916 will be presented in subsequent chapters. Each of these storms was selected in the first place, and later carefully analyzed, by considering the three factors which determine its size, namely, *depth of precipitation, duration,* and *area covered.* The relative importance of these three factors may vary greatly, depending on the other conditions which may enter. On a large watershed, great depths of precipitation over small areas are of little consequence; on a very small watershed, such as is ordinarily considered in the design of sewer systems, the storm area covered can be ignored and the attention confined to the maximum depth and duration of rainfall. It is on these two factors of time and depth that the attention is concentrated in the study of excessive precipitation, as contrasted with the investigation of the 160 great storms in which the third factor, area, occupies an important place.

DETERMINING THE FREQUENCY OF EXCESSIVE PRECIPITATION

The word frequency when applied to meteorological phenomena of irregular occurrence is best defined as the number of times, within a selected period of years, that a particular phenomenon has taken place. Dividing the period by the number of such happenings, the quotient obtained is the average length of time in years during which the phenomenon has happened once. This average number of years is also, though less accurately, spoken of as the frequency of the phenomenon. In the latter sense it is used in this chapter in discussing excessive precipitation of a certain defined intensity, and also in chapter VII in speaking of the occurrence of 160 great storms. Whatever significance may attach to it, the reader is cautioned not to construe it to mean a regular or stated interval of occurrence or recurrence, which is its accepted meaning in certain branches of science.

It is obvious that the value of any frequency determination depends primarily upon the length and amount of reliable records available. Unfortunately, long rainfall records in the United States are rather the exception than the rule. To have confined these studies to their use would have imposed limitations that would have effectually barred large sections of the country from consideration for lack of adequate lengths of records. In order to utilize all existing records that possessed any value, even though they differed materially in length, a method was adopted which may be explained as follows:

Let us assume a number of rainfall stations, say five for convenience of illustration, located within an area possessing uniform rainfall characteristics. At station A complete records have been kept for a period of 70 years; at station B, for 40 years; at station C, for 60 years; at station D, for 80 years; and at station E, for 50 years, the aggregate of the period of record being 300 years. Treating this aggregate as a single record for the area under consideration, we may, by the above definitions of frequency, say that the highest rainfall intensity recorded in the entire period has occurred with a probable frequency of once in 300 years. Likewise the second highest intensity has been equaled or exceeded on an average of once in 150 years, and the third highest rainfall intensity has been equaled or exceeded on an

6

average of once in 100 years. The process is capable of indefinite expansion, being limited only by the amount of data at hand. Thus the sixth highest rainfall intensity would have been equaled or exceeded on an average of once in 50 years; the twelfth highest, once in 25 years; and the twenty-fourth highest, once in 15 years. In this way the individual experiences of the observing stations in the given area are combined to give a weighted average, which may be regarded as the probable average experience for any one point within that area.

To illustrate with an actual case, we may take the quadrangle of the earth's surface bounded by the 39th and 41st parallels and the 83d and 85th meridians, in which the Miami River valley is located, and which for convenience of reference has been designated as 9–E. This quadrangle contains 28 rainfall stations with an aggregate period of record of 713 years, no stations having less than 10 whole years of observations being included. To ascertain what 24-hour rainfall intensity has been equaled or exceeded, on an average, once in 100 years, at any point in this quadrangle, select from the aggregate record the 7 greatest 24-hour intensities, and the least of these is the desired rainfall intensity. Arranged in order of magnitude the figures are as follows:

Greatest 1-day Precipitation Records in Quadrangle 9–E. Aggregate period of record— 713 years

1. Newport Barracks, Ky., May 24–25, 1858 . . . . 6.35 inches
2. Urbana, Ohio, September 18, 1866 . . . . 6.20   "
3. Cincinnati, Ohio, June 17–18, 1868 . . . . 6.00   "
4. Urbana, Ohio, June 15, 1868 . . . . 5.95   '
5. Bellefontaine, Ohio, March 25, 1913 . . . . 5.61   '
6. College Hill, Ohio, June 18, 1875 . . . . 5.50   '
7. North Lewisburg, Ohio, September 27–28, 1884 . . . . 5.40   '
8. Newport Barracks, Ky., August 14, 1850 . . . . 5.40   '

THE PLUVIAL INDEX

From the foregoing table it appears that an intensity of 5.40 inches in 1 day has been equaled or exceeded, and is therefore likely to be equaled or exceeded in future years, on an average, once in 100 years at any point in quadrangle 9–E. This is not to be regarded as an accurately fixed quantity, but rather as an indication of what may be expected. On this account, and for greater convenience of reference, the figure 5.40 has been called the *pluvial index* for quadrangle 9–E, corresponding to a 100-year period and a 24-hour rainfall intensity. How much this pluvial index is likely to vary may be judged from an inspection of the figures preceding and succeeding it in the above table. In this particular case the eighth figure also happens to be 5.40 inches. Arithmetically, this eighth figure would correspond to a period of 713 divided by 8, or nearly 90 years, indicating that in the matter of frequency a rainfall of 5.40 inches on the basis of present records may be equaled or exceeded, on an average, once in from 90 to 100 years.

By a similar process the pluvial index is obtained for the 2 days, 3 days, 4 days, 5 days, and 6 days of greatest precipitation in a 100-year period, for the same quadrangle, as follows:

Greatest 2-day Precipitation Records in Quadrangle 9-E.

1. Richmond, Ind., March 24–25, 1913 .... 9.47 inches
2. Bellefontaine, Ohio, March 25–26, 1913 .... 7.74 "
3. Waynesville, Ohio, October 5–6, 1910 .... 7.68 "
4. Jacksonburg, Ohio, October 5–6, 1910 .... 7.50 '
5. Urbana, Ohio, June 15–16, 1868 .... 7.32 '
6. College Hill, Ohio, September 4–5, 1864 .... 7.25 '
7. Cincinnati, Ohio, March 12–13, 1907 .... 7.19 '
8. Kenton, Ohio, August 1–2, 1875 .... 6.96 '

Greatest 3-day Precipitation Records in Quadrangle 9-E.

1. Richmond, Ind., March 23–25, 1913 ....10.35 inches
2. Bellefontaine, Ohio, March 24–26, 1913 .... 9.26 "
3. Urbana, Ohio, June 15–17, 1868 .... 8.41 "
4. Kenton, Ohio, August 25–27, 1871 .... 8.25 '
5. Marion, Ohio, March 24–26, 1913 .... 8.23 '
6. Camp Denison, Ohio, March 12–14, 1907 .... 7.93
7. Waynesville, Ohio, October 5–6, 1910 .... 7.68 '
8. Dayton, Ohio, March 24–26, 1913 .... 7.67 '

Greatest 4-day Precipitation Records in Quadrangle 9-E.

1. Richmond, Ind., March 23–26, 1913 ....11.11 inches
2. Bellefontaine, Ohio, March 23–26, 1913 ....10.63 "
3. Marion, Ohio, March 23–26, 1913 ....9.61 "
4. Greenville, Ohio, March 23–26, 1913 .... 8.92 '
5. Kenton, Ohio, August 26–29, 1871 .... 8.87 '
6. Upper Sandusky, March 23–26, 1913 .... 8.84 '
7. Urbana, Ohio, June 15–18, 1868 .... 8.71
8. College Hill, Ohio, June 18–21, 1875 .... 8.50 '

Greatest 5-day Precipitation Records in Quadrangle 9-E.

1. Kenton, Ohio, August 25–29, 1871 ....11.37 inches
2. Bellefontaine, Ohio, March 23–27, 1913 ....11.16 "
3. Richmond, Ind., March 23–27, 1913 ....11.15 "
4. Marion, Ohio, March 23–27, 1913 ....10.61 '
5. Upper Sandusky, Ohio, March 23–27, 1913 ....10.41 '
6. Greenville, Ohio, March 23–27, 1913 .... 9.33 '
7. Dayton, Ohio, March 23–27, 1913 .... 8.94 '
8. Newport Barracks, Ky., March 3–7, 1850 .... 8.82 '

Greatest 6-day Precipitation Records in Quadrangle 9-E.

1. Richmond, Ind., March 21–26, 1913............................11.74 inches
2. Kenton, Ohio, August 25–29, 1871............................11.37 "
3. Bellefontaine, Ohio, March 23–27, 1913 (5-day).................11.16 "
4. Marion, Ohio, March 23–27, 1913 (5-day)......................10.61 '
5. Upper Sandusky, Ohio, March 23–27, 1913 (5-day)..............10.41 '
6. Kenton, Ohio, July 28–August 2, 1875......................... 9.96 '
7. Greenville, Ohio, March 21–26, 1913.......................... 9.52 '
8. Dayton, Ohio, March 23–27, 1913 (5-day)...................... 8.94 '

It will be noticed that the great storm of March 1913, which covered a vast area, recurs repeatedly, especially for rainfalls of 3 days' duration and longer. In the list of 1-day precipitation records, March 1913 occurs but once; under the 2-day records it appears twice; and thence increasingly until, under the 5-day and 6-day records, 6 out of 8 are March 1913 storm records. As compared with other storms, that of March 1913 was exceptional in its continued heavy precipitation, although it did not cause extraordinarily high rates of rainfall over short periods. This fact stands out clearly in the foregoing tables. They illustrate that one extensive storm, like that of March 1913, even though its occurrence may not be oftener than once in a century, is likely to dominate the pluvial index for the region affected by it. Since the records of high rates of rainfall utilized in a study of this kind are about equally divided between isolated observations representing local thunderstorms or cloudbursts, and observations of storm rainfall covering large areas, the indications are that an unbalanced condition such as that here illustrated is not likely to occur often.

From the preceding discussion it will be clear that the determination of the pluvial index is subject to certain limitations. As used in this chapter it has been expressed to the nearest tenth of an inch, greater refinement not being considered warranted with the data now available.

Attention is called to the fact that in the table of 3-day precipitation, the 7th item of 7.68 inches at Waynesville covers only two dates, namely, October 5 and 6, 1910. This is because the entire storm duration at that point was recorded on these dates, although the same storm at other points extended over a longer period. This treatment is in accord with the rules for determining maximum accumulated precipitation as explained in the preceding chapter. It is interesting to note that in this case the next figure is 7.67 inches at Dayton for the 3-day period March 24 to 26, 1913, which differs from the preceding by only 0.01 inch. In the list of 6-day precipitation records are four where the rain did not exceed 5 days. These are for Bellefontaine, Marion, Upper Sandusky, and Dayton.

By a similar process it is possible to determine the pluvial index for periods of 50, 25, and 15 years and for 1-day to 6-day intensities, respectively. Thus out of the 713 years of aggregate record, the 14th greatest 24-hour rainfall furnishes the 50-year pluvial index; the 28th, the 25-year pluvial index, and so on. In the latter case, dividing 713 by 25, the quotient 28.5 indicates that the pluvial index should be obtained by averaging the 28th and 29th values. A similar interpolation is required to determine the pluvial index for a 15-year period, the correct value lying between the 48th and 49th figures. Interpolation of this kind need not be resorted to except where its omission would introduce appreciable errors.

Broadly speaking, the foregoing method presupposes two conditions: First, that the rainfall characteristics, especially as regards high rates of precipitation, are essentially uniform at all points within the area of a 2-degree quadrangle of the earth's surface. Second, that there are no permanent or cyclic climatic changes affecting the occurrence of high rates of rainfall.

With regard to the first, it is a fact that in fairly level country, as for instance in Ohio and the middle west, the variations in rainfall conditions within one 2-degree quadrangle are comparatively small. The averaging of rainfall records may there be carried out without sensible error; but, where decided differences in elevation exist within one quadrangle, a very appreciable range of meteorological conditions may have to be averaged. In such a case the resulting pluvial index will not be representative of all points within the quadrangle, but will favor those portions having the highest rates of precipitation. The area of a 2-degree quadrangle in latitude 45 degrees is 14,643 square miles, approximately equivalent to that of a circle having a radius of 69 miles. From this it would seem that serious distortion of the pluvial index caused by the coexistence of both very high and very low areas within such a compass seldom occurs.

As to the second assumption, relating to permanent or cyclic climatic changes, the reader is referred to a discussion of this subject in chapter X, from which it will be clear that the results of the method here adopted are not subject to material error from this source.

ISOPLUVIAL CHARTS

In order to give a clear understanding of the frequency with which high rates of precipitation have occurred over the eastern United States, a series of 24 maps, called *isopluvial charts*, was compiled, figures 13 to 36, showing graphically the pluvial index for each of the 133 two-degree quadrangles east of the 103d meridian.

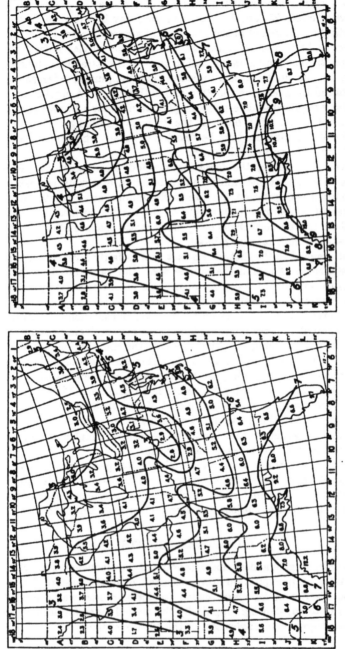

FIG. 14.—ISOPLUVIAL CHART FOR 15-YEAR PERIOD
AND 2-DAY RAINFALL.

FIG. 13.—ISOPLUVIAL CHART FOR 15-YEAR PERIOD
AND 1-DAY RAINFALL.

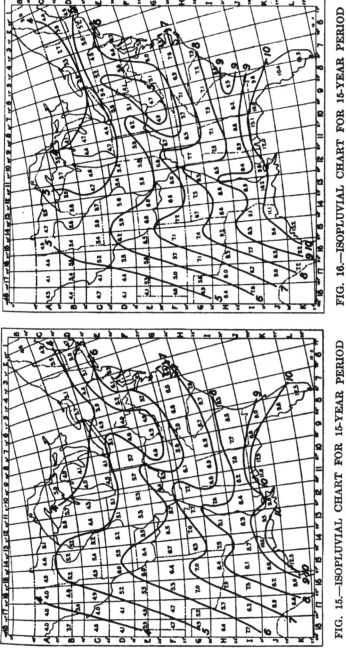

FIG. 16.—ISOPLUVIAL CHART FOR 15-YEAR PERIOD
AND 4-DAY RAINFALL.

FIG. 15.—ISOPLUVIAL CHART FOR 15-YEAR PERIOD
AND 3-DAY RAINFALL.

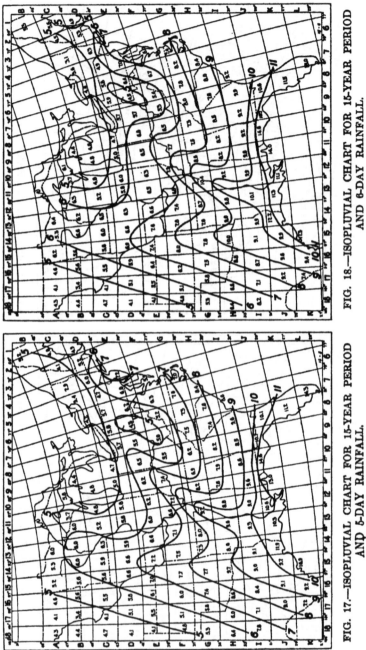

FIG. 18.—ISOPLUVIAL CHART FOR 15-YEAR PERIOD AND 6-DAY RAINFALL.

FIG. 17.—ISOPLUVIAL CHART FOR 15-YEAR PERIOD AND 5-DAY RAINFALL.

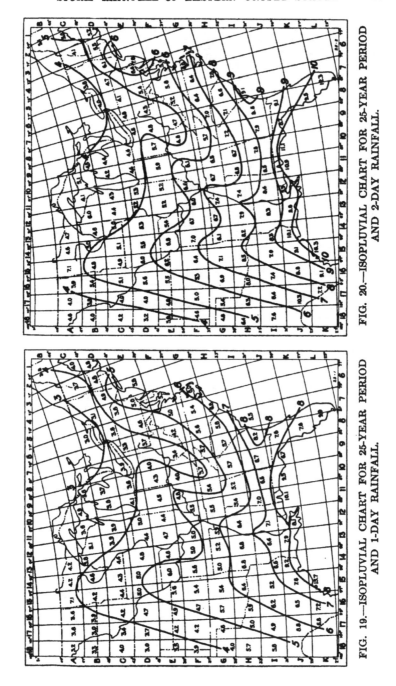

FIG. 20.—ISOPLUVIAL CHART FOR 25-YEAR PERIOD
AND 2-DAY RAINFALL.

FIG. 19.—ISOPLUVIAL CHART FOR 25-YEAR PERIOD
AND 1-DAY RAINFALL.

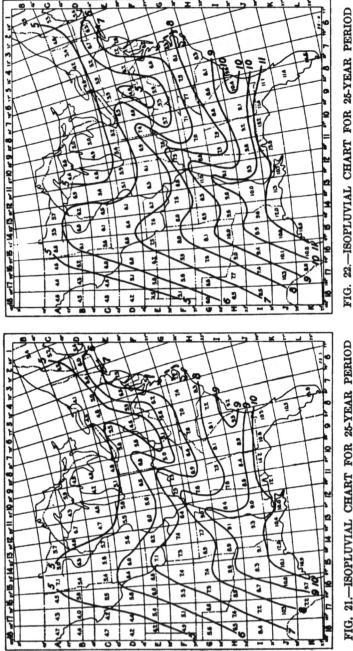

FIG. 22.—ISOPLUVIAL CHART FOR 25-YEAR PERIOD
AND 4-DAY RAINFALL.

FIG. 21.—ISOPLUVIAL CHART FOR 25-YEAR PERIOD
AND 3-DAY RAINFALL.

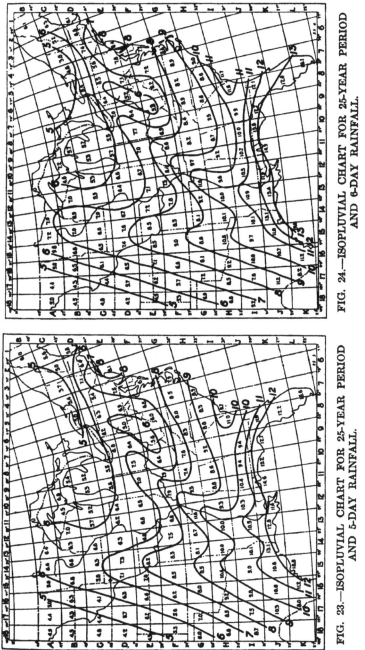

FIG. 24.—ISOPLUVIAL CHART FOR 25-YEAR PERIOD
AND 6-DAY RAINFALL.

FIG. 23.—ISOPLUVIAL CHART FOR 25-YEAR PERIOD
AND 5-DAY RAINFALL.

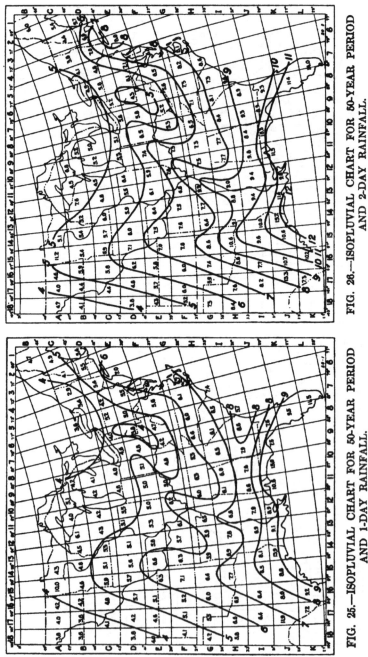

FIG. 26.—ISOPLUVIAL CHART FOR 50-YEAR PERIOD AND 2-DAY RAINFALL.

FIG. 25.—ISOPLUVIAL CHART FOR 50-YEAR PERIOD AND 1-DAY RAINFALL.

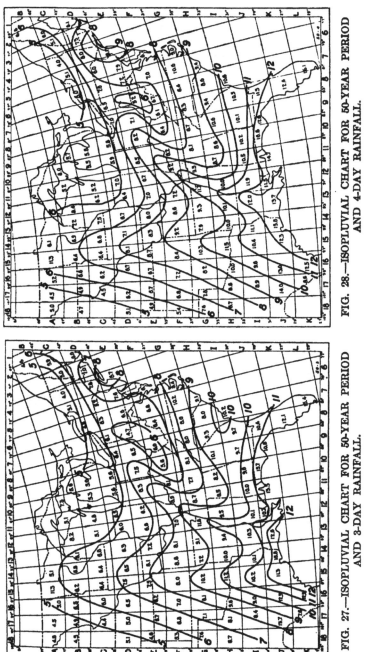

FIG. 28.—ISOPLUVIAL CHART FOR 50-YEAR PERIOD
AND 4-DAY RAINFALL.

FIG. 27.—ISOPLUVIAL CHART FOR 50-YEAR PERIOD
AND 3-DAY RAINFALL.

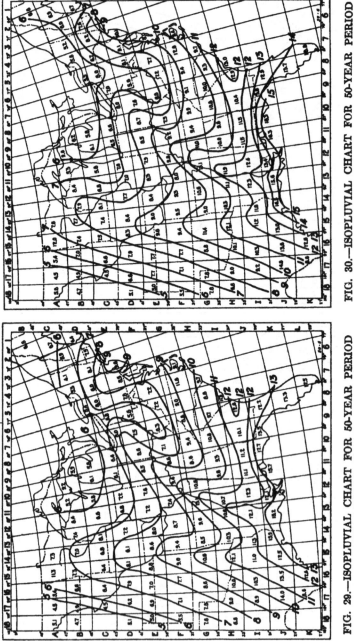

FIG. 30.—ISOPLUVIAL CHART FOR 50-YEAR PERIOD AND 6-DAY RAINFALL.

FIG. 29.—ISOPLUVIAL CHART FOR 50-YEAR PERIOD AND 5-DAY RAINFALL.

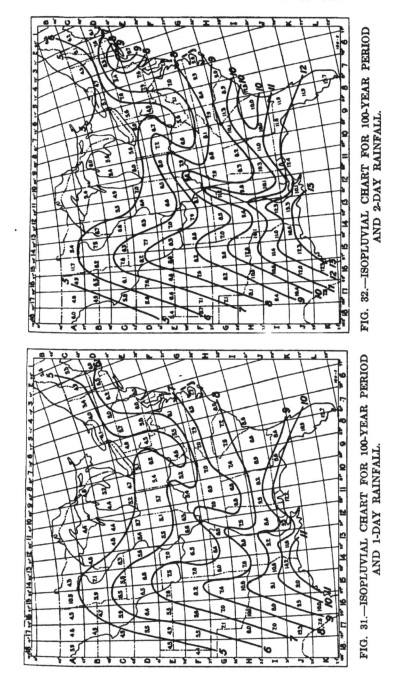

FIG. 32.—ISOPLUVIAL CHART FOR 100-YEAR PERIOD
AND 2-DAY RAINFALL.

FIG. 31.—ISOPLUVIAL CHART FOR 100-YEAR PERIOD
AND 1-DAY RAINFALL.

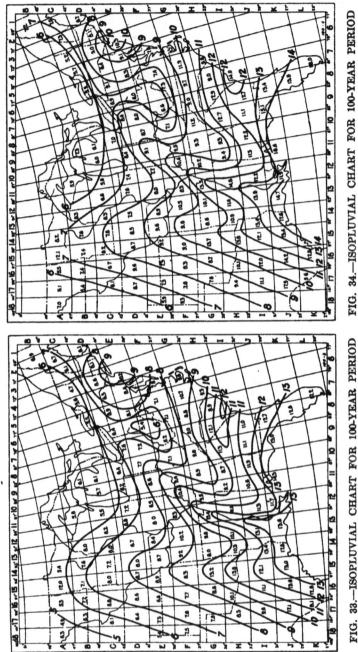

FIG. 34.—ISOPLUVIAL CHART FOR 100-YEAR PERIOD AND 4-DAY RAINFALL.

FIG. 33.—ISOPLUVIAL CHART FOR 100-YEAR PERIOD AND 3-DAY RAINFALL.

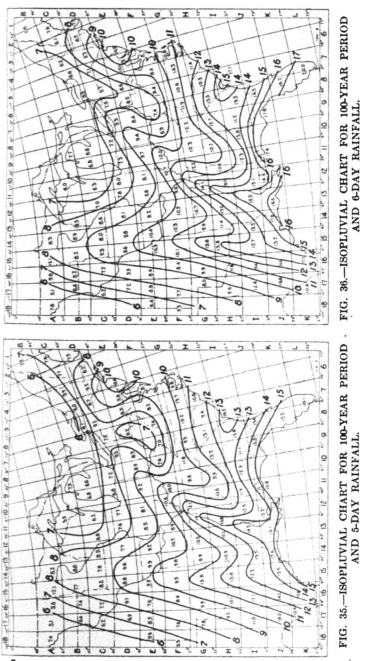

FIG. 36.—ISOPLUVIAL CHART FOR 100-YEAR PERIOD AND 6-DAY RAINFALL.

FIG. 35.—ISOPLUVIAL CHART FOR 100-YEAR PERIOD AND 5-DAY RAINFALL.

7

The data was taken from the excessive precipitation sheets, figure 6. Since many rainfall stations have been established in recent years, the records of which are quite short and would therefore, if included, throw too great weight on the comparatively short period represented by them, it was decided in the compilation of the 100-year and 50-year isopluvial charts to eliminate from consideration all records comprising less than 10 years of complete observations. The periods of record

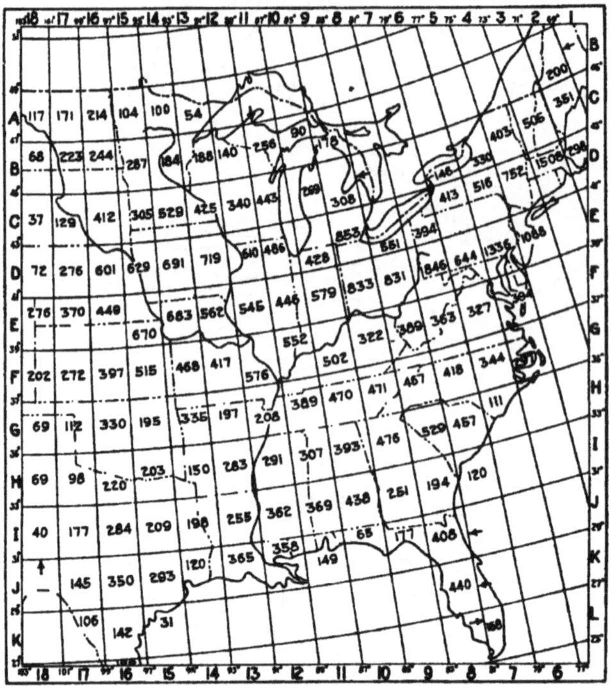

FIG. 37.—AGGREGATE YEARS OF RECORD IN EACH QUADRANGLE.

The figure in each quadrangle represents the sum of the years of record for all those stations in the quadrangle for which excessive precipitation records were copied.

considered are not always continuous, but often consist of two or more shorter periods aggregating or exceeding 10 years. Reference to figures 8 to 12, which show periods of record for rainfall stations in representative quadrangles, will serve to illustrate the necessity for adopting this rule. A similar rule was adopted in the compilation of the 25-year and 15-year isopluvial charts, by which station records less than 5 years in length were omitted.

The aggregate years of record in each quadrangle, at all stations for which excessive precipitation data was gathered, is shown as a number in the center of each quadrangle, in figure 37.

In compiling the data for the summary of excessive precipitation sheets, no rainfall depth was considered whose amount was less than 10 per cent of the mean annual rainfall for a 1-day period, or less than 15 per cent for a period of 2 or more days. In consequence, it happened in several instances that the quotient obtained in dividing the aggregate years of record by a small frequency period such as 15 years, proved to be greater than the total number of rainfall intensities appearing on the excessive precipitation sheets. In such cases the pluvial index was obtained graphically by plotting the previously computed 100, 50, and 25-year pluvial indices and projecting the curve to obtain the 15-year index.

A separate set of isopluvial charts was made up for each of the 4 frequency periods, namely, for 15, 25, 50, and 100 years respectively. There are 6 charts to each set, one each for 1 day, 2 days, 3 days, 4 days, 5 days, and 6 days of accumulated precipitation, making in all 24 charts, see figures 13 to 36.

In the center of each quadrangle was written its pluvial index. Thus in the 15-year, 3-day isopluvial chart shown in figure 15, the number written in each quadrangle denotes the 3-day rainfall intensity which may be expected to occur or be exceeded at any point within that quadrangle on an average of once in 15 years.

In order to bring out the variations in pluvial index for different sections of the United States, lines were drawn across the charts, connecting as nearly as possible all localities having the same pluvial index. These lines, which for convenience of reference have been termed *isopluvial lines*, are shown for successive 1-inch depths of rainfall. To illustrate their significance, the 6-inch line on the chart showing 15-year, 3-day isopluvial lines, figure 15, should be interpreted as indicating all localities in which a rainfall of 6 inches or heavier in 3 days may be expected to occur on an average once in 15 years.

The location and curvature of the isopluvial lines was made the subject of careful study. It was felt that their main function should be to indicate plainly those variations which, from an inspection of all of the 24 charts, appear to be well established and obvious. Great weight was given to the pluvial indices of those quadrangles which have the greatest lengths of records, and conversely little or no weight was attached to local irregularities or apparent inconsistencies in pluvial index traceable to paucity of rainfall data. The 15-year charts were given greater weight than the 100-year charts, as the former are based on nearly 7 times as many records as the latter. For these reasons the location of any one isopluvial line is not necessarily consistent with all pluvial index figures surrounding it. Moreover, some latitude in sketching in the lines resulted from the fact that a pluvial index is, by definition, representative of rainfall conditions at any one point within its quadrangle, the index figure being merely written in the center of the quadrangle for sake of convenience. It is believed that the lines as drawn represent the best possible interpretation which can be put on the data as at present available.

INTERPRETATION OF ISOPLUVIAL CHARTS

It is a matter of common knowledge that the highest rates of rainfall in the eastern United States are most frequent along the Gulf coast, and that heavy rains are frequent over the lower Mississippi Valley, but are rare in the more northern latitudes. These facts are brought out prominently by the isopluvial charts, which show a high pluvial index along the Gulf coast diminishing rapidly toward the interior, except in the Mississippi Valley up which the lines extend like the contours on a topographic map.

Perhaps the most striking feature which all of the charts possess in common is the marked decrease in pluvial index with increase of latitude. If the eastern United States could be imagined to be of uniform elevation, it is probable that the isopluvial lines would run nearly due east and west, evenly spaced, with a slight northeasterly deflection near the Atlantic Ocean caused by increased atmospheric moisture from that source, and conversely a southwesterly trend west of the 97th meridian caused by increasing distance from the Gulf. The reader should not infer from this that high rates of rainfall owe their moisture exclusively to evaporation of ocean waters. Much of it is supplied by evaporation from land surfaces. The isopluvial charts appear to indicate, however, that in the distribution of rainfall intensities in the eastern United States, latitude is the dominating factor.

Difference in altitude is responsible for much variation in pluvial index, a decided decrease being noticeable in mountainous regions. For instance the Appalachian system is responsible for a pronounced southerly deflection in the isopluvial lines. The rapidly increasing elevations of the Great Plains region west of the 97th meridian contribute materially toward decreasing the pluvial index in a westward direction. The influence of the Ozark Mountains is plainly indicated on all of the charts by the strong southward curvature of the lines. In striking contrast are the larger trough-like depressions, like the Mississippi and Ohio River valleys. The low area west of the Ozarks stands out almost as conspicuously, being accentuated by the rapid decrease in pluvial index to the west of it, and by the loops across the low flat lands drained by the Missouri and upper Mississippi Rivers, and the Red River of the north.

The same general tendency, though on a lesser scale, is discernible along the Atlantic coast at large estuaries and river valleys. The manner in which the data are compiled by 2-degree quadrangles does not indicate in sufficient detail the true curvature of the isopluvial lines for such localities, and no attempt has therefore been made to show these influences except in a generalized way. Along the western edge of the charts there is a general paucity of rainfall data, the effect of which is to cause much inconsistency in the pluvial indices of the quadrangles affected. In shaping the most probable courses of the isopluvial lines in this part of the map a wide range of judgment was allowed, and it is to be expected that much adjustment in the lines will be found necessary in the future.

It should not be inferred that excessive precipitation, as represented by the isopluvial charts, has the same normal geographical distribution as the total normal annual rainfall. Because a given region has a high annual rainfall, it does not necessarily follow that it has large and numerous excessive precipitation records. This is clearly shown by a comparison of the isopluvial charts, figures 13 to 36, and the map showing the normal annual rainfall of the United States, figure 5. The annual precipitation of the mountains of western North Carolina, for instance, reaches a maximum of 80 inches, as compared with 40 to 50 inches for the surrounding lowlands; the pluvial indices for this region show no such increase. In fact, they show but a very slight increase on the eastern slope, and a decided decrease on the western slope of the mountains. A similar condition prevails to a lesser extent in the Ozark Mountains of Arkansas and southern Missouri. The comparison just suggested also indicates a very much closer grouping and greater depth of excessive precipitation records along the Mississippi and Ohio Rivers than would be the case if they were proportioned in depth and distribution to the normal annual rainfall.

LIMITATIONS OF ISOPLUVIAL CHARTS

As may be inferred from the preceding remarks it is a foregone conclusion that many of the pluvial index figures shown on the charts will suffer modification as the rainfall records on which they are based grow in length and number. The extent of modification may, in many instances, be forecast with a fair degree of assurance. We may expect it to be small for those quadrangles which already contain many long records, as for example 3–D which has an aggregate period of record of 1508 years, and 5–E which has 1336 years. Slight modifications may also be looked for in quadrangles which, though having a less number of station records, possess a large percentage of long records. For instance, 6–H which has only 4 stations is fairly reliable because of the high average length of record which is nearly 28 years. On the other hand, the index for a quadrangle like 17–G, which comprises parts of Texas and Oklahoma, and has only 11 stations averaging a little over 10 years of record each, cannot be relied upon with any confidence. There are instances among the western quadrangles where rainfall records are so few and short that it became necessary in making up the charts to include many stations with records less than 10 years in length in order to obtain sufficient working data. The pluvial indices for such quadrangles are, therefore, subject to material ultimate correction and should be used with caution.

Aside from the variations in accuracy in different parts of any one chart, attributable to differences in quality and quantity of rainfall records, it is important to note that the individual charts differ from each other in degree of reliability. For it is evident that in the process of determining a 15-year pluvial index, for example, a much larger number of excessive precipitation observations is utilized, thereby assuring correctness of the result within much narrower limits, than is the case in determining a 50-year or 100-year pluvial index from the same data and for the same quadrangle. As the period of recurrence grows longer, the pluvial index becomes progressively less accurate.

Because of these limitations it was deemed advisable to generalize the isopluvial lines. This was done consistently, so that even those sections of the eastern United States where the abundance of rainfall data would have warranted greater detail are treated no differently from those sections which are not so well supplied. A uniform degree of generalization also was maintained on all of the 24 charts. It is believed that this course is justified, and that a greater refinement in the isopluvial lines should not be attempted until rainfall records become sufficiently abundant to enable the determination of pluvial indices for 1-degree quadrangles or even smaller units of earth's surface.

RECORDS OF MOST INTENSE RAINFALL

On the charts, figures 38 to 43, have been assembled the records of rainfall of maximum intensity, which appeared on the excessive precipitation sheets. As in the isopluvial maps, the data is given by quadrangles and for 1, 2, 3, 4, 5 and 6-day periods of rainfall. Thus, in figure 38, the number written in any quadrangle represents the maximum recorded 24-hour rainfall in that quadrangle. These charts, taken in conjunction with figures 13 to 36, show the range of

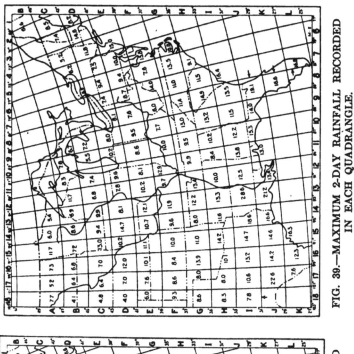

FIG. 39.—MAXIMUM 2-DAY RAINFALL RECORDED
IN EACH QUADRANGLE.

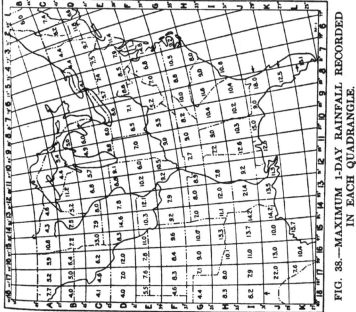

FIG. 38.—MAXIMUM 1-DAY RAINFALL RECORDED
IN EACH QUADRANGLE.

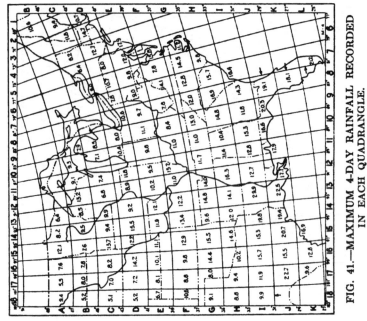

FIG. 41.—MAXIMUM 4-DAY RAINFALL RECORDED
IN EACH QUADRANGLE.

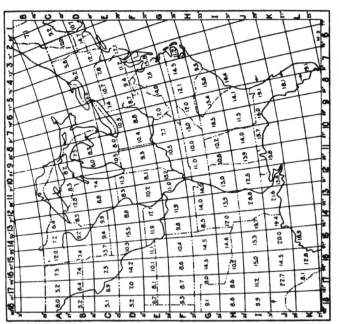

FIG. 40.—MAXIMUM 3-DAY RAINFALL RECORDED
IN EACH QUADRANGLE.

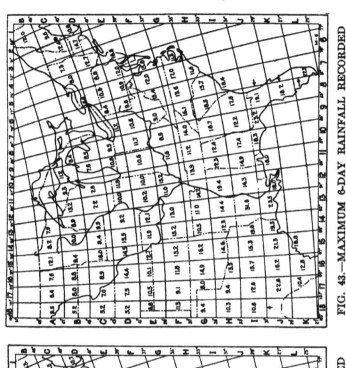

FIG. 43.—MAXIMUM 6-DAY RAINFALL RECORDED
IN EACH QUADRANGLE.

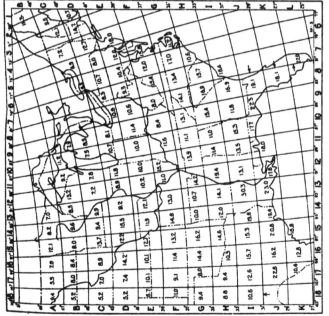

FIG. 42.—MAXIMUM 5-DAY RAINFALL RECORDED
IN EACH QUADRANGLE.

excessive precipitation records used in constructing the isopluvial charts.   •

As stated in chapter IV, the excessive precipitation sheets were made out only for rainfall stations having complete records covering not less than 5 consecutive years, and did not include records subsequent to 1914. Therefore, the data shown in figures 38 to 43 does not, in general, include intense rainfalls at stations having less than 5 years of records, nor those which occurred since December 31, 1914. Occasional exceptions were made as in the case of the great storm of Aug. 17–21, 1915, over Texas and Arkansas, which is responsible for the record of 18.6 inches in 24 hours in quadrangle 14–I, figure 39, and for the values in this quadrangle and some neighboring quadrangles on the succeeding charts. Unfortunately, the charts had been completed at the time of occurrence of the great July 1916 storm in North Carolina, and its record breaking precipitation does not, therefore, appear on them. The records used for platting the storm maps were also consulted, and as these included all station records, regardless of length of period covered, it follows that an occasional intense rainfall figure for a short station record has been used.

No attempt was made to draw isopluvial lines on the charts of most intense rainfall, because of the nature of the data. A careful consideration of the erratic nature of precipitation and the many difficulties which have attended the observing, recording and publishing of rainfall data, as outlined elsewhere in this report, leads to the conclusion that it will require many additional years of rainfall recording before even an approximation to dependable regularity will be discernible in maximum precipitation data when arranged on charts as in figures 38 to 43. It was largely because of this difficulty that the necessity arose for providing the isopluvial charts, which furnish information less liable to subsequent variation.

It is felt, nevertheless, that the charts showing intense rainfall furnish information of considerable value to the practising engineer when interpreted with due regard to the peculiar nature of the data, and when considered in conjunction with the isopluvial index data given in figures 13 to 36. The values given should be looked upon as indicative rather than finite; for they represent essentially temporary maxima, likely to be exceeded in the future. Their applicability is, of course, restricted to small areas.

PRACTICAL APPLICATION OF EXCESSIVE PRECIPITATION DATA

The isopluvial charts furnish a convenient means of determining the pluvial index for any locality within the eastern United States,

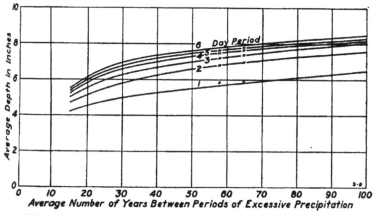

FIG. 44.—FREQUENCY OF EXCESSIVE PRECIPITATION IN
QUADRANGLE 3–D.

The depth shown corresponding to any frequency period is that which will
probably be equaled or exceeded once during that period.

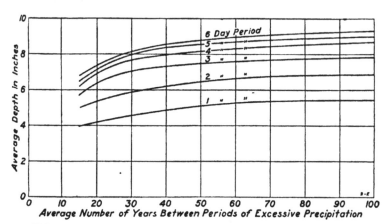

FIG. 45.—FREQUENCY OF EXCESSIVE PRECIPITATION IN
QUADRANGLE 9–E.

The depth shown corresponding to any frequency period is that which will
probably be equaled or exceeded once during that period.

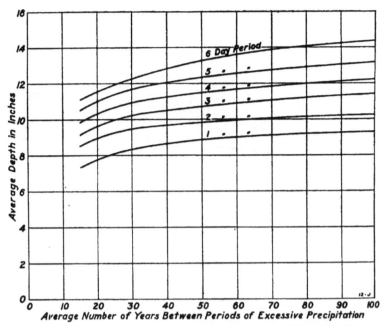

FIG. 46.—FREQUENCY OF EXCESSIVE PRECIPITATION IN
QUADRANGLE 12–J.

The depth shown corresponding to any frequency period is that which will
probably be equaled or exceeded once during that period.

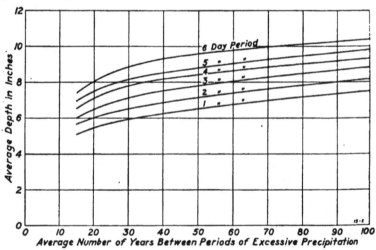

FIG. 47.—FREQUENCY OF EXCESSIVE PRECIPITATION IN
QUADRANGLE 15–E.

The depth shown corresponding to any frequency period is that which will
probably be equaled or exceeded once during that period.

108

by merely interpolating between the isopluvial lines.   In doing this it is necessary to bear in mind the limitations of the charts as above set forth.   Detached mountain spurs and other prominent local topographic features, which though important in themselves yet are so small as to be recognized with difficulty on charts of the size here published, had to be ignored in drawing the isopluvial lines, and allowance should be made for them accordingly.

In applying the pluvial indices to a particular area, it is perhaps desirable first to construct frequency curves, similar to the four shown in figures 44 to 47, for several of the surrounding quadrangles.   The method of constructing these curves from the pluvial indices is readily seen.   The six curves shown for quadrangle 3–D, figure 44, correspond to the 1 to 6-day maximum periods of precipitation.   The curve for the 1-day period is found by platting, to the proper frequency intervals, the pluvial indices 4.2, 4.8, 5.5, and 6.5, taken from the charts for 15, 25, 50, and 100-year frequencies, respectively.   The 2 to 6-day curves are found in a similar way.   This enables the investigator to obtain a clear idea of the additional depths which may be expected, above that for the maximum day, for consecutive periods of 2 to 6 days. These curves also show clearly the manner in which the depths increase with the average period of time between occurrences.

The four quadrangles for which curves are shown in figures 44 to 47, were selected to show the varying characteristics in different sections of the country.   The curves for quadrangle 3–D, figure 44, indicate that in this quadrangle from 65 to 75 per cent of the total rainfall occurs on the maximum day in a 1 to 6-day period of excessive precipitation.   In quadrangle 9–E, figure 45, these percentages are reduced to between 50 and 60.   Similar percentages are determinable for the other two sets of curves.   The curves for quadrangle 12–J, figure 46, indicate the dominance of the storms of long duration to which the Gulf coast is subject.   A significant feature of all the curves is their near approach to a horizontal position at a frequency of 100 years, indicating that the maximum depths of rainfall already recorded will probably not be greatly exceeded in the future.

The isopluvial charts may be used to advantage in the preparation of frequency curves for the design of sewer systems, bridge and culvert openings for small drainage basins, and in the design of dams, levees, and channel improvements where the watersheds involved are not more than a few square miles in area.   Since each pluvial index represents the excessive precipitation at only one station, it will be seen that the rainfall values are applicable to only small areas.   For the larger areas the reader is referred to the results of the time-area-depth investigations given in chapter VIII.

CHAPTER VI.—SELECTING AND SIZING THE 160 GREAT STORMS

The discussion of rainfall statistics in the last two chapters was limited to a consideration of excessive precipitation records at individual stations. This involved the two important rainfall factors of *depth* and *duration*, but ignored a third factor, *area*, which is equally important. The next logical step in the investigation, therefore, is to study a number of large storms as a whole, giving consideration to all three of the factors, time, area, and depth, which determine the size of storms.

This chapter is devoted to a description of the methods used in selecting and determining the relative sizes of 160 great storms. These will be used in subsequent chapters in a discussion of seasonal and geographical distribution, frequency, and cyclic variation of storms. Several of the largest and most important of these 160 storms are also later studied and discussed in much greater detail as to their time-area-depth relations, by means of maps and curves.

PERIOD COVERED

Before proceeding with the actual selection of storms it was necessary to determine what period of years the investigation should cover. It is, of course, highly desirable to have this period as long as possible, and still be sure that the rainfall records during the entire time are sufficiently numerous and well distributed to warrant deductions as to the size, frequency, seasonal distribution, and cyclic variation of the storms which occurred.

Prior to 1843 the records are so few and scattered as to be of negligible value for the objects here in view. From 1843 to 1872 there were still very few rainfall gaging stations, and consequently the chance was remote that the center of a storm area would occur near one of these. It was still more unlikely that any but the greatest storms would cause unusual rainfall records at two or more of these widely separated stations. Only the greatest storm of this period, that of October 3–4, 1869, over Connecticut, was therefore selected for further study and comparison with the greatest storms of recent years in the same region.

Although for the next nineteen years, 1873–1891, the number of rainfall gaging stations was considerably greater, the same handicap persists, that is, there were not sufficient records to be sure of including the smaller storms. For this reason, therefore, detail consideration was limited to the two greatest storms of that period: that of July 27–31, 1887, central over Georgia, and that of May 31–June 1, 1889, central over Pennsylvania.

On July 1, 1891, the Weather Bureau of the United States Department of Agriculture took over the climatological work which for twenty years had been conducted by the Signal Service of the War Department. A large number of additional observing stations were soon established, especially in those parts of the country where but few had existed prior to that time. We can feel sure, for the 25-year period 1892–1916, that not only have we records of all the storms that have occurred which come within the selected limits, but also that the data is sufficient to warrant a study of comparative sizes by means of their time-area-depth relations, and a study of their average frequency and seasonal distribution.

It is to this 25-year period that the greatest amount of study has been given. In the determination of storm frequency, and seasonal and geographical occurrence in the next chapter, no attempt is made to use the less complete data for the 50 years prior to 1892; and in discussing in detail in a later chapter the time, area, and depth storm factors, consideration is limited, for this 50-year period, to only the three great storms just mentioned. We do not believe we are justified, however, in ignoring altogether the data for those early years It has very decided value in supplementing and corroborating the storm experience of the past 25 years, and will be further discussed in a later chapter in connection with the time-area-depth relations of 30 of the most important storms of the past 25 years.

SELECTING THE STORMS

In selecting the more important from the large number of storms which were recorded during the 25-year period, 1892–1916, it was necessary to fix maximum limits of area and duration which were to be treated as being comprised in a single storm. Obviously, it was also necessary to fix minimum limits of depth and area in order to exclude the numerous storms of such small area and depth as to be of little or no consequence in an investigation of this nature. The fixing of these limits was done somewhat arbitrarily, the principal object being so to choose them as to include all storms which could possibly be of interest. The criterion adopted was that each storm selected should have not less than five station records having a 3-day precipitation equaling or exceeding 6 inches.

Of such storms, 160 were found which occurred during the years 1892–1916. A list of these, divided into two groups, northern and southern, is given in tables 4 and 5, giving for each storm the identification number, date, geographical location of principal center, as well as the highest, fifth highest, tenth highest, and twentieth highest 3-day precipitation records. The storms marked with asterisks are the largest and most important and of these a detail study as to their time-area-depth relations is made in a subsequent chapter.

The actual location of all the storms of which there are records, and which come within the limits just described, required a careful and extended search of several sources. Most of them were located by consulting the monthly rainfall chart and the notes on floods in the Monthly Weather Review.* On the chart appear isohyetals showing the precipitation over the entire United States for the current month. Notes in the text generally supplement the chart, and explain in some detail the nature and extent of unusual storms. Since these charts show the amount of rainfall for an entire month, it occasionally happens that although there is a large rainfall over an extensive area it is so distributed throughout the month that the maximum five consecutive days appear as a storm of little or no consequence. In such cases as this, the notes in the text on storms and floods sometimes indicate the character of the precipitation, but it was generally necessary to refer to the daily rainfall records of stations in the storm area to determine whether the precipitation shown by the chart was distributed over an extended period of time, or fell principally as a storm sufficiently intense to come within the limits chosen. Months for which the rainfall chart indicated that no storm of the defined intensity could have occurred were passed by without further search.

DETERMINING THE RELATIVE SIZES OF THE STORMS

It was early observed that the largest storms of the north and east never approach in depth and intensity the maximum Gulf coast and southern Atlantic seaboard storms. To establish an equitable basis for comparison of size all the storms were divided into two groups, the northern group and the southern group. The line of division was somewhat arbitrarily chosen along the north boundaries of North Carolina, Tennessee, Arkansas, and Oklahoma. Of the 160 storms, 47 are in the northern, and 113 in the southern group. This

---

* The Signal Service of the War Department first published the Monthly Weather Review in January 1873 and continued to issue it monthly until 1892, when the Weather Bureau of the U. S. Department of Agriculture was formed to take over the meteorological work of the Signal Service. Since 1892 the Weather Review has been the principal official periodical of the Weather Bureau.

Table 4.—Chronological List of 49 Great Northern Storms.

The greatest, fifth, tenth, and twentieth highest rainfall records in inches for the maximum period of 3 days in each storm are given. The asterisks (*) denote important storms of which the time-area-depth relations were studied in detail by means of maps and curves.

Storm	Date	Center	High Rainfall Records in Inches			
			Max.	5th	10th	20th
a*	1869 Oct. 3- 4	Conn.				
c*	1889 May 31–June 1	Pa.				
4	1893 May 15–17	Ohio	8.40	6.65	5.22	3.40
10*	1894 May 18–22	Pa.	9.20	8.66	7.84	7.10
12	1894 Sept. 18–20	N. J.	9.30	7.29	5.55	4.72
14	1895 Oct. 12–14	Mass.	8.49	7.65	7.33	6.72
15*	1895 Dec. 17–20	Mo.	11.90	9.25	7.09	6.15
18	1896 June 4- 6	Neb.	12.30	6.20	4.85	3.97
21	1896 Sept. 28–30	Va.	6.90	6.00	5.36	4.28
23	1897 Jan. 1- 3	Mo.	9.04	7.75	6.70	5.97
25*	1897 July 12–14	Conn.	10.30	8.59	6.87	5.57
26	1897 July 27–29	N. J.	9.09	6.08	5.06	3.65
31	1898 July 6- 8	Mo.	9.75	6.84	4.30	3.55
33	1898 Aug. 3- 5	Pa.	7.00	6.02	3.87	3.24
39	1898 Sept. 29–Oct. 1	Mo.	13.79	9.44	8.03	5.20
51*	1900 July 14–16	Iowa	13.70	8.20	6.39	4.06
52	1900 Sept. 9–11	Minn.	7.24	6.21	5.75	5.10
65	1902 June 27–29	Ill.	8.10	6.76	6.01	5.31
67	1902 Sept. 21–23	Kans.	8.07	6.61	5.34	4.38
68	1902 Sept. 24–26	Md.	8.80	6.19	5.05	4.30
72*	1903 Aug. 25–28	Iowa	15.46	10.18	8.14	5.80
73	1903 Sept. 12–14	Wis.	6.19	5.81	4.88	3.20
76*	1903 Oct. 8- 9	N. J.	15.00	10.66	9.78	8.30
77	1904 Mar. 24–26	Ill.	7.16	6.52	6.02	4.89
79	1904 Sept. 13–15	N. J.	9.00	7.10	6.47	5.80
83*	1905 June 9–10	Iowa	12.10	7.25	4.80	3.30
84	1905 July 19–21	Mo.	7.60	6.45	4.51	3.18
86*	1905 Sept. 15–19	Mo.	10.50	8.06	7.58	5.55
87	1905 Oct. 16–18	Mo.	7.97	6.47	5.24	3.73
91	1906 Sept. 16–18	Kans.	8.75	6.39	4.57	3.60
94	1907 Jan. 2- 4	Ark.	9.61	7.79	6.82	5.99
97	1907 July 14–16	Iowa	11.10	6.70	5.30	4.41
109*	1909 July 5- 7	Mo.	11.23	8.54	7.55	6.36
110*	1909 July 20–22	Wis.	12.77	8.50	6.45	4.31
111	1909 Nov. 12–14	Kans.	7.18	6.85	6.18	5.30
113	1910 Aug. 28–30	Neb.	8.52	6.50	4.85	2.69
114*	1910 Oct. 4- 6	Ill.	15.18	11.50	10.30	8.90
119	1911 Aug. 29–31	N. J.	7.67	6.65	6.28	5.66
125*	1912 July 20–24	Wis.	11.25	5.27	4.29	2.90
127	1912 Sept. 23–25	Md.	7.75	6.73	5.60	4.56
129	1913 Jan. 6- 8	Tenn.	9.48	7.36	6.45	5.55
130*	1913 Jan. 10–12	Ark.	7.39	6.70	5.74	5.15
132*	1913 Mar. 23–27	Ohio	10.23	8.98	8.75	8.17
142	1914 Sept. 7- 9	Mo.	7.83	6.13	5.14	3.74
143	1914 Sept. 13–15	Iowa	9.22	6.48	5.99	5.28
148	1915 May 26–28	Mo.	10.63	7.10	5.98	5.26
151*	1915 Aug. 17–20	Ark.	14.00	11.73	10.49	8.70
153	1915 Sept. 7- 9	Kans.	10.33	6.30	4.23	3.41
158	1916 Aug. 13–15	Ill.	9.88	6.49	5.15	3.20

Table 5.—Chronological List of 114 Great Southern Storms

The greatest, fifth, tenth, and twentieth highest rainfall records in inches for the maximum period of 3 days in each storm are given. The asterisks (*) denote important storms of which the time-area-depth relations were studied in detail by means of maps and curves.

Storm	Date	Center	High Rainfall Records in Inches			
			Max.	5th	10th	20th
b*	1887 July 27–31	Ga.	13.62	7.45	4.98	4.19
1	1892 Jan. 11–13	Ala.	10.62	7.79	7.10	4.82
2	1892 Apr. 5– 7	Ala.	9.60	6.62	5.10	3.60
3	1892 Apr. 20–22	La.	10.80	7.37	6.30	4.32
5	1893 May 26–28	Ark.	13.22	8.45	6.86	5.70
6	1893 Aug. 26–28	S. C.	13.15	8.32	6.75	4.87
7	1893 Sept. 6– 8	La.	9.49	8.13	7.26	6.08
8	1894 Mar. 18–20	Ark.	6.92	6.35	5.20	3.50
9	1894 Apr. 29–May 1	Tex.	10.25	6.96	6.15	4.97
11	1894 Aug. 4– 6	S. C.	12.60	11.07	9.84	5.42
13*	1894 Sept. 24–26	Fla.	7.52	7.15	6.16	5.22
16	1896 Jan. 31–Feb. 2	Miss.	8.60	7.21	5.42	3.72
17	1896 Apr. 12–14	Miss.	8.09	7.43	6.62	6.20
19	1896 July 6– 8	S. C.	8.30	6.18	5.00	4.15
20	1896 Sept. 19–21	La.	9.00	7.98	5.96	3.62
22	1896 Sept. 25–27	Tex.	11.89	9.84	6.45	4.08
24*	1897 Mar. 22–23	Ga.	7.80	7.11	5.97	4.50
27	1897 Aug. 17–19	La.	12.40	7.04	5.83	4.39
28	1897 Sept. 20–22	Fla.	9.11	7.83	6.03	4.33
29	1897 Dec. 2– 4	Miss.	8.53	7.03	5.52	4.45
30	1898 May 3– 5	Okla.	16.03	6.86	5.41	4.23
32	1898 July 11–13	Fla.	12.32	8.74	6.50	4.15
34	1898 Aug. 27–29	Ga.	10.46	7.48	6.30	4.95
35	1898 Sept. 1– 3	Ga.	9.30	7.02	5.75	4.39
36	1898 Sept. 10–12	La.	8.30	6.30	4.16	3.10
37	1898 Sept. 21–23	N. C.	13.79	9.44	8.03	5.20
38	1898 Sept. 29–Oct. 1	La.	13.02	8.00	6.25	5.10
40	1898 Oct. 2– 4	Ga.	9.17	7.59	6.62	4.74
41	1899 Jan. 4– 6	La.	10.25	7.51	6.75	5.12
42	1899 Mar. 13–15	Ala.	33.00	10.50	7.30	4.62
43*	1899 June 27–July 1	Tex.	19.90	7.70	4.91	3.00
44	1899 Oct. 2– 4	Fla.	8.83	7.00	5.25	4.41
45	1899 Dec. 9–11	Miss.	10.22	7.08	5.21	4.05
46	1900 Jan. 9–11	La.	12.25	7.92	6.41	5.50
47	1900 Feb. 10–12	Ga.	8.80	6.39	5.30	4.23
48	1900 Apr. 5– 7	Tex.	13.90	11.92	10.12	8.48
49*	1900 Apr. 15–17	Miss.	15.97	7.98	4.60	3.17
50	1900 July 13–15	Tex.	10.60	7.65	5.59	3.70
53	1900 Sept. 21–23	Tex.	8.00	6.50	5.50	4.43
54	1901 Jan. 10–12	Miss.	8.02	7.03	6.75	6.00
55	1901 May 20–22	N. C.	15.71	11.35	5.39	3.87
56	1901 June 11–13	Fla.	7.09	6.68	6.40	5.37
57	1901 Aug. 5– 7	N. C.	12.02	8.55	7.04	6.18
58	1901 Aug. 14–16	Ala.	11.44	8.23	7.15	6.03
59	1901 Sept. 16–18	Ga.	8.78	6.85	6.09	5.34
60	1901 Dec. 27–29	Ala.	9.50	7.13	6.20	5.00
61	1902 Feb. 26–28	Ga.	10.20	7.68	5.3	3.5
62	1902 Mar. 15–16	Ga.	10.78	9.52	8.50	7.30
63*	1902 Mar. 26–29	Miss.	10.11	6.52	5.27	3.17
64	1902 June 26–28	Tex.	10.24	7.68	6.91	5.56
66	1902 July 30–Aug. 1	Ark.	14.38	7.25	5.93	4.11
69	1902 Dec. 2– 4	Fla.	11.16	7.67	6.64	4.80
70	1903 May 12–14	Ala.	13.73	7.20	5.75	4.35
71	1903 July 1– 3	Tex.				

Table 5.—*Continued.*

Storm	Date		Center	High Rainfall Records in Inches			
				Max.	5th	10th	20th
74	1903	Sept. 12–14	Fla.	10.00	7.07	5.10	2.95
75	1903	Sept. 13–15	Ga.	12.20	9.00	7.30	5.57
78	1904	June 2– 4	Okla.	11.88	7.00	6.68	5.07
80	1904	Dec. 25–27	La.	10.05	8.92	8.25	6.35
81	1905	Feb. 11–13	Ga.	7.20	6.65	6.11	5.15
82	1905	Apr. 24–26	La.	9.03	6.10	5.07	3.95
85	1905	Aug. 3– 5	Fla.	12.23	6.19	4.66	3.16
88	1906	Mar. 18–20	Miss.	8.45	6.95	6.55	5.90
89	1906	May 22–24	Fla.	12.46	9.00	6.50	3.84
90	1906	Aug. 6– 8	Okla.	9.58	6.73	5.30	4.28
92	1906	Sept. 26–28	Ala.	12.45	9.96	7.26	6.52
93*	1906	Nov. 17–21	Ark.	13.00	10.25	8.90	6.75
95	1907	May 8–10	La.	13.66	8.91	6.74	4.73
96	1907	May 29–31	La.	10.51	8.10	7.35	5.85
98	1907	Sept. 27–29	Fla.	10.40	6.80	6.25	5.10
99	1907	Nov. 18–20	Ark.	7.75	6.65	6.49	5.36
100	1908	May 22–24	Okla.	9.03	8.50	8.28	7.19
101	1908	July 29–31	N. C.	10.73	7.60	4.83	3.07
102*	1908	July 28–Aug. 1	La.	17.62	8.14	6.62	4.11
103*	1908	Aug. 24–26	N. C.	15.58	12.83	9.62	8.30
104	1908	Sept. 24–26	Fla.	7.50	6.56	5.90	4.43
105*	1908	Oct. 20–24	Okla.	13.00	10.18	7.78	6.69
106	1909	May 25–27	Miss.	11.32	8.44	7.39	6.46
107	1909	June 1– 3	Miss.	9.62	6.86	5.77	4.96
108*	1909	June 29–July 3	Fla.	15.85	10.67	8.65	6.50
112	1910	June 12–14	Fla.	8.47	7.21	5.30	4.20
115	1910	Oct. 16–18	Fla.	10.20	9.06	7.78	5.11
116	1911	Apr. 13–15	Ark.	9.33	6.45	5.96	4.71
117	1911	Aug. 13–15	Ark.	10.61	7.17	5.63	4.62
118	1911	Aug. 29–31	Ga.	19.12	8.85	5.70	4.02
120	1911	Dec. 11–13	La.	8.50	6.63	6.00	4.88
121	1912	Mar. 14–16	S. C.	6.45	6.08	5.76	4.95
122	1912	Apr. 15–17	Miss.	8.86	7.02	6.20	5.26
123	1912	Apr. 20–22	Fla.	9.50	7.25	6.31	5.32
124	1912	June 8–10	Fla.	13.85	7.02	5.46	3.67
126	1912	Aug. 9–11	Tex.	13.25	6.19	3.78	2.71
128	1912	Dec. 3– 5	Miss.	7.80	6.78	5.92	5.00
131	1913	Mar. 13–15	Ala.	14.06	8.63	7.68	6.10
133	1913	Sept. 13–15	Ark.	11.43	7.54	6.90	6.28
134	1913	Sept. 25–27	La.	13.33	10.34	8.53	6.52
135*	1913	Oct. 1– 2	Tex.	14.47	10.44	8.51	5.42
136*	1913	Dec. 2– 5	Tex.	14.45	11.00	9.68	7.95
137	1914	Mar. 26–28	La.	12.87	9.00	7.00	6.08
138	1914	Apr. 26–28	Tex.	14.84	7.45	6.46	4.98
139	1914	May 28–30	Tex.	11.92	6.28	4.66	3.21
140	1914	July 13–15	La.	10.17	9.25	6.61	2.30
141	1914	Aug. 6– 8	Tex.	12.33	7.90	6.36	4.78
144*	1914	Oct. 14–15	N. C.	12.00	7.00	5.72	4.25
145	1914	Oct. 23–25	Tex.	15.00	9.10	6.40	4.60
146	1915	Apr. 22–24	Tex.	11.80	8.30	6.69	4.50
147	1915	May 6– 8	Fla.	8.62	6.03	5.47	4.62
149	1915	July 3– 5	La.	10.86	7.44	6.42	5.33
150	1915	Aug. 1– 3	Fla.	16.61	10.11	6.63	4.12
152	1915	Sept. 29–Oct. 1	Miss.	10.40	7.10	5.85	4.85
154	1916	May 1– 3	Tex.	7.90	6.95	5.98	4.95
155	1916	May 21–23	La.	10.05	8.47	6.65	5.30
156*	1916	July 6–10	Ala.	19.69	15.26	12.48	9.91
157*	1916	July 14–16	N. C.	23.68	16.77	14.70	10.70
159	1916	Oct. 16–18	La.	13.08	7.80	5.98	3.95
160	1916	Dec. 20–22	Fla.	7.89	6.01	5.44	4.20

grouping of the storms has certain conspicuous defects, as i
Iowa and Illinois storms in the same class with those of New E
although the sources and characteristics of the two types a
different. Similarly, in the southern group, the storms of Te
the Carolinas are grouped together though they differ mate
type. But aside from these defects the subdivision has prov
of material assistance not only in making an intelligent i
possible of the relative sizes of storms, but also by bringing ou
facts relating to their seasonal distribution.

The relative sizes of the storms were determined by i
graphically the 20 highest 3-day rainfall records in each stor
do this, each storm was given a separate ordinate on a sheet i
section paper, and on this ordinate were platted the highest, i
highest, the tenth highest, and the twentieth highest values
maximum 3-day period of the storm. The storms were the
as to size according to the fifth highest value. Figure 48, pl
this manner, shows the 47 great northern storms and the 1E
southern storms arranged in order of size. The upper an
extremities of the lines representing the storms show, respe
the highest and twentieth highest values; the upper and lowe
show, respectively, the fifth highest and tenth highest value
solid lines represent storms for which maps and time-area-depth
were drawn. The storms can be identified by means of the n
near the lower margin, which are the same as those in tables 4

The fifth highest value was chosen as a basis for compar
size of the storms because the average area represented by five i
is about equal to that of the largest reservoir watershed in the
Valley flood control project. On this basis of comparison two
variable factors, time and area, are kept approximately co
while the third variable, depth, determines the comparative
the storm.

The 3-day maximum period was chosen for several reasons.
intense storms last only 1, 2, or 3 days, and obviously it would
wise to compare the maximum records of such short storm
records representing the total precipitation of 5-day or 6-day
On the other hand, it would not be fair to compare the ma
1-day records of a 6-day storm with the maximum 1-day recor
1-day storm. The 3-day period is considered a fair average
purpose of comparing these types of storm in so far as it is p
to institute such comparisons. Another important reason for ch
the 3-day period is that a storm of this concentration places a ma
burden on the flood protection system that is planned for the
Valley.

As stated, the areas of all the storm centers are made approximately the same for purposes of comparison by using the records of an equal number of stations, namely five, at the center of each storm. This method is fairly reliable when applied to storms in the same section of the United States, but cannot be used so successfully in comparing the areas of storms in different parts of the country, as there are more stations in a given number of square miles in some sections than in others. Fortunately, for the purposes of the Miami Conservancy District, the rainfall gaging stations are much closer together in the northeastern part of the country than in the south and west. In the former section, the greatest storms are generally of much less depth and area than in the latter section. The distribution of rainfall stations tends, therefore, to equalize the number of storms chosen from the different sections.

The above method of comparing storm sizes is of little value when applied to two storms, one of which occurred prior to 1892 and the other subsequent to that time. In such cases the tendency always is to picture the earlier storm smaller than it really was, on account of the greater scarcity of observing stations. This introduces a twofold error. A greater area is represented by an equal number of stations in the earlier than in the later storm, and, since the stations are more widely separated in the earlier storm, there is less probability of the center of the storm having occurred at or near any station, that is, the higher rates of rainfall may not have been recorded at all. It follows from these considerations that a given number of rainfall records, such as the five here arbitrarily chosen, will for early storms necessarily give lower values than for the later storms, and that such values will become progressively smaller the earlier the storm selected. For this reason none of the storms prior to 1892 are shown on the charts in figure 48.

CHAPTER VII.—GEOGRAPHICAL LOCATION SEASONAL DISTRIBUTION, AND FREQUENCY OF THE 160 GREAT STORMS

The geographical and seasonal distribution of the 160 sto discussed in chapter VI are shown by the charts in figure 49 an the maps in figures 50 to 57. The geographical division of the sto into northern and southern groups, made in the preceding cha has been retained in the following discussion. In addition, the sto of each group have been divided, according to the time of the in which they occurred, into four seasonal groups each containin of the storms that occurred in a particular quarter of the year.

The division of the year into quarters was determined from charts shown in figure 49. These charts have, along their horiz axes, 12 spaces to represent the months of the year. Each sto platted as a vertical line, chronologically on the horizontal axis, ing, as in figure 48, the highest, the fifth highest, the tenth hi and the twentieth highest 3-day values. From a study of these it was decided to divide the year into quarters beginning, respect November 1, February 1, May 1, and August 1. These will be re to as the first, second, third, and fourth quarters.

The eight maps, figures 50 to 57, show the storms occurring northern and southern groups during the four quarters, there one map for each quarter for each group. The first four maps the storms in the northern group, and the next four show those southern group. Each storm is represented by its 6-inch isol for the maximum 3-day period of rainfall. It can be identifi merically by means of the table accompanying each map. This also gives the area in square miles enclosed by the 6-inch isoh and the average depth of precipitation in inches over that Where the storm extends beyond the coast line, the area and av depth given are for the land area enclosed by the 6-inch isohye

NORTHERN STORMS

There are 6 storms recorded in the first quarter of the year in northern group. As shown in figure 50, all of these occurred in lower Mississippi and Missouri River valleys. In the second qua only 2 are recorded. One of these, number 132, March 24–26, 1

caused the disastrous floods throughout the Ohio Valley. This storm, as will be seen by referring to figure 51, was very much greater and occurred farther north and east than number 77. The latter storm occurred in southern Illinois not far north of the mouth of the Ohio River. The fact that only two storms of material size have occurred in the northern part of the United States at this season during the past 25 years indicates that storm 132 was phenomenal in time of occurrence as well as in depth of precipitation and area covered.

During the third quarter the storms of this group are located principally in the Mississippi and Missouri River valleys. Two storms of about the average size for this quarter, numbers 10 and 25, and one comparatively small storm, number 26, are located on the Atlantic Coast in eastern Pennsylvania, New Jersey, and the New England states. The location and size of the storms of this quarter are shown in figure 52. One storm, number 4, is located on the south shore of Lake Erie. Most of the storms have comparatively small, kidney shaped areas, were of short duration and had intense rainfall, both of which factors are typical of thunderstorms.

It is during the fourth quarter, August to October, that the greatest storms occur in the northern group. These are massed along the Atlantic Coast and in the lower Mississippi and Missouri River valleys. One storm, number 114, October 4-6, 1910, extends from central Arkansas to central Ohio, as shown in figure 53.

Storm *a*, October 3-4, 1869, belongs to this quarter but is not shown on the map because it did not occur during the years 1892–1916. Its 6-inch isohyetal encloses a large part of central New England, a condition which tends further to establish the principal geographical location of great storms during this quarter of the year.

Although the smaller storms are similar in shape to the storms of the third quarter, the larger ones are shaped more like the storms of the first and second quarters.

SOUTHERN STORMS

Fifteen storms are recorded in the first quarter of the year in the southern group. Thirteen of these are in the lower Mississippi Valley and closely adjoining regions. Two small storms, numbers 69 and 160, are in northeastern Florida. The location and size of the storms in the southern group, occurring in this quarter of the year, are shown in figure 54.

In the second quarter 25 storms are recorded in the southern group. These are shown in figure 55. They are all in the interior of the south, as were all but two of those of the first quarter. This is significant as regards the origin of these storms. It will be recalled

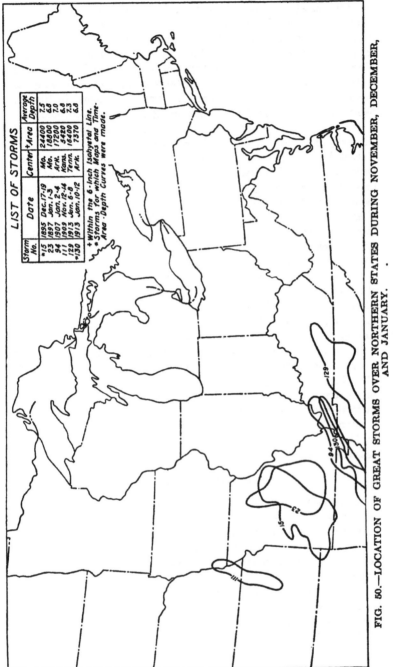

FIG. 50.—LOCATION OF GREAT STORMS OVER NORTHERN STATES DURING NOVEMBER, DECEMBER, AND JANUARY.

LIST OF STORMS

Storm No.	Date		Center	†Area	Average Depth
*15	1895	Dec. 17-19	Mo.	24400	7.5
23	1897	Jan. 1-3	Mo.	18800	6.8
94	1907	Jan. 2-4	Ark.	17200	7.0
111	1909	Nov. 12-14	Kans.	5420	6.6
129	1913	Jan. 6-8	Tenn.	16400	7.3
*130	1913	Jan. 10-12	Ark.	7370	6.8

† Within the 6-inch Isohyetal Line.
* Storms for which Maps and Time-Area-Depth Curves were made.

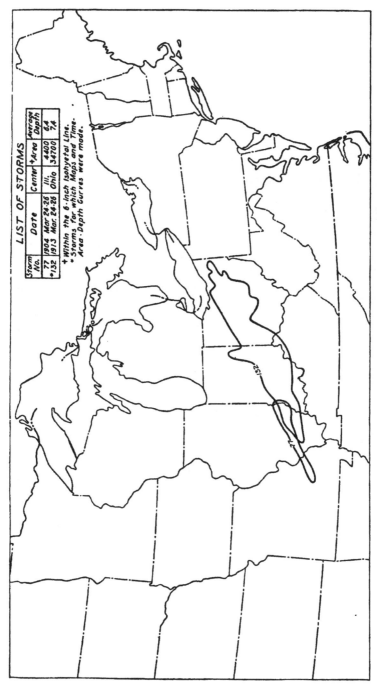

FIG. 51.—LOCATION OF GREAT STORMS OVER NORTHERN STATES DURING FEBRUARY, MARCH, AND APRIL.

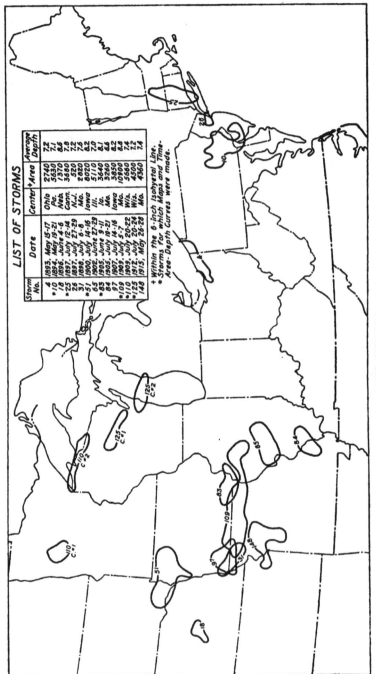

LIST OF STORMS

Storm No.	Date	Center	*Area	Average Depth
4	1893, May 15-17	Ohio	2740	7.2
*10	1894, May 19-21	Pa.	5630	7.1
18	1896, June 4-6	Neb.	1370	8.6
*25	1897, July 12-14	Conn.	3680	7.8
26	1897, July 27-29	N.J.	520	7.5
31	1898, July 6-8	Mo.	2620	7.6
*51	1900, July 14-16	Iowa	6070	8.2
65	1902, June 27-29	Ill.	5110	7.0
*83	1905, June 9-11	Ia.	3640	8.1
84	1905, July 10-21	Mo.	3260	6.6
97	1907, July 14-16	Iowa	3850	8.2
*109	1909, July 5-7	Mo.	10900	6.8
*110	1909, July 20-22	Wis.	5660	7.4
*125	1912, July 20-24	Wis.	4500	7.2
148	1915, May 26-28	Mo.	4360	7.4

+ Within the 6-inch Isohyetal Line.
* Storms for which Maps and Time-Area-Depth Curves were made.

FIG. 52.—LOCATION OF GREAT STORMS OVER NORTHERN STATES DURING MAY, JUNE, AND JULY.

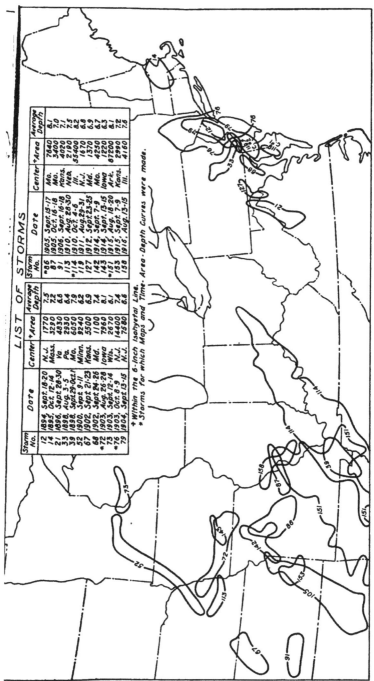

FIG. 53.—LOCATION OF GREAT STORMS OVER NORTHERN STATES DURING AUGUST, SEPTEMBER, AND OCTOBER.

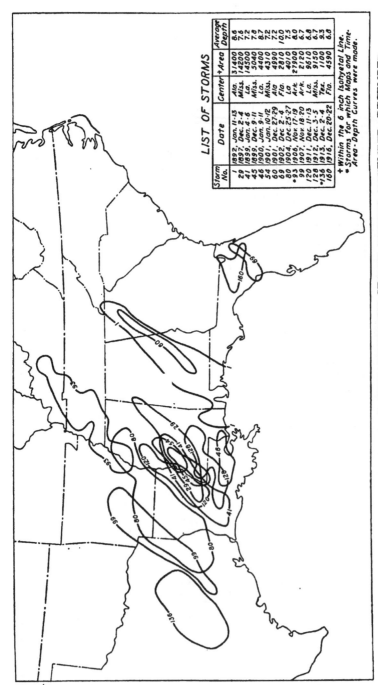

LIST OF STORMS

Storm No.	Date	Center	+Area	Average Depth
1	1892, Jan. 11-13	Ala.	31400	8.6
29	1897, Dec. 2-4	Miss.	14200	7.6
41	1899, Jan. 4-6	La.	14500	7.2
45	1899, Dec. 9-11	Miss.	5040	7.8
46	1900, Jan. 9-11	La.	4400	8.7
54	1901, Jan. 10-12	Miss.	4310	7.2
60	1901, Dec. 27-29	Ala.	4990	7.2
69	1902, Dec. 2-4	Fla.	2810	10.0
80	1904, Dec. 25-27	La.	4010	7.5
*93	1906, Nov. 17-19	Ark.	27900	8.0
99	1907, Nov. 18-20	Ark.	2120	6.7
120	1911, Dec. 11-13	La.	9610	6.8
*128	1912, Dec. 3-5	Miss.	9150	6.7
*136	1913, Dec. 2-4	Tex.	1100	9.3
160	1916, Dec. 20-22	Fla.	4590	6.8

+ Within the 6 inch Isohyetal Line.
* Storms for which Maps and Time-
   Area-Depth Curves were made.

FIG. 54.—LOCATION OF GREAT STORMS OVER SOUTHERN STATES DURING NOVEMBER, DECEMBER, AND JANUARY.

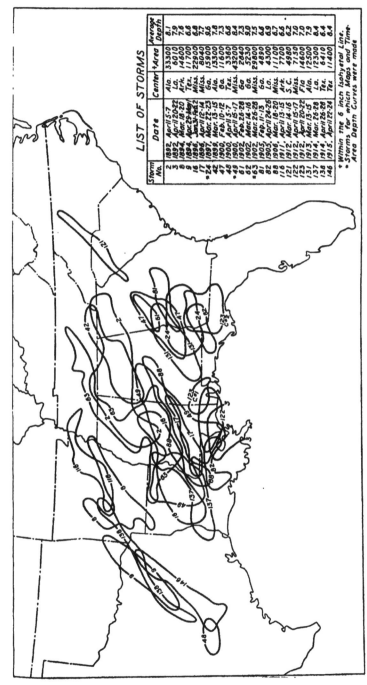

LIST OF STORMS

Storm No.	Date	Center	*Area	Average Depth
2	1892, April 5-7	Ala.	33300	8.1
3	1892, April 20-22	La.	6010	7.9
8	1894, Mar. 18-20	Ark.	14600	7.6
9	1894, Mar. 29-May 1	Tex.	11700	6.6
16	1896, Jan. 31-Feb. 2	Miss.	22900	6.9
17	1896, April 12-14	Miss.	8040	7.7
*24	1897, Mar. 22-23	Ga.	15900	9.6
42	1899, Mar. 13-15	Ala.	15300	7.8
47	1900, Feb. 10-12	Ga.	11600	7.3
48	1900, April 5-7	Tex.	3360	6.6
49	1900, April 15-17	Miss.	43200	8.4
61	1902, Feb. 26-28	Ga.	2640	7.3
62	1902, Mar. 14-16	Ga.	5230	9.0
*63	1902, Mar. 26-28	Miss.	29800	7.5
81	1905, Feb. 11-13	Ga.	4890	6.6
82	1905, April 24-26	La.	4300	6.9
88	1906, Mar. 18-20	Miss.	11100	6.7
116	1911, Mar. 14-16	Ark.	7700	6.6
121	1912, Mar. 14-16	S.C.	4980	6.2
122	1912, April 15-17	Miss.	7150	7.0
123	1912, April 20-22	Fla.	14600	7.9
137	1913, Mar. 13-15	Ala.	12500	7.3
137	1914, Mar. 26-28	La.	12300	8.4
138	1914, April 26-28	Tex.	6410	6.4
146	1915, April 22-24	Tex.	11400	6.4

† Within the 6 inch Isohyetal Line.
* Storms for which Maps and Time-
  Area Depth Curves were made

FIG. 55.—LOCATION OF GREAT STORMS OVER SOUTHERN STATES DURING FEBRUARY, MARCH, AND APRIL.

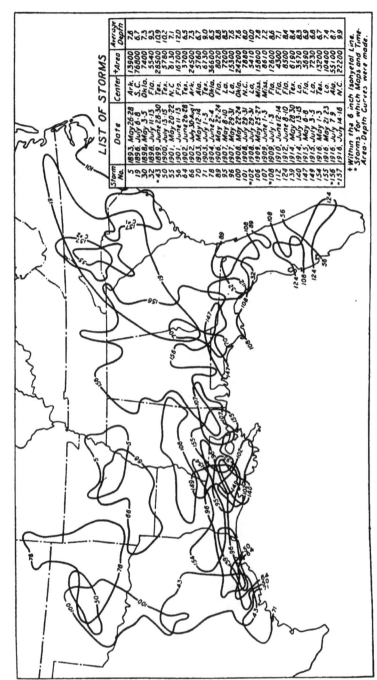

FIG. 56.—LOCATION OF GREAT STORMS OVER SOUTHERN STATES DURING MAY, JUNE, AND JULY.

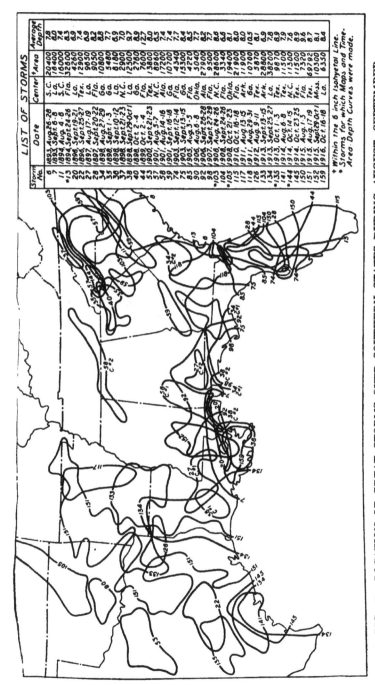

LIST OF STORMS

Storm No.	Date	Center	*Area	Average Depth
6	1893, Aug.26-28	S.C.	20400	7.8
7	1893, Sept.6-8	La.	16400	8.0
11	1894, Aug.4-8	S.C.	10000	7.4
*13	1894, Sept.24-26	Fla.	32500	8.3
20	1896, Sept.19-21	La.	4260	6.9
21	1896, Sept.25-27	Tex.	12900	7.4
27	1897, Aug.17-19	La.	9050	6.9
28	1897, Sept.20-22	Fla.	9050	8.2
34	1898, Aug.27-29	Ga.	10800	8.8
35	1898, Sept.1-3	Ga.	5480	7.7
36	1898, Sept.10-12	La.	8180	6.9
37	1898, Sept.21-23	N.C.	2900	7.0
38	1898, Sept.29-Oct.1	La.	12500	9.7
40	1898, Oct.2-4	Ga.	2760	8.9
44	1899, Oct.2-4	Fla.	13800	12.7
53	1900, Sept.21-23	Tex.	7600	8.0
57	1901, Aug.5-7	N.C.	8960	6.5
58	1901, Sept.16-18	Ala.	17200	7.4
59	1901, Sept.12-14	Ga.	10700	7.7
74	1903, Aug.3-3	Fla.	4340	8.4
75	1903, Sept.13-15	Ga.	13500	8.5
85	1905, Aug.3-3	Fla.	4220	7.7
90	1906, Aug.6-8	Ohio.	5040	8.2
92	1906, Sept.26-28	Ala.	27900	7.7
*98	1907, Sept.27-29	Fla.	9900	8.6
*104	1908, Aug.24-26	N.C.	28500	8.1
108	1908, Sept.21-23	Fla.	13400	8.0
117	1908, Oct.21-23	Ohio.	21900	8.0
115	1911, Aug.13-15	Ark.	21900	8.5
118	1911, Aug.29-31	Ga.	10700	10.3
126	1912, Aug.9-11	Tex.	3870	8.1
133	1913, Sept.13-5	Ark.	28800	6.9
134	1913, Sept.25-27	La.	38200	7.8
*135	1913, Oct.1-3	Tex.	9870	8.9
141	1914, Aug.6-8	Tex.	11500	7.9
*144	1914, Oct.14-15	Ga.	10700	7.6
145	1914, Oct.23-25	Tex.	11500	9.6
150	1915, Aug.17-19	Fla.	7320	8.9
151	1915, Sept.29-Oct.1	Tex.	69792	8.7
152	1915, Sept.29-Oct.1	Miss.	10300	8.1
159	1916, Oct.16-18	La.	6350	8.4

† Within the 6 inch Isohyetal Line.
* Storms for which Maps and Time-Area-Depth Curves were made.

FIG. 57.—LOCATION OF GREAT STORMS OVER SOUTHERN STATES DURING AUGUST, SEPTEMBER, AND OCTOBER.

that the low pressure areas which accompany them originate in the northwest, west, and southwest, and travel eastward. Their paths converge and focus in eastern Texas and Oklahoma and then traverse the country in a northeasterly direction in parallel lines.

In chapter III it was also pointed out that these low pressure areas travel much faster during the winter than during the summer months. These two characteristics of winter lows, namely, rapidity of movement, and parallel direction of travel, result in the long narrow shapes and parallel major axes, which are such striking characteristics of all these storm areas.

It should also be noted that all these storms lie on the ocean side of the paths of lows. This is not true of the storms which occur during the summer months, as may be seen by referring to maps for the third and fourth quarters. This is because the paths of summer lows are not as definite as those for winter lows, and they often remain in practically the same position for several days. A further reason is that many of the storms during the third and fourth quarters are due to thunderstorm conditions.

Another feature of the southern group of the first and second quarters is, that, while the storms do not lie very far inland, they are far enough so that the 6-inch isohyetal does not intersect the coast line. This is a peculiarity that is not true of the storms of the third and fourth quarters, and is probably due to the fact that precipitation is not caused until the warm moist winds reach the higher and much colder plateau a short distance inland. The effect of this inland storm area in the lower Mississippi Valley is plainly evident on the isopluvial charts, figures 13 to 36.

It is during the third quarter, May to July, that the greatest storms generally occur in the southern group. Two of the greatest storms of record in the United States, number 43, June 27–July 1, 1899, over Texas, and number 156, July 6–10, 1916, over Alabama and Georgia, occurred during this quarter. A total of 32 storms are included in this group. These are shown in figure 56, and it is very noticeable that they extend over the entire south from central Texas to the Atlantic Ocean. However, a slight concentration seems to exist along the Gulf Coast. The contrast between the southern storms of long duration, and the short, intense storms of the northern group, during this quarter, is especially remarkable.

Forty-one storms occurred in the southern group during the fourth quarter, August to October, almost a third more than for any other quarter. These are shown in figure 57. It will be noticed that they are scattered over the entire south, as was the case in the third quarter, and that they are slightly concentrated along the Gulf Coast, in Florida, and in North and South Carolina.

In the first half of the year, during the months of November to April, all storms, as here defined, occur in the Mississippi Valley from the Gulf of Mexico to Iowa, Illinois, Indiana, and Ohio. In the second half of the year, May to October, they occur principally along the Atlantic and Gulf coastal regions and in the central Mississippi Valley, the greater number of the interior storms occurring west of the Mississippi River. In both the north and south, the greatest storms occur during the summer months. In the north, they tend to occur most frequently in late summer; and in the south, most frequently in the early summer. Throughout the entire year storms are much more numerous in the south than they are in the north.

FREQUENCY

The frequency of storms, like the frequency of excessive precipitation discussed in a preceding chapter, is defined as the average number of years between occurrences.

For the purpose of discussing the frequency of the 160 storms of the 25-year period, 1892–1916, the division into northern and southern groups will be ignored, and the country considered as a whole. All storms will be looked upon as either winter or summer storms, the winter months being November to April, inclusive, and the summer months May to October, inclusive. This simplifies the discussion, and leaves to the reader the determination of the exact storm season and frequency to which any particular locality is subject. For this purpose the maps, figures 50 to 57, may be used as a guide.

The determination of the frequency of the 160 storms could be still further refined by using the total area subject to these storms, and the average area enclosed by the 6-inch isohyetal, to find the frequency over any given area. But the nature and extent of the data seem hardly to warrant such precision. If this result is desired, however, more confidence can be placed in its determination from the isopluvial charts, figures 13 to 36, described in chapter V.

The investigator should be cautioned not to try to restrict the area, season, or frequency of occurrence too closely. These factors can never be finally and accurately determined for a natural phenomenon. They may be much more definitely fixed after abundant data has been gathered for a long period of years, but until that time a comparatively large "factor of ignorance" must be applied to the results obtainable.

WINTER STORMS

It has been previously pointed out that winter storms are confined almost entirely to the Mississippi Valley. The greatest number

9

occur along the lower Mississippi in Louisiana, Arkansas, Mississippi, Alabama, and western Georgia. The number, frequency, and size of these storms gradually diminishes northward. During the 25-year period in question, 46 storms have occurred (ignoring the 2 small winter storms in Florida), or approximately 2 storms each winter season. Of this number, 38 were in the southern group, quite evenly distributed over western Gerogia, Alabama, Mississippi, Louisiana, Arkansas, and eastern Texas. This means, of course, that a winter storm of the defined intensity and size occurs on an average once every 3 years somewhere in the upper Mississippi Valley, and once or twice each year over some part of the southern states.

SUMMER STORMS

As has been previously shown, summer storms of the size under consideration occur almost exclusively along the Atlantic and Gulf coast, and in the interior region along the Mississippi River. A total of 112 summer storms, over the areas just mentioned, occurred during the 25-year period. This is an average of 4 or 5 summer storms each year somewhere in the United States east of the 103d meridian. Of this number, 12 storms, or an average of 1 storm every 2 years, occurred on the north Atlantic coast. Three of these were during the first half of the season, or an average frequency of 1 storm in about 8 years; and 9 of them occurred in the latter half of the season, giving an average frequency of approximately 1 storm in 3 years.

In the interior region along the Mississippi River there were 26 storms, or an average frequency of one storm each year. Of these, 11 occurred during the first half of the season and 15 during the last half. A previous statement has been made that most of these indicate thunderstorm conditions in shape and area. The effect of the intense local storms is seen on the isopluvial charts, figures 13 to 36, in the shape of the isopluvial lines, and in the great variation of index figures at the centers of adjoining quadrangles.

The remaining 73 summer storms occurred along the southern Atlantic and Gulf coasts, and in Arkansas, eastern Oklahoma, and Texas. This is an average frequency of 3 storms each year. During the first half of the summer season there were 32 storms, and in the latter half 41 storms. The area along the Gulf coast which is subject to these storms is also subject to winter storms. As was previously shown, however, winter storms lie somewhat farther inland than the summer storms. The summer storms of the south are apt to be both intense and long continued, as for example, the July 1916 storms, numbers 156 and 157.

From the foregoing discussion it is apparent that every section of the country is subject at different seasons to two distinct types of storms, namely, short violent storms, and storms of long duration in which the rain falls continuously at a fairly uniform rate. Between these two extremes there are, of course, all gradations. Along the north Atlantic coast, on the high plateau of western Oklahoma, Kansas, Nebraska, and Iowa, and in Wisconsin and Minnesota the short violent storm in which as much as 6 inches falls in 3 days at 5 different stations is the prevailing type. Although the long-continued type of storm does not appear on the maps of these sections, from an examination of the rainfall records it is evidently present, but not intense enough to come within the limits here chosen for selecting storms.

Storms of the short violent type occur in the west most often in midsummer; along the north Atlantic coast they occur most frequently in the late summer, during September and October. Along the Gulf coast the precipitation and area covered by such storms is so much greater than in the west and northeast that the type characteristics are somewhat obscured; and are not apparent from an examination of the maps alone. However, an examination of the daily precipitation records of these regions, for a period of a few years, shows that their principal time of occurrence is midwinter.

The storm type of long duration occurs most frequently in the opposite season, respectively, in each of these regions. These distinctive features are not confined to storm rainfall, but are true for practically the entire precipitation.

CHAPTER VIII.—DESCRIPTIONS AND TIME-AREA-DEPTH RELATIONS OF THE 33 MOST IMPORTANT STORMS

This chapter deals with the relations between area covered and average depth of rainfall in respective periods of 1, 2, 3, 4, and 5 days, for 33 of the largest and most important storms recorded in the United States east of the 103d meridian. A brief description is included for each of the 33 storms, and the method of selecting them from the 160 storms previously discussed is explained.

The chief usefulness of this study lies in its application to different sections of the eastern United States, in determining the most probable distribution and intensity of great storms which may occur in the future. Any such application involves the assumption that the records already available are of sufficient extent and duration to furnish a representative sample of weather conditions for some time to come. Whether the present records are sufficiently so representative may be a debatable question. When the records of another 25 years are available, a similar study of them undoubtedly will furnish results of a higher degree of assurance. In the meantime a searching examination of the methods and results here set forth will, it is felt, convince the most skeptical that a high degree of confidence may be reposed in the conclusions here reached.

OUTLINE OF PROCEDURE FOLLOWED

The work on each of the 33 storms consists of a number of distinct steps. First, the rainfall data pertaining to the storm is assembled. Second, there are determined the 1-day period of greatest average rainfall, the 2-day period of greatest average rainfall, and so on, until the whole duration of the storm is covered.

Third, on a large scale map of the United States, showing all rainfall observing stations, there is platted at each station the figures showing the amount of precipitation for the maximum 1-day of the storm. A similar map is prepared for the maximum 2-day precipitation, and likewise for each successive period until the whole duration of the storm, or until the maximum 5-day period, is covered. This gives for each storm of not more than five days duration a series of as many maps as the number of days included in the storm period. For storms of more than five days the mapping was confined to the five consecutive days of maximum rainfall.

132

Fourth, on each map, using the platted figures like elevations on a topographic map, lines of equal rainfall,—the so-called *isohyetals* or *rainfall contours,*—are carefully drawn. To cover all the days of the storms used for this study, 114 such maps were required. These maps have been combined, so far as seems possible without becoming confused with each other, and are reproduced on 36 charts of one-tenth the original scale, figures 58 to 93, accompanied by a discussion of their individual characteristics.

The fifth step in the study is to measure with a planimeter the area contained within each isohyetal shown on the storm rainfall maps. All these measurements are shown in the tables in the appendix. Sixth, the average depth of rainfall over the area within each isohyetal is calculated. The exact method of carrying out this calculation is explained in a later chapter, and all the results of the calculations are shown in the appendix.

Seventh, on coordinate paper a curve, here designated a *time-area-depth* curve, is platted using the results obtained from each map, platting as coordinates the area in square miles contained within each isohyetal and the average depth of precipitation in inches over such area. Thus, there are as many time-area-depth curves as there are storm precipitation maps. These curves, arranged in two groups according to the geographical location of the storms, and accompanied by discussion of certain storm features which there become apparent, are reproduced in figures 94 to 103.

To assemble the data, draw and measure a map, and derive the corresponding time-area-depth curve required on the average one man's time for about 2½ days. This should be multiplied about 114 times to obtain the results shown in this chapter.

DESCRIPTIONS OF INDIVIDUAL STORMS

In the following pages are given in condensed form, by means of charts and brief notes, the principal features of the more important storms. The rainfall for the successive periods of maximum precipitation is shown for each storm by means of small isohyetal maps reproduced from the maps used in determining the time-area-depth relations. In order to save space each chart shows isohyetal maps for several storms. For instance, the first map, figure 58, shows the maximum 1-day precipitation for three storms, namely, that of December 17–20, 1895, over portions of Oklahoma, Missouri, and Illinois; that of March 22–23, 1897, over portions of Alabama and Georgia; and that of April 15–17, 1900, over portions of Louisiana, Mississippi, and Alabama.

Because of the small scale it was impossible to reproduce all of the isohyetals and maintain clearness. Only the even numbered lines could be shown, and at some of the centers of excessive precipitation some of these had to be eliminated. Thus, the rainfall center in Texas, figure 64, had to be limited above 14 inches to the 20-inch, 25-inch, and 30-inch isohyetals, and a slight distortion of the lines was then

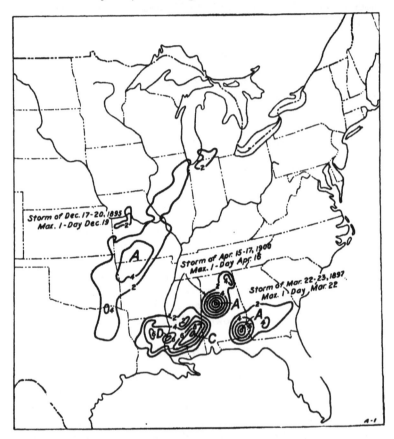

FIG. 58.—RAINFALL MAP FOR MAXIMUM DAY, STORMS OF
DECEMBER 1895, MARCH 1897, AND APRIL 1900.

required to bring them out with clearness. Where it was necessary to omit one or more even numbered isohyetals, the maximum rainfall at the peak is shown just beneath the peak letter. It will be seen from the above statement that low storm peaks, which are shown on the original large scale maps by only one contour, and this an odd numbered one, will not appear on the small scale reproductions.

Figure 59 shows the rainfall distribution during the maximum 2-day period for the storms shown in figure 58, and also for the storm of October 3–4, 1869, in the New England states. The nature of the data available for the latter storm did not permit of platting the maximum 1-day rainfall. It will be noted from an inspection of the

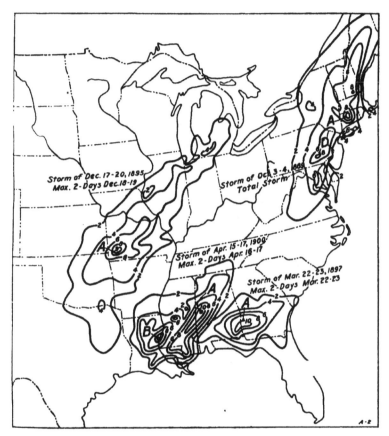

FIG. 59.—RAINFALL MAP FOR MAXIMUM 2 DAYS, STORMS OF OCTOBER 1869, DECEMBER 1895, MARCH 1897, AND APRIL 1900.

dates shown in figures 58 and 59 that the 2-day period of maximum rainfall includes, in addition to the day of maximum precipitation, either the day following or preceding it, depending upon which gives the greater sum. This is in accordance with the rules for selecting such periods.

Figures 60 and 61 show successively the rainfall distribution for the maximum 3-day and 4-day periods for such of the storms shown in figures 58 and 59 as had that many days of excessive precipitation. Figures 62 to 93 show in a similar manner the rest of the 33 great storms. On account of the manner of reproduction it was not found

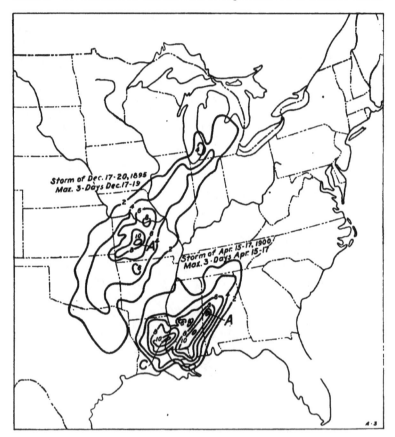

FIG. 60.—RAINFALL MAP FOR MAXIMUM 3 DAYS, STORMS OF
DECEMBER 1895, AND APRIL 1900.

convenient to place the maps in exact chronological order, although that is the general arrangement.

In a few cases the areas covered by adjoining storms overlapped slightly, without, however, causing confusion, the lapping isohyetals being indicated by broken lines. In nearly all cases it was found necessary to generalize the isohyetals and to eliminate some of the lesser rainfall centers, or peaks, because of the small scale of publication. All of the important features, however, have been carefully preserved.

The areas inclosed between isohyetals, and the average rainfall over these areas, will be found listed in the tables in the appendix.

FIG. 61.—RAINFALL MAP FOR MAXIMUM 4 DAYS, STORM OF DECEMBER 1895.

The principal rainfall centers can be identified readily in the tables and on the charts by means of the capital letters. Unless the reader is careful to compare the maps and time-area-depth tables for the same storm period, he may be confused by the fact that a principal storm center does not always bear the same capital letter for different storm periods. Thus a principal center may be lettered A on the

1-day map, and *B* on the 2-day map. It would have been awkward and difficult to maintain uniformity in this respect on account of the shifting, different rates of growth, and the appearance and disappearance of centers in the development of many storms.

Brief notes concerning the principal storms charted are given, describing points of special interest. No uniform treatment has been attempted since the data relating to different storms varies materially with the nature of the storm, the character of the country affected, and the extent of dependable records and reports. Notes pertaining to floods resulting from the storms have been added wherever the information seemed of sufficient interest. It will be noted that, of the 33 storms described, at least 14 are known to have caused maximum recorded floods on one or more rivers.

Storm *a*, October 3–4, 1869

This storm is of peculiar interest in that it is the largest on record in the New England states within the last 50 years, and also because it extended the farthest north of any of the great storms considered in this report. No less than 155 station records of this storm are available. Four of these are from Canada. The map in figure 59 shows the rainfall for the entire storm period which covered parts of two calendar days. No data was available for plotting a 24-hour rainfall map.

The greatest measured rainfall was at Canton, Conn., near Hartford, where the total was 12.35 inches. A number of reports indicate over 8 inches in less than 36 hours. At Springfield, Mass., 8.05 inches fell in 25 hours and 15 minutes; at Middletown, Conn., one gage reader reported 8.90 inches in a little over 37 hours. At the fort at Willet's Point, N. Y., 7.85 inches fell in 25 hours.

This storm resulted in severe floods throughout New England. The Connecticut River experienced the highest stage in a century caused by rainfall alone, and registered 26.3 feet at Hartford. This has been exceeded only by spring floods resulting from combinations of melting snows and rains.

The reader should compare this storm with that of July 12–14, 1897, figures 71 to 73, which occurred in the same region but did not cover so vast an area. The other New England storms for which data are available are of lesser magnitudes. Examples are that of August 7–9, 1874, which affected principally the Connecticut coast region, but was not felt to any great extent in the interior; and that of Feb. 10–14, 1886, over portions of Rhode Island and Connecticut.

The foregoing indicates that storms of great intensity and covering large areas are comparatively rare occurrences in the New England states.

A thorough discussion of this storm, accompanied by records of the precipitation at a large number of stations and a map showing the distribution of the rainfall, was prepared by· James B. Francis and published in the Transactions of the American Society of Civil Engineers, Volume 7, 1878, pp. 224–235. Much of the data here presented for this storm was taken from the article by Mr. Francis.

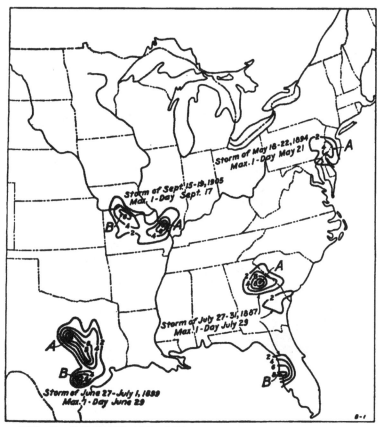

FIG. 62.—RAINFALL MAP FOR MAXIMUM DAY, STORMS OF
JULY 1887, MAY 1894, JUNE 1899, AND SEPTEMBER 1905.

Storm *b*, July 27–31, 1887

The West Indian hurricane which gave rise to this storm, progressed over the Caribbean Sea in a northwesterly course, until in the Gulf of Mexico it passed the 92d meridian. Its center then moved quite suddenly northeastward, reaching the Alabama coast on the

27th, after which its progress became slow and its course erratic. From Alabama it progressed eastward into Georgia, thence westward again across Alabama · into Mississippi where it dissipated. The rainfall which accompanied it was especially heavy in Georgia and

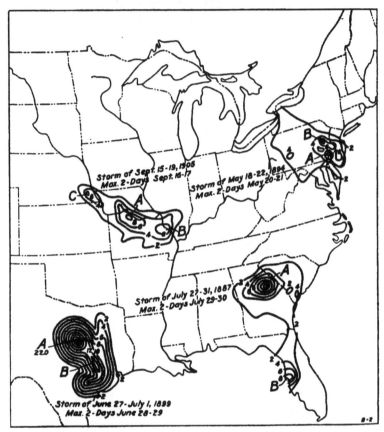

FIG. 63.—RAINFALL MAP FOR MAXIMUM 2 DAYS, STORMS OF JULY 1887, MAY 1894, JUNE 1899, AND SEPTEMBER 1905.

eastern Alabama, see figures 62 to 66. At Union Point, Ga., 16.50 inches fell July 28–30, being the maximum 3-day record for quadrangle 9–H, see figure 40. At Athens, Ga., 12.63 inches fell July 28–31, and at Washington, Ga., 12.41 inches fell during the same period. In Alabama 11.20 inches fell on the 27th and 28th at Opelika.

The rivers of Georgia rose to extremely high stages and inflicted great damage. The Savannah River registered 34.5 feet at Augusta on July 31, and inundated the city. This is within 3.3 feet of the record stage caused by the storm of August 24–26, 1908, figures 76 to 78, described later.

Storm c, May 31–June 1, 1889

This storm at first developed two distinct precipitation centers, figures 71 and 72. The principal one covered the central portions of

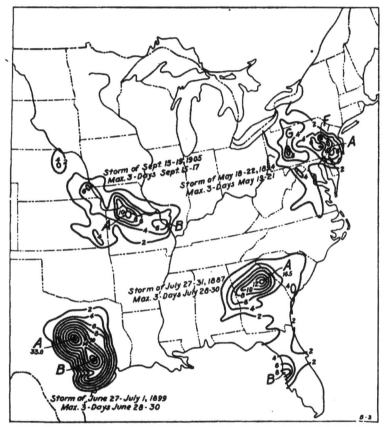

FIG. 64.—RAINFALL MAP FOR MAXIMUM 3 DAYS, STORMS OF JULY 1887, MAY 1894, JUNE 1899, AND SEPTEMBER 1905.

Pennsylvania and Maryland, and the smaller one was on the borderline between Virginia and West Virginia. The actual duration of rainfall did not exceed 36 hours at most points.

Though usually spoken of as the storm that caused the Johnstown disaster, it should not be inferred that its effect was felt to any great extent in the basin of the Allegheny River. With the exception of the Kiskiminetas River, on the headwaters of which Johnstown is located, none of the streams west of the Alleghenies experienced unusual floods. At Pittsburgh the Ohio River exceeded ordinary flood stage by only 2 feet. The storm was more severe east of the mountains,

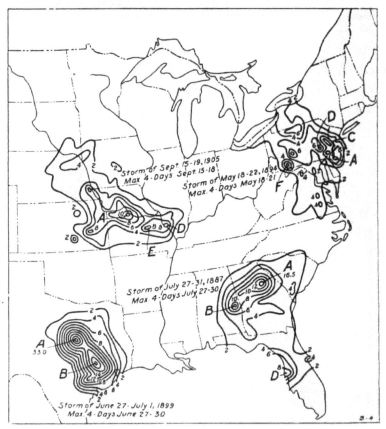

FIG. 65.—RAINFALL MAP FOR MAXIMUM 4 DAYS, STORMS OF
JULY 1887, MAY 1894, JUNE 1899, AND SEPTEMBER 1905.

over the drainage basin of the Susquehanna and Potomac Rivers, causing the greatest flood flow in more than a century on these two streams, and also on the West Branch of the Susquehanna River, and on the Juniata River.

The reader should compare this storm with the one of May 18–22, 1894, figures 62 to 66, which approaches it in intensity though not covering the same area, but which caused the second greatest flood in the main Susquehanna River.

Storm 10, May 18–22, 1894

Heavy rains fell over eastern Pennsylvania and New Jersey on May 19 and 20, followed by still heavier precipitation on the 21st and

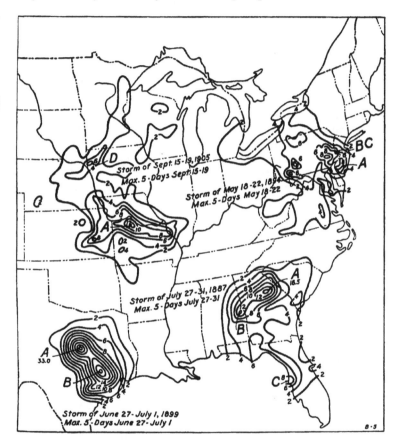

FIG. 66.—RAINFALL MAP FOR MAXIMUM 5 DAYS, STORMS OF JULY 1887, MAY 1894, JUNE 1899, AND SEPTEMBER 1905.

22d, see figures 62 to 66. On the latter two days, centers of heavy rainfall formed in southwestern and in northern Pennsylvania. The area covered by lighter rains during this storm was quite extensive, including parts of Tennessee, Ohio, New York, and New England.

At more than 11 localities in Pennsylvania and New Jersey over 5 inches fell in one day. During the period May 18–22 there fell at Quakertown, Pa., 9.36 inches; at West Chester, Pa., 9.87 inches; at Moorestown, N. J., 9.39 inches. At West Chester 9.03 inches fell on the 20th and 21st.

This storm was responsible for high river stages. The Susquehanna River at Harrisburg, Pa. registered 25.6 feet on May 22, approaching the record stage caused by the storm of 1889.

Storm 13, September 24–26, 1894

This storm was caused by a tropical hurricane the center of which moved in a northwesterly direction over Haiti and Cuba, then changed to a northerly course passing over the Florida peninsula and up the Atlantic coast. It was accompanied all along its path by high rates of precipitation. The storm area covers the Florida peninsula and the coastal regions of Georgia and South Carolina, figures 67 to 69, and probably extended over the ocean for many miles. At 12 stations in Florida the rain exceeded 9 inches during the two days of maximum rain. The greatest depth was at Clermont, Fla., where 12.50 inches was recorded on the 25th and 26th, and a total of 12.76 for the storm. At Federal Point 11.25 inches fell on the 25th and 26th; at Gainesville, Fla., 11.81 inches fell on the same dates; at Grassmere, Fla., 10.89 inches fell on the 25th and 26th; at Hypoluxo, Fla., 10.40 inches fell on the 24th, and 11.98 inches on the 24th and 25th. During the period Sept. 24–26 there fell at Kissimmee, Fla., 11.95 inches; at New Smyrna, Fla., 11.07 inches; and at Jacksonville, Fla., 11.05 inches.

Storm 15, December 17–20, 1895

This storm was remarkable for the evenness of its rainfall distribution, see figures 58 to 61. The precipitation covered an enormous scope of country, extending from northeastern Texas to southern Michigan. The heaviest precipitation occurred over Missouri where the following records were obtained. At Phillipsburg 6.95 inches fell between 5 p. m. of the 17th and 5 p. m. of the 18th, the total fall amounting to 12.20 inches. At Sarcoxie 9.03 inches fell on the 18th and 19th. Over large parts of Lawrence and Callaway counties the rain exceeded 9 inches in depth.

From 5 to 6 inches of rain fell in less than two days over the entire drainage area of the Illinois River, causing a sudden rise but no unusual flood stage. This is of interest inasmuch as some of the great floods in the Illinois River, notably those of May 1892, April 1904, and March 1913, all of which exceeded by far that of December 1895, were caused by less rainfall. The condition of the soil was the governing factor. In 1895 the rain was accompanied by a thaw which took the frost out of the ground and permitted considerable absorption to take place. The 15 months preceding this storm were the driest months of a long period of drouth, which is said to have been the severest in the history of Illinois. The other storms referred to all fell on partly saturated ground, and in some instances found the rivers already swollen as the result of previous rains.

The storm of December, 1895, was responsible for disastrous floods in Missouri. The Osage River attained the highest stages since the memorable flood of 1844. The Sack, Spring, Gasconnade, Meramec, and Cuivre Rivers, and many lesser streams in Missouri also experienced serious floods and inflicted great damage.

Storm 24, March 22–23, 1897

Heavy and general rains prevailed during March, 1897, throughout the south Atlantic states, causing high rates of runoff in all the streams of that section. In some localities rain began on the 2d, and with slight intermission continued until the 23d. The rainfall for the two days, March 22 and 23, was especially heavy and concentrated, see figures 58 and 59. At Newton, Ala., 10.29 inches fell on the 22d, being the greatest 24-hour rainfall on record in quadrangle 10–I, see figure 38; at Fort Gaines and Morgan, Ga., respectively, 11.26 and 11.50 inches fell on the 22d and 23d. The latter is the maximum on record for 2-day rainfall in quadrangle 9–I, figure 39.

As a result of these rains the Chattahoochee River at Eufaula, Ala., on the 24th reached a stage of 50 feet, the highest since March 28, 1888, when 56 feet was recorded. The Alabama River reached 41.5 feet at Selma, Ala., on the 26th.

Storm 25, July 12–14, 1897

This was a New England storm of comparatively short duration, most of the rain falling in from 30 to 36 hours. Its greatest intensities were recorded in western Connecticut and Massachusetts, though lesser rains covered most of the other New England states, see figures 71 to 73. The largest amount of rainfall was recorded, according to Weather Bureau data, at Southington, Conn., where 10.30 inches fell in 33 hours on the 13th and 14th. Three other stations in Connecticut recorded over 9 inches in 3 days. The Connecticut River at Hartford reached 20.8 feet on the 16th, a high though not extraordinary stage. There is some analogy between this storm and that of October 3 and 4, 1869, figure 59.

Storm 43, June 27–July 1, 1899

This storm, according to an account in Monthly Weather Review of July, 1899, appears to have resulted from a semi-tropical hurricane which moved northward from the central portion of the Gulf of Mexico. It caused a high tide at Galveston. The storm passed

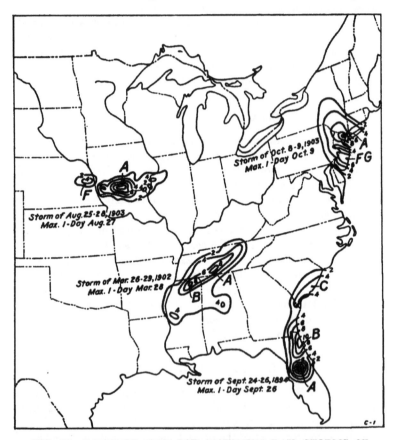

FIG. 67.—RAINFALL MAP FOR MAXIMUM DAY, STORMS OF SEPTEMBER 1894, MARCH 1902, AUGUST 1903, AND OCTOBER 1903.

inland during the night of June 26, its energy greatly diminishing as it progressed. Practically all of the rainfall was confined to the state of Texas, see figures 62 to 66, very heavy rains occurring over the drainage basin of the Brazos River.

During the 72 hours ending 8 a. m., June 30, a depth of rainfall was recorded at Brenham of 19.99 inches, which is the maximum 3-day rainfall record for quadrangle 15–J, see figure 40. During the same period the rainfall at Cuero was 12.86 inches and at Hewitt, 14.95 inches. On the 28th the rain gage at Hearne overflowed at 24 inches, the observer estimating a fall of from 30 to 40 inches during a period

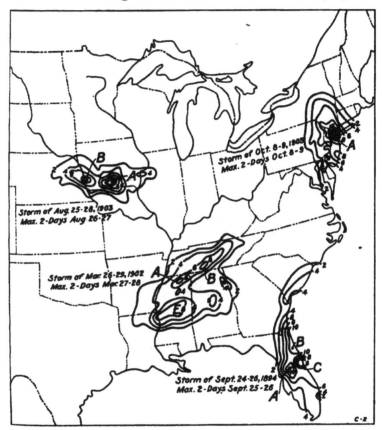

FIG. 68.—RAINFALL MAP FOR MAXIMUM 2 DAYS, STORMS OF SEPTEMBER 1894, MARCH 1902, AUGUST 1903, AND OCTOBER 1903.

of less than 24 hours. At Hallettsville 12.00 inches fell on June 27–29 and at Sugarland 16.00 inches fell on June 27–30.

Precipitation of 33 inches for the maximum period of 3 days was unofficially recorded at Turnersville. Upon inquiry as to the reliability of this record, P. C. Day, Climatologist and Chief of Division of the U. S. Weather Bureau, wrote:

With regard to Turnersville I will state that the Weather Bureau had no observer at that place during the period in question; however, a station was opened there a few months later, and the gentleman who became observer, Mr. James J. M. Smith, reported by special letter as follows: On the morning of June 29, 1899, he measured 11 inches of water in his gage and states that it had run over during the night, amount of loss not known. On the morning of the 30th he again measured 11 inches in his gage, and says that it had again run over, amount of overflow unknown. On the

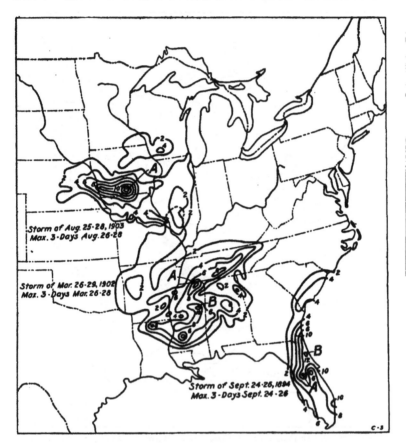

FIG. 69.—RAINFALL MAP FOR MAXIMUM 3 DAYS, STORMS OF
SEPTEMBER 1894, MARCH 1902, AND AUGUST 1903.

morning of July 1 he again measured 11 inches, making a total fall during 90 hours of 33 inches, measured. He advises also that these amounts correspond with those made in various receptacles by others in his vicinity, one case being reported where a tank 12 inches deep and covered with slats 3 inches wide, with openings of ½ inch between slats, was found full of water which had fallen during the 24 hours from 7 a. m. of the 28th of June to the same hour on the 29th.

This storm caused the most extensive floods ever known in the Brazos River, though previous highwater marks were exceeded only in places. The valley of the Brazos was flooded to a great width, varying from 2 miles in McLinnan County to 25 miles near the mouth. The crest of the flood was 17 days in passing from Waco to the mouth,

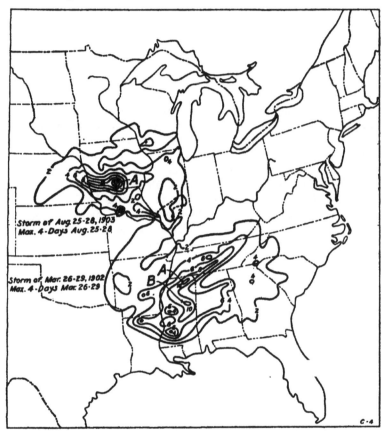

FIG. 70.—RAINFALL MAP FOR MAXIMUM 4 DAYS, STORMS OF MARCH 1902 AND AUGUST 1903.

a distance of about 285 miles. The damage was estimated at over $8,000,000, and 24 lives were lost.

Storm 49, April 15–17, 1900

The area of low barometric pressure which accompanied this storm was central over northern Oklahoma, and moved slowly to the

Lake region, causing extraordinary rains in Louisiana, southern Mississippi, and western Alabama, see figures 58 to 60. In Mississippi the rainfall was in places greater than ever before recorded.

Severe and disastrous floods occurred on the rivers of western Alabama and southern Mississippi. The Alabama, Tombigbee, and Black Warrior Rivers reached extremely high stages. On April 18 the Black Warrior rose to 71.1 feet at Tuscaloosa, Ala., the highest recorded stage at this place. The Pearl and Black Rivers were very high and caused extensive damage and the loss of several lives.

Some of the 24-hour amounts recorded in Mississippi are: Bay St. Louis, 8.77 inches on the 17th; Fayette, 9.25 inches on the 16th and 12.50 on the 15th and 16th; Natchez, 6.75 inches on the 17th and 12.75 on the 16th and 17th; Port Gibson, 7.64 inches on the 17th; Meridian, 6.85 inches. At the latter place 10.57 inches of rain fell from 12:50 p. m., April 15, to 8:10 a. m., April 17, a period of 43 hours and 20 minutes. At Windham 10.80 inches fell on the 16th; at Demopolis, Ala., 9.00 inches fell on the 17th; and at Eutaw, Ala., 12.65 fell on the 16th, with a total of 13.90 on April 15–17. The rainfall at the latter station furnished the records for 1-day, 2-day, and 3-day precipitation in quadrangle 11–I, see figures 38 to 40. Other high records in Ala. were 8.25 inches on the 16th at Greensboro; 11.80 inches on the 17th at Livingston; and 8.75 inches on the 16th at Pushmataha.

Storm 51, July 14–16, 1900

The principal center of excessive precipitation in this storm was in the northwest corner of Iowa, near Primghar, where a total of 13.70 inches fell, of which 13.00 inches fell on the 14th and 15th. This is the record for 2-day precipitation in quadrangle 15–C, see figure 39.

Another less important rainfall center formed in southeastern South Dakota, where at Yankton, 8.07 inches fell on the 14th and 15th, with a total of 9.62 inches for the entire storm period, see figures 71 to 73. Fairly heavy rains fell on the 16th in southeastern Minnesota and portions of Wisconsin. On the whole, the storm is conspicuous among those here listed because of the comparatively small area covered by heavy rainfall.

Storm 63, March 26–29, 1902

As will be seen by reference to figures 67 to 70, this storm had three distinct areas of heavy precipitation, situated approximately in as many states. The largest of these was over Mississippi, the next largest over Tennessee and northern Mississippi, and the smallest was over the center of Alabama. Compared with other storms described in these pages, the amounts of rain which fell during this storm are large but not extraordinary. Rates of runoff, however, were high throughout the area affected because of the previously wet condition of the soil. The Duck River, a tributary of the Tennessee River, experienced the highest stages in half a century, reaching 45.6 feet on the gage at Columbia, Tenn., on March 30. The Cumberland River rose to 65.0 feet at Burnside, Ky., the highest stage on record. The Tennessee River and the rivers of Alabama attained high stages. In Mississippi the Pearl River exceeded all previous records on March 31 with a stage of 36.0 feet at Jackson, and 29.0 feet at Edinburg. At Burnside, Miss. where the highest record is 62 feet, March 31, 1886, and the danger stage 50 feet, the stage on March 30, 1902, was 58.9 feet.

Storm of May 16–31, 1903

This storm is mentioned here as an example of protracted rains, over vast areas, which are capable of bringing about extraordinary floods, without, however, possessing those characteristics which bring them within the definition of great storms as adopted in this report. This storm covered Kansas, Oklahoma, Arkansas, and portions of Nebraska, Iowa, and Missouri. Rain fell for a period of 17 days with occasional interruptions. The precipitation from May 16 to 21 was comparatively light, the heavier rains occurring between the 22d and 31st of May. The heaviest rainfall was reported at Concordia, Kansas, where 3.68 inches fell on the 29th, and 10.59 inches, during the period May 16–31. At several other points amounts in excess of 8 inches for the entire period were recorded. Over the balance of the 200,000 square miles affected, the rain averaged between 3 and 6 inches in 15 days.

The first half of May had also been very wet, and had caused the streams to rise above normal stages. This circumstance had an important bearing on the extraordinary flood conditions which followed the rains during the second half of May. About the 28th all streams in Missouri, Iowa, Nebraska, Kansas, and Oklahoma were out of their banks. The Kansas River reached the highest stages in its history on May 31, with a gage reading of 29.50 feet at Lecompton, and 32.7 feet at Topeka, Kans. On June 1 the Missouri River at Kansas City, Mo., rose to 35.0 feet, the highest stage in its 45-year gage record and only 3 feet lower than the historic flood of June 20, 1844. The Mississippi River at St. Louis on June 10, 1903, crested at 38.0 feet, the third highest record in more than 130 years. Other rivers which reached record stages were the Des Moines River in

Iowa; and the Republican and Smoky Hill Rivers in Kansas, both
tributaries of the Kansas River.

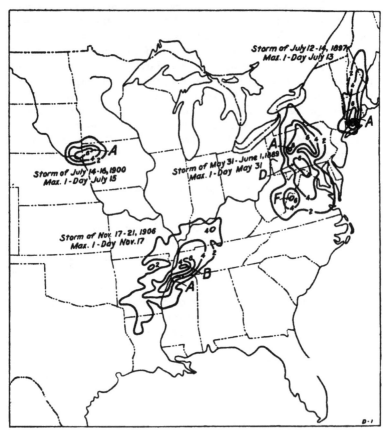

FIG. 71.—RAINFALL MAP FOR MAXIMUM DAY, STORMS OF
MAY 1889, JULY 1897, JULY 1900, AND NOVEMBER 1906.

The existing records relating to the memorable flood of 1844 in
the Mississippi Valley indicate that very probably it was caused by
widespread rainfall similar in character to the precipitation of the
latter half of May, 1903.

Storm 72, August 25–28, 1903

This storm was central in southern Iowa, and extended partly
into eastern Nebraska and northern Missouri. The areas of heavy
precipitation were confined to a comparatively narrow belt, as will

be seen by reference to figures 67 to 70. The storm was remarkable for the intensity of precipitation reported at a number of stations in Iowa. At Afton 9.30 inches fell on the 27th and 10.37 inches on Aug. 25–28; at Woodburn 15.53 inches fell on August 26–28, this being the maximum 3-day rainfall record for quadrangle 14–D, see figure 40. At

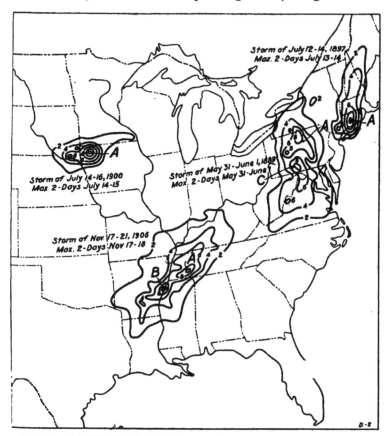

FIG. 72.—RAINFALL MAP FOR MAXIMUM 2 DAYS, STORMS OF MAY 1889, JULY 1897, JULY 1900, AND NOVEMBER 1906.

Corning 11.87 inches fell during the same period, which likewise is the maximum 3-day record for quadrangle 14–E. On the 27th 10.88 inches fell at Hopeville, 11.22 inches at Chariton, and 9.25 inches at Allerton. At Council Bluffs 8.30 inches fell on the 26th and 10.18 inches on the 26th and 27th; these are respectively the maximum 1-day and 2-day records for quadrangle 15–D. The larger streams experienced no unusual flood stages, owing to the peculiar geographic distribution of the rainfall.

Storm 76, October 8–9, 1903

This storm, shown in figures 67 to 68, covered but a small area in the northeastern United States, part of it having extended over the Atlantic Ocean. It was responsible for the greatest flood in 300 years on the Delaware River, caused by rainfall alone, being approached

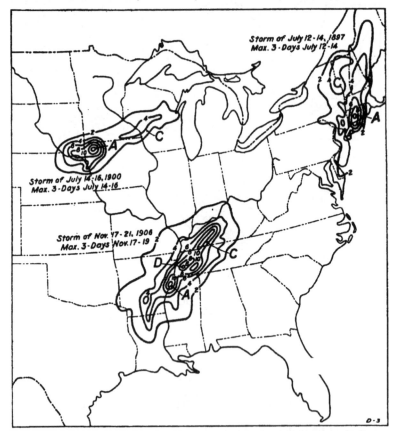

FIG. 73.—RAINFALL MAP FOR MAXIMUM 3 DAYS, STORMS OF
JULY 1897, JULY 1900, AND NOVEMBER 1906.

only by those of October 6, 1786, and January 8, 1841. It also caused the greatest floods on record on the Passaic River in New Jersey, and on the Lackawanna River in Pennsylvania. Along these rivers the damage was particularly severe.

Storm 83, June 9–10, 1905

Compared with the large areas in Iowa, Illinois, and Indiana, which received only from 1 to 2 inches of precipitation, the areas covered by intense rains during this storm were small. Most of the rain fell between 8.30 p. m. of June 9, and 8:30 p. m. of June 10. The center

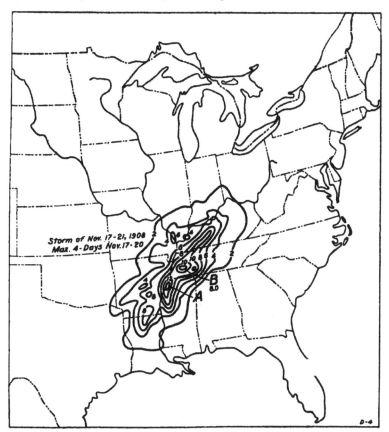

FIG. 74.—RAINFALL MAP FOR MAXIMUM 4 DAYS, STORM OF
NOVEMBER 1906.

of the area of greatest rainfall was at Bonaparte, Iowa, where 12.10 inches fell in about 12 hours. This is the record for 24-hour rainfall in quadrangle 13–E, see figure 38. Although the period of storm rainfall comprised parts of two calendar days, its short duration did not render it advisable to draw a separate map for each day. Figure 76 shows the distribution of rainfall for the entire storm period.

This storm caused the notable flood of June 10 in Devils Creek, a small Iowa stream draining into the Mississippi River below Fort Madison. Its maximum rate of flood discharge appears to have exceeded 500 second feet per square mile from an area of 145 square miles. The Des Moines River at Keosauqua, Iowa, rose nearly 19

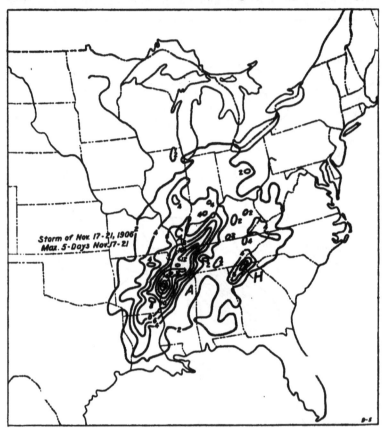

FIG. 75.—RAINFALL MAP FOR MAXIMUM 5 DAYS, STORM OF
NOVEMBER 1906.

feet. The Mississippi River experienced a sudden local rise, registering 18.4 feet at Keokuk on the 10th, which is about 2.5 feet less than the record flood stage of June 6, 1851, at that place.

Storm 86, September 15–19, 1905

Heavy rains fell in Missouri and southern Illinois from September 15 to 17, and were followed by lighter rains on the 18th and 19th, figures 62 to 66.   At Boonville, Mo., 12.98 inches fell in the period September 15 to 19, of which 6.70 inches fell on the 17th.   The 12.98-inch record is the maximum 5-day rainfall for quadrangle 13-F, see figure 42.   At Chester, Ill., 8.06 inches fell in 20.5 hours on the 16th and 17th.

This storm was notable for the unusually quick rises in the streams which were affected by it.   A number of rivers in Missouri experienced destructive floods, among them the Meramec and the Gasconnade. The Missouri River itself rose above danger line from Boonville to Hermann, Mo., and was instrumental in swelling the Mississippi River over 10 feet in 24 hours, causing it to reach a stage of 30.2 feet at St. Louis on September 21.   Although considerable damage was done by inundation along these rivers, no unprecedented stages were reported, and the flood crest in the Mississippi flattened out below Cairo to small proportions.

Storm 93, November 17–21, 1906

This storm caused excessive rainfalls over western Tennessee, northern Mississippi, and eastern Arkansas, from the 16th to the 21st, figures 71 to 75.   At many points the rainfall recorded was considered the heaviest known.   The Tennessee, the Little Tennessee, the Hiwassee, and the Clinch River, all rose to high stages and inflicted much damage.   About twenty lives were lost.

Storm 102, July 28–August 1, 1908

This storm was of tropic origin but was devoid of severity except at a few places.   As will be noted from figures 76 to 80, the area affected did not extend far inland, the principal rains occurring in Louisiana.   This is the more remarkable since the storm lasted 5 days. At Franklin, La., 17.62 inches fell in the 3 days July 29 to 31, of which 9.60 inches fell on July 30.

Storm 103, August 24–26, 1908

Widespread rains occurred over the south Atlantic states on Aug. 24, 25, and 26, 1908.   These were accompanied by heavy downpours over portions of Georgia, and North and South Carolina, see figures 76 to 78.   In 3 days 14.75 inches fell at Carlton, Ga., and 14.14 inches at Greenville, S. C.   At Anderson, S. C., 14.31 inches fell in 3 days, of which 11.65 inches fell on the 25th.   At Monroe, N. C., 15.58 inches fell in 3 days.   In the main, however, the rain was distributed quite evenly.   The damage done to standing crops and to property along streams was incalculable. Heavy rains on the 19th and 21st of August had partly saturated the soil in many places, and the runoff following this storm was, therefore, very great. On many of the south Atlantic coast streams occurred the highest stages at that time on record. Some of these stages have since been exceeded following

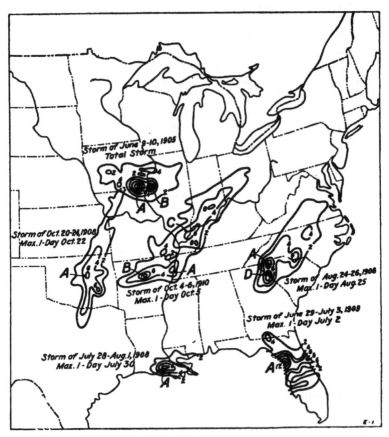

FIG. 76.—RAINFALL MAP FOR MAXIMUM DAY, STORMS OF JUNE 1905, JULY 1908, AUGUST 1908, OCTOBER 1908, JUNE 1909, AND OCTOBER 1910.

the storms of July 1916 in the Appalachian region, described subsequently. The Savannah and Santee Rivers experienced the most destructive floods in their histories, the Savannah at Augusta, Ga., exceeding the great flood of Sept. 11, 1888, by 0.1 foot. The crest stage of 38.8 feet on August 27 is the highest on record in a century.

At Calhoun Falls, S. C., the crest stage was 28.2 feet, also the highest on record.

The Congaree River at Columbia, S. C., attained its record stage on Aug. 27 with 35.8 feet; the Great Pee Dee River at Cheraw, S. C.,

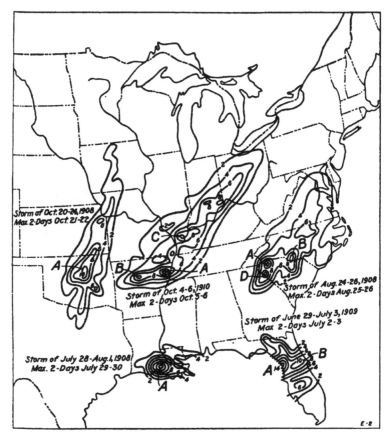

FIG. 77.—RAINFALL MAP FOR MAXIMUM 2 DAYS, STORMS OF JULY 1908, AUGUST 1908, OCTOBER 1908, JUNE 1909, AND OCTOBER 1910.

with 44.3 feet on Aug. 27; the Saluda River at Pelzer, S. C., with 25.6 feet on Aug. 25, and at Chappels, S. C., with 34.7 feet on Aug. 26. In North Carolina the Cape Fear River reached a stage of 68.7 feet at Fayetteville on Aug. 29, the highest ever recorded.

Storm 105, October 20–24, 1908

This storm extended in a northerly direction from southern Oklahoma to Minnesota over a distance of about 700 miles, see figures 76 to 80.   The south to north trend of its axis differs rather strikingly from the usual southwest to northeast direction taken by storms in

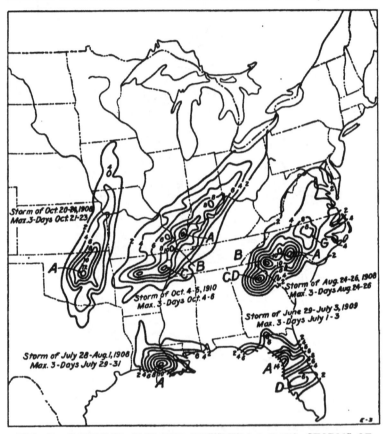

FIG. 78.—RAINFALL MAP FOR MAXIMUM 3 DAYS, STORMS OF
JULY 1908, AUGUST 1908, OCTOBER 1908, JUNE 1909, AND
OCTOBER 1910.

that section of the United States.   This was not due to any progressive movement of the storm center, however, the rains occurring almost simultaneously over the entire storm area.   The greatest rainfall intensities were observed in Oklahoma.   At Meeker, Okla., 16.23 inches fell in 5 days, of which 10.00 inches fell on the 20th.   This establishes the record for heavy precipitation in quadrangle 15–G, see figure 38. In the adjacent quadrangle 16–G this storm also established the maximum record, with 8.95 inches on the 23d, and a total of 14.88 inches in 6 days at Norman, Okla. Four other points in Oklahoma reported over 11 inches in 5 days. Serious floods resulted in Oklahoma, and great damage was done to railroads, bridges, and crops.

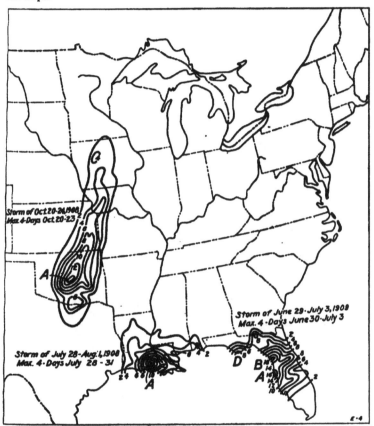

FIG. 79.—RAINFALL MAP FOR MAXIMUM 4 DAYS, STORMS OF JULY 1908, OCTOBER 1908, AND JUNE 1909.

Storm 108, June 29–July 3, 1909

This storm was not caused by any tropical disturbance, but appears to have been the result of a cyclonic movement over the northern part of Florida. Extraordinary rains fell over the central and northern

11

counties of that state, see figures 76 to 80. The greatest recorded
24-hour rainfall was 12.00 inches on July 2 at Rockwell. The total
rainfall recorded at this station in 4 days was 16.77 inches. At
Tarpon Springs, on the west coast, 11.09 inches fell on the 30th and

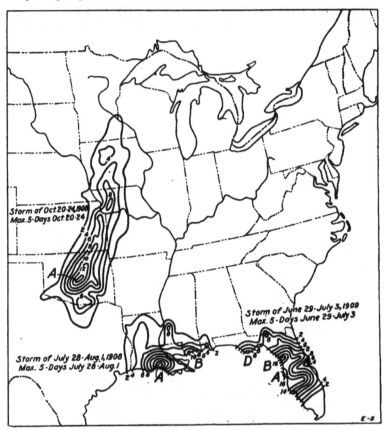

FIG. 80.—RAINFALL MAP FOR MAXIMUM 5 DAYS, STORMS OF
JULY 1908, OCTOBER 1908, AND JUNE 1909.

17.86 inches in 5 days; and at Avon Park, in the interior, 16.06 inches
were recorded in 5 days. No large streams were affected by this
storm.

Storm 109, July 5–7, 1909

Heavy rains occurred on the above dates in eastern Kansas, northern Missouri, southern Iowa, and central Illinois, but the areas affected
by great intensities of precipitation were not extensive, figures 81 to 83.

A small center of very high rainfall formed in central Missouri, near Bagnell, where 11.75 inches fell in 4 days, of which 7.91 inches fell on the 6th, this being the record for 24-hour rainfall in quadrangle 13-F, see figure 38. In northern Missouri a similar downpour, amounting to 11.23 inches in 3 days, was reported at Bethany.

The storm was of short duration but caused, nevertheless, rapid rises in the rivers, having been preceded by wet weather which began on June 29. The Grand River in northwestern Missouri attained unusually high stages, flooding much land. Its discharge affected the Missouri, which at the time was bank full, due to a combination of melting snows and rain, and caused it to inundate vast areas. Other rivers seriously affected were the Neosho and the Marais des Cygnes. This storm contributed materially to the Missouri River floods which inundated the low lying districts of Kansas City and St. Louis.

Storm 110, July 20–22, 1909

This storm covered the region west of Lake Superior as shown in figures 81 to 83, on the 19th to 22d, and produced unusually high rainfall intensities for this section of the United States. At Ironwood, Michigan, on the 21st, 6.72 inches fell, which at that time was the greatest 24-hour rainfall record in Michigan. Destructive floods resulted on a number of streams.

Storm 114, October 4–6, 1910

In intensity as well as in distribution this storm ranks as one of the severest on record in the eastern United States. It covered the lower Ohio River and a portion of the Mississippi River valley, extending from western Arkansas to northeastern Ohio. The distribution of rain for the 3 days of maximum precipitation is shown in figures 76 to 78. It commenced during the night of October 3 and continued until the night of the 6th; on the last two days of this period it extended northeastward into the Great Lakes region. The greatest intensities occurred east of Cairo and in northern Arkansas. Cairo reported 10.95 inches of rain between 5 p. m. of the 3d and noon of the 6th, the greatest 4-day rain recorded in quadrangle 12–F, see figure 41. Marked Tree, Ark., reported 13.99 inches for the three days Oct. 4 to 6, which is the maximum 3-day rainfall in quadrangle 12–G, see figure 40. At Golconda, Ill., nearly 8 inches fell in 24 hours and 15.24 inches fell in 60 hours, which is the 3-day maximum on record in quadrangle 11–F, see figure 40. At Blandville, Ky., 10.1 inches fell in 60 hours. Louisville, Ky., reported 5.06 inches in 24 hours. At Cobden, Ill., 9.70 inches fell on the 4th and 5th, which is the record for 2-day rainfall for quadrangle 12–F, see figure 39.

5–8, were largely responsible for the destructive floods in many streams which followed the rains of January 10–12. High stages were recorded on the Cumberland and Kentucky Rivers. The Miami, Scioto, and Muskingum Rivers were also affected. The greater damage, however, was done in the valley of Green River, Kentucky,

FIG. 82.—RAINFALL MAP FOR MAXIMUM 2 DAYS, STORMS OF JULY 5–7, 1909, JULY 20–22, 1909, JANUARY 1913, OCTOBER 1913, AND OCTOBER 1914.

where large areas were inundated. The crest of the flood on Green River reached 35.5 feet at Lock No. 2 on January 18, being the highest stage on record.

This storm bears some analogy to that of October 4 to 6, 1910, shown in figures 76 to 78, the shapes and extents of the respective storm areas being similar. They occurred, however, at different times of the year; and the soil conditions, in each case, influenced to a large extent the rate of runoff, causing it to be higher in the January flood.

Storm 132, March 23–27, 1913

The unusual severity of this storm as to duration, rainfall intensity, and extent of territory covered, entitle it to be classed as one of the

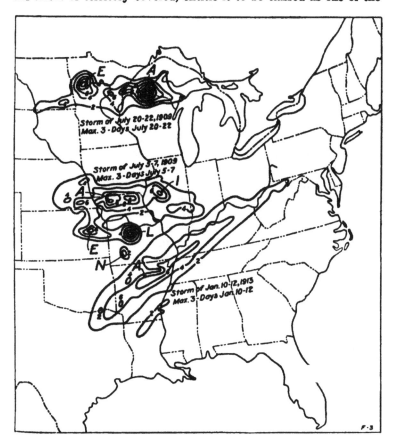

FIG. 83.—RAINFALL MAP FOR MAXIMUM 3 DAYS, STORMS OF JULY 5–7, 1909, JULY 20–22, 1909, AND JANUARY 1913.

greatest storms on record in the eastern United States. In point of flood damage it is without a parallel.

Figures 84 to 88 show the area affected by heavy rains. Outside of the 2-inch isohyetal a vast area was affected by light rains. A prominent feature of these charts is the even distribution of the precipitation and absence of centers of great rainfall intensity. No exceptional 24-hour rates were recorded. On the other hand this storm established the record for continued rainfall at a number of localities. Thus, at Richmond, Ind., 9.47 inches fell on March 24 and 25, and 11.74 inches in 6 days. The rainfall at Richmond during this storm established the record for 2-day, 3-day, 4-day, and 6-day rainfall in quadrangle 9–E. At Bellefontaine and Kenton, Ohio, over 11 inches fell in 5 days. Precipitation amounting to 4 inches or more in 24 hours was reported at a number of localities in Indiana and Ohio.

The unusual duration of heavy precipitation was brought about by the merging of storms accompanying two distinct low pressure areas, which occurred almost simultaneously in a low pressure trough extending from southwest to northeast. One of these areas progressed from Iowa in a northeasterly course across the upper lake region, and the other followed it two days later, moving along a parallel route from central Arkansas toward Lake Erie. The bulk of the precipitation in the lower Ohio basin fell during the 72-hour period, March 24 to 26.

In the Wabash basin and in the lake region the three days of maximum rainfall were March 23, 24, and 25.

Preceding the storm there had been frequent rains, which, though moderate, had brought about a fair state of saturation of the soil. The latter was not frozen at the commencement of the storm. The rates of runoff which resulted were very high, and in the case of some of the basins, were extraordinary. The number of streams which exceeded all previous highwater stages are so numerous that it is practicable to enumerate here only the more important ones, such as the Beaver, Mahoning, Muskingum, Scioto, Olentangy, Little Miami, Miami, White River of Indiana, Wabash, Maumee, and Sandusky. The immediate effect of the floods in the northern tributaries of the Ohio River was to create the highest stages on record in the Ohio River between New Martinsville, W. Va., and Cincinnati, and also at Madison, Ind. At other points the Ohio did not reach the 1884 stage. The rivers flowing into the Ohio from the south were not seriously affected. In this connection a comparison of this storm with that of January, 1913, see figures 81 to 83, is of interest. The latter caused the greater precipitation over the southern tributary basins, and many of these streams attained higher stages than in March, 1913.

In New York state record floods occurred on the Mohawk and upper Hudson Rivers. The damage done by floods in some localities was extraordinary. The total monetary losses caused by the March 1913 storm aggregated in the neighborhood of $200,000,000.

Storm 135, October 1–2, 1913

Heavy rains fell in western Texas during the first four days of October. The area of greatest precipitation was quite compact, as will be seen from figures 81 and 82, which show the fall for Oct. 1 and 2, the two days of maximum precipitation. The heaviest 24-hour record was at San Marcos, Texas, where 13.03 inches fell on Oct. 2. The small isolated rainfall center shown on the charts to the north of the main storm area, was caused by a 24-hour rain of 11.00 inches at Waco, Texas, also on Oct. 2.

High stages occurred on the San Antonio, Guadalupe, Rio Grande, and Colorado Rivers as a result of this storm. The Guadalupe broke all previous flood records, rising to 36.7 feet at Gonzales, Tex. This was exceeded, however, on December 4 of that year by a 38.1-foot stage.

Storm 144, October 14–15, 1914

The heavy rains of October, 1914, in western North Carolina were caused by two storm centers coming in close succession from different directions. A large area including parts of Kentucky, Tennessee, Virginia, North Carolina, South Carolina, and Georgia received light rains. In North Carolina three distinct centers of excessive precipitation formed, as shown in figures 81 and 82. Taken in order from north to south these are at Rock House, Chimney Rock, and Highlands. At Rock House 8.80 inches fell on the 15th and 11.00 inches during the 3 days October 14 to 16. At Chimney Rock 10.00 inches were recorded on the 15th, which is the 1-day maximum for quadrangle 8–G. At Highlands 8.25 inches fell on the 15th and 9.55 inches during the 3 days October 14 to 16. No unusual flood stages resulted from this storm.

Storm 151, August 17–21, 1915

This storm was the continuation of the West Indian hurricane which devastated Galveston, causing great damage to its causeway and to shipping and causing a loss of 275 lives. Its path was similar to that taken by the famous hurricane of September 9, 1900, which nearly destroyed Galveston. Both of these tropical storms moved in a northwesterly direction over the West Indian archipelago in an almost direct course for Galveston. On passing inland their courses curved northward, and then northeasterly across the Mississippi Valley, Great Lakes region, and the St. Lawrence Valley. A comparison of the two tracks shows the 1915 storm to have been south of that of 1900 throughout the gulf region by about 150 miles. Its course over the United States was also south of that of the 1900 hurricane, the turn from northwest to northeast being sharper than in 1900.

The 1915 hurricane passed inland in the early morning of August 17, causing excessive rains over all of eastern Texas and a portion of Louisiana on the 17th, 18th, and 19th. Recurving to the northeast on the 18th it formed a second, large center of precipitation over Arkansas, which extended into Oklahoma, Missouri, Illinois, and

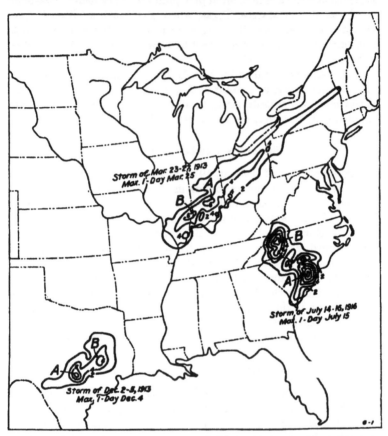

FIG. 84.—RAINFALL MAP FOR MAXIMUM DAY, STORMS OF MARCH 1913, DECEMBER 1913, AND JULY 14–16, 1916.

Tennessee. The larger part of the rain on this area fell on the 19th and 20th.

The storm area consisted, therefore, of two main centers of precipitation, connected by a neck, the greater of these areas covering all of Arkansas and parts of adjoining states, see figures 89 to 92. The rainfall within these areas was exceptionally heavy. At San Augustine in eastern Texas, 19.83 inches were recorded in the 4 days, August 16 to 19, establishing the 2-day, 3-day, and 4-day records for excessive precipitation in quadrangle 14–I.   Other high records in Texas were 17.95 inches at Rockland, 13.70 inches at Galveston, and 19.37 inches at Liberty, all in the 3-day period August 17 to 19.   The rainfall at Liberty set the record for 3-day, 4-day, 5-day, and 6-day rainfall in quadrangle 14–J.

FIG. 85.—RAINFALL MAP FOR MAXIMUM 2 DAYS, STORMS OF MARCH 1913, DECEMBER 1913, AND JULY 14–16, 1916.

High rates were reported in Arkansas as follows: 11.18 inches at Marshall, August 17 to 20; 10.50 inches at Pocahontas, August 18 to 20; 10.90 inches at Okay, August 18 to 20; 10.08 inches at Centerpoint, August 18 and 19; and 14.80 inches at Hardy, August 17 to 20. The latter is the record for excessive precipitation for 2-day, 3-day, and 4-day rainfall in quadrangle 13–G.

Arkansas center.　The time of travel of the storm is clearly seen from the following comparison of dates of different maximum periods in the two centers:

Maximum Periods	Dates of Occurrence	
	Texas Center	Arkansas Center
1 day	August 17	August 19
2 days	"　17–18	"　19–20
3 days	"　17–19	"　18–20
4 days	17–20	17–20

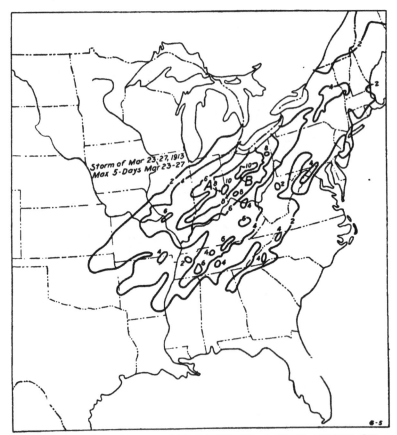

FIG. 88.—RAINFALL MAP FOR MAXIMUM 5 DAYS, STORM OF
MARCH 1913.

For the purpose of study and analysis, a separate set of maps and curves was prepared for each center for the maximum 1-day, 2-day, and 3-day periods of rainfall.　The two 3-day maps overlap somewhat in time and area. Only one map was made for the maximum 4-day period since that period was the same for both centers. As it is interesting to compare the time-area-depth data obtained from the two sets of maps, both sets of data are given in the appendix and both sets of time-area-depth curves are reproduced. The curves for the Arkansas center are in the northern group, figures 94 to 97, and those

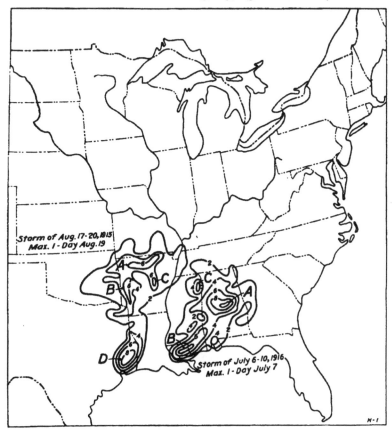

FIG. 89.—RAINFALL MAP FOR MAXIMUM DAY, STORMS OF
AUGUST 1915 AND JULY 6–10, 1916.

for the Texas center are in the southern group, figures 99 to 102. As this is essentially a southern storm, the time-area-depth curves for the Arkansas center are platted only down to the area of the last isohyetal which does not include the Texas center. Only the maps for the maximum periods as determined by the Arkansas center are reproduced, see figures 89 to 92.

Storm 156, July 6–10, 1916

Extraordinary rains fell over Alabama, southeastern Louisiana, and western Georgia during the period July 6–10, as the result of a tropical hurricane which entered the United States just east of the mouth of the Mississippi River on July 5. The area of precipitation, figures 89 to 93, centered over southern Alabama, Mississippi, and

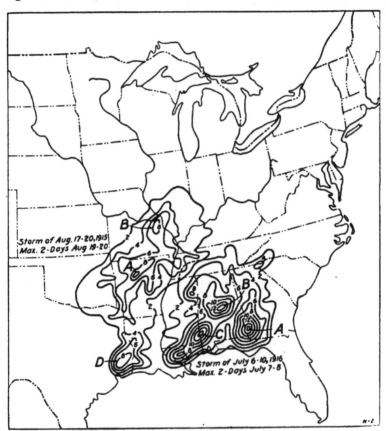

FIG. 90.—RAINFALL MAP FOR MAXIMUM 2 DAYS, STORMS OF AUGUST 1915 AND JULY 6–10, 1916.

Georgia although practically the whole state of Alabama received a total rainfall of 10 inches or more during the five days ending July 10. During this period the area covered by 5 inches or more of rainfall also included the eastern third of Mississippi, the western two-thirds of Georgia, and extended well up into central Tennessee and western

North Carolina. It should be noted that this storm was closely
followed by storm 157, July 14–16, 1916, figures 84 to 87, over practically the same area in western North Carolina. Light rains fell
between the two storms, and consequently the ground was saturated
and most of the streams bank full when the second storm came.

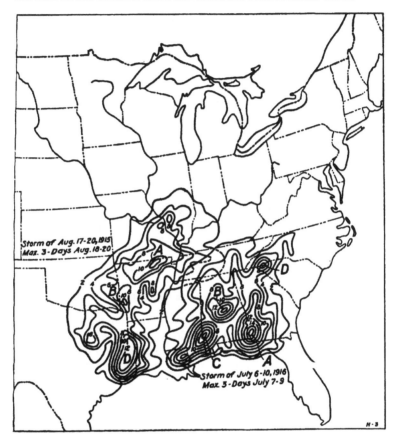

FIG. 91.—RAINFALL MAP FOR MAXIMUM 3 DAYS, STORMS OF
AUGUST 1915 AND JULY 6–10, 1916.

The average rainfall in Alabama for the month of July, determined
from the records at 58 stations, was 16.70 inches, which is 11.19 inches
more than the normal for that month. The notes in the Weather
Bureau publication, Climatological Data, state: "The month of
July, 1916, will go on record as the most disastrous month experienced
in Alabama in the last half century." This same publication estimates

12

the property loss, caused by this storm and the rains which followed it, at 11,000,000 dollars. Some 26,000 people in the state were made destitute and dependent, and 350,000 acres of land in Alabama were submerged by the floods. The tide at Mobile reached 11.6 feet above mean low water, which is about 1.7 feet higher than any previous record.

Some of the most intense precipitation records, in inches, are as follows:

Station	1-Day	Date	2-Day	Date	3-Day	Date
Leakesville, Miss.	10.00	6	16.54	6–7	19.00	6–8
Merrill, Miss.	12.35	6	15.95	6–7	19.95	6–8
Bay Minette, Ala.	12.20	6	17.30	6–7	....	......
Clanton, Ala.	11.23	7	16.88	7–8	17.81	6–8
Eufaula, Ala.	10.86	8	13.20	7–8	14.94	7–9
Evergreen, Ala.	11.95	6	15.27	6–7	18.62	6–8
'Alaga, Ala.	12.70	8	18.32	8–9	19.14	8–10
Blakely, Ga.	9.90	8	17.25	7–8	19.69	7–9
Robertsdale, Ala.	12.50	5	18.08	5–6	19.31	5–7

The Pascagoula River at Merrill, Miss., reached a maximum stage of 27.0 feet at 7 p. m. on the 9th, exceeding the highest previous record by 0.8 foot. The Chattahoochee River at Alaga, Ala., reached a record stage of 44.0 feet at 9 a. m. on the 9th, exceeding the greatest previous record by 3.8 feet. Many other streams closely approached or exceeded the maximum previous recorded stage.

Storm 157, July 14–16, 1916

This storm was produced by a West Indian hurricane or tropical cyclone which approached the south Atlantic seaboard from the southeast, entering South Carolina near Charleston on the morning of July 14. As is commonly the case with hurricanes after they come in contact with land surfaces, this storm rapidly diminished in energy and finally dissipated on July 16, on the eastern slope of the mountains in North Carolina. This illustrates the important influence which high mountain ranges exert in modifying or arresting the progress of cyclonic movements. Statistics published by the Weather Bureau* tend to show that the tracks of cyclones do not as a rule cross the Appalachian Mountains.

Two distinct centers of precipitation formed during this storm, one over South Carolina with a peak of over 16 inches in 2 days, and the other over North Carolina with a peak of over 23 inches in 2 days, see figures 84 to 87. The maximum rainfall at the latter was recorded at two stations, about one mile apart, located on opposite sides of a

* Types of Storms of the United States and their Average Movements. Supplement No. 1, Monthly Weather Review, 1914.

mountain gap, nearly midway between Grandfather Mountain, elevation 5964 feet, and Mount Mitchell, elevation 6711 feet. These stations are Altapass, elevation 2625 feet, on the east slope, where a standard 8-inch rain gage indicated a total fall of 23.77 inches for the 3 days, July 15–17; and Altapass Inn on the west slope, where the total fall was 1.52 inches less. At Altapass 23.22 inches of rain fell between 2 p. m. of July 15 and 2 p. m. of July 16.

FIG. 92.—RAINFALL MAP FOR MAXIMUM 4 DAYS, STORMS OF
AUGUST 1915 AND JULY 6–10, 1916.

To the westward of the mountains the rainfall decreased rapidly as indicated by the isohyetals in figures 84 to 87, the total fall at points 40 miles west of Altapass being less than 1 inch.

Other high rainfall records reported during this storm are 11.05 inches on July 15 at Florence, S. C.; 12.60 inches on July 15 at Kingstree, S. C.; 13.25 inches on July 15 at Effingham, S. C.; 10.65 inches on July 15, at Transon, N. C.; 13.25 inches on July 16 at Blantyre, N. C.; 14.70 inches on July 16 at Brevard, N. C.; and 10.02 inches on the 15th and 10.43 inches on the 16th at Gorge, N. C.

The floods caused by this storm are of peculiar interest. Since the storm progressed in a westerly direction, the lower reaches of the

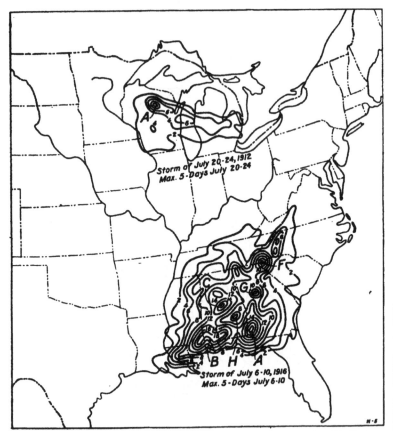

FIG. 93.—RAINFALL MAP FOR MAXIMUM 5 DAYS, STORMS OF JULY 1912 AND JULY 6-10, 1916.

longer Atlantic coast streams were in flood in many cases before their headwaters, resulting in floods less severe than would normally be expected from such rainfall. These floods were, in some measure, attributable to the saturated condition of the soil brought about by previous heavy rains, which covered this region on July 9 to 13. A number of rivers exceeded the flood stages attained during the storm of August 1908. Some of those which broke all previous records were the Catawba River at Mount Holly, N. C., and at Catawba, S. C., with stages of 36.5 and 40.4 feet respectively; the Santee River at Rimini, S. C., 35.8-foot stage; the Wateree River at Camden, S. C., 43.0 feet; and the Broad River at Blairs, S. C., 36.5 feet. West of the Blue Ridge Mountains the upper Yadkin on July 15, reached the highest stages in its record, and the French Broad River at Asheville on July 16 broke all records with a 21-foot stage.

DISCUSSION OF TIME-AREA-DEPTH RELATIONS

There are ten sets of time-area-depth curves, figures 94 to 103, constructed in the manner described on page 133. These are grouped first by region of occurrence, and second by period of maximum precipitation. Thus, each group of five sets comprises the curves for the maximum 1-day, 2-day, 3-day, 4-day, and 5-day periods of about half of the 33 storms just described, there being 17 storms in the northern and 16 in the southern parts of the United States, for which curves are shown. These present the time-area-depth data for each storm in its most useful form. Certain storm features and similarities between storms are not discernible until the time-area-depth curves are drawn and compared.

From a knowledge of the causes of precipitation in interior continental regions it would be expected that the maximum 1-day period of rainfall would be preceded and followed by periods of gradually increasing and decreasing rainfall, especially during the winter months. This is generally found to be true, the maximum 1-day rainfall rarely occurring on the first or last day of the storm period.

Another very noticeable fact is that the depth of rain which falls on the maximum day is almost always more than half the total of the storm period, regardless of the number of days in the latter. This indicates, that much higher rates of precipitation occur in periods of less than a day. The accuracy of this conclusion has been verified at individual stations.

In general, there are no very great differences between the largest storm and the second or third largest. This is more nearly true of the northern group than of the southern, and for the 1-day, 2-day, and 3-day periods than for the 4-day and 5-day periods. The most notable exceptions to this rule are storm 43, for 3, 4, and 5-day periods, and storm 156, for the 4 and 5-day periods, both in the southern group. The highest 3, 4, and 5-day record in storm 43 is 33.0 inches at Turnersville, Texas. The observer was not at the time an Official Weather Bureau observer, although he had a standard gage. The record is open to doubt. The other high records of this phenomenal storm, however, tend to substantiate it.

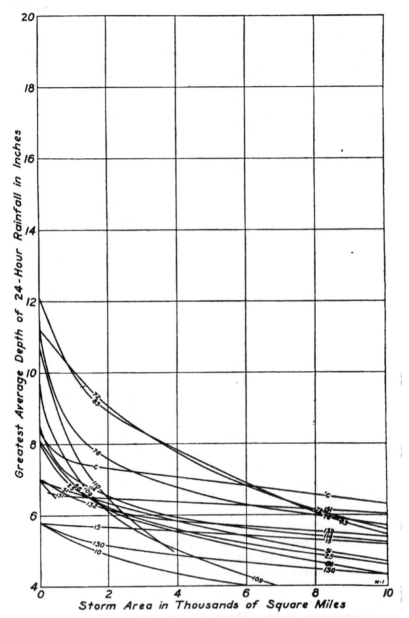

FIG. 94.—TIME-AREA-DEPTH CURVES FOR STORMS OVER
NORTHERN STATES SHOWING GREATEST AVERAGE
DEPTH OF RAINFALL DURING 1 DAY.

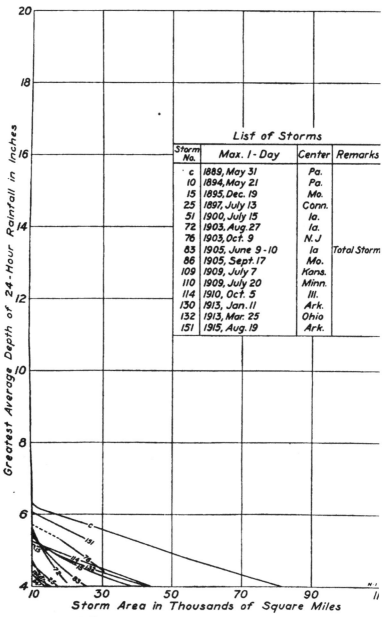

List of Storms

Storm No.	Max. 1-Day	Center	Remarks
c	1889, May 31	Pa.	
10	1894, May 21	Pa.	
15	1895, Dec. 19	Mo.	
25	1897, July 13	Conn.	
51	1900, July 15	Ia.	
72	1903, Aug. 27	Ia.	
76	1903, Oct. 9	N. J	
83	1905, June 9-10	Ia	Total Storm
86	1905, Sept. 17	Mo.	
109	1909, July 7	Kans.	
110	1909, July 20	Minn.	
114	1910, Oct. 5	Ill.	
130	1913, Jan. 11	Ark.	
132	1913, Mar. 25	Ohio	
151	1915, Aug. 19	Ark.	

FIG. 94.—*Continued ; note change in horizontal scale.*

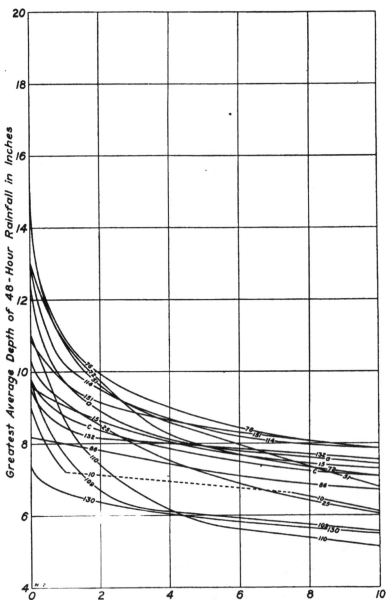

FIG. 95.—TIME-AREA-DEPTH CURVES FOR STORMS OVER
NORTHERN STATES SHOWING GREATEST AVERAGE
DEPTH OF RAINFALL DURING 2 DAYS.

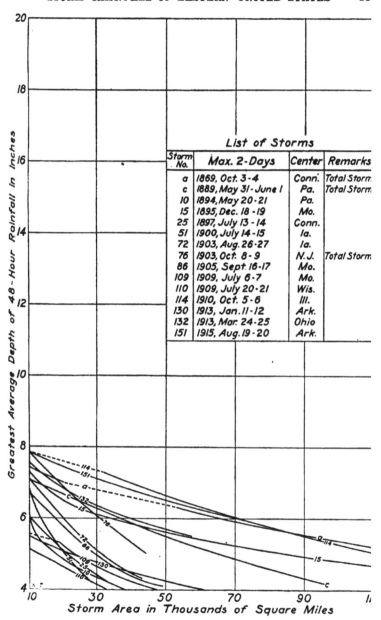

List of Storms

Storm No.	Max. 2-Days	Center	Remarks
a	1869, Oct. 3-4	Conn.	Total Storm
c	1889, May 31-June 1	Pa.	Total Storm
10	1894, May 20-21	Pa.	
15	1895, Dec. 18-19	Mo.	
25	1897, July 13-14	Conn.	
51	1900, July 14-15	Ia.	
72	1903, Aug. 26-27	Ia.	
76	1903, Oct. 8-9	N.J.	Total Storm
86	1905, Sept. 16-17	Mo.	
109	1909, July 6-7	Mo.	
110	1909, July 20-21	Wis.	
114	1910, Oct. 5-6	Ill.	
130	1913, Jan. 11-12	Ark.	
132	1913, Mar. 24-25	Ohio	
151	1915, Aug. 19-20	Ark.	

FIG. 95.—*Continued ; note change in horizontal scale.*

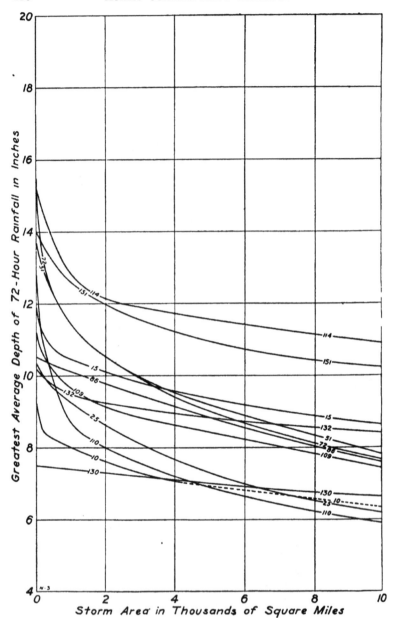

FIG. 96.—TIME-AREA-DEPTH CURVES FOR STORMS OVER
NORTHERN STATES SHOWING GREATEST AVERAGE
DEPTH OF RAINFALL DURING 3 DAYS.

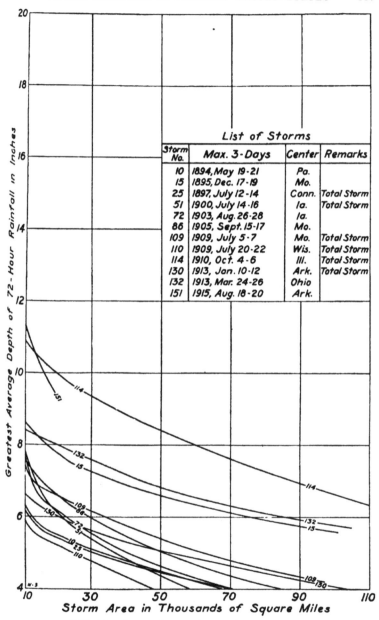

FIG. 96.—*Continued; note change in horizontal scale.*

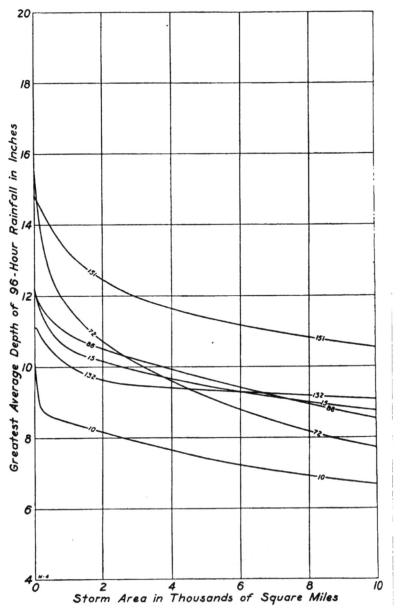

FIG. 97.—TIME-AREA-DEPTH CURVES FOR STORMS OVER
NORTHERN STATES SHOWING GREATEST AVERAGE
DEPTH OF RAINFALL DURING 4 DAYS.

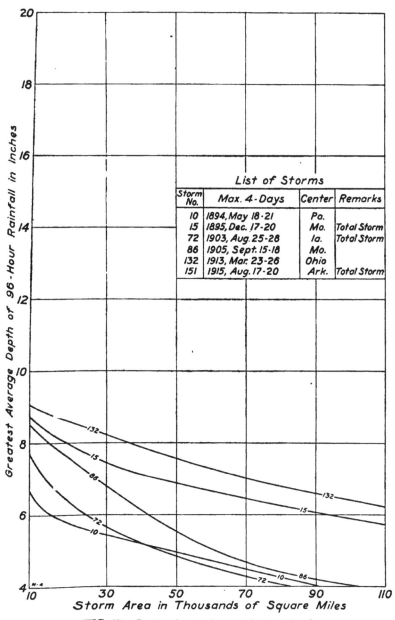

FIG. 97.—*Continued; note change in horizontal scale.*

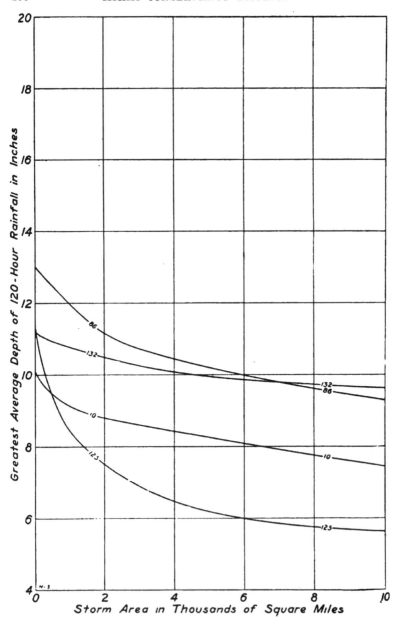

FIG. 98.—TIME-AREA-DEPTH CURVES FOR STORMS OVER
NORTHERN STATES SHOWING GREATEST AVERAGE
DEPTH OF RAINFALL DURING 5 DAYS.

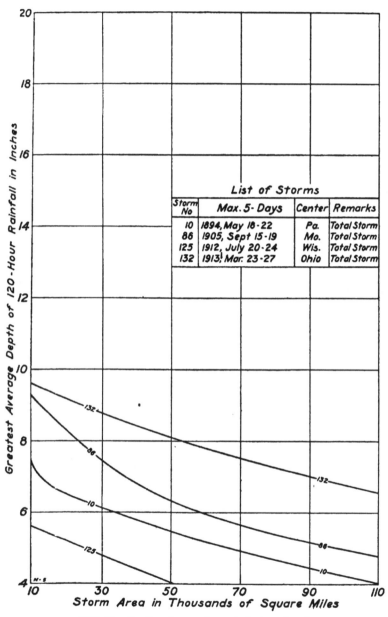

List of Storms

Storm No	Max. 5- Days	Center	Remarks
10	1894, May 18-22	Pa.	Total Storm
86	1905, Sept 15-19	Mo.	Total Storm
125	1912, July 20-24	Wis.	Total Storm
132	1913, Mar. 23-27	Ohio	Total Storm

FIG. 98.—*Continued ; note change in horizontal scale.*

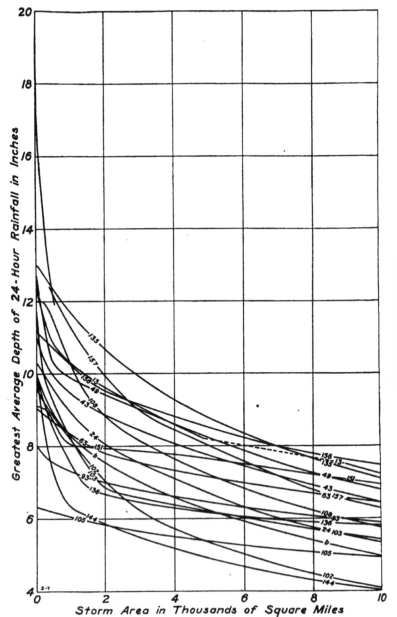

FIG. 99.—TIME-AREA-DEPTH CURVES FOR STORMS OVER
SOUTHERN STATES SHOWING GREATEST AVERAGE
DEPTH OF RAINFALL DURING 1 DAY.

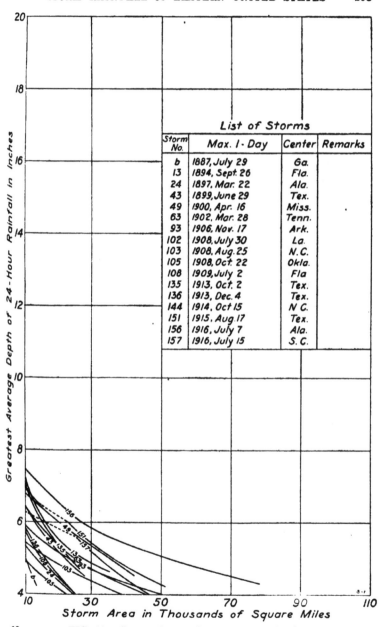

List of Storms

Storm No.	Max. 1-Day	Center	Remarks
b	1887, July 29	Ga.	
13	1894, Sept. 26	Fla.	
24	1897, Mar. 22	Ala.	
43	1899, June 29	Tex.	
49	1900, Apr. 16	Miss.	
63	1902, Mar. 28	Tenn.	
93	1906, Nov. 17	Ark.	
102	1908, July 30	La.	
103	1908, Aug. 25	N.C.	
105	1908, Oct. 22	Okla.	
108	1909, July 2	Fla	
135	1913, Oct. 2	Tex.	
136	1913, Dec. 4	Tex.	
144	1914, Oct 15	N C.	
151	1915, Aug.17	Tex.	
156	1916, July 7	Ala.	
157	1916, July 15	S. C.	

FIG. 99.—*Continued ; note change in horizontal scale.*

13

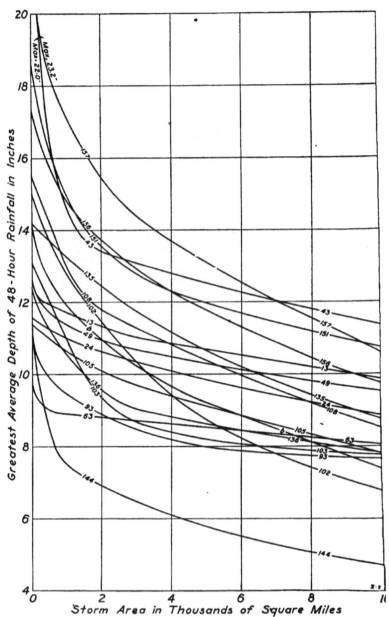

FIG. 100.—TIME-AREA-DEPTH CURVES FOR STORMS OVER
SOUTHERN STATES SHOWING GREATEST AVERAGE
DEPTH OF RAINFALL DURING 2 DAYS.

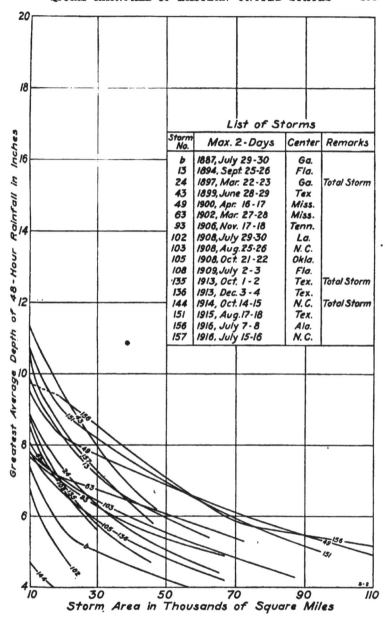

Storm No.	Max. 2-Days	Center	Remarks
b	1887, July 29-30	Ga.	
13	1894, Sept. 25-26	Fla.	
24	1897, Mar. 22-23	Ga.	Total Storm
43	1899, June 28-29	Tex	
49	1900, Apr. 16-17	Miss.	
63	1902, Mar. 27-28	Miss.	
93	1906, Nov. 17-18	Tenn.	
102	1908, July 29-30	La.	
103	1908, Aug. 25-26	N. C.	
105	1908, Oct. 21-22	Okla.	
108	1909, July 2-3	Fla.	
135	1913, Oct. 1-2	Tex.	Total Storm
136	1913, Dec. 3-4	Tex.	
144	1914, Oct. 14-15	N. C.	Total Storm
151	1915, Aug. 17-18	Tex.	
156	1916, July 7-8	Ala.	
157	1916, July 15-16	N. C.	

FIG. 100.—*Continued; note change in horizontal scale.*

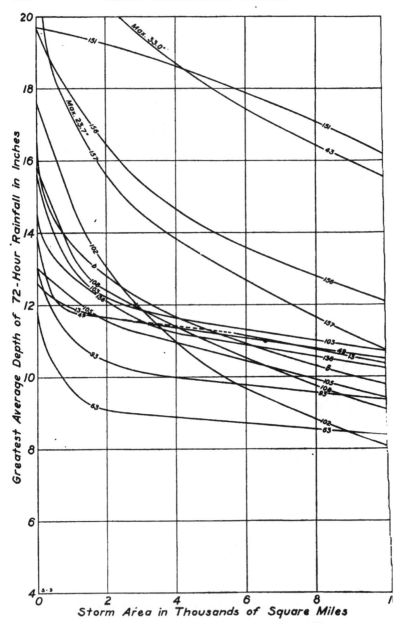

FIG. 101.—TIME-AREA-DEPTH CURVES FOR STORMS OVER
SOUTHERN STATES SHOWING GREATEST AVERAGE
DEPTH OF RAINFALL DURING 3 DAYS.

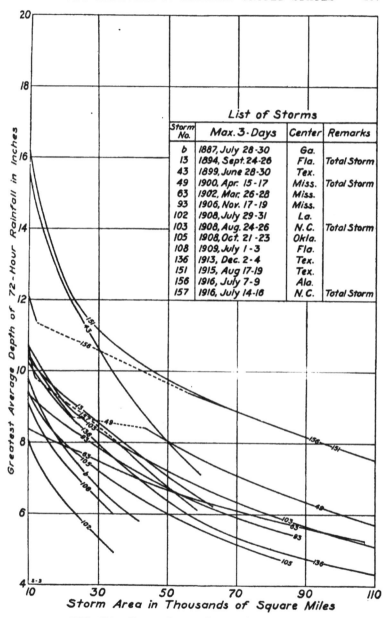

List of Storms

Storm No.	Max. 3-Days	Center	Remarks
b	1887, July 28-30	Ga.	
13	1894, Sept. 24-26	Fla.	Total Storm
43	1899, June 28-30	Tex.	
49	1900, Apr. 15-17	Miss.	Total Storm
63	1902, Mar. 26-28	Miss.	
93	1906, Nov. 17-19	Miss.	
102	1908, July 29-31	La.	
103	1908, Aug. 24-26	N.C.	Total Storm
105	1908, Oct. 21-23	Okla.	
108	1909, July 1-3	Fla.	
136	1913, Dec. 2-4	Tex.	
151	1915, Aug 17-19	Tex.	
156	1916, July 7-9	Ala.	
157	1916, July 14-16	N.C.	Total Storm

FIG. 101.—*Continued; note change in horizontal scale.*

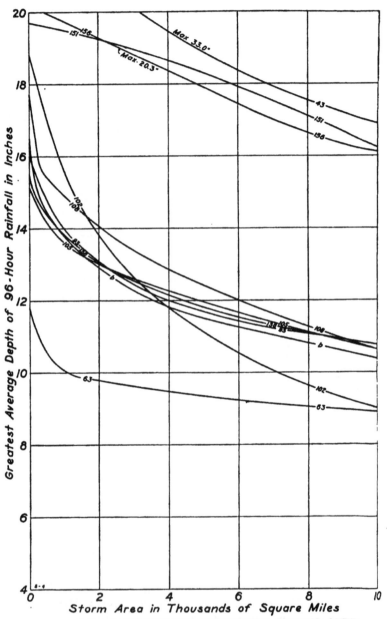

FIG. 102.—TIME-AREA-DEPTH CURVES FOR STORMS OVER
SOUTHERN STATES SHOWING GREATEST AVERAGE
DEPTH OF RAINFALL DURING 4 DAYS.

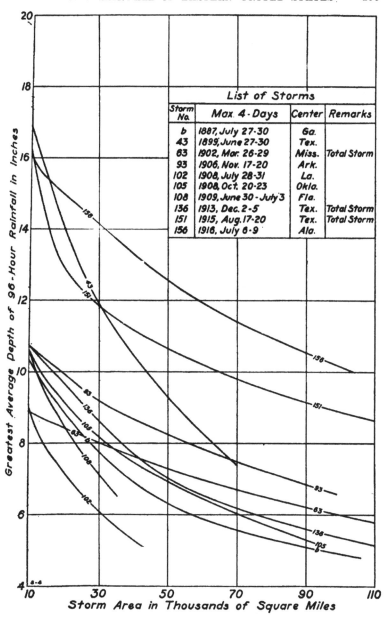

List of Storms

Storm No.	Max. 4 - Days	Center	Remarks
b	1887, July 27-30	Ga.	
43	1899, June 27-30	Tex.	
63	1902, Mar. 26-29	Miss.	Total Storm
93	1906, Nov. 17-20	Ark.	
102	1908, July 28-31	La.	
105	1908, Oct. 20-23	Okla.	
108	1909, June 30 - July 3	Fla.	
136	1913, Dec. 2-5	Tex.	Total Storm
151	1915, Aug.17-20	Tex.	Total Storm
156	1916, July 8-9	Ala.	

FIG. 102.—*Continued ; note change in horizontal scale.*

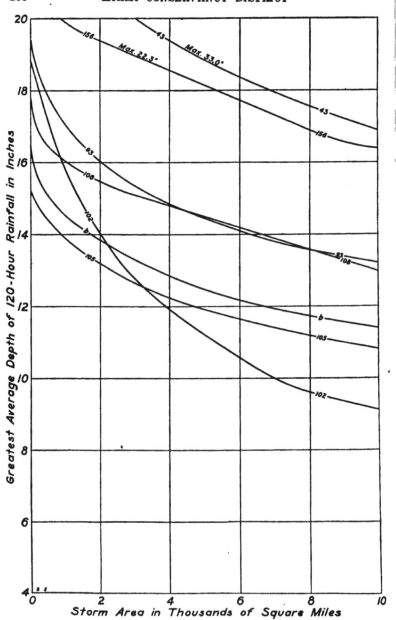

FIG. 103.—TIME-AREA-DEPTH CURVES FOR STORMS OVER
SOUTHERN STATES SHOWING GREATEST AVERAGE
DEPTH OF RAINFALL DURING 5 DAYS.

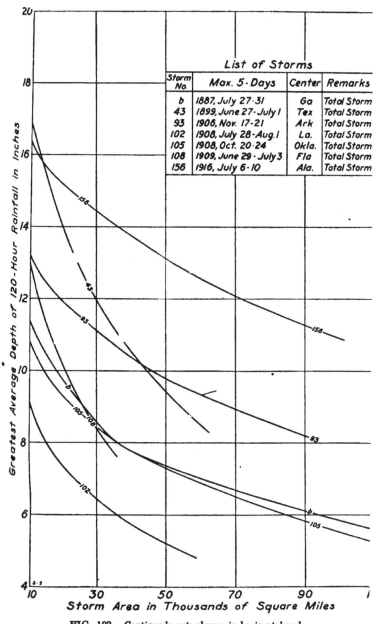

List of Storms

Storm No.	Max. 5·Days	Center	Remarks
b	1887, July 27·31	Ga	Total Storm
43	1899, June 27· July 1	Tex	Total Storm
93	1906, Nov. 17·21	Ark	Total Storm
102	1908, July 28·Aug. 1	La.	Total Storm
105	1908, Oct. 20·24	Okla.	Total Storm
108	1909, June 29 · July 3	Fla	Total Storm
156	1916, July 6· 10	Ala.	Total Storm

FIG. 103.—*Continued; note change in horizontal scale.*

The storm of July 6–10, 1916, number 156, for 4 and 5-day maximum periods is also very exceptional. It is about 50 per cent greater than the next smaller. It is very improbable that this could happen anywhere except close to the ocean or gulf, as such a condition requires the importation to the storm center for several days in succession of immense quantities of air greatly overcharged with water vapor.

The effect of the same factor, distance of land travel, is also apparent in the northern group of storms. Compare the curves, figures 94 to 98, with the maps, figures 50 to 57. Storms 114, 151, 15, and 86 occurred farthest south in the Mississippi Valley, and are the largest for the 3, 4, and 5-day periods. They are less prominent on the 1 and 2-day charts because overshadowed by the short violent storms of the upper Mississippi Valley and the northeast.

USE OF THE TIME-AREA-DEPTH CURVES

Practical use of the time-area-depth curves may be made by first determining the area, A, of any watershed to which the storms are to be applied. Then, by referring to the time-area-depth curves for the region of the United States in which this area is located, the curves for the greatest storms, or for the storms most applicable to the area, as shown by the seasonal maps, figures 50 to 57, give the average depths of precipitation over an area equal to A at the center of the various storms.

This method of using the curves involves a number of assumptions, all of which tend to make the average depth over the area A appear greater than would probably ever occur in a storm identical in size, shape, and characteristics with the one applied. The principal of these assumptions are: (1) that the maximum storm center will occur over area A; (2) that the isohyetal enclosing an area at the principal storm center equal to A will be identical in shape, orientation, and location to area A; (3) if two or more maximum periods of the same storm are used, the further assumption is made that the storm center will not move or change its shape.

It may be found advisable to use the 2-day time-area-depth curve for a different storm than that employed for the 1-day determination. For though the 1-day curve of the one storm may lie higher on the diagram than that for the other, yet the curves for the 2-day period may have their relative positions reversed. In such a case all of the above assumptions would be involved, and the further assumption that the maximum 1-day period of the one storm is followed by a greater 2-day maximum of the other storm. It must not be overlooked, however, that a storm over the area in question, greater than any yet recorded, may materially outweigh all of the above considerations.

In any storm the proportion of rainfall which appears immediately as runoff is largely dependent on the degree of saturation of the soil caused by previous rainfall. Before any direct use can be made of the time-area-depth curve for a given maximum period, therefore, it is necessary to know the amount of rainfall which occurred over a similar area on the preceding days of the storms used as examples. This can be determined indirectly from the time-area-depth curves by taking the difference in height of the curve including the previous days of the storm, and the curve exclusive of them, for the area in question. In considering the effect which a given storm will produce it is necessary to take into account the topography of the area, season of the year, vegetable growth, character of surface soil, its condition, and degree of saturation.

The time-area-depth curves here presented are not adapted for use in the design of sewer systems. In the latter, time units as small as hours and minutes, and units of area as small as acres are essential. Chapter V, on frequency of excessive precipitation, applies more particularly to rainfall over small or negligible areas. In flood control studies, however, units smaller than the day and the square mile, respectively, would be of no practical value.

For the convenience of the reader certain time-area-depth data for each of the 33 storms, taken from the curves, is presented in tabular form in tables 6 and 7. This consists of the depth in inches of the greatest average rainfall on areas of 1, 500, 1000, 2000, 4000, and 6000 square miles, during maximum consecutive periods of 1 to 5 days. Storm 151 is listed in both tables, for the reasons given on page 175, which accounts for there being 17 storms in table 7.

SELECTING THE 33 MOST IMPORTANT STORMS

The entire rainfall investigation, as previously stated, was conducted primarily with a view to its bearing on Miami Valley conditions. In the selection of storms for mapping, it was necessary to limit the work involved by choosing certain restrictions of storm area, depth, and the time during which a storm might be considered to last. From observation and from previous studies of storms made by the Morgan Engineering Company, it was known that heavy precipitation over an appreciable area is rarely continuous for a period of more than five or six days. Several such rainfall periods, however, may occur separated by only one or two days of little or no rainfall.

It is the latter type of storms which cause the greatest floods on rivers having very large drainage areas, where more than three or

Table 6.—Depth in Inches of the Greatest Average Rainfall over Areas of various sizes for 17 important Storms in the Northern part of the United States

No.	Storm Date	Center	Greatest 24-Hour Rainfall Over Area in Square Miles						Greatest 48-Hour Rainfall Over Area in Square Miles					
			1	500	1,000	2,000	4,000	6,000	1	500	1,000	2,000	4,000	6,000
a	1869, October 3–4	Conn.	....	....	....	....	....	....	12.4	10.4	9.7	8.9	8.1	7.8
c	1889, May 31–June 1	Pa.	8.4	7.8	7.5	7.3	7.1	6.8	9.8	9.1	8.8	8.4	7.8	7.5
10	1894, May 18–22	Pa.	5.8	5.5	5.3	4.9	4.3	4.0	9.0	7.8	7.2	7.1	6.9	6.8
15	1895, December 17–20	Mo.	5.8	5.8	5.7	5.7	5.5	5.4	9.6	9.3	9.0	8.6	8.1	7.7
25	1897, July 12–14	Conn.	8.0	7.3	6.8	6.2	5.6	5.2	10.3	9.5	9.1	8.5	7.5	6.9
51	1900, July 14–16	Iowa	7.0	6.8	6.6	6.3	5.8	5.4	13.0	11.7	10.7	9.6	8.6	8.0
72	1903, August 25–28	Iowa	11.2	10.6	10.0	9.1	7.8	6.8	14.7	11.8	10.8	9.7	8.3	7.7
76	1903, October 8–9	N. J.	11.5	9.3	8.4	7.6	6.8	6.3	15.0	11.9	10.9	9.9	9.0	8.4
83	1905, June 9–10	Iowa	12.1	10.9	10.0	8.9	7.8	6.9	....	....	....	....	....	....
88	1905, September 15–19	Mo.	8.1	7.4	6.9	6.2	5.5	5.0	8.2	8.1	8.0	7.8	7.4	7.1
109	1909, July 5–7	Mo.	9.7	8.0	7.2	6.1	5.0	5.0	9.7	8.4	7.7	6.7	6.1	5.9
110	1909, July 20–22	Wis.	10.7	9.0	7.9	6.5	4.9	4.3	11.0	9.7	8.6	7.4	6.2	5.6
114	1910, October 4–6	Ill.	8.5	7.5	7.1	6.5	5.9	5.6	13.0	11.0	10.2	9.5	8.7	8.2
125	1912, July 20–24	Wis.	....	....	....	....	....	....	....	....	....	....	....	....
130	1913, January 10–12	Ark.	5.8	5.6	5.4	5.1	4.9	4.7	7.4	6.8	6.6	6.4	6.0	5.8
132	1913, March 23–27	Ohio	7.0	6.7	6.5	6.2	5.8	5.7	9.5	8.8	8.4	8.2	8.0	7.8
151	1915, August 17–20	Ark.	7.0	6.6	6.5	6.5	6.4	6.2	10.9	10.2	9.7	9.1	8.6	8.3

Table 6.—*Continued.*

Storm No.	Greatest 72-Hour Rainfall Over Area in Square Miles						Greatest 96-Hour Rainfall Over Area in Square Miles						Greatest 120-Hour Rainfall Over Area in Square Miles					
	1	500	1,000	2,000	4,000	6,000	1	500	1,000	2,000	4,000	6,000	1	500	1,000	2,000	4,000	6,000
a																		
c																		
10	9.2	8.2	8.0	7.6	7.1	6.8	10.1	8.6	8.4	8.2	7.6	7.2	10.1	9.5	9.1	8.8	8.4	8.1
15	11.9	10.7	10.5	10.1	9.6	9.2	12.1	11.0	10.5	10.2	9.7	9.3						
25	10.3	9.6	9.3	8.6	7.7	7.0												
51	13.7	12.2	11.4	10.5	9.5	8.9	15.5	12.4	11.7	10.7	9.6	8.8						
72	15.5	12.2	11.4	10.5	9.4	8.7												
76																		
83																		
86	10.5	10.3	10.1	9.8	9.1	8.6	12.2	11.3	10.9	10.5	9.9	9.4	13.0	12.5	12.0	11.2	10.4	10.0
109	11.2	10.1	9.7	9.1	8.6	8.2												
110	12.8	9.8	8.7	8.0	7.2	6.7												
114	15.2	13.7	12.8	12.2	11.7	11.4												
125													11.3	9.4	8.4	7.5	6.5	6.0
130	7.5	7.5	7.4	7.3	7.1	6.9	11.1	10.4	10.0	9.6	9.4	9.3						
132	10.2	9.7	9.5	9.2	9.0	8.7	14.8	13.9	13.2	12.5	11.6	11.2	11.2	10.9	10.8	10.5	10.1	9.9
151	14.0	13.1	12.6	12.0	11.2	10.7												

Table 7.—Depth in Inches of the Greatest Average Rainfall over Areas of various sizes for 17 important Storms in the Southern part of the United States

| No. | Storm | | Greatest 24-Hour Rainfall Over Area in Square Miles | | | | | | Greatest 48-Hour Rainfall Over Area in Square Miles | | | | | |
	Date	Center	1	500	1,000	2,000	4,000	6,000	1	500	1,000	2,000	4,000	6,000
b	1887, July 27–31	Ga.	10.0	8.9	8.3	7.5	6.6	5.9	14.1	12.6	11.9	11.0	9.7	8.8
13	1894, September 24–26	Fla.	12.5	10.8	10.3	9.6	8.6	8.0	12.5	11.9	11.7	11.4	10.9	10.5
24	1897, March 22–23	Ga.	10.3	9.6	8.9	8.1	7.1	6.3	11.6	11.3	11.0	10.6	10.1	9.6
43	1899, June 27–July 1	Texas	11.0	9.9	9.4	8.8	8.0	7.4	22.0	16.6	14.6	13.4	12.8	12.3
49	1900, April 15–17	Miss.	12.7	10.3	9.9	9.4	8.6	7.9	12.6	11.7	11.4	11.0	10.5	10.1
63	1902, March 26–29	Miss.	9.0	8.7	8.3	7.8	7.3	7.0	10.0	9.0	8.9	8.8	8.6	8.4
93	1906, November 17–21	Ark.	8.0	7.5	7.3	7.0	6.6	6.3	11.0	9.9	9.5	8.9	8.2	7.9
102	1906, July 28–August 1	La.	9.6	8.8	8.2	7.0	5.7	4.9	15.0	13.7	12.8	11.5	9.4	8.2
103	1908, August 24–26	N. C.	11.7	9.2	8.2	6.8	6.3	6.0	13.1	12.0	11.0	9.4	8.4	8.0
105	1908, October 20–24	Okla.	6.3	6.2	6.0	5.8	5.5	5.3	11.4	11.0	10.7	10.1	9.4	8.8
108	1909, June 29–July 3	Fla.	12.0	11.5	10.3	8.8	7.6	6.8	15.5	14.2	13.0	11.8	10.8	9.9
135	1913, October 1–2	Texas	13.0	12.3	11.7	10.7	9.2	8.3	14.2	13.7	13.3	12.4	11.1	10.2
136	1913, December 2–5	Texas	10.0	8.5	7.5	6.6	6.3	6.0	12.3	11.2	10.5	9.6	8.8	8.4
144	1913, October 14–15	N. C.	10.0	7.3	6.3	5.8	5.2	4.6	11.3	8.5	7.5	6.9	6.1	5.5
151	1915, August 17–20	Texas	9.1	8.7	8.0	7.9	7.7	7.5	18.6	16.2	15.0	13.6	12.3	11.7
156	1916, July 6–10	Ala.	11.1	10.7	10.2	9.5	8.8	8.2	17.3	15.8	14.8	13.7	12.3	11.3
157	1916, July 14–16	N. C.	19.3	12.2	11.3	9.9	8.4	7.5	23.2	18.4	17.1	15.5	13.7	12.5

Table 7.—*Continued.*

Storm No.	Greatest 72-Hour Rainfall Over Area in Square Miles						Greatest 96-Hour Rainfall Over Area in Square Miles						Greatest 120-Hour Rainfall Over Area in Square Miles					
	1	500	1,000	2,000	4,000	6,000	1	500	1,000	2,000	4,000	6,000	1	500	1,000	2,000	4,000	6,000
b	16.5	14.4	13.7	12.8	11.7	10.9	16.5	14.5	13.8	12.9	11.8	11.2	16.5	15.1	14.5	13.8	12.8	12.2
13	12.6	12.2	11.9	11.7	11.4	11.2												
24																		
43	33.0	25.6	22.1	20.4	18.7	17.5	33.0	26.1	22.4	20.9	19.4	18.4	33.0	26.1	22.4	20.9	19.4	18.4
49	13.9	12.2	11.8	11.7	11.4	11.1												
63	11.8	10.3	9.7	9.1	8.9	8.7	11.8	10.6	10.1	9.8	9.5	9.2						
93	13.0	11.7	11.1	10.5	10.0	9.8	16.0	14.8	14.0	13.1	12.0	11.4						
102	17.6	16.2	14.8	13.1	11.0	9.7	18.8	17.0	15.6	13.8	11.8	10.5	19.4	17.8	17.0	16.0	14.9	14.1
103	15.6	13.9	13.1	12.2	11.0	11.3							18.8	17.1	15.8	14.0	11.9	10.6
105	13.0	12.6	12.2	11.6	11.0	10.4	15.1	14.2	13.6	13.0	12.2	11.7	15.2	14.4	13.9	13.2	12.2	11.6
108	15.8	14.6	13.4	12.4	11.4	10.5	17.7	15.4	14.9	14.1	12.9	12.0	17.8	16.4	16.0	15.5	14.8	14.2
135																		
136	14.5	13.3	12.8	12.1	11.2	10.9	15.5	14.4	13.8	13.0	12.1	11.5						
144																		
151	19.7	19.6	19.5	19.2	18.6	17.9	19.7	19.6	19.5	19.3	18.7	17.9						
156	19.7	18.7	17.8	16.5	14.7	13.6	20.3	20.0	19.8	19.3	18.4	17.4	22.3	20.4	19.9	19.4	18.6	17.7
157	23.7	18.4	17.3	15.6	13.8	12.7												

four days is required for sufficient runoff to collect to cause flood conditions. For an investigation of storms which cause the greatest floods on such rivers a longer rainfall period than three or four days would have to be chosen as the maximum for a storm. A number of storms have been recorded which lasted ten or even fifteen days, but in almost every such instance the intensity during the five consecutive days of maximum rainfall was much less than the intensity of other storms of the same region which lasted only five days or less. It is a well-known fact that these long-continued general storms of late winter and early spring are not those which are the greatest menace to life and property on streams which have small drainage areas. Such storms frequently extend over the entire eastern half of the United States, but in no place is the rainfall very intense. In all streams except the very largest, the time required for rainfall to reach the main channel and run off is so small, in comparison with the duration of the storm, that no serious flood can result. The water simply collects and runs off practically as fast as it falls. On the contrary, the largest streams, such as the Mississippi and Missouri Rivers, experience their greatest floods after such long-continued general storms. When such a storm covers the entire watershed of one of these rivers, the flood crests on all the principal tributaries may reach the main channel at approximately the same time, and cause the greatest flood stages on the lower reaches of the river. A good example of this type of storm is that of May 16–31, 1903, described on page 151.

From a consideration of the area of the Miami watershed and the characteristic length of storms in the United States, it was decided for the purpose of mapping to include the maximum periods of storms up to five days, and for each of these periods the entire continuous storm area included within the 2-inch isohyetal.

The 33 storms selected for detail investigation comprise the 3 greatest storms of the 49-year period, 1843–1891, and 30 of the largest and most important storms taken from the 160 which occurred during the 25-year period, 1892–1916. The latter consist of 15 storms taken from the northern and 15 from the southern group. The 3 early storms were selected as the largest and most important after a careful sifting process, now to be described in more detail.

STORMS WHICH OCCURRED FROM 1843 TO 1872

The search for storm data for the 30-year period 1843–1872 was confined to a few of the largest storms. The object was to learn whether or not any storms during this period were as great or greater than storms of recent years that have occurred in the same sections of the country. It was obviously impossible to study minutely the lesser storms of this early period, on account of scarcity of rainfall records. However, the northeastern part of the United States was somewhat of an exception in this respect, as rainfall records were kept by many scientifically trained men for their own use, or in cooperation with the Smithsonian Institution.

In the middle west and south there were few rainfall gaging stations, and it is necessary to rely very largely on the newspapers and scientific journals for descriptions of rainfall and flood conditions. All of these sources of information were examined, and the time-area-depth relations of a few of the largest storms were analyzed as closely as the character and extent of the data would permit.

In the northern part of the country only one storm furnished sufficient data to exhibit fully its time-area-depth relations. This is storm *a* of October 3–4, 1869, central over Connecticut, a map of which is shown in figure 59, and description given on page 138. The fact that a number of scientists and engineers were making rainfall observations at that time, and yet no other storm was reported which closely approached storm *a* in depth and area covered, is very good proof that this was the largest storm which occurred in the northern part of the country during that period. Doubtless if there were complete data available, so that the true size of other large storms of the period could be determined, some of these would approach storm *a* in magnitude, but it is hardly possible that it could have been materially exceeded.

The time-area-depth curve for storm *a*, figure 95, is exceeded by that of storm 76, Oct. 8–9, 1903, central over New Jersey. The other northeastern storms, in order of size for the 2-day period, are storm 25, July 12–14, 1897, central over Connecticut; storm *c*, May 31–June 1, 1889, central over Pennsylvania; and storm 10, May 18–22, 1894, central over Pennsylvania. The curves for all of these lie below that for storm *a*. It is apparent from this, also, that storm *a* is one of the greatest to which the northeastern states are subject.

Deductions as to the possible maximum size of storms which may have occurred in the northwest and south during this 30-year period must be based on the comparatively few rainfall records available, on newspaper and technical journal reports, and on the comparisons which may be made of this data with the accurate and abundant data of the largest storms of the last 25 years. A comparison of the greatest storms of the past 25 years in Iowa, Illinois, and Ohio, numbers 72, 51, 114, and 132, figure 95, with those in the northeastern states, shows that they vary most at the peak and differ but little for areas of 5,000 to 10,000 square miles. It is, of course, natural that the peak

14

records should vary greatly, even in the same section of the country, as they are for but one or only a very few stations.

The final conclusion is that for the 30-year period 1843–1872 there were no storms in the northeastern states which exceeded that of October 3–4, 1869. The maximum storm which may have occurred in other parts of the United States east of the 103d meridian cannot be so definitely fixed, but from the fairly reliable data available it is safe to say there was no storm which materially exceeded the largest storms of the past 25 years in the same regions.

STORMS WHICH OCCURRED FROM 1873 TO 1891

During the 19 years from 1873 to 1891 the number and distribution of stations leaves little room for doubt that not only were all the greatest storms recorded at a number of stations, but also that the records of northern storms are sufficient to reveal with a considerable degree of accuracy their comparative time-area-depth relations. However, there were large areas in the southern and central western states in which stations were relatively sparse, and consequently it is probable that there are no records for many of the smaller storms. For this reason the investigation for the 19-year period, 1873–1891, was also confined to a comparison of the greatest storms of that period with the greatest storms over the same geographical sections, which have occurred during the last 25 years. After a thorough examination of all the data available, maps and time-area-depth curves were made for two of the greatest storms of the period.

Storm *b*, July 27–31, 1887, over Georgia, is shown in figures 62 to 66; the time-area-depth curves are included with those for the southern group of storms, figures 99 to 103. The curves for this storm are considerably below those for later southern storms, which indicates insufficient data for the earlier period, instead of an increase in the size of storms in late years. However, this storm is high enough in the scale of great southern storms to substantiate the claim that it is one of the largest that has occurred during the period in question. As in the preceding 30-year period, the positive evidence, as shown by records of precipitation, that no greater storm occurred, is corroborated by the negative evidence that there are no river gage, newspaper, or scientific journal records to show that a greater storm did occur during this period.

In the northern and eastern states storm *c*, May 31–June 1, 1889, central over Pennsylvania, is the greatest of the period. The resulting floods caused widespread disaster, that at Johnstown being historic. The maps are shown in figures 71 and 72, and the time-area-depth curves in figures 94 and 95.

There is but little question that storms *a*, *b*, and *c* were among the greatest which occurred during the 49 years from 1843 to 1891. If rainfall gaging stations had been as numerous during this period as during the succeeding 25 years the storms would have been more accurately recorded, and it is possible that they were somewhat greater than is indicated by the data now available.

STORMS WHICH OCCURRED FROM 1892 TO 1916

All storms of magnitude which occurred from 1892 to 1916 are accurately recorded. The number and distribution of rainfall gaging stations for this period make it possible to map these storms and obtain a close comparison of their time-area-depth relations.

How to select the 30 most important storms of this 25-year period from the 160 for which detail precipitation records had been compiled presented no simple problem. It would have been comparatively easy if maps and time-area-depth curves had been prepared for all 160 storms. A great deal of time and labor is required to make these maps and curves, however, and no additional information of value is gained by deducing the time-area-depth relations for a large number of the less important storms. The object was not merely to select for mapping the 30 largest storms from the charts showing relative sizes, as in effect this would have included very few northern storms. Even after dividing the storms into northern and southern groups, it was not thought advisable to map the 15 greatest storms in each group, as this would leave large sections unrepresented. It was finally decided to map four or five of the greatest storms in each group, without regard to geographical position, and then to select the remaining storms to be mapped from different sections of the country, taking the greatest for each section. Those finally selected for mapping are indicated by an asterisk in tables 4 and 5.

The 4 largest storms in the northern group, as shown in the chart of comparative depths of storms, figure 48, were first selected for mapping and time-area-depth study. These are, in order of size, number 151, central over northern Arkansas; number 114, central over southern Illinois; number 76, central over New Jersey, and number 72, central over Iowa. The fifth storm in order of size, number 39, was not mapped because it had no special features, and fell over approximately the same area as storms 151 and 114, both of which exceeded it.

The next 8 storms in order of size were mapped. They are, number 15, central over Missouri, which was platted because of the great area covered and its unusual size for a winter storm; number 132, the great storm of March 23–27, 1913, central over Ohio and

Indiana; number 10, central over Pennsylvania, which is an example of the unusually large storms of long duration on the North Atlantic coast; number 25, central over Connecticut; number 109, central over Missouri; number 110, central over Wisconsin, which is the only storm in that region coming within the selected limits, although another Wisconsin storm was mapped which lies just below those limits; number 51, central over Iowa; and number 86, Sept. 15–19, 1905, central over Missouri. The long major axis of storm 86 had a northwest to southeast direction, which is very unusual for a storm of this size. Another distinctive feature was its long duration at a time of year when short violent storms generally occur.

The next five smaller storms, numbers 94, 23, 14, 129, and 12, were not mapped because they are not of particular interest. Number 83, the next one mapped, is the well-known Devils' Creek storm in Iowa, of June 9–10, 1905. In the area of maximum precipitation the greater part of the rainfall occurred within 12 hours, and all within 24 hours. The maximum record was 12.1 inches.

Only two other storms in the northern group were mapped, number 130, January 10–12, 1913, over northern Arkansas and Kentucky, bearing some resemblance in season, type, and area covered, to storm 132, of March 23–27, 1913, which caused the disastrous flood in the Miami Valley; and number 125, the smallest storm mapped, of July 20–24, 1912, over Wisconsin. This was mapped because of its center depth, which is unusual for that region, and because only one other storm is mapped for that section.

In the southern group, as in the northern, the four greatest storms were mapped. The fifth largest storm, number 56, was central over Florida, was not very unusual in any way, and was not in a position to cause material flood damage. The next greatest storm, number 13, included practically the same area and was platted instead of number 56. Intense precipitation fell over a much larger area in storm 13, as may be seen by comparing the tenth and twentieth values of the two storms.

The next four storms in size, numbers 136, 108, 43, and 135 were platted. The next, number 134, was not platted because it was central over Louisiana, a section most prolific in storms, where great flood damage cannot result.

The two storms following in size, numbers 93 and 105, were platted, and the next two, numbers 150 and 92, were passed by on account of unimportance. Storms 24 and 63 were mapped to help get a representative distribution of storms. Only one other storm was mapped. This is number 102, July 29–31, 1908, over Louisiana. It is of interest chiefly on account of the depth of 17.62 inches at the center.

CHAPTER IX.—MAPPING AND PREPARING TIME-AREA-DEPTH CURVES FOR THE 33 MOST IMPORTANT STORMS

In determining the time-area-depth relations of storms many unforeseeable difficulties and uncertainties were encountered. To determine the best procedure in such cases numerous experiments were tried. In order that other investigators, who may wish to make similar storm studies, may be forewarned of the difficulties and may be able to profit by our experience, this chapter contains a rather detailed description of the various expedients tried and the methods finally adopted.

The relations of storm duration, area, and depth are perhaps the most complex of all that ordinarily receive attention in the discussion of rainfall statistics. Numerous other distinct and independent variable factors enter, such as rate and direction of storm movement, location and distribution of rainfall gaging stations, and time of making observations; and their mutual inter-relations are both elusive and involved. In writing this chapter every attempt has been made to discuss and evaluate each element separately at first, to remove every trace of obscurity or uncertainty in primal meaning, and thus to help the reader to unravel the tangled web of relationships which results from the records of the actual storm phenomena. In spite of this effort, however, this chapter doubtless remains the most difficult in the whole volume and will require the greatest demand upon the attention of the reader. It will be necessary constantly to keep before the mind the nature of each element involved in order to appreciate the discussions of relationships from various aspects.

ASSEMBLING AND COMPUTING DATA FOR MAPPING STORMS

The complete data for each storm selected for mapping was assembled and listed in tabular form as illustrated in figure 104. This form is entirely distinct from the excessive precipitation sheets described in chapter IV, the data for the storms being much more complete. In order to have all the data needed for mapping a storm, the records taken during the storm period at all the stations touched by the storm were copied irrespective of their amounts. This introduced records from many stations where the amount was less than the

213

minimum used in the excessive precipitation sheets. The smallest number of stations for one storm, used in plotting the rainfall maps, was 33 and the greatest number for one storm was 965.

As shown in figure 104, the stations were first listed in alphabetical order, and then opposite each station in the successive columns were copied the daily rainfall records during the storm period considered. The appropriate dates are entered at the tops of the respective columns.

Next, each column was totaled. The day with the greatest total was considered the date of maximum 1-day precipitation. In most of the 33 storms thus studied in detail no doubt exists as to the date of maximum 1-day precipitation. In some cases, however, usually in storms of long duration and covering a large extent of territory, there was some uncertainty as to what should be considered the maximum 1-day period. The date of maximum precipitation over one center of heavy rainfall, for instance, is sometimes different from the date of maximum precipitation over another distinct center of sufficient importance to merit separate consideration.

The period of maximum 2-day precipitation was determined by joining to the date of maximum 1-day precipitation either the day preceding or the day following, using the day on which occurred the greater total sum of precipitation at all the stations. The dates for the 3-day and successive maximum periods were similarly determined.

After the dates were fixed, the total amounts of precipitation at each station for the maximum 1, 2, 3, 4, and 5-day periods were entered in the last columns of the form as illustrated in figure 104.

MAPPING THE STORMS

On outline maps showing the principal geographical features and state boundaries within and adjacent to the area of the storm, each rainfall station within the storm area was located and its precipitation for the period considered was noted beside it. Isohyetal lines were then drawn at 1-inch rainfall intervals, down to and including the 2-inch line.

In appearance these rainstorm maps, figure 105, resemble topographic maps. If data were available to plat rainstorm maps with the same degree of accuracy with which good topographic maps are platted, there could have been no necessity for selecting one from several methods of platting the former. Unfortunately, however, several factors operate to reduce the accuracy of rainstorm maps, making it advisable to devise some means, if possible, to eliminate these inaccuracies. They may be divided into four classes:

MORGAN ENGINEERING COMPANY
MEMPHIS

Subject __Storm of March 23-27, 1913_____ File _____

Project No. _____ Computed by_____ Date _____ Checked by _____ Date _____ Page _____

Station		Daily Precipitation					Max. Precipitation				
		23	24	25	26	27	1	2	3	4	5
Ohio-Continued						-2-					
Camp Dennison		.02	1.92	2.25	2.85	.40 4.17	2.25	5.10	7.02	7.04	7.44
Canal Dover		.30	2.70	1.35	.75	4.05	1.35	2.10	4.80	5.10	5.10
Canton		1.03	2.20	3.00	1.62	.60 5.20	3.00	4.52	6.82	7.85	8.45
Cardington		1.02	2.55	3.00	1.40	5.55	3.00	4.40	6.95	7.97	7.97
Chillicothe	||	.02	1.20	*	4.60		1.20	*	*	*	5.82
Cincinnati	W.B.	2.21	4.15	1.11		6.36	4.15	5.26	7.47	7.47	7.47
Circleville		.15	1.50	1.97	2.29	.37 3.47	1.97	4.26	5.76	5.91	6.28
Clarington		.12	.10	.50	.95	.22 .60	.50	1.45	1.55	1.67	1.89
Columbus	W.B.	.53	2.14	2.89	1.40	.01 5.03	2.89	4.29	6.43	6.96	6.97
Cochocton	φ ||	.09	.20	2.70							
Dayton		.51	2.91	3.28	1.48	.76 6.19	3.28	4.76	7.67	8.18	8.94
Delaware		1.12	2.00	2.46	1.87	T 4.46	2.46	4.33	6.33	7.45	7.45
Demos		.23	.27	1.00	1.85	.72 1.27	1.00	2.85	3.12	3.35	4.07
Dennison	φ	.42	.76	.90							
Frankfort		T	1.20	1.67	2.20	1.42 2.87	1.67	3.87	5.07	5.07	6.49
Gallipolis	||		.14	1.80	1.22	.14	.14	1.94	1.94	1.94	3.16
Garrettsville		1.98	1.03	4.61	.88	.87 5.64	4.61	5.49	6.52	8.50	9.37
Granville		.49	1.43	2.68	2.06	.50 4.11	2.68	4.74	6.27	6.76	7.16
Sratiot		.39	1.29	2.21	3.00	.57 3.50	2.21	5.21	6.50	6.89	7.46
Green		.03	.95	3.20	.46	.98	.95	4.15	4.18	4.18	4.64
Green Hill		.59	1.56	1.54	1.27	.66 3.10	1.54	2.81	4.37	4.96	5.62
Greenville		1.29	1.77	4.45	1.41	.41 6.22	4.45	5.86	7.63	8.92	9.33
Hamilton	||	.32	.70	2.70	*	* 3.40/3.80	2.70	*	*	*	7.52
Haydenville			1.40	.70	1.63	1.40	1.40	2.10	2.10	2.10	3.73
Ironton		.01	.40	2.67	.81	.41	.40	3.07	3.08	3.08	3.89
Kenton		2.00	1.50	3.60	1.20	.35 5.10	3.60	4.80	6.30	8.30	8.65
Killbuck		.70	1.65	3.75	2.00	.70 5.40	3.75	5.75	7.40	8.10	8.80
Kings Mills	||		.69	2.57	4.06	1.22 3.26	2.57	6.63	7.32	7.32	8.54
Lancaster		.38	2.48	1.74	2.48	.42 4.22	1.74	4.22	6.70	7.08	7.50
Mᶜ Connelsville		.21	.24	1.40	1.80	.68 1.64	1.40	3.20	3.44	3.65	4.33
Marietta		.23	.06	.70	1.33	.35 .76	.70	2.03	2.09	2.32	2.67
Marion		1.38	1.97	4.39	1.87	1.00 6.36	4.39	6.26	8.23	9.61	10.61
Milfordton		.58	1.20	2.25	1.62	.60 3.45	2.25	3.87	5.07	5.65	6.25

FIG. 104.—FORM ILLUSTRATING THE METHOD OF COMPILING STORM DATA.

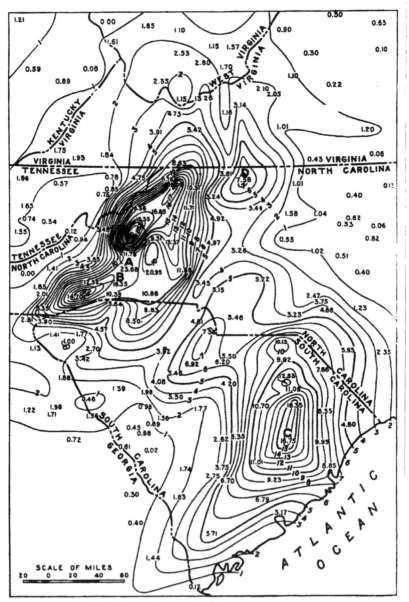

FIG. 105.—MAP ILLUSTRATING METHOD OF PLATTING STORMS.

Isohyetals shown are for 3 days of maximum rainfall during storm of July 13–17, 1916.

SOURCES OF ERROR IN ORIGINAL DATA

Class 1. Arbitrary Location of Observing Stations

Rainfall recording stations, being permanently located, rarely furnish data sufficient to develop precisely the shape of a rainstorm area. A rainstorm map platted from the available records depicts the storm no more accurately than a topographic map platted from random stadia observations would represent actual topography. In comparatively level country the difference between two topographic maps, one from well chosen stadia observations and the other from random stadia observations, would not be great. This also holds for the comparatively regular edges of a rainstorm map. In hilly country, however, the two topographic maps would probably not resemble each other closely; and it is very likely that this can be truly said of two rainstorm maps platted under analogous conditions. In recent years, and particularly in the more densely inhabited parts of the country, stations are comparatively close together, so that storms platted from their records more nearly represent the true conditions. It is noticeable that the closer stations are together, the more irregular appears to be what may be called the topography of the storm. Particularly is this so in the vicinity of the storm centers. For this reason the percentage of error in areas on which great depths fall is larger than in areas covered to less depths. Other things being equal, also, the percentage of error will be greater as the stations are farther apart.

Class 2. Inaccuracy of Records

In some cases the accuracy of the data is questionable. It is subject to two kinds of errors. First, errors due to failure to catch the precipitation properly; and second, personal errors, due to failure to read, record, transcribe, or tabulate correctly. By far the greatest number of observers are carrying on the work voluntarily, and hence it is not probable that the close attention is given to accuracy which is to be expected from paid observers. Occasional inconsistencies in the records were found, indicating errors from these sources, and it was difficult at times to decide whether or not to alter or discard certain data.

Class 3. Arbitrary Division of Time

Another and very important source of inaccuracy in determining the greatest rainfall for a given period is due to the fact that readings at each station are recorded at a stated time each day without regard to the time of beginning or ending of the storm. Under these cir-

cumstances it is almost certain that a record of the greatest rainfall
within a given length of time is seldom obtained. What is recorded
as a 2 days' rainfall may be, and often is, found to be actually a storm
of less than 24 hours' duration. A 24-hour record may likewise cover
only a few hours of rainfall. Errors of this kind obviously are always
in the same direction, and arithmetical averages derived from them,
therefore, cannot be compensating. They are the more serious
because they always indicate a greater length of time for a given
precipitation, and therefore a smaller storm intensity, than the true
one. The effect of this error, however, tends to be less for two days
than for one, less for three days than for two, and in the more serious
storm type extending over a period of several days the error is of little
consequence.

Another factor which tends to reduce this error is the area over
which a given storm occurs. Let us assume that the storm is moving
in an easterly direction and causes a precipitation of 5.2 inches at
station *a* lasting from 11 p. m. on May 18 to 11 p. m. May 19. In its
eastward movement it reaches station *b* and causes a precipitation of
4.5 inches, lasting from 2 p. m. May 19 to 2 p. m. May 20. Although
the precipitation lasts only 24 hours at each station, as stated above,
it is recorded as a 48-hour storm at both stations if the readings are
taken about 6 p. m. But from the time rainfall began at station *a*
until it ceased at station *b* was 39 hours, and the runoff for the storm
area including stations *a* and *b* was in many respects similar to that
which would have been caused had the precipitation at each station
been distributed uniformly throughout the 39 hours.

Class 4. Variations in Time of Observation

Another class of inaccuracies is brought about by the fact that
readings at all stations are not taken simultaneously. Most of the
readings are taken about sundown, but a considerable number are
taken in the morning. A few, the so-called regular Weather Bureau
records, are computed from continuous automatic records and cover
the period from midnight to midnight. It was found that the last
class agrees fairly closely with the evening readings at surrounding
stations. No semblance of agreement could, however, be found to
exist between evening and succeeding morning readings. As the
evening readings are in the preponderance, they necessarily constitute
the controlling data for the maps. The morning readings would have
been discarded altogether had it not been that without them the
remaining material was too scant to be useful.

EXPERIMENTAL MAPS

Before platting the final maps for the 33 storms selected for that purpose, experimental maps were made in order to learn by what method they could be constructed with the greatest facility and accuracy.

To determine the effect of the several sources of inaccuracy enumerated above, experiments were made with the data for two storms. The storm of October 4–6, 1910, was platted on maps by four different methods, here designated the A, B, C, and D methods. The storm of November 17–21, 1906, was platted by two of these, the A and B methods.

By the A method the morning and evening records taken on the same day were treated as though of equal value and as though they had been taken simultaneously. The depths of rainfall for the different periods of maximum accumulated precipitation at each station were taken directly from the computations described on page 214. This method is the simplest of the four and would be the natural way of utilizing the data if corrections for the four sources of error described above could readily be made in the original figures. The only inaccuracies that the A method permits to be corrected, however, are those falling in the second class, that is, inaccuracies of reading and recording, and this only when surrounding stations furnish unmistakable evidence as to the nature of the error.

By the B method the evening records were made to control the lines of equal rainfall within the limits imposed by the morning records of the following day. That is, the isohyetals, although drawn with little regard for the morning records of the calendar day, were not drawn through points shown by the records of the following morning to have had less rain than the lines would indicate. The morning records were shown in the form of a fraction, the numerator in black indicating the reading for the current morning and the denominator in red that of the following morning. This method is, therefore, an attempt to adjust inaccuracies due to some readings being taken in the morning. With these adjustments it becomes perhaps slightly easier to form an opinion as to whether or not certain station records were subject to the inaccuracies noted under class 2 above.

The C method was tried with the hope that it might provide a method of compensating for inaccuracies due to variations in time of observation, for those due to arbitrary division of time, and to some extent for those due to inaccuracy of reading and recording, described above under classes 4, 3, and 2, respectively. By this method a definite date was not selected as that of 1-day maximum accumulated precipitation for all the stations, but the day on which the summation of rainfall for all the stations was greatest (elsewhere considered the date of 1-day maximum accumulated precipitation) was merely taken as a fair *basic date*, and the 1-day maximum accumulated precipitation allowed, at each particular station, to vary 1 day in either direction from this basic date. Thus, in the storm of October 4 to 6, 1910, the date for which the summation for all the stations is greatest, or the basic date, was October 5; but the 1-day maximum accumulated precipitation at Farmersburg, Ohio, was taken as October 4, because the precipitation at Farmersburg on that date was greater than that on either October 5 or 6; the 1-day maximum at Evansville, Ind., was taken as October 5, because greater at Evansville on that date than on October 4 or 6; the 1-day maximum at Farmland, Ind., was taken as October 6, because greater at Farmland on that date than on October 4 or 5. The dates of maximum accumulated precipitation at each particular station were allowed to include the day either preceding or following the basic period, determined by summing up the precipitations at all the stations, provided this gave a greater maximum for that station. The shapes of the time-area-depth curves obtained by the C method did not seem to justify the latter's use. This is mainly due to the fact that it does not make allowance for the progressive movement of a storm. For instance, if the heaviest rainfall of a given storm extends from May 23 to 25 in southeast Missouri, the heaviest rainfall period in Indiana for the same storm may be, and likely will be, from the 24th to the 26th. By the C method it is made to appear that the rain fell simultaneously in the two states.

In the D method, each successive day's record was platted on a separate map and isohyetal lines drawn. The two successive maximum days were combined by laying one map over the other, noting the points of intersection of the isohyetals, and placing a figure at the point indicating the sum of the two lines. Isohyetals for the 2-day period were then drawn through these points. The map so obtained was combined with the map for the day preceding or following, depending on which was the larger, to get points on the isohyetal lines for the 3-day period, and so on.

This would seem to be the most logical method of the four in that it retains on the map of each successive day certain of the irregularities of storm shape of preceding days, which do not appear when an independent map is platted for each maximum period. By this method, however, it is sometimes possible to build up higher storm centers than seem warranted by the data. This feature, perhaps, has a tendency to minimize inaccuracies introduced by the fixed locations of the stations. One disadvantage is that the method is very laborious and time consuming. Another is that the A method must be used for platting the precipitation records for each day of the storm, or else it becomes necessary to readjust the morning records arbitrarily.

CONCLUSIONS FROM EXPERIMENTAL MAPS

Time-area-depth curves were drawn for each of the experimental maps constructed by methods A, B, C, and D, for the storm of October 4–6, 1910. There were four distinct but approximately parallel curves for the storm periods of 2 and 3 days; for the maximum 1-day period the maps and curves constructed by methods A and D would have been identical throughout, hence only one set of maps and curves was drawn for this period. The curves by methods A, B, and D for the maximum 2-day and 3-day periods were found to follow each other closely. Similar comparative curves were drawn from the A and B method experimental maps of the storm of November 17–21, 1906. The C and D method maps and curves were not drawn for this storm. The A and B method curves were not as nearly coincident as for the storm of October 4–6, 1910.

The A method was found to be substantially as accurate, and much simpler than any of the other three. It was, therefore, the method finally adopted for use in constructing maps for the 33 great storms.

The B method was found impractical except where the 1-day maximum occurred on the first day of storm, the 2-day maximum on the first and second days of storm, the 3-day maximum on the first, second, and third days of storm, etc., without making arbitrary adjustments of the records. The C method was discarded because, having failed to compensate for inaccuracies due to splitting storms between days, it has no advantages over the other methods. The D method was discarded because of the great amount of labor involved, and the uncertainty attendant on readjusting the morning records. In fact, for the storm of October 4–6, 1910, in which the D method was compared with others, the curves differ by no great amount from those drawn by the A and B methods.

No way could be devised to allow for inaccuracies due to the arbitrary location of observing stations, except to place more dependence on determinations from storms where a great many station records are available than upon those where few are available. If isohyetals for each day could be drawn and the separate days combined with absolute accuracy, the D method would give results identical with the B method. However, owing to the different times of taking the readings, it is impossible to plot a map which represents accurately the rainfall over a large area for any given exact period of time. This inaccuracy is less for 2 days than for 1 day, less for 3 days than for 2, and so on. Thus in using the D method two maps were combined, the inaccuracy of each of which was as great as that of a single map platted from the combined figures of the two separate maps. The resulting map contains, therefore, not only the errors in its two components, but also the errors in making the combination. Errors of the latter kind were especially introduced when contours did not definitely intersect but lay in the same general direction with a varying distance between them. In such cases it was difficult to determine where to establish points representing their sum. Or, a small peak on one map might lie wholly between two contours on the other map, and it was debatable which of these to add to the peak.

As previously stated, after weighing the advantages and disadvantages of each of the four methods, method A was adopted for subsequent use.

CRITICISM OF USING MAXIMUM STORM PERIODS

In passing, attention should be called to the fact that one of the processes in the time-area-depth study is open to an apparent objection. Reference is made to treating separately the maximum 1-day period, the maximum 2-day period, etc., of the storm, without direct reference to the precipitation of preceding and following days in so far as this might affect the runoff and consequent flood damage of the maximum period.

The disadvantages to which this method of treatment gives rise could have been eliminated in the following manner: Instead of platting a separate map and time-area-depth curve for each maximum period of the storm, a separate map and curve might have been constructed for the first day, the first and second days, the first to the third days, inclusive, the first to the fourth days, and the first to the fifth days of the storm. The data as represented by the time-area-depth curves would then be in a form immediately available for runoff computations.

The latter method has the further advantage of tending to correct the inaccuracies previously mentioned under classes 1 and 4, and offers the same facility for correcting the inaccuracy of class 2 as does the method of taking maximum periods. The inaccuracy, class 1, due to arbitrary location of observing stations, is partially corrected by having, for periods of two or more days, the principal storm features of preceding periods as a guide. Usually the first day of an extended storm is not the day on which the rainfall is a maximum, and consequently the error introduced by the arbitrary location of the stations is not of great importance. Inaccuracies of class 4, those due to variations in time of observation, could be partially corrected by letting the evening records control the lines of equal rainfall within the limits imposed by the morning records of the following day, as described on page 219 for the experimental maps constructed by method B.

In addition to the above advantages, maps for consecutive periods, beginning with the first day of the storm, would show the cumulative depth reached on successive days, and the direction and distance of the storm movement. This information can be ascertained indirectly, of course, from the maps and curves as published.

These objections are not serious, however, and are outweighed by the advantage of having the maximum storm periods the most prominent feature of the results, and the most directly available. This is especially helpful when dealing with small drainage areas, where the runoff collects quickly, and would approximate 100 per cent for the maximum day.

ERRORS DUE TO PERSONAL EQUATION

In platting the isohyetal lines, cases often occur in which either one of two or more courses appears to be correct so far as the available data indicates. The location chosen depends solely upon the judgment of the one doing the platting and may be largely a matter of accident. In order to test the magnitude of this personal equation effect, three maps were platted from the same data by different individuals. The storm of October 20–24, 1908, selected for this trial, shown in figures 76 to 80, was one in which the chance for personal variation was not so great as in some of the others; yet the discrepancy from this source was found to be considerable. The maps were drawn for the total storm rainfall. The relative variation in the small areas around the peak was large, as may be seen from the following table:

Isohyetal	Area in Square Miles Obtained by Three Different Persons		
15 inch	151	55	68
14 "	479	424	301
13 "	1110	958	780
12 "	2220	2190	1370
11 "	4600	4050	2520
10 "	6530	5900	4890
9 "	8870	8210	7800
8 "	12730	13460	12170
7 "	21560	23000	21880
6 "	36660	32010	31710

The proportional variation is much the greatest for the smallest areas, as might be expected, since the peak is dependent usually upon a single maximum record. This might be indicated in general terms by the statement that any isohyetal which does not enclose at least three rainfall records is of only approximate utility. The greater the area enclosed by an isohyetal, the greater will be the number of stations enclosed within it, and the smaller will be the percentage of probable error in drawing the isohyetal.

ASSEMBLING, COMPUTING, AND PLATTING DATA FOR TIME-AREA-DEPTH CURVES

As previously stated, a map was drawn for each period of maximum precipitation in each of the storms. The number of maps required for a single storm varies from one to five, depending on the number of days the storm lasted. Since each of these maps shows the depth and area of precipitation for a definite period of time, we have the three interdependent factors, time, area, and depth, in a definite measurable form. The data in this form, however, is not easily accessible for direct use, or for comparison with similar data for other storms and other periods of the same storm.

The best method which could be devised for combining the three factors, time, area, and depth, in such a manner as to make them comparable for different storms, and for different periods of the same storm, is to plat the data in the form of curves. The data for an entire storm is a group of curves, each of which gives the time-area-depth relations for one maximum period of the storm. Thus there is a curve for each map. All the curves for one storm are not presented together, however, as it is desirable for comparative purposes and for convenience in publication to have on the same chart a number of curves taken from different storms for the same duration of maximum rainfall. These curves are shown in figures 94 to 103.

Since there is a curve corresponding to each map, for a definite period of maximum precipitation, the element of time remains a fixed quantity for one curve. The varying factor of area is platted as abscissas, and the corresponding average depths as ordinates to the curve.

The steps necessary to take these quantities, area and average depth, from a given map and prepare them for platting involved a considerable amount of work. Table 8 gives the computations for the maximum 3-day period of the storm of July 14–16, 1916, a map of which is shown in figure 105. For convenience in reference a capital letter is placed at each center of precipitation. These, and the order of combining them, are shown in column 1 of table 8. Beginning at the greatest center of precipitation, the isohyetals are planimetered, and the area in square inches enclosed is set opposite each. This process is continued at the principal center until the last isohyetal is reached which encloses that center alone. Before planimetering and listing the area enclosed by the next lower isohyetal, the one or more additional centers enclosed by it are treated in the same manner as the principal center. The same process is continued until the 2-inch isohyetal is reached. The isohyetals and the area in square inches enclosed by each are shown in columns 2 and 3 of table 8.

The computations for determining the average depth over the area enclosed by each isohyetal are simple, the average depth of precipitation over the area between any two isohyetals, named in column 5, being considered equal to their arithmetical mean. These average depths are given in column 6. The area in square inches enclosed by each isohyetal, column 3, is reduced to square miles and placed in column 4. Column 7 shows the area between isohyetals, obtained by subtracting from the area enclosed by each isohyetal that enclosed by the next higher. The average depth of precipitation over this interspace is given in column 6, and in column 8 the volume in inch-miles obtained by taking the product of the quantities in columns 6 and 7. The total volume of precipitation in inch-miles enclosed by each isohyetal, as shown in column 9, is the sum of the quantity in column 8 and the total in column 9 for the next higher isohyetal. The average depth of precipitation in inches over the entire area enclosed by any isohyetal is, of course, the quotient obtained by dividing the quantity in column 9 by that in column 4. This is given in column 10.

In the appendix of this volume are given in tabular form the essential time-area-depth quantities for the purpose of checking or recomputing the computations, or for reproducing the curves on a larger scale, if it is desired to group them differently for comparative purposes.

The curve for the maximum 1-day period of storm 157, July 14–16, 1916, is a good example of certain peculiarities met in platting the curves for several other storms. For that reason it is shown alone in figure 106. There were three distinct 1-day rainfall peaks of widely differing characteristics. Two of these occurred on July 15, one each in North and South Carolina, and the third occurred on July 16 in North Carolina over an area differing slightly from that covered by the peak of the preceding day. For areas up to 500 square miles, the greatest average depths of precipitation for a 24-hour period occurred in the North Carolina center on July 16; for areas from 500 to 3,000 square miles, in South Carolina on July 15; and for areas greater than 3,000 square miles in North Carolina on July 15.

15

Table 8.—Form of Time-Area-Depth Computations, Storm of July 14–16, 1916.
Maximum 3-day Period, July 14–16

Center	Isohyetal	Area Enclosed		Between Isohyetals	Average Depth	Net Area	Volume	Total Volume	Average Depth
	·Inch	Square Inches	Square Miles	Inches	Inches	Square Miles	Inch-Miles	Inch-Miles	Inches
1	2	3	4	5	6	7	8	9	10
A	22	0.004	5	23.7–22	22.8	5	110	110	22.8
	21	0.03	41	22–21	21.5	36	770	880	21.5
	20	0.06	82	21–20	20.5	41	840	1,720	21.0
	19	0.09	120	20–19	19.5	38	740	2,460	20.5
	18	0.15	210	19–18	18.5	90	1,660	4,120	19.6
	17	0.32	440	18–17	17.5	230	4,020	8,140	18.5
	16	0.60	820	17–16	16.5	380	6,270	14,410	17.6
	15	0.84	1,150	16–15	15.5	330	5,120	19,530	17.0
	14	1.07	1,460	15–14	14.5	310	4,500	24,030	16.5
	13	1.34	1,840	14–13	13.5	380	5,130	29,160	15.8
	12	1.69	2,310	13–12	12.5	470	5,870	35,030	15.2
B	15	0.03	41	16.4–15	15.7	41	640	640	15.7
	14	0.09	120	15–14	14.5	79	1,150	1,790	14.9
	13	0.19	260	14–13	13.5	140	1,890	3,680	14.2
	12	0.32	440	13–12	12.5	180	2,250	5,930	13.5
AB	11	2.87	3,930	12–11	11.5	1,180	13,600	54,560	13.9
	10	4.10	5,610	11–10	10.5	1,680	17,600	72,160	12.9
	9	4.77	6,530	10– 9	9.5	920	8,740	80,900	12.4
	8	5.64	7,720	9– 8	8.5	1,190	10,100	91,000	11.8
	7	6.54	8,960	8– 7	7.5	1,240	9,300	100,300	11.2
	6	7.57	10,400	7– 6	6.5	1,440	9,360	109,660	10.6
	5	8.86	12,100	6– 5	5.5	1,700	9,350	119,010	9.8
C	16	0.08	110	16.8–16	16.4	110	1,800	1,800	16.4
	15	0.26	360	16–15	15.5	250	3,880	5,680	15.8
	14	0.55	750	15–14	14.5	390	5,660	11,340	15.1
	13	0.94	1,290	14–13	13.5	540	7,290	18,630	14.4
	12	1.31	1,790	13–12	12.5	500	6,250	24,880	13.9
	11	1.97	2,700	12–11	11.5	910	10,500	35,380	13.1
	10	2.73	3,740	11–10	10.5	1,040	10,900	46,280	12.4
	9	3.85	5,270	10– 9	9.5	1,530	14,500	60,780	11.5
	8	4.90	6,710	9– 8	8.5	1,440	12,200	72,980	10.9
	7	6.33	8,660	8– 7	7.5	1,950	14,600	87,580	10.1
	6	8.65	11,800	7– 6	6.5	3,140	20,400	107,980	9.2
	5	12.00	16,400	6– 5	5.5	4,600	25,300	133,280	8.1
ABC	4	26.02	35,600	5– 4	4.5	7,100	32,000	284,290	8.0
D	7	0.035	48	7.6– 7	7.3	48	350	350	7.3
	6	0.25	340	7– 6	6.5	292	1,900	2,250	6.6
	5	0.64	880	6– 5	5.5	540	2,970	5,220	5.9
	4	1.25	1,710	5– 4	4.5	830	3,740	8,960	5.2
A to D	3	33.21	45,500	4– 3	3.5	8,190	28,700	321,950	7.1
	2	42.72	58,500	3– 2	2.5	13,000	32,500	354,450	6.1

Note.—Blackfaced figures show maximum precipitation at rainfall centers.

It is probable that each curve would suffer some modification if it
were possible to free the rainfall observations from the errors arising

from arbitrary division of time commented on elsewhere in this chapter. For instance, the maximum 24-hour record at Altapass, North Carolina, was 22.22 inches between 2 p. m. of July 15 and 2 p. m. of July 16. But the time-area-depth curve for July 16 shows a maximum, nearly 3 inches smaller, although Altapass was the center of that storm peak. This is due to the fact that the maximum 1-day rainfall at Altapass, based on regular evening readings, is 19.32 inches for July 16, and this, therefore, became the upper limit of the curve.

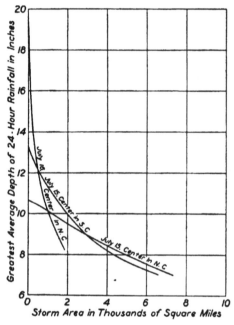

FIG. 106.—TIME-AREA-DEPTH CURVES FOR STORM OF JULY 14–17, 1916.

This storm occurred over the Carolinas. The curves show the greatest average depth of rainfall during the maximum 1-day period.

To what extent the remainder of the curve is subject to errors from this source can not be ascertained, because of lack of data. The indications are that, as the area increases, the averages obtained from the increased number of station records tends to decrease such errors.

An examination of the three curves referred to shows that no single curve furnishes a complete answer. The curve adopted for the maximum 1-day period, and shown in the group curves, is the envelope of these three curves. In the group curves, where cases of this kind occur, and the angle between the two curves at the point of intersection is sharp, no attempt is made to join them in a smooth curve.

A different kind of break in the curves is occasionally caused by a very great increase in the area without a corresponding reduction in average depth of rainfall. This occurs at the point where the base of the greatest storm peak joins the main body of the storm. The curve for storm 10, figure 95, is a good example of such a break. The 6-inch isohyetal, enclosing an area of 1020 square miles over which there was an average rainfall of 7.2 inches, is the lowest which encloses center A alone. The 5-inch isohyetal, enclosing an area of 7460 square miles, over which there was an average depth of rainfall of 6.2 inches, includes centers A to E. Any curved line between these two points is misleading, hence they are connected by a straight dotted line.

The limitations inherent in rainfall data, as described in the preceding pages, together with the manner of compiling the time-area-depth curves, cause the latter to indicate depths of rainfall slightly less than the actual. This is true especially of intense rains occurring over small areas, as illustrated in the case of the July 1916 storm over the Carolinas, but is not so material in storms which have a gradual and even distribution as for instance the March 1913 storm in Ohio.

Continuing this line of reasoning, it follows that errors of this kind are to be looked for principally in the upper portions of the time-area-depth curves, while the lower portions would be little affected. Since the former relate to comparatively small areas of minor interest in connection with flood control problems, such inaccuracies are of little moment.

DEGREE OF ACCURACY ATTAINED

The character of the operations in each step was allowed to govern the extent of detail with which they were checked for accuracy. Assembling the data for platting the maps consisted in listing the stations in the storm area, and placing opposite each the amount of rainfall recorded on the several days of the storm. When ordinary care is exercised but few errors will occur, and each of these will affect only the record at one station for one day of the storm. This would, in general, have a negligible influence on the final maps and curves. For these reasons it was not considered necessary or advisable to check the data completely, as this would have required an additional amount of time equal to about half of that consumed in compiling the data.

The second step, that of determining the maximum consecutive periods of the storm, can be sufficiently checked by inspection, and this was done. It was very important that the stations be accurately located on the outline map used in mapping the storms, as the mislocation of a single station would introduce an error, perhaps a serious one, into all the storm maps on which that station occurred. The maps described on page 214 were used, and as is there stated, the location of each station was carefully checked to insure its accuracy. After platting the figures showing the amount of rainfall at each station, these were not checked individually, as an error of location or amount could affect only the one map, and any serious error of this character could be found by comparing the several maps of the same storm. This comparative check was performed for the purpose just mentioned and to check the proper location of the isohyetal lines. The proper scaling of the maps was assured by planimetering an area, reading the result and then scaling the same area again without setting the instrument at zero. The second reading corresponds to twice the area scaled, and if half the second reading differs materially from the first another trial is made.

Each operation in making the computations and platting the time-area-depth curves was individually checked. The areas in square miles, in columns 4 and 9, table 8, and the volume of rainfall in inch-miles in column 8, are computed to 2 significant places up to 1000, and to 3 significant figures beyond that point. The resulting average depth of rainfall in column 10 is computed to the nearest tenth of an inch.

CHAPTER X.—VARIATION IN MEAN ANNUAL RAINFALL OVER EASTERN UNITED STATES

One of the chief benefits which may come from the study of past meteorological events is the ability to apply the knowledge so gained in forecasting future conditions. This forecasting cannot be done intelligently unless proper allowance is made for the chance variations which may occur, and for regular periodic or progressive changes, if there are such. For this reason, studies have been made of the variations in monthly and annual rainfall over certain areas for the years 1888 to 1916. The amount of data collected prior to 1888 is not sufficient to warrant its use in these investigations. Studies were made, first for a comparatively large area including the entire portion of the United States east of the 103d meridian, together with the parts of North and South Dakota, Nebraska, and Texas that lie west of this meridian; and second, for a comparatively small interior area comprising the states of Illinois, Indiana, Ohio, Kentucky, and West Virginia. The larger area covers 1,837,500 square miles and the smaller one, 198,900 square miles.

DATA USED

The investigation of storm rainfall, described in the preceding chapters, is confined to a study of excessive precipitation at single stations, or of storms equal to or exceeding a well defined minimum. However, to determine the maximum amount of variation which may be expected and to ascertain if there is any evidence of cyclic and progressive changes, it was necessary to consider additional data. The following material was utilized in the studies described in this chapter.

1. All precipitation records in the United States east of the 103d meridian made during the 29-year period 1888–1916.

2. Records of the 160 storms, discussed in the preceding chapters, which occurred over the same area during the 25-year period 1892–1916.

3. The records of excessive precipitation over the same area, during the 44-year period 1871–1914.

The records of the 160 storms, and excessive precipitation records for the years subsequent to 1888 are, of course, included in the total precipitation record referred to in 1. The object of using them separately, as previously stated, is to determine the effect of storms and of excessive precipitation in producing the annual variations of the total rainfall.

RESULTS OF STUDY

It seems reasonable to suppose that by averaging the annual rainfall over that portion of the United States east of the 103d meridian, local or sectional variations such as might exist for the smaller area would in a large measure be compensated, and the resulting annual totals would show but little variation from year to year. Essentially this was found, except that the variations for the larger area were slightly larger than had been anticipated.

Figure 107 shows the average rainfall over the larger area for each month and year of the 29-year period considered, while figure 108 shows similar quantities for the smaller area. The small diagram included within each figure shows the average monthly rainfall obtained by averaging the 29 values for each month. These quantities together with corresponding quantities for the Dayton cooperative station and for the state of Ohio are given in table 9. They show that on the average the rainfall is considerably greater during the summer than it is during the winter.

Table 9.—Average Monthly Rainfall on Four Different Areas for the Period 1888 to 1916

Month	Dayton	Ohio	*5 Interior States	Eastern U. S.
January.................	3.41	3.18	3.30	2.42
February.................	2.65	2.69	2.80	2.48
March..................	3.83	3.55	3.74	2.90
April...................	2.98	3.02	3.39	3.09
May....................	3.69	3.71	4.01	3.87
June...................	3.89	3.83	4.03	4.02
July....................	3.26	3.94	3.96	3.98
August.................	3.19	3.29	3.46	3.69
September...............	2.85	2.78	3.04	3.19
October.................	2.62	2.51	2.44	2.50
November...............	2.84	2.67	2.91	2.26
December...............	2.62	2.71	2.83	2.44
Mean Annual............	37.83	37.88	39.91	36.85
Mean Monthly...........	3.15	3.16	3.33	3.07

The values of average rainfall shown in figures 107 and 108 were obtained in the following manner. From the Weather Bureau publications entitled Climatological Data and Monthly Weather Review there was obtained for each year of the period 1888–1916 inclusive,

* Illinois, Indiana, Ohio, Kentucky, and West Virginia.

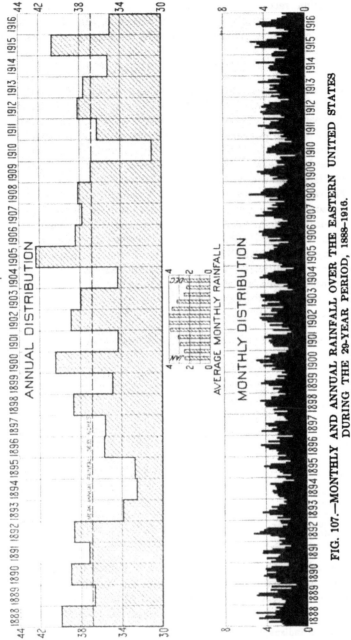

FIG. 107.—MONTHLY AND ANNUAL RAINFALL OVER THE EASTERN UNITED STATES
DURING THE 29-YEAR PERIOD, 1888–1916.

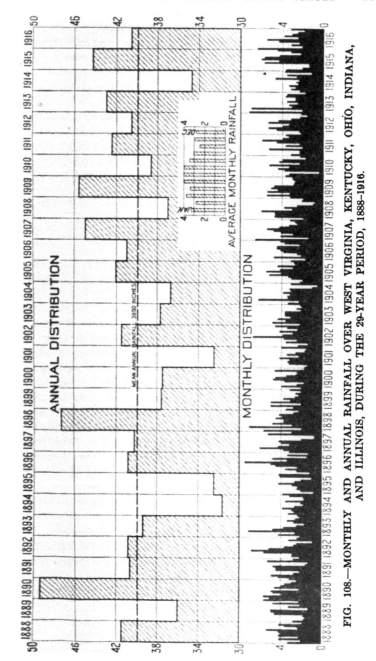

FIG. 108.—MONTHLY AND ANNUAL RAINFALL OVER WEST VIRGINIA, KENTUCKY, OHIO, INDIANA, AND ILLINOIS, DURING THE 29-YEAR PERIOD, 1888–1916.

the average monthly rainfall for each state. By taking as weights the area of each state in square miles, it was possible to compute the average depth of rainfall for a given month over any desired section of the United States by adding the weighted means in inch-miles for all the states within such section, and dividing the sum by the aggregate area of the section in square miles. The total rainfall each year for the section was then obtained by adding the averages, so computed, for the twelve calendar months.

Figures 107 and 108 show that the variations in annual rain'all for the larger area are approximately two-thirds as great as those for the smaller area. The average yearly departure from the mean annual rainfall for the two areas studied, and also for Dayton and for Ohio, for the period considered, was found to be as follows:

> Dayton..................................13.86 per cent
> Ohio...................................... 9.08 "   "
> Five interior states................... 8.27 "   "
> Eastern United States................. 5.84 "

This indicates, as seems reasonable, that the larger the area considered the smaller is the variation in total annual precipitation over that area.

Figures 107 and 108 possess many features in common. Comparing them year by year, the two diagrams agree in 72 per cent of the cases as to annual totals being greater or less than the mean annual rainfall. The great period of drouth which occurred in 1894 and 1895 is represented in figure 107 by deficiencies of 5½ and 4½ inches, and in figure 108 by deficiencies of 8 and 7½ inches, respectively.

In a few instances there are marked discrepancies between the two diagrams, which must be attributed to decided differences in precipitation over large areas. Thus, for the year 1900, the rainfall over the eastern half of the United States averaged 3½ inches in excess of the 29-year mean, as against a deficiency of 2½ inches for the smaller area. In 1909 the smaller area averaged an excess of 5¾ inches while the rainfall over the larger area averaged normal. The greatest excess shown in figure 108 is 9¼ inches for 1890, as against only 2 inches for the same year in figure 107. In making these comparisons it should, of course, be borne in mind that the fluctuations for the larger area tend to be less than for the smaller area, the former having not only a smaller mean annual rainfall, but also the greater tendency for compensating extreme local variations.

A striking feature of both diagrams is the absence of any evidence as to cyclic or progressive variation. Although it may be argued that the 29-year period is too short to furnish conclusive evidence regarding this matter, it seems reasonable to suppose that if any important cyclic variation does exist some indications, at least, would be present in one or both of these diagrams.

In order to show graphically the extent of the departure of the maximum rainfall from the mean for different continuous periods, four diagrams were constructed, figures 109 to 112. The first two figures apply to the larger area; the latter two, to the smaller area. Figure 109 shows the maximum departure in inches from the monthly means for continuous periods varying in length from 1 to 6 months. Thus the greatest excess of precipitation for any month during the 29 years over the mean for that month, was in February, 1903, when the average over the 37 states was 4.30 inches, an excess of 1.82 inches. The maximum deficiency for any one month was in March, 1910, when the average precipitation over 37 states was 1.99 inches below the mean. In the upper portion of the diagram, the monthly precipitation for the 6 months, January to June, 1911, shows an average monthly deficiency of only 0.68 inch below the 29-year means for those months. Similarly, the 6-month period, September, 1902, to February, 1903, has an average monthly excess of 0.81 inch over the means for the period.

An inspection of the chart shows that the principal excesses and deficiencies generally occur during the spring and early summer months, illustrating a tendency for the depth of rainfall to be more erratic during these months than at other times of the year.

The second diagram, figure 110, shows the maximum variation for continuous periods of 1, 2, 3, 4, 8, 12, 16, 20, 24, and 28 years, respectively, from the 29-year mean. The departures are expressed as percentages of the 29-year mean.

Figures 111 and 112 are similar to figures 109 and 110. Figure 111 shows that the maximum departures from the normal rainfall also occur during the spring and early summer months in the case of the smaller area. In practically all cases the actual values of the departures are seen to be considerably greater for the smaller area than for the larger area, as would be expected.

EFFECT OF STORMS AND EXCESSIVE PRECIPITATION

It is conceivable that the variations in the number of great storms occurring each year might be responsible for a large part of the fluctuations in the total annual rainfall. With this in mind table 10 was prepared, showing the chronological distribution of the 160 storms previously discussed. Table 2 is analogous, showing the chronological distribution of rainfalls in which there were one or more records of excessive precipitation. A comparison of these two tables with the

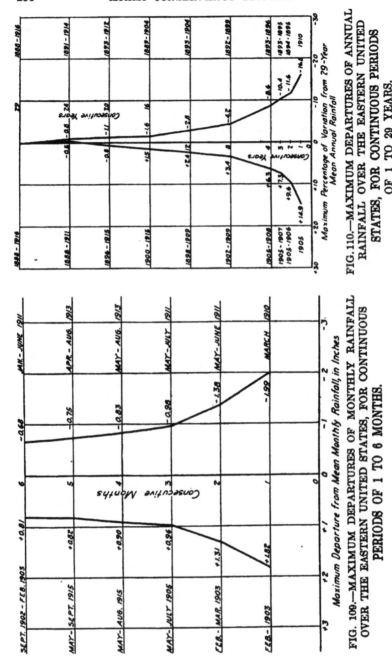

FIG. 109.—MAXIMUM DEPARTURES OF MONTHLY RAINFALL
OVER THE EASTERN UNITED STATES, FOR CONTINUOUS
PERIODS OF 1 TO 6 MONTHS.

FIG. 110.—MAXIMUM DEPARTURES OF ANNUAL
RAINFALL OVER THE EASTERN UNITED
STATES, FOR CONTINUOUS PERIODS
OF 1 TO 29 YEARS.

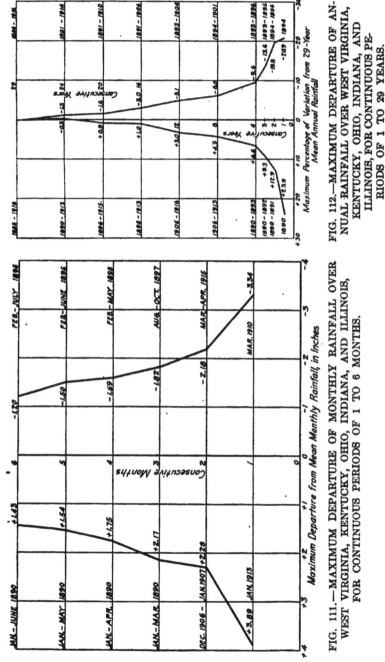

FIG. 112.—MAXIMUM DEPARTURE OF ANNUAL RAINFALL OVER WEST VIRGINIA, KENTUCKY, OHIO, INDIANA, AND ILLINOIS, FOR CONTINUOUS PERIODS OF 1 TO 29 YEARS.

FIG. 111.—MAXIMUM DEPARTURE OF MONTHLY RAINFALL OVER WEST VIRGINIA, KENTUCKY, OHIO, INDIANA, AND ILLINOIS, FOR CONTINUOUS PERIODS OF 1 TO 6 MONTHS.

chart showing the total annual rainfall in the eastern United States, figure 107, indicates that the number of large storms and the amount of excessive precipitation occurring each year have a decided influence on the amount of annual rainfall.

Table 10.—Number of Great Northern and Southern Storms occurring each year, and the Yearly Departure from Normal, during the 25-year period 1892–1916

Year	Southern Storms	Northern Storms	Total	Departure from Normal
1892.........	3	0	3	− 2
93.........	3	1	4	− 1
94.........	4	2	6	0
95.........	0	2	2	− 3
96.........	5	2	7	0
97.........	4	3	7	0
98.........	8	3	11	+ 4
99.........	5	0	5	0
1900.........	6	2	8	+ 1
01.........	7	0	7	0
02.........	6	3	9	+ 2
03.........	4	3	7	0
04.........	2	2	4	− 1
05.........	3	4	7	0
06.........	5	1	6	0
07.........	4	2	6	0
08.........	6	0	6	0
09.........	3	3	6	0
10.........	2	2	4	− 1
11.........	4	1	5	0
12.........	6	2	8	+ 1
13.........	5	3	8	+ 1
14.........	7	2	9	+ 2
15.........	5	3	8	+ 1
16.........	6	1	7	0
Total.......	113	47	160	

Since 160 storms occurred in the 25 years, 1892–1916, the average number each year is about 6. On account of unavoidable inexactness of the data, let us assume that any year is normal in which 5, 6, or 7 storms occurred. The storms are listed by years in table 10. This table shows the number of storms in the northern and southern groups, and the departure from the normal for each year. A comparison of this annual departure, with the departure of the annual rainfall for the same area, figure 107, year for year, shows the influence of storms in causing the variation of the total rainfall. It has already been pointed out that this variation from the 29-year mean over the eastern part of the United States is quite small except for two or three years. The same thing is true of the variation in number of storms. The excess and deficiency, respectively, in the annual rainfall and in the annual number of storms, generally occur in the same year. The

years of maximum departure, however, do not occur simultaneously in the two sets of data. The years 1892 and 1895 are minima for storms; the total precipitation for the former is above normal, and for the latter it is third from the minimum. The year 1898, though having the maximum number of storms, is but little above normal in total rainfall, while the year 1905, though being a maximum for total precipitation, has a normal number of storms, and 1910, the year of minimum total rainfall, has a deficiency of only one storm.

It is clear from this, as might be expected, that the storm rainfall and the non-storm rainfall vary, each independently of the other. Sometimes the two departures are in opposite directions and tend to compensate, producing normal total precipitation; and sometimes they are in the same direction, producing a greater total departure. Although the non-storm precipitation has not been isolated and studied separately, its annual variation apparently is less than that of the storm rainfall, since the latter stamps its principal features on the variation of the total rainfall.

This relation between storm and non-storm precipitation could be determined much more accurately if maps and time-area-depth curves were made for all the storms. From the curves, the total inch-miles of water in each storm could be determined, and thus the proportion of storm rainfall to total rainfall, and the annual variation of each, could be found quite accurately. This process, however, would involve a much greater amount of work than the results would probably warrant.

Figure 113 gives a clear idea of the sequence of the 160 storms we are discussing. This shows the 47 storms in the northern group, and the 113 storms in the southern group, arranged chronologically. The date of occurrence, and the greatest, fifth, tenth, and twentieth greatest 3-day rainfall records are shown graphically. The storms are numbered so that they may be identified by consulting tables 4 and 5.

Perhaps the most noticeable fact brought out by this chart is the manner in which the storms in one group supplement those in the other. For instance, during years in which the storm rainfall was definitely below normal in the north, the number of storms occurring in the south was either normal or above normal. Conversely, during years in which the number of northern storms was greater than normal, the storm rainfall in the south was either normal or deficient. This tends to confirm the conclusion reached from a study of figures 107 and 108; namely, that there is no cyclic or progressive variation in rainfall.

The evidence of the records of excessive precipitation supplements that of the 160 large storms. Table 2, page 77, gives the number of rainfalls recorded in the United States east of the 103d meridian causing excessive precipitation at 1 station, at 2 to 5 stations, and at 6 or more stations, in each year from 1871 to 1914, inclusive. On account of the lesser limits chosen for these rainfalls they are much more numerous than the storms just described. It is instructive to compare the number of such rainfalls with the total annual rainfall, year for year, in the same manner in which the storms and the total annual rainfall were compared. Since these numerous rainfalls represent a much larger percentage of the total annual precipitation than do the 160 great storms, it is natural that they should vary in number from year to year much in the same manner as does the total precipitation. This they do, as may be seen by comparing column 5 of table 2, with figure 107.

There is as little evidence of cyclic variation or permanent change in the amount of annual rainfall to be found in records of excessive precipitation, as in the records of storms and of total precipitation.

CHAPTER XI.—EUROPEAN STORM RAINFALL

At about the time the investigation of storm conditions in the eastern United States was begun, an examination of the most pertinent and easily available French and German rainfall and storm literature also was commenced. This research of foreign literature was not undertaken in the belief that any relation necessarily exists between the storm rainfall of the two continents, and, therefore, an exhaustive study, such as that made of storms of the eastern United States, was not made of European storms. The much longer European rainfall and flood records, however, might be expected to throw additional light on the origin and cause of great rainstorms and floods, as well as on the relative size and frequency of ordinary great floods as compared with the maximum of a very long period of years.

Although this investigation deals primarily with storms, it is advisable to include some consideration of flood runoff of European rivers, because the runoff records are available for a much longer period of years than are the rainfall records. The material here given is necessarily fragmentary, as it is in the nature of brief summaries of the work of a number of different authorities. These summaries have been placed in what seems to be the most logical order, and proper acknowledgments accorded. Care has been used to transcribe the data accurately, but only that part of the large amount of translated material which is directly pertinent to this investigation is here presented.

PRECIPITATION IN RIVER BASINS OF NORTH GERMANY*

Western Germany, influenced by the ocean, has more uniform climatic conditions than the interior of the continent. This is because the cyclonal rains, coming from the ocean, reach the interior much more feebly and irregularly than they do the coast country; and also, because the local summer rains in the interior make up a greater percentage of the rainfall than they do in the coast country.

Considering the rainfall by seasons, the percentage of annual rainfall which occurs during the months of December, January, and February, decreases from west to east. On the west coast it is 24

* Die Niederschläge in den Norddeutschen Stromgebieten. By Prof. Dr. G. Hellmann. Berlin, 1906. Dietrich Reimer. In three volumes. Vol. I, text; Vols. II and III, tables.

per cent, and in the eastern part of Germany, or inland country, it is 16 per cent of the mean annual rainfall. The percentage which occurs during March, April, and May, varies from 18 per cent at the west coast to 24 per cent inland, the relations being reversed. This difference is more marked during June, July, and August. The percentage of the annual rainfall then occurring varies from 30 per cent at the west to 40 per cent at the east. In September, October, and November the distribution is again like that of winter; namely, the decrease is from 30 per cent at the coast to 22 per cent inland. The north German inland country, therefore, has relatively heavy rainfall in spring and summer; the coast region, on the other hand, in autumn and winter. Comparing the cold half of the year, October to March, with the warm half, April to September, the coast country at the west receives about 50 per cent of its rainfall in each half, while the inland country has 15 per cent more in the warm half of the year than in the cold.

Leaving the Swiss stations out of consideration, the maximum monthly rainfall advances apparently regularly from south to north. The south and southwest, the upper Elbe and a part of the upper Rhine basins, have the maximum in June; the middle and northern part, the major portion of the area here studied, in July; a small inland strip of northern Hanover and Schleswig-Holstein, in August; the northwest coast around the Elbe mouth, the west coast of the Baltic (Schleswig-Holstein), the east coast of the Baltic in East Prussia around the mouth of the Memel, the extreme southwest of Germany on the basin of the upper and middle Mosel, and the part of the Jura Mountains in Switzerland draining into the Rhine, have the maximum in October.

The month of least average rainfall likewise advances from south to north. January is the month of least average rainfall for the region furthest from the coast, including the entire southern part except the Swiss stations. A great area northeast of this, including the upper and middle basins of the Memel, Pregel, and Weichsel, as well as the right tributaries of the Oder, and two small districts in western Germany, have February as the month of minimum rainfall. Practically all of western Germany, as well as most of the coast of the Baltic, have April as the minimum. From this it is evident that the time of occurrence of extremes is exactly 6 months apart for wide areas.

Besides the principal maximum and minimum, there are evident in annual curves of rainfall still further secondary extremes. These occur in March, October, and December, and are confined to the western and northwestern part of Germany, that is, to the lands near the coast. They are caused by the action of great cyclonal rainfalls covering wide areas and occurring chiefly in the cold season. Such storms have often caused floods in the Rhine, Ems, and Weser basins.

Stations at low elevations show less variation of annual rainfall than those located at high elevations. The smallest mean monthly rainfall in the low country amounts to from 0.98 to 1.58 inches. In the driest regions it goes as low as 0.79 inch, as at Prag, where the mean annual is 17.6 inches; and apparently somewhat lower at other points. At the stations lying nearer the ocean and receiving greater rainfall, it frequently rises to over 1.58 inches, as 1.73 inches at Trier, mean annual, 26.8 inches; 1.77 inches at Kleve, mean annual 30.4 inches; and 2.16 inches at Aachen, mean annual 33.4 inches. In the mountains the records naturally vary much more. Only one of the stations studied did not have a minimum mean monthly rainfall of less than 3.94 inches.

In the level country of middle and north Germany, the greatest monthly rainfall fluctuates between 2.56 and 3.74 inches. In the driest sections it remains somewhat lower. Posen, with a mean annual rainfall of 19.4 inches, has a maximum monthly rainfall of 2.44 inches; Prag, with a mean annual of 17.65 inches has a maximum monthly rainfall of 2.48 inches. At many inland stations, where the summer rains are more pronounced than in the regions of ocean climate, this increases to 4 inches or more. Limberg, for instance, has a mean annual rainfall of 27.7 inches, and a maximum monthly rainfall of 4.1 inches.

In the mountainous country of north Germany, Klaustal, on the plateau of the upper Harz mountains, which has a mean annual rainfall of 53 inches, has the greatest mean monthly rainfall of 5.86 inches. The Swiss stations in the upper Rhine basin have a much higher mean monthly rainfall, as for instance Altstätten with 6.5 inches, and a mean annual of 49.8 inches; St. Beatenberg, 6.9 inches, mean annual 57 inches; St. Gallen, 7.36 inches, mean annual 56.5 inches; and Einsiedeln, 7.8 inches, mean annual 62 inches. Other Swiss stations run up to above 11.8 inches. It can therefore be taken that the mean monthly rainfall of the north German river basins varies between 0.79 and 12.6 inches.

About 4000 stations, 2348 of them in north Germany, were studied for this work. Their total records covered an aggregate of approximately 30,000 years. Only about 30 per cent of the 4000 had records extending through a period of 10 years or more. The records belong to different periods, and are scattered irregularly over the area considered.

Analysis shows that continuous rainfall records of not less than

30 to 40 years are required to give a reliable figure for the mean annual. To determine the mean monthly rainfall within 2 per cent, 60 year records are necessary.

RELATION BETWEEN THE MAXIMUM 24-HOUR RAINFALL AND THE MONTHLY AND ANNUAL RAINFALL IN GERMANY

At a number of representative stations a quite uniform relation was found to exist between the average maximum 24-hour rainfall for a given month and the average rainfall for that month. The former is generally about 25 or 30 per cent of the latter. At stations where the annual precipitation is small this percentage may increase to about 32 per cent; at stations where the annual precipitation is large, it may decrease to about 22 per cent. The percentage is somewhat smaller during the winter months of October to March than during the summer months of April to September. In the winter it averages about 25, and in the summer about 29 per cent.

A similar relation exists between the greatest 24-hour rainfall which has ever been recorded during a given month and the average rainfall for that month. The former generally is from 65 to 80 per cent of the latter. Here, also, the percentage is greater at stations where the annual rainfall is small than where it is large. And, too, the percentage is smaller in the winter than in the summer months, averaging about 70 per cent in the former and 80 per cent in the latter.

The records at 28 representative stations were used to determine the relation of the average greatest 24-hour rainfall per year, at a given station, to the mean annual rainfall at that station. The stations were located principally in north Germany, and each had a complete record of 35 or more years. The determination of this relation at the individual stations showed that the average greatest 24-hour rainfall per year is relatively larger at stations having little rainfall than at stations having much rainfall.

By analyzing the results an empirical formula was deduced for determining the average greatest 24-hour rainfall per year at any station in the region to which the formula is applicable.

Let M be the average greatest 24-hour rainfall per year at a given station; H, the mean annual rainfall at that place; and P, the percentage ratio of the former to the latter. Then, for that station

$$P = \frac{100M}{H}$$

By plotting on coordinate paper the actual values of P and H obtained at the 28 representative stations, the manner in which P de-

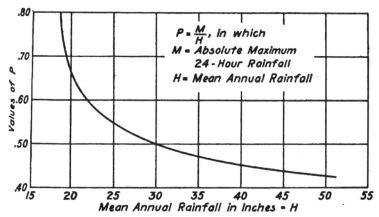

FIG. 114.—CURVE SHOWING RELATION AT INDIVIDUAL STATIONS
BETWEEN THE AVERAGE GREATEST ANNUAL 24-HOUR
RAINFALL AND THE MEAN ANNUAL RAINFALL.

The curve is based on the records at 68 stations in Germany and Switzerland.

creases with an increasing value of H was clearly shown. In this
way was found the desired general equation for determining the average
greatest 24-hour rainfall at any station, when the mean annual rainfall is known.

Between P and H there exists an equation of the form

$$P = a + \frac{b}{H}.$$

The constants a and b are found to have the values, $a = 2.11$ and
$b = 84.17$, when H is expressed in inches. Therefore,

$$P = 2.11 + \frac{84.17}{H}$$

$$M = 0.8417 + 0.0211H$$

Naturally these have no physical meaning, but are to be looked
upon only as empirical formulas limited by the amount and character
of the data upon which they are based. The following values obtained
by the use of these equations may make clearer the relation between
H, P, and M:

H Inches	P Per Cent.	M Inches	H Inches	P Per Cent.	M Inches
15	7.78	1.17	40	4.21	1.69
20	6.32	1.26	45	3.98	1.79
25	5.47	1.37	50	3.79	1.90
30	4.91	1.47	55	3.64	2.00
35	4.51	1.58	60	3.51	2.11

Only a few of the 28 stations used in determining the foregoing equations have a large annual rainfall. In south Germany and Switzerland, 40 additional representative stations were therefore chosen, each of which had a record of 24-hour maxima for at least 21 years. With these, the average greatest 24-hour rainfall per year in per cent of the corresponding mean annual was again calculated for each station. The 68 values of P and the corresponding values of H were plotted on coordinate paper, and a curve was sketched in averaging the points. The curve obtained is shown in figure 114. Attempts to establish equations similar to those above were not satisfactory because of the irregularity of the plotted points. This irregularity is ascribed to the fact that the records were shorter, the stations more scattered, and the records of less uniform value than the 28 records used in the preceding analysis.

MAXIMUM 24-HOUR RAINFALLS OF 3.94 INCHES (100 MM) OR MORE

The stations having long-term records are too few to suffice for determining the frequency and extent of the extreme greatest rains. Such rains occur either as local thunderstorms, for the determination of which the stations having long records are too far apart; or as general rains, whose extent and greatest amount cannot be accurately determined by such stations. The establishment of a close net of rainfall stations has made possible a more accurate insight into the conditions of maximum rainfall. Some results of studies of the occurrence of extreme maximum 24-hour rainfall, based on records at 445 stations, will here be given. Only maxima of at least 3.94 inches (100 mm) will be considered.

Considering, first, the bulk of the maximum 24-hour measurements, it is plain that the stations in the level country are to be differentiated from those in the mountains. The extreme maximum values occur among the latter, as follows:

Sudeten* { North side	9.42	inches
South side	13.7	"
Erzgebirge	5.86	"
Harz Mts.	6.15	"
Thüringerwald	5.40	"
Schwarzwald	6.90	"
Vogesen	5.40	"
Alps (Rhine Basin) { Austria	7.60	"
Switzerland	11.00	"

* Mountains between Silesia and Bohemia, southeastern Germany.

The highest figure given for the Alps may readily have been exceeded, since stations in the high country are not numerous. Except for the Sudeten and the Alps, the other mountain rainfalls are not unusually high and are often exceeded by maxima in the level country. These level country maximum 24-hour rainfalls occur during the most intensive rains. Between the maxima occurring in the mountains and in the flat country there is the general difference that the mountain rains generally fall as widespread rains lasting the entire 24 hours, while the low country maxima fall in a short time, often in a few hours, and are generally accompanied by thunderstorms. An hour's rain of the former would fall in 5 minutes in the latter. The former are widespread, and cause extensive floods; the latter are mostly local, but cause relatively large damage within limited areas.

The greatest 24-hour rainfalls in the level country occur in the continental and dry areas, where a rainfall of 5.5 to 5.9 inches in 24 hours is not uncommon; while in wet northwest Germany and in the coast regions, the limiting value of 4.72 inches has not yet been exceeded.

Following are some values of absolute maximum 24-hour rainfalls, in inches, at representative stations in Germany:

Station	Mean Annual Rainfall	24-Hour Rainfall	Ratio
Inland Regions:			
Welikij Dwor, Gouv. Wilna..........	....	5.50	....
Rominten, Ostpreussen.............	....	5.67	
Kurwien, Ostpreussen...............	25.0	5.67	.23
Wildgarten, Westpreussen...........	....	6.06	....
Berlin, Brandenburg................	22.9	6.54	.28
Sommerfeld, Brandenburg...........	....	5.86	
Triebel, Brandenburg................	....	5.63	....
Schlanstedt, Sachsen................	21.25	6.02	.28
Pirna, Kgr. Sachsen.................	26.8	6.15	.23
Coast Regions:			
Kolberg, Pommern..................	22.8	4.01	.18
Schwerin, Mecklenburg.............	23.8	4.65	.20
Kiel, Schleswig-Holstein.............	26.8	3.94	.15
Ahlden, Hanover............•......	22.9	4.60	.20
Nieder Warsburg, Westfalen.........	24.6	4.40	.18

The highest maxima in the drier regions amount to about 20 to 30 per cent of the corresponding mean annual rainfall; but in the moister regions, to only 15 to 19 per cent. The law previously found, namely, that the average 24-hour maximum does not increase in proportion to the annual rainfall, is, therefore, confirmed by the absolute maximum 24-hour rainfall. This influence of inland position upon absolute maximum 24-hour rainfall is well shown on a map with a small dot for each station where a maximum 24-hour rainfall of over 3.94 inches has occurred. Very few dots occur in or near the coast regions; they are very numerous n central and southern Germany; more occur on the upper third of the Elbe Basin than anywhere else; and there are practically none in the rainy western mountainous country of north Germany. By far the greater number of maximum 24-hour rainfalls occur in the summer months of June, July, and August. A study shows that differences exist between the extreme west and the middle and eastern areas.

Considering, first, only the individual or local maxima which occur at one or two stations and which, as already remarked, belong exclusively to the low country, the following distribution in per cent is obtained:

Location	April	May	June	July	Aug.	Sept.	Oct.
North Germany (per cent)......	2	13	22	32	17	13	1
Austria (per cent).............	......	11	39	20	28	2	......

Such a table is not given for the low country stations of south Germany and Switzerland, because very few of the stations in the low country have a maximum 24-hour rainfall of 3.94 inches or more.

If the 24-hour maxima are arranged according to drainage basins, the following table shows their percentage of occurrence in the various seasons.

River Basin	Winter	Spring	Summer	Autumn
Weichsel.................	....	4	87	9
Oder....................	....	10	80	10
Elbe....................	....	18	74	8
Rhine...................	27	6	33	34

The floods caused by heavy rains must therefore occur in the warm half of the year on the easterly streams, Weichsel, Oder and Elbe. On the Rhine, the cold half of the year has a greater percentage of such floods. This would stand out still more on the Rhine if the maxima of local character were eliminated, especially on the upper Rhine, so far as it is represented by the stations of Switzerland, Austria, Bavaria and Alsace-Lorraine. The greatest floods caused by rainfall occur in the warm half of the year on the Memel, Pregel, and Weser rivers, just as on the Weichsel, Oder, and Elbe.

HEAVY LOCAL RAINS OF SHORT DURATION

The following table shows the average intensity, in inches per minute, and frequency of heavy rainfalls of short duration. It is the average for north Germany for the period 1891–1902.

Duration (minutes) Intensity (inches per minute)	4.4	11.8	25.6	41.5	58.0	1 h. 30 m.	2 h. 37 m.	4 h. 31 m.
	0.071	0.048	0.037	0.030	0.024	0.018	0.013	0.008
Number..	91	357	346	167	185	319	123	151

It will be noted that the intensity diminishes with the duration, as would naturally be expected.

The most noteworthy cloudbursts in North Germany in 1891–1902 were:

Station	Date	Duration	Inches of Rain-fall
Waltershausen in Thüringen.........	Aug. 14, 1884	1 h.	2.96
Neustadt on the Haardt............	Sept. 7, 1886	1 h.	3.86
Schwerin in Mecklenburg...........	May 11, 1890	1 h. m.	4.37
Bobersburg in Schlesweig..........	June 21, 1895	2 h. 35	5.05
Wildgarten in Westpreussen........	Aug. 1, 1896	1 h. 40 m.	5.27
Kemnits in Oberlausits.............	July 17, 1887	2 h.	6.30
Görlsdorf in Brandenburg..........	June 12, 1889	2 h. 15 m.	5.20
Berlin...........................	Apr. 14, 1902	3 h. 30 m.	5.63

STORMS AND FLOODS IN EUROPEAN RIVER BASINS

Storm and flood data pertaining to a few European rivers has been gathered from a number of sources. An exhaustive study of the storms and floods in each river basin has not been attempted, although the material originally translated was much more comprehensive than the short summaries here published.

The data given is not equally comprehensive for all the basins, because the records and studies of some were much more accessible and complete than for others. As far as the available data would permit, however, the following information and records are presented for each river: (a) The location, area, topographic, and hydrographic features of the river basins; (b) the normal and extreme rainfall conditions; (c) the principal floods of the past, their relative sizes, and the depths of rainfall and peculiarities of the storms which caused them.

Floods in the Elbe River*

The Elbe rises in Bohemia, northern Austria, flows northward and northwestward into Germany at Tetschen, Bohemia, past Dresden, Magdeburg, and Hamburg, and into the North Sea at Cuxhaven.

* Reports of the Austrian Hydrographic Bureau, and "Der Elbstrom, sein Stromgebiet und seine wichtigsten Nebenflüsse."

The Moldau and Little Elbe, both rising in northern Austria, join to form the main Elbe about 23 miles below Prague at Melnik, Bohemia. The drainage area of the Moldau at its junction with the Elbe at Melnik is 10,850 square miles, the drainage area of the Little Elbe at that place is 5,290 square miles, and the drainage area of the total Elbe at the German border at Tetschen, Bohemia, is 19,730 square miles.

The mean annual rainfall over the Moldau basin, as determined from the records for the years 1876–1900, is 29.05 inches; that of the Little Elbe for the same period of years is 28.25 inches; and that of the total Elbe above Tetschen is 27.10 inches.

The three greatest floods of recent years on the upper Elbe and its tributaries were those of September 1–4, 1890, July 27–31, 1897, and September 7–13, 1899. Measurements of the discharge of the Moldau, the Little Elbe, and the Elbe were made during each of these floods. Measurements of the Moldau were made at Prague, a short distance above its junction with the Little Elbe, where it has a drainage area of 10,400 square miles; measurements on the Little Elbe were made at Brandeis, where it has a drainage area of 5180 square miles; and measurements of the total Elbe were made at Tetschen. Following are the results of the measurements, the gage heights, the average rainfall, and the total runoff for each basin during each flood.

	Average Rainfall in Inches	Max. Discharge in Second Feet	Max. Gage Height in Feet	Total Runoff in Per Cent.
Moldau at Prague, Bohemia:				
September 1–4, 1890.....	4.27	140,000	16.6	57
July 27–31, 1897........	5.02	69,000	11.0	25*
September 7–13, 1899....	4.01	69,800	....	19
Little Elbe at Brandeis, Bohemia:				
September 1–4, 1890.....	2.48	16,580	9.1	37
July 27–31, 1897........	4.65	25,400	10.1	29
September 7–13, 1899....	2.86	11,780	....	17
Total Elbe at Tetschen, Bohemia:				
September 1–4, 1890.....	3.35	157,000	27.8	53
July 27–31, 1897........	4.67	94,200	18.7	26*
September 7–13, 1899....	3.37	76,200	....	19.5

The greatest recorded rates of discharge of the two tributaries and of the total Elbe at Tetschen, are as follows: On the Moldau at Prague, March 29, 1845, there was a discharge of 159,000 second feet; on the Little Elbe at Brandeis, March 9, 1891, there was a discharge of 38,830 second feet; and on the total Elbe at Tetschen, March 31, 1845, there was a discharge of 197,700 second feet.

The above table indicates that the amount of rainfall in a storm is not an accurate index of the size of flood which it produces. The

* Average annual percentage is 27 for Elbe above German border.

fact that the 1890 flood, with less rainfall, was greater than the flood of 1897, is explained by the Austrian Hydrographic Bureau as follows:

Some increase in the rate of runoff was due to the harvests being all gathered before September, 1890, while they were still standing in July and August, 1897. The main cause of the difference in percentage of runoff, however, was the excessive rainfall during August, 1890. The average for the month over the entire area above the German border was 5.57 inches, as against the 15-year average of 2.96 inches. Thus, the flood rainfall of September 1–4, 1890, found the streams already with considerable flow, and the ground thoroughly saturated. The 1897 flood rainfall on July 27–31, on the other hand, was preceded by 26 days below the average in rainfall, since an average of only 2.36 inches of rain fell during the first 26 days of July, 1897, as against a 15-year average for July of 3.50 inches. The distribution of the rainfall over the drainage area, and the temperature conditions, in 1890, were also much more conducive to a high rate of runoff than in 1897.

The results of a comparison of the flood periods of July-August, 1897, and September, 1899, are summarized by the Austrian Hydrographic Bureau as follows:

(a) The volume of rainfall which fell at the time of the flood of 1897 was much more considerable than that of the year 1899, the former amounting, for the total basin, to about 1.4 times the latter.

(b) Similarly, the volume of runoff in 1897 far exceeded that in 1899, in that in the former the volume for the total basin was 1.8 times, for the Moldau basin 1.6 times, and for the Little Elbe basin 2.7 times greater than in the latter.

(c) In 1897 higher maxima were reached generally. Only on the Moldau and Wattawa rivers did the flood wave of 1899 reach a greater height.

(d) Finally, in 1899, as a result of the smaller rainfall, lower runoff coefficients than in 1897 were obtained, which phenomenon is especially noticeable on the Little Elbe basin above Melnik.

Floods in the Oder River[*]

The Oder River rises in northern Austria and enters Germany in the province of Silesia. The tributaries coming into the Oder on the

[*] Führer durch die Sammelausstellung aus dem Gebiete des Wasserbaues Brussels, 1910.
Centralblatt der Bauverwaltung, 1894, pp. 345–350.
Jahrbuch für die Gewässerkunde, 1901.
Handbuch der Ingenieur Wissenschaften III, I Band, 4 Auflage.
Der Oderstrom, sein Stromgebiet und seine wichtigsten Nebenflüsse.

left or west bank have a steep fall. They come down from the mountains of the Austrian boundary, which stretch their foothills far toward the Oder, leaving only a narrow strip of level country along the river. Their drainage basins receive a very heavy rainfall, especially in the upper portions. In the comparatively narrow strip of flat land along the river the rainfall is low, often less than 20 inches of mean annual rainfall; in the mountains it is 40 inches or more, sometimes as high as 55 inches. In the low country the average daily rainfall in storms is 1.4 to 1.6 inches, with a maximum of 3.3 inches, while in the mountainous country it is 3.1 inches on the average, with a maximum of 8 inches. At Neuwiese, a station near the source of the Görlitzer Neisse tributary, a maximum 24-hour rainfall of 13.6 inches was recorded on March 29, 1897. The total rainfall at this station during the storm of July 26–31, 1897, was 17.8 inches.

The following table shows the rainfall in inches at 13 representative stations on the Oder basin. The number of years of record end with the year 1890.

Station	Years of Record	Maximum Annual Rainfall	Minimum Annual Rainfall	Mean Annual Rainfall	Maximum 24-Hour Rainfall
Breslau...................	82	26.0	6.1	22.2	4.4
Ratibor...................	34	31.4	14.3	24.1	3.5
Kreuzburg.................	18	30.8	15.0	22.0	....
Zechen-Guhrau.............	36	32.6	14.8	22.0	2.6
Wang.....................	26	73.5	22.6	47.2	6.1*
Eichberg.................	32	35.5	21.0	26.8	3.8
Frankfurt on Oder..........	43	26.8	13.8	20.3	3.7
Posen....................	40	27.3	11.3	19.6	3.3
Stettin..................	43	25.8	12.6	20.2	3.3
Pammin..................	24	24.7	19.5	26.6	3.0
Prenzlau.................	33	27.3	12.2	17.8	2.5
Hinrichshagen.............	28	24.5	12.2	18.6	2.7
Lübbenow.................	32	26.5	15.8	19.7	3.3

The cloudburst of August 2–3, 1888, on the basins of the upper Queis, Katzbach, and Bober caused the greatest floods in these streams since 1804. The Queis, Katzbach, and Bober are tributaries on the left bank of the Oder in Silesia. The storm of 1888 began on the evening of August 2 and lasted from 15 to 18 hours. The mean annual rainfall and the average precipitation during this cloudburst, as determined from a number of station records on each basin, are as follows:

Basin	Mean Annual Rainfall	Rainfall August 2–3, 1888	Per cent of Mean Annual
Queis...............	32.62	5.64	17.3
Katzbach...........	25.08	4.19	16.7
Bober..............	31.78	3.90	12.3

* This occurred on July 17, 1882; on July 30, 1897, there was a rainfall at this station of 8.66 inches.

Another great summer flood occurred June 14–23, 1894, on the upper part of the Oder and Weichsel River basins. The average rainfall over these basins, for the maximum period of 24 hours, June 16, as determined from the records of 14 representative stations, was 2.23 inches; the average for the maximum 3-day period, June 15–17, as determined from the records of 5 of these stations, was 5.83 inches; and the average for the whole storm period, June 14–23, at 4 stations, was 9.60 inches.

The storm fell on saturated ground. There had been intermittent rains for 14 days preceding, most of which had run off. The rate of rainfall on the upper Olsa basin, an upper tributary of the Oder, above Teschen, having a drainage area of 185 square miles, was 14,120 cubic feet per second on June 16, as estimated from the records at two stations. One of these stations, Istehna, reported 3.06 inches; the other, Jablunkau, reported 3.48 inches. These are two of the highest records for the maximum day. The discharge at the maximum stage on June 17 was 18,800 second feet, or 101.5 second feet per square mile, showing that the rainfall on the mountains must have been considerably higher than that recorded at the stations, which lie in the valleys, and that practically all the rainfall ran off. This occurred in a basin, the greater part of which was covered with heavy woods and the soil of which under ordinary conditions is not impermeable.

A third great flood of recent years on the Oder basin was that caused by the storm of July 27–31, 1897. This flood did great damage and in the province of Silesia 28 lives were lost. Extensive protective measures were carried out as a result of this flood. The average rainfall over the drainage areas of the principal tributaries of the upper Oder during the maximum 24-hour period, July 29–30, and during the total 4 days of rainfall, July 27–31, 1897, was as follows:

Tributary	Drainage Area Square Miles	Maximum 24-Hour Rainfall	Total Rainfall July 27–31
Lomnitz	45.2	6.61	9.96
Zacken	105	4.56	7.05
Kemnitzbach	44.2	4.33	6.97
Zippelbach	17.5	5.67	6.90
Bober	793.4	3.31	4.94
Queis	388.6	3.79	5.56

The maximum discharge of the Bober was 67,750 second feet, or 85.5 second feet per square mile; that of the Queis was 42,400 second feet, or 109.0 second feet per square mile.

Floods in the Danube River*

The Danube is formed by the union of the two mountain streams, Breg and Brigach, which rise in the Black Forest in the Grand Duchy of Baden, Germany. These streams join at Donau-Eschingen, in Baden, and, from there, the Danube flows through Wurtemberg and Bavaria and enters Austria at Passau, about 200 miles above Vienna. The distance from the source of the Danube to Vienna is about 500 miles, and its total length is about 1735 miles. The drainage area above Vienna is about 39,400 square miles, and the total drainage area is about 315,500 square miles.

The annual average rainfall over the Danube basin above Vienna for the 14-year period 1896–1909 was as follows:

1896........47.1 inches		1903........41.5 inches
1897........45.4 "		1904........38.3 "
1898........38.7 "		1905........42.5 "
1899........44.0 '		1906........44.9 '
1900........40.8 '		1907........40.0 '
1901........39.2 '		1908........34.8 '
1902........40.8 "		1909........42.0 '

Maximum......................47.1 inches
Mean.........................41.6 "
Minimum......................34.8 "

The Hydrographic Bureau of Austria, from whose records much of the material here is taken, has made an exhaustive study of floods and flood conditions on the Danube at Vienna.

Records of floods on the Danube extend back to the year 1000, during which time there have been repeated disastrous floods, causing great loss of life and property. The greatest was in 1501, the second greatest in 1787, and the third greatest in 1899.

The greatest flood in the Danube ever recorded, that of 1501, had a maximum discharge at Vienna of 494,200 second feet, or 12.6 second feet per square mile. The next greatest, that of 1787, discharged 416,540 second feet, or 10.6 second feet per square mile. The third greatest, that of 1899, had a maximum discharge of 370,650 second feet, or 9.5 second feet per square mile at Vienna. A very large flood in the Danube also occurred July 16–31, 1897. In this flood the maximum discharge at Vienna was 337,100 second feet, or 8.6 second feet per square mile.

The following amounts in inches for maximum 1-day and 2-day

---

* La Danube dans la Basse Autriche, by M. Armand. Annales des Ponts et Chaussées, 1910, V.

Reports of the Austrian Hydrographic Bureau.

periods and for the entire storm of 7 days, at stations reporting greatest rainfall, will give the reader an idea of the precipitation in the storm of September 8–14, 1899:

Period	Station	Sept. Storm	Mean Annual	Year 1899	Sept., 1899	Mean for Sept.
Max. 1 day............	Mühlau.........	11.3	61.1	78.9	22.5	5.35
Max. 2 days............	Langbathsee.....	17.0	......	98.6	27.0	......
Total 7 days............	Alt-Aussee*......	26.2	82.5	105.6	31.2	7.72

The tabulation below shows the average amount of rainfall over the Danube watershed above Vienna during three large storms, and the discharge at Vienna for these and the two greatest previous floods. For the storm of July, 1897, the records at 371 stations were used in determining the average rainfall; for the storm of September, 1899, the records at 472 stations were used; there is nothing to indicate the number of records used in determining the average for the storm of August, 1890.

Storm	Max. 1-Day Rainfall, Inches	Total Storm Rainfall, Inches	Max. Disch. Sec.-Ft.
Causing flood of 1501.........	....	....	494,200
Causing flood of 1787.........	....	....	416,540
August 24 to Sept. 5, 1890.....	1.48	7.11	....
July 16–31, 1897.............	1.65	5.90	337,100
September 8–14, 1899.........	2.37	6.71	370,650

A study of rainfall and runoff shows that the greatest recorded flood may some day be exceeded. For example, the great flood of 1899 occurred in September when the Danube is low. The same rainfall could occur in June or July, when the river is high due to the melting of snow and glaciers. Also, part of the precipitation at the time of the flood of 1899 was in the form of snow in the mountains. If this had all been rain, the resulting flood would have been greater. The soil was capable of more absorption in September than if the flood had occurred earlier in the year. The rainfall could have had a longer duration and local heavy downpours could have increased its effect in raising the river at Vienna, as shown by past rainfall records.

In short, despite the records of over 900 years at Vienna, the government experts are convinced, after studies with the most complete and long term records of rainfall and gage heights, that provision should be made for a greater flood than has ever yet been experienced. This was also borne out by a study of the possible combinations of

---

* The precipitation for the maximum day at Alt-Aussee was 9.58 inches.

various tributary floods, which showed that combinations much more conducive to maximum stages in the Danube were liable to occur.

These studies also showed that, as might be expected, the most favorable conditions for a great flood were when the rainstorm travelled in the direction of flow of the main river. A lighter storm travelling thus might cause a greater rise than a heavier rainfall going upstream or across the basin.

The Hydrographic Bureau of Austria concluded that, on the average, great floods may be expected about once a decade, sometimes twice. Extraordinary floods, like 1899, 1787, and 1501, due to widespread, excessive, extended rainfall, or to snow melting combined with heavy rains, occur seldom, but may occur several times during a century. A record of 100 years is not long enough to fix the maximum flood. A record of 500 years or more is scarcely sufficient. For example, the 1899 flood, the maximum of the 19th century, might have been taken as the maximum to determine the channel capacity. It was the maximum for over a hundred years, yet the flood of 1501 had a 33 per cent greater maximum discharge at Vienna.

Floods in the Seine River*

The river Seine of northern France flows west through Paris and empties into the English Channel. Its principal tributaries above Paris are the Aube, Yonne, and Marne. The mouth of the latter is but a short distance above the city. The Oise empties into the Seine from the north a short distance below Paris. The total area of the Seine basin is 30,400 square miles; the drainage area of the main stream and its tributaries above the mouth of the Oise is 16,730 square miles.

The regimen of the Seine basin is regular. The highest altitudes at the headwaters reach 2600 feet. About one-third of the drainage area is impermeable strata; the rest permeable. Rains falling on the impermeable strata always produce the highest floods at Paris. The floods caused by the tributaries flowing over the permeable strata take considerably longer in reaching Paris. The Seine, consequently, remains in flood for several days. If during that period another rainfall occurs, a higher flood follows. The river may continue rising, therefore, during whole months, from the effects of successive rains. Floods on the other great rivers of France, the Rhone, Garonne, and

* The information and data here given was taken principally from three sources: The Report of the Flood Commission of Paris, 1910; Les Inondations du Bassin de la Seine; and Observations faites sur la Seine à Paris pendant la Crue de Janvier-Fevrier-Mars, 1910. M. Arana. (Annales des Ponts et Chaussées, 1911, VI, p. 600.)

Loire, on the contrary, are generally the result of one rainfall. Their rise is rapid and high, but of short duration.

The* annual rainfall over the Seine basin during the 37-year period, 1871–1907, determined from the average of the records at 130 stations, was as follows:

Year	Annual Rainfall	Year	Annual Rainfall	Year	Annual Rainfall	Year	Annual Rainfall
1871	.23.9	1881	.25.3	1891	.27.8	1901	.26.4
1872	.34.6	1882	.34.0	1892	.27.2	1902	.26.5
1873	.26.5	1883	.29.0	1893	.23.4	1903	.27.0
1874	.21.5	1884	.24.0	1894	.25.3	1904	.25.1
1875	.28.2	1885	.30.4	1895	.26.8	1905	.28.2
1876	.27.6	1886	.31.8	1896	.30.4	1906	.27.9
1877	.32.4	1887	.24.2	1897	.27.0	1907	.25.0
1878	.34.4	1888	.28.0	1898	.23.2	Mean	.27.5
1879	.28.9	1889	.26.7	1899	.26.1	Max.	.34.6
1880	.29.7	1890	.26.2	1900	.25.7	Min.	.21.5

The average annual rainfall at various stations in the upper region of the basin varies considerably from the general average of all the upper region stations. Thus the maximum average annual rainfall at any station is 69.4 inches, and the minimum is only 22.4 inches. About 46 per cent of the rain falls between November 1 and April 30, and 54 per cent between May 1 and October 31. The greatest rainfall is in the upper mountainous part of the basin, the next greatest toward the sea, and the least amount in the intermediate region. Thus the region around Paris receives the smallest amount of rainfall in France, the annual amount being only 22.9 inches.

The most abundant and persistent rains rarely cause a flood in the Seine during the warm season. In the cold season, the floods are more frequent; but even then they are rare enough to class the Seine among the quiet rivers. During the 18th and 19th centuries there were at Paris 33 floods of 16 to 20-foot stages, 11 above 20 feet, and 3 disastrous floods above 23 feet. During the same two centuries there were only 10 floods during the warm season, which reached stages of 10 to 17 feet. The rainfall statistics show that although heavy rainfalls are required to cause great floods on the Seine, they do not necessarily produce them.

The table on the following page gives the rainfall in inches at 15 representative stations on the Seine watershed above Paris, which caused four great floods in the past half century.

Studies of rainfall on the Seine basin show that floods of the cold season, November 1 to May 1, of any considerable height at Paris, have always been preceded by a warm season with more than the

* Supplément au Manuel Hydrologique du Bassin de la Seine, 1909.

17

normal rainfall of 14.7 inches in the basin above. This was the case
in the flood of January 28, 1910. The rainfall in the warm season of
1909 was 17.3 inches. This was also the case in the floods of 1896–
1897, when the warm season rainfall was 19.6 inches. Two floods
followed in November, 1896, and February, 1897, with stages of 17.4
feet and 18.4 feet, respectively, at Paris. It also requires abundant
rain in December and January, to bring about a flood.

Station	Mean Annual Rainfall	Term of Record	Sept. 22–24, 1866	Feb. 12– Mar. 15, 1876	Nov. 8– Dec. 9, 1882	Dec. 21, 1882–Jan. 3, 1883
Les Settons.....	68.9	1861–1880	5.95	31.80	19.40	13.86
Pannetiere......	35.0	1861–1880	4.76	9.06	10.30	3.50
Saulien.........	34.3	1861–1880	5.27	11.90	9.34	5.07
Penilly.........	29.8	1861–1880	3.82	6.54	5.59	2.40
Montbard......	27.8	1861–1880	4.10	7.05	6.94	3.90
Chanceaux......	38.8	1861–1880	4.88	10.00	8.30	5.40
Chantillon......	28.8	1861–1880	3.98	6.15	7.88	2.84
Chaumont......	37.6	1865–1880	3.23	9.05	10.60	5.75
Vassy..........	33.1	1865–1880	3.38	8.98	8.70	3.50
Bar-le-Duc......	37.4	1861–1880	2.40	9.90	10.80	4.40
La Neuville.....	35.0	1871–1880	2.48	7.55	8.58	3.30
Montmont......	29.6	1866–1880	....	7.00	8.70	2.91
Sainte-Menehould....	29.4	1871–1880	1.77	7.80	6.57	2.20
Hirson..........	29.5	1859–1878	2.72	8.31	7.40	4.05
Paris..........	22.9	1867–1880	1.38	2.99	5.79	1.34

The following are the greatest floods which have occurred at
Paris. The table indicates that great floods on the Seine are not so
frequent as on many other rivers. The stages are at la Tournelle
bridge, Paris:

Year	Stage		Year	Stage
1649	25.2 feet		1740	26.0 feet
1651	25.7 "		1764	24.0 "
1658	28.9 "		1802	24.4 "
1690	24.8 "		1836	21.0 "
1711	25.0 "		1876	21.3 "
			1910	27.6 "

This shows that the maximum flood at Paris occurred in 1658 and
reached a stage of 28.9 feet at the Pont de la Tournelle. Some records
show that a greater flood occurred in 1611 or 1615, with a stage of
about 30 feet. The 1658 record is called the maximum, however.
The 1910 flood, which reached a stage of 27.6 feet, was next in size
to the 1658 flood. It is claimed by some authorities, however, that
other floods intervened which had greater discharges, though recorded
of less height on the gage because the channel of the Seine in Paris
had a greater carrying capacity than in 1910.

The rains which brought about the flood of January–February–March, 1910, reaching a maximum on January 28, occurred in three distinct rainfall periods. The rainfall in inches during each of these periods, obtained by taking the average of the records at 15 representative stations, was as follows: During the period November 28 to December 9, 1909, the average rainfall was 2.8 inches; December 15–31, 1909, it was 2.4 inches; and January 9–27, 1910 it was 6.0 inches. During the maximum 4 days of the last period, January 18–21, 1910, the average rainfall at these 15 stations was 3.1 inches. The maximum rainfall for this 4-day period occurred on the basin of the Yonne River, a tributary which enters the Seine above Paris. At the 3 stations of maximum rainfall on this basin the amounts were 6.1, 7.5, and 7.8 inches. The maximum 1-day rainfall also occurred at these 3 stations, on January 19, being 2.6, 3.0, and 3.0 inches, respectively.

The river at Paris began rising on January 18 and reached its maximum, 27.6 feet at la Tournelle, Paris, on January 28, where it remained practically constant 12 hours. The crest at the Pont d' Austerlitz was 28.3 feet. It then fell slowly until February 7, when it was still 6 feet above the lower wharves along the Seine, which are covered at a stage of 8.7 feet. It then began to rise again, and on February 12 reached a second crest of 17.9 feet at Austerlitz. It stayed 17 hours at this level, fell to 17.5 feet at Austerlitz on the 14th, and then rose to a third crest of 18.4 feet on the 17th, where it stayed for 8 hours. It then fell gradually to a stage of 14.4 feet on February 22, and again rose gradually to the fourth crest of 18.4 feet on March 1. The river slowly receded from this stage until March 15, when it exposed the lower wharves.

The flow of the Seine at Paris in the 1910 flood has been variously estimated. The generally accepted figure is 88,500* second feet, or 5.26 second feet per square mile of drainage area. This was for a gage height of 27.6 feet.at la Tournelle, Paris, which is the greatest height reached since 1658, when there was a stage of 28.9 feet at la Tournelle. The discharge during the 1658 flood was practically the same.

Floods in the Upper Loire River†

The part of the Loire considered as the Upper Loire is that above the department of Saone-et-Loire, somewhat southeast of the center of France. It drains an area of 3090 square miles, stretching from south to north a distance of 180 miles, lying between the basins of the Ardèche and Allier Rivers.

* Annales des Ponts et Chaussées, 1911, VI, p. 600.

† Les Crues de la Loire Supérieure, by M. Jollois. Annales des Ponts et Chaussées, 1881, 1st Semestre, p. 273.

From the source to Saint-Just, a distance of 113 miles, the Loire is a deep-bedded torrent, widening out in only four small submersible valleys at Brives, Saint-Vincent, Vorey, and Bas. The last is the greatest, and it is only about three miles long. At Saint-Just, the Loire enters the plain of Forez, which it traverses for 28 miles. Then it enters the gorge of Pinay, which is 19 miles in length; and, finally it runs through the plain of Roanne for about 20 miles.

The average seasonal and annual rainfall, in inches, on different parts of the Upper Loire River basin is tabulated below. These figures were determined by averaging the records made during the 10-year period, 1861–1870, at 4 stations on the right bank of the Upper Loire, at 5 stations on the left bank of the Upper Loire, at 5 stations on the right bank of the Forez, at 2 stations each on the left bank of the Forez and on the right and left banks of the Roanne.

Average Seasonal and Annual Rainfall, in Inches, in Upper Loire Basin

	Winter	Spring	Summer	Autumn	Year
Upper Loire:					
Right bank...............	9.63	11.40	9.25	15.82	46.10
Left bank...............	4.41	7.74	7.63	9.43	29.21
Le Forez:					
Right bank...............	5.18	8.44	8.87	9.91	32.40
Left bank...............	3.78	6.00	6.24	7.48	23.50
Roanne:					
Right bank...............	6.88	8.62	9.03	9.57	34.10
Left bank...............	5.53	7.13	8.41	9.58	30.65

The author recognizes four storm types and gives certain data for the greatest storm of each type. They are:

1. Storms produced by very heavy rainfall on the Mezenc Mountains in the Upper Loire basin, with little or no rainfall on the rest of the basin. This rainfall is generally brought by south winds.

The most remarkable storm of this type was that of October 8–9, 1878, which moved from south to north, diminishing in intensity. The average distribution of the rainfall over the watershed was as follows:

Mezenc Mountain region.........................7.88 inches
Upper basin (exclusive of Mezenc Mountains)........3.85 "
Forez basin..2.12 "
Roanne basin......................................1.68 '

Of the 7.88 inches which fell in the region of the Mezenc Mountains, 6.78 inches fell on October 8. The maximum rainfall fell on the 8th throughout the basin except at one station, where it occurred on the 7th.

2. Storms produced by a general rain all over the basin, but heavier in the upper part. This type of rainfall is generally accompanied by a south wind, turning to southwest.

The greatest storm of this type was that of September 24–25, 1866. The average distribution of rainfall over the watershed was as follows:

```
Upper Loire basin.................................5.5 inches
Forez basin......................................3.8   "
Roanne basin.....................................5.5   "
```

The 24th was the day of greatest rainfall, except at two stations, at one of which the maximum occurred on the 23d and at the other on the 25th.

3. Storms produced by general rain over all the basin, but greater below the mouth of the Forez than above. They are generally accompanied by winds from the southwest to west.

The greatest flood produced by this type of storm was that of October 18–20, 1872, in which the maximum rainfall occurred in the Roanne basin. The fore part of the month was very rainy, and the ground was thoroughly saturated. The average distribution of rainfall over the watershed was as follows:

```
Upper Loire basin.................................1.63 inches
Forez basin......................................4.00   "
Roanne basin.....................................4.47   "
```

The maximum rainfall in the Upper Loire and Forez basins occurred on the 18th; in the Roanne basin, on the 19th.

4. Storms produced by light rains in the upper basin and heavy rains in the basins of the lower tributaries, accompanied by much runoff of melted snow. Storms of this type are generally accompanied by southwest to northwest winds.

The greatest recorded storm of this type is that of April 8–9, 1879. At the beginning of April a heavy layer of snow covered all the Loire basin, when a warm wind, accompanied by rain, crossed the basin from west to east. The flood was caused mainly by melting snow. The average distribution of rainfall in inches on April 7 and 8 was as follows:

	April 7	April 8	Total
Upper Loire basin..........	1.15	0.81	1.96
Forez basin................	0.50	1.03	1.53
Roanne basin	0.00	1.16	1.16

The floods of the first three classes, which are the most common, attain their maximum the same day as the maximum rainfall. Rainfall observations for the purpose of flood prediction, therefore, are not of much value. The rapidity of propagation of the flood peak is greater the higher the flood.

Except at a few points, the flood wave remains within the banks in the upper portion of the stream.

Floods in the Garonne River*

The Garonne is in southern France, and flows westward into the Atlantic. Above Toulouse, and below, on the left bank, the river receives the drainage from the Pyrenees; on the right bank, below Toulouse, it is fed by tributaries from the central table-land of France and from the Cevennes and other mountain ranges.

At Toulouse the area drained by the Garonne is only about 2400 square miles, which is approximately one-seventh the area of the Seine basin above Paris, yet the maximum flow of the Garonne at Toulouse is six times that of the Seine at Paris. The maximum floods occur in the spring and early summer, and are caused by rain and melting snow. Two of the greatest floods of the Garonne occurred in June and September, 1875.

The flood of June 21–24, 1875, was caused by 4 days' rainfall, averaging about 6.7 inches in depth at 11 representative stations, accompanied by some melting snow from the mountains. The maximum 24-hour rainfall at these stations averaged about 3.4 inches in depth. The rainfall on the upper basin of the Tarn, one of the Garonne tributaries, was very great. At .4 typical stations, the records were 7.7, 7.7, 7.6, and 8.8 inches. The corresponding maximum 24-hour readings were 4.0, 3.7, 4.0, and 5.8 inches. The discharge of the Tarn resulting at Palisse on June 23 was 229,500 second feet. In this flood, 540 lives were lost, part of Toulouse was flooded and destroyed, and a number of villages were submerged.

During the 5-day period, September 9–13, 1875, the records show a remarkable rainfall covering the part of France between the Loire basin, the Pyrenees, and the Mediterranean coast. The average rainfall at 23 selected stations for the five days was 14.73 inches, and the 24-hour maximum rainfall averaged for these stations was 6.63 inches. The widespread and record-breaking rain caused the greatest floods known in that part of France. The rain and floods had been greater in mountainous parts, but, in the plains, this is the maximum on record. At Saint-Gervais there was recorded 12.6 inches during

* Annales des Ponts et Chaussées.

this 5-day period. During the 4-day period, October 14-17, 1874, at this station, however, there fell 34.8 inches of rainfall.

One of the stations whose record was used in determining the above averages, that of Bleymard, at the summit of the Cevennes, received 22.8 inches of rainfall during the 5-day period, 15.7 inches of which fell on the 13th. At Vialas, at the source of the Ardèche River, there was 20.0 inches during the 5-day period, 7.92 inches of which occurred on September 13th. This gave a flood of 18.4 feet at Salavas, whereas the greatest flood known on the Ardèche occurred in October, 1827, reached 55.8 feet on this gage, and was produced by a rainfall of 31.2 inches in 24 hours at Joyeuse, a station in the basin.

The flood of the Garonne at Toulouse in September, 1875, was very disastrous to life and property. Over 200 dead bodies were found after the flood, and many other lives were lost. The damage was done at the suburb of St. Cyprien, across the river from Toulouse. Toulouse is on higher ground and is protected by massive quays, which have been built out into the river, encroaching on the channel. St. Cyprien is on low ground on the opposite bank, a few feet above normal river stage, and occupies a triangular tract at a bend in the river, across which is the natural path of overflow. The central and most thickly populated part of St. Cyprien, in 1875, lay at a lower level than the raised bank of the river skirting this triangle. A masonry arch bridge and several low dams obstructed the channel, in addition to the quays. The flood swept across the suburb and did enormous damage. The remedies proposed were to construct a diversion channel back of St. Cyprien, across the bend; to replace the bridge by one of greater waterway; to widen the river at narrow points; and to change the fixed to movable dams. The diversion channel was estimated to cost $1,500,000.

Floods in the Durance River[*]

The Durance is different from the other French rivers. The impermeability of its basin, the great number and the elevation of its mountains, its meteorology so entirely different from that of the rest of France, and finally the steep slope, not only of the river but also of its tributaries, help to explain the fact that it is the most extraordinarily torrential of the great French rivers.

The Durance, located in the southeast of France, drains an area of 5730 square miles and empties into the lower course of the Rhone from the east. It is about 190 miles in length, in the course of which

[*] La Durance. Régime, Crues et Inondations, by M. E. Imbeaux. Annales des Ponts et Chaussées, 1892, 1st Semestre.

it falls from an elevation at its source of 6760 feet to one of 43 feet above sea level at its mouth.

The rainfall records are very incomplete, covering only a short period. Prior to 1882 there were not enough rainfall observing stations to determine with accuracy the annual average rainfall over the basin. For the years 1882–1888, however, this average can be very well determined and is given below.

Annual Rainfall, in Inches, in Durance Basin

Basin	1882	1883	1884	1885	1886	1887	1888	Mean
Above Mirabeau........	32.1	27.9	21.0	34.0	45.6	29.7	34.6	32.2
Below Mirabeau........	23.4	20.0	20.0	29.6	32.3	26.1	29.0	25.8
Entire basin...........	30.4	26.2	20.8	33.1	42.8	29.0	33.1	30.8

A comparison of the records at a few stations, where readings extended back from 30 to 57 years, showed that the 1882–1888 period was one of more than usual rainfall, and the average for the basin is adopted by the author as 27.6 inches, which is less than the mean annual for France of 30.3 inches.

The number of days of rainfall per year, however, is less than in any other part of France. From 1882 to 1889 this number averaged only 68. These days are not divided equally between the seasons. They average 13 in summer, 14 in winter, 16 in spring, and 19 in autumn, with a maximum of 9 days per month, in November, and a minimum of 3 days per month, in August.

The rainfall is not divided equally between the seasons, or between the months. October and November are the wettest months, the mean rainfall for these months for 1882–1888 being 4.2 inches and 4.43 inches, respectively, or about one-seventh the annual rainfall. The driest month is August, with a mean of 1.62 inches, or about one-twentieth of the mean annual. January, February, and March are also dry months. The result is that winter is the driest of the seasons, with a mean of 5.25 inches, or 17 per cent of the mean annual rainfall. Spring and summer have about 23 per cent of the mean annual. Autumn is the wettest season, with an average of 11.25 inches, or 37 per cent of the mean annual. The great rains of autumn are those that produce the highest floods. The geographical distribution of the rainfall over the basin of the Durance is favorable for producing floods, because the heaviest rainfall is upon the steepest and most impermeable parts.

The record of floods is of much greater length than that of rainfall. Prior to 1832, however, a complete record of floods is not available, and there are no records of gage heights or discharges. Partial records back to 1226 show numerous floods, generally occurring in the autumn.

From 1832 to 1890 there were 188 floods. Of these, 46 were great enough to cause gage heights of 13 to 16 feet or over at Mirabeau, 54 miles above the mouth of the stream and above which the Durance has a drainage area of 4150 square miles. Of the 46 floods, 7 reached 16.5-foot stages at Mirabeau. There were as many as 9 floods in a single year, as in 1848 and 1856; and there were only 9 years during the whole period which were exempt from floods reaching at least a 10-foot stage at Mirabeau.

The 7 exceptional floods (16.5-foot or higher gage records at Mirabeau) occurred in 1836, 1843, 1863, 1882, and 3 in 1886. The greatest flood was the last of the three in 1886, which occurred on November 10–11, causing a discharge at Mirabeau of about 237,000 second feet, or 57 second feet per square mile. This was practically equalled by the flood of 1843. Such floods cause approximately 20-foot stages in the Durance at Mirabeau. The shifting character of the material in the stream bed prevents a closer comparison of different floods by means of gage heights. The flood of October 28, 1882, had a maximum discharge at Mirabeau of 203,000 second feet; the flood of October 26–27, 1886, a maximum of 176,700 second feet; and the flood of November 8, 1886, a maximum of 191,000 second feet.

The Durance floods generally occur either in spring, from March 15th to June 15th, or in the fall, from September 15th to December 15th. Spring floods are rarely very high. Fall floods are often high and dangerous. Due to the distribution and form of the precipitation, winter and summer are almost totally free from floods. Floods of the Durance seldom coincide with those of the Rhone, owing to the very complex regimen of the latter as compared with that of the former.

This paper, by the engineer in charge of the studies on the Durance at the time, outlines the results of the work done by order of the government to establish a basis for flood prevention measures. The government ordered this study made because of the great damage done by the three floods of 1886.

Floods in the Tiber River*

The drainage area of the Tiber is 6455 square miles. The mean annual rainfall over the basin is 34.8 inches, and there is no appreciable snow fall.

* From the Proceedings of the Institution of Civil Engineers.
The Works on the Tiber, by M. Ronna. Bulletin de la Société d'Encouragement pour l'Industrie Nationale. November, 1898.
Deutsche Bauzeitung, 1893, p. 99.
Ueber die Tiberregulierung, Wochenschrift des Oesterreichischen Ingenieure und Architekten Verein. 1877, p. 114.

The first flood recorded at Rome was in 413 B. C.   The greatest flood ever recorded was that of 1598, A. D., when the maximum discharge was 106,000 second feet, or 16.4 second feet per square mile.   The second greatest flood was the disastrous flood of 1870, when the maximum stage was 8 feet lower than in 1598.   However, the actual discharge during this flood was but little less than that of the flood of 1598, the great difference in stage being due to obstructions and encroachments in the channel.   The stage in 1598 was 64.2 feet, in 1870 it was 56.5 feet.   The 1870 flood was 12 feet deep in the streets along the quays, and covered about one half the inhabited area of the city.   The water surface slope was 2.89 feet per mile in the 6-mile course through the city.   In the upper part it was only 0.8 foot per mile, but in the lower part it was 5.81 feet per mile.   The obstruction at one of the bridges caused a difference in level of 22 inches.

CHAPTER XII.—APPLICATION TO MIAMI VALLEY OF STUDY OF GREAT STORMS

In preparing plans for flood prevention in the Miami Valley one of the first problems requiring definite decision was the size of the maximum flood against which protection should be provided. The engineer at once recognizes that the size of the maximum possible flood cannot be determined directly from the size of the maximum possible rainstorm, vital as this may be, since the greatest rainstorms do not necessarily cause the greatest floods. Moreover, the flood prevention works of the Miami Valley on occasion will submerge large areas of valuable agricultural land. As the frequency and depth of this flooding will have a marked effect on the value of these lands, the design of the works was materially affected by the desire to secure flood protection with a minimum damage to the land above the dams. The ability to make accurate estimates of the frequency of floods of any given magnitude therefore is of great value.

In determining the size of the greatest flood within the bounds of probability, which may occur over a given area, there are three primary conditions to be considered. They are:

(1) Topographic conditions affecting the rate of runoff, such as: size, shape, and slope of drainage area; slope and condition of channel of main stream and tributaries; number and position of tributaries; and surface cover of watershed.

(2) Geographic location of the drainage area with regard to direction and path of storm movement, sources of moisture, distance from ocean bodies, and location of mountain ranges.

(3) Past records of great storms and corresponding stream stages, so far as they may throw light on the maximum storm which may occur in the future. The time-area-depth relations, season of occurrence, frequency, and geographic location are the principal features of former storms to be examined in determining their bearing on any flood problem. The study of former storms should not be confined to those which have occurred over the given drainage basin, especially if the past record is for comparatively few years, since greater storms may have chanced to occur over other areas which are comparable in nearly every respect. This raises the difficult question as to just what other areas are comparable in the matter of the greatest storms to which they are subject.

SIZE OF FLOOD SELECTED AS BASIS OF DESIGN

In working out the plans for protecting the Miami Valley against floods, all three of the factors just mentioned were thoroughly investigated. The second and third factors, geographic location, and past records of storms, are the subjects treated in this volume. After mature consideration of all of these factors the Official Plan for flood protection was designed to provide against a hypothetical storm which would cause a maximum flood runoff almost 40 per cent in excess of that of the storm of March 23–27, 1913, the latter having caused the greatest rate of runoff during the 100 years of record for the Miami River. The reasons for adopting this basis will now be discussed in considerable detail.

GEOGRAPHIC LOCATION OF MIAMI VALLEY IN ITS RELATION TO STORMS

As has been previously stated, the general direction of storm movement in the United States is from west to east. The low barometric centers which accompany storms generally originate in the northwest, west, and southwest. Their most frequented tracks converge and focus in eastern Texas, Oklahoma, and Arkansas, and then either extend up the Mississippi and Ohio Valleys or pass off the south Atlantic coast in Georgia or South Carolina. Occasionally a storm travels northwest across the gulf, then changes its direction to north or northeast through Texas, Oklahoma, Missouri, Illinois, Indiana, and Ohio. This is the characteristic path of West Indian hurricanes. The location of the 160 largest storms which occurred in the eastern part of the United States during the 25-year period 1892–1916 can be clearly seen by referring to figures 50 to 57.

Heavy, long-continued rains in the United States obtain their supply of water largely from the Gulf of Mexico and the Atlantic Ocean. A large part of the rain of the Miami Valley doubtless comes from the ocean and gulf, and on account of the long distance of land travel, and the higher latitude, the greatest rainfall in southwestern Ohio cannot be nearly as heavy as may occur over the south Atlantic and Gulf states. In traveling such long distances over land the air, laden with moisture when it leaves the ocean or gulf, is subject to large variations of temperature, the average of which is gradually decreasing with increasing latitude. This causes the air to lose much of its moisture. The effect of distance of land travel and increasing latitude in decreasing the depth of storm rainfall is plainly evident in all of the investigations described in chapters V to VIII.

Not all the rainfall of the northern interior part of the United States, however, comes directly from the ocean and gulf. A large part of the rainfall of southwestern Ohio, especially during the summer months, is water that has evaporated from the forests and fields to the south of it. The forests and fields of Missouri and Arkansas, for instance, will evaporate water at the rate of one-eighth of an inch, and perhaps one-fourth of an inch a day, as long as the supply is plentiful. As just stated, this is largely the source of water supply in the frequent thunderstorms of the summer months. Such storms, however, while they may have considerable depths over limited areas, are rarely if ever the cause of serious flood damage in the Miami River. From this it appears that after all, the sources of moisture of destructive storms in the Miami Valley are the ocean and gulf, and the maximum intensity of rainfall in such storms is subject to the limitations imposed by latitude and distance of land travel.

Another geographic feature that tends to reduce the maximum rainstorm which can occur in the Miami Valley is the location of the Appalachian Mountains to the east and southeast. The effect of high mountains in preventing the passage of great quantities of water vapor is discussed in chapter III.

From the foregoing discussion it will be seen that the Miami Valley is located in one of the principal paths of storms. It is so shielded from the two primary sources of moisture, however, that the otherwise possible maximum storm is greatly reduced.

GREAT STORMS OF THE PAST AS APPLIED TO MIAMI VALLEY CONDITIONS

The three factors, time, area, and depth are necessary in determining the absolute and relative sizes of storms. These factors were determined for 33 of the largest and most important storms which have occurred in the eastern part of the United States during the past 75 years, and the results are given and discussed in chapter VIII. In deciding which of these storms are applicable to Miami Valley conditions it is necessary to consider their geographical distribution; and to determine which of those applicable would cause the greatest floods if duplicated in the Miami Valley, it is necessary to consider their seasonal distribution. These subjects are discussed in chapter VII.

Of the 33 great storms for which detail investigations were made, we may at once eliminate from consideration the 16 in the southern group. The investigation of all the factors which influence the size of storms in the northern interior of the United States furnishes conclusive evidence that these southern storms cannot be duplicated in the Miami Valley. Of the 17 storms in the northern group, 5 were east of the Appalachian Mountains, and could not be duplicated in the Miami Valley on account of the barrier these mountains furnish against the passage of great quantities of water vapor. The remaining 12 storms in the northern group are those which, within the limits of possibility, may be considered applicable to the Miami watershed, although only one of them, that of March 23–27, 1913, caused a considerable flood in the Miami River. A list of these 12 storms, with the center and date of occurrence of each, is given in table 11. There will also be found tabulated the maximum average depth of rainfall in each storm over an area of 4000 square miles, expressed as a percentage of the maximum average depth of the storm of March 23–27, 1913, over an equal area, for 1 to 5-day periods. The numbers in this table are computed from depths taken from the time-area-depth curves, figures 94 to 98.

This table shows which storms exceeded that of March, 1913. For the maximum period of one day, storms 72 and 83 materially exceeded storm 132. Both were Iowa summer storms. Storm 72 lasted 4 days, but except for the first day it was practically the same size as storm 132 over an area of 4000 square miles at the center; number 83, the Devils Creek storm of June 9–10, 1905, fell principally in 12 hours, and altogether in 24 hours. For the 1-day maximum period those two storms exceeded that of 1913 by 33 and 35 per cent, respectively. It was early recognized, however, that a 1-day maximum rainfall is of too short duration to use as a basis for designing flood protection works for the Miami Valley. A storm of longer duration, although the rainfall on the maximum day might be considerably smaller, could easily cause a more serious flood.

The 2-day maximum average depth of the 1913 storm over an area of 4000 square miles is exceeded in only four cases, and the greatest of these, storm 114, October 4–6, 1910, over Illinois, exceeds it by only 9 per cent. Since the 2-day maximum period is still too short to cause the most destructive floods in the Miami Valley, we may pass this without further comment.

In the investigation of local conditions in the Miami Valley it was found that the maximum 3-day period in a storm would place the greatest burden on a system of protection works. Hence it is the maximum 3-day period, more than any other, which is used as a basis of comparing the sizes of the 12 great storms applicable to the Miami Watershed. In table 11, storms 114 and 151 stand out preeminently for the 3-day period. These two storms have several features in common. They differ but little in average depth, area covered, geographical location, and season of occurrence. Both occurred during the summer and early fall, when atmospheric conditions are most conducive to maximum rates and depths of rainfall, but, fortunately, when the percentage of runoff is generally least. The centers of both of them were far south of the Miami Watershed, and it is very questionable indeed if such storms can occur over the latter area. It is still more doubtful if such storms could occur during the winter months, when the large percentage and rates of runoff cause the greatest floods.

Table 11.—Comparative Intensities of 12 Great Storms Applicable to the Miami Valley, for Maximum Periods of 1 to 5 Days

The greatest average depth of rainfall over an area of 4000 square miles is expressed as a per cent of the depth for the corresponding period and area of the storm of March 23–27, 1913.

Storm	Date	Center	Periods of Maximum Rainfall				
			1-Day	2-Day	3-Day	4-Day	5-Day
15	Dec. 17–20, 1895	Mo.	95	100	107	103	......
51	July 14–16, 1900	Iowa	100	108	106	......	......
72	Aug. 25–28, 1903	Iowa	133	104	105	102	......
83	June 9–10, 1905	Iowa	135	......	......	......	......
86	Sept. 15–19, 1905	Mo.	95	93	101	105	103
109	July 5– 7, 1909	Mo.	86	76	96	......	......
110	July 20–22, 1909	Wis.	83	76	80	......	......
114	Oct. 4– 6, 1910	Ill.	102	109	130	......	......
125	July 20–24, 1912	Wis.	......	......	......	......	64
130	Jan. 10–12, 1913	Ark.	83	75	79	......	......
132	Mar. 23–27, 1913	Ohio	100	100	100	100	100
151	Aug. 17–20, 1915	Ark.	110	108	125	123	......

Of the 12 storms listed, all but 2 occurred during the summer months. One of the 2 winter storms, number 130, had an average 3-day maximum depth only 79 per cent as great as that of March 1913. The other winter storm, number 15, December 17–20, 1895, is 7 per cent in excess of storm 132 for the 3-day maximum period and 4000 square mile area. The two are quite similar in average depth and area covered for other maximum periods. This storm is also in the southern Missouri-Illinois-Arkansas storm area.

The only one of the 12 storms which materially exceeds number 132 for the 4-day period is storm 151, which has already been discussed for its 3-day maximum depth. It needs no further mention here, since it exceeded storm 132 for the 4-day period by 23 per cent, as compared to 25 per cent for the 3-day period.

Only 3 of the 12 storms lasted 5 days. These are: number 86, September 15–19, 1905, which had an average depth 3 per cent greater than storm 132; number 125, which had an average depth only 64 per cent as great as storm 132; and number 132, March 23–27, 1913.

From the foregoing analysis of the greatest storms of the past 75 years which are applicable to Miami Valley conditions, it is seen that storm 132 was much farther north than either of the two storms that greatly exceeded it in depth. Only two other storms occurred during the winter months, both of which were much farther south in the Mississippi Valley, and only one of these, number 15, was comparable in size to storm 132. It is apparent, then, that storm 132, March 23–27, 1913, was unusually great for the region of its occurrence, and this is given additional emphasis by the fact that it occurred in late winter. Certainly, from the storm evidence of the past 75 years there is no indication that it will ever be greatly exceeded in the region of the Miami Valley.

RELATION OF GREAT STORMS TO MAXIMUM POSSIBLE

If it were necessary to depend wholly on the records of storms which have occurred in the United States, it might be thought possible for moderately great storms to occur over a period of a few hundred years, and then to find, as an exception, a storm three or four times as great. Theoretically that is very improbable, simply because water vapor in sufficient quantities cannot be transported from the ocean or gulf fast and long enough to cause such exceptional storms. As stated in chapter XI, however, records were collected of the stages of rivers in Europe for long periods of time, and these furnish fairly conclusive proof that such great exceptional storms actually do not occur. On the Danube at Vienna, for instance, we have records since about the year 1000 A. D.; fairly accurate records are available for stages of floods in the Tiber at Rome for more than 2,000 years; and records have been made of floods on the Seine at Paris for a long period of years.

The greatest flood stage on the Danube at Vienna, during the 900-year period for which records are available, was reached in the year 1501. This was about 30 to 35 per cent greater than the maximum flood of the past 100 years, that of 1899, which was the third greatest flood of the entire period. The second greatest flood on the Danube at Vienna occurred in 1787, and was about 85 per cent as large as the greatest flood recorded in 900 years. This is some indication of what may be expected in the Miami Valley during the course of a thousand years, as to how much the most extreme flood may exceed the greatest flood recorded in a 75 or 100-year period. The Danube record has been kept with as great accuracy as that of any river in Europe.

In Paris accurate records of the heights of floods are available for the past three hundred years. In 1611 the river reached a stage of thirty feet, the highest stage ever recorded. In 1910 it reached a stage only two and one half feet less; in 1658 it reached a stage about a foot less than the 1611 stage; and there have been a number of floods of only three or four feet less than the maximum of 1611. The salient fact is that the great floods have always remained within three or four feet of each other, and there has been no phenomenal flood that brought down half again or twice as much water as other great floods.

Records of the Tiber at Rome have been kept with some degree of accuracy since 413 B. C. The greatest flood which ever occurred at Rome was in 1598, A. D., but in 1870 there was a flood of about the same stage, and in the last few years there have been other floods but little smaller. Here again we find that the maximum flood of a thousand or two thousand years is closely approached by other large floods which occur with much greater frequency.

In the United States records are not available for a sufficient length of time to determine with confidence just what relation the maximum recorded storms bear to the greatest storms that might occur in one or two thousand years. It is necessary, therefore, to provide a factor of safety larger than would be required if our records extended through several hundred years.

REASONS FOR CHOOSING AS BASIS OF DESIGN, A FLOOD 40 PER CENT GREATER THAN THAT OF MARCH 1913

After making the extensive investigation of storms in the eastern United States, it is believed that the March, 1913, flood is one of the great floods of centuries in the Miami Valley. In the course of three or four hundred years, however, a flood 15 or 20 per cent greater may occur. We do not believe a flood will ever occur which is more than 20 or 25 per cent in excess of that of March 1913. There is a factor of ignorance, however, against which we must provide, and the only way to do this is arbitrarily to increase the size of the maximum flood provided for. If longer records were available a closer estimate could be made, but in planning works on which the protection of the Miami Valley depends, it is necessary to go beyond human judgment. This has been done on all the other phases of the design, and we believe it would not be good engineering practice to stop at our judgment on this phase. We must be able to say that the engineering works are absolutely safe in every respect. For this reason provision is made for a flood nearly 40 per cent greater than that of March 1913. This is 15 or 20 per cent in excess of what is believed to be the greatest possible flood that will ever occur.

18

APPENDIX

TIME–AREA–DEPTH DATA

APPENDIX.—TIME–AREA–DEPTH DATA

The tables give the essential time-area-depth data for the 33 storms for which maps and curves were drawn. The storms are arranged chronologically; and in each case, the computations for the maximum 1-day period are given first, those for the maximum 2-day period, second, and so on until all the computations are given.

The first column gives the periods of maximum rainfall; the second column gives the rainfall centers. These centers are indicated by the letters used to designate them on the storm maps, figures 58 to 93. The sequence of these letters for a given period of a given storm indicates the manner in which the different centers were combined. Attention is called to the fact that a given letter does not necessarily identify the same center for different periods of rainfall of a given storm. In each case the letters apply only to the centers indicated on the corresponding map.

The third column shows the isohyetal depths for which the enclosed areas and average depths of rainfall are given in the fourth and fifth columns, respectively. The abbreviations *Max.* and *Min.* in the third column indicate whether the center is a peak or depression area; and the figures opposite, in the fifth column, give the maximum or minimum depths of precipitation at the centers.

The area in square miles, in the fourth column, is computed to 2 significant figures for quantities up to 1000, and to 3 significant figures for greater amounts. The average depths of rainfall are computed to the nearest tenth of an inch.

Not all of the areas and corresponding average depths of rainfall were used in platting the time-area-depth curves. Only those corresponding to the largest peaks were used. Occasionally, as for the maximum 1-day period of the storm of July 14–16, 1916, the curve represents data taken from two or more peaks. For convenience in reference, the figures in the fifth column used in constructing the curves are shown in bold-faced type.

276

STORM a, OCTOBER 3-4, 1869

Period of Maximum Rainfall	Rainfall Center	Isohyetal-Inches	Area in Square Miles	Average Depth in Inches	Period of Maximum Rainfall	Rainfall Center	Isohyetal-Inches	Area in Square Miles	Average Depth in Inches
2 days Oct. 3–4	A	Max.		12.4	2 days Oct. 3–4	E	Max.		6.6
		12	14	12.2			6	900	6.3
		11	110	11.6		F	Max.		8.0
		10	320	10.8			7	270	7.5
		9	770	10.0			6	770	6.9
		8	1,550	9.2		G	Min.		4.0
		7	3,260	8.3			5	96	4.5
	B	Max.		8.1		A to G	5	53,000	6.3
		8	82	8.1			4	81,400	5.7
		7	3,680	7.5		H	Max.		5.8
	C	Max.		8.0			5	180	5.4
		7	1,140	7.5			4	1,220	4.6
	ABC	6	20,300	7.0		I	Min.		2.3
	D	Max.		8.1			3	920	2.6
		8	140	8.1		A to I	3	122,000	5.0
		7	3,660	7.5			2	185,000	4.1
		6	7,840	7.0					

STORM b, JULY 27-31, 1887

Period of Maximum Rainfall	Rainfall Center	Isohyetal-Inches	Area in Square Miles	Average Depth in Inches	Period of Maximum Rainfall	Rainfall Center	Isohyetal-Inches	Area in Square Miles	Average Depth in Inches
1 day July 29	A	Max.		10.0	2 days July 29–30	A	3	21,200	5.5
		10	14	10.0		B	Max.		8.6
		9	150	9.5			8	550	8.3
		8	520	8.8			7	1,150	7.9
		7	1,270	8.0			6	1,890	7.3
		6	2,400	7.3			5	2,300	7.0
		5	3,930	6.6			4	3,190	6.3
		4	6,300	5.8			3	4,750	5.4
		3	9,360	5.1		C	Max.		4.6
		2	13,800	4.2			4	230	4.3
	B	Max.		8.1			3	1,070	3.7
		8	260	8.1		D	Min.		1.7
		7	1,000	7.7			2	360	1.8
		6	1,810	7.1		A to D	2	66,300	3.7
		5	2,560	6.6	3 days July 28–30	A	Max.		16.5
		4	2,960	6.4			16	14	16.3
		3	3,750	5.8			15	82	15.6
		2	4,890	5.0			14	300	14.8
	C	Max.		2.2			13	740	14.0
		2	5,890	2.1			12	1,380	13.3
	D	Max.		2.9			11	2,310	12.6
		2	200	2.5			10	3,890	11.7
2 days July 29–30	A	Max.		14.1			9	5,800	11.0
		14	14	14.1			8	8,040	10.3
		13	110	13.6			7	10,900	9.6
		12	380	12.8			6	15,300	8.7
		11	890	12.1			5	20,400	7.9
		10	1,440	11.5			4	27,300	7.0
		9	2,190	10.8			3	41,600	5.8
		8	3,370	10.0		B	Max.		8.6
		7	4,740	9.3			8	620	8.3
		6	6,710	8.6			7	1,270	7.9
		5	9,350	7.6			6	1,920	7.4
		4	13,900	6.6			5	2,660	6.9

STORM b, JULY 27–31, 1887 (Continued)

Period of Maximum Rainfall	Rainfall Center	Isohyetal-Inches	Area in Square Miles	Average Depth in Inches	Period of Maximum Rainfall	Rainfall Center	Isohyetal-Inches	Area in Square Miles	Average Depth in Inches
3 days July 28–30	B	4	3,760	6.2	4 days July 27–30	F	Min.		2.0
		3	10,100	4.5			3	12,900	2.5
	C	Max.		5.1		A to F	3	106,000	4.8
		5	55	5.1			2	131,000	4.4
		4	640	4.6	5 days July 27–31	A	Max.		16.5
		3	1,510	4.0			16	27	16.3
	ABC	2	96,400	4.2			15	190	15.6
4 days July 27–30	A	Max.		16.5			14	700	14.8
		16	14	16.3			13	1,560	14.1
		15	96	15.6			12	2,810	13.4
		14	300	14.8			11	4,350	12.7
		13	780	14.0		B	Max.		12.4
		12	1,490	13.3			12	96	12.2
		11	2,520	12.6			11	360	11.7
		10	4,270	11.7		AB	10	8,090	11.7
	B	Max.		12.1			9	12,200	11.0
		12	27	12.1			8	15,800	10.4
		11	160	11.6			7	19,900	9.8
		10	410	10.9			6	25,800	9.1
	AB	9	8,020	10.8			5	33,300	8.2
		8	11,900	10.0		C	Max.		8.6
		7	16,100	9.4			8	970	8.3
		6	21,100	8.7			7	1,850	7.9
		5	27,400	8.0		D	Max.		7.7
	C	Max.		5.6			7	140	7.4
		5	710	5.3		CD	6	7,900	6.8
	ABC	4	43,000	6.7			5	15,100	6.2
	D	Max.		8.6		A to D	4	72,400	6.6
		8	1,100	8.3		E	Max.		5.1
		7	1,740	8.0			5	140	5.1
		6	2,600	7.5			4	1,560	4.5
		5	3,920	6.8		F	Min.		2.5
		4	5,740	6.1			3	3,370	2.7
	E	Max.		5.1		A to F	3	117,000	5.4
		5	96	5.1			2	149,000	4.8
		4	880	4.6					

STORM c, MAY 31–JUNE 1, 1889

Period of Maximum Rainfall	Rainfall Center	Isohyetal-Inches	Area in Square Miles	Average Depth in Inches	Period of Maximum Rainfall	Rainfall Center	Isohyetal-Inches	Area in Square Miles	Average Depth in Inches
1 day May 31	A	Max.		8.4	1 day May 31	A to E	4	29,800	5.6
		8	96	8.2		F	Max.		6.3
		7	330	7.7			6	140	6.2
	B	Max.		7.7			5	1,920	5.5
		7	1,000	7.5			4	4,700	4.9
	C	Max.		7.6		A to F	3	53,600	4.8
		7	190	7.3			2	81,000	4.0
	ABC	6	6,160	6.8	2 days May 31–June 1	A	Max.		9.8
		5	11,100	6.2			9	220	9.4
	D	Max.		7.1			8	1,260	8.6
		7	1,440	7.1			7	3,720	7.9
		6	3,660	6.7		B	Max.		8.6
	E	Max.		6.6			8	110	8.3
		6	41	6.4			7	620	7.6
	DE	5	7,910	6.1		AB	6	8,310	7.2

STORM c, MAY 31–JUNE 1, 1889 (Continued)

Period of Maximum Rainfall	Rainfall Center	Isohyetal-Inches	Area in Square Miles	Average Depth in Inches	Period of Maximum Rainfall	Rainfall Center	Isohyetal-Inches	Area in Square Miles	Average Depth in Inches
2 days	C	Max.		8.3	2 days	E	Max.		6.2
May		8	410	8.2	May		6	230	6.1
31–		7	2,740	7.6	31–	A to E	5	30,600	6.3
June 1		6	6,160	7.0	June 1		4	43,600	5.7
	D	Min.		3.6			3	62,200	5.1
		4	110	3.8			2	95,400	4.2
		5	290	4.2					

STORM 10, MAY 18–22, 1894

Period of Maximum Rainfall	Rainfall Center	Isohyetal-Inches	Area in Square Miles	Average Depth in Inches	Period of Maximum Rainfall	Rainfall Center	Isohyetal-Inches	Area in Square Miles	Average Depth in Inches
1 day	A	Max.		5.8	3 days	D	Max.		7.7
May		5	780	5.4	May		7	82	7.3
21		4	2,180	4.8	19–21	A to D	6	4,230	7.0
		3	6,020	4.0		E	Min.		4.6
		2	11,400	3.3			5	120	4.8
2 days	A	Max.		9.0		F	Max.		8.2
May		9	40	9.0			8	27	8.1
20–21		8	140	8.6			7	660	7.5
		7	530	7.8			6	1,400	7.0
		6	1,020	7.2		A to F	5	9,980	6.3
	B	Max.		8.9			4	15,500	5.7
		8	48	8.4		G	Max.		9.0
		7	810	7.6			9	27	8.9
		6	1,370	7.1			8	120	8.6
	C	Max.		7.7			7	370	7.9
		7	260	7.5			6	860	7.1
		6	930	6.8			5	1,930	6.2
	D	Max.		7.9			4	11,900	4.8
		7	57	7.5		H	Max.		4.6
		6	540	6.6			4	680	4.3
	E	Min.		4.3		A to H	3	49,200	4.5
		5	380	4.7		I	Max.		3.9
	A to E	5	7,460	6.2			3	1,030	3.4
		4	10,800	5.7	4 days	A to I	2	74,800	3.8
		3	24,500	4.5	May	A	Max.		10.1
	F	Max.		3.4	18–21		10	14	10.0
		3	980	3.2			9	69	9.6
	G	Max.		3.4			8	250	8.8
		3	370	3.2		B	Max.		9.4
	H	Max.		6.0			9	20	9.2
		5	270	5.5			8	250	8.6
		4	530	5.0		C	Max.		9.0
		3	960	4.3			9	27	9.0
	I	Max.		3.3			8	410	8.5
		3	250	3.2		ABC	7	3,500	7.8
	A to I	2	56,600	3.4			6	5,460	7.3
3 days	A	Max.		9.2		D	Max.		8.2
May		9	14	9.1			8	68	8.1
19–21		8	160	8.5			7	850	7.6
	B	Max.		9.0			6	1,520	7.1
		8	82	8.5		A to D	5	10,900	6.6
	C	Max.		8.7		E	Max.		9.8
		8	200	8.3			9	20	9.4
	ABC	7	1,720	7.7			8	164	8.6

STORM 10, MAY 18-22, 1894 (Continued)

Period of Maximum Rainfall	Rainfall Center	Isohyetal-Inches	Area in Square Miles	Average Depth in Inches	Period of Maximum Rainfall	Rainfall Center	Isohyetal-Inches	Area in Square Miles	Average Depth in Inches
4 days May 18-21	E	7	580	7.8	5 days May 18-22	D	Max.		8.7
		6	1,040	7.2			8	190	8.4
		5	2,920	6.1		E	Min.		5.5
	A to E	4	27,400	5.5			6	41	5.8
	F	Max.		8.1			7	200	6.4
		8	14	8.0		A to E	7	6,580	8.0
		7	110	7.5			6	9,650	7.5
		6	290	6.9			5	14,500	6.8
		5	550	6.2		F	Max.		9.0
		4	1,070	5.4			9	27	9.0
	G	Max.		7.9			8	82	8.6
		7	490	7.4			7	410	7.7
		6	1,750	6.8			6	900	7.1
		5	5,170	5.9			5	2,340	6.1
	H	Max.		5.1		G	Max.		9.9
		5	370	5.0			9	27	9.4
	GH	4	10,400	5.2			8	160	8.7
	I	Max.		5.0			7	450	7.9
		4	250	4.5			6	1,030	7.1
	A to I	3	57,000	4.8			5	2,030	6.3
	J	Min.		1.8		H	Max.		7.9
		2	380	1.9			7	940	7.4
	K	Max.		4.3			6	2,990	6.8
		4	370	4.2			5	7,230	6.0
		3	2,850	3.6		I	Max.		5.2
	A to K	2	94,900	3.9			5	410	5.1
5 days May 18-22	A	Max.		10.1		A to I	4	45,800	5.7
		10	14	10.0		J	Min.		2.8
		9	120	9.5			3	140	2.9
		8	820	8.6		A to J	3	61,400	5.1
	B	Max.		9.8		K	Max.		4.4
		9	140	9.4			4	370	4.2
	C	Max.		9.7			3	2,850	3.6
		9	580	9.4		A to K	2	112,000	4.0
	BC	8	1,600	8.9					

STORM 13, SEPTEMBER 24-26, 1894

Period of Maximum Rainfall	Rainfall Center	Isohyetal-Inches	Area in Square Miles	Average Depth in Inches	Period of Maximum Rainfall	Rainfall Center	Isohyetal-Inches	Area in Square Miles	Average Depth in Inches
1 day Sept. 26	A	Max.		12.5	1 day Sept. 26	AB	4	12,600	6.6
		12	59	12.2			3	16,400	5.9
		11	180	11.8		C	Max.		6.0
		10	360	11.1			5	3,060	5.5
		9	620	10.4			4	7,060	4.9
		8	910	9.8			3	10,800	4.4
		7	1,300	9.1		ABC	2	35,600	4.6
		6	1,720	8.5	2 days Sept. 25-26	A	Max.		12.5
	B	Max.		11.3			12	130	12.2
		11	74	11.2			11	560	11.7
		10	750	10.5			10	1,210	11.0
		9	1,370	10.1		B	Max.		12.0
		8	2,210	9.5			11	1,580	11.5
		7	3,520	8.8			10	2,940	11.0
		6	4,830	8.2		AB	9	6,810	10.4
	AB	5	8,960	7.5			8	9,680	9.9

STORM 13, SEPTEMBER 24-26, 1894 (Continued)

Period of Maximum Rainfall	Rainfall Center	Isohyetal-Inches	Area in Square Miles	Average Depth in Inches	Period of Maximum Rainfall	Rainfall Center	Isohyetal-Inches	Area in Square Miles	Average Depth in Inches
2 days Sept. 25-26	C	Min.		2.9	3 days Sept. 24-26	AB	10	5,650	11.2
		3	14	3.0			9	8,790	10.6
		4	110	3.4		C	Max.		12.0
		5	250	4.0			11	140	11.5
		6	490	4.8			10	490	10.8
		7	770	5.4			9	1,040	10.1
	ABC	7	13,800	9.0		ABC	8	15,700	9.8
		6	17,800	8.5		D	Min.		3.5
		5	25,500	7.6			4	27	3.8
		4	35,300	6.7			5	120	4.4
	D	Max.		6.0			6	250	4.9
		5	4,510	5.5			7	520	5.8
		4	7,000	5.2		A to D	7	24,400	8.9
	A to D	3	54,000	5.8			6	32,600	8.3
		2	62,400	5.4			5	37,600	7.9
3 days Sept. 24-26	A	Max.		12.5			4	40,900	7.6
		12	260	12.2		E	Max.		6.0
		11	1,100	11.7			5	3,440	5.5
	B	Max.		12.6			4	7,120	5.0
		12	260	12.3		A to E	3	55,000	6.8
		11	2,330	11.6			2	62,800	6.2

STORM 15, DECEMBER 17-20, 1895

Period of Maximum Rainfall	Rainfall Center	Isohyetal-Inches	Area in Square Miles	Average Depth in Inches	Period of Maximum Rainfall	Rainfall Center	Isohyetal-Inches	Area in Square Miles	Average Depth in Inches
1 day Dec. 19	A	Max.		5.8	2 days Dec. 18-19	G	Max.		3.2
		5	6,440	5.4			3	180	3.1
		4	15,800	4.9		A to G	2	212,000	3.7
		3	34,900	4.1	3 days Dec. 17-19	A	Max.		11.9
	B	Max.		4.2			11	82	11.4
		4	490	4.1			10	660	10.6
		3	900	3.8			9	3,390	9.7
	C	Max.		3.7			8	5,790	9.2
		3	3,000	3.4		B	Max.		9.2
	ABC	2	89,800	3.2			9	120	9.1
2 days Dec. 18-19	A	Max.		9.6			8	880	8.2
		9	460	9.3		AB	7	13,400	8.3
		8	1,770	8.7			6	24,400	7.5
		7	4,150	8.0		C	Max.		6.9
		6	13,100	7.0			6	1,100	6.4
		5	27,100	6.2		ABC	5	61,600	6.3
	B	Max.		6.2		D	Max.		5.4
		6	55	6.1			5	490	5.2
		5	2,000	5.5		A to D	4	101,000	5.6
	C	Max.		5.6		E	Max.		6.0
		5	1,110	5.3			5	470	5.5
	ABC	4	73,000	5.2			4	1,110	4.9
	D	Max.		4.2		F	Max.		4.5
		4	960	4.1			4	160	4.2
	E	Max.		4.4		A to F	3	145,000	5.0
		4	940	4.2			2	235,000	4.0
	A to E	3	129,000	4.5	4 days Dec. 17-20	A	Max.		12.2
	F	Min.		1.1			12	21	12.1
		2	1,230	1.6			11	210	11.5
							10	630	10.8

STORM 15, DECEMBER 17-20, 1895 (Continued)

Period of Maximum Rainfall	Rainfall Center	Isohyetal-Inches	Area in Square Miles	Average Depth in Inches	Period of Maximum Rainfall	Rainfall Center	Isohyetal-Inches	Area in Square Miles	Average Depth in Inches
4 days Dec. 17-20	A	9	3,820	9.7	4 days Dec. 17-20	D	6	580	6.2
		8	6,650	9.2		A to D	5	77,600	6.3
	B	Max.		9.3			4	122,000	5.6
		9	120	9.2		E	Min.		1.8
		8	820	8.6			2	380	1.9
	AB	7	15,100	8.3			3	1,560	2.4
		6	31,000	7.4		F	Min.		2.4
	C	Max.		6.7			3	530	2.7
		6	1,770	6.4		A to F	3	174,000	4.9
	D	Max.		6.3			2	254,000	4.1

STORM 24, MARCH 22-23, 1897

Period of Maximum Rainfall	Rainfall Center	Isohyetal-Inches	Area in Square Miles	Average Depth in Inches	Period of Maximum Rainfall	Rainfall Center	Isohyetal-Inches	Area in Square Miles	Average Depth in Inches
1 day March 22	A	Max.		10.3	2 days March 22-23	A	Max.		11.5
		10	120	10.1			11	410	11.3
		9	410	9.7			10	1,830	10.7
		8	890	9.0			9	4,270	10.0
		7	1,560	8.4			8	6,910	9.4
		6	3,100	7.5			7	9,760	8.9
		5	4,610	6.8			6	13,900	8.2
		4	6,600	6.1			5	18,600	7.5
	B	Max.		6.0			4	30,100	6.3
		5	55	5.4			3	38,400	5.7
		4	990	4.6			2	65,100	4.4
	AB	3	18,700	4.5					
		2	27,800	3.8					

STORM 25, JULY 12-14, 1897

Period of Maximum Rainfall	Rainfall Center	Isohyetal-Inches	Area in Square Miles	Average Depth in Inches	Period of Maximum Rainfall	Rainfall Center	Isohyetal-Inches	Area in Square Miles	Average Depth in Inches
1 day July 13	A	Max.		8.1	2 days Ju.y 13-14	B	5	560	5.7
		8	55	8.1		AB	4	15,200	5.5
		7	230	7.7			3	25,000	4.7
		6	1,050	6.8			2	33,300	4.2
		5	2,410	6.1	3 days July 12-14	A	Max.		10.3
		4	7,270	5.0			10	60	10.2
		3	12,500	4.4			9	560	9.6
	B	Max.		3.2			8	1,500	8.9
		3	530	3.1			7	2,410	8.4
	C	Max.		3.8			6	3,680	7.8
		3	190	3.4			5	8,550	6.4
	ABC	2	22,600	3.6		B	Max.		6.8
2 days July 13-14	A	Max.		10.3			6	250	6.4
		10	41	10.2			5	1,080	5.7
		9	480	9.6		AB	4	20,200	5.4
		8	1,460	8.8		C	Max.		5.5
		7	2,260	8.4			5	860	5.2
		6	3,560	7.7			4	8,020	4.6
		5	6,710	6.7		AC	3	51,700	4.4
	B	Max.		6.8			2	74,200	3.9
		6	110	6.4					

STORM 43, JUNE 27–JULY 1, 1899

Period of Maximum Rainfall	Rainfall Center	Isohyetal-Inches	Area in Square Miles	Average Depth in Inches
1 day June 29	A	Max.		11.0
		10	190	10.5
		9	440	10.0
		8	2,140	8.8
		7	3,220	8.4
		6	5,120	7.7
		5	8,260	6.8
		4	12,300	6.1
	B	Max.		8.1
		8	120	8.1
		7	380	7.7
		6	590	7.3
		5	1,120	6.5
		4	1,970	5.6
	AB	3	20,500	5.2
		2	28,600	4.5
2 days June 28–29	A	Max.		22.0
		20	68	21.0
		15	300	18.3
		14	380	17.5
	B	Max.		14.6
		14	530	14.3
	AB	13	1,970	13.4
		12	3,720	13.0
		11	5,350	12.5
		10	7,100	12.0
		9	10,800	11.2
		8	14,600	10.5
		7	16,800	10.1
		6	19,300	9.6
		5	22,400	9.0
		4	31,200	7.8
		3	39,800	6.8
		2	47,200	6.2
3 days June 28–30	A	Max.		33.0
		30	82	31.5
		25	220	29.0
		20	550	25.1
		19	620	24.5
		18	680	23.9
		17	740	23.4
		16	790	22.9
	B	Max.		20.0
		20	110	20.0
		19	550	19.6
		18	1,120	19.0
		17	1,970	18.4
		16	2,960	17.8
	AB	15	5,150	18.0
		14	6,050	17.4
		13	7,800	16.6
		12	9,020	16.0
		11	10,100	15.5

Period of Maximum Rainfall	Rainfall Center	Isohyetal-Inches	Area in Square Miles	Average Depth in Inches
3 days June 28–30	AB	10	12,800	14.5
		9	16,200	13.4
		8	20,500	12.4
		7	24,200	11.6
		6	28,500	10.9
		5	33,100	10.1
		4	39,800	9.2
		3	47,800	8.2
		2	59,400	7.1
4 days June 27–30	A	Max.		33.0
		30	96	31.5
		25	250	29.1
		20	580	25.3
		19	670	24.5
		18	740	24.0
		17	780	23.6
		16	890	22.7
	B	Max.		20.1
		20	110	20.0
		19	630	19.6
		18	1,620	19.0
		17	3,110	18.3
		16	4,790	17.6
	AB	15	7,610	17.7
		14	9,240	17.1
		13	11,500	16.4
		12	13,300	15.9
		11	15,200	15.3
		10	17,500	14.7
		9	19,800	14.1
		8	22,500	13.4
		7	28,700	12.1
		6	35,100	11.1
		5	42,000	10.2
		4	48,100	9.5
		3	55,600	8.7
		2	69,800	7.4
5 days June 27– July 1	A	Max.		33.0
		30	96	31.5
		25	250	29.1
		20	580	25.3
		19	670	24.5
		18	740	24.0
		17	780	23.6
		16	890	22.7
	B	Max.		20.8
		20	110	20.4
		19	670	19.6
		18	1,640	19.0
		17	3,310	18.2
		16	5,200	17.6
	AB	15	7,640	17.7
		14	9,270	17.2

STORM 43, JUNE 27–JULY 1, 1899 (Continued)

Period of Maximum Rainfall	Rainfall Center	Isohyetal-Inches	Area in Square Miles	Average Depth in Inches	Period of Maximum Rainfall	Rainfall Center	Isohyetal-Inches	Area in Square Miles	Average Depth in Inches
5 days	AB	13	11,600	16.4	5 days	AB	7	29,400	12.1
June		12	13,500	15.9	June		6	36,200	11.0
27–		11	15,400	15.3	27–		5	44,300	10.0
July		10	17,500	14.8	July		4	53,100	9.1
1		9	20,000	14.1	1		3	62,500	8.3
		8	22,600	13.5			2	77,000	7.2

STORM 49, APRIL 15–17, 1900

Period of Maximum Rainfall	Rainfall Center	Isohyetal-Inches	Area in Square Miles	Average Depth in Inches	Period of Maximum Rainfall	Rainfall Center	Isohyetal-Inches	Area in Square Miles	Average Depth in Inches
1 day	A	Max.		12.7	1 day	J	Min.		1.4
April		12	55	12.4	April		2	850	1.7
16		11	140	11.8	16	C to J	2	42,400	4.5
		10	250	11.2	2 days	A	Max.		12.7
		9	440	10.4	April		12	82	12.4
		8	680	9.8	16–17		11	590	11.6
		7	890	9.2			10	2,810	10.7
		6	1,230	8.5			9	6,110	10.1
		5	1,630	7.8			8	10,800	9.4
		4	2,380	6.7			7	14,700	8.9
		3	4,050	5.4			6	18,000	8.4
	B	Max.		5.0		B	Max.		12.8
		5	82	5.0			12	68	12.4
		4	810	4.6			11	160	11.8
		3	1,780	4.0			10	520	10.9
	AB	2	9,010	4.1			9	990	10.2
	C	Max.		10.8			8	3,040	9.1
		10	370	10.4			7	4,790	8.5
		9	1,160	9.8		C	Max.	140	7.4
		8	2,870	9.0			7	140	7.2
		7	4,160	8.6		D	Max.		8.0
		6	5,570	8.0			7	270	7.5
		5	7,430	7.4		BCD	6	10,500	7.4
	D	Max.		9.3		E	Max.		9.5
		9	120	9.2			9	68	9.3
		8	370	8.7			8	200	8.7
		7	680	8.2			7	370	8.2
		6	1,120	7.5			6	730	7.3
	E	Max.		6.4		A to E	5	42,400	7.3
		6	680	6.2		F	Max.		5.5
	DE	5	6,840	5.9			5	2,330	5.3
	F	Min.		2.3		G	Max.		5.2
		3	150	2.6			5	55	5.1
		4	410	3.2		A to G	4	60,600	6.5
	G	Min.		2.5		H	Min.		1.7
		3	68	2.7			2	96	1.8
		4	150	3.1			3	510	2.4
	C to G	4	23,700	5.8		I	Min.		2.9
	H	Min.		2.3			3	82	2.9
		3	270	2.6		A to I	3	85,400	5.6
	I	Max.		4.2			2	110,000	4.9
		4	140	4.1	3 days	A	Max.		13.9
	C to I	3	33,000	5.1	April		13	82	13.5
					15–17		12	250	12.8

STORM 49, APRIL 15–17, 1900 (Continued)

Period of Maximum Rainfall	Rainfall Center	Isohyetal-Inches	Area in Square Miles	Average Depth in Inches	Period of Maximum Rainfall	Rainfall Center	Isohyetal-Inches	Area in Square Miles	Average Depth in Inches
3 days April 15–17	A	11	680	12.0	3 days April 15–17	D	7	850	8.0
	B	Max.		12.8		E	Min.		3.9
		12	96	12.4			4	27	3.9
		11	1,940	11.5			5	140	4.5
	AB	10	6,480	11.0			6	530	5.2
		9	9,840	10.5		F	Min.		5.1
		8	13,300	10.0			6	150	5.5
		7	18,000	9.3		A to F	6	43,200	8.4
	C	Max.		13.8		G	Max.		6.4
		13	68	13.4			6	68	6.2
		12	360	12.7		A to G	5	54,900	7.8
		11	1,370	11.8		H	Max.		5.7
		10	2,920	11.1			5	2,400	5.4
		9	4,870	10.4		A to H	4	69,100	7.1
		8	8,520	9.6		I	Min.		2.9
		7	11,900	9.0			3	68	2.9
	D	Max.		9.5		A to I	3	94,400	6.2
		9	82	9.3			2	121,000	5.4
		8	370	8.7					

STORM 51, JULY 14–16, 1900

Period of Maximum Rainfall	Rainfall Center	Isohyetal-Inches	Area in Square Miles	Average Depth in Inches	Period of Maximum Rainfall	Rainfall Center	Isohyetal-Inches	Area in Square Miles	Average Depth in Inches
1 day July 15	A	Max.		7.0	2 days July 14–15	ABC	4	10,400	6.7
		6	1,370	6.5			3	15,900	5.6
		5	3,830	5.9			2	28,400	4.2
		4	6,430	5.3	3 days July 14–16	A	Max.		13.7
		3	9,310	4.8			13	68	13.4
		2	14,300	4.0			12	270	12.7
2 days July 14–15	A	Max.		13.0			11	550	12.1
		13	27	13.0			10	1,160	11.2
		12	150	12.6			9	1,920	10.6
		11	370	12.0		B	Max.		9.6
		10	680	11.3			9	96	9.3
		9	1,100	10.6		AB	8	4,220	9.4
		8	1,780	9.8			7	6,020	8.9
	B	Max.		8.1			6	8,020	8.2
		8	140	8.1			5	11,000	7.5
	AB	7	4,380	8.4			4	16,400	6.5
		6	6,160	7.9		C	Max.		5.5
		5	7,830	7.4			5	1,460	5.3
	C	Min.		3.9			4	5,020	4.7
		4	110	3.9		ABC	3	36,800	5.0
							2	60,200	3.9

STORM 63, MARCH 26–29, 1902

Period of Maximum Rainfall	Rainfall Center	Isohyetal-Inches	Area in Square Miles	Average Depth in Inches	Period of Maximum Rainfall	Rainfall Center	Isohyetal-Inches	Area in Square Miles	Average Depth in Inches
1 day March 28	A	Max.		9.0	1 day March 28	AB	6	5,820	7.0
		9	68	9.0			5	11,000	6.3
		8	660	8.6			4	17,600	5.6
		7	1,680	7.9			3	22,000	5.2
	B	Max.		8.6		C	Max.		4.0
		8	140	8.3			4	140	4.0
		7	530	7.7			3	1,060	3.6

STORM 63, MARCH 26-29, 1902 (Continued)

Period of Maximum Rainfall	Rainfall Center	Isohyetal-Inches	Area in Square Miles	Average Depth in Inches	Period of Maximum Rainfall	Rainfall Center	Isohyetal-Inches	Area in Square Miles	Average Depth in Inches
1 day March 28	D	Max.		4.8	3 days March 26-28	F	Max.		8.3
		4	700	4.4			8	68	8.2
		3	1,710	3.9			7	410	7.6
	A to D	2	58,100	3.6		A to F	6	29,800	7.5
2 days March 27-28	A	Max.		9.7			5	44,800	6.8
		9	200	9.5		G	Max.		7.0
		8	450	9.0			6	1,640	6.5
		7	880	8.2			5	3,980	5.9
	B	Max.		9.0		A to G	4	73,900	6.0
		9	96	9.0			3	107,000	5.2
		8	730	8.6		H	Max.		6.7
		7	2,260	7.8			6	120	6.4
	AB	6	7,150	7.1			5	370	5.8
		5	12,000	6.5			4	1,100	4.9
	C	Min.		3.3			3	2,400	4.2
		4	370	3.7		A to H	2	194,000	4.0
	D	Max.		7.0	4 days March 26-29	A	Max.		11.8
		7	250	7.0			11	100	11.4
		6	1,190	6.6			10	440	10.7
		5	3,010	5.9			9	1,290	9.9
	E	Max.		8.5		B	Max.		10.8
		8	5,240	8.5			10	250	10.4
		7	9,420	8.1			9	940	9.7
		6	13,700	7.6		C	Max.		10.0
		5	17,300	7.2			9	440	9.5
	A to E	4	57,700	5.8		D	Max.		9.7
		3	72,300	5.3			9	250	9.4
	F	Max.		6.2		E	Max.		9.4
		6	18	6.1			9	340	9.2
		5	420	5.5		A to E	8	10,400	8.9
		4	1,350	4.8		F	Max.		9.5
		3	2,860	4.1			9	960	9.2
	A to F	2	105,000	4.5			8	2,180	8.8
3 days March 26-28	A	Max.		11.8		A to F	7	25,500	8.2
		11	75	11.4		G	Max.		8.3
		10	310	10.7			7	300	7.7
		9	750	10.0		A to G	6	46,700	7.4
		8	1,680	9.2			5	67,000	6.8
	B	Max.		10.8		H	Max.		6.3
		10	190	10.4			6	68	6.2
		9	630	9.8			5	480	5.6
	C	Max.		9.4		A to H	4	90,200	6.2
		9	340	9.2			3	129,000	5.4
	D	Max.		9.5		I	Max.		4.5
		9	610	9.2			4	360	4.2
	BCD	8	4,930	8.8			3	3,580	3.6
	A to D	7	14,400	8.1		J	Max.		4.1
	E	Max.		10.0			4	190	4.0
		9	440	9.5			3	2,230	3.6
		8	1,430	8.8		A to J	2	268,000	3.9
		7	3,130	8.1					

STORM 72, AUGUST 25–28, 1903

Period of Maximum Rainfall	Rainfall Center	Isohyetal-Inches	Area in Square Miles	Average Depth in Inches	Period of Maximum Rainfall	Rainfall Center	Isohyetal-Inches	Area in Square Miles	Average Depth in Inches
1 day Aug. 27	A	Max.		11.2	3 days Aug. 26–28	A	Max.		15.5
		11	110	11.1			15	10	15.2
		10	520	10.6			14	35	14.4
		9	920	10.1			13	95	13.8
		8	1,510	9.5			12	180	13.3
		7	2,460	8.7			11	470	12.2
		6	3,290	8.2			10	1,280	11.1
	B	Max.		6.2			9	2,040	10.5
		6	96	6.1			8	3,390	9.7
	AB	5	4,860	7.3			7	5,210	8.9
		4	6,570	6.6			6	7,960	8.1
	C	Max.		4.8			5	12,800	7.1
		4	190	4.4			4	24,300	5.9
	D	Max.		4.6		B	Max.		6.4
		4	330	4.3			6	20	6.2
	A to D	3	12,400	5.1			5	100	5.7
	E	Min.		0.5			4	260	5.0
		1	55	0.7		C	Max.		4.3
		2	250	1.3			4	1,420	4.2
	A to E	2	20,000	4.1		ABC	3	44,900	4.8
	F	Max.		5.3		D	Max.		4.5
		5	300	5.2			4	40	4.2
		4	890	4.7			3	220	3.6
		3	2,120	4.0		E	Min.		1.5
		2	3,460	3.4			2	9,240	1.7
2 days Aug. 26–27	A	Max.		14.7		A to E	2	104,000	3.4
		14	14	14.4		F	Max.		4.4
		13	55	13.6			4	620	4.2
		12	120	13.0			3	5,470	3.6
		11	410	11.9			2	17,700	2.8
		10	710	11.3	4 days Aug. 25–28	A	Max.		15.5
		9	1,100	10.7			15	20	15.2
		8	1,780	9.9			14	65	14.8
		7	2,810	9.0			13	160	14.0
		6	3,640	8.4			12	250	13.4
	B	Max.		10.2			11	590	12.3
		10	27	10.1			10	1,380	11.3
		9	140	9.6			9	2,020	10.7
		8	410	8.9			8	3,540	9.8
		7	1,570	7.9			7	5,220	9.1
		6	3,220	7.2			6	8,200	8.1
	AB	5	10,300	7.1			5	13,800	7.1
	C	Max.		5.5			4	27,600	5.8
		5	55	5.3		B	Max.		6.6
	ABC	4	15,500	6.5			6	30	6.3
	D	Max.		5.0			5	130	5.7
		5	55	5.0			4	320	5.0
		4	660	4.5		C	Max.		4.3
	E	Min.		2.5			4	1,820	4.2
		3	200	2.7		ABC	3	51,800	4.8
	F	Max.		4.8		D	Max.		6.0
		4	41	4.4			5	95	5.5
	A to F	3	29,200	5.1			4	210	5.0
		2	43,200	4.3			3	560	4.0

STORM 72, AUGUST 25–28, 1903 (Continued)

Period of Maximum Rainfall	Rainfall Center	Isohyetal-Inches	Area in Square Miles	Average Depth in Inches	Period of Maximum Rainfall	Rainfall Center	Isohyetal-Inches	Area in Square Miles	Average Depth in Inches
4 days Aug. 25–28	E	Max.		4.4	4 days Aug. 25–28	F	Min.		1.0
		4	1,130	4.2			2	9,180	1.5
		3	8,040	3.5		A to F	2	144,000	3.3

STORM 76, OCTOBER 8–9, 1903

Period of Maximum Rainfall	Rainfall Center	Isohyetal-Inches	Area in Square Miles	Average Depth in Inches	Period of Maximum Rainfall	Rainfall Center	Isohyetal-Inches	Area in Square Miles	Average Depth in Inches
1 day Oct. 9	A	Max.		11.5	2 days Oct. 8–9	A	12	140	13.1
		11	55	11.3			11	330	12.2
		10	82	11.0			10	680	11.3
		9	140	10.4			9	1,300	10.4
		8	640	8.9			8	3,420	9.2
		7	780	8.7			7	5,300	8.6
	B	Max.		7.4		B	Max.		9.8
		7	96	7.2			9	82	9.4
	AB	6	3,290	7.0			8	360	8.7
	C	Max.		6.9		C	Min.		6.2
		6	190	6.5			7	27	6.6
	ABC	5	5,300	6.5		BC	7	1,660	7.8
	D	Min.		3.4		D	Min.		5.2
		4	41	3.7			6	27	5.6
	E	Min.		3.6		E	Max.		8.3
		4	480	3.8			8	14	8.2
	F	Max.		8.0			7	41	7.6
		7	590	7.5		A to E	6	14,400	7.4
	G	Min.		5.7		F	Min.		4.6
		6	41	5.8			5	96	4.8
	FG	6	1,660	6.5		G	Min.		3.1
		5	2,740	6.3			4	14	3.5
	H	Min.		3.6			5	55	4.2
		4	41	3.8		H	Max.		6.3
	A to H	4	17,900	5.3			6	120	6.2
		3	24,200	4.9		A to H	5	20,900	6.8
		2	30,500	4.4			4	27,000	6.3
2 days Oct. 8–9	A	Max.		15.0			3	36,400	5.6
		14	27	14.5			2	44,000	5.0
		13	68	13.8					

STORM 83, JUNE 9–10, 1905

Period of Maximum Rainfall	Rainfall Center	Isohyetal-Inches	Area in Square Miles	Average Depth in Inches	Period of Maximum Rainfall	Rainfall Center	Isohyetal-Inches	Area in Square Miles	Average Depth in Inches
1 day	A	Max.		12.1	1 day	AB	6	3,640	8.1
		12	41	12.0			5	4,740	7.5
		11	190	11.6			4	6,200	6.8
		10	410	11.0			3	9,040	5.7
		9	780	10.3		C	Max.		5.7
		8	1,180	9.7			5	140	5.5
		7	1,720	9.0			4	1,010	4.6
	B	Max.		11.0			3	3,150	3.9
		10	27	10.4		D	Max.		3.8
		9	160	9.6			3	1,850	3.5
		8	420	8.9		A to D	2	34,000	3.5
		7	750	8.3					

STORM 86, SEPTEMBER 15–19, 1905

Period of Maximum Rainfall	Rainfall Center	Isohyetal-Inches	Area in Square Miles	Average Depth in Inches	Period of Maximum Rainfall	Rainfall Center	Isohyetal-Inches	Area in Square Miles	Average Depth in Inches
1 day Sept. 17	A	Max.		8.1	4 days Sept. 15–18	A	Max.		12.2
		8	41	8.0			12	27	12.1
		7	370	7.6			11	250	11.6
		6	930	6.9			10	1,480	10.7
		5	2,110	6.1			9	3,860	10.0
		4	3,740	5.4			8	5,940	9.4
		3	6,820	4.6			7	8,760	8.8
	B	Max.		6.7		B	Max.		8.9
		6	790	6.4			8	140	8.4
		5	2,830	5.7			7	780	7.7
		4	5,420	5.2		C	Max.		8.2
		3	8,960	4.5			8	41	8.1
	AB	2	27,700	3.6			7	220	7.6
2 days Sept. 16–17	A	Max.		8.2		ABC	6	18,400	7.7
		8	440	8.1		D	Max.		8.1
		7	3,200	7.6			8	250	8.0
		6	5,230	7.2		E	Max.		8.5
	B	Max.		8.0			8	290	8.2
		7	1,030	7.5		DE	7	3,040	7.6
		6	2,260	7.0			6	4,740	7.2
	AB	5	14,400	6.3		A to E	5	34,200	6.6
		4	19,700	5.8			4	47,000	5.8
		3	26,900	5.2			3	69,200	4.7
	C	Max.		6.2		F	Min.		1.2
		6	55	6.1			2	600	1.6
		5	380	5.6		G	Max.		5.7
	D	Max.		5.4			5	600	5.4
		5	410	5.2		H	Max.		5.4
	CD	4	2,710	4.5			5	200	5.2
		3	4,380	3.3		GH	4	4,520	4.6
	A to D	2	49,400	4.1			3	9,320	4.1
3 days Sept. 15–17	A	Max.		10.5		A to G	2	123,000	3.8
		10	700	10.2	5 days Sept. 15–19	A	Max.		13.0
		9	1,930	9.8			12	490	12.5
		8	4,070	9.1			11	990	12.0
		7	5,480	8.7			10	1,780	11.3
		6	7,640	8.1			9	5,050	10.2
	B	Max.		8.1			8	9,910	9.3
		8	96	8.0			7	15,400	8.7
		7	1,030	7.5		B	Max.		8.9
		6	2,260	7.0			8	120	8.4
	AB	5	18,700	6.7			7	820	7.6
	C	Max.		6.2		C	Max.		8.2
		6	82	6.1			8	14	8.2
		5	2,260	5.5			7	290	7.5
	ABC	4	32,800	5.8		ABC	6	27,600	7.8
		3	48,600	5.1			5	39,500	7.1
	D	Max.		4.3			4	53,100	6.4
		4	1,110	4.2		D	Max.		6.6
		3	4,640	3.7			6	290	6.3
	E	Max.		3.3			5	1,040	5.7
		3	1,440	3.2		E	Max.		5.5
	A to E	2	84,700	4.0			5	370	5.2

STORM 86, SEPTEMBER 15-19, 1905 (Continued)

Period of Maximum Rainfall	Rainfall Center	Isohyetal-Inches	Area in Square Miles	Average Depth in Inches
5 days Sept. 15-19	DE	4	5,460	4.8
	F	Min.		1.4
		2	410	1.7
		3	1,360	2.3
	G	Max.		4.8
5 days Sept. 15-19	G	4	330	4.4
	A to G	3	99,800	5.1
	H	Max.		4.0
		3	1,620	3.5
	A to H	2	168,000	4.1

STORM 93, NOVEMBER 17-21, 1906

Period of Maximum Rainfall	Rainfall Center	Isohyetal-Inches	Area in Square Miles	Average Depth in Inches
1 day Nov. 17	A	Max.		7.9
		7	550	7.5
		6	3,900	6.6
		5	8,320	6.0
	B	Min.		3.4
		4	360	3.7
	C	Max.		5.1
		5	110	5.1
	ABC	4	21,400	5.1
		3	34,900	4.5
	D	Max.		3.5
		3	270	3.2
	E	Max.		4.8
		4	320	4.4
		3	1,310	3.7
	A to E	2	77,600	3.4
2 days Nov. 17-18	A	Max.		10.6
		10	120	10.3
		9	880	9.6
		8	1,630	9.1
		7	4,790	8.0
	B	Max.		11.0
		10	96	10.5
		9	460	9.7
		8	1,100	9.0
		7	2,370	8.2
	AB	6	15,200	7.3
	C	Min.		4.3
		5	164	4.6
	ABC	5	25,100	6.6
	D	Min.		3.0
		4	260	3.5
	A to D	4	41,100	5.7
		3	66,900	4.9
		2	124,000	3.8
3 days Nov. 17-19	A	Max.		13.0
		12	140	12.5
		11	460	11.8
		10	960	11.1
	B	Max.		11.7
		11	140	11.5
		10	360	10.9
	AB	9	3,910	10.0
	C	Max.		10.4
		10	1,090	10.5
3 days Nov. 17-19	C	9	2,300	10.0
	D	Min.		5.2
		6	41	5.6
		7	250	6.3
		8	1,410	7.3
	A to D	8	13,900	9.1
		7	19,200	8.6
	E	Max.		8.0
		7	410	7.5
	A to E	6	27,900	8.0
		5	37,400	7.4
	F	Min.		3.0
		4	460	3.5
	A to F	4	56,800	6.4
		3	86,300	5.4
		2	132,000	4.4
4 days Nov. 17-20	A	Max.		16.0
		15	180	15.5
		14	380	15.0
		13	820	14.2
		12	1,590	13.4
		11	2,710	12.6
	B	Min.		7.0
		8	380	7.5
		9	1,190	8.2
		10	1,900	8.7
	C	Max.		13.0
		12	82	12.5
		11	360	11.7
	ABC	10	10,700	10.7
		9	16,000	10.3
		8	21,700	9.8
		7	29,900	9.2
	D	Max.		9.4
		8	370	8.5
		7	960	7.9
	E	Max.		10.0
		9	380	9.5
		8	2,400	8.6
		7	4,600	8.1
	A to E	6	50,300	8.3
	F	Max.		7.0
		6	730	6.5
	A to F	5	72,000	7.4
	G	Min.		3.2

STORM 93, NOVEMBER 17-21, 1906 (Continued)

Period of Maximum Rainfall	Rainfall Center	Isohyetal-Inches	Area in Square Miles	Average Depth in Inches	Period of Maximum Rainfall	Rainfall Center	Isohyetal-Inches	Area in Square Miles	Average Depth in Inches
4 days Nov. 17-20	G	4	420	3.6	5 days Nov. 17-21	E	Max.		9.2
	H	Min.		3.3			9	480	9.1
		4	330	3.6		A to E	8	39,800	10.4
	A to H	4	98,700	6.6		F	Max.		8.0
		3	139,000	5.7			8	140	8.0
		2	197,000	4.8		A to F	7	53,900	9.6
5 days Nov. 17-21	A	Max.		19.4			6	70,500	8.9
		19	41	19.2			5	95,300	8.0
		18	160	18.7			4	137,000	7.0
		17	470	17.9		G	Min.		0.1
		16	880	17.2			1	120	0.6
		15	1,490	16.5			2	340	1.2
		14	2,410	15.7			3	670	1.8
		13	4,070	14.8		A to G	3	202,400	5.8
		12	6,260	14.0		H	Max.		8.6
	B	Min.		9.2			8	380	8.3
		10	140	9.6			7	990	7.8
		11	640	10.3			6	2,220	7.1
	C	Max.		12.4			5	3,790	6.4
		12	1,100	12.2			4	6,760	5.6
	ABC	11	14,700	12.6		I	Max.		5.1
		10	19,800	12.0			5	230	5.0
		9	25,500	11.5			4	520	4.7
	D	Min.		6.9		HI	3	28,500	4.0
		7	140	7.0		A to I	2	378,000	4.4
		8	520	7.4					

STORM 102, JULY 28-AUGUST 1, 1908

Period of Maximum Rainfall	Rainfall Center	Isohyetal-Inches	Area in Square Miles	Average Depth in Inches	Period of Maximum Rainfall	Rainfall Center	Isohyetal-Inches	Area in Square Miles	Average Depth in Inches
1 day July 30	A	Max.		9.6	3 days July 29-31	A	Max.		17.6
		9	160	9.3			17	150	17.3
		8	530	8.8			16	250	17.0
		7	890	8.3			15	410	16.8
		6	1,410	7.6			14	580	16.0
		5	2,000	7.0			13	790	15.3
		4	3,480	5.9			12	1,100	14.6
		3	6,650	4.8			11	1,560	13.7
		2	12,400	3.7			10	2,330	12.6
2 days July 29-30	A	Max.		14.1			9	2,890	12.0
		14	150	14.0			8	3,630	11.3
		13	440	13.7			7	5,410	10.0
		12	660	13.4			6	7,090	9.2
		11	1,030	12.7			5	9,160	8.4
		10	1,640	11.9		B	Max.		7.2
		9	2,010	11.5			7	190	7.0
		8	2,420	11.0			6	750	6.6
		7	3,010	10.3			5	2,410	5.9
		6	4,120	9.3		A-B	4	17,700	6.7
		5	7,430	7.6			3	25,100	5.8
		4	9,530	6.9			2	34,300	4.9
		3	14,600	5.7	4 days July 28-31	A	Max.		18.8
		2	22,600	4.6			18	120	18.4
							17	230	18.0

STORM 102, JULY 28–AUGUST 1, 1908 (Continued)

Period of Maximum Rainfall	Rainfall Center	Isohyetal-Inches	Area in Square Miles	Average Depth in Inches	Period of Maximum Rainfall	Rainfall Center	Isohyetal-Inches	Area in Square Miles	Average Depth in Inches
4 days July 28–31	A	16	330	17.6	5 days July 28– Aug. 1	A	14	860	16.1
		15	550	16.8			13	1,190	15.4
		14	780	16.1			12	1,550	14.7
		13	1,000	15.7			11	2,080	13.9
		12	1,360	14.9			10	2,760	13.0
		11	1,850	14.0			9	3,890	12.0
		10	2,740	12.9			8	5,020	11.2
		9	3,700	12.0			7	5,780	10.7
		8	4,870	11.2		B	Max.		11.0
		7	6,890	10.1			10	41	10.5
	B	Max.		7.4			9	190	9.7
		7	230	7.2			8	290	9.3
	AB	6	11,000	8.8			7	920	8.1
		5	15,700	7.8		AB	6	12,800	8.5
		4	24,500	6.6			5	16,800	7.8
		3	34,700	5.7		C	Max.		7.9
		2	44,200	5.0			7	14	7.8
5 days July 28– Aug. 1	A	Max.		18.8			6	120	6.7
		18	120	18.4			5	380	5.7
		17	260	17.9		ABC	4	31,800	6.3
		16	410	17.4			3	43,800	5.5
		15	580	16.9			2	58,700	4.8

STORM 103, AUGUST 24–26, 1908

Period of Maximum Rainfall	Rainfall Center	Isohyetal-Inches	Area in Square Miles	Average Depth in Inches	Period of Maximum Rainfall	Rainfall Center	Isohyetal-Inches	Area in Square Miles	Average Depth in Inches
1 day Aug. 25	A	Max.		11.1	1 day Aug. 25	F	5	590	5.3
		11	41	11.1			4	1,640	4.8
		10	68	10.9			3	4,480	4.0
		9	110	10.4		G	Max.		3.4
		8	200	9.5			3	900	3.2
		7	300	8.9		A to G	2	49,000	3.6
		6	550	7.8	2 days Aug. 25–26	A	Max.		13.1
	B	Max.		7.2			13	68	13.1
		7	230	7.1			12	250	12.6
		6	750	6.7			11	490	12.1
	C	Max.		6.2			10	820	11.4
		6	360	6.1			9	1,030	11.1
	ABC	5	6,650	5.9			8	1,260	10.6
		4	10,600	5.4			7	1,510	10.1
	D	Max.		11.7		B	Max.		11.5
		11	41	11.4			11	96	11.3
		10	96	10.9			10	380	10.7
		9	220	10.1			9	1,100	9.9
		8	340	9.6			8	1,880	9.3
	E	Max.		9.6			7	2,790	8.7
		9	68	9.3		C	Max.		7.9
		8	110	9.0			7	550	7.5
	DE	7	780	8.6		ABC	6	9,090	7.8
		6	1,150	7.9		D	Max.		12.8
		5	1,750	7.1			12	41	12.4
		4	2,530	6.3			11	110	11.8
	A to E	3	19,100	4.9			10	230	11.1
	F	Max.		5.6			9	330	10.6

STORM 103, AUGUST 24–26, 1908 (Continued)

Period of Maximum Rainfall	Rainfall Center	Isohyetal-Inches	Area in Square Miles	Average Depth in Inches	Period of Maximum Rainfall	Rainfall Center	Isohyetal-Inches	Area in Square Miles	Average Depth in Inches
2 days Aug. 25–26	E	Max.		9.8	3 days Aug. 24–26	C	13	180	13.8
		9	200	9.4			12	380	13.1
	DE	8	1,200	9.3			11	620	12.5
		7	1,370	9.1		D	Max.		14.3
		6	1,850	8.4			14	14	14.2
	F	Max.		6.6			13	55	13.6
		6	96	6.3			12	96	13.1
	G	Min.		4.3			11	200	12.2
		5	55	4.6		CD	10	1,510	11.6
	H	Max.		6.1		A to D	9	9,450	10.8
		6	68	6.1		E	Max.		10.5
	I	Max.		6.5			10	250	10.3
		6	3,070	6.3			9	750	9.8
	A to I	5	27,100	6.6		A to E	8	13,400	10.2
		4	41,100	5.9			7	16,600	9.7
		3	62,200	5.1			6	20,400	9.1
		2	87,000	4.3		F	Min.		4.2
3 days Aug. 24–26	A	Max.		15.6			5	450	4.6
		15	82	15.3		G	Max.		9.2
		14	200	14.8			9	590	9.1
		13	410	14.1			8	2,270	8.6
		12	820	13.3			7	5,370	8.0
		11	1,410	12.6			6	8,210	7.4
	B	Max.		14.1		H	Max.		6.4
		14	68	14.1			6	68	6.2
		13	200	13.7		A to H	5	42,700	7.6
		12	600	12.9		I	Max.		5.6
		11	960	12.4			5	2,630	7.4
	AB	10	4,590	11.5		A to I	4	61,600	6.7
	C	Max.		14.8			3	86,300	5.8
		14	68	14.4			2	112,000	5.0

STORM 105, OCTOBER 20–24, 1908

Period of Maximum Rainfall	Rainfall Center	Isohyetal-Inches	Area in Square Miles	Average Depth in Inches	Period of Maximum Rainfall	Rainfall Center	Isohyetal-Inches	Area in Square Miles	Average Depth in Inches
1 day Oct. 22	A	Max.		6.3	2 days Oct. 21–22	B	7	220	6.7
		6	410	6.2		AB	7	7,120	8.6
		5	3,240	5.6			6	9,200	8.1
	B	Max.		6.2		C	Max.		7.0
		6	41	6.1			7	68	7.0
		5	190	5.6			6	340	6.6
	C	Max.		5.2		ABC	5	16,600	7.0
		5	580	5.1		D	Max.		6.2
	ABC	4	10,300	4.9			6	55	6.1
		3	16,600	4.4			5	230	5.6
	D	Max.		3.3		E	Max.		5.4
		3	1,360	3.1			5	780	5.3
	A to D	2	34,200	3.4		A to E	4	29,700	5.9
2 days Oct. 21–22	A	Max.		11.4			3	43,900	5.1
		11	230	11.2			2	66,800	4.2
		10	960	10.7	3 days Oct. 21–23	A	Max.		13.0
		9	1,970	10.1			13	55	13.0
		8	4,520	9.2			12	520	12.6
	B	Min.		6.4			11	1,530	11.8

STORM 105, OCTOBER 20–24, 1908 (Continued)

Period of Maximum Rainfall	Rainfall Center	Isohyetal-Inches	Area in Square Miles	Average Depth in Inches	Period of Maximum Rainfall	Rainfall Center	Isohyetal-Inches	Area in Square Miles	Average Depth in Inches
3 days Oct. 21–23	A	10	3,760	11.0	4 days Oct. 20–23	D	Max.		7.6
		9	5,480	10.6			7	960	7.4
		8	8,100	9.9			6	2,330	6.9
	B	Min.		6.1		A to D	5	41,000	7.5
		7	180	6.4		E	Max.		5.8
	AB	7	12,100	9.1			5	160	5.5
	C	Min.		5.2		A to E	4	52,900	6.8
		6	68	5.6		F	Max.		4.3
	D	Max.		7.4			4	770	4.2
		7	110	7.2		A to F	3	69,300	6.0
	A to D	6	19,400	8.1			2	92,700	5.2
		5	31,000	7.1	5 days Oct. 20–24	A	Max.		15.2
	E	Max.		5.1			15	60	15.1
		5	68	5.0			14	360	14.6
	A to E	4	43,400	6.4			13	820	14.0
	F	Max.		4.7			12	2,100	13.2
		4	300	4.4			11	3,960	12.2
	A to F	3	58,800	5.6			10	5,900	11.7
		2	82,600	4.7			9	8,460	11.1
4 days Oct. 20–23	A	Max.		15.1			8	14,000	10.1
		15	41	15.0		B	Max.		8.7
		14	270	14.6			8	520	8.4
		13	680	13.9		AB	7	22,000	9.1
		12	2,080	13.0			6	31,300	8.4
		11	3,970	12.3		C	Max.		7.6
		10	5,760	11.7			7	920	7.3
		9	8,160	11.1			6	2,370	6.8
		8	12,500	10.2		ABC	5	47,800	7.4
		7	18,000	9.4			4	62,400	6.8
	B	Max.		7.7		D	Max.		5.2
		7	460	7.4			5	500	5.1
	C	Max.		7.1			4	1,600	4.7
		7	96	7.0		A to D	3	90,100	5.8
	ABC	6	26,600	8.4			2	131,000	4.7

STORM 108, JUNE 29–JULY 3, 1909

Period of Maximum Rainfall	Rainfall Center	Isohyetal-Inches	Area in Square Miles	Average Depth in Inches	Period of Maximum Rainfall	Rainfall Center	Isohyetal-Inches	Area in Square Miles	Average Depth in Inches
1 day July 2	A	Max.		12.0	1 day July 2	C	4	580	4.7
		12	200	12.0		ABC	3	16,600	4.8
		11	360	11.8			2	24,400	4.1
		10	510	11.4	2 days July 2–3	A	Max.		15.5
		9	740	10.8			15	180	15.3
		8	1,050	10.1			14	320	15.0
		7	1,560	9.3			13	470	14.5
	B	Max.		8.8			12	640	13.9
		8	150	8.4			11	820	13.4
		7	440	7.8			10	1,160	12.6
	AB	6	3,660	7.8		B	Max.		11.8
		5	5,000	7.2			11	450	11.4
		4	7,990	6.2			10	1,260	10.8
	C	Max.		5.5		AB	9	3,970	10.8
		5	120	5.3			8	5,160	10.3

STORM 108, JUNE 29–JULY 3, 1909 (Continued)

Period of Maximum Rainfall	Rainfall Center	Isohyetal-Inches	Area in Square Miles	Average Depth in Inches
2 days July 2-3	AB	7	6,780	9.7
		6	8,760	8.9
		5	11,000	8.2
	C	Max.		7.3
		7	160	7.2
		6	1,190	6.6
		5	3,260	5.9
	D	Max.		5.5
		5	120	5.3
	A to D	4	19,700	6.8
		3	26,200	6.0
		2	29,700	5.6
3 days July 1-3	A	Max.		15.8
		15	200	15.4
		14	370	15.0
		13	550	14.5
		12	820	13.9
		11	1,140	13.2
	B	Max.		12.0
		11	920	11.5
	AB	10	3,560	11.6
		9	4,710	11.1
		8	6,010	10.5
		7	7,870	9.8
		6	10,100	9.1
	C	Max.		6.2
		6	41	6.1
	D	Max.		7.8
		7	600	7.4
		6	2,460	6.7
	A to D	5	20,200	7.4
		4	24,300	7.0
	E	Max.		7.9
		7	340	7.5
		6	660	7.0
		5	1,030	6.5
		4	1,480	6.0
	A to E	3	30,600	6.4
		2	34,200	6.0
4 days June 30- July 3	A	Max.		17.7
		17	55	17.4
		16	110	17.0
		15	180	16.4
		14	300	15.6
	B	Max.		16.7
		16	160	16.4
		15	300	16.0
		14	480	15.4
	AB	13	1,370	14.6

Period of Maximum Rainfall	Rainfall Center	Isohyetal-Inches	Area in Square Miles	Average Depth in Inches
4 days June 30- July 3	AB	12	2,600	13.6
		11	3,970	12.9
		10	5,410	12.2
		9	6,900	11.7
		8	9,420	10.8
		7	12,200	10.0
		6	15,000	9.4
	C	Max.		6.4
		6	41	6.2
	ABC	5	21,800	8.2
		4	25,900	7.6
	D	Max.		8.0
		8	180	8.0
		7	480	7.7
		6	750	7.2
		5	1,120	6.7
		4	1,510	6.1
	A to D	3	31,900	7.0
		2	35,700	6.5
5 days June 29- July 3	A	Max.		17.8
		17	68	17.4
		16	250	16.8
		15	1,560	15.7
		14	2,610	15.2
	B	Max.		16.8
		16	160	16.4
		15	340	15.9
		14	480	15.5
	AB	13	4,380	14.7
		12	6,670	14.0
		11	8,350	13.5
		10	9,720	13.0
		9	11,200	12.6
		8	13,200	12.0
		7	15,500	11.3
		6	18,300	10.6
	C	Max.		6.4
		6	68	6.2
	ABC	5	25,100	9.2
		4	28,600	8.6
	D	Max.		8.5
		8	230	8.3
		7	480	7.9
		6	780	7.3
		5	1,150	6.8
		4	1,550	6.2
	A to D	3	32,800	8.1
		2	36,000	7.6

STORM 109, JULY 5–7, 1909

Period of Maximum Rainfall	Rainfall Center	Isohyetal-Inches	Area in Square Miles	Average Depth in Inches	Period of Maximum Rainfall	Rainfall Center	Isohyetal-Inches	Area in Square Miles	Average Depth in Inches
1 day July 7	A	Max.		9.7	2 days July 6–7	H	5	55	5.4
		9	41	9.4		A to H	4	19,100	5.3
		8	190	8.7		I	Max.		5.0
		7	520	8.0			4	300	4.5
		6	900	7.4		A to I	3	30,800	4.6
		5	1,440	6.7		J	Max.		8.4
		4	2,260	5.9			8	27	8.2
	B	Max.		6.5			7	150	7.5
		6	27	6.3			6	490	6.8
		5	110	5.7			5	1,030	6.1
		4	330	4.9			4	1,750	5.4
	AB	3	4,290	4.9			3	2,820	4.7
	C	Max.		3.7		K	Max.		4.4
		3	68	3.4			4	330	4.2
	ABC	2	7,560	3.8			3	1,000	3.7
	D	Max.		5.8		A to K	2	67,400	3.6
		5	96	5.4		L	Max.		5.8
		4	320	4.8			5	110	5.4
		3	1,010	3.9			4	380	4.8
		2	1,820	3.3			3	1,010	4.0
	E	Max.		4.6			2	1,780	3.3
		4	480	4.3		M	Max.		4.0
		3	2,190	3.7			3	670	3.5
		2	6,740	2.9			2	4,180	2.7
	F	Max.		2.9		N	Max.		2.6
		2	5,590	2.5			2	340	2.3
	G	Max.		2.9	3 days July 5–7	A	Max.		11.2
		2	1,930	2.5			11	27	11.1
2 days July 6–7	A	Max.		9.7			10	260	10.4
		9	55	9.4			9	980	9.8
		8	330	8.7			8	2,310	9.0
		7	690	8.1		B	Max.		9.0
		6	1,210	7.4			8	730	8.5
		5	2,300	6.5		AB	7	5,540	8.3
	B	Max.		8.4			6	8,520	7.7
		8	110	8.2		C	Max.		7.2
		7	770	7.6			7	14	7.1
		6	1,450	7.1			6	110	6.5
	C	Max.		6.3		D	Max.		6.3
		6	68	6.2			6	27	6.2
	D	Max.		6.4		A to D	5	13,500	6.9
		6	160	6.2		E	Max.		9.7
	BCD	5	5,110	6.0			9	82	9.4
	E	Max.		5.3			8	270	8.8
		5	96	5.2			7	710	8.0
	F	Max.		6.4			6	1,250	7.4
		6	41	6.2			5	2,450	6.5
		5	160	5.7		F	Max.		6.5
	G	Max.		7.5			6	82	6.3
		7	55	7.3			5	190	5.9
		6	550	6.6		G	Max.		6.8
		5	1,480	5.9			6	410	6.4
	H	Max.		5.7			5	890	5.9

STORM 109, JULY 5-7, 1909 (Continued)

Period of Maximum Rainfall	Rainfall Center	Isohyetal-Inches	Area in Square Miles	Average Depth in Inches	Period of Maximum Rainfall	Rainfall Center	Isohyetal-Inches	Area in Square Miles	Average Depth in Inches
3 days July 5-7	H	Max.		6.1	3 days July 5-7	L	10	41	10.7
		6	27	6.1			9	150	9.9
		5	560	5.5			8	410	9.0
	I	Max.		8.7			7	740	8.3
		8	290	8.4			6	1,200	7.6
		7	1,140	7.7			5	1,710	7.0
		6	2,400	7.1			4	2,330	6.3
		5	5,000	6.3			3	3,330	5.5
	A to I	4	34,500	6.0		M	Max.		4.9
	J	Max.		5.3			4	480	4.5
		5	82	5.2			3	2,460	3.7
		4	2,930	4.5		A to M	2	96,400	4.2
	K	Max.		4.4		N	Max.		5.8
		4	410	4.2			5	110	5.4
	A to K	3	57,700	5.1			4	400	4.8
	L	Max.		11.2			3	860	4.1
		11	14	11.1			2	1,290	3.6

STORM 110, JULY 20-22, 1909

Period of Maximum Rainfall	Rainfall Center	Isohyetal-Inches	Area in Square Miles	Average Depth in Inches	Period of Maximum Rainfall	Rainfall Center	Isohyetal-Inches	Area in Square Miles	Average Depth in Inches
1 day July 21	A	Max.		10.7	3 days July 20-22	A	Max.		12.8
		10	150	10.4			12	14	12.4
		9	270	10.0			11	41	11.7
		8	400	9.6			10	96	11.0
		7	580	8.9			9	180	10.3
		6	930	8.0			8	450	9.2
		5	1,490	7.1		B	Max.		8.5
		4	2,520	6.0			8	82	8.3
		3	3,260	5.4		AB	7	2,200	7.9
		2	3,900	5.0			6	3,500	7.4
	B	Max.		5.5		C	Max.		6.4
		5	110	5.3			6	250	6.2
		4	300	4.8		ABC	5	6,200	6.6
		3	590	4.2		D	Max.		5.2
		2	990	3.5			5	230	5.1
2 days July 20-21	A	Max.		11.0		A to D	4	11,700	5.6
		10	230	10.5		E	Max.		11.1
		9	370	10.1			11	41	11.1
		8	530	9.6			10	230	10.7
		7	890	8.8			9	380	10.2
		6	1,590	7.8			8	530	9.7
		5	2,970	6.7			7	890	8.8
		4	4,110	6.1			6	2,160	7.4
	B	Max.		7.8			5	3,160	6.8
		7	160	7.4			4	4,560	6.1
		6	1,480	6.6		F	Max.		4.2
		5	3,300	6.0			4	580	4.1
		4	5,870	5.3		A to F	3	31,400	4.7
	AB	3	24,800	4.4			2	50,000	3.9
		2	39,900	3.7					

STORM 114, OCTOBER 4–6, 1910

Period of Maximum Rainfall	Rainfall Center	Isohyetal-Inches	Area in Square Miles	Average Depth in Inches	Period of Maximum Rainfall	Rainfall Center	Isohyetal-Inches	Area in Square Miles	Average Depth in Inches
1 day Oct. 5	A	Max.		8.1	2 days Oct. 5–6	A to G	6	32,100	7.3
		7	160	7.5		H	Min.		4.5
		6	480	6.8			5	200	4.7
		5	1,100	6.1		A to H	5	50,200	6.6
	B	Max.		8.5		J	Min.		2.9
		8	68	8.3			3	300	2.9
		7	330	7.7			4	1,880	3.4
		6	710	7.0		K	Min.		3.5
		5	1,410	6.3			4	55	3.7
	AB	4	7,280	5.1		A to K	4	72,000	6.0
		3	11,500	4.5			3	95,000	5.4
	C	Max.		8.0			2	131,000	4.6
		7	400	7.6	3 days Oct. 4–6	A	Max.		15.2
		6	1,510	6.8			15	41	15.1
		5	3,650	6.0			14	180	14.7
	D	Max.		7.1			13	290	14.2
		7	41	7.1			12	810	13.1
		6	180	6.6			11	1,440	12.4
		5	630	5.8		B	Max.		13.3
	E	Max.		6.0			13	55	13.2
		5	670	5.5			12	320	12.6
	F	Min.		3.7			11	900	11.9
		4	270	3.8		C	Max.		14.0
	C to F	4	19,100	4.9			13	68	13.5
	G	Max.		4.3			12	250	12.8
		4	440	4.2			11	710	11.9
	C to G	3	34,900	4.3		ABC	10	8,240	11.1
	H	Min.		1.9			9	13,900	10.5
		2	970	1.9		D	Max.		11.5
	A to H	2	76,000	2.9			11	96	11.3
2 days Oct. 5–6	A	Max.		12.9			10	230	10.8
		12	68	12.5			9	550	10.1
		11	230	11.8		E	Min.		6.5
		10	410	11.2			7	96	6.7
		9	790	10.4			8	530	7.3
	B	Max.		11.0		A to E	8	26,500	9.5
		10	68	10.5		F	Min.		6.3
		9	260	9.7			7	55	6.6
	AB	8	2,600	9.2		A to F	7	38,600	8.9
		7	5,400	8.3		G	Min.		4.1
	C	Max.		10.2			5	140	4.5
		10	250	10.2			6	440	5.2
		9	730	9.8		A to G	6	55,400	8.2
		8	2,980	8.8		H	Min.		4.6
		7	6,630	8.1			5	190	4.8
	D	Max.		8.1		A to H	5	70,100	7.6
		8	55	8.1		J	Max.		5.1
		7	600	7.6			5	140	5.1
	E	Max.		8.3		A to J	4	91,200	6.9
		8	360	8.2			3	125,000	6.0
	F	Max.		8.3		K	Max.		4.4
		8	270	8.2			4	110	4.2
	EF	7	3,790	7.6			3	740	3.6
	G	Min.		5.6		A to K	2	168,000	5.1
		6	96	5.8					

STORM 125, JULY 20–24, 1912

Period of Maximum Rainfall	Rainfall Center	Isohyetal-Inches	Area in Square Miles	Average Depth in Inches	Period of Maximum Rainfall	Rainfall Center	Isohyetal-Inches	Area in Square Miles	Average Depth in Inches
5 days July 20–24	A	Max.		11.3	5 days July 20–24	C	5	4,130	6.2
		11	27	11.2		D	Max.		6.2
		10	120	10.7			6	41	6.1
		9	300	10.0			5	640	5.5
		8	520	9.4		E	Max.		5.2
		7	860	8.6			5	68	5.1
	B	Max.		7.6		A to E	4	19,000	5.3
		7	180	7.3		F	Max.		4.2
	AB	6	2,180	7.4			4	260	4.1
		5	3,640	6.6		G	Min.		2.5
	C	Max.		7.8			3	150	2.7
		7	860	7.4		A to G	3	36,700	4.5
		6	2,220	6.9			2	62,800	3.6

STORM 130, JANUARY 10–12, 1913

Period of Maximum Rainfall	Rainfall Center	Isohyetal-Inches	Area in Square Miles	Average Depth in Inches	Period of Maximum Rainfall	Rainfall Center	Isohyetal-Inches	Area in Square Miles	Average Depth in Inches
1 day Jan. 11	A	Max.		5.8	2 days Jan. 11–12	D	4	320	4.1
		5	930	5.4		A to D	3	56,500	4.1
		4	6,490	4.6		E	Min.		0.8
		3	13,200	4.1			1	96	0.9
	B	Max.		4.8			2	200	1.2
		4	220	4.4		F	Max.		3.8
	C	Max.		4.4			3	1,150	3.4
		4	140	4.2		G	Min.		1.9
	BC	3	2,600	3.5			2	55	1.9
	ABC	2	33,600	3.2		A to G	2	89,800	3.5
	D	Max.		5.4	3 days Jan. 10–12	A	Max.		7.5
		5	680	5.2			7	2,140	7.3
		4	2,460	4.7		B	Max.		7.4
	E	Max.		4.2			7	190	7.2
		4	190	4.1		AB	6	7,370	6.8
	DE	3	14,300	3.7		C	Max.		6.1
	F	Min.		0.0			6	250	6.1
		1	27	0.5		ABC	5	23,000	5.9
		2	120	1.2			4	43,900	5.2
	DEF	2	27,000	3.1		D	Max.		5.7
	G	Max.		2.4			5	68	5.4
		2	1,100	2.2			4	200	4.8
2 days Jan. 11–12	A	Max.		7.4		E	Min.		1.4
		7	96	7.2			2	55	1.7
		6	1,550	6.5			3	230	2.3
		5	7,410	5.7		A to E	3	81,800	4.4
	B	Max.		6.7		F	Min.		0.8
		6	250	6.4			1	55	0.9
		5	2,010	5.6			2	140	1.2
	C	Min.		3.5		A to F	2	124,000	3.8
		4	140	3.7		G	Max.		2.9
	ABC	4	22,100	5.0			2	4,740	2.5
	D	Max.		4.2					

STORM 132, MARCH 23–27, 1913

Period of Maximum Rainfall	Rainfall Center	Isohyetal-Inches	Area in Square Miles	Average Depth in Inches	Period of Maximum Rainfall	Rainfall Center	Isohyetal-Inches	Area in Square Miles	Average Depth in Inches
1 day March 25	A	Max.		7.0	1 day March 25	B	Max.		6.2
		7	96	7.0			6	740	6.1
		6	820	6.6		AB	5	4,930	5.8

STORM 132, MARCH 23–27, 1913 (Continued)

Period of Maximum Rainfall	Rainfall Center	Isohyetal-Inches	Area in Square Miles	Average Depth in Inches
1 day March 25	C	Max.		5.7
		5	680	5.4
	D	Max.		5.6
		5	200	5.3
	E	Max.		5.3
		5	370	5.2
	A to E	4	16,000	5.0
	F	Max.		4.6
		4	140	4.3
	G	Max.		4.2
		4	82	4.1
	H	Max.		4.2
		4	120	4.1
	J	Min.		0.5
		1	120	0.7
		2	440	1.3
		3	900	1.9
	K	Max.		5.3
		5	120	5.2
		4	560	4.6
	A to K	3	43,000	4.0
		2	87,300	3.3
2 days March 24–25	A	Max.		9.5
		9	120	9.3
		8	590	8.7
	B	Max.		8.5
		8	1,100	8.4
	C	Max.		8.4
		8	300	8.3
	ABC	7	6,890	7.8
	D	Max.		7.1
		7	250	7.1
	E	Max.		7.2
		7	200	7.1
	A to E	6	16,600	7.0
		5	28,700	6.4
	F	Max.		5.6
		5	120	5.3
	G	Max.		6.4
		6	82	6.2
		5	410	5.6
	H	Max.		6.1
		6	82	6.1
		5	820	5.6
	J	Max.		5.1
		5	82	5.1
	A to J	4	57,600	5.5
	K	Max.		4.0
		4	340	4.0
	L	Max.		5.1
		5	110	5.1
		4	550	4.2
	M	Max.		5.1
		5	82	5.1

Period of Maximum Rainfall	Rainfall Center	Isohyetal-Inches	Area in Square Miles	Average Depth in Inches
2 days March 24–25	M	4	380	4.6
	A to M	3	105,000	4.6
		2	181,000	3.7
3 days March 24–26	A	Max.		10.2
		10	120	10.1
		9	620	9.6
	B	Max.		9.3
		9	96	9.2
	C	Max.		9.0
		9	120	9.0
	ABC	8	6,980	8.6
	D	Max.		9.3
		9	200	9.2
		8	1,680	8.6
	E	Max.		8.5
		8	68	8.3
	A to E	7	20,300	8.0
	F	Min.		5.1
		6	410	5.5
	A to F	6	34,700	7.4
	G	Max.		6.4
		6	68	6.2
	A to G	5	55,800	6.6
	H	Max.		6.5
		6	68	6.3
		5	270	5.7
	I	Max.		6.9
		6	120	6.5
		5	680	5.7
	J	Max.		7.1
		7	120	7.1
		6	740	6.6
	K	Max.		6.6
		6	41	6.3
	JK	5	5,480	5.7
	L	Max.		5.2
		5	160	5.1
	M	Max.		7.3
		7	110	7.2
		6	860	6.6
		5	2,410	5.9
	A to M	4	105,000	5.7
	N	Max.		5.7
		5	260	5.4
		4	1,620	4.6
	O	Max.		5.3
		5	96	5.2
		4	700	4.6
	A to O	3	195,000	4.7
	P	Max.		4.7
		4	590	4.4
		3	3,490	3.6
	A to P	2	291,000	4.0

STORM 132, MARCH 23-27, 1913 (Continued)

Period of Maximum Rainfall	Rainfall Center	Isohyetal-Inches	Area in Square Miles	Average Depth in Inches	Period of Maximum Rainfall	Rainfall Center	Isohyetal-Inches	Area in Square Miles	Average Depth in Inches
4 days March 23-26	A	Max.		11.1	4 days March 23-26	T	3	3,720	3.7
		11	41	11.1		U	Max.		4.5
		10	380	10.6			4	41	4.2
		9	1,560	9.8			3	310	3.6
	B	Max.		10.6		A to U	2	328,000	4.2
		10	160	10.3	5 days March 23-27	A	Max.		11.2
		9	2,000	9.6			11	68	11.1
	C	Max.		9.4			10	680	10.5
		9	140	9.2			9	3,350	9.7
	D	Max.		9.0		B	Max.		11.2
		9	110	9.0			11	140	11.1
	A to D	8	17,200	8.7			10	1,990	10.5
	E	Max.		8.5			9	5,430	9.9
		8	140	8.3		C	Max.		9.9
	A to E	7	31,100	8.2			9	1,570	9.5
	F	Max.		8.0		D	Max.		9.6
		7	550	7.5			9	510	9.3
	G	Min.		5.7		A to D	8	22,400	9.1
		6	96	5.8		E	Max.		8.5
	H	Max.		7.1			8	140	8.3
		7	250	7.1		F	Min.		6.3
	A to H	6	52,300	7.5			7	250	6.6
	J	Max.		6.4		G	Max.		9.4
		6	68	6.2			9	55	9.2
	A to J	5	75,000	6.9			8	320	8.6
	K	Max.		6.3		A to G	7	39,600	8.4
		6	420	6.2		H	Max.		8.0
		5	2,900	5.6			7	1,030	7.5
	L	Max.		5.4		A to H	6	60,900	7.8
		5	340	5.2		J	Max.		6.3
	M	Min.		3.6			6	550	6.2
		4	140	3.8		K	Max.		6.8
	N	Min.		3.9			6	110	6.4
		4	96	3.9		A to K	5	90,600	7.0
	O	Max.		6.9		L	Max.		7.1
		6	140	6.5			7	140	7.1
		5	750	5.7			6	1,030	6.6
	P	Max.		7.1		M	Max.		6.9
		7	140	7.1			6	1,050	6.5
		6	770	6.6		N	Min.		3.0
	Q	Max.		6.6			4	300	3.5
		6	41	6.3			5	1,440	4.3
	PQ	5	5,540	5.7		O	Max.		6.7
	R	Max.		6.5			6	82	6.4
		6	96	6.3		P	Max.		6.2
		5	270	5.8			6	200	6.1
	A to R	4	137,000	5.9		Q	Max.		7.1
	S	Max.		5.8			7	140	7.1
		5	560	5.4			6	960	6.6
		4	2,740	4.7		L to Q	5	17,900	5.6
	A to S	3	212,000	5.1		A to Q	4	177,000	6.3
	T	Max.		4.7		R	Max.		5.8
		4	700	4.4			5	560	5.4

STORM 132, MARCH 23-27, 1913 (Continued)

Period of Maximum Rainfall	Rainfall Center	Isohyetal-Inches	Area in Square Miles	Average Depth in Inches	Period of Maximum Rainfall	Rainfall Center	Isohyetal-Inches	Area in Square Miles	Average Depth in Inches
5 days March 23-27	R S	4	2,680	4.7	5 days March 23-27	T U	3	1,460	4.0
		Max.		5.4			Max.		4.8
		5	1,900	5.2			4	1,150	4.4
		4	7,580	4.7			3	8,170	3.6
	A to S	3	300,000	5.2		V	Max.		4.5
	T	Max.		6.5			4	150	4.2
		6	55	6.2			3	4,800	3.5
		5	250	5.6		A to V	2	492,000	4.2
		4	480	5.1					

STORM 135, OCTOBER 1-2, 1913

Period of Maximum Rainfall	Rainfall Center	Isohyetal-Inches	Area in Square Miles	Average Depth in Inches	Period of Maximum Rainfall	Rainfall Center	Isohyetal-Inches	Area in Square Miles	Average Depth in Inches
1 day Oct. 2	A	Max.		13.0	2 days Oct. 1-2	A	Max.		14.2
		13	55	13.0			14	68	14.1
		12	340	12.6			13	640	13.6
		11	770	12.0			12	1,230	13.1
		10	1,270	11.4			11	1,900	12.5
		9	1,970	10.7			10	2,740	11.9
		8	2,700	10.1			9	3,710	11.3
		7	3,940	9.3			8	5,590	10.3
		6	5,610	8.5			7	7,280	9.7
		5	8,490	7.5			6	9,280	9.0
		4	11,900	6.6		B	Max.		8.6
	B	Max.		7.5			8	68	8.4
		7	55	7.3			7	250	7.8
		6	230	6.7			6	590	7.0
		5	520	6.0		AB	5	13,800	7.9
		4	860	5.4			4	20,900	6.8
	AB	3	19,400	5.5			3	30,100	5.8
		2	27,000	4.7		C	Max.		11.2
	C	Max.		11.0			10	82	10.6
		10	41	10.5			9	140	10.1
		9	82	10.0			8	230	9.5
		8	120	9.5			7	300	9.0
		7	180	8.8			6	400	8.4
		6	250	8.2			5	550	7.6
		5	400	7.2			4	850	6.5
		4	810	5.8			3	1,260	5.5
		3	1,230	5.0		ABC	2	45,600	4.7
		2	1,940	4.1					

STORM 136, DECEMBER 2-5, 1913

Period of Maximum Rainfall	Rainfall Center	Isohyetal-Inches	Area in Square Miles	Average Depth in Inches	Period of Maximum Rainfall	Rainfall Center	Isohyetal-Inches	Area in Square Miles	Average Depth in Inches
1 day Dec. 4	A	Max.		10.0	1 day Dec. 4	AB	4	8,960	5.8
		9	95	9.5			3	14,200	5.0
		8	280	8.9			2	33,400	3.3
		7	620	8.1	2 days Dec. 3-4	A	Max.		12.5
		6	1,070	7.4			12	55	12.2
		5	1,890	6.1			11	280	11.6
	B	Max.		7.5			10	660	11.0
		7	240	7.2			9	1,200	10.3
		6	1,020	6.7			8	1,920	9.6
		5	2,180	6.0			7	8,570	8.0

STORM 136, DECEMBER 2-5, 1913 (Continued)

Period of Maximum Rainfall	Rainfall Center	Isohyetal-Inches	Area in Square Miles	Average Depth in Inches	Period of Maximum Rainfall	Rainfall Center	Isohyetal-Inches	Area in Square Miles	Average Depth in Inches
2 days Dec. 3-4	A	6	13,800	7.4	3 days Dec. 2-4	E	4	1,300	4.4
		5	20,100	6.8			3	5,180	3.7
		4	27,400	6.2		A to E	2	144,000	3.9
		3	40,400	5.3	4 days Dec. 2-5	A	Max.		15.5
		2	67,000	4.2			15	41	15.2
3 days Dec. 2-4	A	Max.		14.5			14	330	14.6
		14	55	14.2			13	900	13.9
		13	230	13.7			12	1,780	13.2
	B	Max.		13.6			11	3,590	12.3
		13	55	13.3			10	5,340	11.7
	AB	12	930	12.9		B	Max.		11.0
		11	1,810	12.3			10	140	10.5
		10	3,150	11.5		AB	9	13,000	10.4
	C	Max.		11.1			8	18,700	9.8
		10	220	10.5			7	22,300	9.5
	ABC	9	8,820	10.4			6	27,300	8.9
	D	Max.		11.0			5	34,100	8.2
		10	96	10.5			4	48,500	7.1
		9	440	9.7		C	Max.		5.3
	A to D	8	14,900	9.5			5	82	5.1
		7	19,000	9.1			4	2,160	4.5
		6	22,700	8.7		ABC	3	117,000	5.0
		5	28,900	8.0		D	Max.		4.3
		4	37,700	7.2			3	18,900	3.6
		3	55,100	6.0		A to D	2	284,000	3.6
	E	Max.		4.9					

STORM 144, OCTOBER 14-15, 1914

Period of Maximum Rainfall	Rainfall Center	Isohyetal-Inches	Area in Square Miles	Average Depth in Inches	Period of Maximum Rainfall	Rainfall Center	Isohyetal-Inches	Area in Square Miles	Average Depth in Inches
1 day Oct. 15	A	Max.		10.0	1 day Oct. 15	E	4	360	5.6
		9	41	9.5			3	600	4.8
		8	140	8.8			2	1,050	3.8
		7	260	8.2	2 days Oct. 14-15	A	Max.		11.5
		6	440	7.5			11	14	11.3
		5	680	6.8			10	55	10.7
	B	Max.		8.3			9	160	9.9
		8	27	8.3			8	300	9.3
		7	110	7.6			7	440	8.7
		6	250	7.0			6	730	7.8
		5	920	5.9		B	Max.		9.5
	AB	4	3,810	5.2			9	27	9.3
	C	Max.		5.2			8	190	8.6
		5	27	5.1			7	410	8.0
		4	140	4.7			6	770	7.3
	ABC	3	6,630	4.5		C	Min.		3.6
	D	Max.		3.3			4	14	3.8
		3	1,100	3.2			5	68	4.3
	A to D	2	16,300	3.4		ABC	5	3,140	6.4
	E	Max.		8.8			4	5,200	5.7
		8	27	8.4		D	Max.		6.4
		7	55	8.0			6	14	6.2
		6	110	7.3			5	96	5.6
		5	200	6.4			4	230	5.0

STORM 144, OCTOBER 14–15, 1914 (Continued)

Period of Maximum Rainfall	Rainfall Center	Isohyetal-Inches	Area in Square Miles	Average Depth in Inches	Period of Maximum Rainfall	Rainfall Center	Isohyetal-Inches	Area in Square Miles	Average Depth in Inches
2 days Oct. 14–15	A to D	3	11,500	4.5	2 days Oct. 14–15	E	6	230	7.5
	E	Max.		11.0			5	340	6.9
		10	14	10.5			4	550	6.0
		9	27	10.0			3	920	5.0
		8	68	9.1		A to E	2	25,000	3.5
		7	120	8.4					

STORM 151, AUGUST 17–20, 1915

Period of Maximum Rainfall	Rainfall Center	Isohyetal-Inches	Area in Square Miles	Average Depth in Inches	Period of Maximum Rainfall	Rainfall Center	Isohyetal-Inches	Area in Square Miles	Average Depth in Inches
1 day Aug. 17 Texas center	D	Max.		8.0	2 days Aug. 17–18 Texas center	IK	6	8,080	7.1
		8	1,100	8.0		DJIK	5	38,600	7.6
		7	5,320	7.6			4	52,300	6.8
		6	7,320	7.3			3	66,000	6.1
	I	Max.		7.9			2	94,700	5.0
		7	270	7.5	3 days Aug. 17–19 Texas center	D	Max.		19.7
		6	920	6.8			19	1,830	19.3
	DI	5	14,000	6.5			18	2,900	19.0
	J	Max.		9.1			17	4,180	18.5
		9	160	9.1			16	5,720	18.0
		8	420	8.7			15	6,840	17.6
		7	780	8.2			14	7,580	17.3
		6	1,670	7.3			13	8,530	16.9
		5	2,810	6.6			12	9,430	16.4
	DIJ	4	23,400	6.0			11	10,700	15.8
		3	31,300	5.3			10	12,300	15.1
		2	51,600	4.2			9	14,600	14.3
2 days Aug. 17–18 Texas center	D	Max.		18.6			8	17,800	13.2
		18	55	18.3			7	21,000	12.4
		17	140	17.8		A	Max.		13.0
		16	270	17.2			13	160	13.0
		15	410	16.6			12	410	12.7
		14	630	15.9			11	620	12.3
		13	890	15.2			10	1,180	11.4
		12	1,570	14.0			9	2,450	10.4
		11	2,750	12.9			8	5,800	9.3
		10	6,410	11.5		B	Max.		10.6
		9	8,780	11.0			10	260	10.3
		8	11,300	10.4			9	620	9.8
		7	13,400	10.0			8	3,290	8.8
	J	Max.		9.1		AB	7	16,700	8.4
		9	300	9.1		I	Max.		10.5
		8	930	8.7			10	96	10.2
		7	1,530	8.2			9	250	9.8
	DJ	6	19,700	9.0			8	3,540	8.6
	I	Max.		10.0			7	7,260	8.0
		10	55	10.0		L	Max.		7.1
		9	300	9.6			7	230	7.0
		8	620	9.0		DABIL	6	69,800	8.9
		7	1,100	8.4			5	93,900	8.0
	K	Max.		7.9			4	124,000	7.2
		7	3,020	7.5			3	163,000	6.3
							2	211,000	5.4

STORM 151, AUGUST 17–20, 1915 (Continued)

Period of Maximum Rainfall	Rainfall Center	Isohyetal-Inches	Area in Square Miles	Average Depth in Inches	Period of Maximum Rainfall	Rainfall Center	Isohyetal-Inches	Area in Square Miles	Average Depth in Inches
1 day	A	Max.		6.7	3 days	B	10	270	10.0
Aug.		6	3,420	6.4	Aug.		9	1,620	9.6
19	B	Max.		6.2	18–20		8	3,670	9.0
Ark.		6	2,000	6.1	Ark.	C	Max.		8.7
center	C	Max.		7.0	center		8	630	8.3
		6	480	6.5		D	Max.		12.3
	ABC	5	16,000	5.8			12	4,960	12.1
		4	25,000	5.3			11	6,460	12.0
	D	Max.		9.1			10	8,320	11.6
		9	960	9.0			9	9,990	11.3
		8	2,450	8.7			8	11,600	10.9
		7	4,300	8.2		E	Max.		8.4
		6	5,780	7.8			8	340	8.2
		5	7,210	7.3		A to E	7	55,100	9.0
		4	9,160	6.7		F	Max.		8.7
	A to D	3	60,600	4.8			8	1,440	8.3
	E	Max.		3.7			7	2,330	8.0
		3	1,490	3.4		G	Max.		8.2
	F	Min.		0.9			8	160	8.1
		2	770	1.4			7	410	7.7
	G	Min.		0.3		A to G	6	87,200	8.3
		2	3,510	1.2			5	116,000	7.5
	A to G	2	103,000	3.8			4	147,000	6.8
2 days	A	Max.		10.9			3	196,000	6.0
Aug.		10	270	10.5			2	255,000	5.2
19–20		9	820	9.8	4 days	A	Max.		14.8
Ark.		8	2,880	8.9	Aug.		14	250	14.4
center		7	9,900	7.9	17–20		13	500	13.9
	B	Max.		8.3	Total		12	960	13.3
		8	1,520	8.1	storm		11	2,670	12.1
		7	4,960	7.7			10	5,370	11.3
	C	Max.		7.9			9	11,200	10.4
		7	380	7.5		B	Max.		9.8
	ABC	6	31,500	7.1			9	580	9.4
		5	49,600	6.5		C	Max.		9.1
		4	74,100	5.9			9	330	9.0
	D	Max.		9.1		ABC	8	25,300	9.4
		9	1,070	9.0		D	Max.		19.7
		8	2,380	8.7			19	1,850	19.3
		7	4,760	8.1			18	2,930	19.0
		6	7,480	7.5			17	4,180	18.6
		5	9,480	7.1			16	5,790	18.0
		4	13,200	6.4			15	6,840	17.6
	A to D	3	135,000	5.1			14	7,580	17.3
		2	183,000	4.4			13	8,530	16.9
3 days	A	Max.		14.0			12	9,530	16.4
Aug.		13	270	13.5			11	11,000	15.8
18–20		12	790	12.8			10	12,900	15.0
Ark.		11	1,620	12.2			9	15,900	13.9
center		10	4,820	11.0			8	19,400	13.0
		9	10,500	10.2		E	Max.		10.6
		8	19,800	9.4			10	260	10.3
	B	Max.		10.1			9	1,570	9.6

STORM 151, AUGUST 17-20, 1915 (Continued)

Period of Maximum Rainfall	Rainfall Center	Isohyetal-Inches	Area in Square Miles	Average Depth in Inches	Period of Maximum Rainfall	Rainfall Center	Isohyetal-Inches	Area in Square Miles	Average Depth in Inches
4 days	E	8	4,660	8.9	4 days	G	7	410	7.7
Aug.	A to E	7	66,700	9.9	Aug.	H	Max.		7.3
17-20	F	Max.		10.5	17-20		7	250	7.1
Total		10	68	10.3	Total	A to H	6	107,000	8.7
storm		9	270	9.7	storm		5	139,000	8.0
		8	3,370	8.6			4	178,000	7.2
		7	7,500	8.0			3	233,000	6.3
	G	Max.		8.2			2	293,000	5.6
		8	120	8.1					

STORM 156, JULY 6-10, 1916

Period of Maximum Rainfall	Rainfall Center	Isohyetal-Inches	Area in Square Miles	Average Depth in Inches	Period of Maximum Rainfall	Rainfall Center	Isohyetal-Inches	Area in Square Miles	Average Depth in Inches
1 day	A	Max.		11.2	2 days	B	15	250	15.7
July		11	110	11.1	July		14	540	15.0
7		10	500	10.6	7-8		13	880	14.5
		9	1,380	9.9			12	1,280	13.8
		8	2,850	9.2			11	1,880	13.1
		7	5,100	8.4			10	3,140	12.0
		6	8,010	7.8			9	5,010	11.1
	B	Max.		9.9			8	9,270	9.9
		9	220	9.4		C	Max.		16.0
		8	660	8.8			16	20	16.0
		7	1,240	8.2			15	110	15.6
		6	3,490	7.1			14	310	14.9
	AB	5	17,800	6.9			13	600	14.2
		4	25,600	6.1			12	1,020	13.5
	C	Max.		6.2			11	1,460	12.9
		6	360	6.1			10	1,980	12.3
		5	1,010	5.7			9	4,050	10.9
		4	2,100	5.1			8	5,210	10.3
	D	Max.		5.1		BC	7	18,900	9.4
		5	440	5.0		D	Max.		8.2
		4	1,090	4.7			8	240	8.1
	A to D	3	67,300	4.6			7	2,030	7.6
	E	Max.		4.8		BCD	6	34,300	8.2
		4	960	4.4		A to D	5	70,200	5.8
		3	1,920	3.9			4	88,000	5.6
	A to E	2	78,300	4.3		E	Max.		5.1
2 days	A	Max.		17.3			5	160	5.0
July		17	55	17.2			4	780	4.6
7-8		16	140	16.9		F	Min.		1.4
		15	380	16.0			2	500	1.7
		14	640	15.4			3	1,000	2.1
		13	1,300	14.4		A to F	3	120,000	5.0
		12	1,930	13.8			2	144,000	4.6
		11	2,760	13.1	3 days	A	Max.		19.7
		10	3,900	12.3	July		19	230	19.4
		9	5,110	11.7	7-9		18	420	19.0
		8	6,900	10.9			17	650	18.4
		7	9,440	9.9			16	990	17.8
		6	12,400	9.1			15	1,670	16.9
	B	Max.		16.9			14	2,160	16.3
		16	55	16.4			13	2,720	15.6

STORM 156, JULY 6–10, 1916 (Continued)

Period of Maximum Rainfall	Rainfall Center	Isohyetal-Inches	Area in Square Miles	Average Depth in Inches
3 days July 7–9	A	12	3,720	14.9
		11	5,690	13.7
		10	8,030	12.8
		9	10,200	12.1
		8	12,500	11.4
	B	Max.		17.6
		17	55	17.3
		16	260	16.7
		15	620	16.0
		14	1,000	15.4
		13	1,460	14.8
		12	2,070	14.1
		11	3,240	13.2
		10	5,940	11.9
		9	8,740	11.2
		8	16,800	9.9
	C	Max.		17.4
		17	75	17.2
		16	200	16.7
		15	500	16.0
		14	740	15.5
		13	1,040	14.9
		12	1,340	14.4
		11	1,700	13.8
		10	2,280	12.9
		9	3,890	11.5
		8	5,260	10.7
	D	Max.		12.5
		12	20	12.2
		11	50	11.8
		10	200	10.8
		9	380	10.2
		8	620	9.5
	A to D	7	56,000	9.4
		6	75,400	8.7
		5	93,800	8.0
		4	114,000	7.4
	E	Max.		5.9
		5	90	5.4
		4	600	4.6
	A to E	3	146,000	6.6
		2	182,000	5.8
4 days July 6–9	A	Max.		20.3
		20	80	20.2
		19	260	19.7
		18	450	19.2
		17	740	18.5
		16	1,100	17.9
		15	1,910	16.9
		14	2,680	16.7
		13	3,530	15.6
	B	Max.		20.0
		19	1,640	19.5

Period of Maximum Rainfall	Rainfall Center	Isohyetal-Inches	Area in Square Miles	Average Depth in Inches
4 days July 6–9	B	18	2,500	19.2
		17	3,410	18.7
		16	4,450	18.2
		15	5,440	17.7
		14	7,240	16.9
	C	Max.		18.6
		18	50	18.3
		17	210	17.7
		16	520	17.0
		15	1,080	16.2
		14	1,790	15.5
	BC	13	13,300	15.6
	D	Max.		13.2
		13	200	13.1
	E	Max.		13.5
		13	90	13.2
	A to E	12	25,200	14.6
		11	33,200	13.8
		10	41,600	13.2
		9	52,300	12.4
	F	Min.		3.8
		4	150	3.9
		5	440	4.3
		6	850	4.9
		7	1,420	6.2
		8	2,480	6.4
	A to F	8	69,300	11.4
	G	Max.		12.5
		12	35	12.2
		11	95	11.8
		10	200	11.1
		9	410	10.3
		8	660	9.6
	H	Max.		9.6
		9	500	9.3
		8	1,840	8.7
	A to H	7	89,000	10.6
		6	104,000	10.0
		5	116,000	9.5
		4	136,000	8.2
		3	162,000	8.0
		2	194,000	7.0
5 days July 6–10	A	Max.		22.3
		22	23	22.2
		21	120	21.7
		20	280	20.8
		19	530	20.3
		18	840	19.6
		17	1,200	19.0
		16	1,620	18.5
		15	2,320	17.5
		14	3,000	16.8
		13	4,090	15.9

STORM 156, JULY 6–10, 1916 (Continued)

Period of Maximum Rainfall	Rainfall Center	Isohyetal-Inches	Area in Square Miles	Average Depth in Inches	Period of Maximum Rainfall	Rainfall Center	Isohyetal-Inches	Area in Square Miles	Average Depth in Inches
5 days July 6–10	B	Max.		20.0	5 days July 6–10	F	11	580	12.8
		19	1,710	19.5			10	840	12.1
		18	2,580	19.2			9	1,240	11.2
		17	3,560	18.8		G	Max.		14.3
		16	4,870	18.1			14	40	14.2
		15	6,200	17.6			13	160	13.7
		14	8,180	16.8			12	340	13.0
	C	Max.		18.8			11	640	12.3
		18	100	18.4			10	1,060	11.6
		17	320	17.8			9	1,730	10.8
		16	680	17.1		H	Min.		4.4
		15	1,260	16.4			5	180	4.7
		14	2,190	15.6			6	400	5.1
	BC	13	14,400	15.7			7	900	5.9
	D	Max.		14.4			8	1,980	6.8
		14	700	14.2		A to H	8	86,700	11.4
		13	2,360	13.7			7	100,000	10.8
	A to D	12	29,300	14.7			6	116,000	10.2
		11	37,300	14.0		I	Max.		9.2
	E	Max.		13.8			9	30	9.1
		13	100	13.4			8	400	8.5
		12	450	12.7			7	740	8.0
		11	1,350	11.9		J	Max.		8.4
	A to E	10	50,000	13.1			8	160	8.2
		9	63,300	12.4			7	600	7.7
	F	Max.		15.9		IJ	6	2,240	7.3
		15	25	15.4		A to J	5	135,000	9.6
		14	90	14.8			4	154,000	9.0
		13	220	14.0			3	181,000	8.2
		12	390	13.4			2	230,000	6.8

STORM 157, JULY 14–16, 1916

Period of Maximum Rainfall	Rainfall Center	Isohyetal-Inches	Area in Square Miles	Average Depth in Inches	Period of Maximum Rainfall	Rainfall Center	Isohyetal-Inches	Area in Square Miles	Average Depth in Inches
1 day July 15	A	Max.		13.2	1 day July 15	B	9	1,530	9.7
		13	41	13.1			8	2,480	9.3
		12	260	12.6			7	3,520	8.7
		11	580	12.0			6	4,640	8.1
		10	860	11.5			5	5,970	7.6
		9	1,270	10.8			4	7,350	7.0
		8	1,790	10.2		ABCD	3	25,900	5.6
		7	2,380	9.5			2	36,000	4.7
		6	3,080	8.8	1 day July 16	A	Max.		19.3
	C	Max.		9.4			19	3	19.2
		9	14	9.2			18	14	18.7
		8	110	8.6			17	27	18.1
		7	290	7.9			16	41	17.6
		6	660	7.2			15	55	17.0
	D	Max.		6.6			14	96	16.0
		6	190	6.3			13	150	15.1
	ACD	5	6,560	7.3			12	190	14.6
		4	10,900	6.1			11	230	14.0
	B	Max.		10.6			10	310	13.1
		10	700	10.3			9	490	12.2

STORM 157, JULY 14–16, 1916 (Continued)

Period of Maximum Rainfall	Rainfall Center	Isohyetal-Inches	Area in Square Miles	Average Depth in Inches	Period of Maximum Rainfall	Rainfall Center	Isohyetal-Inches	Area in Square Miles	Average Depth in Inches
1 day July 16	A	8	740	10.9	2 days July 15–16	C	13	230	13.4
		7	1,070	9.9			12	420	13.0
		6	1,420	9.0			11	670	12.4
		5	1,850	8.2			10	960	11.9
	B	Max.		14.7			9	1,530	11.0
		14	14	14.3			8	2,040	10.4
		13	68	13.7			7	2,630	9.7
		12	150	13.0			6	3,380	9.0
		11	270	12.3		D	Max.		9.8
		10	440	11.6			9	27	9.4
		9	640	11.0			8	140	8.7
		8	960	10.1			7	420	7.9
		7	1,340	9.4			6	840	7.2
		6	1,760	8.7		E	Max.		7.4
		5	2,260	8.0			7	27	7.2
	AB	4	5,560	7.2			6	530	6.5
	C	Max.		6.4		CDE	5	7,350	7.4
		6	27	6.2		A to E	4	26,800	7.6
		5	230	5.6		F	Max.		7.4
		4	730	4.8			7	86	7.2
	ABC	3	9,000	5.9			6	340	6.7
		2	13,900	4.7			5	880	6.0
	D	Max.		5.8			4	1,750	5.2
		5	68	5.4		A to F	3	37,700	6.5
		4	410	4.7			2	46,200	5.8
		3	1,030	4.0	3 days July 14–16	A	Max.		23.7
		2	2,160	3.2			22	5	22.8
2 days July 15–16	A	Max.		23.2			21	41	21.5
		22	4	22.5			20	82	21.0
		21	27	21.6			19	120	20.5
		20	82	20.9			18	210	19.6
		19	120	20.4			17	440	18.5
		18	210	19.6			16	820	17.6
		17	410	18.5			15	1,150	17.0
		16	750	17.6			14	1,460	16.5
		15	1,070	17.0			13	1,840	15.8
		14	1,380	16.4			12	2,310	15.2
		13	1,750	15.8		B	Max.		16.4
		12	2,140	15.2			15	41	15.7
	B	Max.		16.1			14	120	14.9
		15	41	15.5			13	260	14.2
		14	120	14.8			12	440	13.5
		13	220	14.2		AB	11	3,930	13.9
		12	410	13.4			10	5,610	12.9
	AB	11	3,780	13.8			9	6,530	12.4
		10	5,450	12.9			8	7,720	11.8
		9	6,390	12.3			7	8,960	11.2
		8	7,410	11.8			6	10,400	10.6
		7	8,720	11.2			5	12,100	9.8
		6	10,200	10.5		C	Max.		16.8
		5	11,800	9.8			16	110	16.4
	C	Max.		13.8			15	360	15.8

STORM 157, JULY 14–16, 1916 (Continued)

Period of Maximum Rainfall	Rainfall Center	Isohyetal-Inches	Area in Square Miles	Average Depth in Inches	Period of Maximum Rainfall	Rainfall Center	Isohyetal-Inches	Area in Square Miles	Average Depth in Inches
3 days	C	14	750	15.1	3 days	C	5	16,400	8.1
July		13	1,290	14.4	July	ABC	4	35,600	8.0
14–16		12	1,790	13.9	14–16	D	Max.		7.6
		11	2,700	13.1			7	48	7.3
		10	3,740	12.4			6	340	6.6
		9	5,270	11.5			5	880	5.9
		8	6,710	10.9			4	1,710	5.2
		7	8,660	10.1		A to D	3	45,500	7.1
		6	11,800	9.2			2	58,500	6.1

STATE OF OHIO

THE MIAMI CONSERVANCY DISTRICT

Contract Forms and Specifications

BY
THE ENGINEERING STAFF OF THE DISTRICT

TECHNICAL REPORTS
Part VI

DAYTON, OHIO
1918

PREFATORY NOTE

This volume is the sixth of a series of Technical Reports issued in connection with the planning and execution of the notable system of flood protection works now being built in the Miami Valley.

The Miami Valley, which forms a part of the large interior plain of the central United States and comprises about 4,000 square miles of gently rolling topography in southwestern Ohio, is one of the leading industrial centers of the country. From the great flood of March, 1913, which destroyed in this valley alone over 360 lives and probably more than $100,000,000 worth of property, there resulted an energetic movement to prevent a recurrence of such a disaster. This movement developed gradually into a great cooperative enterprise for the protection of the entire valley by one comprehensive project. The Miami Conservancy District, established in June, 1915, under the newly enacted Conservancy Act of Ohio, became the agency for securing this protection. On account of the size and character of the undertaking, the plans of the district have been developed with more than usual care.

A Report of the Chief Engineer, submitting a plan for the protection of the district from flood damage, was printed March, 1916, in three volumes of about 200 pages each. Volume I contains a synopsis of the data on which the plan is based, a description of its development, and a statement of the plan in detail. Volume II contains a legal description of all lands affected by the plan. Volume III contains the contract forms, specifications, and estimates of quantities and cost.

After various slight modifications, the Report of the Chief Engineer was adopted by the board of directors as the Official Plan of the district, and was republished in May, 1916. This plan for flood protection contemplates the building of five earth dams across the valleys of the Miami River and its tributaries to form retarding basins, and the improvement of several miles of river channel within the half dozen largest cities of the valley. It is estimated that the dams will contain nearly 9,000,000 cubic yards of earth; that their outlet structures will contain nearly 200,000 cubic yards of concrete; that the river channel improvements will involve the excavation of nearly 5,000,000 cubic yards; and that the whole project will cost about $25,000,000.

5

In order to plan the project intelligently, many thorough investigations and researches had to be carried out, the results of which have proved of great value to the district and will also, it is believed, be of widespread use to the whole engineering profession. To make the results of these studies available to the residents of the State and to the technical world at large, it is planned to publish a series of Technical Reports containing all data of permanent value relating to the history, investigations, design, and construction of the flood prevention works.

The following list shows the reports prepared by the engineering staff of the district which have already been published, the number of pages and number of illustrations in each part, and the price, post free, for which each part may be purchased.

Part I.—The Miami Valley and the 1913 flood; 125 pages, 44 illustrations, 50 cents.

Part II.—History of the Miami flood control project, 196 pages, 41 illustrations, 50 cents.

Part III.—Theory of the hydraulic jump and backwater curves. Experimental investigation of the hydraulic jump as a means of dissipating energy; 111 pages, 88 illustrations, 50 cents.

Part IV.—Calculation of flow in open channels; 283 pages, 79 illustrations, 75 cents.

Part V.—Storm rainfall of eastern United States; 310 pages, 114 illustrations, 75 cents.

Part VI.—Contract forms and specifications; 192 pages, 3 folding plates, 50 cents.

Atlas of Selected Contract and Information Drawings to accompany Part VI; 139 plates, 11 by 15 inches, $1.50.

The following parts of the technical reports are in the course of preparation:

General design of Miami flood control project.

Rainfall and runoff in the Miami Valley.

Laws relating to flood prevention work.

Flood prevention works in other localities.

Earth dams.

Construction of protection system.

The object of issuing this particular Volume, Part VI of the Technical Reports, is to put into permanently accessible form the legal documents and specifications prepared in the fall of 1917 for letting the contracts for construction. These papers were prepared with unusual care, and contain some novel features which, it is believed, will be of general interest and use to the engineering profession.

However, the work is not being done by contract. Because of the unusual conditions resulting from the entrance of the United States into the war, and because of the excessive demand upon construction organizations, contractors generally did not bid on the work except on a cost plus basis and under conditions which would leave the burden of every risk upon the district. In view of this condition the district undertook to carry on the work with its own forces. The contract forms and specifications herein reproduced are being used in the case of certain minor parts of the work, on which it was advisable to let contracts. While doing the work by force account, the district is following these specifications.

ARTHUR E. MORGAN,
Chief Engineer.

Dayton, Ohio,
September, 1918.

CONTENTS

Page

Officers of The Miami Conservancy District 4
Prefatory note 5
Contents.. 8
List of Plates 12

INTRODUCTION
Information supplied to prospective bidders 13
Clarity of language 14
Form of bond 14
Special characteristics of specifications 15
Use of forms and specifications 17
Index to specifications 17
Acknowledgments 17

ADVERTISEMENTS—Exhibit A
Dams and appurtenances 18
Dayton and Hamilton 20
Notice to subcontractor 22

INFORMATION FOR BIDDERS
List of proposed contracts 24
General information 26
List of contract items 37
Estimates of quantities 39

PROPOSAL—GERMANTOWN DAM AND APPURTENANCES
Introductory 67
Schedule of prices 68
Conditions 70
List of contract drawings 70
Experience and references 71
Plant and equipment 71

AGREEMENT—Exhibit B
Blank form 73
Schedule of unit prices and estimated quantities 75
List of contract drawings 79

BOND—Exhibit C
Blank form 81

SPECIFICATIONS
Explanatory note 84

Section   **General Conditions**
0.1   Definitions 86
0.2   Sureties 87
0.3   Time and order of completion 88
0.4   Extension of time 89
0.5   District to furnish right-of-way 90

8

Section		Page
0.6	Inspection and right of access	90
0.7	To remedy defective work	91
0.8	Retaining imperfect work	91
0.9	Engineer cannot waive obligations	92
0.10	To direct work	92
0.11	To provide for emergencies	93
0.12	To modify methods and equipment	94
0.13	To furnish lines and grades	94
0.14	To determine quantities and measurements	95
0.15	To define terms and explain plans	95
0.16	Completeness of specifications, estimates, and drawings	96
0.17	Items of work	97
0.18	Changes and alterations	97
0.19	Extra work	97
0.20	Progress estimates	99
0.21	Final payment	99
0.22	Payment by check	100
0.23	Delayed payments	100
0.24	Payment only in accordance with Contract	101
0.25	Money retained for defects and damages	101
0.26	Claims for damages	101
0.27	Remedies cumualtive	101
0.28	Acceptance shall not constitute waiver	101
0.29	Collateral works	102
0.30	Persons interested in Contract	102
0.31	Personal attention of Contractor	102
0.32	Contractor's address	102
0.33	Agents, superintendents, and foremen	103
0.34	Compliance with laws	103
0.35	Character of employees	103
0.36	To maintain communications	103
0.37	To guard against accidents	104
0.38	Contractor responsible for claims	104
0.39	Hindrances and delays	105
0.40	Night work	105
0.41	Delivery of materials	105
0.42	Infringements of patents	106
0.43	Protection against claims for labor and material	106
0.44	Assignment	106
0.45	Removal of equipment	107
0.46	Suspension of work if Contract is violated	107
0.47	Board for Engineer and assistants	108
0.48	Police and sanitary regulations	109
0.49	Intoxicants	109
0.50	Camps	109
0.51	Sanitary conveniences	109
0.52	Garbage and camp refuse	110
0.53	Doctors, hospitals	110
0.54	Medical supervision	110
0.55	Medical report	111
0.56	Examinations	111
0.57	No direct compensation	111

Section	General Specifications	Page
0.58	Water supply	112
0.59	Pumps and drains	112
0.60	Borrow pits	112
0.61	Protection of trees	113
0.62	Materials obtained from construction	113
0.63	Assistance in making tests	114
0.64	Protection of existing structures	114
0.65	Excavation—classification	114
0.66	Excavation—work to be done	115
0.67	Excavation—limits of	115
0.68	Excavation—cleaning surface for inspection	116
0.69	Excavation—use of explosives	116
0.70	Excavation—finishing work	116
0.71	Excavation—payment	116
0.72	Embankment—work to be done	117
0.73	Embankment—Class A, dams constructed by hydraulic monitor and dredge pump method	118
0.74	Embankment—Class B, dams constructed by semi-hydraulic method	119
0.75	Embankment—Class C, dams constructed by sprinkling and rolling method	120
0.76	Embankment—Class D, levees constructed with material from river channel excavation	121
0.77	Embankment—Class E, levees constructed with material from borrow pits	122
0.78	Embankment—Class F, miscellaneous	122
0.79	Embankment—preparation of foundation	122
0.80	Embankment—materials to be used	122
0.81	Embankment—finishing work	123
0.82	Embankment—payment	123
0.83	Backfilling—description	124
0.84	Backfilling—material to be used	124
0.85	Concrete—description	125
0.86	Concrete—cement	125
0.87	Concrete—fine aggregate	126
0.88	Concrete—coarse aggregate	126
0.89	Concrete—water	126
0.90	Concrete—large stones	127
0.91	Concrete—samples	127
0.92	Concrete—preparation of foundations	127
0.93	Concrete—mixing	127
0.94	Concrete—forms	128
0.95	Concrete—placing	128
0.96	Concrete—construction joints	129
0.97	Concrete—contraction joints	129
0.98	Concrete—finishing surfaces	129
0.99	Concrete—cold weather precautions	130
0.100	Concrete—protection of	130
0.101	Concrete—replacing faulty work	130
0.102	Concrete—placing metal work	131

Section   Page
0.103 Concrete—measurement for payment 131
0.104 Concrete—prices to include 131
0.105 Mortar 132

Detail Specifications

To save space the detail specifications for some items, as indicated by the asterisks below, are not printed in this volume. The omitted items include: First, all the highway specifications, which are based upon the standard specifications of the Ohio State Highway Department, and which are, therefore, readily accessible elsewhere; and second, numerous items of only minor importance in these contracts, for which the specifications are, therefore, relatively brief.

Item   Page
1   Stream control at dams 133
2   Clearing and grubbing 135
3   Removal of buildings ....*
4   Trimming and shaping slopes of river channels 136
5   Excavation, soil stripping 137
6, 7, 8, 6a, 7a, and 8a   Excavation for new channels in channel improvements, for cutoff trenches and for stream channels........138
9, 10, and 11   Excavation for river channel improvements 140
12 Excavation—overhaul for river channel improvements 141
13, 14, 15, 13a, 14a, and 15a   Excavation for outlet works and spillways at dams 142
16, 17, and 18   Excavation for minor drainage systems 143
19, 20, and 21   Excavation for street and highway cuts and fills.......144
22, 23, 24, 22a, 23a, 24a   Excavation for retaining walls, bridge piers and flood gate structures, except at dams 147
25, 26, 27, 25a, 26a, and 27a   Excavation and backfilling of trenches, except at dams 148
28, 29, and 30   Excavation, miscellaneous 150
31, 32, and 33   Embankment for main dams 151
34 and 35   Embankment for levees 153
36 Embankment, miscellaneous 154
37 Backfilling 155
38 Surface dressing and grassing 156
41 and 41a   Concrete in outlet works, spillways, retaining walls, bridge piers, and similar structures ...-...........................157
42 Concrete in conduit linings at dams 158
43 and 43a   Concrete in aprons, paving at dams and similar flat shapes.159
44 and 44a   Concrete in monolithic slope revetment 160
45 Concrete in flexible slab revetment or channel lining 161
46 and 46a   Concrete in culverts, bulkheads, gate chambers and similar structures, except at dams...*
47 Concrete in manholes, catch basins, sewers, and similar shapes ....*
48 Stone masonry 162
50 Cement 164
51 Reinforcing steel 166
52 Vitrified brick lining ....*
53 Drilling foundations or masonry 167
        * Not included herein.

Item		Page
54	Grouting foundations or masonry	167
55	Furnishing vitrified round or channel pipe	*
56	Laying vitrified round or channel pipe	*
57	Dry rubble paving	169
58	Riprap	170
59	Gutter paving	171
60	Crushed rock or gravel	*
61, 62, 63, 64x, 65, and 66.	Paving for streets and highways	*
67	Guard rails	172
68	Wire fences	173
70	Protective boom and accessories	175
71	Timber and lumber, miscellaneous	176
72 and 72x	Timber piles	177
73	Wood sheet piling	179
74	Steel sheet piling	180
75	Furnishing cast iron pipe	*
76	Laying cast iron pipe	*
77	Flood gates	
78	Steel highway bridges	
81	Raising, removing, and re-erecting existing steel bridges	
82	Miscellaneous cast iron, wrought iron, and steel	
83	Bronze	*
84	Wire rope	*
85	Cleaning up	182
86	Pumping stations	*
87	Pumping equipment	*
88	Removal and re-erection of existing pumping and power stations	*
89	Sheeting left in place	183
90	Extra work	183
Index to specifications		184

* Not included herein.

PLATES ACCOMPANYING INFORMATION FOR BIDDERS

(Inserted in back of volume.)

General map of the Miami Valley.
Diagram showing rainfall at Dayton, Ohio, 1893 to 1916.
Diagram showing river stages at Dayton, Ohio, 1893 to 1916.

CONTRACT FORMS AND SPECIFICATIONS

INTRODUCTION

This volume shows the published announcements of proposed lettings, the advance information printed for dissemination among prospective bidders, and the contract forms, including general specifications and the more important detail specifications, all as prepared in advance of and in preparation for the letting of contracts for the construction of flood protection works in the Miami Valley. There has been prepared to accompany this volume an Atlas of Selected Contract and Information Drawings, including 139 printed drawings on pages eleven by fifteen inches in size. The drawings have been selected to show: First, the extent and nature of the preparation of the plans preparatory to the letting of contracts; second, the general methods followed for the solution of the innumerable problems in design of details, arising during the preparation of the plans.

Great care was taken in the preparation of this material to make its arrangement and wording as nearly perfect as possible. Most of the matter was electrotyped so that it might be in convenient form for use in printing a large number of different contracts, and it is here reprinted for distribution among engineers in the belief that the profession in general and engineers engaged upon similar projects in particular will find it useful in their practice. Before taking up the details of these forms it seems desirable to describe briefly some of the ends sought for in their preparation, and to point out certain valuable characteristics which it is believed distinguish this material from some of that in use elsewhere.

INFORMATION SUPPLIED TO PROSPECTIVE BIDDERS

It is an axiom in the letting of contracts that the more the uncertainties in the execution of the work can be eliminated, and the plainer the terms and conditions under which the work is to be executed can be set forth, the lower will be the bids received and the less will be the chance that controversies will subsequently arise which may hamper or delay the completion of the work.

Effort was made on this project, therefore, to determine in advance those physical and commercial conditions of importance to a contractor under which the work is to be done. This information is set forth under the heading, Information for Bidders, and it is believed that engineers will here find a suggestive compendium of the information needed by a contractor. This information includes general geographic information, weather and runoff data, boring and test pit records, maps and descriptions of the work, with estimates of quantities, standard prices of some indispensable supplies, together with available kinds and sources of various materials.

To give the fullest possible opportunity to local bidders, who might not in some cases be possessed of great resources, the proposed work was divided into as large a number of separate contracts as seemed compatible with rapid progress and efficient execution. This policy resulted in the long list of separate contracts as shown on page 24. At the same time, in order not to discriminate unduly against the larger contractor, provision was made, as shown in the form of proposal, page 70, to receive bids upon different contracts in combination. By this means a contractor possessed of large means and equipment could bid upon the work in large units without incurring the danger of being awarded only small or isolated parts of the work.

CLARITY OF LANGUAGE

The contract forms and the specifications are vital to the legal as well as the engineering integrity of a contract. While preserving all the legal requirements, a thorough effort was made in all the contract forms to reduce the language to the most perfect simplicity and clarity. To this end every phrase was scrutinized, all superfluous material was deleted, and in many cases ancient legal verbiage was replaced by the more usual language of every-day life. The careful reader will note many changes from hoary legal forms that are in more or less general use. No harm can come from the changes, it is believed, while there is a positive gain from the greater directness and intelligibility secured by the substitutions. This procedure was followed not only in the wording of the agreement and bond, but throughout the entire preparation of the specifications, and in the case of the latter leads to one of their most useful characteristics.

FORM OF BOND

The primary object of a bond is to furnish a guarantee for the satisfactory completion of a contract at the prices and within the time stipulated therein. Anything short of that is a makeshift. Any conditions or reservations introduced in the bond tending to limit that security decrease its value as a guarantee of the completion of the work.

The form of bond adopted is designed to provide the fullest possible protection to the district without imposing upon the sureties any unreasonable obligations.   It requires the surety to indemnify the district against the contractor's

(1) non-payment of bills for labor and materials;

(2) non-satisfaction of claims for damage to life or property;

(3) violation of laws, ordinances, etc.;

(4) infringement of patents;

(5) non-satisfaction of claims originating from any operations under the contract;

(6) failure to faithfully perform the contract in all particulars, and within the time specified.

It specifically provides that the district may order alterations in the work, or materials to be furnished, and may grant extensions of time, without first obtaining the consent of the sureties, except that such changes must not alter the general character of the work as a whole, nor cause the total amount to be paid the contractor to be increased by more than 25 per cent.

It reserves to the surety the right to take over and complete the contract in case of failure of the contractor.

There are no requirements in this form of bond that may not be found in existing standard forms in use by leading bonding companies. It is believed, however, that few other forms contain all of its provisions, while on the other hand there have been excluded from it certain objectionable clauses, sometimes met with in standard forms, which under certain conditions may detract from the usefulness of the bond from the standpoint of the owner.

The district pays for the bond—indirectly it is true, but fully, nevertheless—and it pays for full and adequate protection.   The surety should assume the full risk for which it is to be paid.   It is important, therefore, that the bond shall make the desired protection possible, and not fail of its purpose through the insertion of unnecessary conditions tending to lessen the responsibility of the surety.

SPECIAL CHARACTERISTICS OF SPECIFICATIONS

As explained under the heading Specifications, on page 29, and in the explanatory note, on page 84, just preceding the printed specifications, the latter are divided into three portions, General Conditions, General Specifications, and Detailed Specifications. The sections of each portion are numbered consecutively in such a manner as to form a self-indexing system. The first two portions are printed in full in this volume, but to economize space, of the detailed specifications, only those especially characteristic of this project are here included. The highway specifications are omitted because they are based upon the standard specifications of the Ohio State Highway Department, and hence are readily accessible elsewhere. Other items omitted are of only minor importance in this project and hence are relatively brief. The table of contents lists all the sections of the specifications, at the same time indicating those sections not included in this volume.

Throughout the specifications, care is taken to bind the district to as careful observance of terms as the contractor. The contractor is safeguarded against arbitrary action on the part of the engineer. Wherever possible, the customary phrases "approved," "satisfactory," "as directed," "to the engineer's satisfaction," are eliminated, being replaced by definite description of the results to be attained or the methods to be used. But in emergencies where the success of the work or the welfare of the public may be at stake, the engineer is given full authority to act, with power to enforce his directions. In various sections the language is so drawn as to prescribe and insure the engineer's fairness.

The contractor's risks are reduced as much as possible not only by the large amount of advance information furnished to bidders, and by furnishing a flood and weather warning service during construction, but also by the district's assuming certain risks during construction which may be beyond the control of the contractor. Thus, items 6.2 and 9.2 provide that if objectionable materials wash into certain excavations through no fault of the contractor, he shall be paid for their removal. Similarly, the contractor is not held responsible for damages caused by floods during construction to lands lying above the permanent dams. The attempt is made throughout the specifications to prescribe only that reasonable normal quality of work which would naturally be expected in work of standard quality for the different purposes required. Thus the contractor is relieved of the danger of petty oppression upon the part of inspectors or the engineer.

Rate of progress upon the main contracts of this project is a vital element of the design, and is covered in section 0.3 in connection with the contract drawings.

Engineers will find of particular interest on account of their novel features sections: 0.8, Retaining imperfect work; 0.19, Extra work; 0.21 to 0.23, Final payment; 0.26, Filing claims by contractor for damages; 0.65, Classification of excavation; and 0.85 to 0.105, Relating to concrete.

USE OF FORMS AND SPECIFICATIONS

For each of the important contracts there was printed and bound together: the appropriate Advertisement—Exhibit A; the Agreement—Exhibit B; the Bond—Exhibit C; and from the Specifications—Exhibit D, the General Conditions, the General Specifications, and such of the Detail Specifications as related to that particular contract. To facilitate such separated use each item of the Detail Specifications begins a new page just as they are printed in this volume. The contracts for all the dams as printed were identical, containing blanks that were appropriately filled in afterwards for each individual dam. The same was true for all the contracts for local flood protection.

INDEX TO SPECIFICATIONS

This volume is supplied with a carefully compiled and unusually complete index to the specifications. This will be found invaluable in any extensive use of the specifications. Each of the printed contracts described above is supplied with a similar index, which experience has shown to be exceedingly useful to the engineers in charge during construction.

ACKNOWLEDGMENTS

The specifications as finally adopted represent largely the work of Arthur E. Morgan, Chief Engineer, and Charles H. Paul, Assistant Chief Engineer; H. S. R. McCurdy also had an active part in their preparation; but the unusual variety of previous experience represented by the large number of engineers engaged upon this project enabled special first-hand knowledge to be utilized on all sections of the specifications. Among those contributing in an important way to the final result are Daniel W. Mead, consulting engineer; J. H. Kimball, O. N. Floyd, G. H. Matthes, Walter M. Smith and A. B. Mayhew. The specifications and forms were reviewed from a legal viewpoint by the attorneys for the district, O. B. Brown, and John A. McMahon, of Dayton, Ohio, and E. J. B. Schubring, of Madison, Wis.

ADVERTISEMENTS

FLOOD CONTROL WORKS, DAMS AND APPURTENANCES

OFFICE OF THE BOARD OF DIRECTORS
THE MIAMI CONSERVANCY DISTRICT

Dayton, Ohio, September 15, 1917.

Sealed proposals will be received at the office of the Secretary, Board of Directors, The Miami Conservancy District, Dayton, Ohio, until 2 o'clock P. M. November 15, 1917, for the construction of dams and appurtenances as follows:

Contract No. 1.—Germantown Dam and appurtenances, including Road No. 1, involving approximately the following principal quantities: Excavation, 200,000 cu. yds.; embankment, 850,000 cu. yds.; concrete, 20,000 cu. yds.; paving and riprap, 1,000 cu. yds.; iron and steel, 120 tons.

Contract No. 2.—Englewood Dam and appurtenances, including Road No. 3, Road No. 4, and Road No. 5, involving approximately the following principal quantities: Excavation, 375,000 cu. yds.; embankment, 3,500,000 cu. yds., ; concrete, 38,000 cu. yds ; paving and riprap, 2,000 cu. yds.; iron and steel, 180 tons.

Contract No. 3.—Lockington Dam and Appurtenances, including Road No. 8 and Road No. 9, involving approximately the following principal quantities: Excavation, 200,000 cu. yds.; embankment, 1,000,000 cu. yds.; concrete, 37,000 cu. yds.; paving and riprap, 1,000 cu. yds.; iron and steel, 50 tons.

Contract No. 4.—Taylorsville Dam and appurtenances, including Road No. 12 and Road No. 13, involving approximately the following principal quantities: Excavation, 750,000 cu. yds.; embankment, 1,100,000 cu. yds.; concrete, 57,000 cu. yds.; paving and riprap, 5,000 cu. yds.; iron and steel, 400 tons.

Contract No. 5.—Huffman Dam and appurtenances, including Road No. 16 to Station 60+50, and Road No. 17, involving approximately the following principal quantities: Excavation, 300,000 cu. yds.; embankment, 1,400,000 cu. yds.; concrete, 45,000 cu. yds.; paving and riprap, 1,500 cu. yds.; iron and steel, 400 tons.

All of which works are a part of the system of flood control of The Miami Conservancy District, to be carried out under authority of the Conservancy Act of Ohio, and are in accordance with the Official Plan of the District.

Proposals must be on the blank forms furnished by the Board, and must be accompanied by a certified check for not less than 5 per cent of the aggregate amount of the bid, figured on the basis of the estimated quantities and the unit prices bid, but which in no case need exceed $50,000; such check to be drawn to the order of the Treasurer of The Miami Conservancy District, as a guarantee that the bidder, if awarded a contract, will, within 10 days after the contract is delivered to him for that purpose, execute the same, and furnish surety bond for the faithful performance of the contract, in the sum of 40 per cent of the contract price; said contract and bond to be on the standard forms which have been adopted by the Board.

If any bidder, to whom an award has been made, shall fail to execute the contract or to furnish satisfactory bond within the time hereinbefore specified, or as extended by the Board, the award shall thereupon become void, in which case the proceeds of the certified check shall become the property of the District, and the contract may be awarded to the next lowest or best bidder; and such next lowest or best bidder shall thereupon assume the contract, as if he were the party to whom the award was first made.

Each bidder must, in his proposal, present satisfactory evidence that he has been engaged in constructing works of the general character covered by his proposal, and that he is fully prepared, and has the necessary capital, to begin the work promptly, and to conduct it as required by the contract and specifications. Proposals not containing such evidence will not be recognized as bids.

The right is reserved to reject any or all bids, and to waive any technical defects, as the interests of the District may require.

Drawings, specifications, proposal blanks, and other information may be obtained on application to the Chief Engineer, The Miami Conservancy District, Dayton, Ohio, at whose office drawings, boring samples, and other data may be inspected.

EZRA M. KUHNS,
Secretary.

THE MIAMI CONSERVANCY DISTRICT
DAYTON, OHIO

NOTICE TO SUBCONTRACTORS

The larger contracts to be let contain numerous smaller pieces of work for which the main contractor may not be equipped, and which he may desire to sublet. The Chief Engineer of the Conservancy District would be pleased to hear from any persons or firms who would be in position to take any of these subcontracts, and will put them in touch with those who will bid on the larger contracts as a whole. On the reverse side of this sheet is a blank form which may be filled out by those interested. Some of the kinds of work which may be included in these subcontracts are:

Clearing and grubbing trees from base of dam.
Soil stripping from base of dams, levees, and borrow pits.
Surface dressing and grassing of dams.
Seeding or sodding of levees.
Trimming slopes of river channels.
Removal of buildings.
Excavation for minor drainage systems.
Construction of levees.
Excavation and fills for highway construction.
Rolled gravel paving for highways.
Water bound macadam paving for highways.
Street paving; bituminous, macadam, brick, concrete, etc.
Guard rails for highways.
Construction of wire fences.
Laying vitrified pipe.
Gutter paving.
Stone masonry.
Stone paving.
Riprap.
Small concrete structures.
Road culverts.
Concrete slope revetment for river banks.
Concrete blocks for slope protection.
Setting flood gates in canals and sewers.

22

INFORMATION CONCERNING SUBCONTRACTORS

To be filled out by persons or firms interested in subcontracts under the main contractors, and sent to the Chief Engineer of The Miami Conservancy District, Dayton, Ohio.

Name ....

Address ....

Date ....

How long in contracting business............................

Kind of work handled..

....

....

Where was work located......................................

....

....

What size jobs have you handled..............................

....

.. ..

....

What equipment have you.....................................

....

....

....

....

In what kind of contracts are you particularly interested........

. ...

...., ...

....

In what sized contracts are you particulary interested ....

....

....

Remarks: ....

INFORMATION FOR BIDDERS

(Published in September, 1917.)

LIST OF PROPOSED CONTRACTS

Contract Number	Designation of Contract
1.	Germantown Dam and Appurtenances, including Road No. 1.
2	Englewood Dam and Appurtenances, including Roads Nos. 3, 4, and 5.
3	Lockington Dam and Appurtenances, including Roads Nos. 8 and 9.
4	Taylorsville Dam and Appurtenances, including Roads Nos. 12 and 13.
5	Huffman Dam and Appurtenances, including Roads Nos. 16 (part) and No. 17.
6	Road No. 2 (Germantown Basin.)
7	Road No. 6 (Englewood Basin.)
8	Road No. 7 (Englewood Basin.)
9	Road No. 10 (Lockington Basin.)
10	Road No. 11 (Lockington Basin.)
11	Road No. 14 (Taylorsville Basin.)
12	Road No. 15 (Taylorsville Basin.)
13	Road No. 16 (part not included in Contract No. 5) (Huffman Basin.)
14	Unassigned.
15	Unassigned.
16	Steel Highway Bridge at Germantown Spillway.
17	Steel Highway Bridge at Englewood Spillway.
18	Steel Highway Bridge at Lockington Spillway.
19	Steel Highway Bridge at Mill Creek.
20	Steel Highway Bridge at Taylorsville Spillway.
21	Steel Highway Bridge at Huffman Spillway.
22	Unassigned.
23	Bridge Piers and Abutments (For bridges to be moved.)
24	Raising Bridge over Brush Creek.
25	Moving Bridge over Stillwater River.
26	Unassigned.
27	Tippecanoe City Local Protection—General.
28	Tippecanoe City Local Protection—Drainage Works.
29	Tippecanoe City Local Protection — Sewage Pumping Station.

Contract
Number

30 Tippecanoe City Local Protection—Equipment for Sewage Pumping Station.

31 Tippecanoe City Local Protection—Substructure for Water and Electric Light Station.

32 Tippecanoe City Local Protection—Superstructure for Water and Electric Light Station.

33 Tippecanoe City Local Protection—Moving equipment for Water and Electric Light Station.

34 Fairfield Local Protection.

35 Piqua Local Protection—Above Ash Street.

36 Piqua Local Protection—Below Ash Street.

37 Troy Local Protection—Miami River.

38 Troy Local Protection—Pearson Levee.

39 Troy Local Protection—East above Adams Street.

40 Troy Local Protection—West above Adams Street.

41 Dayton Local Protection—Above Island Park.

42 Dayton Local Protection—Island Park to Washington Street.

43 Dayton Local Protection—Washington Street to Stewart Street.

44 Dayton Local Protection—Stewart Street to Broadway.

45 Dayton Local Protection—Mad River.

46 Dayton Local Protection—Wolf Creek.

47 Unassigned.

48 Unassigned.

49 West Carrollton Local Protection.

50 Miamisburg Local Protection—East Side.

51 Miamisburg Local Protection—West Side.

52 Miamisburg Local Protection—Drainage Works.

53 Franklin Local Protection—East Side.

54 Franklin Local Protection—West Side.

55 Middletown Local Protection—General.

56 Middletown Local Protection — Clearing above Poast Town Bridge.

57 Unassigned.

58 Hamilton Local Protection—Miami River.

59 Hamilton Local Protection—Four Mile Creek.

60 Hamilton Local Protection—Drainage Works.

61 to 99—Unassigned.

101, 102, etc.—Removal of Buildings.

GENERAL INFORMATION

The Board of Directors of The Miami Conservancy District invites proposals for the construction of Flood Control Works, in accordance with the accompanying Advertisements, which indicate the general character of the work; the time and place of opening bids; the requirements as to certified check, and bond; and the necessity of experience and ability on the part of the bidder.

Drawings and specifications may be examined at the office of the Chief Engineer, Conservancy Building, Dayton, Ohio, at any time during office hours. A set of Drawings, Specifications, and duplicate Proposal Blanks, for any contract, will be furnished to any prospective bidder upon receipt of a deposit of $5.00, which deposit will be returned upon return of the Drawings and Specifications in good condition within five days after proposals are opened.

Proposals

Proposals must be submitted on the form provided for that purpose. They must not be changed in form, and no alterations or interlineations shall be made therein. Should the bidder decide to explain or to qualify his bid he should do so in a supplemental statement attached to the proposal; but any proposal may be rejected which contains explanations or qualifications which change or modify the character or conditions of the proposal as printed, or which make it not comparable with other proposals.

Proposals will be received from the same bidder on more than one contract, but a proposal on only a part of a contract will not be considered. In case more than one contract is awarded to the same bidder, each such contract will be executed separately and, unless other arrangements are made between the parties, will be handled as a separate contract throughout the progress of the work.

In executing a proposal it is important that every blank space be filled. Proposals must be signed and enclosed in a sealed envelope, which shall be endorsed with the name and address of the bidder and the title of the contract, and addressed to the Miami Conservancy District, Dayton, Ohio. If the proposal is made by an individual it shall be signed with his usual business signature and his address shall be given; if it is made by a firm it shall be signed with the copartnership name by a member of the firm, and the name and address of each member shall be given; if it is made by a corporation it shall be signed by a duly authorized officer, with the corporate name attested by the corporate seal, and the business address of the corporation shall be given. No telegraphic proposal will be considered.

Proposals will be opened and read publicly at the time given in the Advertisement, and bidders are invited to be present.

The withdrawal of a proposal at any time before the time set for opening will be permitted.

Certified Check

A certified check, of the amount specified in the Advertisement, shall be enclosed with the proposal. Certified checks will be returned to bidders to whom award is not made, within three days after the execution of the contract or the rejection of all bids. Interest at the rate of six per cent per annum will be paid for each day that such checks are held after the 30th day from date of opening of the proposals. Certified checks of a bidder to whom award is made will be returned within three days after the execution of the contract with him, provided award does not become void for the reasons given in the Advertisement.

Surety Bond

One or more surety bonds, the aggregate amount of which shall be 40 per cent of the contract price, must be furnished at the time each contract is executed.

Local Conditions

It is the desire of the District to give the bidders all assistance possible in familiarizing themselves with the work to be done, and with local conditions that might affect its execution. Rainfall, run-off, and hydrographic data have been assembled and are made a part hereof. Boring and test pit records are shown on the Drawings, and samples obtained from the borings and test pits are available for inspection.

This data has been assembled with care, but its accuracy is not guaranteed. It may or may not be accurate as to any particular details, and is given only as the best information the District has available. Bidders shall satisfy themselves as to local conditions affecting the work, and no information concerning local conditions derived from the data mentioned above, from specifications or drawings, or from the engineer or his assistants, will relieve a contractor from any risk or from fulfilling any or all of the terms of his contract. Each bidder or his representative should visit the site of the work and familiarize himself with local conditions; failure to do so may be considered sufficient cause for the rejection of a proposal.

Ohio Laws

Bidders should become familiar with Ohio Laws regarding hours of labor, handling of contagious disease in camps, storage and use of explosives, claims for payment, damages, etc., and other laws affecting construction work as well as local ordinances and regulations.

General Map

A general map of the district is attached hereto, showing the relative locations of the various principal divisions of the work, and of the principal roads, railroads, trolley lines, cities and towns, as well as other information of general interest. The engineer, when requested to do so, will endeavor to furnish guides, familiar with the work, to show prospective bidders over the work at the expense of the bidder.

Drawings

Drawings appurtenant to each contract are classified as Information Drawings, Contract Drawings, and Supplemental Drawings.

Information Drawings show information of value to a bidder in making up his bid, such as a general map, boring records, etc., or to a contractor in laying out his camp sites, borrow pits, etc., such as a topographic map of the territory adjacent to the work. The data shown on these drawings has been obtained by the engineer for various uses outside of the actual requirements of the contract. It may or may not be accurate as to any particular details, and is given only as the best information the District has available. Its correctness is not guaranteed and it is not to be considered as a part of the contract.

Contract Drawings show the general plan of the work and such details as are necessary to enable the bidder to make an intelligent proposal. They show locations, types, and details of structures, and other general conditions that would influence the amount of a bid, but they do not necessarily show all minor details, nor give all dimensions that would be needed for a working drawing. For instance, in showing the cross section of a spillway weir the principal dimensions are given on the Contract Drawings, but not necessarily all the radii of curvature for the crest or the ogee face. Again, the general arrangement of steel reinforcement is shown on the Contract Drawings, but minor dimensions, cutting and bending details, etc., may be left to be shown on Supplemental Drawings. It is the intention to show on the Contract Drawings enough information to define clearly the character and scope of the work, without encumbering them with so many details as to endanger their legibility or clearness.

Supplemental Drawings will be issued from time to time during the progress of the work, giving additional dimensions or details needed or desired in carrying out the plans. They will supplement the Contract Drawings with such detailed information as is necessary for the final execution of the work.

Specifications

The specifications fall into three parts; first, the General Conditions, Sections 0.1 to 0.57, governing the relations of the contractor to the District, to the public, and to his employees; second, the General Specifications, Sections 0.58 to 0.105, defining and classifying the most commonly used materials, and indicating in general how the work shall be done; and third, the Detail Specifications, Sections 1.1 to 90.2. The Detail Specifications by supplementing the Agreement, General Conditions, General Specifications, the Estimates, and the Drawings, define in detail how each kind of work shall be done, and how it will be measured for payment.

The different kinds of construction work to be done and of materials to be furnished are classified into items, an item meaning a certain kind of work to be done, or material to be furnished, in a certain definite manner, regardless of its location. Each item is designated by a name and number, the numbers running from 1 to 90; as Item 2, Clearing and Grubbing; or Item 51, Reinforcing Steel. All sections relating to any item bear the number of the item as an index number before the decimal point. For instance, Item 50 covers Cement. Section 50.1 gives the requirements as to brands, Section 50.2 specifies the manner of delivery and storage, Section 50.3 covers inspection, tests, etc., and Section 50.4 specifies the methods of determining quantities of cement for payment.

All item numbers with letter "a" affixed refer to work below mean low water level as indicated on the drawings. For instance, Items 41 and 41a include concrete in outlet works, spillways, retaining walls, bridge piers, and other similar structures, above and below mean low water, respectively. The same item may be met on various parts of the project. Wherever in the estimates or on the drawings "Item 41" is indicated, regardless of whether it occurs in channel improvement at Hamilton, or in the construction of the Englewood Dam, or in some other construction, the work to be done is the placing of concrete in accordance with the Specifications under Item 41, supplemented in all cases by the General Specifications, which apply to all items. Item 41a refers to the same kind of construction as Item 41, except that the work is located below mean low water, as shown on the drawings.

In brief, the General Conditions and General Specifications relate, in so far as they are applicable, to all work. The estimates and the drawings indicate under what item or items any particular part of the work is to be done, and the Specifications for that item give the necessary further directions for doing that particular work.

Estimates of Quantities

The estimates are classified by items, but are also grouped in the divisions into which the work naturally falls. Thus, the estimates for each dam are grouped separately, the work in each case being classified according to items. In the same manner, the estimates for each local flood control improvement are listed separately. Within these natural groups, the individual features are also grouped together in the estimates. Thus, in the estimates for local flood control work at Hamilton, the required amount of work and material is listed by items separately for flood gates, levees, sewers, or other constructions.

While the specifications for doing the work are uniform for any item, wherever it may occur, the unit price will not necessarily be the same. For instance, Clearing and Grubbing under Item 2 might cost more at the Germantown damsite than at the Huffman damsite, though the specifications are identical, because of the heavier timber at one place than at the other. The contractor will bid by reference to items, but the entire work will be divided into contracts of such size and nature that the same unit price can reasonably apply to all work under any one item in a single contract. For instance, if the construction of the Germantown Dam were to be let as one contract, the contractor would bid only one unit price for Clearing and Grubbing the damsite, in his bid on that particular contract.

Progress

Special attention of bidders is directed to Article 2 of the Agreement, and to paragraphs 2 and 3 of Section 0.3 of the Specifications entitled Time and Order of Completion. A complete and well designed construction plant, an effective organization, and maintenance of the required rate of progress will be insisted upon.

Embankment

Three different methods of building embankments for main dams are described in the Specifications. The proposal provides for alternate bids, depending on the bidder's choice of methods of construction. Except for a minimum amount of Class C embankment which will be required at some dams, the contractor may, if he so desires, substitute one class of embankment for another, provided that the written consent of the engineer be secured, and provided further that such substitution shall involve no additional cost to the District. But if a substitution is made by order of the engineer, requiring a class of embankment of higher unit price, payment will then be made at the unit price bid for the class of embankment required; except that in case the contractor, by failing to carry out the terms of his contract as to rate of progress or otherwise, shall put the safety of the work, or the safety of life or property, in jeopardy; and if for the sake of safety it becomes necessary under such conditions, in the opinion of the engineer, to substitute for one class of embankment another class on which the unit contract price is higher; in such cases the engineer may order such substitution and the contractor shall be paid the unit price of the class contemplated in the contract, and not the unit price of the class made necessary by his failure to carry out the terms of his contract.

Right-of-Way

Right-of-way necessary for camp sites at dams, for borrow pits, and for spoil banks, will be furnished by the District as provided in Section 0.5 of the Specifications.

Progress Payments

Progress payments, as provided in Section 0.20 of the Specifications will be made approximately once a month during the active progress of the work.

Experience and Fitness

It is essential that bidders present full evidence that they have had the experience and are prepared financially to properly handle the work, as required in the Advertisement.

Natural Gas

A supply main of the Union Natural Gas Corporation, and a branch line to Dayton, from that main, pass within one-half mile of

the Huffman damsite. Natural gas is available at practically all of
the cities and towns where work under these contracts is contemplated.

Coal

The principal points from which the coal supply for the Miami
Valley is obtained are in Ohio, West Virginia and Kentucky.
Freight rates from Ohio points were, at the time of this writing,
from about 90c to $1.10 per ton; from points in West Virginia and
Kentucky the freight rates were from about $1.25 to $1.35 per ton.

Electric Power

The Englewood, Taylorsville, and Huffman damsites are within
the territory served by the Dayton Power and Light Co., and that
company has signified its willingness to furnish power at the damsites, in accordance with its regular Demand Power Schedule (see
next page), it being understood that the current will be sold as
"High Tension Energy," and the contractor will furnish his own
step-down equipment. The company would expect a minimum return of $40.00 per year per horse power contracted for, and the con-
tract period must be long enough so that the Company would get
the cost of the lines, etc., (approximately $2,000 per mile), out of the
job. This Demand Power Schedule for 500 kilowatts at 60 per cent
load factor calls for a rate of approximately 1¼c per kilowatt hour.

For the Lockington Dam, power may be obtained at the Lockington sub-station of the Western Ohio Electric Co. at a rate of
about 2c per kilowatt hour, the contractor to furnish his own step-
down equipment and distribution lines.

Power for the Germantown Dam might be furnished by the Ohio
Electric Railway Co. If that company is not in position to supply
it, the Dayton Power and Light Co. will probably be willing to
make satisfactory arrangements to handle the business.

Bidders may obtain complete and up-to-date information on the
power situation by consulting the Power Engineer of the Dayton
Power and Light Co.; the Superintendent of Power and Light of
the Western Ohio Electric Co.; or the General Manager of the
Ohio Electric Railway Co.

Rate Schedule of Dayton Power and Light Co.

The Demand Power Rate is as follows:

Hours use of Demand	Rate	Hours use of Demand	Rate	Hours use of Demand	Rate
1 to 50	$ .0680	161 to 165	$ .0330	271 to 275	$ .0260
51 " 55	.0598	166 " 170	.0325	276 " 280	.0259
56 " 60	.0566	171 " 175	.0320	281 " 285	.0258
61 " 65	.0534	176 " 180	.0315	286 " 290	.0257
66 " 70	.0502	181 " 185	.0310	291 " 295	.0256
71 " 75	.0470	186 " 190	.0305	296 " 300	.0255
76 " 80	.0455	191 " 195	.0300	301 " 310	.0254
81 " 85	.0440	196 " 200	.0295	311 " 320	.0253
86 " 90	.0425	201 " 205	.0292	321 " 330	.0252
91 " 95	.0410	206 " 210	.0289	331 " 340	.0251
96 " 100	.0395	211 " 215	.0286	341 " 350	.0250
101 " 105	.0390	216 " 220	.0283	351 " 360	.0249
106 " 110	.0385	221 " 225	.0280	361 " 370	.0248
111 " 115	.0380	226 " 230	.0278	371 " 380	.0247
116 " 120	.0375	231 " 235	.0276	381 " 390	.0246
121 " 125	.0370	236 " 240	.0274	391 " 400	.0245
126 " 130	.0365	241 " 245	.0272	401 " 425	.0244
131 " 135	.0360	246 " 250	.0270	426 " 450	.0243
136 " 140	.0355	251 " 255	.0268	451 " 475	.0242
141 " 145	.0350	256 " 260	.0266	476 " 500	.0241
146 " 150	.0345	261 " 265	.0264	501 and above	.0240
151 " 155	.0340	266 " 270	.0262		
156 " 160	.0335				

From monthly bills as computed on the Demand Power Rate quantity discounts will be allowed, according to the size of the gross bill, as follows:

When bill is less than	$ 10.00	No quantity discount
" " " from $ 10.01 to	25.00	Discount will be 1 %
" " " " 25.01 "	50.00	" " 2 %
" " " " 50.01 "	75.00	" " 3½%
" " " " 75.01 "	100.00	" " 5½%
" " " " 100.01 "	125.00	" " 8 %
" " " " 125.01 "	150.00	" " 10½%
" " " " 150.01 "	175.00	" " 12½%
" " " " 175.01 "	200.00	" " 13½%
" " " " 200.01 "	225.00	" " 15 %
" " " " 225.01 "	250.00	" " 16 %
" " " " 250.01 "	275.00	" " 17 %
" " " " 275.01 "	300.00	" " 18 %
" " " " 300.01 "	325.00	" " 19 %
" " " " 325.01 "	350.00	" " 20 %
" " " " 350.01 "	375.00	" " 21 %
" " " " 375.01 "	400.00	" " 22 %
" " " " 400.01 "	450.00	" " 23 %
" " " " 450.01 "	500.00	" " 24 %
" " " " 500.01 "	550.00	" " 25½%
" " " " 550.01 "	600.00	" " 26½%
" " " " 600.01 "	650.00	" " 27½%
" " " " 650.01 "	700.00	" " 28½%
" " " " 700.01 "	750.00	" " 29½%
" " " " 750.01 "	800.00	" " 30½%
" " " " 800.01 "	850.00	" " 31½%
" " " " 850.01 "	900.00	" " 32½%
" " " " 900.01 "	950.00	" " 33½%
" " " " 950.01 "	1000.00	" " 35 %
" " " " 1000.01 "	1050.00	" " 36 %
" " " " 1050.01 "	1100.00	" " 37 %
" " " " 1100.01 "	1200.00	" " 38 %
" " " " 1200.01 "	1300.00	" " 38½%
" " " " 1300.01 "	1400.00	" " 39 %
" " " " 1400.01 "	1500.00	" " 39½%
" " " " 1500.01 "	1750.00	" " 40 %
" " " " 1750.01 "	2000.00	" " 40½%
" " " " 2000.01 "	2250.00	" " 41 %
" " " " 2250.01 "	2500.00	" " 41½%
" " " " 2500.01 "	2750.00	" " 42 %
" " " " 2750.01 "	3000.00	" " 42½%
" " " " 3000.01 "	3250.00	" " 43 %
" " " " 3250.01 "	3500.00	" " 43½%
" " " " 3500.01 "	3750.00	" " 44 %
" " " " 3750.01 "	4000.00	" " 44½%
" " " " 4000.01 "	4250.00	" " 45 %
" " " " 4250.01 "	4500.00	" " 45½%
" " " " 4500.01 "	4750.00	" " 46 %
" " " " 4750.01 "	5000.00	" " 46½%
" " " " 5000.01 and above		" " 47 %

The discounts shown above are based upon quantity, and are considered as a part of the rate schedule. An additional discount of 5% will be allowed after deducting the quantity discount for payment within discount days.

Note.—The foregoing schedule has been amended by adding two mills ($0.002) net per kilowatt hour, effective June 15, 1917.

33

Telephone Service

Telephone service can be obtained at reasonable cost from the telephone companies doing business within the District. It is probable that the contractor's construction cost for connections at the damsites will be only for the branch to his camp from the nearest company line, which is not far in any case. The service charge will probably be about $12 to $15 per month.

Freight Service

The railroads and electric lines reaching the various sections of work are shown on the general map. Railroad freight can be switched from the Cincinnati, Hamilton & Dayton Railroad to the Dayton, Covington & Piqua Electric Railway at West Milton; from the Cincinnati, Hamilton & Dayton Railroad to the Western Ohio Electric Railway at Wapakoneta; from the Cincinnati, Hamilton & Dayton Railroad to the Dayton & Troy Electric Railway at Tippecanoe City; from the Dayton & Troy Electric Railway to the Western Ohio Electric Railway at Piqua; and from the Cleveland, Cincinnati, Chicago & St. Louis Railway to the Ohio Electric Railway at Dwyer Station (south of Dayton).

Source of Supply—Sand and Gravel for Concrete

All of the sand and gravel deposits mentioned below, except the commercial pits, are owned and controlled by the District, and are available to the contractor to the extent necessary to his work. The limits of borrow pit areas are shown on the drawings.

Germantown Dam.—Sand and gravel in sufficient quantity and of acceptable quality for the work at Germantown are found in the islands in Twin Creek, about 2,000 feet upstream from the dam. Near the south side of the creek the available supply is only a few feet deep, but on the north side borings penetrated to a depth of 20 feet without reaching the bottom of acceptable material. Separated upon the ¼-inch mesh screen the proportion of gravel in the mixture is somewhat in excess of that of the sand.

About two miles southeast of Germantown, with a siding connection to the Cincinnati Northern R. R., is a commercial gravel pit furnishing satisfactory washed and screened concrete aggregates.

Englewood Dam.—Two distinct sand and gravel deposits occur in the vicinity of the Englewood damsite, one in the Stillwater River, downstream from the National Road and about 4,000 feet upstream from the dam, and the other in a small ridge about 2,000 feet upstream from the dam and 1,000 feet east of the river.

The river deposit comprises an island about 1,200 feet long and 250 feet wide. Borings from 12 to 20 feet deep have not reached the bottom of acceptable material. Separated upon the ¼-inch screen, the proportion of sand is somewhat greater than that of gravel. There is probably sufficient material here to satisfy the requirements.

The bank deposit forms a narrow, winding ridge about 1,500 feet long and rising 15 or 20 feet above the surrounding river plain. Test pits and borings in this material show that more or less clay is mixed with the sand and gravel and indicate that thorough washing will be required. The proportion of sand is greatly in excess of that of gravel.

Lockington Dam.—Two principal gravel deposits have been developed for the work at Lockington, one in the bed of Loramie Creek, extending about 1,000 feet upstream from the dam, and the other in the area adjacent to Mill Creek on the Bailey Farm, about half a mile north of the damsite.

The Loramie Creek deposit averages about eight feet deep. While sufficient sand for concrete requirements may perhaps be obtained at this place, the amount of coarse aggregate will be insufficient.

At Mill Creek the sand and gravel bed extends for a distance of at least 800 feet along the stream. This deposit contains sufficient material for the concrete requirements, but the proportion of fine aggregate, as separated on the ¼-inch screen, is at least equal to that of the coarse aggregate. As more or less clay is distributed throughout the material, thorough washing will be necessary to produce sufficiently clean aggregates.

Taylorsville Dam.—An ample supply of acceptable sand and gravel may be found in the river bed, extending downstream from the Taylorsville dam. Separated upon the ¼-inch screen, the average of a number of samples shows the material to consist of sand and gravel in about equal proportions.

Huffman Dam.—Acceptable sand and gravel in sufficient quantity are found in the areas of shallow flowage in the river bed west of the right-angle bend just upstream from the Huffman damsite. Analyses upon a ¼-inch screen show the proportion of gravel to be somewhat in excess of the sand.

Conduit Lining.—Coarse aggregate for special concrete required in conduit linings at dams (Item 42) cannot be obtained locally.

Bids on similar material were received by the City of Dayton in November, 1916, from the following firms: Henkel & Sullivan, Cincinnati; Arabia Granite Co., Cincinnati; Harris Granite Quarries Co., Salisbury, N. C.

Local Improvements.—Generally speaking, sand and gravel of suitable quality may be found within easy reach of the work at the various cities and towns where local channel improvements are contemplated.

Stone for Stone Masonry, Rubble Paving, and Riprap

It is possible that some stone for Item 48, Stone Masonry, Item 57, Dry Rubble Paving, and Item 58, Riprap, can be obtained from the best of the Brassfield limestone, a ledge deposit lying under comparatively shallow cover at the higher elevations near each of the damsites. For Item 57 some of the thickest of the hard limestone layers in the Cincinnatian formation may also be available. For Item 58, the harder boulders, where of sufficient size, will be acceptable. With these exceptions, all rock for the foregoing items must be obtained from quarries producing stone of acceptable quality, size, and shape. There are several such quarries in commercial operation within reasonable transportation distance by rail of the work and in addition, satisfactory ledges exist, not on property of the District, where the contractor can probably acquire quarrying privileges.

Flood Warnings

Within the limits of the drainage area there are about 30 U. S. Weather Bureau stations, and 17 river gages maintained by the District. Observers at all these stations send in reports promptly by telephone or telegraph, to the hydrographer of the District, whenever a rainfall of more than one inch in 24 hours occurs, or whenever there is any important change in river conditions. Thus the District keeps in close touch with the situation at all times and inasmuch as special attention and study have been given this subject, fairly accurate predictions of flood conditions can be made several hours in advance.

It is the present intention to continue this service during the construction period, to collect and utilize promptly the best information obtainable as to flood conditions, and to make the predictions available to contractors upon request at any time. These predictions will cover not only the comparatively rare big floods, but also those of moderate size that might necessitate the protection of contractor's equipment located in the river bottoms.

LIST OF CONTRACT ITEMS

Item

1	Stream control at dams
2	Clearing and Grubbing
3	Removal of buildings
4	Trimming and shaping slopes of river channels
5	Excavation, soil stripping
6 and 6a	Excavation, Class 1, for new channels and channel improvements, for cut-off trenches and for stream control channels
7 and 7a	Excavation, Class 2, for same
8 and 8a	Excavation, Class 3, for same
9	Excavation, Class 1, for river channel improvements
10	Excavation, Class 2, for same
11	Excavation, Class 3, for same
12	Excavation—Overhaul for river channel improvements
13	Excavation, Class 1, for outlet works and spillways at dams
14	Excavation, Class 2, for same
15	Excavation, Class 3, for same
16	Excavation, Class 1, for minor drainage systems
17	Excavation, Class 2, for same
18	Excavation, Class 3, for same
19	Excavation, Class 1, for street and highway cuts and fills
20	Excavation, Class 2, for same
21	Excavation, Class 3, for same
22 and 22a	Excavation, Class 1, for retaining walls, and bridge piers, and flood gate structures, except at dams
23 and 23a	Excavation, Class 2, for same
24 and 24a	Excavation, Class 3, for same
25 and 25a	Excavation, Class 1, and backfilling for trenches
26 and 26a	Excavation, Class 2. and backfilling for trenches
27 and 27a	Excavation, Class 3, and backfilling for trenches
28	Excavation, Class 1, miscellaneous
29	Excavation, Class 2, miscellaneous
30	Excavation, Class 3, miscellaneous
31	Embankment, Class A, for main dams constructed by hydraulic monitor and dredge pump method
32	Embankment, Class B, for main dams constructed by semihydraulic method
33	Embankment, Class C, for main dams constructed by sprinkling and rolling method
34	Embankment, Class D, for levees built from material taken from river channel excavation
35	Embankment, Class E, for levees built from materials taken from borrow pits
36	Embankment, Class F, miscellaneous
37	Backfilling
38	Surface dressing and grassing
41 and 41a	Concrete in outlet works, spillways, retaining walls, bridge piers, and similar structures
42	Concrete in conduit linings at dams
43 and 43a	Conrete in aprons, paving at dams, and similar flat shapes

Item
44 and 44a	Concrete in monolithic slope revetment
45	Concrete in flexible slab revetment or channel lining
46 and 46a	Concrete in culverts, bulkheads, gate chambers, and similar structures, except at dams
47	Concrete in manholes, catch basins, sewers, and similar shapes
48	Stone masonry
50	Cement
51	Reinforcing steel
52	Vitrified brick lining
53	Drilling foundations or masonry
54	Grouting foundations or masonry
55	Furnishing vitrified round or channel pipe
56	Laying vitrified round or channel pipe
57	Dry rubble paving
58	Riprap
59	Gutter paving
60	Crushed rock or gravel
61	Gravel paving for streets and highways
62	Water-bound macadam paving for same
63	Bituminous-bound macadam paving for same
64	Brick paving on concrete foundation, for same
64x	Brick paving on rolled foundation, for same
65	Concrete paving for same
66	Other paving not here mentioned which may be required for replacing existing pavements
67	Guard rails
68	Wire fences
70	Protective boom and accessories
71	Timber and lumber, miscellaneous
72	Timber piles 20 feet, or less, in length
72x	Timber piles more than 20 feet in length
73	Wood sheet piling
74	Steel sheet piling
75	Furnishing cast iron pipe
76	Laying cast iron pipe
77	Flood gates
77x	Furnishing flood gates and accessories that are to be installed by the District
77y	Setting in place flood gates and accessories that are to be furnished by the District
78	Steel highway bridges
81	Raising, removing, and re-erecting existing steel bridges
82	Miscellaneous cast iron, wrought iron, and steel
83	Bronze
84	Wire rope
85	Cleaning up
86	Pumping stations
87	Pumping equipment
88	Removal and re-erection of existing pumping and power stations
89	Sheeting left in place
90	Extra work

ESTIMATES OF QUANTITIES

Note.—An explanation of the system of item numbers used in the following estimates will be found under the heads of Specifications, page 29, and Estimates of Quantities, page 30. A complete list of contract items is given on pages 37 and 38.

Contract No. 1—Germantown Dam and Appurtenances, including Road No. 1.

Stream Control at Dam	Item	Estimated Quantity
Risk, damage, construction and removal of cofferdams outside the limits of the dams, etc......	1	1 job
Cut-off Trench at Dam		
Earth excavation, Class 1, above mean low water	6	12,000 cu. yds.
Earth excavation, Class 1, below mean low water	6a	5,000 " "
Mixed excavation, Class 2, above mean low water	7	500 " "
Rock excavation, Class 3, above mean low water	8	200 " "
Rock excavation, Class 3, below mean low water	8a	100 " "
Wood sheet piling ⎫	73	38,000 ft. B. M.
Steel sheet piling ⎬ Alternatives........... ⎰	74	110 tons
Main Dam		
Clearing and grubbing under dam.............	2	4 acres
Clearing and grubbing in borrow pits..........	2	3 "
Excavation, soil stripping...................	5	25,000 cu. yds.
Excavation, Class 1, for minor drainage systems.	16	1,000 " "
Embankment, Class A, hydraulic fill..........	31 ⎫	
Embankment, Class B, semi-hydraulic fill.......	32 ⎬	700,000 " "
Embankment, Class C, rolled fill.,............	33	130,000 " "
Surface dressing and grassing................	38	28,000 " "
Furnishing vitrified pipe.....................	55	5,000 lin. ft.
Furnishing vitrified pipe specials..............	55	10 pieces
Laying vitrified pipe.........................	56	5,020 lin. ft.
Gutter paving...............................	59	1,100 sq. yds.
Concrete in gutter system....................	41	25 cu. yds.
Cement....................................	50	250 bbls.
Spillway		
Earth excavation, Class 1....................	13	32,000 cu. yds.
Mixed excavation, Class 2....................	14	800 " "
Rock excavation, Class 3....................	15	100 " "
Earth excavation, Class 1, for minor drainage systems..................................	16	20 "
Concrete in slope facing......................	41	500 "
Concrete in paving and floor..................	43	500 " "
Cement....................................	50	1,400 bbls.
Gutter paving..............................	59	20 sq. yds.
Outlet Works		
Clearing and grubbing.......................	2	3 acres
Earth excavation, Class 1, above mean low water	13	92,000 cu. yds.
Earth excavation, Class 1, below mean low water	13a	9,500 " "
Mixed excavation, Class 2, above mean low water	14	18,000 " "
Mixed excavation, Class 2, below mean low water	14a	26,000 " "
Rock excavation, Class 3, above mean low water	15	400 "
Rock excavation, Class 3, below mean low water	15a	200 "
Mixed excavation, Class 2, for minor drainage systems..................................	17	100 "
Concrete in side and head walls of entrance and outlet above mean low water................	41	2,800 "
Concrete in conduits above mean low water.....	41	2,300 "
Concrete in conduits below mean low water.....	41a	5,000 "

	Item	Estimated Quantity
Outlet works (Continued)		
Concrete in side and head walls of entrance and outlet and submerged weirs below mean low water...	41a	3,700 cu. yds.
Concrete in conduit linings....................	42	1,400 " "
Concrete in paving and floors above mean low water...	43	100 " "
Concrete in paving and floors below mean low water...	43a	2,600 " "
Cement..	50	27,500 bbls.
Reinforcing steel...................·.......___..	51	35,000 lbs.
Backfilling around structures..................	37	10,000 cu. yds.
Embankment, Class F, miscellaneous...........	36	8,000 " "
Furnishing vitrified pipe......................	55	900 lin. ft.
Furnishing vitrified pipe specials..............	55	20 pieces
Laying vitrified pipe..........................	56	940 lin. ft.
Protective boom and accessories, lumber........	70	22,000 ft. B. M.
Protective boom and accessories, concrete.......	41	300 cu. yds.
Protective boom and accessories, miscellaneous cast iron, wrought iron and steel.............	82	9,500 lbs.
Protective boom and accessories, reinforcing steel	51	3,000 "
Drilling foundations or masonry...............	53	300 lin. ft.
Miscellaneous cast iron, wrought iron, and steel (including pipe)...........................	82	2,000 lbs.
Dry rubble paving............................	57	600 cu. yds.
Riprap.......................................	58	400 " "
Crushed rock or gravel........................	60	6,200 " "
General		
Cleaning up.................................	85	1 job
Road No. 1		
Clearing and grubbing..................	2	1 acre
Grading		
Excavation from cut, Class 1................	19	8,400 cu. yds.
Excavation from cut, Class 3................	21	100 " "
Culverts		
Eight 12-inch vitrified pipe culverts		
Two 18-inch vitrified pipe culverts		
Two 2-ft. x 2-ft. concrete box culverts		
Ten 12-inch iron pipe culverts		
Excavation, Class 1......................	19	230 cu. yds.
Backfilling.............................	37	100 " "
Concrete...............................	46	80 " "
Cement................................	50	134 bbls.
Reinforcing steel.......................	51	1,440 lbs.
Vitrified pipe		
12-inch.............................	55	260 lin. ft.
18-inch.............................	55	60 " "
Corrugated iron pipe, 12-inch..............	82	1,500 lbs.
Concrete slab bridges.		
Two 15-ft. spans		
Excavation, Class 1......................	22	60 cu. yds.
Excavation, Class 2......................	23	10 " "
Backfilling.............................	37	20 " "
Concrete in substructure...............	41	90 "
Concrete in superstructure..............	46	48 " "
Cement................................	50	200 bbls.
Reinforcing steel.......................	51	7,000 lbs.
Iron pipe railing.......................	82	780 "
Road surfacing		
Unrolled gravel..........................	60	1,400 cu. yds.
Guard rails.................................	67	3,200 lin. ft.
Wire fences		
Fence.....................................	68	8,100 " "
Miscellaneous lumber.....................	71	200 ft. B. M.
Miscellaneous iron and steel................	82	900 lbs.

Contract No. 2—Englewood Dam and Appurtenances, including Roads Nos. 3, 4 and 5.

Stream Control at Dam	Item	Estimated Quantity
Risk, damage, and work not otherwise paid for ..	1	1 job
Earth excavation, Class 1, above mean low water	6	8,000 cu. yds.
Earth excavation, Class 1, below mean low water	6a	2,000 " "

Cut-off Trench at Dam

Earth excavation, Class 1, above mean low water	6	35,000 " "
Earth excavation, Class 1, below mean low water	6a	2,000 "
Mixed excavation, Class 2, above mean low water	7	1,000 "
Mixed excavation, Class 2, below mean low water	7a	100 "
Rock excavation, Class 3, above mean low water	8	100 "
Rock excavation, Class 3, below mean low water	8a	10 " "
Steel sheet piling ⎫	74	130 tons
Wood sheet piling ⎭ Alternatives............ ⎰	73	45,000 ft. B. M.

Main Dam

Clearing and grubbing under dam.............	2	3 acres
Clearing and grubbing in borrow pits...........	2	25 "
Soil stripping.............................	5	75,000 cu. yds
Earth excavation, Class 1, for minor drainage systems...........................	16	3,500 " "
Embankment, Class A, hydraulic fill...........	31 ⎫	
Embankment, Class B, semi-hydraulic fill......	32 ⎬	2,970,000 " "
Embankment, Class C, rolled fill...............	33	530,000 " "
Surface dressing and grassing.................	38	85,000 " "
Furnishing vitrified pipe.....................	55	22,000 lin. ft.
Furnishing vitrified pipe specials..............	55	10 pieces
Laying vitrified pipe........................	56	22,020 lin. ft.
Gutter paving.	59	2,000 sq. yds.
Concrete in gutter system....................	41	150 cu. yds.
Cement...................................	50	600 bbls.

Spillways

Clearing and grubbing......................	2	1 acre
Earth excavation, Class 1, main spillway......	13	110,000 cu. yds.
Earth excavation, Class 1, temporary spillway...	13	45,000 " "
Mixed excavation, Class 2....................	14	7,000 " "
Embankment, Class F, miscellaneous..........	36	8,000 "
Concrete in weir............................	41	3,400 "
Concrete in side walls.......................	41	9,800 "
Concrete in floor paving.....................	43	3,000 "
Cement...............................	50	22,000 bbls.
Backfilling around structures......	37	1,800 cu. yds.
Surface dressing and grassing...............	38	2,800 " "
Furnishing vitrified pipe.....................	55	700 lin. ft.
Furnishing vitrified pipe specials..............	55	20 pieces
Laying vitrified pipe........................	56	740 lin. ft.
Dry rubble paving..........................	57	500 cu. yds.
Crushed stone or gravel......................	60	400 " "

Outlet Works

Earth excavation, Class 1, above mean low water	13	8,000 "
Earth excavation, Class 1, below mean low water	13a	3,000 "
Mixed excavation, Class 2, above mean low water	14	17,000 " "
Mixed excavation, Class 2, below mean low water	14a	26,000 " "
Concrete in walls, entrance and outlet above mean low water..................................	41	2,150 "
Concrete in conduits above mean low water.....	41	3,050 "
Concrete in side walls and head walls, entrance and outlet, and submerged weirs below mean low water...............................	41a	1,200 " "

Outlet Works (Continued)	Item	Estimated Quantity
Concrete in conduits below mean low water.....	41a	8,900 cu. yds.
Concrete in conduit linings....................	42	2,000 " "
Concrete in floor of entrance and outlet below mean low water..........................	43a	2,900 " "
Cement.......................................	50	33,000 bbls.
Reinforcing steel in partition walls.............	51	31,000 lbs.
Protective boom and accessories, concrete in piers	41	600 cu. yds.
Protective boom and accessories, lumber........	70	34,000 ft. B. M.
Protective boom and accessories, miscellaneous cast iron, wrought iron and steel.............	82	13,000 lbs.
Protective boom and accessories, reinforcing steel	51	4,000 "
Backfilling around structures..................	37	28,000 cu. yds.
Furnishing vitrified pipe......................	55	600 lin. ft.
Furnishing vitrified pipe specials...............	55	20 pieces
Laying vitrified pipe..........................	56	640 lin. ft.
Dry rubble paving............................	57	1,000 cu. yds.
Riprap.......................................	58	300 " "
Crushed stone or gravel.......................	60	6,000 " "

General

Removal of buildings.........................	3	1 job
Cleaning up..................................	85	1 "

Road No. 3

Clearing and grubbing........................	2	1 acre
Grading		
Excavation from cut, Class 1..............	19	1,000 cu. yds.
Excavation from cut, Class 2..............	20	200 " "
Excavation from cut, Class 3..............	21	100 " "
Excavation from borrow, Class 1............	19	4,200 "
Culverts		
Three 12-Inch vitrified pipe culverts		
Three 12-Inch iron pipe culverts		
Excavation, Class 1.....................	19	50 "
Backfilling............................	37	30 "
Concrete...............................	46	14 " "
Cement................................	50	24 bbls.
Reinforcing steel......................	51	150 lbs.
Vitrified pipe, 12-inch.................	55	80 lin. ft.
Corrugated iron pipe, 12-inch............	82	450 lbs.
Road surfacing		
Rolled gravel..........................	61	2,880 sq. yds.
Guard rails..................................	67	410 lin. ft.
Wire fences		
Fence.................................	68	1,320 " "
Miscellaneous lumber......................	71	100 ft. B. M.
Miscellaneous iron and steel................	82	450 lbs.

Road No. 4

Clearing and grubbing........................	2	1 acre
Grading		
Excavation from cut, Class 1..............	19	3,300 cu. yds.
Excavation from cut, Class 2..............	20	200 " "
Excavation from cut, Class 3..............	21	100 " "
Culverts		
Three 12-inch vitrified pipe culverts		
One 18-inch vitrified pipe culvert		
One 2-ft. x 2-ft. concrete box culvert		
Four 12-inch iron pipes		
Excavation, Class 1.....................	19	100 "
Backfilling............................	37	50 "
Concrete...............................	46	33 " "
Cement................................	50	57 bbls.

Road No. 4 (Continued)	Item	Estimated Quantity
Reinforcing steel.........................	51	800 lbs.
Vitrified pipe		
12-inch............................	55	80 lin. ft.
18-inch............................	55	26 " "
Corrugated iron pipe, 12-inch..............	82	600 lbs.
Road surfacing		
Rolled gravel............................	61	15,000 sq. yds.
Guard rails..............................	67	9,260 lin. ft.
Wire fences		
Fence..................................	68	1,000 " "
Miscellaneous lumber......................	71	100 ft. B. M.
Miscellaneous iron and steel................	82	300 lbs.

Road No. 5		
Clearing and grubbing.......................	2	0.5 acre
Grading		
Excavation from cut, Class 1................	19	1,800 cu. yds.
Excavation from borrow, Class 1.............	19	900 " "
Excavation from cut, Class 3................	21	100 " "
Culverts		
Four 12-inch vitrified pipe culverts		
One 18-inch vitrified pipe culvert		
Four 12-inch iron pipe culverts		
Excavation, Class 1.......................	19	90 "
Backfilling..............................	37	40 " ..
Concrete...............................	46	24 " "
Cement.................................	50	40 bbls.
Reinforcing steel.........................	51	240 lbs.
Vitrified pipe		
12-inch............................	55	90 lin. ft.
18-inch............................	55	22 " "
Corrugated iron pipe, 12-inch..............	82	600 lbs.
Road surfacing		
Unrolled gravel..........................	60	560 cu. yds.
Guard rails..............................	67	250 lin. ft.
Wire fences		
Fence..................................	68	2,000 lin. ft.
Miscellaneous lumber......................	71	200 ft. B. M.
Miscellaneous iron and steel................	82	300 lbs.

Contract No. 3—Lockington Dam and Appurtenances, including Roads Nos. 8 and 9.

Stream Control at Dam	Item	Estimated Quantity
Risk, damage, construction and removal of coffer-dams outside the limits of the dam, etc.......	1.	1 job

Cut-off Trench at Dam		
Earth excavation, Class 1, above mean low water	6	32,000 cu. yds.
Earth excavation, Class 1, below mean low water	6a	14,000 " "
Rock excavation, Class 3, above mean low water	8	300 " "
Rock excavation, Class 3, below mean low water	8a	200 " "
Wood sheet piling } Alternatives............. {	73	80,000 ft. B. M
Steel sheet piling }	74	180 tons

Main Dam		
Clearing and grubbing.......................	2	10 acres
Soil stripping used in dams...................	5	35,000 cu. yds.
Embankment, Class A, hydraulic fill..........	31 }	
Embankment, Class B, semi-hydraulic fill.......	32 }	685,000 " "

Main Dam (Continued)

	Item	Estimated Quantity
Embankment, Class C, rolled fill	33	265,000 cu. yds.
Surface dressing and grassing	38	34,000 " "
Earth excavation, Class 1, for minor drainage systems	16	1,000 " "
Furnishing vitrified pipe	55	8,400 lin. ft.
Furnishing vitrified pipe specials	55	10 pieces
Laying vitrified pipe	56	8,420 lin. ft.
Gutter paving	59	1,000 sq. yds.
Concrete in gutter system	41	60 cu. yds.
Cement	50	300 bbls.

Outlet Works and Spillway

	Item	Estimated Quantity
Earth excavation, Class 1, above mean low water	13	`92,000 cu. yds.
Earth excavation, Class 1, below mean low water	13a	36,000 " "
Rock excavation, Class 3, below mean low water	15a	8,000 " "
Grouting of foundations or masonry	54	200 " "
Drilling of foundations or masonry	53	500 lin. ft.
Miscellaneous cast iron, wrought iron, and steel (pipe for grouting)	82	2,000 lbs.
Concrete in spillway weir above mean low water	41	3,100 cu. yds.
Concrete in walls above mean low water	41	20,400 " "
Concrete in spillway below mean low water	41a	50 " "
Concrete in walls and submerged weirs below mean low water	41a	8,600 "
Concrete in conduit linings	42	200 "
Concrete in paving and floors below mean low water	43a	3,800 " "
Cement	50	50,000 bbls.
Reinforcing steel in partition wall	51	30,000 lbs.
Reinforcing steel in bases of retaining walls	51	40,000 "
Furnishing vitrified pipe	55	900 lin. ft.
Furnishing vitrified pipe specials	55	20 pieces
Laying vitrified pipe	56	940 lin. ft.
Embankment, Class F, miscellaneous fill	36	2,000 cu. yds.
Backfilling around structures	37	1,000 " "
Dry rubble paving	57	600 " "
Riprap	58	400 " "
Crushed rock or gravel	60	200 " "
Protective boom and accessories, lumber	70	29,000 ft. B. M.
Protective boom and accessories, concrete	41	500 cu. yds.
Protective boom and accessories, reinforcing steel	51	3,000 lbs.
Protective boom and accessories, miscellaneous cast iron, wrought iron, and steel	82	11,000 "

General

	Item	Estimated Quantity
Removal of buildings	3	1 job
Cleaning up	85	1 "

Road No. 8

	Item	Estimated Quantity
Clearing and grubbing	2	1 acre
Grading		
Excavation from cut, Class 1	19	1,800 cu. yds.
Excavation from borrow, Class 1	19	200 " "
Culverts		

Culverts
Three 12-inch vitrified pipe culverts
One 18-inch vitrified pipe culvert
One 3-ft. x 2-ft. concrete box culvert

Road No. 8 (Continued)	Item	Estimated Quantity
Eight 12-inch iron pipe culverts		
Excavation, Class 1	19	100 cu. yds.
Backfilling	37	70 " "
Concrete	46	30 " "
Cement	50	50 bbls.
Reinforcing steel	51	900 lbs.
Vitrified pipe		
12-inch	55	70 lin. ft.
18-inch	55	22 " "
Corrugated iron pipe, 12-inch	82	1,200 lbs.
Road surfacing		
Unrolled gravel	60	1,450 cu. yds.
Guard rails	67	11,700 lin. ft.
Wire fences		
Fence	68	1,700 " "
Miscellaneous lumber	71	200 ft. B. M.
Miscellaneous iron and steel	82	600 lbs.

Road No. 9		
Clearing and grubbing	2	1 acre
Grading		
Excavation from cut, Class 1	19	10,400 cu. yds.
Excavation from cut, Class 3	21	100 " "
Culverts		
Seven 12-inch vitrified pipe culverts		
Two 18-inch vitrified pipe culverts		
Four 3-ft. x 2-ft. concrete box culverts		
Eighteen 12-inch iron pipe culverts		
Excavation, Class 1	19	260 cu. yds.
Backfilling	37	90 " "
Concrete	46	100 " "
Cement	50	170 bbls.
Reinforcing steel	51	3,850 lbs.
Vitrified pipe		
12-inch	55	190 lin. ft.
18-inch	55	50 " "
Corrugated iron pipe, 12-inch	82	2,700 lbs.
Bridges		
One 10-ft. span concrete slab bridge		
Excavation, Class 1	22	170 cu. yds.
Backfilling	37	80 " "
Concrete in substructure	41	115 " "
Concrete in superstructure	46	11 " "
Cement	50	170 bbls.
Reinforcing steel	51	1,100 lbs.
Iron pipe railing	82	350 "
Substructure for steel bridge		
Excavation, Class 1	22	180 cu. yds.
Backfilling	37	60 " "
Concrete	41	130 " "
Cement	50	170 bbls.
Reinforcing steel	51	100 lbs.
Road surfacing		
Unrolled gravel	60	1,540 cu. yds.
Guard rails	67	1,560 lin. ft.
Wire fences		
Fence	68	3,860 " "
Miscellaneous lumber	71	600 ft. B. M.
Miscellaneous iron and steel	82	600 lbs.

Contract No. 4—Taylorsville Dam and Appurtenances, including Roads Nos. 12 and 13.

Stream Control at Dam	Item	Estimated Quantity
Risk, damage, placing and removing 12,000 cu. yds. of cofferdam outside the limits of the dam, etc.	1	1 job
Cut-off Trench at Dam		
Earth excavation, Class 1, above mean low water	6	14,000 cu. yds.
Earth excavation, Class 1, below mean low water	6a	6,000 " "
Earth excavation, Class 2, below mean low water	7a	100 " "
Rock excavation, Class 3, above mean low water	8	100 " "
Rock excavation, Class 3, below mean low water	8a	100 " "
Wood sheet piling } Alternatives	73	120,000 ft. B. M.
Steel sheet piling }	74	315 tons
Main Dam		
Clearing and grubbing	2	0.5 acre
Soil stripping	5	30,000 cu. yds.
Embankment, Class A, hydraulic fill	31 }	
Embankment, Class B, semi-hydraulic fill	32 }	920,000 " "
Embankment, Class C, rolled fill	33	120,000 " "
Surface dressing and grassing	38	35,000 " "
Earth excavation, Class 1, for minor drainage systems	16	1,200 " "
Furnishing vitrified pipe	55	8,100 lin. ft.
Furnishing vitrified pipe specials	55	10 pieces
Laying vitrified pipe	56	8,120 lin. ft.
Gutter paving	59	1,200 sq. yds.
Concrete in slope gutters	41	70 cu. yds.
Cement	50	300 bbls.
Outlet Works and Spillway		
Clearing and grubbing	2	18 acres
Earth excavation, Class 1, above mean low water	13	452,000 cu. yds.
Earth excavation, Class 1, below mean low water	13a	80,000 " "
Mixed excavation, Class 2, above mean low water	14	70,000 " "
Mixed excavation, Class 2, below mean low water	14a	105,000 " "
Rock excavation, Class 3, above mean low water	15	1,000 "
Rock excavation, Class 3, below mean low water	15a	300 "
Mixed excavation, Class 2, for minor drainage systems	17	400 "
Earth excavation, Class 1, for minor drainage systems	16	300 "
Miscellaneous cast iron, wrought iron, and steel (pipe)	82	4,000 lbs.
Concrete in side and partition walls above mean low water	41	18,000 cu. yds.
Concrete in side and partition walls and submerged weirs below mean low water	41a	16,800 " "
Concrete in spillway weir above mean low water	41	5,700 "
Concrete in spillway weir below mean low water	41a	300 "
Concrete in conduit linings	42	700 "
Concrete in floor and paving	43	100 "
Concrete in floor of entrance and outlet below mean low water	43a	14,500 " "
Cement	50	76,000 bbls.
Reinforcing steel in partition wall	51	81,000 lbs.
Reinforcing steel in footings of retaining walls	51	80,000 "
Furnishing vitrified pipe	55	3,700 lin. ft.
Furnishing vitrified pipe specials	55	60 pieces
Laying vitrified pipe	56	3,820 lin. ft.
Embankment, Class F, miscellaneous fill	36	110,000 cu. yds.
Backfilling around structures	37	1,000 " "
Dry rubble paving	57	2,000 " "

Outlet **Works and Spillway** (Continued)	Item	Estimated Quantity
Riprap.	58	3,000 cu. yds.
Crushed rock or gravel.	60	600 " "
Protective boom and accessories, concrete in piers	41	900 " "
Protective boom and accessories, reinforcing steel	51	8,000 lbs.
Protective boom and accessories, lumber.	70	32,000 ft. B. M.
Protective boom and accessories, miscellaneous cast iron, wrought iron, and steel.	82	12,000 lbs.
General		
Removal of buildings.	3	1 job
Cleaning up.	85	1 "
Road No. 12		
Clearing and grubbing.	2	2 acres
Grading		
Excavation from cut, Class 1.	19	2,900 cu. yds.
Excavation from borrow, Class 1.	19	100 " "
Excavation from cut, Class 3.	21	50 " "
Culverts		
Five 12-inch vitrified pipe culverts		
One 18-inch vitrified pipe culvert		
One 5-ft. x 3½-ft. concrete box culvert		
Four 12-inch iron pipe culverts		
Excavation, Class 1.	19	140 cu. yds.
Backfilling.	37	60 " "
Concrete.	46	48 " "
Cement.	50	78 bbls.
Reinforcing steel.	51	1,300 lbs.
Vitrified pipe		
12-inch.	55	120 lin. ft.
18-inch.	55	22 " "
Corrugated iron pipe, 12-inch.	82	600 lbs.
Road surfacing		
Unrolled gravel.	60	550 cu. yds.
Wire fences		
Fence.	68	7,000 lin. ft.
Miscellaneous lumber.	71	200 ft. B. M.
Miscellaneous iron and steel.	82	600 lbs.
Road No. 13		
Clearing and grubbing.	2	1.5 acres
Grading		
Excavation from cut, Class 1 .	19	2,100 cu. yds.
Excavation from borrow, Class 1.	19	800 " "
Excavation from cut, Class 3.	21	50 " "
Culverts		
One 12-inch vitrified pipe culvert		
Two 18-inch vitrified pipe culverts		
One 2-ft. x 2-ft. concrete box culvert		
Three 3-ft. x 2-ft. concrete box culverts		
Ten 12-inch iron pipe culverts		
Excavation, Class 1.	19	140 cu. yds.
Backfilling.	37	70 " "
Concrete.	46	75 " "
Cement.	50	130 bbls.
Reinforcing steel.	51	3,760 lbs.
Vitrified pipe		
12-inch.	55	40 lin. ft.
18-inch.	55	44 " "
Corrugated iron pipe, 12-inch.	82	1,500 lbs.
Road surfacing		
Unrolled gravel.	60	1,200 cu. yds.
Guard rails.	67	5,410 lin. ft.
Wire fences		
Fence.	68	4,100 " "
Miscellaneous lumber.	71	200 ft. B M.
Miscellaneous iron and steel.	82	900 lbs.

Contract No. 5—Huffman Dam and Appurtenances, including Roads Nos. 16 and 17.

Stream Control at Dam	Item	Estimated Quantity
Risk, damage, placing about 10,500 cu. yds. of cofferdam outside the limits of the dam, removal of 9,500 cu. yds. of said cofferdam, etc........	1	1 job
Earth excavation in diversion channel, Class 1, above mean low water.....................	6	17,000 cu. yds.
Earth excavation in diversion channel, Class 1, below mean low water.................. .	6a	7,000 " "

Cut-off Trench at Dam		
Earth excavation, Class 1, above mean low water	6	17,000 " "
Earth excavation, Class 1, below mean low water	6a	8,000 "
Mixed excavation, Class 2, above mean low water	7	300 "
Mixed excavation, Class 2, below mean low water	7a	100 "
Rock excavation, Class 3, above mean low water	8	100 "
Rock excavation, Class 3, below mean low water	8a	100 " "
Wood sheet piling } Alternatives........ .... {	73	122,000 ft. B. M.
Steel sheet piling }	74	350 tons

Main Dam		
Clearing and grubbing under dam.............	2	4 acres
Clearing and grubbing in borrow pits..........	2	10 "
Soil strippings.............................	5	42,000 cu. yds.
Earth excavations, Class 1, for minor drainage systems...................................	16	1,000 " "
Embankment, Class A, hydraulic fill..........	31 }	
Embankment, Class B, semi-hydraulic fill......	32 }	1,160,000 " "
Embankment, Class C, rolled fill..............	33	245,000 " "
Surface dressing and grassing.................	38	40,000 " "
Furnishing vitrified pipe.....................	55	10,500 lin. ft.
Furnishing vitrified pipe specials......	55	10 pieces
Laying vitrified pipe........................	56	10,520 lin. ft.
Gutter paving...	59	600 sq. yds.
Concrete in gutter system...................	41 .	75 cu. yds.
Cement...................................	50	200 bbls.

Outlet Works and Spillway		
Clearing and grubbing.......................	2	2 acres
Earth excavation, Class 1, above mean low water	13	30,000 cu. yds.
Earth excavation, Class 1, below mean low water	13a	18,000 " "
Mixed excavation, Class 2, above mean low water	14	52,000 " "
Mixed excavation, Class 2, below mean low water	14a	88,000 " "
Rock excavation, Class 3, above mean low water	15	200 "
Rock excavation, Class 3, below mean low water	15a	100 "
Earth excavation, Class 1, for minor drainage systems...................................	16	200 "
Mixed excavation, Class 2, for minor drainage systems...................................	17	100 "
Embankment, Class F, miscellaneous..........	36	19,000 "
Backfill around structures....................	37	1,000 "
Concrete in walls above mean low water.......	41	14,800 "
Concrete in walls and submerged weirs below mean low water...................	41a	12,200 "
Concrete in spillway above mean low water.....	41	4,200 "
Concrete in spillway below mean low water......	41a	200 "
Concrete in conduit linings..................	42	470 "
Concrete in floor and paving above mean low water...................................	43	100 "
Concrete in floor and paving, entrance and outlet, below mean low water....................	43a	10,600 " "
Cement...................................	50	54,000 bbls.

Outlet Works and Spillway (Continued)	Item	Estimated Quantity
Reinforcing steel in partition wall..............	51	80,000 lbs.
Reinforcing steel in bases of retaining walls......	51	70,000 "
Protective boom and accessories, concrete in piers	41	700 cu. yds.
Protective boom and accessories, lumber........	70	35,000 ft. B. M.
Protective boom and accessories, reinforcing steel	51	6,500 lbs.
Protective boom and accessories, miscellaneous cast iron, wrought iron, and steel.............	82	14,500 "
Miscellaneous cast iron, wrought iron, and steel..	82	4,000 "
Furnishing vitrified pipe.....................	55	3,100 lin. ft.
Furnishing vitrified pipe specials...............	55	50 pieces
Laying vitrified pipe.........................	56	3,200 lin. ft.
Dry rubble paving..........................	57	800 cu. yds.
Riprap....................................	58	500 " "
Crushed rock or gravel.......................	60	800 " "

General

Removal of buildings...:.....................	3	1 job
Cleaning up................................	85	1 "

Road No. 16

Clearing and grubbing.......................	2	1 acre
Grading		
Excavation from cut, Class 1................	19	11,000 cu. yds.
Excavation from borrow, Class 1.............	19	1,000 " "
Excavation from cut, Class 2...............	20	200 " "
Excavation from cut, Class 3...............	21	100 " "
Culverts		
Four 12-inch vitrified pipe culverts		
Six 18-inch vitrified pipe culverts		
Four 2-ft. x 2-ft. concrete box culverts		
Two 3-ft. x 3-ft. concrete box culverts		
One 5-ft. x 3½-ft. concrete box culvert		
Eight 12-inch iron pipe culverts		
Excavation, Class 1.....................	19	400 cu. yds.
Backfilling.........................	37	170 " "
Concrete............................	46	190 " "
Cement.............................	50	320 bbls.
Reinforcing steel......................	51	6,300 lbs.
Vitrified pipe		
12-inch.............................	55	130 lin. ft.
18-inch.............................	55	260 " "
Corrugated iron pipe, 12-inch.............	82	1,500 lbs.
Road surfacing		
Rolled gravel............................	61	11,240 sq. yds.
Guard rails................................	67	3,000 lin. ft.
Wire fences		
Fence....................................	68	12,600 " "
Miscellaneous lumber.......................	71	300 ft. B. M.
Miscellaneous iron and steel.................	82	1,200 lbs.

Road No. 17

Clearing and grubbing.......................	2	0.5 acre
Grading		
Excavation from cut, Class 1................	19	3,000 cu. yds.
Excavation from borrow, Class 1.............	19	500 " "
Excavation from cut, Class 3...............	21	50 " "
Culverts		
One 18-inch vitrified pipe culvert		
Two 12-inch iron pipe culverts		
Excavation, Class 1.........................	19	30 cu. yds.
Backfilling............................	37	15 " "
Concrete............................	46	12 " "

	Item	Estimated Quantity
Road No. 17 (Continued)		
Cement....................................	50	20 bbls.
Reinforcing steel.........................	51	70 lbs.
Vitrified pipe, 18-inch...................	55	56 lin. ft.
Corrugated iron pipe, 12-inch.............	82	300 lbs.
Road surfacing		
Rolled gravel.............................	61	8,300 sq. yds.
Guard rails...............................	67	7,500 lin. ft.
Wire fences		
Fence.....................................	68	2,040 " "
Miscellaneous lumber......................	71	200 ft. B. M.
Miscellaneous iron and steel..............	82	300 lbs.

Contract No. 6—Road No. 2 (Germantown Basin).

	Item	Estimated Quantity
Clearing and grubbing.....................	2	0.5 acres
Grading		
Excavation from borrow, Class 1...........	19	20,550 cu. yds.
Excavation from borrow, Class 2...........	20	2,200 " "
Culverts		
One 3-ft. x 2-ft. concrete box culvert		
Excavation, Class 1.......................	19	50 cu. yds.
Backfilling...............................	37	20 " "
Concrete..................................	46	30 " "
Cement....................................	50	50 bbls.
Reinforcing steel.........................	51	2,000 lbs.
Road surfacing		
Unrolled gravel...........................	60	200 cu. yds.
Guard rails...............................	67	1,900 lin. ft.
Wire fences...............................	68	1,600 " "

Contract No. 7—Road No. 6 (Englewood Basin).

	Item	Estimated Quantity
Excavation from borrow, Class 1...........	19	2,000 cu. yds.
Road surfacing, unrolled gravel...........	60	260 " "

Contract No. 8—Road No. 7 (Englewood Basin).

	Item	Estimated Quantity
Clearing and grubbing.....................	2	1 acre
Grading		
Excavation from cut, Class 1..............	19	6,400 cu. yds.
Excavation from borrow, Class 1...........	19	26,500 " "
Excavation from cut, Class 2..............	20	700 " "
Excavation from cut, Class 3..............	21	300 "
Culverts		
One 3-ft. x 2-ft. concrete box culvert		
Excavation, Class 1.......................	19	40 "
Backfilling...............................	37	20 " "
Concrete..................................	46	40 " "
Cement....................................	50	70 bbls.
Reinforcing steel.........................	51	2,700 lbs.
Four 12-inch vitrified pipe culverts		
Excavation, Class 1.......................	19	40 cu. yds.
Backfilling...............................	37	20 " "
Concrete..................................	46	20 " "
Cement....................................	50	30 bbls.
Reinforcing steel.........................	51	200 lbs.
Vitrified pipe............................	55	120 lin. ft.

Road No. 8 (Continued)	Item	Estimated Quantity
One 18-inch vitrified pipe culvert		
Excavation, Class 1	19	20 cu. yds.
Backfilling	37	10 " "
Concrete	46	10 " "
Cement	50	20 bbls.
Reinforcing steel	51	100 lbs.
Vitrified pipe	55	40 lin. ft.
Iron pipe culverts	82	600 lbs.
Road surfacing		
Unrolled gravel	60	1,100 cu. yds.
Guard rails	67	2,200 lin. ft.
Wire fences		
Fence	68	9,400 lin. ft.
Miscellaneous lumber	71	200 ft. B. M.
Miscellaneous iron and steel	82	600 lbs.

Contract No. 9—Road No. 10 (Lockington Basin).

	Item	Estimated Quantity
Clearing and grubbing	2	0.5 acre
Grading		
Excavation from cut, Class 1	19	300 cu. yds.
Excavation from borrow, Class 1	19	26,700 " "
Culverts		
One 4-ft. x 3-ft. concrete box culvert		
Excavation, Class 1	19	50 "
Backfilling	37	30 " "
Concrete	46	30 " "
Cement	50	55 bbls.
Reinforcing steel	51	1,600 lbs.
One 3-ft. x 2-ft. concrete box culvert		
Excavation, Class 1	19	50 cu. yds.
Backfilling	37	30 " "
Concrete	46	20 " "
Cement	50	35 bbls.
Reinforcing steel	51	1,300 lbs.
Iron pipe culverts	82	300 "
Road surfacing		
Unrolled gravel	60	400 cu. yds.
Guard rails	67	2,600 lin. ft.
Wire fences		
Fence	68	1,000 " "
Miscellaneous iron and steel	82	300 lbs.

Contract No. 10—Road No. 11 (Lockington Basin).

	Item	Estimated Quantity
Clearing and grubbing	2	4.5 acres
Grading		
Excavation from cut, Class 1	19	10,000 cu. yds.
Excavation from cut, Class 2	20	300 " "
Excavation from cut, Class 3	21	200 " "
Culverts		
One 4-ft. x 3-ft. concrete box culvert		
Excavation, Class 1	19	30 cu. yds.
Backfilling	37	20 " "
Concrete	46	15 " "
Cement	50	25 bbls.
Reinforcing steel	51	700 lbs.

Road No. 11 (Continued)	Item	Estimated Quantity
Eight 18-inch vitrified pipe culverts		
Excavation, Class 1	19	120 cu. yds.
Backfilling	37	100 " "
Concrete	46	55 " "
Cement	50	105 bbls.
Reinforcing steel	51	600 lbs.
Vitrified pipe	55	200 lin. ft.
Iron pipe culverts	82	1,700 lbs.
Concrete slab bridge		
Excavation, Class 1	22	50 cu. yds.
Excavation, Class 2	23	10 " "
Excavation, Class 3	24	20 " "
Backfilling	37	38 "
Concrete in substructure	41	60 "
Concrete in superstructure	46	40 " "
Cement	50	160 bbls.
Reinforcing steel	51	4,000 lbs.
Pipe railing	82	450 "
Road surfacing		
Unrolled gravel	60	1,600 cu. yds.
Guard rails	67	900 lin. ft.
Wire fences		
Fence	68	16,800 " "
Miscellaneous lumber	71	500 ft. B. M.
Miscellaneous iron and steel	82	1,650 lbs.

Contract No. 11—Road No. 14 (Taylorsville Basin).

	Item	Estimated Quantity
Clearing and grubbing	2	0.5 acre
Grading		
Excavation from borrow, Class 1	19	34,000 cu. yds.
Culverts		
One 4-ft. x 3-ft. concrete box culvert		
Excavation, Class 1	19	100 "
Backfilling	37	50 "
Concrete	46	30 " "
Cement	50	50 bbls.
Reinforcing steel	51	1,800 lbs.
Road surfacing		
Rolled gravel	61	2,300 sq. yds.
Guard rails	67	2,400 lin. ft.
Wire fences	68	2,400 " "

Contract No. 12—Road No. 15 (Taylorsville Basin).

	Item	Estimated Quantity
Clearing and grubbing	2	0.5 acre
Grading		
Excavation from borrow, Class 1	19	15,200 cu. yds.
Culverts		
One 4-ft. x 3-ft. concrete box culvert		
Excavation, Class 1	19	100 "
Backfilling	37	50 "
Concrete	46	30 " "
Cement	50	50 bbls.
Reinforcing steel	51	1,800 lbs.
Road surfacing		
Unrolled gravel	60	200 cu. yds.
Guard rails	67	2,200 lin. ft.
Wire fences	68	2,200 " "

Contract No. 13—Road No. 16 (part not included in Contract No. 5)
(Huffman Basin)

	Item	Estimated Quantity
Clearing and grubbing......................	2	0.5 acre
Grading		
Excavation from cut, Class 1...............	19	3,500 cu. yds.
Excavation from cut, Class 3...............	21	100 " "
Culverts		
One 18-inch vitrified pipe culvert		
One 2-ft. x 2-ft. concrete box culvert		
One 3-ft. x 2-ft. concrete box culvert		
One 6-ft. x 4-ft. concrete box culvert		
Excavation, Class 1.....................	19	100 cu. yds.
Backfilling...........................	37	40 " "
Concrete..............................	46	80 " "
Cement...............................	50	130 bbls.
Reinforcing steel......................	51	4,000 lbs.
Vitrified pipe, 18-inch................	55	22 lin. ft.
Iron pipe culverts.......................	82	1,500 lbs.
Concrete slab bridges		
Excavation, Class 1....................	22	130 cu. yds.
Excavation, Class 2....................	23	20 " "
Excavation, Class 3....................	24	20 " "
Backfilling...........................	37	40 " "
Concrete in substructure...............	41	30 " "
Concrete in superstructure.............	46	20 " "
Cement...............................	50	80 bbls.
Reinforcing steel.....................	51	1,300 lbs.
Road surfacing		
Rolled gravel.........................	61	10,500 sq. yds.
Wire fences		
Fence................................	68	12,600 lin. ft.
Miscellaneous lumber..................	71	200 ft. B. M.
Miscellaneous iron and steel..........	82	1,500 lbs.

Note.—Contracts Nos. 14 and 15 are unassigned. Nos. 16 to 21 are steel highway bridge contracts. Nos. 22 to 26 are either unassigned or small miscellaneous contracts.

Contract No. 27—Tippecanoe City Local Protection—General.

Levee Construction

Soil stripping...............................	5	10,000 cu. yds.
Embankment, Class E.......................	35	145,000 " "
Surface dressing and grassing.................	38	15,000 " "

Retaining Walls

Excavation, Class 1..........................	22	
Excavation, Class 2..........................	23	2,830 "
Excavation, Class 3..........................	24	
Concrete....................................	41	1,550 " "
Cement.....................................	50	2,070 bbls.
Reinforcing steel...........................	51	1,000 lbs.
Miscellaneous iron and steel............	82	5,000 "

Street Modifications

Excavation, Class 1..........................	19	
Excavation, Class 2..........................	20	9,400 cu. yds.
Excavation, Class 3..........................	21	
Macadam paving.............................	62	2,900 sq. yds.
Brick paving................................	64	2,500 " "
Cement.....................................	50	525 bbls.

General

Clearing and grubbing.......................	2	1 acre
Cleaning up................................	85	1 job

Contract No. 28—Tippecanoe City Local Protection—Drainage.

	Item	Estimated Quantity
Excavation, Class 1, minor drainage...........	16 ⎫	
Excavation, Class 2, minor drainage...........	17 ⎬	10,250 cu. yds.
Excavation, Class 3, minor drainage...........	18 ⎭	
Excavation, Class 1, and backfilling for trenches.	25 ⎫	
Excavation, Class 2, and backfilling for trenches.	26 ⎬	19,400 " "
Excavation, Class 3, and backfilling for trenches.	27 ⎭	
Concrete apron............................	43	50 "
Concrete in culverts, etc.....................	46	220 "
Concrete in sewers, manholes, etc.............	47	2,800 " "
Cement....................................	50	4,100 bbls.
Reinforcing steel............................	51	40,000 lbs.
Furnishing vitrified pipe.....................	55	List price, $806
Laying vitrified pipe.........................	56	9,000 inch-ft.
Furnishing cast iron pipe.....................	75	3 tons
Laying cast iron pipe........................	76	580 inch-ft.
Flood gates.................................	77	11,000 lbs.
Miscellaneous cast iron and steel.............	82	20,000 "

Note.—Contracts Nos. 29 to 33 pertain to a Sewage Pumping Station and to a Waterworks and Electric Light Station.

Contract No. 34—Fairfield Local Protection.

	Item	
Clearing and grubbing........................	2	3 acres
Removal of buildings.........................	3	1 job
Excavation, soil stripping.....................	5	6,500 cu. yds.
Embankment, Class E, from borrow pits........	35	65,000 " "
Surface dressing and grassing.................	38	6,000 " "

Drainage System

	Item	
Excavation, Class 1, open ditch................	16 ⎫	
Excavation, Class 2, open ditch...............	17 ⎬	57 "
Excavation, Class 3, open ditch...............	18 ⎭	
Excavation, Class 1, for street and highway cuts and fills.................................	19 ⎫	
Excavation, Class 2, for street and highway cuts and fills.................................	20 ⎬	57 "
Excavation, Class 3, for street and highway cuts and fills.................................	21 ⎭	
Excavation, Class 1, and backfilling for trenches.	25 ⎫	
Excavation, Class 2, and backfilling for trenches.	26 ⎬	65 "
Excavation, Class 3, and backfilling for trenches.	27 ⎭	
Concrete in culverts, gate chambers, etc........	46	200 " "
Cement....................................	50	350 bbls.
Reinforcing steel............................	51	5,000 lbs.
Furnishing vitrified pipe.....................	55	List price, $68
Laying vitrified pipe.........................	56	720 inch-ft.
Furnishing cast iron pipe.....................	75	1 ton
Laying cast iron pipe........................	76	220 inch-ft.
Flood gates.................................	77	10,000 lbs.
Miscellaneous cast iron, wrought iron, and steel..	82	1,000 "

Street Modifications

	Item	
Excavation, Class 1, for street and highway cuts and fills.................................	19 ⎫	
Excavation, Class 2, for street and highway cuts and fills.................................	20 ⎬	8,660 cu. yds.
Excavation, Class 3, for street and highway cuts and fills.................................	21 ⎭	
Gravel paving for streets and highways.........	61	10,000 sq. yds.
Guard rails.................................	67	5,000 lin. ft.

General	Item	Estimated Quantity
Excavation, Class 2, for retaining walls.........	23	20 cu. yds.
Excavation, Class 1, miscellaneous.............	28	
Excavation, Class 2, miscellaneous.............	29	57 " "
Excavation, Class 3, miscellaneous.............	30	
Concrete in retaining walls...................	41	10 " "
Cleaning up..............................	85	1 job

Contracts Nos. 35 and 36.—Piqua Local Protection.

Note.—The local protection work at Piqua is divided into two contracts numbered 35 and 36. Contract No. 35 will contain approximately 74,000 cu. yds. of channel excavation, 75,000 cu. yds. of levee embankment, and 75 cu. yds. of concrete. Contract No. 36 will contain approximately 110,000 cu. yds. of channel excavation, 114,000 cu. yds. of levee embankment, and 2,500 cu. yds. of concrete.

Improvement of River Channel

Trimming and shaping slopes of river channels...	4	4,000 sq. yds.
Soil stripping..............................	5	3,000 cu. yds.
Excavation, Class 1..........................	9	
Excavation, Class 2..........................	10	184,200 " "
Excavation, Class 3..........................	11	
Overhaul per 100 feet........................	12	100,000 " "

Levee Construction

Soil stripping..............................	5	16,500 cu. yds.
Embankments, Class D (from river excavation)..	34	182,000 " "
Embankment, Class E (from borrow pits).......	35	7,000 " "
Surface dressing and grassing.................	38	16,300 "

Revetment

Excavation, Class 1, below mean low water.....	22a	
Excavation, Class 2, below mean low water......	23a	850 "
Excavation, Class 3, below mean low water......	24a	
Concrete.................................44, 44a		1,400 " "
Cement..................................	50	1,750 bbls.
Reinforcing steel...........................	51	16,000 lbs.

Retaining Walls

Excavation, Class 1..........................	22	
Excavation, Class 2..........................	23	900 cu. yds.
Excavation, Class 3..........................	24	
Backfilling.................................	37	50 " "
Concrete..................................	41	1,150 " "
Cement..................................	50	1,500 bbls.
Timber piles...............................	72	500 lin. ft.

Flood Gate Structures

Excavation, Class 2..........................	23	
Excavation, Class 2, below mean low water......	23a	50 cu. yds.
Concrete..................................	46	
Concrete, below mean low water..............	46a	50 " "
Cement..................................	50	75 bbls.
Timber and lumber, miscellaneous.............	71	1,000 ft. B. M.
Miscellaneous cast iron, wrought iron, and steel..	82	2,000 lbs.

Drainage System	Item	Estimated Quantity
Excavation, Class 1	16	
Excavation, Class 2	17	160 cu. yds.
Excavation, Class 3	18	
Excavation and backfilling for trenches	25	100 " "
Concrete in culverts	46	30 " "
Cement	50	40 bbls.
Furnishing vitrified pipe	55	List price, $100
Laying vitrified pipe	56	1,200 inch-ft.
Furnishing cast iron pipe	75	2 tons
Laying cast iron pipe	76	600 inch-ft.
Flood gates	77	1,000 lbs.
Miscellaneous cast iron, wrought iron, and steel	82	2,000 "

Street Modifications

	Item	
Excavation, Class 1	19	1,000 cu. yds.
Gravel paving for streets and highways	61	2,500 sq. yds.
Guard rails	67	1,000 lin. ft.

General

	Item	
Clearing and grubbing	2	17 acres
Cleaning up	85	1 job

Contracts Nos. 37 and 40—Troy Local Protection.

Note.—This work is to be divided into four contracts numbered 37, 38, 39 and 40, of which Contract No. 37 contains the main work. Contract No. 38 contains 10,000 cu. yds. of levee embankment; Contract No. 39 contains 4,500 cu. yds. of levee embankment; Contract No. 40 contains 57,500 cu. yds. of levee embankment together with the work listed under the headings of Raising Bridges and Street Modifications.

Improvement of River Channel

	Item	
Trimming and shaping slopes of river channels	4	1,100 sq. yds.
Excavation, Class 1	6	
Excavation, Class 2	7	
Excavation, Class 3	8	
Excavation, Class 1, below mean low water	6a	88,600 cu. yds.
Excavation, Class 2, below mean low water	7a	
Excavation, Class 3, below mean low water	8a	
Excavation, Class 1	9	
Excavation, Class 2	10	167,900 " "
Excavation, Class 3	11	
Overhaul, per 100 feet	12	10,000 " "
Steel sheet piling	74	80 tons

Levee Construction

	Item	
Soil stripping	5	18,040 cu. yds.
Embankment, Class D	34	109,100 " "
Embankment, Class E	35	36,850 " "
Surface dressing and grassing	38	18,480 " "

Revetment

	Item	
Concrete in monolithic slope revetment	44	970 " "
Concrete in flexible slab revetment	45	280 " "
Cement	50	1,675 bbls.
Reinforcing steel	51	55,300 lbs.
Timber piles	72	5,000 lin. ft.
Miscellaneous cast iron	82	5,600 lbs.

Retaining Walls	Item	Estimated Quantity
Excavation, Class 1	22	
Excavation, Class 2.	23	620 cu. yds.
Excavation, Class 3	24	
Backfilling	37	100 " "
Concrete	41	530 " "
Stone masonry	48	10 " "
Cement	50	690 bbls.
Reinforcing steel	51	1,000 lbs.
Crushed rock and gravel	60	10 cu. yds.

Flood Gate Structures		
Excavation, Class 2	23	
Excavation, Class 2, below mean low water	23a	100 "
Concrete	46	
Concrete below mean low water	46a	105 " "
Cement	50	135 bbls.
Flood gates	77	9,000 lbs.

Drainage System		
Excavation, Class 1	16	
Excavation, Class 2	17	1,330 cu. yds.
Excavation, Class 3	18	
Excavation, Class 1, and backfilling for trenches	25	
Excavation, Class 2, and backfilling for trenches	26	920 "
Excavation, Class 3, and backfilling for trenches	27	
Concrete in apron, etc.	43	25 "
Concrete in culverts and gate chambers	46	150 " "
Concrete in sewer manholes and catch basins	47	42 " "
Cement	50	275 bbls.
Furnishing vitrified pipe	55	List price, $1,000
Laying vitrified pipe	56	10,700 inch-ft.
Furnishing cast iron pipe	75	5 tons
Laying cast iron pipe	76	750 inch-ft.
Flood gates	77	9,600 lbs.
Miscellaneous iron and steel (manhole covers)	82	4,500 "

Bridge Protection		
Steel sheet piling	74	170 tons

Raising Bridges		
Excavation, Class 2	23	200 cu. yds.
Concrete	41	100 " "
Cement	50	120 " "
Raising bridge	81	1 job

Street Modifications		
Excavation, Class 1	19	6,500 cu. yds.
Cement	50	640 bbls.
Water-bound macadam paving	62	4,800 sq. yds.
Concrete paving	65	3,000 " "
Guard rails	67	200 lin. ft.
Miscellaneous steel	82	43,000 lbs.

General		
Clearing and grubbing	2	15 acres
Cleaning up	85	1 job

Contracts Nos. 41 to 46—Dayton Local Protection.

Note.—The Dayton work will be divided into five contracts. Contract No. 41 is essentially a levee contract containing approximately 95,000 cu. yds. of levee embankment. Contract No. 45 is also a levee contract containing approximately 35,000 cu. yds. of levee embankment. Contract No. 46 contains approximately 75,000 cu. yds. of channel excavation, and 23,000 cu. yds. of levee embankment. Contract No. 42 is the contract for the main river improvement, and contains approximately 970,000 cu. yds. of channel excavation, 80,000 cu. yds. of levee embankment, and 25,000 cu. yds. of concrete. Contract No. 43 contains 462,000 cu. yds. of channel excavation, 205,000 cu. yds. of levee embankment, and 3,500 cu. yds. of concrete. Contract No. 44 is unassigned.

Improvement of River Channel	Item	Estimated Quantity
Trimming and shaping slopes	4	43,800 sq. yds.
Soil stripping	5	22,500 cu. yds.
Excavation, Class 1	6	
Excavation, Class 2	7	110,100 " "
Excavation, Class 3	8	
Excavation, Class 1	9	
Excavation, Class 2	10	1,653,000 " "
Excavation, Class 3	11	
Overhaul, per 100 feet	12	4,250,000 " "
Surface dressing and grassing	38	20,000 " "
Steel sheet piling	74	84 tons

Levee Construction		
Soil stripping	5	33,000 cu. yds.
Embankment, Class D, (from river excavation)	34	367,500 " "
Embankment, Class E, (from borrow pits)	35	167,000 " "
Surface dressing and grassing	38	68,000 " "

Revetment		
Concrete	44	
Concrete below mean low water	44a	8,500 " "
Concrete in flexible slab revetment	45	3,000 " "
Reinforcing steel	51	330,000 lbs.
Timber piles	72	15,000 lin. ft.
Steel sheet piling	74	84 tons
Miscellaneous iron and steel	82	45,000 lbs.
Cement	50	12,500 bbls.

Retaining Walls		
Excavation, Class 1	22	
Excavation, Class 2	23	
Excavation, Class 3	24	
Excavation, Class 1, below mean low water	22a	45,880 cu. yds.
Excavation, Class 2, below mean low water	23a	
Excavation, Class 3, below mean low water	24a	
Backfilling	37	500 " "
Concrete	41	
Concrete below mean low water	41a	20,000 " "
Cement	50	32,000 bbls.
Reinforcing steel	51	135,000 lbs.
Furnishing vitrified pipe	55	List price, $1,000
Laying vitrified pipe	56	18,000 inch-ft.
Crushed rock or gravel	60	500 cu. yds.
Timber piles	72	40,000 lin. ft.
Miscellaneous iron and steel	82	15,000 lbs.
Stone masonry	48	2,000 cu. yds.

Flood Gate Structures

	Item	Estimated Quantity
Excavation, Class 1	22	
Excavation, Class 2	23	
Excavation, Class 3	24	
Excavation, Class 1, below mean low water	22a	140 cu. yds.
Excavation, Class 2, below mean low water	23a	
Excavation, Class 3, below mean low water	24a	
Concrete	46	
Concrete below mean low water	46a	200 " "
Cement	50	275 bbls.
Reinforcing steel	51	4,000 lbs.
Steel sheet piling	74	12 tons
Flood gates	77	44,000 lbs.
Miscellaneous iron and steel	82	3,000 "

Drainage System

	Item	Estimated Quantity
Excavation, Class 1	16	
Excavation, Class 2	17	30 cu. yds.
Excavation, Class 3	18	
Excavation, Class 1, and backfilling for trenches	25	
Excavation, Class 2, and backfilling for trenches	26	
Excavation, Class 3, and backfilling for trenches	27	
Excavation, Class 1, for same, below mean low water	25a	640 "
Excavation, Class 2, for same, below mean low water	26a	
Excavation, Class 3, for same, below mean low water	27a	
Concrete in culverts, etc	46	10 "
Concrete in sewers, manholes, and catch basins	47	150 " "
Cement	50	200 bbls.
Furnishing vitrified pipe	55	List price, $200
Laying vitrified pipe	56	2,400 inch-ft.

Bridge Protection

	Item	Estimated Quantity
Concrete	41	25 cu. yds.
Cement	50	40 bbls.
Steel sheet piling	74	500 tons

Street Modifications

	Item	Estimated Quantity
Excavation, Class 1	19	
Excavation, Class 2	20	5,000 cu. yds.
Excavation, Class 3	21	
Gravel paving	61	2,000 sq. yds.
Macadam paving	62	2,500 " "
Cement	50	35 bbls.

General

	Item	Estimated Quantity
Excavation, Class 1	28	
Excavation, Class 2	29	120 cu. yds.
Excavation, Class 3	30	
Clearing and grubbing	85	40 acres

Note.—Contracts Nos. 47 and 48 are unassigned.

Contract No. 49.—West Carrollton Local Protection.

Levee Construction	Item	Estimated Quantity
Soil stripping	5	2,650 cu. yds.
Embankment, Class E, (from borrow)	35	28,700 " "
Surface dressing and grassing	38	4,600 " "

Retaining Walls

Excavation, Class 1	22 ⎫	
Excavation, Class 2	23 ⎬	295 "
Excavation, Class 3	24 ⎭	
Backfilling	37	10 "
Concrete	41	260 " "
Cement	50	360 bbls.
Miscellaneous iron and steel	82	2,500 lbs.

Drainage System

Excavation, Class 1, and backfilling for trenches.	25 ⎫	
Excavation, Class 2, and backfilling for trenches.	26 ⎬	60 cu. yds.
Excavation, Class 3, and backfilling for trenches.	27 ⎭	
Concrete in culverts, etc.	46	25 " "
Concrete in manholes, etc.	47	10 " "
Cement	50	50 bbls.
Vitrified pipe	55	List price, $195
Laying vitrified pipe	56	1,440 inch-ft.
Furnishing cast iron pipe	75	2 tons
Laying cast iron pipe	76	600 inch-ft.
Flood gates	77	2,500 lbs.
Miscellaneous cast iron, wrought iron and steel	82	1,000 "

Raising Bridges

Raising existing bridges	81	1 job

Street Modifications

Excavation for cuts and fills, Class 1	19	2,000 cu. yds.
Gravel paving	61	4,200 sq. yds.
Guard rails	67	1,000 lin. ft.

General

Clearing and grubbing	2	1 acre
Cleaning up	85	1 job

Contracts Nos. 50, 51, and 52—Miamisburg Local Protection.

The work at Miamisburg will be divided into three contracts. Contract No. 50 will comprise all work, except drainage, on the east side of the river, and Contract No. 51 will comprise the work on the west side of the river. The respective amounts of embankment in these contracts is indicated below. The drainage work will be done under Contract No. 52.

Improvement of River Channel

Trimming and shaping slopes	4	3,600 sq. yds.
Excavation, Class 1	9 ⎫	
Excavation, Class 2	10 ⎬	2,150 cu. yds.
Excavation, Class 3	11 ⎭	

Levee Construction	Item	Estimated Quantity
Soil stripping, east side of river................	5	14,400 cu. yds.
Soil stripping, west side of river...............	5	4,500 " "
Embankment, Class E, from borrow pit, east side of river..................................	35	171,800 " "
Embankment, Class E, from borrow pit, west side of river..................................	35	50,500 "
Surface dressing and grassing, east side of river..	38	7,825 " "
Surface dressing and grassing, west side of river..	38	2,500 "

Revetment

Concrete..................................	44 ⎫	
Concrete below mean low water...............	44a ⎬	750 " "
Cement..................................	50	1,120 bbls.
Reinforcing steel.............................	51	12,500 lbs.
Timber piles................................	72	1,150 lin. ft.
Miscellaneous iron and steel..................	82	100 lbs.

Retaining Walls

Excavation, Class 1..........................	22 ⎫	
Excavation, Class 2..........................	23 ⎪	
Excavation, Class 3..........................	24 ⎪	
Excavation, Class 1, below mean low water.....	22a ⎬	670 cu. yds.
Excavation, Class 2, below mean low water.....	23a ⎪	
Excavation, Class 3, below mean low water.....	24a ⎭	
Concrete..................................	41 ⎫	
Concrete below mean low water...............	41a ⎬	1,150 " "
Cement..................................	50	1,720 bbls.
Reinforcing steel.............................	51	28,000 lbs.
Furnishing vitrified pipe......................	55	List price, $120
Laying vitrified pipe.........................	56	1,800 inch-ft.
Timber piles................................	72	100 lin. ft.
Miscellaneous iron and steel..................	82	500 lbs.

Flood Gate Structures

Excavation, Class 1..........................	22 ⎫	
Excavation, Class 2..........................	23 ⎪	
Excavation, Class 3..........................	24 ⎪	
Excavation, Class 1, below mean low water.....	22a ⎬	470 cu. yds.
Excavation, Class 2, below mean low water.....	23a ⎪	
Excavation, Class 3, below mean low water.....	24a ⎭	
Concrete..................................	46	650 " "
Cement..................................	50	1,050 bbls.
Reinforcing steel.............................	51	7,000 lbs.
Timber piles................................	72	100 lin. ft.
Steel sheet piling............................	74	13 tons
Flood gates................................	77	40,600 lbs.
Miscellaneous iron and steel..................	82	2,000 "

Drainage System

Excavation, Class 1..........................	16 ⎫	
Excavation, Class 2..........................	17 ⎬	1,000 cu. yds.
Excavation, Class 3..........................	18 ⎭	
Excavation, Class 1, and backfilling for trenches.	25 ⎫	
Excavation, Class 2, and backfilling for trenches.	26 ⎬	4,010 "
Excavation, Class 3, and backfilling for trenches.	27 ⎭	
Concrete aprons.............................	43	10 "
Concrete in culverts and gate chambers.........	46	140 "
Concrete in sewers, manholes, and catch basins..	47	660 " "
Cement..................................	50	1,230 bbls.
Reinforcing steel.............................	51	15,000 lbs.
Furnishing vitrified pipe......................	55	List price, $1710

Drainage System (Continued)	Item	Estimated Quantity
Laying vitrified pipe	56	16,200 inch-ft.
Timber piles	72	100 lin. ft.
Furnishing cast iron pipe	75	5.6 tons
Laying cast iron pipe	76	1,400 inch-ft.
Flood gates	77	35,500 lbs.
Miscellaneous iron and steel	82	8,000 "

Street Modifications		
Excavation	19	25,000 cu. yds.
Gravel paving	61	3,830 sq. yds.
Water-bound macadam paving	62	600 " "
Brick paving on concrete foundation	64	800 " "
Guard rails	67	200 lin. ft.

General		
Clearing and grubbing	2	13 acres
Cleaning up	85	1 job

Contracts Nos. 53 and 54—Franklin Local Protection.

The work at Franklin will be divided into two contracts for the two sides of the river. The principal quantities in each are indicated below.

Improvement of River Channel		
Trimming slopes	4	3,050 sq. yds.
Excavation, Class 1	9	
Excavation, Class 2	10	100,600 cu. yds.
Excavation, Class 3	11	
Surface dressing and grassing	38	850 " "

Levee Construction		
Soil stripping, east side of river	5	10,400 "
Soil stripping, west side of river	5	200 "
Embankment, Class D, east side of river	34	52,000 "
Embankment, Class E, east side of river	35	11,300 "
Embankment, Class E, west side of river	35	60,850 "
Surface dressing and grassing, east side of river	38	5,750 "
Surface dressing and grassing, west side of river	38	5,500 "

Revetment		
Concrete	44	
Concrete below mean low water	44a	830 " "
Cement	50	1,340 bbls.
Reinforcing steel	51	21,700 lbs.
Timber piles	72	5,500 lin. ft.
Miscellaneous iron and steel	82	500 lbs.

Retaining Walls		
Excavation, Class 1	22	
Excavation, Class 2	23	
Excavation, Class 3	24	
Excavation, Class 1, below mean low water	22a	4,900 cu. yds.
Excavation, Class 2, below mean low water	23a	
Excavation, Class 3, below mean low water	24a	
Backfilling	37	50 "
Concrete	41	
Concrete below mean low water	41a	2,750 " "
Cement	50	3,950 bbls.

Retaining Walls (Continued)	Item	Estimated Quantity
Reinforcing steel	51	8,500 lbs.
Furnishing vitrified pipe	55	List price, $150
Laying vitrified pipe	56	3,000 inch-ft.
Timber piles	72	9,350 lin. ft.
Miscellaneous iron and steel	82	500 lbs.

Flood Gate Structures

Excavation, Class 1	22	
Excavation, Class 2	23	
Excavation, Class 3	24	
Excavation, Class 1, below mean low water	22a	450 cu. yds.
Excavation, Class 2, below mean low water	23a	
Excavation, Class 3, below mean low water	24a	
Concrete	46	
Concrete below mean low water	46a	850 " "
Cement	50	1,280 bbls.
Reinforcing steel	51	7,000 lbs.
Timber piles	72	200 lin. ft.
Steel sheet piling	74	15 tons
Flood gates	77	28,500 lbs.
Miscellaneous iron and steel	82	3,500 "

Drainage System

Excavation, Class 1, and backfilling for trenches	25	
Excavation, Class 2, and backfilling for trenches	26	190 cu. yds.
Excavation, Class 3, and backfilling for trenches	27	
Concrete aprons	43	5 " "
Concrete culverts and gate chambers	46	70 " "
Concrete sewers, manholes, and catch basins	47	15 " "
Cement	50	145 bbls.
Reinforcing steel	51	1,500 lbs.
Furnishing vitrified pipe	55	List price, $1137.50
Laying vitrified pipe	56	8,400 inch-ft.
Furnishing cast iron pipe	75	3 tons
Laying cast iron pipe	76	600 inch-ft.
Flood gates	77	10,000 lbs.
Miscellaneous iron and steel	82	4,000 " "

Street Modifications

Excavation for street fills	19	21,700 cu. yds.
Gravel paving	61	2,900 sq. yds.
Water-bound macadam paving	62	700 " "
Guard rails	67	500 lin. ft.

General

Clearing and grubbing	2	7 acres
Cleaning up	85	1 job

Contract No. 55.—Middletown Local Protection.

Improvement of River Channel	Item	Estimated Quantity
Excavation, Class 1	6	
Excavation, Class 2	7	
Excavation, Class 1, below mean low water	6a	65,000 cu. yds.
Excavation, Class 2, below mean low water	7a	

Levee Construction

Excavation, soil stripping	5	14,100 " "
Embankment, Class D, (from cut)	34	50,000 " "
Embankment, Class E, (from borrow)	35	120,000 " "
Surface dressing and grassing	38	14,500 " "

Revetment

Concrete	44	
Concrete below mean low water	44a	25 "

Retaining Walls

Excavation, Class 1	22	
Excavation, Class 2	23	550 "
Excavation, Class 2, below mean low water	23a	
Concrete	41	450 " "
Cement	50	600 bbls.

Flood Gate Structures

Excavation, Class 1	22	
Excavation, Class 1, below mean low water	22a	500 cu. yds.
Concrete	46	
Concrete below mean low water	46a	330 " "
Cement	50	430 bbls.
Reinforcing steel	51	9,000 lbs.
Steel sheet piling	74	9 tons
Flood gates	77	32,500 lbs.
Hand rails (miscellaneous iron and steel)	82	2,000 "

Drainage System

Excavation, Class 1	16	
Excavation, Class 2	17	30 cu. yds.
Excavation, Class 3	18	
Excavation and backfilling, Class 1	25	
Excavation and backfilling, Class 2	26	270 "
Excavation and backfilling, Class 3	27	
Concrete in culverts, etc.	46	70 " "
Cement	50	110 bbls.
Reinforcing steel	51	2,000 lbs.
Furnishing vitrified pipe	55	List price, $264
Laying vitrified pipe	56	2,760 inch-ft.
Furnishing cast iron pipe	75	2 tons
Laying cast iron pipe	76	500 inch-ft.
Flood gates	77	17,000 lbs.
Miscellaneous cast iron, wrought iron, and steel	82	4,000 "

Street Modifications

Excavation, Class 1	19	200 cu. yds.
Gravel paving	61	250 sq. yds.

General

Clearing and grubbing	2	1 acre
Cleaning up	85	1 job

Note.—Contract No. 57 is unassigned.

Contracts Nos. 58, 59, and 60—Hamilton Local Protection.

Note.—The work at Hamilton will be done under Contract No. 58, except that under Items 6 to 8a, New Channel Excavation, and that under Drainage System, which will be done under Contracts Nos. 59 and 60 respectively.

Improvement of River Channel	Item	Estimated Quantity
Trimming and shaping slopes of river channels...	4	156,800 sq. yds.
Soil stripping...	5	2,000 cu. yds.
Excavation, Class 1...	6	
Excavation, Class 2...	7	
Excavation, Class 3...	8	
Excavation, Class 1, below mean low water....	6a	84,030 " "
Excavation, Class 2, below mean low water....	7a	
Excavation, Class 3, below mean low water....	8a	
Excavation, Class 1...	9	
Excavation, Class 2...	10	1,950,000 " "
Excavation, Class 3...	11	
Overhaul, per 100 feet...	12	9,750,000 " "
Surface dressing and grassing...	38	23,500 " "
Steel sheet piling...	74	100 tons

Levee Construction

	Item	
Clearing and grubbing...	2	2 acres
Removal of buildings...	3	1 job
Soil stripping...	5	11,000 cu. yds.
Embankment, Class D, from river excavation...	34	280,000 " "
Embankment, Class E, from borrow pits...	35	13,000 " "
Surface dressing and grassing...	38	33,000 " "

Revetment

	Item	
Concrete...	44	
Concrete below mean low water...	44a	9,800 " "
Concrete in flexible slab revetment...	45	9,000 " "
Cement...	50	27,000 bbls.
Reinforcing steel...	51	1,146,300 lbs.
Timber piles...	72	44,000 lin. ft.
Steel sheet piling...	74	300 tons
Miscellaneous iron and steel...	82	188,300 lbs.

Retaining Walls

	Item	
Excavation, Class 1...	22	
Excavation, Class 2...	23	
Excavation, Class 3...	24	
Excavation, Class 1, below mean low water...	22a	28,870 cu. yds.
Excavation, Class 2, below mean low water...	23a	
Excavation, Class 3, below mean low water...	24a	
Concrete...	41	
Concrete below mean low water...	41a	12,730 " "
Cement...	50	19,700 bbls.
Reinforcing steel...	51	254,600 lbs.
Furnishing vitrified pipe...	55	List price, $60
Laying vitrified pipe...	56	1,200 inch-ft.
Timber piles...	72	20,800 lin. ft.
Miscellaneous iron and steel...	82	4,000 lbs.

Drainage System	Item	Estimated Quantity
Excavation, Class 1	16 ⎫	
Excavation, Class 2.	17 ⎬ 1,250 cu. yds.	
Excavation, Class 3.	18 ⎭	
Excavation, Class 1, and backfilling for trenches.	25 ⎫	
Excavation, Class 2, and backfilling for trenches.	26	
Excavation, Class 3, and backfilling for trenches.	27	
Excavation, Class 1, for same, below mean low water.	25a ⎬ 32,645 " "	
Excavation, Class 2, for same, below mean low water.	26a	
Excavation, Class 3, for same, below mean low water.	27a ⎭	
Concrete aprons.	43	10 " "
Concrete culverts and gate chambers.	46 ⎫	
Concrete culverts and gate chambers below mean low water.	46a ⎬ 205 " "	
Concrete sewers, manholes, and catch basins. . . .	47	2,740 " "
Cement.	50	4,500 bbls.
Reinforcing steel.	51	59,000 lbs.
Vitrified brick lining.	52	25 cu. yds.
Furnishing vitrified pipe.	55	List price, $2200
Laying vitrified pipe.	56	21,600 inch-ft.
Bituminous-bound paving.	63	760 sq. yds.
Asphalt paving.	66	1,700 " "
Timber and lumber.	71	1,000 ft. B. M.
Timber piles.	72	3,000 lin. ft.
Furnishing cast iron pipe.	75	10 tons
Laying cast iron pipe.	76	2,200 inch-ft.
Flood gates. . . . . :	77	40,000 lbs.
Miscellaneous iron and steel.	82	20,000 "

Bridge Protection

	Item	Quantity
Excavation.	23	30 cu. yds.
Concrete.	41	25 " "
Cement.	50	30 bbls.
Steel sheet piling.	74	210 tons

Street Modifications

	Item	Quantity
Earth excavation for fills.	19	7,000 cu. yds.
Paving, water-bound macadam.	62	2,800 sq. yds.
Paving brick on concrete foundation. . .	64	670 " "
Guard rails.	67	400 lin. ft.

General

	Item	Quantity
Clearing and grubbing.	2	40 acres
Cleaning up.	85	1 job

PROPOSAL

FOR THE

CONSTRUCTION OF GERMANTOWN DAM AND APPURTENANCES

(Contract No. 1)

........................1917

To the Board of Directors,
The Miami Conservancy District,
Dayton, Ohio.

Pursuant to the advertisement of your Board dated September 15, 1917, for bids for the construction of dams and appurtenances, (Contract No. 1, Germantown Dam and Appurtenances, including Road No. 1) as designated in said Advertisement and in the specifications and contract drawings of the District; the undersigned bidder herewith proposes to do all work, perform all services, and furnish all material (unless otherwise definitely provided in the specifications or on the contract drawings), within the time, at the rate, and in the manner required, at the prices named in the following schedule; all in accordance with the specifications for dams and appurtenances adopted by the District; and agrees, if awarded this contract, to execute a valid contract and bond, which contract shall include the agreement, bond, advertisement, specifications, and drawings, in forms prepared by the District, and as set forth in the above mentioned form of agreement.

The bidder further agrees that in case of his failure to execute such contract with necessary bond, within the time fixed in the Advertisment, the check accompanying this proposal, and the money payable thereon, shall be and remain the property of the Miami Conservancy District, as liquidated damages for such failure.

The bidder further agrees that, in case the contract is awarded to him, he will begin work within the time specified in Article 2 of the Agreement; that the end of the period of preparation described in Article 2 of the Agreement shall be April 1, 1918, but not less than ninety days after the execution of the contract; that he will maintain the rate of progress outlined in the specifications and contract drawings, and that the work shall be completed on or before September 1, 1921.

SCHEDULE OF PRICES

Item	Classification	Estimated Quantity	Unit of Quantity	Unit Price
1	Stream control.....................	1	job	......
2	Clearing and grubbing..............	11	acre	....
5	Excavation, soil stripping............	25,000	cu. yd.	....
6	" Class 1, for cut-off trench and stream control channel...................	12,000	" "	....
7	" Class 2, for same........	400	" "	.....
8	" Class 3, for same........,.	200	" "	.....
6–a	" Class 1, for same, below mean low water........	5,000	" "	....
7–a	" Class 2, for same, below mean low water........	100	" "	....
8–a	" Class 3, for same, below mean low water........	100	" "	....
13	" Class 1, for outlet works and spillways..........	125,00	" "	....
14	" Class 2, for same........	19,000	" "	....
15	" Class 3, for same........	500	" "	....
13–a	" Class 1, for same, below mean low water........	10,000	" "	....
14–a	" Class 2, for same, below mean low water........	26,000	" "	,...
15–a	" Class 3, for same, below mean low water........	200	" "	...
16	" Class 1, for minor drainage systems..............	1,000	" "	....
17	" Class 2, for same........	100	" "	....'
18	" Class 3, for same........	50	" "	.....
19	" Class 1, for street and highway cuts and fills.......	8,600	" "	....
20	" Class 2, for same........	100	" "	.....
21	" Class 3, for same........	100	" "	.....
22	" Class 1, for retaining walls, bridge piers, etc., except at dams..............	100	" "	....
23	Excavation, Class 2, for same........	10	" "	.....
24	" Class 3, for same........	10	" "	.....
28	" Class 1, miscellaneous.....	2,000	" "	.....
29	" Class 2, miscellaneous.....	1,000	" "	.....
30	" Class 3, miscellaneous.....	300	" "	.....
*{31	Embankment, Class A.............	700,000	" "	....
{33	" Class C..............	130,000	" "	....
*{32	" Class B..............	700,000	" "	....
{33	" Class C..............	130,000	" "	....
*33	" Class C..............	830,000	" "	....
36	" Class F, (Miscellaneous fill).................	8,000	" "	......
37	Backfilling........................	10,000	" "	....
38	Surface dressing and grassing........	28,000	" "	....

*Alternatives. See specifications, items 0.72, 31.1, 32.1, and 33.1. The bidder is required to fill in prices on only one of these three alternate methods of building the dam; he may bid on more than one if he so desires, but in comparing bids no preference will be given to any one method.

Item	Classification	Estimated Quantity	Unit of Quantity	Unit Price
41	Concrete, in outlet works, spillways, retaining walls, bridge piers, and similar structures.....	6,000	cu. yd.	....
41–a	" in same, below mean low water...................	8,700	" "	....
42	" in conduit linings at dams...	1,400	" "	....
43	" in aprons, paving at dams, and similar flat shapes....	600	" "	....
43–a	" in same, below mean low water...................	2,600	" "	....
46	" in culverts, bulkheads, gate-chambers, and similar structures, except at dams.....	130	" "	....
50	Cement...........................	29,500	bbl.	....
51	Reinforcing steel....................	50,000	lb.	....
53	Drilling foundations or masonry......	500	lin. ft.	....
54	Grouting foundations or masonry......	100	cu. yd.	....
55	Furnishing vitrified round or channel pipe................... Note.—The unit prices on this item will be obtained by deducting the given percentage from the price list given on the next page at the bottom of the schedule.	$2,600	list price	% off
56	Laying vitrified round or channel pipe.. Note.—The quantity in inch-feet will be obtained by multiplying the length in feet of the pipe laid, by its nominal diameter in inches.	60,000	inch-ft.	....
57	Dry rubble paving..................	600	cu. yd.	....
58	Riprap.........................	400	" "	....
59	Gutter paving.....................	1,100	sq. yd.	....
60	Crushed rock or gravel............ ...	7,600	cu. yd.	....
67	Guard rails........................	3,200	lin. ft.	....
68	Wire fence.......	8,100	" "	....
70	Protective boom and accessories, lumber	22	M.ft.B.M.	....
71	Timber and lumber, miscellaneous.....	· 1	" " "	....
72	Timber piles 20 ft. or less in length....	None	lin. ft.	....
72–x	Timber piles more than 20 ft. long.....	None	" " "	....
73	Wood sheet piling...................	40	M.ft.B.M.	....
74	Steel sheet piling..	110	ton	....
82	Miscellaneous cast iron, wrought iron, and steel.......................	15,000	. lb.	....
85	Cleaning up.......................	1	job	....
89	Sheeting left in place.....	1	M.ft.B.M.	....
90	Extra work, at actual necessary cost plus 15%, as defined in Section 0.19 of the Specifications.			

CONDITIONS

The bidder shall cross out two of the three conditions following, and it is agreed that unless he does so the proposal is not conditional upon the award of any other contract.

(1) The undersigned is bidding on other contracts and makes this proposal subject to the condition that no other contract will be awarded him in addition to this one except Contracts Nos.......

....
(Last three words may be crossed off, if not required.)

(2) The undersigned is bidding on other contracts, and makes this proposal subject to the condition that the following contracts will be awarded to him in addition to this one. Contracts Nos......

....

(3) This proposal is not conditional upon the award of any other contract.

Note.—If the bidder wishes to make different unit prices, depending on whether or not more than one contract is awarded to him, he may submit more than one proposal on the same contract, using Condition No. 3 in the proposal that is not conditional upon the award of any other contract, and using Condition No. 2 in the proposals that are conditional upon the award of other contracts.

LIST OF CONTRACT DRAWINGS

The Contract Drawings to be listed in Article 6 of the Agreement, as forming a part of the contract for this work are as follows:

Drawing Number		Accession Number
1	General Plan of Dam and Appurtenances	1740
2	Cross Section of Dam and Drainage Details	1571
3	Spillway	1725
4	Outlet Works	1730
5	Outlet Works Details	1541
6	Road No. 1. Plan and Profile, Station 0+00 to Station 26+00	1557
7	Road No. 1. Plan and Profile, Station 26+00 to Station 68+50.8	1385
8	Road No. 1. Plan and Profile, Station 78+98.4 to Station 95+60	1579
9	Piers for Protective Boom	1577
10	Protective Boom Details	1600
11	Retaining Walls, Typical Sections	1769
12	Highways, Typical Sections	1605
13	Highways, Pipe Culverts	1610

14 Highways, Concrete Culverts, 2 ft.×2 ft. and 3 ft.×3 ft. 1615
15 Highways, Concrete Culverts, 3 ft.×2 ft. and 4 ft.×3 ft. 1620
16 Highways, Concrete Culverts, 5 ft.×3½ ft. & 6 ft.×4 ft. 1625
17 Highways, Concrete Slab Bridge...................1585
18 Highways, Handrails 1630
19 Highways, Wire Fences and Guard Rails...........1635
20 Stream Control, Sheet 1..........................1733
21 Stream Control, Sheet 2..........................1784

EXPERIENCE AND REFERENCES

The bidder declares that he has been engaged in the construction of works similar in character to that covered by this proposal for.............years, and that he has built the following works:

Nature of Work	Locality
....	....
....	....
....	....
....	....
....	....

and he further declares that he is fully prepared and has the necessary capital to begin the work promptly and to conduct is as required by the contract and specifications. He refers to the following persons who are competent to advise as to his financial standing:

Name	Address
....	....
....	....
....	....
....	....

PLANT AND EQUIPMENT

The bidder states that he has the following equipment available for immediate use on this contract, and that he proposes, at the times indicated, to add to his equipment the items listed on the next page.

Equipment Available for Immediate Use

Equipment to be Acquired

Date	Item of Equipment
....	....
....	....
....	....
....	....
....	....
....	....
....	....
....	....
....	....
....	....

The undersigned bidder declares that the only persons or parties interested in this proposal as principals are named herein; that this proposal is made without collusion with any other person, firm or corporation; that no member of the Board of Directors of the Miami Conservancy District or other officer of the District, or any person in the employ of the Board of Directors, is directly or indirectly interested as contracting party, partner, stock holder, surety or otherwise in the performance of the contract; that he or his agent has carefully examined the location of the proposed work, the proposed form of contract and the drawings for said contract, and is informed as to local conditions affecting the proposed work.

This proposal is conditioned on the award being made, and the work being financed and ready to proceed, on or before January 15, 1918, unless by the agreement of both parties this proposal is held binding for a longer period.

Bidder's Signature		Address
....	[Seal]	....
....	[Seal]	....
....	[Seal]	....

[SEAL]
 Attest:

....
                    Secretary.

AGREEMENT

(Note.—Bidder will not fill in blanks on this form. Proposals are to be made on the separate form furnished for that purpose.)

THIS AGREEMENT, made and entered into this_____ day of_____, in the year One Thousand Nine Hundred and _____ by and between The Miami Conservancy District, a corporation organized and existing under the Conservancy Act of Ohio, passed February 6, 1914, acting through its Board of Directors, by virtue of the power vested in it by said Act,

party of the first part, and_____

_____

_____

of the City of_____County of_____,

and State of_____, hereinafter designated as the **Contractor,** party of the second part,

WITNESSETH: That the parties to these presents each in consideration of the undertakings, promises, and agreements on the part of the other herein contained, have undertaken, promised, and agreed, and do hereby undertake, promise, and agree, the party of the first part for itself, its successors, and assigns, and the part____

of the second part for_____and_____heirs, executors, administrators, and assigns, as follows:

Article 1.—In consideration of the payments to be made as hereinafter provided, and of the performance by the party of the first part of all of the matters and things by it to be performed as herein provided, the Contractor, party of the second part, agrees, at his own sole cost and expense, to perform all the labor and services, and furnish all the materials, plant, and equipment necessary to complete, and to complete in good, substantial, workmanlike, and approved manner, within the time hereinafter specified, and in accordance with the terms, conditions and provisions of this Contract and of the instructions, orders and directions of the Engineer made in accordance with this Contract, the following work to-wit:_____

....

....

Article 2.—The Contractor further agrees to begin work within 45 days from the date of execution hereof, and to prosecute the same with speed and diligence so as to insure the completion of the work on or before...

....

It is hereby agreed that the period from the date of execution of this Agreement to...............................hereafter called the **End of Period of Preparation,** shall be utilized by the Contractor in assembling plant and equipment, organizing forces, and in getting the work under way so that thereafter he may maintain the rate of progress prescribed, and complete the work within the time specified in this Contract.

The maintenance of the required rate of progress on this Contract, and its completion within the specified time, being to an exceptional degree necessary for the complete success of the work, on account of the urgent need for relief and the very serious disadvantage of a delayed construction period, the Contractor agrees to take all precautions in preparation and management which may be necessary to insure the rate of progress and the time of completion required by this Contract.

Article 3.—The Miami Conservancy District, party of the first part, agrees to pay, and the Contractor, party of the second part, agrees to accept, as full compensation, satisfaction and discharge, for all work done and all materials furnished, whether mentioned in the following schedule or not, and for all costs and expenses incurred and damages sustained, and for each and every matter, thing or act performed, furnished or suffered, in the full and complete performance and completion of the work of this Contract in accordance with the terms, conditions, and provisions thereof and of the instructions, orders and directions of the Engineer thereunder, except **Extra Work** which shall be paid for as provided in Section 0.19 of the Specifications, and except as in this Contract otherwise specifically provided, a sum equal to the amount of the actual work done and materials furnished, as determined by the Engineer, under each Item in the following Schedule multiplied by the Unit Price applicable to each such Item, as set forth in the following Schedule. to-wit:

SCHEDULE OF UNIT PRICES AND ESTIMATED QUANTITIES

Note.—Bidders will not fill in blanks on this form. Use proposal blanks which are furnished as separate sheets.

(Items as here printed are those included in contracts for construction of dams and appurtenances. Appropriate changes were made in other contracts.)

Item	Classification	Estimated Quantity	Unit of Quantity	Unit Price
1	Stream control.....................	....	job	....
2	Clearing and grubbing..............	....	acre	....
3	Removal of buildings...............	....	job	....
5	Excavation, soil stripping............	....	cu. yd.	....
6	" Class 1, for cut-off trench and stream control channel..................	....		...
7	" Class 2, for same.........	.....		...
8	" Class 3, for same........	....		...
6-a	" Class 1, for same, below mean low water........	....		...
7-a	" Class 2, for same, below mean low water........	....		...
8-a	" Class 3, for same, below mean low water........	.....		...
13	" Class 1, for outlet works and spillways.........	....		...
14	" Class 2, for same........	....		...
15	" Class 3, for same........	....		...
13-a	" Class 1, for same, below mean low water........	....		..
14-a	" Class 2, for same, below mean low water........	....		..
15-a	" Class 3, for same, below mean low water........	....		..
16	" Class 1, for minor drainage systems..............	....		...
17	" Class 2, for same.......	....		...
18	" Class 3, for same........	....		...
19	" Class 1, for street and highway cuts and fills......	....	" "	...
20	" Class 2, for same........	....		....
21	" Class 3, for same........	....		....
22	" Class 1, for retaining walls, bridge piers, etc., except at dams..............	....	" "	....

Item	Classification	Estimated Quantity	Unit of Quantity	Unit Price
23	Excavation, Class 2, for same......	....	cu. yd.	....
24	" Class 3, for same......	....	" "	....
28	" Class 1, miscellaneous....	....	" "	....
29	" Class 2, miscellaneous...	....	'	....
30	" Class 3, miscellaneous..	....	'	....
31	Embankment, Class A.........	....	'	....
32	" Class B.........	...'	.. '	....
33	" Class C.........	....	'	....
36	" Class F, (Miscellaneous fill).........	....	'	....
37	Backfilling.........	....	'	....
38	Surface dressing and grassing.........	....	'	....
41	Concrete, in outlet works, spillways, retaining walls, bridge piers, and similar structures....	....	'	...
41-a	" in same, below mean low water.........	....	'	....
42	" in conduit linings at dams...	....	'	....
43	" in aprons, paving at dams, and similar flat shapes....	....	'	....
43-a	" in same, below mean low water.........		'	....
46	" in culverts, bulkheads, gate-chambers, and similar structures, except at dams.....	....	" "	...
50	Cement.........	....	bbl.	....
51	Reinforcing steel.........	....	lb.	....
53	Drilling foundations or masonry......	....	lin. ft.	....
54	Grouting foundations or masonry.....	....	cu. yd.	....
55	Furnishing vitrified round or channel pipe......... Note.—The unit prices on this item will be obtained by deducting the given percentage from the price list given on the next page at the bottom of the schedule.	....	list price	% off
56	Laying vitrified round or channel pipe.. Note.—The quantity in inch-feet will be obtained by multiplying the length in feet of the pipe laid, by its nominal diameter in inches.	....	inch-ft.	......
57	Dry rubble paving.........	....	cu. yd.	....
58	Riprap......	....	" "	....
59	Gutter paving.........	....	sq. yd.	....
60	Crushed rock or gravel.........	....	cu. yd.	....

Item	Classification	Estimated Quantity	Unit of Quantity	Unit Price
61	Rolled gravel paving for highways.....	....	sq. yd.	....
62	Water-bound macadam paving for highways.........................	....	" "	....
67	Guard rails............................	....	lin. ft.	....
68	Wire fence........................	....	" "	...
70	Protective boom and accessories, lumber	....	M.ft.B.M.	......
71	Timber and lumber, miscellaneous.....	....	" " "	....
72	Timber piles 20 ft. or less in length....	....	lin. ft.	....
72-x	Timber piles more than 20 ft. long.....	.. ..	" "	....
73	Wood sheet piling......................	....	M.ft.B.M.	......
74	Steel sheet piling.....................	....	ton	....
82	Miscellaneous cast iron, wrought iron, and steel...........................	....	lb.	....
85	Cleaning up....................	....	job	......
89	Sheeting left in place.............	....	M.ft.B.M.	......
90	Extra work, at actual necessary cost plus 15%, as defined in Section 0.19 of the Specifications.	•		

PRICE LIST OF VITRIFIED PIPE

Adopted by Eastern Manufacturers, May 2, 1912.

Calibre of Pipe Inches	Straight Pipe Per Foot	Elbows and Curves Each	Slants 1 ft. or Less Per Foot Long Side	Increasers and Reducers Each	Channel Pipe Per Foot	Branches 1 or 2 ft. long with Inlets up to 12 in. Each	Branches 2½ or 3 ft. long with Inlets up to 12 in. Each	Branches 2½ or 3 ft. long with Inlets 15 in. or Larger—Each	Stoppers or Plugs Each
3 & 4	$0.25	$0.75	$0.75	$1.00	$0.15	$1.00	$1.25	......	$0.08
5 & 6	.40	1.20	1.20	1.60	.24	1.60	2.00	......	.13
8	.55	1.65	1.65	2.20	.33	2.20	2.75	......	.19
9 & 10	.80	2.40	2.40	3.20	.48	3.20	4.00	......	.27
12	1.00	3.00	3.00	4.00	.60	4.00	5.00	......	.33
15	1.35	4.05	4.05	5.40	.81	5.40	6.75	$4.95	.45
18	1.90	5.70	5.70	7.60	1.14	7.60	9.50	13.30	......
20	2.25	6.75	6.75	9.00	1.35	9.00	11.25	15.75	......
21 & 22	3.00	9.00	9.00	12.00	1.80	12.00	15.00	21.00	......
24	3.25	9.75	9.75	13.00	1.95	13.00	16.25	22.75	......
27	4.50	13.50	13.50	18.00	2.70	......	22.50	27.00	......
30	5.50	16.50	16.50	22.00	3.30	......	27.50	33.00	......
33	6.25	18.75	18.75	25.00	3.75	......	31.25	37.50	......
36	7.00	21.00	21.00	28.00	4.20	......	35.00	42.00	......

Article 4.—In case of default in completing the whole work to be done under this Contract within the time herein specified, including such extensions as may have been granted, the Contractor hereby agrees to pay to the party of the first part as liquidated damages for such default; First, a sum sufficient to compensate said first party for the cost and expense of employing engineers, inspectors, and employees to the extent that their services are reasonably required during the period of default by the work of this Contract; and Second, a sum equal to one-half of one per cent on all moneys that have been paid the Contractor under this Contract for each calendar month or part thereof that the completion of the whole work under this Contract is delayed. The party of the first part shall have the right to deduct such liquidated damages from any moneys due or to become due the Contractor, and the amount, if any, still owing after such deduction shall be paid on demand by the Contractor or his Surety. Payment of such liquidated damages shall not relieve the Contractor or his Sureties from any other obligation under this Contract.

Article 5.—If the Contractor shall fail to comply with any of the terms, conditions, provisions, or stipulations of this Contract according to the true intent and meaning thereof, then the party of the first part may avail itself of any or all remedies provided in that behalf in the Contract, and shall have the right and power to proceed in accordance with the provisions thereof.

Article 6.—It is hereby agreed by the parties to this Agreement that the following exhibits attached thereto and made parts thereof shall constitute integral parts of said Agreement, the whole to be collectively known and referred to as the **Contract:**

Advertisement	Exhibit A
Agreement	Exhibit B
Bond	Exhibit C
Specifications	Exhibit D

which specifications include:

(a) General Conditions
(b) General Specifications
(c) Detail Specifications

Contract Drawings   Exhibit E
(As listed on next page.)

Contract Drawings for Contract No ....

No.	Title	Accession Number

Article 7.—The Contractor agrees to furnish a bond or bonds in such form and in such amount and with such Sureties, as is hereinafter provided, and as will be satisfactory to the Board, conditioned upon the faithful and complete performance and carrying out of this Contract.

IN WITNESS WHEREOF: The Miami Conservancy District, party of the first part, its Board of Directors having duly approved this Contract by resolution passed on the_____day of_____, 191___, has hereunto affixed its corporate name and the name of the President of its Board of Directors, and has hereto attached its corporate seal attested by the Secretary of its Board of Directors; and_____

_____

_____, party of the second part, has hereunto affixed his their name___ and seal___ its corporate name by its President thereunto duly authorized, and has hereto attached its corporate seal attested by its Secretary, in **duplicate**, the day and year first above written.

[SEAL]   **THE MIAMI CONSERVANCY DISTRICT**

President, Board of Directors.
PARTY OF THE FIRST PART

Attest:

Secretary, Board of Directors.

By_____
President.

_____---[SEAL]

_____[SEAL]

[SEAL]   _____[SEAL]

Attest: PARTY OF THE SECOND PART

Secretary.

Witnesses.

Exhibit C

BOND

(Note.— Bidders will not fill in blanks on this form.)

KNOW ALL MEN BY THESE PRESENTS, That we_____

----

of_____, County of_____ and State

of_____, as **Principal**, hereinafter called the **Contractor,**

and _____

----

organized under the laws of the State of_____ and
duly authorized to transact business within the State of Ohio, as
Surety, hereinafter called the **Surety,** are held and firmly bound un-
to THE MIAMI CONSERVANCY DISTRICT, a corporation or-
ganized under the laws of the State of Ohio, with its principal place
of business located at Dayton, County of Montgomery, and State of
Ohio, as **Obligee,** hereinafter called the **District,** in the sum of_____

----

dollars ($_____) lawful money of the United States, for the

payment whereof to the District the Contractor binds_____

_____, _____ heirs, executors, ad-
ministrators, successors, and assigns, and the Surety binds itself,
its successors, and assigns, firmly by these presents.

WHEREAS the Contractor and the District have entered into

a written contract, dated the_____ day of_____

_____191___, hereinafter called the **Contract,**
a copy of which is attached hereto and made a part hereof, for_____

THE CONDITION OF THIS OBLIGATION IS SUCH that if the said Contractor shall pay all persons, firms or corporations who perform labor or furnish equipment, supplies and materials for use in the work thereunder, and shall satisfy all claims against the District, for damages to life, limb or property that may be caused by the acts of, or negligence of, the Contractor or any of his agents or employees, and shall satisfy all suits or claims brought against the District arising from the violation of any law, ordinance, regulation, order or decree on the part of the Contractor or any of his agents or employees; or from any infringement, or alleged infringement of patents in the work under said Contract; or howsoever originating from any of the operations under said Contract; and shall fully indemnify and save harmless the District from all cost and damage which it may suffer by reason of failure so to do, and shall fully reimburse and repay the District all outlay and expense which the District may incur in making good any such default, and in all other particulars shall faithfully perform the Contract on his part according to all the terms, covenants, and conditions thereof and within the time specified therein, then this obligation shall be void; otherwise to remain in full force and effect.

PROVIDED,

First—That should the Contractor fail to comply with the provisions of the Contract to such an extent that the Contract shall be forfeited, the Surety shall have the right to assume the Contract and proceed to perform or sublet the same, as therein provided. And the Surety shall, in that event, be subrogated to all the rights and interests of the Contractor arising out of the Contract, and be entitled to hold and use all of the equipment and properties of the Contractor which may be necessary for the completion of the Contract; and all moneys which may be due the Contractor at the time of his default or which may thereafter become due said Contractor, under or by virtue of the Contract, shall become due and payable to the Surety as the work progresses, subject to all of the terms of the Contract.

Second—That any alterations which may be made in the terms of the Contract, or in the work or materials to be furnished thereunder; or the granting by the District of any extension of time; or any forbearance or action on the part of either the District or the Contractor toward the other under said Contract; shall not in any way release the Contractor and the Surety, or either of them, their heirs, executors, administrators, successors, or assigns, from their liability hereunder; notice to the Surety of any such alteration, extension, forbearance, or action, being hereby waived; provided that the written consent of the Surety shall first be obtained if any alteration be required which shall alter the general character of the work as a whole, or which shall increase the total amount to be paid to the Contractor by more than 25 per cent.

Third—No right of action shall accrue hereunder to or for the use or benefit of any one other than the District, and the District's rights hereunder may not be assigned without the written consent of the Surety.

Signed and sealed, this------------------------------- day of

---------------------------19----.

Witnesses:

----------------------   --------------------------[SEAL]

----------------------   --------------------------[SEAL]

----------------------   --------------------------[SEAL]
                                                      PRINCIPAL.
[Seal]

                                        President.
                                                SURETY.
Attest:

    Secretary.

SPECIFICATIONS

EXPLANATORY NOTE

The specifications fall into three parts; first, the General Conditions, Sections 0.1 to 0.57, governing the relations of the contractor to the District, to the public, and to his employees; second, the General Specifications, Sections 0.58 to 0.105, defining and classifying the most commonly used materials, and indicating in general how the work shall be done; and third, the Detail Specifications, Sections 1.1 to 90.2. The Detail Specifications by supplementing the Agreement, General Conditions, General Specifications, the Estimates, and the Drawings, define in detail how each kind of work shall be done, and how it will be measured for payment.

The different kinds of construction work to be done and of materials to be furnished are classified into items, an item meaning a certain kind of work to be done, or material to be furnished, in a certain definite manner, regardless of its location. Each item is designated by a name and number, the numbers running from 1 to 90; as Item 2, Clearing and Grubbing; or Item 51, Reinforcing Steel. All sections relating to any item bear the number of the item as an index number before the decimal point. For instance, Item 50 covers Cement. Section 50.1 gives the requirements as to brands, Section 50.2 specifies the manner of delivery and storage, Section 50.3 covers inspection, tests, etc., and Section 50.4 specifies the methods of determining quantities of cement for payment.

All item numbers with letter "a" affixed refer to work below mean low water level as indicated on the drawings. For instance, Items 41 and 41a include concrete in outlet works, spillways, retaining walls, bridge piers, and other similar structures, above and below mean low water, respectively. The same item may be met on various parts of the project. Wherever in the estimates or on the drawings "Item 41" is indicated, regardless of whether it occurs in channel improvement at Hamilton, or in the construction of the Englewood Dam, or in some other construction, the work to be done is the placing of concrete in accordance with the Specifications under Item 41, supplemented in all cases by the General Specifications, which apply to all items. Item 41a refers to the same kind of construction as Item 41, except that the work is located below mean low water, as shown on the drawings.

In brief, the General Conditions and General Specifications relate, in so far as they are applicable, to all work. The estimates and the drawings indicate under what item or items any particular part of the work is to be done, and the Specifications for that item give the necessary further directions for doing that particular work.

GENERAL CONDITIONS

0.1 Definitions.—Wherever the words herein defined, or pronouns used in their stead, occur in this Contract and Specifications, they shall have the meanings here given:

The word **District** or the expressions **Party of the First Part** or **First Party** shall mean the Corporation of The Miami Conservancy District.

The word **Board** shall mean the Board of Directors of The Miami Conservancy District, or any agency or officer duly authorized to act for the District in the execution of the work required by this Contract.

The word **Directors** shall mean the individual members of the Board of Directors acting in their official capacity.

The word **Engineer** shall mean the Chief Engineer of The Miami Conservancy District, or his properly authorized agents, engineers, assistants, inspectors, and superintendents, acting severally within the scope of the particular duties entrusted to them.

The word **Contractor** or the expressions **Party of the Second Part** or **Second Party** shall mean the person, persons, partnership, or corporation entering into this Contract for the performance of the work required by it, and the legal representatives of said party, or the agent appointed to act for said party in the performance of the work.

The words **Surety** or **Sureties** shall mean the bondsmen or party or parties who have made secure the fulfillment of the Contract by a Bond, and whose signatures are attached to said Bond.

The word **Contract** shall mean, collectively, all of the covenants, terms, and stipulations contained in the various portions of this Contract, to-wit:

Advertisement, Agreement, Bond, Specifications, and Contract Drawings.

The word **Specifications** shall mean, collectively, all of the terms and stipulations contained in those portions of the Contract known as the General Conditions, General Specifications, and Detail Specifications.

The word **Drawings** shall mean, collectively, all of the drawings attached to the Contract and made part thereof, and also such supplementary drawings as the Engineer may issue from time to time in order to elucidate said contract drawings or for showing details which are not shown thereon, or for the purpose of showing changes in the work as authorized under Section 0.18 of the Specifications entitled **Changes and Alterations.**

The words **Contract Price** shall mean either the unit price or unit prices named in the Agreement, or the total of all payments under the Contract at the unit price or unit prices, as the case may be.

Wherever in this Contract the words **Directed, Required, Permitted, Ordered, Instructed, Designated, Considered Necessary, Prescribed,** or words of like import are used, it shall be understood that the direction, requirement, permission, order, instruction, designation, or prescription, etc., of the Engineer is intended; and similarly, the words **Approved, Acceptable, Satisfactory,** or words of like import, shall mean approved by, or acceptable or satisfactory to, the Engineer, unless another meaning is plainly intended.

Wherever figures are given in this Contract after the word **Elevation** or an abbreviation of it, or where figures representing elevations are given, they shall mean distance in feet above the United States Geological Survey Sea Level Datum, as determined from various bench marks placed by said Survey in and adjoining the lands of The Miami Conservancy District.

0.2 Sureties.—With the execution and delivery of this Contract the Contractor shall give security for the faithful performance of the Contract by filing with the Board one or more Surety bonds in the form annexed hereto, the aggregate amount of which shall be not less than 40 per cent of the estimated total contract price. Each bond must be signed by the Contractor and the Sureties. The Sureties, and the amount in which each will qualify, must be to the reasonable satisfaction of the Board.

Should any Surety upon the Contract be deemed unsatisfactory at any time by the Board, notice will be given the Contractor to that effect, and the Contractor shall forthwith substitute a new Surety or Sureties satisfactory to the Board. And no further payment shall be deemed due or shall be made under this Contract, until the new Surety or Sureties shall qualify and be accepted by the Board.

At the expiration of each year after the date of Contract, the liability of the Sureties under the bonds may be reduced to the extent to which, in the opinion of the Board, the need for Surety is reduced, until such liability shall amount to 25 per cent of the original liability, at which amount the bonds shall remain in full force and virtue until the completion of the entire work.

0.3 Time and Order of Completion.—The Contractor agrees that the work shall be commenced and carried on at such points, and in such order of precedence, and at such times and seasons as may be directed by the Engineer, in accordance with Section 0.10 of the Specifications. The Engineer shall have the right to have the work discontinued, for such time as may be necessary, in whole or in part, should the condition of the weather, or of flood, or other contingency make it desirable so to do, in order that the work shall be well and properly executed. Extension of time may be granted the Contractor for discontinuance of work so required, as provided in Section 0.4 of the Specifications, entitled **Extension of Time.**

The Contractor further agrees that he will begin work not later than at the time specified in the Agreement, and will progress therewith at such a rate that the work shall be completed in accordance with the Contract. (It is further agreed that the rate of progress shall be at least such that the work accomplished up to the end of each six months period,—the first period commencing at the end of the period of preparation as specified in the Agreement,—shall have the same proportion to the whole work, as the time consumed since the end of the period of preparation has to the entire time allowed for completing the Contract after the period of preparation. For the application of this rule, the period from December 1 to March 31 inclusive, shall be considered as equal to 2 months, and the period from April 1 to November 30, inclusive, shall be considered as equal to 10 months.) The foregoing provision in parenthesis will be used only on contracts extending over a period of 12 months or more from the end of the period of preparation. It shall not be construed to limit or restrict any special construction program that may be outlined or specified on the contract drawings as, for example, the required progress for stream control at dams. Determination as to whether this rate of progress is being maintained shall be made by comparing the value, at the contract price, of the work done as shown in the progress estimates, with the total estimated contract price. Any failure to maintain the required rate of progress, after taking into consideration extensions of time that have been granted, shall be a breach of contract, just as would be a failure to complete the entire work within the specified time.

The Board shall have the right, at its discretion, to extend the time for the completion of the work beyond the time stated in this Contract, for reasons set forth in Section 0.4, entitled **Extension of Time,** but such extension, if so granted, shall waive no other obligations of the Contractor or of the Sureties, and if the time for the completion of the work be extended by the Board, then in such case, the District shall be fully authorized and empowered to make such deductions from the final estimate of the amount due the Contractor, as are stipulated in Article 4 of the Agreement, for each calendar day that the Contractor shall be in default for the completion of the work beyond the date to which the time of completion shall have been extended by the Board. The contractor may be permitted or required to continue and finish the work or any part thereof after the time fixed by the Contract for completion, or as it may have been extended, but such action shall in no wise operate as a waiver on the part of the District of its right to collect the liquidated damages agreed upon in case of such delay, or of any of its rights under this Contract.

0.4 Extension of Time.—Delays due to causes beyond the control of the Contractor other than such as reasonably would be expected to occur in connection with or during the performance of the work, may entitle the Contractor to an extension of time for completing the work sufficient to compensate for such delay. No extension of time shall be granted, however, unless the Contractor shall immediately, but in any case within 15 days from the initiation of the delay, notify the Engineer in writing of such delay, and of the time of beginning and the cause of the same, and unless he shall within 15 days after the expiration of such delay notify the Engineer in writing of the extension of time claimed on account thereof,—and then only to the extent, if any, allowed by the Engineer. To allow or to require completion after the time specified will not constitute an extension of time. No extension of time shall operate to release the Surety from any of its obligations.

The Contractor declares that he has familiarized himself with weather, river, and local conditions and other circumstances which may, or are likely to, affect the performance and completion of the work, and that he has carefully examined the data and information pertinent thereto collected by the Engineer and on file in his office, and agrees that, taking these conditions and circumstances into account, he will provide adequate equipment and prosecute the work in such manner, and with such diligence, that the same will be completed within the time specified herein, or as the same may be extended, even though the most adverse conditions which reasonably could be expected to occur during the period of construction do prevail during the performance of the work. It is understood, however, that as to river and weather conditions the Contractor shall provide against the most adverse conditions and circumstances which reasonably are to be expected to occur on the average within a 15-year period, as shown by past records. '

When the work of the District is enjoined by legal proceedings which prevent the Contractor from prosecuting any of the work of this Contract, an extension of time may be granted sufficient, in the opinion of the Engineer, to compensate for such delay. But no injunction as to a part of the Contractor's work shall entitle him to an extension of time, unless, in the opinion of the Engineer, such injunction unavoidably delays the completion of the whole Contract.

0.5 District to Furnish Right-of-Way.—The Board will furnish to the Contractor all right-of-way, which in the opinion of the Engineer, is necessary for carrying on the work, for borrow pits or spoil banks, for camp sites at dams, and for securing access to the site of the work. In case of serious interference to the work by delay in furnishing such right-of-way, the Contractor shall be allowed an extension of time equivalent to the time lost by unavoidable delay in the completion of the Contract because of the failure to furnish the right-of-way on time.

In case of serious delay and loss to the Contractor because of failure of the District to furnish right-of-way which in the opinion of the Engineer is necessary for the work, the Contractor shall be compensated for such loss.

0.6 Inspection and Right of Access.—The Board contemplates, and the Contractor hereby agrees to, a thorough and minute inspection by the Engineer, or by any of his agents or by any agents which the Board may appoint for such purpose, of all work and material furnished under this Contract, in order to ascertain whether all workmanship or materials are in strict accordance with the requirements of this Contract. The Contractor is entitled to such inspection upon application to the Board or to the Engineer, provided, that when the Contractor requests the inspection of materials not yet delivered to the site of the work, he shall pay the reasonable traveling expenses of the Engineer or agent in making such inspection, and he shall not require the continued presence of the Engineer or agent at the place of manufacture or shipment, for the purpose of making intermittent or occasional inspections.

The Contractor shall furnish to the Board, the Engineer, or any of their agents, access at all times to the work and to the premises used by the Contractor, and shall provide them every reasonable facility, including ladders, steps, scaffolds or platforms, as may be desirable, for the purpose of inspection, even to the extent of discontinuing portions of the work temporarily, or of uncovering or taking down portions of finished work. In case of taking down or uncovering work, should the work thus exposed or examined prove satisfactory, the uncovering or taking down, and replacing or making good, of the parts disturbed, shall be paid for as Extra Work as provided in Section 0.19 of the Specifications, entitled **Extra Work;** but should the work examined prove unsatisfactory, the uncovering, taking down, replacing and making good shall be at the expense of the Contractor. The Contractor shall make no charge for temporary discontinuance of work for purposes of inspection.

The Contractor agrees to make suitable provision for the unrestricted inspection by any authorized engineer or agent at places of manufacture, of any materials being made or prepared for use under this Contract.

The Contractor shall regard and obey the directions and instructions of any authorized engineer or agent with reference to correcting any defective work or replacing any materials found to be not in accordance with the Specifications and Drawings, and in case of dispute the Contractor may appeal to the Chief Engineer, whose decision shall be final; but pending such decision the instructions of said engineer or agent shall be followed, and the Contractor shall make no claim for damages or delay on this account, except as provided in Section 0.39, of the Specifications, entitled **Hindrances and Delays.**

The Contractor shall, whenever so requested, give the Engineer access to the proper invoices, bills of lading, and other records, and shall, without charge therefor, provide scales with a capacity of not less than 300 pounds, and assistance for weighing or measuring any of the materials.

0.7 To Remedy Defective Work.—If the work, or any portion thereof, shall be damaged in any way, or if defects not readily detected by inspection shall develop before the final completion and acceptance of the whole work, the Contractor shall forthwith make good without compensation such damage or defect, in a manner satisfactory to the Engineer. Any materials brought upon the ground for use in the work, which shall be condemned by the Engineer as unsuitable, or not in conformity with the Contract, shall be immediately discarded and removed by the Contractor to a satisfactory distance from the work.

If the Contractor shall fail to replace any defective or damaged work or material after reasonable notice, the Engineer may cause such work or materials to be replaced, and the expense thereof shall be borne by the Contractor.

0.8 Retaining Imperfect Work.—If the Contractor shall execute any part of the work defectively, and if the imperfection, in the opinion of the Engineer shall not be of such magnitude or importance as to necessitate, or be of such nature as to make impracticable or dangerous or undesirable, the removal and reconstruction of the imperfect part, then the Engineer shall, with the written approval of the Board, have the right to make such deduction as may be just and reasonable, from the amounts due or to become due the Contractor, instead of requiring the imperfect part to be removed and reconstructed.

0.9 Engineer Cannot Waive Obligations.—It is expressly agreed that neither the Engineer, nor any of his assistants or agents shall have any power to waive the obligations of this Contract for the furnishing by the Contractor of good and suitable material and for his performing good work as herein described. Failure or omission on the part of the Engineer, or any of his assistants or agents, to condemn defective or inferior work or material, shall not imply acceptance of the work, or release of the Contractor from obligation to at once tear out, remove and properly replace the same without compensation, and at his own cost and expense, at any time upon the discovery of said defective work and material, prior to the final acceptance of the entire Contract and the release of the Contractor by the Board, notwithstanding that such work or such material may have been estimated for payment, or payments may have been made on the same. Neither shall such failure or omission, nor any acceptance by the Engineer or by the Board, be construed as barring the Board, at any subsequent time, from recovery of damages, and of such a sum of money as may be needed to remove and to build anew all portions of the work in which fraud was practiced or improper work or material hidden.

0.10 To Direct Work.—It is mutually agreed that the Engineer shall have the right to direct the manner in which all work under this Contract is to be conducted, in so far as may be necessary to secure the safe and proper progress and quality of the work.

Upon all questions concerning the execution of the work, and the interpretation of the Drawings and Specifications, and on the determination of quantities and cost, the decision of the Chief Engineer shall be final and binding on both parties, and his estimates and decisions shall be a condition precedent to the right of the Contractor to receive any money under this Contract.

Whenever, in his opinion it is necessary to do so, in order to insure the safe and proper completion of the Contract, he shall determine the order of precedence, and the time and seasons at which any portion or portions of the work shall be commenced and carried on.

He shall especially direct the manner of conducting the work when it is in locations where the party of the first part is doing other work either by contract, or by its own force, or where it is necessary to have constructed or reconstructed railroads, traction lines, telephone or telegraph lines, highways or other works affected by the improvement, in order that conflict may be avoided and the work on this Contract be harmonized with that on other contracts or with other work being done in connection with, or growing out of, any operations of the District.

0.11 To Provide for Emergencies.—It is understood by all parties to this Contract that unusual conditions may arise on the work which will require that immediate and unusual provisions be made to protect the public from danger of loss or damage due directly or indirectly to the prosecution of the work, and that it is part of the service required of the Contractor to make such provisions.

The Contractor shall use such foresight and shall take such steps and precautions as may be necessary, to protect the public from danger of damage or loss of life or property, which would result from the interruption of public water supply or other public service, or from the failure of partly completed work.

Whenever, in the opinion of the Engineer, an emergency exists, against which the Contractor has not taken sufficient precaution for the safety of the public or the protection of the works to be constructed under this Contract, or of adjacent structures or property which may be injured by processes of construction on account of such neglect; and whenever, in the opinion of said Engineer, immediate action shall be considered necessary in order to protect public or private, personal or property interests liable to loss or damage on account of the operations under this Contract, then, and in that event, the Engineer, upon giving notice to the Contractor, may provide suitable protection to said interests by causing such work to be done and material to be furnished as, in the opinion of the Engineer, may seem reasonable and necessary.

The cost and expense of said work and material shall be borne by the Contractor, and if he shall not pay said cost and expense upon presentation of the bills therefor duly certified by the Engineer, then said cost and expense shall be deducted from any amounts due or which may become due said Contractor. In case the Board shall decide that all or part of the expense incurred in meeting any emergency is such as for any reason cannot be justly charged to the Contractor, it may compensate the Contractor for all or part of the work done and material furnished in meeting such emergency.

0.12 **To Modify Methods and Equipment.**—Except where otherwise directly specified in the Contract, the Contractor shall design, lay out, and be responsible for the methods and equipment used in fulfilling the Contract; but such methods and equipment, when required, shall have the approval of the Engineer. Whenever required, the Contractor shall furnish to the Engineer for his information, bills of materials, descriptions and copies of drawings showing in reasonable detail the materials and construction of any· construction plants, false work, structures, parts of any structures, or appliances, to be furnished or built under this Contract, for which complete detail drawings are not to be issued by the Engineer. If at any time the Contractor's methods or equipment appear to the Engineer to be unsafe, inefficient, or inadequate for securing the safety of the workmen, the quality of work, or the rate of progress required, he may order the Contractor to increase their safety and efficiency, or to improve their character, and the Contractor shall comply with such orders. If at any time the Contractor's working force, in the opinion of the Engineer, shall be inadequate for securing the necessary progress, as herein stipulated, the Contractor shall, if so directed, increase the force or equipment to such an extent as to give reasonable assurance of compliance with the schedule of progress; but the failure of the Engineer to make such demand shall not relieve the Contractor of his obligation to secure the quality, the safe conducting of the work, and the rate of progress required by the Contract; and the Contractor alone shall be responsible for the safety, efficiency and adequacy of his plant, appliances and methods.

0.13 **To Furnish Lines and Grades.**—All lines and grades will be given by the Engineer, but the Contractor shall provide such materials as are not normally part of an engineering equipment, and give such assistance as reasonably may be required by the Engineer to enable measurements and inspections to be made. He shall not be required, except for brief intervals, to furnish men or material to do the work which would naturally belong to members of a surveying or inspection party. It is the intention not to delay the work for the giving of lines or grades, but if necessary, working operations shall be suspended for such reasonable time as the Engineer may require for this purpose. No special compensation shall be made for the cost to the Contractor, of any of the work or delay occasioned by giving lines and grades, by making other necessary measurements, or by inspection; but such costs, it is agreed, shall be included in the Unit Prices stipulated for the appropriate items of construction. The Contractor shall keep the Engineer informed, a reasonable time in advance, of the times and places at which he intends to do work, in order that lines and grades may be furnished and necessary measurements for record and payment may be made with the minimum of inconvenience to the Engineer or of delay to the Contractor.

All marks and stakes must be carefully preserved by the Contractor, and in case of their unnecessary destruction by him, or any of his employees, such stakes will be replaced by the Engineer at the Contractor's expense.

0.14   To Determine Quantities and Measurements.—The Engineer shall make all measurements, and determine all quantities and amounts of work and materials done or furnished under this Contract.

Unless specifically so stated in detail in the Contract or Specifications, no extra measurements, or measurements according to local custom of any kind, shall be allowed in measuring the work under this Contract; but only the length, area, solid contents, number, weight, or time, in standard units, as the case may be, shall be considered.

It is stipulated and agreed that the planimeter shall be considered an instrument of precision adapted to the measurement of areas.

0.15   To Define Terms and Explain Plans.—The various parts of the Contract are intended to be explanatory of each other, but should any discrepancy appear, or any misunderstanding arise, as to the import of anything contained in either, the explanation of the Chief Engineer shall be final and binding. Correction of any error or omission in the Drawings and Specifications may be made by the Engineer, when such correction is necessary to bring out clearly the intention which is indicated by a reasonable interpretation of the Drawings and Specifications as a whole.

Whenever, in the Specifications, or in Drawings which are a part of this Contract or which may be furnished to the Contractor for directing his work, the terms or descriptions of various qualities of workmanship, material, structures, processes, plant, or other features of the Contract, are described in general terms, the meaning and fulfillment of which must depend upon individual judgment; then, in all such cases, the question of the fulfillment of such

Specifications or requirements shall be decided by the Chief Engineer, and said material shall be furnished, said work shall be done, and said structure, process, plant, or feature shall be constructed, furnished or carried on in full and complete accordance with his interpretation of the same, and to his full satisfaction and approval.

0.16 Completeness of Specifications, Estimates, and Drawings. —The Specifications and Drawings, taken in connection with the estimates and other provisions of this Contract, are intended to describe and show the work required to be done, and the material required to be furnished. The quantities of the various classes of work to be done and materials to be furnished under this Contract, are presented for the purpose of comparing on a uniform basis the proposals offered for the work under this Contract. It is recognized to be impracticable to determine beforehand with accuracy the quantity entering into the construction.

The Specifications and Drawings are to be taken, therefore, as indicating the approximate amount of work, its approximate nature and position, and the method of construction, insofar as the same are determined in advance, and neither the Board nor any agent or officer thereof, guarantees that the actual amount of work will correspond with the quantities estimated by the Engineer; and the Contractor hereby agrees that he will make no claim for anticipated profits, or for losses, because of any difference between the quantities of the various classes of work actually done, or materials actually furnished, and the estimated quantities stated in the Agreement. The work is intended to be constructed in accordance with the best practice and with due regard for the safety of the structures; and in the event of any doubt as to the meaning of any portion of the Contract, Specifications, Drawings, Supplementary Drawings, or instructions of the Engineer, the same shall be understood to call for the best types of construction, both as to materials and workmanship, which reasonably can be interpreted.

Wherever, in the Specifications or in the Drawings, standard specifications or other publications are referred to, copies thereof may be found, available for reference, at the office of the Chief Engineer.

Any work or material not herein specified or shown on the Drawings, but which by fair implication would be included in any items in the Contract, shall be done or furnished by the Contractor without additional charge therefor.

0.17 Items of Work.—The division into Items has been made to enable the Contractor to bid on the different portions of the work in accordance with his estimate of their unit cost, so that in the event of any increase or decrease in the quantities of any particular kind of work, the actual quantities executed may be paid for at the Unit Price for that particular kind of work.

0.18 Changes and Alterations.—The Board reserves the right to make such alterations, eliminations, and additions as it may elect in the line, grade, form, location, dimensions, plan or material of the work herein contemplated, or any part thereof, either before or after the commencement of construction.

If such changes diminish the quantity of work to be done, they shall not constitute a claim for damages or anticipated profits on the work that may be dispensed with; provided, that if said first party shall make such changes or alterations as shall make useless any work already done or material already furnished or used in said work, the Engineer shall make reasonable allowance therefor, which action shall be binding on both parties. If they increase the amount of work, such increase shall be paid for according to the quantities actually done, and at the Unit Prices established for such work under this Contract. Extra work shall be paid for as hereinafter provided.

0.19 Extra Work.—If, during the performance of this Contract, it shall become necessary or desirable for the proper completion of the work hereunder to order additional work done or materials furnished, which in the opinion of the Engineer are not susceptible of classification under the Schedule of Unit Prices, the Contractor shall, if ordered in writing by the Engineer, do and perform such work and furnish such materials; and he shall be paid therefor the actual and necessary net cost, as determined by the Engineer, plus 15 per cent thereof. Such actual net cost shall cover all labor and materials necessary for the performance of the extra work, including any extraordinary expenses incurred directly on account thereof, the wages of foremen, and the expense attached to Contractor's liability insurance covering the labor so employed; but in making payment to the Contractor for such extra work, no allowance shall be made for overhead charges, general superintendence, general expenses, contingencies, or use and depreciation of, and wear and tear upon plant. Charge for extra work shall not include the maintenance of the Contractor's camp, except in case such camp be maintained primarily to carry on extra work.

The Contractor shall have no further claim in excess of the foregoing, and this method of payment shall not apply to the performance of any work or the furnishing of any material which in part or in whole is, in the opinion of the Engineer, susceptible of classification under the Schedule of Unit Prices. In case any work or material shall be required to be done or furnished under the provisions of this section, the Contractor shall at the end of each day during the progress thereof furnish to the Engineer daily time slips showing the name or number of each workman employed thereon, with the time worked, the character of work he is doing and the wages paid or to be paid to him, and also a daily memorandum of the material used on such extra work, showing the amount and character of such material, from whom purchased, and the amount paid or to be paid therefor. If required by the Engineer, the Contractor shall produce any books, vouchers, records, and memoranda showing the work and materials actually paid for and the actual prices therefor. Such daily time slips and memoranda shall not be binding upon the District, except in so far as they are reasonable, accurate, and pertinent, and if any question or dispute should arise as to the correct cost of such extra work or material, the determination of the Engineer upon such question shall be final and conclusive.

It is further agreed that if at any time the number of men employed by the Contractor on work done under this section of the Contract, or the character of such men or the plant and equipment so employed, are not, in the opinion of the Engineer, the best adapted for the satisfactory prosecution of the work, the Engineer shall so notify the Contractor in writing, whereupon the latter shall increase or decrease the number of men, or substitute different men, or make such changes in plant and equipment, as shall be ordered by the Engineer. Extra work so ordered shall constitute a part of the work to be done under this Contract, and all and singular the provisions and conditions of this Contract and of the Bond accompanying it shall apply to the said extra work as if the same were specified in the Contract.

As a condition precedent to the right to receive any money for extra work and material furnished under this Contract, the Contractor shall deliver to the Board, before the 15th day of the month following a month in which any such extra work has been done or extra material furnished, an itemized bill of the cost of such materials or work, and unless it is so filed the claim for an extra shall be deemed waived.

0.20 Progress Estimates.—In order to assist the Contractor to prosecute the work advantageously, the Engineer shall, from time to time, during the active progress of the work, approximately once a month, make a determination of all work done and materials incorporated in the work by the Contractor up to that time, and a progress estimate, in writing, showing: the value of such work and materials under and according to the terms of this Contract; any other amounts due the Contractor; all deductions made in accordance with the provisions of the Contract; then, from the balance, a deduction of 10 per cent of such balance, or a larger percentage, if in the opinion of the Engineer the protection of the District so requires; then, from the remainder, a deduction of the total amount of all previous payments; and finally, the amount due the Contractor under such progress estimate. Such progress estimates shall not be required to be made by strict measurements, but they may be made either by measurement or by approximation. Progress estimates may, at any time, be omitted if, in the opinion of the Engineer, the protection of the District so requires.

Estimates of value of not to exceed 90 per cent of their cost will be included in the body of the estimate for material intended for use in the work, delivered at the site of the work, provided that if such material shall greatly exceed the amount required for the succeeding month's consumption or use, or if material is delivered so far in advance that it may suffer damage or depreciation before being used, no estimate or allowance for the excess shall be made, if the Board so elects.

In case work is nearly suspended or in case only unimportant progress is being made, the Engineer may, at his discretion, make progress estimates at greater intervals than once a month.

Upon such progress estimate being made and certified in writing to the Board, the District shall, within 10 days after the date of the estimate, pay to the Contractor the amount due him under such estimate; provided, however, that the District may at all times reserve and retain from such amount, in addition to the 10 per cent heretofore mentioned, any sum or sums which, by the terms hereof, or of any law of the State of Ohio, it is or may be authorized or required to reserve or retain.

0.21 Final Payment.—Whenever, in the opinion of the Engineer, the work covered by this Contract has been completed, he shall so certify in writing to the Board, and shall submit a final estimate, showing the total amount of work done by the Contractor, and its value under and according to the terms of this Contract;

any other amounts due the Contractor; all deductions made in accordance with the provisions of the Contract; the total of all previous payments; and the amount due the Contractor under such final estimate. On or before the expiration of 30 days after date of the acceptance of the work by the Board, the District shall pay to the Contractor the amount due him on the final estimate. Provided, however, that before he shall be entitled to payment of such amount, the Contractor shall execute and file with the Board, a release, in proper form, of all claims against the District on account of this Contract, except for the Contractor's equity in the amounts kept or retained under the terms of this Contract; and except for the interest, if any, due on the final estimate, as provided hereinafter; and except any other claims that have theretofore been filed in accordance with the provisions of the Contract, which are listed and itemized in detail in a statement attached to and made a part of such release, giving reasons for, nature of, and amount of each claim so listed. All prior estimates upon which payment may have been made, shall be superseded by the final estimate.

0.22 **Payment by Check.**—Payment may be made by check signed by the properly authorized officers of the District. The Contractor shall designate some bank or other agent within the limits of the Miami Conservancy District, authorized by him to receive payments on progress estimates, unless otherwise agreed upon between the Contractor and the Board.

0.23 **Delayed Payments.**—Should any payment due the Contractor on any estimate be delayed, through fault of the District, beyond the time stipulated, such delay shall not constitute a breach of contract or be the basis for a claim for damages, but the District shall pay the Contractor interest on such amount at the rate of 6 per cent per annum for the period of such delay. The term for which interest will be paid shall be reckoned, in the case of progress estimates, from the 10th day after date of the estimate, to the date of payment of the estimate; and in case of the final estimate, from the 30th day after the acceptance, to the date of payment of the final estimate. The date of payment of any estimate shall be considered the day on which the payment is made or offered as evidenced by the records of the Board's office. If interest shall become due on any progress estimate, the amount thereof, as determined by the Board shall be added to a succeeding estimate. If interest shall become due on the final estimate, it shall be paid on a supplementary voucher prepared by the Board; provided, however, that the Contractor shall not be entitled to interest on any sum or sums which by the terms hereof the District may be authorized to reserve or retain.

0.24 Payment Only in Accordance with Contract.—The Contractor shall not demand, nor be entitled to receive, payment for the work or materials, or any portion thereof, except in the manner set forth in this Contract, and after the Engineer shall have given a certificate for such payment.

0.25 Money Retained for Defects and Damages.—The Contractor shall pay to the District all expenses, losses and damages, as determined by the Engineer, incurred in consequence of any defect, omission, or mistake of the Contractor or of his employees, or the making good thereof, and the District may apply any moneys which otherwise would be payable at any time hereunder, to the payment thereof.

0.26 Claims for Damages.—It is agreed that if the Contractor shall claim compensation for any alleged damage by reason of the acts or omissions of the Board, or its agents, he shall, within 10 days after the sustaining of such damage, make a written statement to the Engineer of the nature of the alleged damage. On or before the last day of the month succeeding that in which any such damage is claimed to have been sustained, the Contractor shall file with the Engineer an itemized statement of the details and amount of such damage, and upon request of the Engineer shall give him access to all books of account, receipts, vouchers, bills of lading and other books or papers containing any evidence of the amount of such damage. Unless such statement shall be filed as thus required, his claim for compensation shall be forfeited and invalidated, and he shall not be entitled to payment on account of any such damage.

0.27 Remedies Cumulative.—Any remedy provided in this Contract shall be taken and construed as cumulative, that is, additional to each and every other remedy herein provided.

0.28 Acceptance Shall Not Constitute Waiver.—No order, measurement, determination, or certificate by the Engineer, or order by the Board for payment of money, or payment for, or acceptance of the whole or any part of the work by the Engineer or the Board, or extension of time, or possession taken by the Board or its employees, shall operate as a waiver of any portion of this Contract or of any power herein provided, except as provided for in Section 0.8

of these Specifications, entitled **Retaining Imperfect Work**; nor shall any waiver of any breach of this Contract be held to be a waiver of any other or subsequent breach.

0.29 Collateral Works.—The Board reserves the right to have such agent or agents as it may elect enter the property or location on which the works herein contracted for are to be constructed or installed, for the purpose of constructing or installing such collateral works as said first party may desire, or for the construction or reconstruction of railroads, traction lines, telephone and telegraph lines, highways or other works affected by the improvement. Such collateral works will be constructed or installed with as little hindrance or interference as possible with the Contractor. The party of the second part hereby agrees not to interfere with, or prevent the performance of, any collateral work by the agent or agents of the party of the first part.

0.30 Persons Interested in Contract.—The Contractor hereby declares that no other person or corporation has any interest hereunder as Contractor.

This Contract shall be void if any member of the Board of Directors of The Miami Conservancy District, or other officer of the District, or any person in the employ of the Board of Directors, or of the District, is, or shall become, directly or indirectly, interested as contracting party, partner, stockholder, surety or otherwise, in the performance of the Contract, or in the supplies, work or business to which it relates, or in any portion of the profits thereof; provided, however, that incidental and unintentional interest, such as ownership of stock or interest in a railroad, telephone, electric traction or other company which may, as an incident to its usual business, furnish services or supplies to the Contractor, shall not be included within the meaning of this provision.

0.31 Personal Attention of Contractor.—The Contractor shall give his personal attention constantly to the faithful prosecution of the work, and shall be present, either in person or by a duly authorized representative on the site of the work, continually during its progress. He shall maintain an office on or adjacent to the site of the work, and shall at all times keep in said office a complete copy of the Specifications and Drawings.

0.32 Contractor's Address.—Both the address given in the bid or proposal upon which this Contract is founded, and the Contractor's office at or near the site of the work are hereby designated as places to either of which notices, letters and other communications to the Contractor shall be mailed or delivered. The delivering at either of the above named places of any notice, letter or other communication from the Board or its Agents to the Contractor shall be deemed sufficient service thereof upon the Contractor, and the date of said service shall be the date of such delivery. The first named address may be changed at any time by notice from the Contractor to the Board. Nothing herein contained shall be deemed to preclude or render inoperative the service of any notice, letter, or other communication upon the Contractor personally.

0.33 Agents, Superintendents and Foremen.—When the Contractor is not present on any part of the work where it may be desired to give directions, orders may be given by the Engineer and shall be received and obeyed by the Superintendent or Foreman who may have charge of the particular part of the work in reference to which orders are given.

0.34 Compliance with Laws.—The Contractor shall keep himself fully informed of all laws, ordinances, and regulations in any manner affecting those engaged or employed in the work, or the materials and appliances used in the work, or in any way affecting the conduct of the work, and of all orders and decrees of bodies or tribunals having jurisdiction or authority over the same. He shall at all times himself observe and comply with, and shall cause his agents and employees to observe and comply with, such existing and future laws, ordinances, regulations, orders and decrees; and shall protect the District against any claim or liability arising from or based upon the violation of any such law, ordinance, regulation, order or decree, whether by himself or his employees.

0.35 Character of Employees.—The Contractor shall employ only competent, skillful, faithful and orderly men to do the work, and whenever the Engineer shall notify the Contractor, in writing, that any man on the work is, in his opinion, incompetent, unfaithful, disorderly or otherwise unsatisfactory, the Contractor shall discharge such man from the work, and shall not again employ him, except with the written consent of the Engineer.

0.36 To Maintain Communications. — The Contractor shall build and maintain such temporary bridges, roads, railroads, telegraph and telephone lines and other means of communications as are necessary and not otherwise provided for, where such communication is interfered with on the work of his Contract, and shall provide for convenient access to the various parts of the work and to adjacent private property which may be affected by the work. He shall provide such temporary fences or guards as may be necessary either to keep live stock on adjoining property from entering the lands occupied by the works, or to make roads and other communications safe by night as well as by day.

In case, in the opinion of the Engineer, such temporary works are dangerous or insufficient, the Contractor shall bring them to the condition of safety or sufficiency required by the Engineer. He shall not disturb, close or obstruct any existing highways or other communications until he has obtained permits therefor from the Engineer.

The Contractor shall be responsible for the sufficiency and safety of all such temporary works, and shall be responsible for all damage resulting from their insufficiency, either in construction, maintenance or operation.

0.37 To Guard Against Accidents.—The Contractor, at all times throughout the performance of this Contract, shall take all precautions necessary to effectually prevent any accident in any place affected by his operations in consequence of the work being done under this Contract, and shall, to this end, put up and maintain suitable and sufficient barriers, signs, lights, or other necessary protection.

The Contractor shall save harmless the District from any suits or claims of every name or description brought against it, for and on account of any injury or damage to person or property, received or sustained by any person or persons, by or from the Contractor or any duly authorized sub-contractor or any agent, employee or workman, by or on account of work done under this Contract, or any extensions or additions thereto, whether caused by negligence or not, or by or in consequence of any negligence in guarding the same, or any material or explosives used or to be used for the same, or by or on account of any material, implement, appliance or machine used in its construction; or by or on account of any accident or of any act or omission of the Contractor or of any duly authorized sub-contractor or any agent, employee or workman.

The Contractor agrees that so much of the money due him under this Contract as shall be considered necessary by the Board may be retained until all suits or claims for damages as aforesaid have been settled and evidence to the effect has been furnished to the Board.

0.38 Contractor Responsible for Claims.—The Contractor shall assume the defense of, and save harmless, the District from all claims of any kind arising from his operations in the performance of the Contract. But he shall not be held responsible for damage which is inevitable or necessary because of the nature of the improvement, and which does not result in any way from his manner of doing the work, such as flooding of lands above the permanent dams.

0.39 Hindrances and Delays.—The risks and uncertainties in connection with the work are assumed by the Contractor as a part of this Contract, and are compensated for in the contract price for the work. The Contractor, except as otherwise definitely specified in this Contract, shall bear all loss or damage for hindrances or delays, from any cause, during the progress of any portion of the work embraced in this Contract, and also all loss or damage arising out of the nature of the work to be done, or from the action of the elements, inclement weather and floods, or from any unforeseen and unexpected conditions or circumstances encountered in connection with the work, or from any other cause whatever; and except as otherwise definitely specified in this Contract, no charge other than that included in the contract price for the work, shall be made by the Contractor against the District for such loss or damage.

Should the work be stopped by order of the party of the first part for any cause other than those authorized in this Contract, then and in that event, such expense as, in the opinion of the Engineer, is caused to the Contractor thereby, other than the legitimate cost of carrying on this Contract, shall be paid by the party of the first part.

0.40 Night Work.—Unless otherwise specified or ordered by the Engineer, work may be done by night as well as by day, and night work may be required if necessary in exigencies, or to complete work on which night work is feasible. But no night work of any kind shall be done without the knowledge of the Engineer.

Where night work is in progress sufficient lights shall be provided to safeguard the workmen, and the public, and to afford adequate facilities for properly placing and inspecting the materials.

For night work the Contractor shall receive no extra compensation, but the compensation for such work and all expenses incident thereto shall be considered as included in the contract price.

0.41 Delivery of Materials.—Materials to be used for work under this Contract shall be delivered sufficiently in advance of their proposed use to prevent delays, and they shall be delivered approximately in the order required.

The Engineer shall be notified a reasonable time in advance of the proposed manufacture of metal-work, so that arrangements may be made for inspection.

0.42. Infringements of Patents.—The Contractor shall be held responsible for any claims made against the District for any infringements of patents by the use of patented articles, or methods, used by him in the construction and completion of the work, or any patented process connected with the work agreed to be performed under the Contract, or of any patented materials used upon the said work, and shall save harmless the District from all costs, expenses and damages which the District shall be obliged to pay by reason of any infringement or alleged infringement of patents used in the construction and completion of the work.

0.43 Protection Against Claims for Labor and Material.—The Contractor agrees that he will save harmless the District from all claims against it for material furnished or work done under this Contract.

It is further agreed by said Contractor that he shall, if so requested, furnish the Board with satisfactory evidence that all persons who have done work or furnished material under this Contract have been duly paid for such work or material, and in case such evidence is demanded and not furnished as aforesaid, such amount as may in the opinion of said Board be necessary to meet the claim of the persons aforesaid may be retained from the money due said party of the second part under this Contract until satisfactory evidence be furnished that all liabilities have been fully discharged.

When required by the laws of Ohio, moneys due the Contractor may be retained for protection against claims.

0.44 Assignment.—The Contractor shall not assign, transfer, convey, sublet or otherwise dispose of this Contract, or his right, title or interest in or to the same or any part thereof, without the previous consent in writing of the Board. If the Contractor shall, without such previous written consent, assign, transfer, convey, sublet, or otherwise dispose of this Contract, or of his right, title, or interest therein, to any other person, company or other corporation, or by bankruptcy, voluntary or involuntary, or by assignment under the insolvency laws of any state; this Contract may at the option of the Board be revoked and annulled, and the District shall thereupon be relieved and discharged from any and all liability and obligations growing out of the same to the Contractor, and to his assignee, trustee, or transferee; and no right under this Contract, or to any money to become due hereunder shall be asserted, excepting as provided herein, against the District, in law or equity by reason of any so-called assignment of this Contract, or any part thereof, or of any moneys to become due hereunder, unless authorized as aforesaid by the written consent of the Board.

0.45 Removal of Equipment.—The Contractor shall not sell, assign, mortgage, hypothecate, or remove equipment or materials which have been installed and which may be necessary for the completion of the Contract, without the written consent of the Engineer.

0.46 Suspension of Work if Contract is Violated.—If the work to be done under this Contract shall be abandoned by the Contractor, or if this Contract shall be assigned, or placed in bankruptcy, or the work sublet by him, otherwise than as herein specified, or if at any time the Engineer shall be of the opinion, and shall so certify in writing to the Board, that the performance of the Contract is unnecessarily or unreasonably delayed, or that the Contractor is violating any of the conditions or agreements of this Contract, or is executing the same in bad faith or not in accordance with the terms thereof, or is not making such progress in the execution of the work as to indicate its completion within the time specified in this Contract, or within the time to which the completion of the Contract may have been extended by the Board, the Board may notify the Contractor to discontinue all work or any part thereof, under this Contract, by a written notice to be served upon the Contractor as hereinbefore provided, and a copy of which notice shall be given to his Surety, or the authorized agent for the latter; within two weeks from the date of such notice, the Contractor shall discontinue the work, or such part therof as the Board shall designate, whereupon the Surety may, at its option, assume this Contract, or that portion thereof on which the Board has ordered the Contractor to discontinue work, and proceed to perform the same, and may, with the written consent of the Board, sublet the work or portion of the work so taken over; provided, however, that the Surety shall exercise its option, if at all, within two weeks after written notice to discontinue work has been served upon the Contractor and upon the Surety or its authorized agent. The Surety, in such event, shall take the Contractor's place in all respects, and shall be paid by the party of the first part for all work performed by it in accordance with the terms of this Contract; and if the Surety under the provisions hereof shall assume said entire Contract, all moneys remaining due the Contractor at the time of his default, shall thereupon become due and payable to the Surety as the work progresses, subject to all of the terms of this Contract.

In case the Surety does not, within the hereinbefore specified time, exercise its right and option to assume this Contract or that portion thereof on which the Board has ordered the Contractor to discontinue work, then the Board shall have the power to complete by contract or otherwise, as it may determine, the work herein described, or such part thereof as it may deem necessary, and the Contractor agrees that the Board shall have the right to take possession of and use any of the materials, plant, tools, equipment, supplies and property of every kind provided by the Contractor for the purpose of his work, and to procure other tools, equipment and materials for the completion of the same, and to charge to the Contractor the expense of said contracts, labor, materials, tools and equipment and expenses incident thereto. The expense so charged shall be deducted by the District out of such moneys as may be due or may at any time thereafter become due the Contractor under and by virtue of this Contract, or any part thereof. The Board shall not be required to obtain the lowest figures for the work of completing the Contract, but the expense to be deducted shall be the actual cost of such work. In case such expense is less than the sum which would have been payable under this Contract if the same had been completed by the Contractor, then the Contractor shall be entitled to receive the difference; and in case such expense shall exceed the amount which would have been payable under the Contract if the same had been completed by the Contractor, then the Contractor shall pay the amount of such excess to the District on notice from the Board of the excess so due; but such excess shall not exceed the amount due, under this Contract, at the time the Contractor is notified to discontinue said work, or any part thereof. plus the amount of the bond or bonds executed by the Contractor for the performance of this Contract. When any particular part of the work is being carried on by the Board, by contract or otherwise, under the provisions of this section, the Contractor shall continue the remainder of the work in conformity with the terms of this Contract, and in such manner as in nowise to hinder or interfere with the persons or workmen employed, as above provided, by the Board.

0.47. **Board for Engineer and Assistants.**—In cases where the Contractor maintains a camp for lodging and feeding employees. he shall, when requested, furnish board or lodging to the Engineer and his assistants, and shall charge the District therefor at the rate of 25c per meal or lodging, or $5.00 per man per week. He shall not furnish such meals and lodging gratis, but shall submit monthly itemized statements to the Engineer of such meals and lodging furnished, for which payment shall be made by the District.

0.48 Police and Sanitary Regulations.—The Contractor and his employees shall promptly and fully carry out the police and sanitary regulations as hereinafter described or as may from time to time be prescribed by the Engineer, to the end that proper work shall be done, good order shall prevail, and the health of employees and of the people using water from the drainage areas, and of the local communities affected by the operations under this Contract, may be conserved and safeguarded. The Contractor shall summarily dismiss and shall not again engage, except with the written consent of the Engineer, any employee who violates the police or sanitary regulations.

0.49 Intoxicants.—The Contractor shall not sell, nor shall he permit or suffer the introduction or use of, intoxicating liquor upon the works embraced in this Contract, or upon any of the grounds occupied or controlled by him.

0.50 Camps.—The Contractor shall if necessary provide suitable and satisfactory buildings for the housing, feeding and sanitary necessities of the men. Such quarters shall provide an amount and arrangement of space per man which will be sufficient and suitable for the maintenance of cleanliness, decency and health. Fly screens shall be provided wherever and whenever necessary. Plans for such quarters shall be submitted to and approved by the Engineer before the quarters are built. The Contractor shall also provide suitable stabling for the animals employed upon the work. All such buildings shall be located only at places approved by the Engineer. The stables shall be at an approved distance from the quarters for men.

The Contractor shall provide at convenient points an ample supply of drinking water of proper quality, and ample shower baths and clothes-washing facilities with hot and cold water, for his employees.

In case of temporary camps or work in or near cities where camps need not be maintained, or of sanitary conveniences away from camp, the provisions concerning camps, drinking water, sanitary conveniences and garbage may be modified to suit the case, upon the written consent of the Engineer.

0.51 Sanitary Conveniences.—Buildings for the sanitary necessities of all persons employed on the work, beginning with the first person employed, shall be constructed and maintained by the Contractor, in the number, manner and places approved or ordered. Unless other types are submitted by the Contractor and approved by the Engineer, these conveniences shall be of an approved incinerator type, except that closets having watertight removable receptacles may be used in special cases if and as permitted. All persons connected with the works shall be obliged to use these conveniences. Satisfactory precautions shall be taken to render the interior of the incinerators and other closets inaccessible to flies. All excreta shall be incinerated daily. The Contractor shall rigorously prohibit the committing of nuisances.

0.52 Garbage and Camp Refuse.—Garbage, both liquid and solid shall be promptly and satisfactorily removed by the Contractor from the buildings, and immediately placed in approved tight receptacles of sufficient capacity for about one day's ordinary production. At least once each day all such garbage shall be incinerated or otherwise thoroughly and satisfactorily disposed of; provided that the Contractor may if he prefers, provide running water in kitchens and other buildings, and dispose of liquid waste and sewage by means of approved drains and cesspools.

The Contractor shall clean of refuse the camp stable, and adjoining grounds at least twice each week, particular care being exercised to prevent conditions which lead to the breeding of flies, mosquitos or other disease-carrying insects or vermin.

In case the Contractor shall neglect or refuse to keep the premises in a clean and sanitary condition, the Engineer may have the work done which in his opinion is necessary to accomplish this purpose, and the cost thereof shall be charged to the Contractor.

0.53 Doctors, Hospitals.—The Contractor shall retain the services of one or more acceptable qualified medical and surgical practitioners, who shall have the care of his employees, shall inspect their dwellings, the stables and the sanitaries as often as necessary or required, and shall supply medical attendance and medicine to the employees whenever needed. The Contractor shall make satisfactory arrangements for hospital service for the proper care of sick or injured employees, and shall give the Engineer satisfactory assurance that the above medical and hospital arrangements have been made. At such places as directed, all articles necessary for giving First Aid to the Injured shall be provided.

0.54 Medical Supervision.—The medical supervision of the Contractor over his employees shall extend to the physical examination of all applicants for employment, in order to prevent persons having communicable diseases which would make them a menace to other employees, from being connected with the work, and the

Contractor shall employ only persons shown by such examinations to be free from such communicable diseases. Whenever, in the opinion of the Engineer, it is necessary for the protection of the public health or the health of the employees, the Contractor shall remove from the work to a hospital, or shall remove permanently from the work or from any camp, any employee whose presence is believed to endanger the health of other persons.

0.55 Medical Report.—Once each week, or more frequently if required, the Contractor shall give the Engineer, in such detail as may be prescribed from time to time, a written report signed by a physician in regular attendance, setting forth clearly the health conditions of the camp or camps and of the employees. If any case of communicable disease, or any case of doubtful diagnosis is discovered, it shall be reported at once to the proper authority and to the Engineer, and the Contractor shall take immediate steps to safeguard the interests of his employees, and the public in general.

0.56 Examinations.—The Engineer shall have the right, in order to determine whether the requirements of this Contract as to sanitary matters are being complied with, to enter and inspect, in person or by agent, any camp or building or any part of the work, and to cause any employee to be examined or to be vaccinated or otherwise treated; also to inspect the drinking water and food supplied to the employees.

0.57 No Direct Compensation.—No direct payment will be made for any work or materials required to meet the sanitary requirements hereinbefore specified, but compensation therefor shall be considered as having been included in the Unit Prices stipulated in the Agreement.

GENERAL SPECIFICATIONS

0.58 Water Supply.—The Contractor shall provide at convenient points an ample supply of water of proper quality for all the operations required under this Contract. A proper piping system shall be installed, maintained, and extended from time to time, to distribute this water to the various portions of the work and plant where it is needed. Whenever necessary, the water shall be under sufficient pressure to give an effective stream from a nozzle for cleaning rock or other surfaces on or against which concrete or masonry is to be built; for washing stones to be used in concrete; for sprinkling, for puddling, and for other purposes. Hose connections and hose, water-casks or other sufficient means, shall be provided also for fighting fires in the more important temporary structures and construction plants, and responsible persons shall be instructed as to the operation of such fire apparatus, so as to prevent or minimize loss of time from fire. These provisions may be waived or modified on any contract by the Engineer to the extent to which in his judgment the limited importance or the nature of the work may justify.

0.59 Pumps and Drains.—The Contractor shall provide all the necessary pumps, temporary pipes, drains, ditches, and other means for removing water from the excavations or other parts of the work, or for preventing the slopes of excavations or embankments from sliding or caving, and he shall, where necessary, satisfactorily remove the water. All compensation for, and all expense incidental to, the fulfillment of the provisions of this section, unless otherwise specifically provided, shall be considered as having been included in the prices stipulated in the Agreement.

0.60 Borrow Pits.—The Contractor may, without charge therefor, open gravel pits, sand pits, or other borrow pits upon lands controlled by the District, but no materials obtained therefrom shall be estimated for payment under any excavation item, except as otherwise specifically provided herein. The location of all borrow pits shall be subject to the approval of the Engineer. It is the intention that all borrow pits shall be left at the termination of work under this Contract in as sightly a condition as practicable. To this end borrow pits shall be shaped and graded as shown on the Drawings, or as ordered. All stumps, boulders or debris, shall be satisfactorily disposed of. If ordered, the surface shall be covered with soil of designated thickness and grassed. With the exception of the surface dressing and grassing, if ordered, which will be paid for under Item 38, and clearing and grubbing, if ordered, which will be paid for under Item 2, no direct payment shall be made for any of the above work, but compensation shall be considered as having been included under the item for which the materials are used.

Where permitted, excavations which are not to be lined with concrete may be enlarged beyond the prescribed limits to obtain materials for construction or for other purposes of the Contractor, but such enlargement of the excavations shall not be paid for under any of the excavation items, except as otherwise specifically provided herein; nor shall any direct payment be made for excavations of whatever nature, not a prescribed part of the permanent structures, which may be made for any purposes of the Contract in connection with the plant, equipment or operations incidental to construction.

0.61 Protection of Trees.—Trees, shrubs, and other obstructons shall be removed from areas other than the sites of the proposed structures only to the extent directed or permitted. Where required, trees or shrubs which are not to be removed under this Contract shall be satisfactorily protected by the Contractor from injury.

0.62 Materials Obtained from Construction.—Materials from the excavations, or from clearing or stripping or other operations, which are suitable and required, shall be used in the construction, as and where directed or permitted, and shall be paid for according to the class of work in which they are used, in addition to the price paid for original removal; except in the case of embankment material for streets and highways, and of material for refilling excavations where such refill is not designated to be paid for as embankment or back-filling. Materials which are not suitable, or for any other reason are not used in the work, shall be satisfactorily disposed of. Earth and rock not used in the construction shall be deposited in spoil banks, which shall be located only at such places as are designated by the Engineer. Spoil banks shall be shaped and graded as directed and may be covered with soil of designated thickness and grassed, the intention being to leave the finished work in as sightly a condition as practicable. Compensation for depositing and grading materials in spoil banks shall be considered as having been included in the prices bid for the excavation, except as provided in Item 12, Excavation—Overhaul for River Channel Improvements. Where it is ordered that spoil banks be covered with top soil and grassed, payment for this work shall be made under Item 38, Surface Dressing and Grassing.

0.63   Assistance in Making Tests.—The Contractor shall render such aid as may be required for making tests of the bearing power of earth, and of the character of foundations; for digging test pits; and for making any other excavations for that purpose which may be directed. Such assistance shall be paid for under Section 0.19 of the Specifications, entitled Extra Work.

0.64   Protection of Existing Structures.—The Contractor shall carefully protect from injury any existing buildings, foundations, fills, land, pavements, pipes, conduits, sewers, drains, cisterns, wells, railway tracks or other works, property or structures that may be liable to injury by the work covered by this Contract, except in so far as the work of the Contract requires their modification or removal. He shall take all precautions necessary for such protection, and shall be fully responsible for, and shall make good, any injury to such works, property or structures that may occur by reason of his operations.

0.65   Excavation—Classification.—Excavation, as indicated on the Drawings or as determined by the Engineer, shall be classified as follows:

Class 1—Earth.—Earth, as a name for excavated material shall be used to include, (except where material is classified as Class 2. Mixed Excavation) all glacial deposit, whether cemented or not, except solid boulders one-half cubic yard or more in volume; it shall include all alluvial deposits, and disintegrated material lying on top of ledge rock, and all material of whatever nature not properly included under Classes 2 or 3. It shall not include the Cincinnatian Formation, which consists of layers of clay or shale below the glacial material, in which are imbedded thin horizontal layers of limestone from a fraction of an inch to a few inches in thickness. Neither shall it include earth or any material which has been classified on the Drawings as Class 2. Mixed Excavation.

Class 2—Mixed Excavation.—The term Mixed Excavation, as a name for excavated material, shall be used to include:

The material known geologically as the Cincinnatian Formation, which lies under the glacial deposits, and consists of thin, approximately horizontal layers of limestone, from a fraction of an inch to a few inches thick, alternated with and imbedded in clay or shale, and which may or may not require blasting for its removal; and.

Mixtures of any or all materials which, in the opinion of the Engineer, it is impracticable to classify as Class 1, Earth or Class 3, Rock, either because of the character of the material, or because of the method used in its removal.

Class 3—Rock.—Rock, as a name for excavated material, shall be used to include the solid ledge rock formation, whenever found, (except when classified as Class 2, Mixed Excavation) which is preglacial, and which can be removed properly only by means of explosives, barring or wedging, or by some other recognized method for quarrying solid rock. It shall include solid boulders of a half cubic yard or more in volume. It shall also include existing concrete, or masonry with mortar joints, which may be removed under instructions. It shall not include thin layers of limestone imbedded in clay or shale, nor cemented gravel, nor any material classified as Class 2, Mixed Excavation.

0.66 Excavation—Work to be Done.—The necessary excavations shall be made for stripping foundations for embankments, for cut-off trenches, pipe trenches, discharge conduits and appurtenances, spillways, channels, drainage systems, permanent street and highway relocations and alterations, channel improvements, and for any other purposes necessary for the completion of all work contemplated under this Contract.

0.67 Excavation—Limits of.—The Drawings for the various structures show, as nearly as was practicable to determine beforehand, the depths, widths, and slopes for the proposed excavations; but all such limits are estimated only, and will be finally determined from the nature of the materials encountered as the work progresses. No excavation outside the prescribed limits will be paid for except when so specifically provided. All material which shall have been loosened outside the prescribed lines shall be removed at the expense of the Contractor, unless allowed to remain by special permission of the Engineer.

Where concrete or masonry is to be built without forms against the sides or bottom of an excavation, especial care shall be taken to preserve the remaining material in an undisturbed condition. In such cases the material shall not be disturbed by machinery or by powder within 6 inches of the prescribed limits, unless with the Engineer's written permission. The remainder shall be removed by hand work, only a short time before placing the concrete or masonry. Should any material be removed beyond the prescribed limits, or be so deteriorated or loosened as, in the opinion of the Engineer, to require removal, the resulting spaces shall be refilled at the Contractor's expense with concrete or other material as directed.

Should it be found that a foundation on which concrete is to be placed deteriorates with exposure, special care shall be taken not to excavate it until immediately before covering it with concrete.

0.68 Excavation—Cleaning Surface for Inspection.—From time to time during the excavation of Rock or Mixed Excavation, whenever so directed, the surface of the material shall be cleaned, so that the area designated can be inspected to determine its quality.

0.69 Excavation—Use of Explosives.—The use of explosives in a manner which might disturb or endanger the stability, safety, or quality of the work will not be allowed. Explosives shall be kept on hand by the Contractor in sufficient quantities to avoid delay on the work. Such explosives shall be stored, handled, and used as prescribed by the laws and regulations of the State of Ohio and subdivisions thereof. The attention of the Contractor is directed to the Act of April 27, 1908, Section 5906, General Code of Ohio, which requires that he shall file a bond in the amount of $5,000 with the County Commissioners of each and every county in which the Contractor shall store or handle explosives in quantities exceeding 100 pounds.

0.70 Excavation—Finishing Work.—The slopes of all excavations which are to be dressed and grassed as hereinafter provided, or which are to remain permanently exposed, shall be finished to the prescribed lines in a careful and workmanlike manner. Slopes of all rock excavations which are to remain permanently exposed shall be trimmed to a reasonably uniform surface with maximum projections from the general surface of not more than one foot, unless otherwise permitted by the Engineer.

0.71 Excavation—Payment.—The excavation to be paid for shall include the quantity in cubic yards excavated in accordance with the Drawings, Specifications and instructions of the Engineer, measured in excavation unless otherwise specified under the respective items of excavation.

The unit prices stipulated for excavation items shall include the entire cost of making the excavations, and of refilling those portions that may be required, where such refilling is not designated for payment as backfilling or as embankment. These unit prices shall cover the cost of the maintenance of the excavations during construction, including damming, pumping, bailing, draining, unwatering and other work necessary; provided, however, that when excavations are made for concrete or other structures the Engineer shall decide, in case the question arises, to what extent the maintenance of the excavation pertains to the excavation item or to the concrete or other item. Sheeting left in place by written order of the Engineer, will be paid for as provided under Item 89; otherwise the cost of placing and removal of sheeting and bracing shall be included in the price bid for excavation. These unit prices shall also include the cost of all other operations incidental to the excavation, and of the disposal of the materials excavated, except as provided in Section 0.62 of the Specifications, entitled Materials Obtained from Construction.

No payment will be made for excavation as **below mean low water level,** under Items 6a, 7a, 8a, etc., unless the estimated elevation of mean low water is shown and so designated, on the Drawings; in which case the estimated elevation shall be used as the dividing line, regardless of where the water level may be at the time the work is done. In cases where this estimated elevation is not shown, excavation of a given Class, whether above or below mean low water level, will be paid for at the unit price stipulated for excavation not below mean low water level.

0.72 **Embankment—Work to Be Done.**—Embankment of the Class specified shall be constructed where shown on the Drawings or where required by the Engineer. Sections 0.73, 0.74, and 0.75 of the Specifications, entitled Embankment—Class A, B, and C, respectively, describe three methods of constructing embankments in the main dams. The proposal provides for alternate bids on embankment depending on the bidder's choice of methods of construction. A minimum amount of Class C embankment will be required at some dams, as is indicated in the schedule of quantities and shown on the Drawings. Except for this minimum amount of Class C embankment, the Contractor may, if he so desires, substitute one class of embankment for another provided that the written approval of the Engineer first be given, and provided further that such substitution shall involve no additional cost to the District; that is, if Class B embankment is to be substituted for Class A, at the Contractor's request, and the unit price of Class B is greater than that of Class A, the material will be paid for at the Class A price. On the other hand, if a substitution is made by order of the Engineer (rather than with his permission), requiring a class of embankment of higher unit price, payment will then be made at the unit price bid for the class of embankment required, except that in case the Contractor, by failing to carry out the terms of the Contract as to rate of progress or otherwise, shall in the opinion of the Engineer put the safety of the work, or the safety of life or property, in jeopardy; and if for the sake of safety it becomes necessary under such conditions, in the opinion of the Engineer, to substitute for one class of embankment another class on which the unit contract price is higher; in such cases the Engineer may order such substitution and the Contractor shall be paid the unit price of the class contemplated in the Contract, and not the unit price of the class made necessary by his failure to carry out the terms of the Contract.

0.73   Embankment—Class A.—The term Class A Embankment shall be used to include embankment for main dams, constructed by **Hydraulic Monitor and Dredge Pump Method:** This method consists of washing down or otherwise delivering material to a dredge pump, which in turn delivers it under pressure through a discharge pipe onto the embankment. In washing down materials, properly constructed nozzles shall be used, with a sufficient flow of water, both as to velocity and quantity, to loosen the materials at the borrow pits, and sufficient additional water to insure the transportation from borrow pits to sump. The dredge pumps, pipes, flumes, sluices, and other transporting equipment shall be carefully installed to meet all requirements for the proper placing of the materials in the embankment.

The outer slopes of the dam shall be carried up in advance of the central portion. The width of these outer levees, and the depth and width of the settling pond, will largely control the gradation of materials, and shall be subject at all times to the direction of the Engineer. Care shall be taken to place in the outer levees material which will conform to the requirements for the outer portion of the dam, and to satisfactorily compact such material. The material for the main body of the embankment shall be delivered near the outer edge of the embankment slopes, and so manipulated that the coarser material will remain near the outer slopes, and the finer materials will be carried toward the center, the finer materials being deposited in a settling pond so as to form an impervious core. It is of the utmost importance to avoid stratificaton within this core. The delivery and manipulation of the materials shall be regulated so as to obtain the best distribution of material, and to keep the water surface of the settling pond as nearly uniform in width as possible. The bottom of settling ponds shall be kept approximately level longitudinally to prevent undue flow of silt to one end.

Sluicing operations shall be conducted at all times in such a manner as to give a puddled core of as uniform width and slopes as the nature of the material used will permit, and to raise both sides of the embankment evenly. Embankment operations shall, unless otherwise permitted in writing, be commenced in the deeper portion at the river section or at the stream control gap, and shall be carried as nearly level longitudinally as possible. Should the use of a series of settling ponds be permitted, they shall be of convenient lengths, separated by temporary cross dams of sufficient thickness to withstand the pressure against them, built of selected material and puddled by hand, or by some equally efficient method, to the satisfaction of the Engineer. As the sluiced material is washed against the temporary cross dam and raised in the embankment, the stratification between it and the cross dam shall be broken by the use of shovels or other suitable tools.

Drainage of surplus water from a higher to a lower settling pond may be permitted if suitable means for such drainage is provided. Whenever the waste water carries silt from one settling pond to another, the outlet shall be immediately raised or sluicing operations in that pond discontinued until such time as this condition has been remedied. The Contractor shall take care of all excess water in the settling pools by means of wells, built up in sections, discharging at approved points, or by pumping, siphoning, or other approved methods. Such devices shall provide for the removal, at the desired elevation, of the comparatively clean water at the surface of the settling pool, while affording ample opportunity for the finer particles to settle properly. Before the completion of all the work, any outflow device within the dam shall be satisfactorily filled with concrete or other acceptable material unless otherwise ordered. No outflow device which will remain within the dam shall be installed until written permission for such installation has been obtained from the Engineer; and such permission will not be given until evidence satisfactory to the Engineer, has been presented to show that adequate provisions can and will be made for completely and satisfactorily filling such device wherever necessary, before the completion of the work. The cost of all labor and all materials required under the provisions in this section, shall be included in the price stipulated in the Agreement.

Variations of this method during construction, which will accomplish equally effective results, will be considered.

0.74 Embankment—Class B.—The term Class B embankment shall be used to include embankment for main dams, constructed by the **Semi-Hydraulic Method**, as follows: The material shall be obtained from borrow pits, or from excavations, transported in cars or by other suitable means, dumped at the outer edge of the embankment slopes, and sluiced toward the center of the dam. In depositing the material in the dam, care shall be taken where necessary, to place the coarser material in the downstream portion of the dam and the finer material in the upstream portion. The sluicing shall be carried on in such a manner that the coarser material will remain near the outer slopes of the embankment, and the finer material will be carried toward the center, the finest material being deposited in a settling pond so as to form an impervious core. It is of the utmost importance to avoid stratification within this core. The quantity and pressure of water, and the direction and point of application of the stream, shall be regulated according to the nature of the material so as to obtain the best distribution of material, and to keep the water surface of the settling pond as nearly uniform in width as possible. The material of the upstream slope shall not be washed clean, as it is desired to have a certain percentage of clay and sand retained near the upstream slope. The bottoms of settling ponds shall be kept approximately level longitudinally to prevent undue flow of silt to one end.

Sluicing shall be done at all times in such manner as to give a puddled core of as uniform width and slopes as the nature of the material used will permit, and to raise both sides of the embankment evenly, and the quantity or direction of the sluicing stream shall be changed as required to obtain this result. The material shall be so placed on the slopes and sluiced or puddled in such a manner that it will be compacted by water as close as practicable to the outer edge of the finished slopes. Embankment operations shall, unless otherwise permitted in writing, be commenced in the deeper portion at the river section, or at the stream control gap, and shall be carried as nearly level longitudinally as possible. Should the use of a series of settling ponds be permitted, they shall be of convenient lengths, separated by temporary cross dams of sufficient thickness to withstand the pressure against them, built of selected material and puddled by hand, or by some equally efficient method, to the satisfaction of the Engineer. As the sluiced material is washed against the temporary cross dam and raised in the embankment, the stratification between it and the cross dam shall be broken by the use of shovels or other suitable tools. The Contractor shall take care of all excess water in a satisfactory manner.

The cost of all labor and materials required under the provisions of this section shall be included in the price stipulated in the Agreement.

0.75 Embankment—Class C.—The term Class C Embankment shall be used to include embankment for main dams, constructed by the **Sprinkling and Rolling Method,** as follows: The embankment shall be built as nearly as possible in horizontal layers, each layer not exceeding 6 inches in thickness when acceptably compacted. Unless the earth is sufficiently moist when spread, each layer shall be wet so as to secure the desired compacting, and if required, the top of the embankment shall be wet just before a layer of earth is spread. The wetting shall be done in a manner which will avoid the formation of pools of water, and will secure the uniform moistening of all portions of embankment. Compacting shall be accomplished by not less than 12-ton, 3-wheeled steam rollers, or traction engines, with grooved, banded, or corrugated rolls. The roller shall pass over every part of each layer that can be traversed by the roller, as many times as may be necessary to thoroughly compact it. The distributing, wetting, and rolling shall be done to the satisfaction of the Engineer. Any places that become boggy or springy shall be corrected to the satisfaction of the Engineer, or as much material as necessary shall be removed and the hole refilled in a satisfactory manner. Embankment operations shall, unless otherwise permitted in writing, be commenced in the deeper portion at the river section or at the stream control gap. Unless otherwise permitted in writing, the embankment shall be carried up approximately level, and in any case the top of unfinished embankment shall not have a slope steeper than 10 to 1 in any direction. Portions of the embankment which are small in area, portions next to walls, and other portions which the rollers cannot reach for any reason, shall be formed of selected materials compacted by means of extra heavy tampers used energetically, or if permitted, by depositing the earth through water in such a manner that all of it shall be thoroughly saturated, or by other means which will secure a degree of compacting at least equivalent to that obtained by rolling as specified. All stones over 4 inches in diameter shall be removed from the portion of the embankment required to be water tight. Stones permitted to remain shall be separated one from another by earth, and in no case shall they be allowed to collect in nests. Stones and cobbles removed from the impervious portions of the embankment shall be placed on the slopes in the manner and to the extent directed. The cost of all labor and materials required under the provisions of this section shall be included in the price stipulated in the Agreement.

0.76 Embankment—Class D.—The term Class D Embankment shall be used to include embankments for levees built from material taken from river channel excavation and constructed as follows:

The embankment shall be carried up in approximately horizontal layers, extending entirely through the fill. Where the material is deposited in a wet condition, the layers shall be not more than 3 feet in thickness, but where the material is deposited in other than a wet condition the layers shall not exceed 1 foot in thickness, and some means satisfactory to the Engineer shall be used for compacting it. Such precautions shall be taken as will preserve the slopes and prevent improper segregation of materials. The object is to secure a compact, stable, reasonably impervious embankment, conforming to the prescribed lines. To allow for shrinkage and settlement, sufficient overfill not to exceed 10 per cent, shall be made by the Contractor at the time of depositing the material. Where embankment is joined to a fill previously placed, or to a natural bank, a satisfactory bond shall be secured between the old and new materials by breaking the old surface or otherwise.

0.77 Embankment—Class E.—The term Class E Embankment shall be used to include embankment for levees built from material taken from borrow pits. It shall be constructed according to the specifications for Class D Embankment, as set forth in Section 0.76 of the Specifications.

0.78. Embankment—Class F.—The term Class F Embankment includes such miscellaneous embankments or fills as are so designated on the Drawings or by the Engineer, in which no special requirements are made as to compacting.

0.79 Embankment—Preparation of Foundation.—Unless otherwise directed the sites of embankments shall be prepared as regards clearing and grubbing, removal of buildings and soil stripping, as specified under Items 2, 3, and 5, to the end that all perishable or otherwise objectionable matter shall be removed. When, in the opinion of the Engineer, sufficient soil has been removed, and material has been uncovered to provide a suitable foundation for the purposes intended, the surface of the material so uncovered shall be picked, plowed, or otherwise satisfactorily roughened, as directed, to make a bond with the embankment material.

Springs encountered in the base of the embankment shall be satisfactorily controlled by plugging, draining, or other approved methods. When advisable, springs shall be led into the outlet channel or away from the embankments, as directed. Approved excavation and materials for this purpose shall be paid for under the appropriate items.

The provisions of this section may be waived or modified as may seem desirable or permissible, in cases where Class F Embankment is specified.

0.80 Embankment—Materials to be Used.—Embankments shall be made of acceptable materials from the excavations, or from borrow pits located on land provided by the District. Such materials shall be carefully selected and shall contain a sufficient proportion of fine particles to render the embankment reasonably impervious to water. The maximum size, quantity, and placing of large stones, shall be as allowed by the Engineer. The material shall be sufficiently stable not to slide or slough at the prescribed slopes, under a condition of saturation. Where feasible to do so, all material from the excavations which are suitable for the purpose shall be placed in the embankment. No frozen material shall be placed in any portion of the embankment by any method whatsoever, nor shall any material be placed on a frozen surface, and operations shall be suspended when, in the opinion of the Engineer, ice conditions are such as to be objectionable. Care shall be taken that no earth not acceptable for embankment purposes, or any roots, brush or other perishable material be incorporated within the embankment. All timber, posts, trestles, staging, etc., shall be removed to the extent directed by the Engineer.

Where Class A, Class B or Class C Embankment is specified, it is the intention that the outer portion of the dam shall be composed of coarse material which shall be permanently stable under a condition of complete saturation. Any material which has a tendency to slough, slide or wash when subjected to any condition which may reasonably be expected to occur, will not be acceptable. The center portion of the dam shall be composed of material containing sufficient clay or other finely divided particles to be highly impervious to water. To obtain these results the Contractor shall use only carefully selected materials. Where the material from one borrow pit does not contain the requisite coarse and fine particles in the desired proportions, other sources shall be utilized, and if necessary to obtain satisfactory results, more than one class of material shall be deposited at the same time in the same or different portions of the embankment.

The provisions of this section shall not apply to Class F Embankment, and may be waived or modified as to Class D Embankment.

0.81 Embankment—Finishing Work.—The slopes and tops of all embankments that are to be dressed and grassed, or which are to remain exposed, shall be finished in a careful and workmanlike manner.

0.82 Embankment—Payment.—The embankment to be paid for shall include the quantity in cubic yards deposited in accordance with the Drawings, Specifications and instructions of the Engineer, measured in embankment unless otherwise specified under the respective items for embankment, and properly placed within the following limits: The prescribed neat limits for embankments, the contour of the natural ground or as left by the stripping, the prescribed limits for excavations the refilling of which is not included for payment under the provisions of Item 37, Backfilling, and the outside neat lines of any structures included within the embankment.

The unit prices stipulated for embankment items shall cover the entire cost of constructing the embankments, including the cost of preparing foundations (except as provided in this section), excavating from borrow pits, pumping, sluicing, transporting, and grading of materials, placing, puddling, draining, compacting, trimming slopes, refilling settlements and depressions, and all labor and materials necessary for doing the work in accordance with the requirements of the Contract, as well as the cost of maintaining embankments until the completion of the Contract. But such prices shall not include the cost of clearing and grubbing, removal of buildings, soil stripping, surface dressing and grassing, which shall be paid for under Items 2, 3, 5, and 38 respectively.

0.83 Backfilling—Description.—The term Backfilling shall be used to include refilling of trenches, and filling around structures in cases where the methods specified for building embankments in which the structures are located are not sufficient, or practicable, to secure satisfactory results. It is the intention to require backfilling around certain structures not only for refilling the excavation, but also, in some cases, for securing a compact, water-tight fill against the structures above the ground line. It may be required in trenches, where settlement subsequent to the first filling is undesirable, or where water-tightness is a requirement. The material shall be selected and placed with care. It may be deposited through water if practicable; otherwise it shall be deposited in layers about 6 inches thick, each layer being sprinkled as required, and well tamped, or rolled, before the next layer is placed.

Trenches in which pipes or conduits are laid, or in which sewers are built, shall be backfilled by tamping in 6-inch layers to a plane 2 feet above the top of the pipe, conduit, or sewer, above which plane the backfilling may be done by depositing the material through water if practicable; otherwise the tamping in layers shall be continued, or, if practicable and permitted, the layers shall be compacted by rolling.

Wherever backfilling is required, either in the Specifications, in the Drawings, or by the Engineer, it is evident that special care is necessary to accomplish the result desired, and the work shall be done carefully, and thoroughly, and strictly in accordance with instructions.

0.84 Backfilling—Material to be Used.—Materials used for Backfilling shall be carefully selected from the materials from the excavations, or from borrow pits if the former are not sufficient or suitable. The materials shall be free from roots, brush, or any other perishable matter or material which will cause unduly prolonged settlement. Except as allowed by the Engineer frozen material shall not be used, nor shall backfilling be placed on or against a frozen surface. The maximum size, quantity and placing of large stones, shall be as allowed by the Engineer. Payments where allowed for backfilling will be made under Item 37 of the Detail Specifications.

0.85 Concrete—Description.—All concrete shall be composed of cement, fine aggregates, coarse aggregates, and water of the qualities hereinafter specified, and mixed in proportions to be prescribed by the Engineer. As the cement is to be paid for separately as a distinct item, the proportions of all the ingredients may be varied by the Engineer to suit the requirements of the various structures, and the Contractor shall make no claim for extra payment because of such variations in mixture, the contract price for each item being understood to apply to any mixture which may be specified. It is acknowledged that different local sands and gravels give widely different strength tests, and for this reason, and to accommodate the mixture to the detail requirements of the work, the Engineer shall order the proportions modified from time to time as may be necessary to obtain the best results.

The object is to secure dense, impervious concrete, except in cases where the chief requirement is weight rather than imperviousness; and the Contractor shall use only such materials, satisfactorily proportioned, mixed, and placed, as the Engineer may direct. From time to time tests will be made both of the materials in advance of construction, and also of the concrete actually being placed in the work. The Contractor shall furnish the cement and aggregates or concrete for all such tests, and shall give such assistance as may be required in facilitating tests. Except in special cases the proportions by volume are expected to range between 1 of cement, 2 of fine aggregate, and 4 of coarse aggregate for the richer mixtures; and 1 of cement, 3 of fine aggregate, and 6 of coarse aggregate for the leaner mixtures. In the construction of heavy walls, a rich mixture at the exposed face, and a leaner mixture at the back or in the body of the wall, may be required.

Large stones may be imbedded in concrete at the option of the Contractor whenever, in the opinion of the Engineer, the introduction thereof will be no detriment to the quality of the work.

0.86 Concrete—Cement.—Cement shall mean Portland cement, as defined by the Standard specifications for cement of the American

Society for Testing Materials revised to 1909, and published in the 1915 Year Book of that Society.

The cost of cement shall not be included in the unit prices bid for concrete, but will be paid for separately under Item 50, Cement.

0.87 Concrete—Fine Aggregate.—Fine aggregate or sand, shall be composed of grains from hard, tough, durable rocks, free from soft, decayed, friable or soluble material. It shall be clean, and shall not contain a sufficient quantity of organic silt, clay or other finely-divided matter to render it unsuitable. If objectionable quantities of silt or other matter adhere to the sand grains, the sand shall be satisfactorily washed. The size of the grains shall be acceptably graded from fine to coarse particles with no grains larger than will pass a one-fourth inch mesh screen. Sand, when mixed into mortar in the proportion by weight of 1 part cement and 3 parts sand may be rejected if it does not develop within 28 days, a compressive strength at least equal to the strength of a similar mortar made of the same cement and standard Ottawa sand, tested at the same age. The use of limestone screenings will not be allowed.

0.88 Concrete—Coarse Aggregate.—Coarse aggregate shall consist of clean, hard, durable, insoluble, broken stone, or gravel, of such size as to pass through a screen having 2½-inch round holes, and be retained on a ¼-inch mesh screen. It shall be acceptably graded in size between the limits prescribed. Coarse aggregate, as here defined, shall be used in all classes of concrete structures except the concrete linings of conduits for dams, and the concrete used in flexible slab revetment, in which cases coarse aggregate of smaller size shall be used, as specified under the appropriate items. Fragments of approximately the same size, or of flat, elongated shapes, or improperly graded, will not be satisfactory. Such screening, grading and washing shall be resorted to as will produce material of acceptable quality. **Run of crusher** or **run of bank**, i. e. gravel or stone containing a mixture of both the fine and coarse aggregates, shall not be used without screening.

In general the local gravels will be deemed satisfactory for use as coarse aggregate if properly screened and washed, but all such gravel shall be subject to test and approval by the Engineer under the provisions of Section 0.91 of the Specifications, entitled Samples. Few or none of the local ledge rocks are satisfactory for use as coarse aggregate.

0.89 Concrete—Water.—The water used in mixing concrete shall be reasonably clean and free from oil, acid, strong alkalies, or vegetable matter.

0.90 Concrete—Large Stones.—Where permitted, clean, sound, non-oil-bearing stones of acceptable size may be imbedded in the concrete provided they can be thoroughly incorporated in the mass and surrounded by an acceptable thickness of concrete. Such stones at the moment of placing in the concrete shall be clean, wet, and free from frost, and shall be well bedded by joggling. Projecting parts of such stones shall be cleaned and wet again, if required, before being covered with concrete. Large stones shall not be placed without first having been inspected by the Engineer.

0.91 Concrete—Samples.—Samples of both fine and coarse aggregate which the Contractor proposes to use shall be submitted to the Engineer for approval a sufficient time in advance of use to allow for the necessary tests for determining the suitability of the material for the use contemplated, and for determining the exact proportions required under the various items.

0.92 Concrete—Preparation of Foundations.—Springs encountered in the foundations shall be plugged, piped or otherwise satisfactorily disposed of. No concrete shall be placed until the foundation is prepared to the satisfaction of the Engineer. If upon rock, the surface shall be scrupulously freed from all dirt, gravel, scale, loose fragments or other objectionable substances. Streams of steam, air, or water under sufficient pressure, and wire brushes or other effective means shall be used to accomplish this cleaning. Steam jets or hot water shall be used to throughly remove snow, ice or frost if any be found upon the foundation when it is desired to lay concrete.

Earth foundations, if required, shall be thoroughly compacted by rolling or other approved methods, and in no cases shall the Contractor deposit concrete on newly placed fills or on such as, in the opinion of the Engineer, are likely to shrink or settle in a manner which may injure the concrete structure. Earth foundations shall be thoroughly and satisfactorily wetted immediately prior to placing the concrete thereon. If frozen, the soil shall be thawed out in a satisfactory manner, and then thoroughly compacted even though this may have been done previous to freezing.

0.93 Concrete—Mixing.—All operations incidental to mixing concrete shall proceed with sufficient dispatch to insure the bonding together of the successive batches as a true monolith. Concrete shall be mixed in approved mechanical batch mixers, in batches of suitable size, except that, when permitted, it may be mixed by hand in a thorough and satisfactory manner. In determining proportions of ingredients, 100 pounds of cement shall be considered 1 cubic foot. The coarse and fine aggregates shall be measured separately, and uncompacted, in approved measuring boxes. Suitable means shall be provided for controlling and accurately measuring the water. The entire batch shall remain in the mixer not less than 60 seconds and longer if necessary to secure a thoroughly satisfactory mix. Except in special cases authorized by the Engineer, concrete which has its initial set before placing in the work shall not be re-tempered, but shall be at once discarded and removed.

0.94 Concrete—Forms.—The Contractor shall provide and maintain in good condition all necessary forms, molds and centers for shaping the concrete. Such forms shall be true to the required shapes and sizes, properly braced, and strong enough to withstand, without springing or warping, all operations incidental to placing the concrete. They shall be made mortar-tight, and the faces in contact with the concrete shall be satisfactorily smooth and clean. To prevent adhesion to the concrete, the contact surfaces of all forms, whether of wood or steel, shall, where required, be coated with soap, mineral oil, or other suitable substance. Wherever required, forms shall be thoroughly wet before placing concrete, so as to prevent injurious drying of the surface of the concrete. Small rods, or wires, to hold the forms in place may be imbedded in the concrete, provided proper means be used to remove the end portions within the concrete, approximately 2 inches from an exposed face. All holes left after removal of the rod or wire ends shall be immediately and completely filled with cement mortar. Forms shall not be removed until the Engineer has approved such removal, but the Contractor alone shall be responsible for all injury to concrete due to their premature removal. Forms shall be removed with great care, so as to avoid injury to the concrete. Forms unsatisfactory in any respect shall not be used, and, if condemned, shall be removed immediately from the work.

0.95 Concrete—Placing.—Provision shall be made for the rapid transportation of fresh concrete from the point of mixing to the work. Care shall be taken in conveying and depositing concrete to avoid methods which tend to produce a segregation of the component parts.

Concrete shall be deposited in approximately level layers, the thickness of which shall be no greater than will readily permit proper compacting, and such manipulation as will remove entrained air and surplus water, and produce, from the accepted materials, the most compact, dense and impervious concrete practicable. The concrete shall be well spaded against the forms.

Concrete shall not be deposited in water without explicit permission, and then only in strict accordance with directions. Care shall be taken that no water interferes in any way with the proper placing of concrete, and the Contractor shall not allow water to rise on any concrete until, in the opinion of the Engineer, it has set sufficiently. All laitance or other substance shall be removed from the surface of concrete at such times, depending upon weather conditions and rate of hardening, as the Engineer may direct. Where new concrete is joined to old, or to rock, the surface of the old concrete or rock shall be thoroughly cleaned, using wire brushes and a jet of water from a hose if necessary, and the surface shall be clean and wet at the moment the fresh concrete is placed. If required, surfaces of concrete previously placed shall be satisfactorily roughened. A wash of neat cement grout shall be scrubbed into the contact surface of the old concrete or the rock with steel brooms immediately before placing concrete and shall be thoroughly worked into all crevices and depressions. When required, a coat of mortar shall be spread over the contact surfaces thus prepared. Provisions for bonding new concrete to old shall be made by the use of forms of such styles and dimensions as to properly form the concrete joints into steps, dovetails, grooves, or other ordered shapes. Where required large stones, dowels, or steel reinforcing rods shall be embedded in concrete for bonding new work to old.

0.96 Concrete—Construction Joints.—Wherever required the work shall be so prosecuted that construction joints will occur only at designated places, and when so directed, the Contractor shall complete by continuous depositing of concrete, sections of the work comprised between such joints.

0.97 Concrete—Contraction Joints.—Contraction joints, expansion joints, and slip-joints shall be formed where and as directed in concrete structures. In general, such joints shall be made by building smooth surfaces, acceptably coated if required with some approved substance to render them inadhesive, and shall be grooved or otherwise shaped as directed, or shall have metal strips imbedded therein.

0.98 Concrete—Finishing Surfaces.—Unless otherwise specified, all concrete surfaces which are to remain permanently exposed shall be given a neat appearance, by removing in an approved manner all rough edges and projections. Honey-combed sections shall either be pointed up with cement mortar or taken out and the spaces refilled with concrete or mortar, as the Engineer may direct. Recesses formed by the removal of defective concrete shall be refilled with concrete as specified in Section 0.101 of the Specifications entitled Concrete—Replacing Faulty Work. Patching and plastering, or washing with cement grout, shall be done only under the specific orders and directions of the Engineer.

Exposed surfaces which are not cast against forms shall be brought to proper lines and grades and shall be given a suitable finish by means of troweling or floating.

0.99 Concrete—Cold Weather Precautions.—Concrete shall not be mixed or deposited during freezing weather without the specific direction or permission of the Engineer. When so directed or permitted, approved precautions shall be taken for removing snow, ice, or frost from the materials and from the surfaces upon which the concrete is to be placed. The water and aggregates shall be heated, and satisfactory provisions made for keeping the fresh concrete from freezing, including adequate coverings, steam pipes or other appliances and materials. Such protection of concrete shall continue, until in the opinion of the Engineer there will be no further danger from freezing. The laying of concrete may, however, be prohibited at any time when, in the judgment of the Engineer, the conditions are unsuitable or the proper precautions are not being taken.

0.100 Concrete—Protection of.—The Contractor shall not permit walking or working over or upon finished surfaces of concrete until sufficiently hardened. Concrete shall be kept moist for at least 4 weeks or until covered with earth, and during this time shall be protected from freezing. Every precaution shall be taken to prevent concrete from drying until there is no danger of cracking or crazing due to lack of moisture. To this end concrete surfaces, unless covered with wet canvas or other equally effective material, shall be sprinkled with sufficient frequency to preclude any possibility of drying out. Alternate drying and wetting shall be particularly guarded against. Great care shall be exercised at all times to prevent injury to concrete surfaces which will be exposed in the finished work. The Contractor may at any time prior to the final acceptance of the work be required to clean all exposed faces of concrete.

0.101 Concrete—Replacing Faulty Work.—If, upon removing the forms, molds, or centers, any voids or other imperfections be found, such faults shall be corrected by the Contractor without additional cost to the District, by patching or other methods, as directed, even to the extent of taking down and replacing unsatisfactory concrete; or if any forms shall fail or become displaced, and as a result the section of concrete contained therein shall, in the opinion of the Engineer, be injured or out of line or be subjected to premature stresses, the Contractor, without additional cost to the District, shall satisfactorily replace so much of said section of concrete as the Engineer may deem necessary, even to the extent of rebuilding the entire section.

The Contractor shall have no claim for payment for cement used in concrete which is rejected as faulty or unsatisfactory.

0.102 Concrete—Placing Metal Work.—There shall be embedded in, or set in, or attached to the concrete wherever directed, reinforcing steel, structural steel shapes, castings, piping or other metal objects shown on the Drawings, required by these Specifications or ordered. The concrete shall be packed tightly around metal work to prevent leakage, and to secure perfect adhesion, and all necessary precautions shall be taken to prevent these objects from being loosened, displaced, broken or deformed.

0.103 Concrete—Measurement for Payment.—The quantities of concrete to be paid for shall be the number of cubic yards deposited in place in accordance with the Drawings, Specifications, and instructions of the Engineer. Whenever the structures are of such types that concrete is to be built against the sides of any excavation, the concrete shall be measured as if the excavation were made exactly to the prescribed lines. Any rubble or brick masonry used by permission of the Engineer for convenience in setting pipes or other objects, or for other purposes in connection with concrete structures, and any mortar used with approval in the laying of any such masonry, shall be measured and paid for as of the class of concrete in which it is used.

0.104 Concrete—Prices to Include.—Except for cement, reinforcing steel and other metal work, as provided elsewhere, the unit prices for the different items of concrete shall cover the entire cost of construction thereof, which shall include: the cost of materials for fine and coarse aggregates, and of large stones; the cost of all mixing, conveying, placing of concrete and large stones; the cost of damming, bulkheading, pumping, bailing; the cost of preparing foundations, piping or otherwise caring for water, and removing laitance; the cost of grout wash or mortar beds used in building upon old work or on rock, and of coatings for contraction joints; the cost of protecting concrete against frost, floods and injury from other sources; the cost of erecting and removing forms, centers, molds, bracing, landing platforms and scaffolds, and of the lumber, steel, nails, wire and other material used therein; the cost of finishing, patching, and wetting; the cost of removing and replacing defective work; and any other work and materials necessary to complete the concrete construction as required by this Contract.

But such unit prices shall not include the cost of furnishing cement, the cost of furnishing and placing reinforcing steel, piping, conduits, structural steel, castings and other metal work, which shall be paid for under the items provided therefor.

0.105 Mortar.—Mortar shall consist of Portland cement, fine aggregates and water, mixed in such proportions as the Engineer may direct. The cement, fine aggregates and water shall conform to the specifications for these materials for concrete, except that the fine aggregate in mortar used for filling joints of pipes or brick masonry shall have no grains larger than will pass a ⅛-inch mesh screen.

Payment for mortar will not be made separately, but will be included in the unit price of the concrete, masonry, pipe, or other work in which it is used; except that the cement will be paid for as a separate item,

DETAIL SPECIFICATIONS

To save space the detail specifications for some items are not printed in this volume. The omitted items include: first, all highway specifications, which are based upon the standard specifictions of the Ohio State Highway Department, and which are, therefore, readily accessible elsewhere: and second, numerous items of only minor importance in these contracts for which the specifications are, therefore, relatively brief. All the items are, however, listed in the table of contents and index.

Item 1

ˡ Stream Control at Dams

1.1 Description.—Under this item shall be included all work and material and all uses of the resources of the Contractor which may be necessary in order to adapt the work of constructing the dams and appurtenances to the safe control of the flow of the stream during construction, and until the entire work of constructing the dams and appurtenances shall have been fully completed, the object being to secure the safety of life and property below the dams, and to prevent damage or delay in completion which might result from inadequate steps being taken.

1.2 Methods of Equipment.—The contract drawings suggest a method of stream control, which the Contractor may follow, and which he will be required to follow unless he shall, with the approval of the Engineer, adopt some other method or some variation of this method. No proposal for different or amended methods of stream control will be considered which is not accompanied by complete drawings and description, or which entails a total expense to the District greater than the method of stream control shown in the contract drawings or otherwise described in the Contract. It is specifically agreed, however, that the adoption by the Contractor of the method of stream control suggested on the contract drawings; or the approval by the Engineer of an elaboration or revision of that method, or of a new method, suggested by the Contractor; or the approval by the Engineer of any other matters pertaining to stream control, shall not in any degree relieve the Contractor from full liability and responsibility on all matters pertaining to stream control; but such approval shall cover only the right of the Contractor to proceed with the work.

1.3 Payment.—For all operations, materials, labor and equipment except as specified hereinafter, and for all the resources of the Contractor used for, or incidental to, the operations under this item, the Contractor shall receive the lump sum stipulated in Item 1. No payment under any other item of the Contract shall be made for cofferdams, slope protection, embankment of a temporary nature or outside the prescribed lines for the dam, or for any other work or materials incidental to the control of the flow of the stream not specifically designated on the Drawings for payment under other items, compensation for all such incidental items being considered as included in the lump sum stipulated for Item 1; excepting that excavating slopes of embankment to make satisfactory bond between old and new work shall be paid for under Item 6, Excavation. The price for Item 1 shall be considered full compensation to the Contractor for all risks and damages to the work under this Contract resulting from the flow of the stream and he shall make good without cost to the District all injury to the work of whatever nature resulting therefrom. The Contractor shall have no claim or charge against the District because of any unforeseen expense in building or rebuilding work resulting from the necessities or results of stream control operations.

The lump sum stipulated in Item 1 shall be estimated for payment when the dam is built to its full height, and when, in the opinion of the Engineer, all damage to the dam or appurtenances resulting from the flow of the stream has been made good and no further work under this item will be required.

Item 2

Clearing and Grubbing

2.1 Description.—Such portions of the sites of embankments, highway rights-of-way, borrow pits, excavations to be used for embankment material, and other areas, as the Engineer may designate, shall be cleared of all trees, stumps, bushes, and other similar encumbrances. Roots, one inch or more in diameter, shall be grubbed as directed.

2.2 Disposal of Materials.—Materials obtained from the clearing and grubbing operations, shall be the property of the Contractor unless otherwise specified, and must be removed from the vicinity of the work or otherwise satisfactorily disposed of.

2.3 Payment.—The quantity to be paid for under this item shall be the area in acres prescribed or designated by the Engineer, within the limits of the work, from which all encumbrances as herein provided shall have been satisfactorily removed. In the case of isolated trees the area to be paid for shall be that covered by the overhang of the branches. Should any growth occur between the time of clearing and the time of placing of embankment, the Contractor shall satisfactorily remove such growth, but payment shall be made once only for any given area, regardless of how many times it may be found necessary to go over such area to render its condition satisfactory at the time of use. The unit price stipulated for this item shall include the cost of clearing the area of all encumbrances as specified, of disposing of all materials, and all expense incidental thereto, except as hereinbefore provided under Section 0.62 of the Specifications, entitled Materials Obtained from Construction.

Item 4

Trimming and Shaping Slopes of River Channels

4.1 Description.—This item shall include the performance of such work, in addition to the channel excavation, as may be necessary to give the channel slope or bank a true shape conforming to the lines and grades shown on the Drawings or given by the Engineer. All irregularities shall be corrected, and the surface left even and smooth without projections. Wherever revetment is specified, the surface here described shall mean the sub-grade. Wherever surface dressing and grassing is required the sub-grade shall be trimmed as herein specified immediately before the placing of the top soil.

The channel excavation at the slopes shall be so made that the natural ground shall not be disturbed back of the line of finished slope, or the line of sub-grade in case revetment or top soil is required.

4.2 Payment.—The quantity to be paid for under this item shall be the area in square yards, included within the limits shown on the Drawings or designated by the Engineer. The price stipulated for this item shall include the cost of all operations and materials necessary for the trimming and shaping of the slopes, and the disposal of surplus materials, and all expense incidental thereto, except as provided in Section 0.62 of the Specifications, entitled **Materials Obtained from Construction.**

Item 5

Excavation, Soil Stripping

5.1 Description.—Soil stripping shall include the excavation within the area to be covered by embankment, of all top soil, peat, grass, weeds, muck, boulders, loose rock, perishable or otherwise objectionable matter, to such depths as may be ordered, in order that the earth embankment may be suitably bonded to a satisfactory natural foundation. Where such material overlies material to be excavated, stripping shall not be estimated for payment as a separate item, but the surface material shall be measured with the excavation directly underlying.

No payment shall be made for the stripping of borrow pits, but the cost of such stripping shall be included as a part of the contract price for furnishing and placing the other material taken from the borrow pit.

5.2 Payment.—Measurement for payment under this item shall include the quantity of material in cubic yards, measured in excavation, actually removed from within the limits shown on the Drawings or prescribed by the Engineer. The price bid shall include the cost of all operations and materials incidental to the excavation and disposal of the materials excavated, except as hereinbefore provided under Section 0.62 of the Specifications entitled Materials Obtained from Construction.

Items 6, 7, 8, 6a, 7a, and 8a

Excavation for New Channels in Channel Improvements

6.1   **Description.**—These items shall include all excavation
7.1   for new channels in channel improvements; for cut-off
8.1   trenches at embankments; and for stream control chan-
6a.1   nels; classified as described in Section 0.65 of the Specifi-
7a.1   cations entitled Excavation—Classification.

8a.1   Item 6 shall include such excavation that may be
classified properly as Class 1—Earth.

Item 7 shall include such excavation that may be classified properly as Class 2—Mixed Excavation.

Item 8 shall include such excavation that may be classified properly as Class 3—Rock.

In some cases there has been shown and so designated on the
Drawings, the estimated elevation of mean low water at points
where the excavations listed above are to be made. In such cases,
and only in such cases, Items 6a, 7a, and 8a shall include the excavation, below such estimated elevation, that may properly be classified,
respectively, as Class 1—Earth; Class 2—Mixed Excavation; and
Class 3—Rock. Where no such estimated elevation is shown on
the Drawings all excavation whether above or below mean low
water, will be included under Items 6, 7, and 8, and not under Items
6a, 7a, or 8a.

Excavations shall have such side slopes as are shown on the
Drawings or designated by the Engineer.

6.2   **Maintenance During Construction.**—Should material
7.2   wash into a cut-off trench before the placing of embank-
8.2   ment, or stony gullies due to wash be formed, such ma-
6a.2   terial and stones shall be carefully removed, if directed,
7a.2   and the Contractor shall be paid the contract price per
8a.2   cubic yard for material so removed.

The excavations for channel improvements and for
stream control channels shall be maintained in good order by the
Contractor at his own expense, and if such maintenance involves the
removal of material from outside the prescribed limits, no payment
shall be made the Contractor for the handling of such outside material.

6.3
7.3
8.3
6a.3
7a.3
8a.3

Payment.—Payment for these items will be made in accordance with Section 0.71, of the Specifications, entitled Excavation—Payment, except that in cases of excavation for Channel Improvement, where it is not practicable to measure the material in excavation, the quantities may be determined from spoil banks or embankments in which the material has been placed, making deduction of 5 per cent for shrinkage.

Items 9, 10, and 11

Excavation for River Channel Improvements

9.1   **Description.**—These items shall include all river chan-
10.1   nel excavation for channel improvements (except excava-
11.1   tion for new channels, and for structures as provided for
elsewhere) classified as described in Section 0.65 of the
Specifications, entitled Excavation—Classification.

Item 9 shall include all such excavation that may be classified
properly as Class 1—Earth.

Item 10 shall include all such excavation that may be classified
properly as Class 2—Mixed Excavation.

Item 11 shall include all such excavation that may be classified
properly as Class 3—Rock.

Excavation under Items 9, 10, or 11, will be classified without regard to its location above or below the elevation of mean low water.

Finished work shall present a reasonably smooth and uniform
surface, and shall nowhere project above grade or within the limits
of the prescribed section.

9.2   **Payment.**—Payment for these items will be made in
10.2   accordance with Section 0.71 of the Specifications, en-
11.2   titled Excavation—Payment, with the following excep-
     tions:

This material will be measured in the spoil banks or the embankments in which it is placed, with a deduction of 5 per cent for
shrinkage, as it is not practicable to measure it in excavation. Deduction will be made for material excavated more than 0.3 foot below grade or more than 0.3 foot beyond prescribed limits.

Any material that washes into the cut through no fault of the
Contractor shall be removed by him at the unit price per yard stipulated for these items.

Required haul beyond the limit of free haul will be paid for under
Item 12, Excavation—Overhaul for River Channel Improvements.

Item 12

Excavation—Overhaul for River Channel Improvements

12.1   Description.—This item shall include the transporting beyond the limit of free haul, which will be 2000 linear feet unless otherwise specified, of material excavated for river channel improvements,—under Items 9, 10, or 11 only,—and deposited in spoil banks, embankments, or other places. It shall not include the transporting of any material except under the Engineer's specific orders.

12.2   Payment.—The overhaul to be paid for under this item shall be the quantity in cubic yards per 100 feet transported beyond the free haul limit. The length of haul shall be understood to mean the distance from the center of mass of the material in place before excavation to the center of mass of the same material as deposited; the locations of either center of mass will be estimated from the best data available, when determinations by actual measurement are not practicable.

Items 13, 14, 15, 13a, 14a, and 15a

Excavation for Outlet Works and Spillways at Dams

13.1   **Description.**—These items shall include all excava-
14.1   tions for outlet works and spillways at Dams, including
15.1   approach and discharge channels, classified as described
13a.1   in Section 0.65 of the Specifications, entitled Excavation
14a.1   —Classification.

15a.1   Item 13 shall include such excavation that may be
classified properly as Class 1—Earth.

Item 14 shall include such excavation that may be classified
properly as Class 2—Mixed Excavation.

Item 15 shall include such excavation that may be classified
properly as Class 3—Rock.

In some cases there has been shown and so designated on the
Drawings, the estimated elevation of mean low water at points
where these excavations are to be made. In such cases, and only in
such cases, Items 13a, 14a, and 15a, shall include the excavation,
below such estimated elevation, that may properly be classified,
respectively, as Class 1—Earth; Class 2—Mixed Excavation; and
Class 3—Rock. Where no such estimated elevation is shown on
the Drawings, all excavations whether above or below mean low
water will be included under Items 13, 14, and 15, and **not** under
Items 13a, 14a, or 15a.

13.2   **Sheeting and Bracing.**—Satisfactory sheeting and
14.2 · bracing shall be used, unless omitted by written permis-
15.2   sion of the Engineer, to hold the sides of the excavation
13a.2   for the outlet conduits, and to prevent damage from
14a.2   seepage into the cut. All sheeting and bracing shall be
15a.2   removed as the backfill is placed, unless otherwise or-
            dered in writing by the Engineer. The trench shall at
all times be kept satisfactorily unwatered in accordance with Section 0.59 of the Specifications, entitled Pumps and Drains, except
that after the excavation is completed and while concrete is being
laid the unwatering of excavation shall be considered as part of the
work required under Concrete, Items 41a and 43a.

13.3   **Payment.**—Payment for these items shall be made in
14.3   accordance with Section 0.71 of the Specifications, en-
15.3   titled Excavation—Payment.
13a.3
14a.3
15a.3

Items 16, 17, and 18

Excavation for Minor Drainage Systems

16.1   **Description.**—These items shall include all excavation
17.1·   in connection with the minor surface drainage systems
18.1   such as are used for disposing of springs in embankment
foundations, or for draining low sections of land within
the retarding basins, or for handling surface water on the berms of
embankments, or for similar purposes; classified in accordance with
Section 0.65 of the Specifications, entitled Excavation—Classification.

Item 16 shall include all such excavation that may be classified
properly as Class 1—Earth.

Item 17 shall include all such excavation that may be classified
properly as Class 2—Mixed Excavation.

Item 18 shall include all such excavation that may be classified
properly as Class 3—Rock.

16.2   **Payment.**—Payment for these items will be made in
17.2   accordance with Section 0.71 of the Specifications, en-
18.2   titled Excavation—Payment.

Items 19, 20, and 21

Excavation for Street and Highway Cuts and Fills

19.1   **Description.**—These items shall include the excava-
20.1   tion of material from street and highway cuts, and from
21.1   borrow pits, trenches for culverts, ditches, drains, and
from other points as required for street and highway construction or alteration, not including bridge piers and abutments:
the placing of such material in the highway fill as directed; and the
excavation and removal of such material as is unsuitable for road
foundations. The materials excavated under these items which are
suitable shall be used in the construction to the extent required, but
payment will be made only for the original removal.

This excavation shall be classified in accordance with Section
0.65 of the Specifications, entitled Excavation—Classification.

Item 19 shall include all such excavation that may be classified
properly as Class 1—Earth.

Item 20 shall include all such excavation that may be classified
properly as Class 2—Mixed Excavation.

Item 21 shall include all such excavation that may be classified
properly as Class 3—Rock.

19.2   **Disposal of Material.**—All suitable material from the
20.2   excavation shall be used as far as practicable to build em-
21.2   bankments and to make refills in excavations. Surplus excavation and waste material shall be used to widen
embankments or shall be deposited in such places and in such manner as may be suitable.

19.3   **Borrow Pits.**—Wherever the material from the
20.3   excavation suitable for use is insufficient in quan-
21.3   tity to form the necessary embankments and refills, the
Engineer will designate where additional Class 1 material
may be procured. Unless otherwise directed by the Engineer a
berm of at least four feet shall be left between the outside of the embankment and the edge of a borrow pit, with provision for side slopes
of 1½ to 1 to the bottom of the borrow pit. Unless the Engineer
gives the Contractor specific written orders to excavate material
other than earth from borrow pits, all material obtained from borrow pits will be paid for under Item 19, as earth excavation, regardless of its actual character.

19.4   **Construction of Embankments and Refills.**—The ma-
20.4   terials used in street and highway embankments and re-
21.4   fills shall be free from all matter which is perishable or of
a nature to cause unduly prolonged settlement. Embankments shall be constructed to conform to the lines and grades shown on the Drawings or prescribed by the Engineer, and such overfill. not to exceed 10 per cent, shall be placed as may be required by the Engineer to allow for settlement. Where directed by the Engineer embankments shall be built in horizontal layers approximately five feet in thickness, each layer being built to the full width. Backfilling around culverts, bridge abutments or other structures shall be thoroughly compacted by puddling, tamping or other approved methods as directed.

19.5   **Maintenance During Construction.**—The Contractor
20.5   shall keep the surface of the roads constructed under this
21.5   Contract in good condition, during the life of the Contract. When so ordered, he shall fill any depressions caused by settlement or by wear and tear due to use, and he shall be paid therefor at the unit prices stipulated for the respective items entering into such repairs.

19.6   **Readjusting Grades.**—Attention is called to the un-
20.6   certainty of ledge rock elevations, which, owing to the
21.6   nature of the work to be done, cannot be determined in advance. In order to avoid shallow rock excavation, or to obtain sufficient suitable materials for embankments and refills, or for any other purposes, the grades shown on the Drawings may be changed or modified at any time by the Engineer, to obtain the desired results.

19.7   **Payment.**—Unless otherwise shown on the Drawings
20.7   or specifically ordered by the Engineer, work under these
21.7   items shall be paid for in accordance with Section 0.71 of the Specifications, entitled Excavation—Payment, and as follows:

First. Clearing and Grubbing will be paid for under Item 2.

Second. All excavations in which concrete structures are to be constructed will be measured for payment to lines 1 foot outside of the foundations of the structure and to such slopes as will stand without sliding; provided that when the character of the material cut into is such that it can be trimmed to the required lines of the structure, and the concrete placed against the sides of the excavation without the use of intervening forms, payment will not be made for excavation outside of the required limits of the concrete.

Third. All excavations for culverts or drains in which circular pipes are to be placed, shall be measured for payment to the required elevation of the lowest part of the pipe, to a bottom width 2 feet greater than the nominal diameter of the pipe and to such side slopes as will stand without sliding.

Fourth. All excavation not included in the two preceding paragraphs shall be measured for payment as excavation in place before removal, to include the total amount of excavation required for embankments and refills as well as that which by the direction of the Engineer is excavated and wasted.

Fifth. Backfilling will be paid for under Item 37.

With the exceptions noted above the unit prices stipulated under these items shall include the cost of all labor and materials, and all other expense necessary for completing the excavation, hauling it as far as is necessary, and properly placing and depositing it in the embankments and refills, or otherwise disposing of it in a satisfactory manner,

Items 22, 23, 24, 22a, 23a, and 24a

Excavation for Retaining Walls, Bridge Piers, and Flood Gate Structures

22.1	**Description.**—These items shall include all excava-
23.1	tions for retaining walls, bridge piers and abutments,
24.1	flood gate structures and similar structures except at
22a.1	Dams, where such excavation will be included in that
23a.1	for outlet works and spillways. The excavation will be
24a.1	classified in accordance with Section 0.65 of the Specifi-

cations, entitled Excavation—Classification.

Item 22 shall include such excavation that may be classified properly as Class 1—Earth.

Item 23 shall include such excavation that may be classified properly as Class 2—Mixed Excavation.

Item 24 shall include such excavation that may be classified properly as Class 3—Rock.

In some cases there has been shown and so designated on the Drawings, the estimated elevation of mean low water at points where these excavations are to be made. In such cases, and only in such cases. Items 22a, 23a, and 24a shall include the excavation, below such estimated elevation, that may properly be classified, respectively, as Class 1—Earth; Class 2—Mixed Excavation; and Class 3—Rock. Where no such estimated elevation is shown on the Drawings, all excavations, whether above or below mean low water, will be included under Items 22, 23, and 24, and not under Items 22a, 23a, or 24a.

22.2	**Sheeting and Bracing.**—Satisfactory sheeting and
23.2	bracing shall be used, unless omitted by written permis-
24.2	sion of the Engineer, to hold the sides of the excavation
22a.2	and to prevent damage from seepage into the cut. All
23a.2	sheeting and bracing shall be removed as the backfill is
24a.2	placed, unless otherwise ordered in writing by the Engi-

neer. The excavation shall at all times be kept satisfactorily unwatered in accordance with Section 0.59 of the Specifications, entitled Pumps and Drains. All voids left or caused by withdrawal of sheeting shall be filled immediately with suitable material and tamped or puddled as required.

22.3	**Payment.**—Payment for these items will be made in
23.3	accordance with Section 0.71 of the Specifications, en-
24.3	titled Excavation—Payment.
22a.3	Unless otherwise shown or prescibed the excava-
23a.3	tions will be measured to vertical planes 1 foot outside
24a.3	of the required limits of the foundation of the structure.

Items 25, 26, and 27; 25a, 26a, and 27a

Excavation and Backfilling of Trenches

25.1   **Description.**—These items shall include all excava-
26.1   tion and backfilling of trenches for pipes, sewers, con-
27.1   duits, levee culverts and similar structures, except at
25a.1   Dams, where such excavation is included in the items
26a.1   for Excavation for Minor Surface Drainage Systems;
27a.1   and except in connection with highway construction
in which case it is included in the items for Street and
Highway Cuts and Fills. This excavation shall be classified as provided in Section 0.65 of the Specifications, entitled Excavation—
Classification.

Item 25 shall include such excavation that may be classified
properly as Class 1—Earth.

Item 26 shall include such excavation that may be classified
properly as Class 2—Mixed Excavation.

Item 27 shall include such excavation that may be classified
properly as Class 3—Rock.

In some cases there has been shown and so designated on the
Drawings, the estimated elevation of mean low water at points
where these excavations are to be made. In such cases, and only in
such cases, Items 25a, 26a, and 27a shall include the excavation, below such estimated elevation, that may properly be classified, respectively, as Class 1—Earth; Class 2—Mixed Excavation; and
Class 3—Rock. Where no such estimated elevation is shown on
the Drawings, all excavation whether above or below mean low
water will be included under Items 25, 26, and 27, and not under
Items 25a, 26a, or 27a.

All excavation shall be made in open cut, and no tunneling will
be allowed except by written permission or order from the Engineer.

25.2   **Sheeting and Bracing.**—Satisfactory sheeting and
26.2   bracing shall be used, unless omitted by written per-
27.2   mission of the Engineer, to prevent any movement
25a.2   which would injure the new or existing structures, and
26a.2   to prevent damage from seepage into the trench. Where
27a.2   used it shall be kept driven ahead of the excavation
whenever practicable. If, in the opinion of the Engineer, additional bracing is required, it shall be placed under his
orders at the expense of the Contractor. All sheeting and bracing

shall be removed as the backfill is placed, unless otherwise ordered in writing by the Engineer. The excavation shall be kept satisfactorily unwatered in accordance with Section 0.59 of the Specifications, entitled Pumps and Drains. All voids left or caused by withdrawal of sheeting shall be filled immediately with suitable material, tamped or puddled as required.

25.3	**Payment.**—Payment for these items shall be made
26.3	in accordance with Section 0.71 of the Specifications, en-
27.3	titled Excavation—Payment. Unless otherwise shown
25a.3	or prescribed the width of trench will be measured as
26a.3	2 feet greater than the nominal diameter of the pipe
27a.3	to be laid therein, or as 2 feet more than the base

width of the concrete or masonry sewer or structure; the sides of the trench will be considered as being vertical; the depth and length shall be the required depth and length of the structure. The unit prices stipulated for these items shall also include the cost of backfilling the trenches as prescribed in Sections 0.83 and 0.84 of the Specifications, entitled Backfilling.

Items 28, 29, and 30

Excavation, Miscellaneous

28.1   **Description.**—These items shall include all excavation
29.1   of whatever nature, shown on the Drawings or ordered
30.1   by the Engineer, not included in Items 1 to 27 inclusive,
classified in accordance with Section 0.65 of the Specifications, entitled Excavation—Classification.

Item 28 shall include all such excavation that may be classified
properly as Class 1—Earth.

Item 29 shall include all such excavation that may be classified
properly as Class 2—Mixed Excavation.

Item 30 shall include all such excavation that may be classified
properly as Class 3—Rock.

28.2   **Payment.**—Payment for these items will be made in
29.2   accordance with Section 0.71 of the Specifications, en-
30.2   titled Excavation—Payment.

Items 31, 32, and 33

Embankment for Main Dams

31.1   **Description.**—These items shall include all embank-
32.1   ments for main dams as shown on the Drawings or re-
33.1   quired by the Engineer.

Item 31 shall include all such embankment that may be classified properly as Class A Embankment, as described in Section 0.73 of the Specifications.

Item 32 shall include all such embankment that may be classified properly as Class B Embankment, as described in Section 0.74 of the Specifications.

Item 33 shall include all such embankment that may be classified properly as Class C Embankment as described in Section 0.75 of the Specifications.

The proposal provides for alternate bids on embankment depending on the bidder's choice of methods of construction. A minimum amount of Class C embankment will be required at some dams, as is indicated in the schedule of quantities and shown on the Drawings. Except for this minimum amount of Class C embankment, the Contractor may, if he so desires, substitute one class of embankment for another, provided that the written approval of the Engineer first be given, and provided further that such substitution shall involve no additional cost to the District; that is, if Class B embankment is to be substituted for Class A at the Contractor's request, and the unit price of Class B is greater than that of Class A, the material will be paid for at the Class A price. On the other hand, if a substitution is made by order of the Engineer (rather than with his permission), requiring a class of embankment of higher unit price, payment will then be made at the unit price bid for the class of embankment required, except that in case the Contractor, by failing to carry out the terms of the Contract as to rate of progress or otherwise, shall in the opinion of the Engineer put the safety of the work, or the safety of life or property, in jeopardy; and if for the sake of safety it becomes necessary under such conditions, in the opinion of the Engineer, to substitute for one class of embankment another class on which the unit contract price is higher; in such cases the Engineer may order such substitution, and the Contractor shall be paid the unit price of the class contemplated in the Contract, and not the unit price of the class made necessary by his failure to carry out the terms of the Contract.

31.2   **Allowance for Shrinkage and Settlement.**—Sufficient
32.2   overfill shall be placed to provide for shrinkage and set-
33.2   tlement. To this end, the top and the slopes of the dam
shall be built to lines determined by adding not more
than one per cent to the net vertical depth of fill required at any
point. - In calculating shrinkage and settlement at the time of trimming and dressing the surface, it will be considered that shrinkage
will continue at a uniform rate for one year after the completion of
the fill to any given elevation. At the time of trimming, the required excess to provide for shrinkage at any given point shall be
varied according to the age of the fill so that no excess will be required on a fill which has been in place for one year.

While slight depressions on the slopes may be made to conform
to the required lines by subsequent filling, such procedure is not
looked upon with favor, and the Contractor shall carefully conduct
his operations in such manner that the material originally placed
shall conform approximately to the limits prescribed. Any loss of
section by displacements, slides or slips, either inward or outward,
which may occur before final acceptance of the work, shall be satisfactorily corrected by the Contractor, without expense to the District.

31.3   **Trimming Slopes.**—As the last work to be done under
32.3   these items, the Contractor shall carefully trim the slopes
33.3   and top of the embankment to the required gross lines.
and depressions, if any, shall be brought to grade with
acceptable material satisfactorily compacted. If required, the slopes
shall then be dressed with top soil of the thickness prescribed, and
grassed, as provided under Item 38, Surface Dressing and Grassing.
In case excess material has been placed in the fill, the Contractor
will not be required to remove it, except in so far as is necessary to
leave the berms in good condition, to leave the slopes and surface
uniform, and to prevent interference or lack of conformity with conduits, spillways, or other parts of the work.

31.4   **Payment.**—Payment under these items will be made
32.4   in accordance with Section 0.82 of the Specifications, en-
33.4   titled Embankment—Payment.

Items 34 and 35

Embankment for Levees

34.1   **Description.**—These items shall include all embank-
35.1   ments for levees as shown on the Drawings or required
by the Engineer.

Item 34 shall include all such embankment that may be classified
properly as Class D Embankment, as described in Section 0.76 of
the Specifications.

Item 35 shall include all such embankment that may be classified
properly as Class E Embankment, as described in Section 0.77 of
the Specifications.

34.2   **Allowance for Shrinkage and Settlement.**—Sufficient
35.2   overfill shall be placed to provide for shrinkage and settlement, depending upon the character of materials used
as well as the manner of placing. In no case, however, will the
overfill required on this account exceed 10 per cent. Any loss of
section by displacements, slides or slips, which may occur before
final acceptance of the work, shall be satisfactorily replaced by the
Contractor, without expense to the District.

34.3   **Trimming Slopes.**—Before the completion of the work
35.3   under these items the Contractor shall trim the slopes
and top of the embankment to the required lines and
grades, in a workmanlike manner. If required, the slopes shall then
be dressed with top soil of the thickness prescribed, and grassed, as
provided under Item 38, Surface Dressing and Grassing. In case
excess material has been placed in the fill, the Contractor will not
be required to remove it, except in so far as is necessary to leave the
berms in good condition and to leave the slopes and surface uniform.

34.4   **Payment.**—Payment under these items will be made
35.4   in accordance with Section 0.82 of the Specifications, entitled Embankment—Payment, except that the unit
prices, under Item 34, Embankment Class D, shall not include the
cost of excavation, which will be included for payment under other
items. Under Item 35, however, the unit prices shall include the
cost of excavation from borrow pits.

Item 36

Embankment, Miscellaneous

36.1   Description.—This item shall include all embankments as shown on the Drawings or required by the Engineer, that may be classified properly as Class F Embankment as described in Section 0.78 of the Specifications.

36.2   Allowance for Shrinkage and Settlement.—Sufficient over-fill not exceeding 15 per cent shall be placed to allow for settlement, depending upon the character of materials used and the manner of placing.

36.3   Payment.—Payment under this item will be made in accordance with Section 0.82 of the Specifications, entitled Embankment—Payment.

Item 37

Backfilling

37.1   Description.—Backfilling shall be in accordance with Sections 0.83 and 0.84 of the Specifications. This item shall include only such backfilling as is specified in the Estimates or Drawings as a separate item, or as may be ordered as a separate item by the Engineer.

37.2   Payment.—The backfilling to be paid for under this item shall include the quantity, in cubic yards, measured or estimated in place, between the outside neat lines of the structure and the prescribed lines for payment for excavation, or as shown on the Drawings or designated by the Engineer. The unit prices stipulated for backfilling under this item shall include the cost of selecting, placing, and compacting the material; of obtaining material from borrow pits, if necessary, and hauling it to the site; of furnishing and using water for sprinkling, wetting or puddling, of protecting pipes, conduits and other structures or property; and of all work or material incidental to placing the backfilling and leaving it in a satisfactory condition at the completion of the Contract.

Item 38

Surface Dressing and Grassing

38.1 Description.—The surface of embankments, refills, exposed slopes of excavations, borrow pits, spoil banks, concrete surfaces, and any other exposed surfaces, wherever directed, shall be dressed with top soil of the thickness prescribed, and grassed. In general, top soil shall be not less than 6 inches, nor more than 24 inches, in thickness.

38.2 Materials.—Surface dressing shall consist of fertile loam soil, unless the use of clay mixed with fertilizer is permitted by the Engineer. Suitable materials obtained from the excavations and stripping operations may be used for surface dressing under the provisions of Section 0.62 of the Specifications, entitled Materials Obtained from Construction, and any additional top soil that may be required may be obtained from points designated by the Engineer.

The seed shall be first quality grass seed of a suitable mixture for the purpose; and if required, suitable seed for a nurse crop also shall be provided.

38.3 Methods.—The surface soil shall be brought into good condition for a seed bed before the seed is sown. It shall be sown with the quantity and kind of seed prescribed. If in the interval between placing the top soil and sowing, weeds shall grow on the surface to be grassed, the Contractor shall remove the weeds without additional compensation. Any sloughing or washing of the top soil which may occur shall be made good, and any portions of the seeded area which are not thoroughly covered with grass shall be reprepared and reseeded, without additional cost to the District, to the end that all surfaces prepared under this item shall have a satisfactory smooth covering of top soil and grass at the termination of this Contract.

38.4 Payment.—The surface dressing and grassing to be paid for under this item shall include the quantity of soil, in cubic yards, measured in place, actually deposited and satisfactorily grassed in accordance with the Drawings or as required by the Engineer. The unit price stipulated under this item shall include the obtaining, transporting, and placing of soil, the fertilizing and seeding, and all labor and materials incidental to securing a satisfactory stand of grass. If material excavated in stripping the foundation for embankment, or in other prescribed excavations, is used for dressing, it will be paid for both as excavation and as surface dressing.

Note—There are no Items 39 or 40.

Items 41 and 41a

Concrete in Retaining Walls, Bridge Piers, and Similar Structures

41.1 **Description.**—These items shall include concrete
41a.1 requiring accurately shaped forms, in outlet works,
spillways, retaining walls, bridge piers, and similar
structures, in which may or may not be imbedded steel reinforcement or other metal work; it may be proportioned by volume from about 1 of cement, 2 of sand, and 4 of coarse aggregate, to about 1 of cement, 3 of sand and 6 of coarse aggregate, as the Engineer may direct. These items shall not include concrete in the lining of conduits, which is covered by Item 42, or that in floors or aprons of outlet works, spillways, etc., which is covered by Items 43 and 43a.

Under Item 41a will be included all concrete requiring accurately shaped forms, in outlet works, spillways, retaining walls and similar structures, placed below the plane of estimated mean low water shown and so designated in the Drawings. Provided, however, that whenever for any structure no such plane is indicated in the Drawings, all concrete, whether above or below mean low water level, will be included under Item 41 and not under Item 41a.

41.2 **Payment.**—Payment under these items shall be made
41a.2 in accordance with Sections 0.103 and 0.104 of the Specifications, entitled Concrete—Measurement for Payment, and Concrete—Prices to Include.

Item 42

Concrete in Conduit Linings at Dams

42.1 Description.—The lining of the outlet conduits, except as otherwise indicated in the Drawings, shall consist of a special grade of concrete, mixed in proportions by volume, of approximately 1 of cement, 1½ of sand, and 2½ of coarse aggregate, the latter to be composed of fragments of granite, trap, or other suitable igneous rock approved by the Engineer. Such coarse aggregate shall pass through a screen having 1-inch round holes, and be retained on a ¼-inch mesh screen. No large stones shall be placed in this class of concrete. As maximum possible density and exceptionally smooth and uniform interior surfaces are essential for the conduit linings, more than ordinary care shall be exercised in building, finishing and securing forms, and in proportioning, mixing and placing the concrete, and the Contractor shall employ none but skilled labor for this work. Both the fine and the coarse aggregates shall be especially well graded. To secure the proper gradation it may be necessary to separate the coarse aggregate into not more than three sizes, and thoroughly remix in such proportions as directed. After placing in the forms, the concrete shall be so manipulated by spading and joggling as to expel entrained air, not only from the concrete, but also from between the face of concrete and the forms. No plastering or patching of concrete surfaces forming the waterway shall be done unless expressly permitted, and if so permitted, shall be done in strict accordance with directions. No thin patching or plastering will be permitted, but recesses shall be cut of such shape as to retain the patches, and of such depth as to insure their permanency. If required, anchor bolts shall be set in drilled holes, and these and wire mesh or other suitable device embedded in the patch. The Contractor shall supply carborundum brick or emery wheels, if necessary, and shall dress all inequalities on the surface of this concrete.

42.2 Payment.—Payments under this item shall be made in accordance with Sections 0.103 and 0.104 of the Specifications, entitled Concrete—Measurement for Payment, and Concrete—Prices to Include.

Items 43 and 43a

Concrete in Aprons, and Similar Flat Shapes

43.1
43a.1 Description.—These items shall include concrete in aprons, floors, paving at dams, and similar flat shapes requiring either no forms at all, or in places, crude or inexpensive forms only, and whether or not containing reinforcing steel. It may be proportioned by volume from about 1 of cement, 2 of sand and 4 of coarse aggregate, to about 1 of cement, $2\frac{1}{2}$ of sand and 5 of coarse aggregate, as the Engineer may direct.

Under Item 43a will be included all concrete in aprons, paving at dams, and similar flat shapes placed below the plane of estimated mean low water shown and so designated in the Drawings. Provided, however, that whenever, for any structure no such plane is indicated in the Drawings, all concrete, whether above or below mean low water level, will be included under Item 43, and not under Item 43a.

43.2
43a.2 Payment.—Payment under these items shall be made in accordance with Sections 0.103 and 0.104 of the Specifications, entitled Concrete—Measurement for Payment, and Concrete—Prices to Include.

Items 44 and 44a

Concrete in Monolithic Slope Revetment

44.1   **Description.**—These items shall include concrete in
44a.1   monolithic slope revetment on river banks, and in concrete footing walls and copings bordering revetment,
which may contain steel reinforcement. It may be proportioned by
volume, from about 1 of cement, 2 of sand and 4 of coarse aggregate, to about 1 of cement, 2½ of sand and 5 of coarse aggregate,
as the Engineer may direct. The Contractor shall dress all surfaces
of embankments on which concrete revetment is to be placed, to
true planes and to the slopes shown in the Drawings, and the Specifications relating to preparation of foundations shall govern the
operations of the Contractor prior to placing concrete. Concrete
shall be deposited behind forms and in panels extending up and
down the embankment, but in pouring, concrete shall not be permitted to flow down the embankment except through chutes. The
lagging composing the face forms shall be securely fastened, to the
approval of the Engineer, and shall be laid parallel with the direction of flow of the stream.

Under Item 44a will be included all concrete in monolithic slope
revetment placed below the plane of estimated mean low water
shown and so designated in the Drawings. Provided, however, that
whenever, for any structure no such plane is indicated in the Drawings, all concrete, whether above or below mean low water level,
will be included under Item 44, and not under Item 44a.

44.2   **Payment.**—Payment under these items shall be
44a.2   made in accordance with Sections 0.103 and 0.104 of the
Specifications, entitled Concrete—Measurement for Payment, and Concrete—Prices to Include.

Item 45

Concrete in Flexible Slab Revetment or Channel Lining

45.1 Description.—Concrete blocks for flexible slope revetment or channel lining, shall be cast in approved molds (which shall not be loosened in less than 48 hours, nor entirely removed until directed), in a place affording ample storage facilities, and unless otherwise directed, shall be aged for not less than 28 days before being placed in the work. The concrete will contain reinforcing steel or other metal and may be proportioned, by volume, from about 1 of cement, 2 of sand, and 3 of coarse aggregate, to about 1 of cement, 2½ of sand, and 5 of coarse aggregate, as the Engineer may direct. The maximum size of the coarse aggregate shall be such as will pass through a screen having 2-inch round holes. Steam curing may be resorted to, provided the process and appliances shall be satisfactory. The time required for curing by such steaming process will be determined by the Engineer from actual tests.

The blocks shall be assembled and secured in place in the manner shown in the Drawings.

45.2 Payments.—Except that the quantity to be paid for shall be measured in the blocks, payments under this item shall be made in accordance with Sections 0.103 and 0.104 of the Specifications, entitled Concrete—Measurement for Payment, and Concrete—Prices to Include.

Wire rope, if shown on the Drawings as a part of the permanent installation, will be paid for under Item 84.

Item 48

Stone Masonry

48.1 Description.—This item shall include the building of stone masonry as shown on the Drawings or as required by the Engineer. Unless ctherwise shown on the Drawings, this masonry shall be of the class known as pitch-faced, range, squared-stone masonry, with backing of concrete, or of large roughly bedded stones, as the Contractor may elect. In case the courses are not of the same thickness, they shall decrease in thickness uniformly, from bottom to top.

Stones shall be hard and durable, free from seams or other imperfections, of approved size, quality, and shape, and in no case shall have less bed than rise. They shall be clean and wet at the time of laying; they shall be well bonded, laid on natural beds, and solidly settled into place in a full bed of mortar. They shall not be laid in freezing weather except under special permission and directions from the Engineer.

Face joints shall be not more than 1 inch thick, and shall be finished flush, as the stones are laid. Mortar for the joints shall satisfy the requirements of Section 0.105 of the Specifications, entitled Mortar. The maximum size of grains in fine aggregate shall be such as will pass a ¼-inch mesh screen.

At least one-fifth of the face of the wall (and of the back if stones are used for backing) shall be headers; those in the face shall be so arranged that a header shall not be laid over a vertical joint, nor shall a vertical joint occur over a header. No header shall be less in length than three times the rise of the course. The whole structure shall be bonded together in accordance with good practice.

All concrete used for backing of stone masonry shall satisfy the requirements of Sections 0.85 to 0.102 of the Specifications, entitled Concrete. Such concrete, however, will not be measured for payment as concrete, but as stone masonry, as provided hereinafter.

Stone masonry shall be kept moist for at least 2 weeks after being laid, and shall be protected from freezing until thoroughly dried out.

48.2 Payment.—The quantity of stone masonry to be paid for under this item shall be the number of cubic yards, measured in place, laid in accordance with the Drawings, Specifications and instructions of the Engineer, including whatever concrete may be used for backing or interior. The unit price stipulated for this item shall include the cost of all labor and materials incidental to completing the construction to the satisfaction of the Engineer, except the cost of cement, reinforcing steel, and other metal work, as provided elsewhere in the Specifications and except the cost of materials obtained from construction as provided in Section 0.62.

Note.—There is no Item 49.

Item 50

Cement

50.1 Brands.—Before delivery of cement is made the brand shall receive the approval of the Engineer. Preference shall be given to cements which, by their records, show uniformity of composition and tendency to maintain high strength of mortar with increased age.

50.2 Delivery and Storage.—Cement shall be delivered in cloth sacks or other strong, well-made packages, each plainly marked with the manufacturer's brand. The weight of cement contained in such packages shall be 94 pounds net, unless some other weight be adopted as a commercial standard during the life of this Contract. Packages in broken or damaged condition shall be rejected, or accepted only as fractional packages. All cement shall be dry, free from lumps, caking and water marks. Pending shipment and in transit cement shall be kept under proper seal.

At all times while cement is required, the Contractor shall have at the site of the work an abundant supply, sufficiently in advance of use to allow satisfactory tests to be made, and shall carefully guard against possible shortage on account of rejection, irregular delivery, or any other cause. The Contractor shall provide storage facilities in an approved location. Store houses shall be weathertight, shall have tight floors a suitable distance above the ground, shall be large enough to admit of keeping on hand a sufficient supply to prevent delays or interruptions to the work by irregular delivery, and shall have sufficient floor space for storing each carload separately, and affording convenient access thereto for sampling, counting of packages and removal. Suitable, accurate scales shall be provided by the Contractor for weighing, and occasionally, as the Engineer may direct, packages of cement shall be weighed by the Contractor to determine whether or not they contain full measure.

The Contractor shall employ a competent storekeeper who can speak and understand the English language, who shall have charge of the storehouse, and shall keep suitable records of the delivery and use of all cement. Copies of these records shall be furnished the engineer at the close of each day's work, showing in such detail as he may reasonably require, the quantity used in each part of the work during the day.

50.3 Inspection, Tests and Requirements.—The cement will be subjected to thorough inspection and tests at the place of manufacture or on the work. It shall conform in all respects to the Standard Specifications for cement, of the American Society for Testing Materials, as published in the 1915 Year Book of that Society, and in addition, standard 1 to 3 mortar briquettes shall show an increase in tensile strength of not less than 20 per cent from the test at 7 days to the test at 28 days. Cement will not be accepted if tests indicate that it does not possess reasonably uniform composition. The Contractor shall notify the Engineer a sufficient time in advance of manufacture or shipment to allow for proper sampling, and unless this is done, or if additional tests are necessary, the Contractor shall rehandle the cement in the storehouse for the purpose of obtaining samples as directed. An agent of the Board shall have the right at all times, of inspecting the raw materials and process of manufacture at the cement works.

Cement will be accepted which has passed satisfactorily all the tests as required herein, including the 28-day tensile test. After any brand of cement shall have sufficiently proved its worth in actual practice upon the work, the Engineer may, should he deem it advisable, allow its use without waiting for the result of the 28-day test. Cement kept in storage may be subjected to retesting.

If the tests prove any cement unsatisfactory which has been delivered at the site of the work, such cement shall be at once plainly marked for identification, and promptly removed from the work and its vicinity.

50.4 Payment.—Unless specifically excepted, all cement actually used in accordance with the Specifications shall be paid for at the unit price for this item, which shall include the cost of the cement, and all expense incidental to delivering it upon the work, to weighing as directed, and to putting it into the concrete, or masonry, or other parts of the work in which it is to be used. For purposes of payment, a barrel of cement shall be considered the equivalent of 376 pounds net of cement.

Cement used in concrete for filling unauthorized excavations, that used in work which is rejected as faulty or unsatisfactory, cement that is wasted, or used in excess of requirements, and all cement used by the Contractor for purposes other than those for which payment is provided in the Contract, shall not be paid for under this item, but the cost thereof shall be considered as having been included in the price stipulated for the appropriate items.

Item 51

Reinforcing Steel

51.1 Description.—Steel reinforcement for concrete shall be free from mill scale, and shall strictly fulfill all the requirements and be subject to all the tests for the structural steel grade given in the Standard Specifications for Billet-Steel Concrete Reinforcement Bars, as published in the 1915 Year Book of the American Society for Testing Materials. Bars shall be twisted or otherwise deformed, and of the sizes and weights shown in the Drawings or ordered, and shall be secured in the positions required in an approved manner, so as to withstand, without displacement, the pouring and spading of concrete, until completely embedded.

51.2 Protection of Steel.—Steel shall be effectively protected from damage at all times, and if rust shall form, all loose rust scales shall be cleaned off with wire brushes or other implements. Any mortar, oil, grease, paint or dirt, which shall adhere to the reinforcement, shall be removed prior to imbedding the steel in the concrete. Ends of rods which are to be left projecting for a considerable time shall be painted with cement grout. Where reinforcing bars project from the concrete, precautions shall be taken to prevent the bars from being struck or jarred in such a way as to injure the bond between steel and concrete.

51.3 Payment.—The quantity of reinforcing steel to be paid for shall be the number of pounds placed in accordance with the Drawings or orders. The unit price stipulated for this item shall include the cost of metal, the royalty, if any, the cost of transporting, cutting, bending, placing, fastening in position, cleaning, protecting, and all other labor and materials connected therewith. Waste material, due to the fact that the lengths furnished are longer than necessary, and wires, clips, or other devices used for securing the metal, shall not be included for payment. The quantity paid for shall include, however, extra metal in authorized laps wherever the lengths required would be unreasonably long for single bars. In determining quantities, except as the Engineer may require test weighing on scales, commercial unit weights shall be used.

Item 53

Drilling Foundations or Masonry

53.1 Description.—Under this item the Contractor shall, if ordered, drill holes in rock, concrete, or masonry for grouting, for inspection, for drainage, or for other purposes, at the places and to the depths to be designated by the Engineer. The diameter of such holes shall be not less than 2 inches, unless otherwise ordered. Holes shall not be smaller in diameter than the size ordered, and not more than $\frac{1}{2}$ inch larger, and shall be drilled by means of diamond, shot, or other suitable rotating type of drill, capable of boring a smooth hole without disturbing the adjacent rock walls. Churn drills or other types of drills operating by means of concussion shall not be used for this purpose, unless special written permission from the Engineer is given for such use.

53.2 Payment.—The quantity to be paid for under this item shall be the actual number of linear feet drilled in accordance with orders. The unit price stipulated for this item shall include all expense necessary for drilling the holes as required.

Item 54

Grouting Foundations or Masonry

54.1 Description.—Under this item the Contractor shall place pipes for grouting or drainage operations, and shall force cement grout into such pipes or into holes drilled under Item 53, to the extent required to test the foundation or to close any seams in the rock, concrete, or masonry that may be found to exist.

54.2 Pipes.—Under this item the Contractor shall, if ordered, set in the holes drilled under Item 53, and at such other places as may be directed, pipes for grouting or for tests. The pipes shall be standard weight merchant pipe, not greater than $2\frac{1}{2}$ inches in nominal diameter, fitted with standard couplings. Any other standard fittings shall be supplied, and all cutting, threading, cold bending and fitting shall be done necessary to place the pipe complete as ordered. Water-tight joints shall be made between the pipes and the sides of holes drilled in rock or masonry, with an approved filler, and by an acceptable method.

54.3 Grout.—Under this item the Contractor shall also, if ordered, force grout composed of cement and water, or of cement, sand and water, in proportions to be prescribed, into seams in any ledge rock, or any other places as directed. Sand for grout shall be clean, and shall be of such fineness that 100 per cent will pass a sieve having 64 openings per square inch, and 45 per cent will pass a sieve having 1,600 openings per square inch.

54.4 Mixing and Placing Grout.—The apparatus for mixing and placing grout shall be of an acceptable type, equal in efficiency to a machine having for its essential part an air-tight chamber in which the grout is effectively stirred, and from which it is forced into the work by air under any required pressure up to 80 pounds per square inch. Grouting shall be so conducted that the Engineer will be satisfied that the desired filling has taken place. Holes shall be made, if ordered, to test the efficacy of the filling. If it be discovered that any voids have not been thoroughly filled by the first application of grout, the process shall be repeated until satisfactory results are obtained.

54.5 Payment.—The quantity of grout to be paid for under this item shall be the actual number of cubic yards mixed in accordance with directions, measured in its liquid state before being forced into the work. The Contractor shall provide suitable means for convenient measuring. The unit price stipulated for this item shall include all labor and materials necessary for doing the grouting, as directed, except that cement and pipe will be paid for under the items provided therefor.

Item 57

Dry Rubble Paving

57.1 Description.—Dry Rubble Paving shall consist of sound, durable stones having established weathering qualities, imbedded in sound, durable crushed rock or gravel which shall be not larger than 2½ inches maximum size. Stone blocks shall have roughly squared top faces. None of the softer varieties of limestone found in this locality are satisfactory for Dry Rubble Paving.

57.2 Method of Laying.—All paving shall be of the thickness and extent shown on the Drawings or ordered. At least one-third of the area of the paving shall be composed of stones having a depth equal to the thickness of the paving, and the majority of the stones shall have a depth of at least two-thirds the thickness of the paving. No stone having a depth less than one-half the thickness of the paving shall be used. The stones of the various depths shall be well distributed. Stones shall be placed by hand, close together, and thoroughly imbedded in the crushed rock or gravel foundation. All voids shall be filled with crushed rock or gravel, and the joints of the face of the paving filled with tightly-driven spalls.

57.3 Payment.—The quantity of Dry Rubble Paving to be paid for under this item shall be the number of cubic yards actually placed within the limits prescribed. The crushed rock or gravel used in this connection shall be included for payment under this item. The price stipulated for this item shall include the cost of all work and materials, and all expenses incidental to preparing the foundation and constructing the paving.

Item 58
Rip-Rap

58.1 Description.—Rip-Rap of required thickness and composed of tough, durable stones of the sizes to be prescribed, shall be placed as designated in the Drawings or ordered. Stones placed on or in the slopes of embankments, as a result of the grading of embankment material, will not be classed as rip-rap. The Contractor will not be required to place rip-rap by hand, with the exception of such hand work as is necessary to conform to the specified lines. None of the softer limestone formations found in this locality are acceptable for rip-rap, and only such of the harder limestone as quarry to acceptable shapes and sizes, and the larger boulders, will be satisfactory.

58.2 Payment.—The quantity of rip-rap to be paid for under this item shall be the volume in cubic yards actually placed within the prescribed limits. The unit price stipulated for this item shall include the cost of all work incidental to furnishing and placing the rip-rap, and leaving it in satisfactory condition.

Item 59
Gutter Paving

59.1 Description.—Cobble gutters shall be laid along the berms on the slopes of the Dams, and at other places as shown on the Drawings or ordered. The stones used shall be tough, durable, rounded cobbles of the size shown on the Drawings, except that, if permitted by the Engineer, durable quarry stone of acceptable sizes and shapes may be used. The stones shall be laid with the longest dimensions vertical, and the largest stones shall be selected and placed along the outside of the gutter. All the stones shall be bedded in cement mortar to the extent and in the manner shown on the Drawings.

59.2 Payment.—The quantities to be paid for under this item shall be the number of square yards in place as shown on the Drawings or designated by the Engineer. The unit price stipulated for this item shall include the cost of all labor and materials and all other expense necessary for completing the cobble gutters, except cement which will be paid for under the item provided therefor.

Item 67
Guard Rails

67.1 Description.—Where indicated on the Drawings or ordered by the Engineer, the Contractor shall furnish and erect wooden guard-rails of the type shown on the Drawings.

67.2 Materials.—Posts shall be of sound, seasoned red cedar, white oak, tamarack, chestnut, or other satisfactory wood, entirely stripped of bark or skin, not less than 6½ feet long, and not less than 6 inches in diameter at any section. Posts shall be straight, sloped on top, and notched for the side rails as shown on the Drawings. They shall be set plumb, not more than 8 feet apart between centers, and set not less than 3 feet into the ground, with the sides toward the highway on a true line or as directed. The lower part of each post, to a point about 3 feet from the top shall be charred as directed.

Rails shall be of seasoned spruce, shortleaf yellow pine or other satisfactory wood, surfaced on 4 sides, of a grade equal to that specified under Item 71, Timber and Lumber Miscellaneous, and of a length equal to twice the distance between adjoining posts, approximately 16 feet.

67.3 Erection.—Post holes shall be excavated sufficiently large to properly admit the post and leave room for setting. Refilling shall be of proper material and thoroughly tamped. The top rail shall be set on a slope, and the side rail shall be notched into the post. Both rails shall be neatly joined and securely nailed, and shall show a true alignment when the guard rail is completed.

67.4 Painting.—When the lumber is dry and weather conditions are suitable, the guard rails shall be painted as follows: Before being assembled the notches in the posts, the tops of the posts, and all parts of the rails that will be inaccessible after erection shall be painted with two coats, and after erection the rails shall be given three coats, of best white lead and linseed oil.

67.5 Payment.—The quantity of guard rail to be paid for under this item shall be the number of linear feet actually built in accordance with the Drawings and directions, as measured along the structure. The unit price stipulated under this item shall include the cost of furnishing and erecting all posts and rails, of excavating and refilling post holes, the cost of painting, charring, spikes, and nails, and all expense incidental to completing the work in a satisfactory manner.

Item 68

Wire Fences

68.1 Description.—Wire fences of the types shown on the Drawings and described herein shall be furnished and erected wherever ordered.

68.2 Materials.—In general, posts shall be of sound, seasoned red cedar, white oak, tamarack, chestnut, or other satisfactory wood, entirely stripped of bark or skin. They shall be reasonably straight and roughly roofed on top; and shall be set plumb and at the locations ordered. The lower part of each post, to a height which will be about 6 inches above the surface of the ground when set, shall be charred as directed. In general, posts shall be spaced about 16 feet apart except at corner posts or straining posts, at either side of gates, and at other places as required, where they shall be spaced about 8 feet apart as shown on the Drawings. Line posts shall be not less than 5 inches in diameter at any section, and shall be set into earth not less than 2½ feet, or into rock not less than 1½ feet. Straining and corner posts shall be not less than 8 inches in diameter at any section, and shall be set into earth not less than 3½ feet, or into rock not less than 2 feet, unless otherwise ordered. Other types of posts of equivalent durability, strength and appearance may be submitted for approval as substitutes for the wooden posts specified.

At corner and straining posts, at gates. and at other places required, wooden braces and No. 9 gage wire guys shall be furnished and placed as shown on the Drawings or directed.

The fencing shall consist of galvanized steel wires of the number, size and approximate spacing shown on the Drawings or described herein, and, unless otherwise shown on the Drawings, shall be equal in all respects to the fencing made by the Page Woven Wire Fence Co., or by the American Steel & Wire Co., having the top and bottom wires No. 9 gage, and all other wires No. 11 gage. Wire gage numbers used are those of the U. S. Steel Wire Gage, and the gage specified is that of the wire before galvanizing. Unless otherwise indicated on the Drawings, either type of fencing may be furnished, as the Contractor may elect.

The Contractor shall furnish and set in the fences, wherever required, gates of approved design, similar to that shown on the Drawings, with hinges and fastenings complete.

68.3 Erection.—Postholes shall be excavated sufficiently large to properly admit the post, and leave room for setting. Where rock is encountered, the posts shall be set in holes excavated by blasting or other satisfactory method, or shall be secured by other methods permitted or directed. Refilling shall be of proper material and thoroughly tamped. The fencing shall be attached to the posts by means of galvanized staples, or by other approved means. During erection the fencing shall be subjected to suitable tension, and when completed, shall be satisfactorily taut, true to line, and fully in conformity with the Drawings and with good practice.

68.4 Payment.—The quantity to be paid for under Item 68 shall be the number of linear feet of fence actually built in accordance with the Drawings or orders, as measured along the structure, but not including gate openings, the length of each of which shall be the distance between centers of gate posts. The price stipulated under this item shall include the cost of furnishing and erecting all posts, braces, wire guys and wire fencing, the cost of excavating and refilling post holes, of charring posts, and of hardware, and all expense incidental to completing the work in a satisfactory manner. But the cost of metal work required for gates, hinges and gate fastenings, shall be paid for under Item 82, Miscellaneous Cast Iron, Wrought Iron and Steel; and that of lumber required for gates, other than that required for gate posts, shall be paid for under Item 71, Timber and Lumber, Miscellaneous.

Note.—There is no Item 69.

Item 70

Protective Boom and Accessories

70.1 Description.—The Contractor shall furnish and place protective booms, including chains, bolts, and other accessories as shown on the Drawings or ordered.

70.2 Lumber.—Timber and lumber of various sizes shall be furnished and built to the shapes shown on the Drawings, to form the timber portion of the protective boom. All timber and lumber to be paid for under this item shall be spruce, Douglas fir, or other satisfactory wood, of a grade equal to that specified under Item 71, Timber and Lumber, Miscellaneous. When fairly wet, its weight shall not exceed 45 pounds per cubic foot. Before being assembled, and while dry, all timber and lumber shall be painted with two coats of best white lead and linseed oil, and a third coat shall be put on after erection and when danger of marring is past.

70.3 Wrought Iron.—Under this item wrought iron chains, bolts, steel plates, and other accessories required in the construction, or for the anchorages, of the protective boom shall be furnished and placed as shown on the Drawings or ordered. All wrought iron shall be of an acceptable quality for the purpose contemplated. The chains shall conform to the Standard Specifications for Iron and Steel Chain of Class B, Iron Chain, published in the 1915 Year Book of the American Society of Testing Materials.

70.4 Payment.—The quantity of timber and lumber to be paid for under this item shall include the number of thousand feet board measure actually placed in accordance with the Drawings or orders. Should any round timber be used it shall be measured for payment as square timber of the largest size, omitting fractions of an inch, which can be inscribed in the small end of the log. The unit price stipulated for this item shall include the cost of all labor and materials necessary for furnishing, working, painting and placing the protective boom, and all expense incidental thereto, except the cost of wrought iron and steel which will be paid for under Item 82, Miscellaneous Cast Iron, Wrought Iron and Steel.

Item 71

Timber and Lumber, Miscellaneous

71.1 Description.—Timber and lumber of various sizes not otherwise provided for may be ordered and shall be furnished by the Contractor, worked and built into place by him for drains, sills, platforms or grillages, and for other similar purposes.

It shall be of the grade known commercially as No. 1 or No. 1 Common, as specified in the standard rules for grading timber and lumber adopted by the Northern Pine Manufacturers Association, the Southern Pine Association, and the National Hardwood Lumber Association in effect January 1, 1916.

Unless noted specifically in the Drawings or Specifications, the Contractor will not be required to furnish any hardwood lumber under this item.

71.2 Payment.—The quantities of timber and lumber to be paid for under this item shall include the number of thousand feet board measure actually placed in accordance with the Drawings, Specifications and orders and not included under other items. Should any round timber be used it shall be estimated as square timber of the largest size, omitting fractions of an inch, which can be inscribed in the small end of the log. Any second-hand lumber which has been notched for previous use shall be estimated for payment as of the size of its minimum cross section. The unit price stipulated for this item shall include the cost of all bolts, spikes and other fastenings and of all other materials, and all expense incidental to furnishing, working and placing the timber and lumber satisfactorily.

No timber or lumber shall be paid for under this item which is not specifically ordered, and this item shall not be interpreted to include the lumber used for forms, molds, and centers for concrete or other masonry, for sheeting and bracing, for scaffolds or braces, for sluices or trestles, or for other temporary or construction purposes, the cost of which is to be included in the unit price stipulated for the work in connection with which it is used; unless the Engineer shall specifically order in writing that such material be permanently left in place, in which case it shall be paid for as provided under Item 89.

Items 72 and 72x

Timber Piles

72.1   **Description.**—The Contractor shall furnish and drive
72x.1 piles of the quality herein specified in the positions
shown in the Drawings or ordered. Such piles shall be
of the lengths prescribed by the Engineer, and for the purpose of
determining the lengths to be required, test piles of lengths to be
ordered, shall be driven at the times, locations, and to the 'depths
required by the Engineer.

Item 72 shall include all such piles 20 feet, or less, in length.

Item 72x shall include all such piles more than 20 feet in length.

72.2   **Quality.**—Piles shall be of spruce or hard wood, and
72x.2 shall be cut from growing trees when the sap is down;
they shall be close grained and solid, free from cracks,
shakes, large or unsound knots, decay, or any other defects which
would impair their strength or durability. Piles shall be cut above
the ground swell of trees, shall be uniformly tapering, and a straight
line from center of butt to center of tip shall lie wholly within the
pile. Before driving, all piles shall be peeled.

72.3   **Size and Length.**—For lengths not exceeding 30 feet,
72x.3 piles shall have a diameter at tip of not less than 9
inches; for piles longer than 30 feet the minimum diameter shall be 8 inches. In all cases the diameter of the butt at the
cut-off shall be at least 12 inches. Piles must be of such lengths as
to insure sound wood, free from the effects of driving, when cut off
at the required elevation. No splicing will be permitted.

72.4   **Driving.**—It is expected that piles will be driven to
72x.4 the depths shown on the Drawings, but modification
may be made by the Engineer in accordance with the
conditions developed. A steam-hammer is preferred, but a drop-
hammer may be used, provided the weight of the hammer and height
of fall are satisfactory. Where directed, a water jet shall be used to
assist in driving. The small end shall be pointed, and the butt end
cut off square, as directed. Care shall be taken not to damage the
piles by excessive hammering. If directed, a suitable cap or follower shall be used. All piles broken, split, or otherwise injured,
and piles driven too low, out of position, or in any respect improperly driven shall be satisfactorily replaced at the Contractor's expense.

72.5   **Cutting to Grade.**—Piles shall be cut off at the re-
72x.5   quired elevations. Where timber caps or grillages are
          to be used care shall be taken to insure a level cut, free
from sloping or curved surfaces, to the end that a smooth, level
bearing shall be provided at the exact elevation.

72.6   **Payment.**—The amount of piling to be included for
72x.6   payment under these items shall be the number of linear
          feet actually in place in accordance with the Drawings
and directions, measured below the cut-off, and in addition, one
quarter of the quantity measured above the cut-off, no allowance
being made for piles that are damaged, broken, or otherwise unfit
for use. Test piles when ordered, shall be paid for under these
items when driven to the satisfaction of the Engineer. Payment
for test piles shall be made for the full length ordered by the Engineer, whether cut off or not. The unit price stipulated for these
items shall include the entire cost of all labor and materials necessary for furnishing. placing, and cutting off the piles in a satisfactory manner, and all expense incidental thereto.

Item 73

Wood Sheet Piling

73.1 Description.—The Contractor shall furnish and drive wood sheet piling of the quality herein specified, in the positions shown in the Drawings or ordered. Such piles shall be of the lengths prescribed by the Engineer, and for the purpose of determining the lengths to be required, test piles, of lengths to be prescribed, shall be driven at the times, locations, and to the depths required by the Engineer.

73.2 Materials.—Sheet piles shall be of spruce, or longleaf yellow pine, of a grade equal to that specified under Item 71, Timber and Lumber, Miscellaneous. Sheet piles shall be tongued and grooved, or splined and grooved, and of the dimensions shown on the Drawings or ordered, but shall not be less than 4 inches, nor more than 6 inches in thickness. Where 6-inch sheet piles are required, each pile shall consist of three 2-inch planks spiked or bolted together as shown on the Drawings.

73.3 Driving.—It is expected that sheet piles will be driven to the depths shown on the Drawings, but modifications may be made by the Engineer in accordance with the conditions developed. Where directed, sheet piles shall be driven by the aid of a water jet and with a suitable cap or follower. All piles broken, split, or otherwise injured, and piles driven too low, out of position, or in any respect improperly driven, shall be satisfactorily replaced at the Contractor's expense. Piles shall be sawed off at the required elevations, and the pile head after sawing shall be sound and level.

73.4 Payment.—The amount of sheet piling to be paid for under this item shall be the number of thousand feet board measure of sheet piling actually in place in accordance with the Drawings and directions, measured below the cut-off, and in addition, one quarter of the quantity measured above cut-off, no allowance being made for piles that are damaged, broken or otherwise unfit for use. Test piles, when ordered, shall be paid for under this item when driven to the satisfaction of the Engineer. Payment for test piles shall be made for the full length ordered by the Engineer, whether cut off or not. The unit price stipulated for this item shall include the entire cost of all labor and materials necessary for furnishing, placing and cutting off the piles in a satisfactory manner, and all expense incidental thereto.

Item 74 ·

Steel Sheet Piling

74.1 Description.—Steel Sheet piling of the lengths shown on the Drawings shall be furnished and driven by the Contractor. Test piles, of lengths to be prescribed, shall be driven at the times and locations and to the depths required by the Engineer.

74.2 Requirements.—The piling shall be open hearth steel, with an ultimate tensile strength, in pounds per square inch, of not less than 55,000 and not more than 70,000. Test specimens shall bend cold, without cracking, 180 degrees around a pin of diameter equal to the thickness of the specimen. Steel sheet piles shall be of some standard make of integral rolled section, shall have a positive interlock capable of withstanding a pull of not less than 9,500 pounds per linear inch, and shall be so designed as to permit a change in direction of at least 15 degrees either way at the joint. Unless otherwise specified on the Drawings, piles shall have a web thickness of not less than $\frac{3}{8}$ inch and shall weigh not less than 35 pounds per square foot when interlocked. Where shown on the Drawings, corners shall be formed by the intersection of a special member. Each pile shall have a hole punched near one end for handling.

74.3 Tests.—Mill tests and inspection will be made under the supervision of the Engineer, but the Contractor shall, at his own expense, arrange for a testing machine at the mill or shop, and the test pieces for breaking, together with all facilities for conducting the prescribed tests.

74.4 Driving.—Sheet piles shall be carefully driven to line and to the required depths. Where necessary to procure the proper penetration, a water jet shall be used. The methods and equipment shall be such as not to injure the piles unnecessarily. For this purpose suitable drive caps shall be used. Unless otherwise permitted. a steam hammer shall be employed for driving. Should the use of drop hammer be allowed, the Contractor shall employ only operators who are skilled in work of this kind, and the weight of hammer and height of fall at all times shall be subject to the approval of the Engineer. Unless otherwise permitted, the piling shall be assembled, before driving, in sections not less than 10 feet long, and no pile in any section shall be driven more than 5 feet lower than the general level of that section, nor shall the level of sheet piling in any section vary more than 5 feet from the general level of the adjacent section. To provide for this method of procedure the hammer shall be properly offset from the leads, and other provisions made to permit ready movement of the pile driver. All piles which are damaged, and piles out of position, or in any respect improperly driven, shall be satisfactorily replaced at the Contractor's expense.

74.5 Payment.—The amount of steel sheet piling to be paid for under this item shall be the number of tons of 2000 pounds actually driven in accordance with the Drawings and Specifications. Test piles, when ordered, shall be paid for under this item, when driven to the satisfaction of the Engineer. Payment for test piles shall be made for the full length ordered. The unit price stipulated for this item shall include the entire cost of all labor and materials necessary for furnishing and driving the piles in a satisfactory manner, and all expense incidental thereto. If piles are required to be cut off, such work shall be paid for as Extra Work, under Section 0.19 of the Specifications.

Item 85

Cleaning Up

85.1 Work to be Done.—Before the completion of the work included in this Contract the Contractor shall completely remove and satisfactorily dispose of all temporary works, to the extent directed. He shall tear down and dispose of all temporary buildings, trestles or stagings constructed by him; shall remove or grade to the extent directed all embankments or cofferdams made for construction purposes; shall satisfactorily fill excavations as directed; shall remove all plant and equipment; shall remove the rails and ties and other woodwork of any railway built by him; shall satisfactorily dispose of all rubbish resulting from the operations under this Contract, and shall do all work necessary to restore the territory embraced within the zone of his operations to a condition at least as sightly as at the beginning of work under this Contract. Surface dressing and grassing shall be placed, if directed, to cover unsightly areas, and, where ordered, shall be paid for under Item 38.

85.2 Payment.—For all labor, materials and miscellaneous work required to leave the grounds in an acceptable condition, with the exception of surface dressing and grassing, which shall be paid for under Item 38. the Contractor shall receive the lump sum stipulated for this item, which shall be included for payment in the final estimate.

Item 89

Sheeting Left in Place

89.1 Description.—This item shall include all sheeting or sheet piling placed by the Contractor as an incidental to work under some other item or items of the Contract, and afterwards left in place by written order of the Engineer. This item shall not include sheet piling driven under Item 73, full payment for which is provided under that item.

89.2 Payment.—The quantity of sheeting to be paid for under this item shall be the number of thousand feet board measure left in place, and included under this item, by written orders of the Engineer. Such sheeting will be paid for at the rate of $15.00 per thousand feet board measure, unless a lower unit price is submitted in the Proposal, and named in the Agreement. This unit price shall include the cost of all bolts, spikes, nails and other materials entering into the construction of such sheeting or sheet piling, and shall be in full settlement for all expense incidental thereto.

Item 90

Extra Work

90.1 Description.—This item shall include all Extra Work performed in connection with this Contract, as provided for under Section 0.19 of the Specifications.

90.2 Payment.—Payment for work done under this item shall be the actual necessary cost plus 15 per cent, as prescribed in Section 0.19 of the Specifications, entitled Extra Work.

INDEX TO SPECIFICATIONS

A
Item

Abandonment of work by Contractor........0.46
Absorption of water by vitrified brick........52.2
Acceptance not implied by failure to condemn
  defective work or material..........0.9, 0.26
Acceptance shall not constitute waiver.......0.28
Access to invoices, bills of lading, etc., to be
  furnished Engineer upon request......
  ....0.6, 0.19, 0.26, 50.2
Access to works for inspection......0.6, 0.36, 50.3
Accidents, Contractor to guard against......0.37
Acts or omissions of Board or its agents......
  ....0.5, 0.9, 0.11, 0.12, 0.15, 0.26, 0.39
Additional work or materials ordered, when
  paid for as extra work.................0.19
Adequacy of plant, equipment and methods..
  ....0.4, 0.12, 0.13, 1.2
Adverse conditions, responsibility for effect of
  ....0.4, 0.11, 0.12, 0.37, 0.39, 1.3
Adverse weather conditions as affecting rate
  of progress.....................0.4, 0.39, 1.1
Agents of Contractor..Bond, 0.22, 0.29, 0.33, 0.37
Agents of District.........0.1, 0.6, 0.6, 0.9, 0.26
Aggregates, concrete....................0.87
Aggregates for road paving..60.2, 61.2, 61.3, 61.6
Agreement............................Agreement
Alluvial deposits, classification of.........0.65
Alterations and changes....Bond, 0.16, 0.18,0.19
Anchor rods, bolts, tie rods, steel.........82.1
Approval of Contract by Board, formal.Agreement
Aprons, concrete in.....................43.1
Assignment of contract requires consent.0.44, 0.46
Assignment of interest in equipment or ma-
  terials requires consent of Engineer.....0.45
Assumption of Contract by Board...........0.46
Assumption of Contract by Sureties....Bond, 0.46

B

Backfilling (see also Refilling)............
  ....0.62, 0.71, 0.84, 1.94, 37.1, 37.2
Bank, agency for receiving payment........0.22
Bank run (sand and gravel)......0.88, 60.1, 60.2
Bankruptcy of Contractor..........0.44, 0.46
Barriers (see also Fences; Guard rails)......0.37
Batch mixing..........................0.93
Baths for employees at camps...........0.50
Beginning, time of...............Agreement, 0.3
Berms......................16.1, 19.3, 31.3
Bills for damages...................0.25, 0.26
  for extra work.....................0.19
  for meals and lodging for Engineers....0.47
  of lading, invoices, etc., to be furnished
  Engineer upon request...0.6, 0.19, 0.26, 50.2
  of material to be furnished by Contractor
  showing materials for construction plant. 0.12
Bituminous-bound macadam pavement..6.11, 63.7
  filler for joints in pavement..........64.7, 65.7
  road materials.....................61.3, 63.7
Blasting, (see also Laws).........0.37, 0.65, 0.69
Block, stone, paving...................57.1
Blocks, brick paving................61.3, 64.7
  concrete, for revetment.............45.1
  stone, paving......................57.1, 57.2
  vitrified...........................52.1
Board, (of Directors) defined...........0.1
Board for Engineer and assistants at camps..0.47
Bolts, anchor, etc...................42.1, 70.3
  nails, etc., in lumber and sheeting...71.2, 89.2
  plates, etc., for booms.............70.3, 83.1
Bond, (surety) applies to extra work...Bond, 0.19
  conditions and form of ...Agreement, Bond, 0.2
  may be reduced....................0.2
  not forfeited by changes, etc.......Bond, 0.19
  to be filed with every County for storing
  and handling explosives.............0.69
Bonding new concrete to old, or to rock....
  ....0.90, 0.95, 0.104
Bonding old and new embankments.....0.76, 1.3
Booms and accessories................70.2
Borrow pits................0.5, 0.60, 5.1, 19.1
Boulders, classification of. 0.65
Boulders, stumps, etc., in borrow pits.......
  ....0.60, 0.62, 5.1

C
Item

Breach of Contract....
  .....Agreement, Bond, 0.3, 0.23, 0.28, 0.46
Brick laying..........................52.6
  lining for sewers...................52.1
  pavement on concrete foundation....
  ....61.1, 61.3, 64.6 to 65.6
  pavement on rolled foundation......
  ....61.1, 64.7 to 65.6
  pavement to be closed to traffic after completion........................64.7
  paving blocks..................61.3, 64.7
  vitrified..........................52.1
Bridge for finishing concrete pavement......65.6
Bridge piers...................22.1, 41.1
Bridges, highways, etc., to be maintained....0.36
Bridges, raising, re-erecting and removing...
  ....81.1, 81.2
Broken stone for bedding rubble paving..57.1, 57.2
  for concrete.......................0.88
  for paving.........................60.1, 60.2
Bronze..........................77.2, 77.3
Buildings, foundations, etc., protection of...0.64
Buildings, removal of...............3.1, 3.2, 3.3
Bulkheading.....................0.71, 0.104

C

Camp construction and maintenance.......
  ....0.50, 0.51, 0.53, 0.54
Camp maintenance not part of extra work
  cost..............................0.19
Camps, right-of-way for, at dams to be fur-
  nished by District...................0.5
Carborundum brick for finishing conduit
  linings............................42.1
Cast iron flood gates..................77.2
Cast iron pipes.....................75.1, 75.2
Cast iron, wrought iron and steel......82.1, 82.3
Catch basins, concrete in..............47.1
Catch basins, gratings for.............82.1
Cement, definition of................0.85, 50.1
  delivery and storage of.............50.2
  grout...0.95, 0.98, 0.104, 51.2, 54.1, 54.3, 54.4
  how paid for......................0.86, 50.4
  in rejected concrete not paid for........0.101
  requirements, specifications, tests......
  ....0.85, 0.93, 50.2, 50.3
  weight of...............0.93, 50.2, 50.4
Cemented materials, classification of........0.65
Cesspools at camps...................0.52
Chains, iron, for booms.............70.3, 76.1
Changes and alterations.....Bond, 0.16, 0.18, 0.19
Changes ordered in plant, force.......0.12, 0.19
Channel lining, concrete...............45.1
  pipe, vitrified Agreement, 55.1, 55.2, 56.1, 56.2
  slopes, trimming and shaping........
  ....0.70, 4.1, 4.2, 9.1, 9.2
Character of employees...............0.35
Check, payment by...................0.22
Cincinnatian formation, classification of......0.65
Claims against District, release of, at time of
  final payment......................0.21
  arising from violation of laws......Bond, 0.34
  arising from infringement of patents..Bond, 0.42
  Contractor responsible for all.......Bond, 0.38
  for anticipated profits or for losses due to
  differences in quantities............0.16
  for damages cannot be made for delay in
  payments.........................0.23
  for damages suffered by Contractor, how
  filed.............................0.26
  for damages to life, limb or property..Bond, 0.37
  for extension of time, how made.........0.4
  for extra work, how made and settled......0.19
  for labor and material, protection against..
  ....Bond, 0.43
  for unforeseen expense incurred in stream
  control...........................1.3
Classification of Embankment..........
  ....0.73, 0.74, 0.75, 0.76, 0.77, 0.78
  of excavation......................0.65
  of excavation above and below water level...
  ....6.1, 16.1, 21.5

Item

Clay, classification of...................... 0.65
Cleaning concrete forms................. ....0.94
  exposed faces of concrete................0.100
  foundations for concrete...............0.92
Cleaning metal work..............51.2, 76.1
  premises................0.50, 0.51, 0.52, 85.1
  stones to be imbedded in concrete....0.58, 0.90
  surface for inspection of excavation....... 0.70
Clearing and grubbing.................2.1, 2.2
Cobble stone paving for gutters at dams..... 59.1
Cofferdams, no payment for.............0.71, 1.3
  removal of............................ 85.1
  responsibility for.....................
    ....Agreement, 0.71, 0.104, 1.2, 85.1
Cold weather precautions...............
    ....0.80, 0.84, 0.92, 0.99, 61.7
Collateral works not to be interfered with...
    ....0.10, 0.29
Communications to Contractor, where to be
  sent.................,............... 0.32
Compacting backfilling................... 0.83
  concrete.............................0.95
  embankments............0.73, 0.75, 0.76, 0.82
  foundations for concrete.................. 0.92
  pavements.....................61.3, 61.6
Compensation, (see also Payment).....Agreement
Compensation for delays...........0.4, 0.5, 0.39
Completeness of specifications, estimates,
  and drawings........................0.16
Completion of contract by Surety.....Bond, 0.46
  of work, as affected by weather, river and
    local conditions...................... 0.4
  of work by District................... 0.46
  of work, default in........Agreement, 0.3, 0.46
  time and order of........Agreement, 0.3
Compliance with laws..............Bond, 0.34
Concrete aggregates for...........0.87, 0.88
  backing for masonry walls............. 48.1
  below water level............38.3, 43.1, 46.1
  bonding new to old or to rock..0.90, 0.95, 0.104
  cement used in................ ...0.85, 0.101
  cold weather precautions.0.84, 0.92, 0.99, 0.104
  construction joints in..............0.96, 0.104
  contraction joints in..............0.97, 0.104
  curb for pavement.....................64x.7
  depositing in water..................... 0.95
  description of.......................... 0.85
  expansion joints..................0.97, 0.104
  finishing surfaces of.....0.95, 0.102, 0.105, 42.1
  forms for.............................. 0.94
  foundation for brick pavement......61.3, 64.6
  in aprons............................ 43.1
  in bulkheads.......................... 46.1
  in catch basins........................ 47.1
  in conduit linings at dams.............. 42.1
  in copings for revetment................ 45.1
  in culverts........................... 46.1
  in flexible slab revetment.............. 45.1
  in floors............................ 43.1
  in footing walls for revetment.......... 44.1
  in gate chambers...................... 46.1
  in manholes.......................... 47.1
  in outlet works...................... 41.1
  in paving at dams..................... 43.1
  in piers............................. 41.1
  in retaining walls..................... 41.1
  in sewers........................... 47.1
  in spillways.......................... 41.1
  joining new work to old, or to rock.
    ....0.90, 0.95, 0.104
  large stones in.........................
    .....0.58, 0.84, 0.90, 0.95, 0.104, 42.1
  materials for...................0.90, 42.1
  may contain rubble or brick masonry.....0.103
  mixing........................0.93, 42.1
  not to be placed on new fills............ 0.92
  pavement.............61.1, 61.2, 61.3, 65.6
  pavement over trench excavations 65.6
  pavement to be closed to traffic while hard-
    ening............................ 65.6
  payment for.........................0.103
  placing..............................0.95
  placing metal work in.................0.102
  plum rock in 0.58, 0.85, 0.90, 0.95, 42.1
  preparation of foundations for..0.92, 0.95, 0.104
  proportioning (see also respective items)... 0.85
  protection of................. ..0.99, 0.100

Item

Concrete rejected, cement not paid for......0.101
  replacing faulty work.....................0.101
  retempering, not permitted................ 0.93
  samples of.....................0.85, 0.91
  tests.........................0.85, 0.91
  water for............................ 0.89
Condemned materials to be removed from
  work.............................. 0.7
Conducting collateral works...........0.10, 0.29
Conducting work, manner of.............. 0.3
Conduit linings at dams, concrete in........ 42.1
Consent, (see Written Consent)
Construction joints in concrete.....0.96, 0.104
Construction plant (see Equipment)
Contagious diseases.................0.54, 0.55
Contract, defined....................... 0.1
  drawings, list of................Agreement
  exhibits composing...............Agreement
  failure to comply with...Agreement, Bond, 0.46
  price, defined..........Agreement, 0.1, 0.3
  provisions of, cannot be waived by Engineer
    ....0.9, 0.28
Contraction joints in concrete........0.97, 0.104
Contractor, defined 0.1
Contractor's address.....................0.32
  bond.................Agreement, Bond, 0.2
  representative.............0.22, 0.31, 0.33
Control of stream discharge....0.3, 0.74, 1.1, 1.3
Controlling water in mixing concrete....... 0.93
Controlling springs in foundations......0.79, 16.1
Core, impervious, in dams................0.73
Correcting defective work........ ..0.6, 0.7
County Commissioners require bond for stor-
  ing and handling explosives 0.69
Crushed rock or gravel for concrete.......... 0.88
  for pavement.......................
    ....60.1, 60.2, 61.1, 61.2, 61.3, 61.6, 61.7
  for rubble paving...................57.1, 57.2
Crushed rock or gravel for paving......60.1, 61.1
Crusher, run of.................0.88, 60.1, 60.2
Culverts, concrete in..................... 46.1
Curb, concrete, for pavement.............. 65.6
Curb, sandstone, for pavement............. 61.3
Curing concrete blocks for revetment....... 45.1
Cut, (see Excavation)

D

Damage, caused by failure of District to furnish right-of-way................ .... 0.5
  claims cannot be made for delay in payment 0.23
  from delays, nature of work, weather, floods,
    etc., included in contract price......... 0.39
  inevitable, responsibility for............ 0.38
  resulting from insufficiency of temporary
    works, Contractor responsible for 0.36
Damaged work, replacing of..... ..0.6, 0.7, 0.9
Damages and defects, money retained for.... 0.25
Damages, liquidated..............Agreement, 0.3
  recovery of, not debarred to Board by
    failure to condemn defective work....... 0.9
  suffered by Contractor... .... 0.26
  suffered by District through negligence of
    Contractor.....................Bond, 0.25
Dams (see also Embankment)
  constructed by hydraulic method 0.73
  constructed by semi-hydraulic method ... 0.74
  constructed by sprinkling and rolling
    method............................ 0.75
  embankment for................31.1, 83.1
  methods of constructing...............
    0.71, 0.72, 0.73, 0.74, 0.75, 31.1, 31.2, 31.3
  stream control at....0.75, 1.1, 1.2, 1.3, 6.1, 6.2
Decision and estimates of Chief Engineer precedent to payment to Contractor....... 0.10
Decision of Chief Engineer, in case of dispute,
  causing delay not a basis for damage
  claim.............................. 0.6
  in case of dispute, Contractor to follow
    temporary instructions....... .... 0.6
  final as to interpretation of drawings or
    specifications...................... 0.6
Decrease or increase of plant, equipment or
  force may be ordered by Engineer.......
    ....0.12, 0.19
Decrease or increase of quantities.Bond, 0.16, 0.17
Deductions for defects and damages........ 0.25

Item

Deduction for District's expense of protecting public in emergencies.................. 0.11
for expense to District in completing contract................................0.46
for imperfect work..................0.7, 0.9
for liquidated damages............ Agreement
for non-completion 0.3
for protection against claims.............. 0.43
from progress estimates.............. ... 0.20
until suits and claims are settled ... 0.37
Default,in completion..Agreement, Bond, 0.3, 0.46
Defective materials, removal of.0.7, 0.9, 0.62, 50.3
Defective work, correcting...........
...................0.6, 0.9, 0.25, 0.101, 0.104
Defective work or materials, payment for, does not relieve Contractor............. 0.9
Defects and damages, money retained for.... 0.25
Definition of terms and explanation of plans..0.15
Definitions.............................. 0.1
Deformed steel bars..................72.4, 73.2
Delayed payments......................... 0.23
Delays and hindrances.....................
..............Agreement, 0.3, 0.4, 0.5, 0.39
Delays, beyond control of Contractor....0.4, 0.39
caused by failure of District to furnish right-of-way........................ 0.5
caused by legal proceedings.............. 0.4
compensation for................0.4, 0.5, 0.39
damage resulting from................. 0.39
expiration of (extension of time).......... 0.4
in delivery of materials................ 0.41
in payments on progress estimates not breach of Contract.................. 0.21
in progress of work....Agreement, 0.3, 0.5, 0.39
notification of, in writing (extension of time)........................ 0.4
occasioned by giving lines and grades, or making measurements, not chargeable to District..................0.13, 0.14
pending decision by Chief Engineer in case of dispute not a basis for damage claim.. 0.6
unreasonable (progress)................ 0.46
Delivery of materials.............0.41, 50.2
Depreciation of plant not to be included in cost of extra work.................. 0.19
Deterioration of foundation surfaces......... 0.67
Direction of work by Engineer....0.3, 0.10, 0.28
Directors, defined........................ 0.1
Discontinuance of work if Contract is violated 0.46
Discontinuance of work on account of weather, etc......................... 0.3
Disintegrated rock, classification of......... 0.65
Disorderly employees to be discharged....... 0.35
Displacements in embankments, correction of. 31.2
Displacement of forms................0.101
Disposal of materials obtained from construction..................0.62, 2.3, 3.1, 3.2, 3.3 5.1
Disposal of materials resulting from removal of buildings................3.1, 3.2
Dispute, in case of, Contractor to follow instructions given pending decision by Chief Engineer.................... 0.6
Disputes as to cost of extra work, how settled. 0.19
Disputes as to interpretation of drawings or specifications........................ 0.10
District, defined......................... 0.1
Ditches, drains, pumps, etc.............. 0.59
Doctors, medical attendance.....0.53, 0.54, 0.55
Dowels, in concrete...................... 0.95
Drainage systems, (surface drainage) excavation for........................... 16.1
Draining springs in foundations...0.79, 0.92, 16.1
Drains and pumps........0.59, 0.71, 0.82, 0.104
Drains for sewage disposal at camps........ 0.52
Drains, iron weeper..................... 82.3
Drawings, accompanying Contract, list of....
...................................Agreement
defined.....................Agreement, 0.1
estimates, and specifications, completeness of................................ 0.16
to be furnished by Contractor showing details of construction plant........ 0.19
to be furnished by Contractor showing details of flood gates............. 77.1
where of doubtful meaning shall be interpreted to call for best type of construction. 0.16

Item

Drawings and specifications, may be corrected by Engineer when not clear............ 0.15
indicate approximate quantities only...... 0.16
interpretation of by Chief Engineer final
..........................0.10, 0.15
Dredging, (see Channel excavation)
Dressing and grassing.
....0.60, 0.70, 34.3, 38.1, 38.2, 38.3, 38.4..
Drilling foundations or masonry for grouting
....................54.1, 54.2, 54.3, 54.4
Dynamite (see Explosives)

E

Earth, classification..................... 0.65
Earth covering for brick pavement.......... 64.7
Elevation, defined........................ 0.1
Eliminations, additions or alterations.......
........................Bond, 0.16, 0.18, 0.19
Embankment, bonding new work to old......
..........................0.76, 0.79, 1.3
classification, construction, etc..0.72, 0.73, 0.74, 0.75, 0.76, 0.77, 0.78, 0.79, 0.80, 0.81, 0.82, 34.1
constructed by hydraulic method.....0.73, 31.1
constructed by semi-hydraulic method.0.74, 31.1
constructed by sprinkling and rolling method......................0.75, 31.2
finishing slopes...............0.81, 0.82, 34.3
for highways, streets, etc............... 19.4
for levees.....0.76, 0.77, 34.1, 34.2, 34.3, 34.4
joining new work to old..............0.76, 1.3
materials to be used................0.76, 0.80
miscellaneous................0.78, 36.1, 36.2
nature of work to be done................ 0.72
payment for........................... 0.82
preparation of foundation for............. 0.79
removal of trestles from................. 0.80
shrinkage allowance in...................
............0.76, 6.3, 19.7, 36.2, 37.2
substitution of one class for another...0.71, 31.1
Emergencies, Contractor to provide for...... 0.11
Emergencies, Engineer's right to act during...
..........................0.11, 0.71, 31.2
Employees, safety of..................... 0.12
Employees, to be of good character......... 0.35
End of period of preparation......Agreement, 0.3
Engineer, defined........................ 0.1
Equipment and methods, Contractor to design and be responsible for....0.4, 0.12, 1.2
modification of, may be required by Engineer........................0.12, 0.19
must be safe, efficient and adequate.0.4, 0.12, 1.2
Equipment, mortgaging or removal of, requires consent...................... 0.45
Estimate of quantities...............Agreement
Estimate of rate of progress, how computed.. 0.3
Estimated quantities are approximate and not guaranteed..................... 0.16
Estimates, progress and final..........0.20, 0.21
Examination, medical, of employees......0.54, 0 56
Excavated materials, disposal of...0.60, 0.71, 19.1
Excavation, below water level.0.71, 9.1, 13.2, 22.2
classification of........................ 0.65
cleaning surface for inspection............ 0.66
finishing work.......................... 0.70
for bridge piers........................ 22.1
for cut-off trenches..................... 6.3
for outlet works........................ 12.1
for retaining walls...................... 22.1
for river channel improvements.....9.1, 9.2, 12.1, 12.2
for spillways........................... 13.1
for stream control channels............... 6.1
for streets and highways................. 19.1
for surface drainage systems............. 16.1
limits of....................0.67, 6.1, 19.7
may be enlarged to secure material........ 0.60
miscellaneous.......................... 28.1
nature of work to be done................ 0.66
overbreak.............................. 0.67
payment...........................0.71, 19.4
soil stripping.......................... 5.1
use of explosives............0.64, 0.67, 0.69
wet.......................0.71, 6.1, 13.2, 22.2
Execution of work, manner of, subject to decision of Chief Engineer. 0.10, 0.19
Exhibits composing Contract.........Agreement

Item

Expansion joints in brick pavement.........64.7
Expansion joints in concrete..............0.97
Expense caused by orders to stop work may
  be paid by District...............0.43
Explanatory note to specifications...Specifications
Explosives (see also Laws).......0.37, 0.65, 0.69
Extension of time, claims for, etc..........0.3, 0.4
  does not impair bond.............Bond, 0.3
  for delays in furnishing right-of-way......0.5
  no waiver of obligations on part of Contractor or Sureties................0.3, 0.28
Extra or special measurements not allowed..0.14
Extra work, specification.........0.19, 90.1, 90.2
  caused by tearing down work for inspection,
   how paid for.......................0.6
  claims for, how made and settled..........0.19
  does not impair bond..............Bond, 0.19

F

Factory inspection of materials............
  ....0.6, 0.41, 50.3, 51.1, 75.2, 77.2
Failure of Engineer to demand increase in
  force or equipment shall not relieve Contractor of obligation as to rate of progress. 0.12
  of temporary or partly completed works,
   Contractor responsible for........0.11, 0.36
  to comply with terms of Contract.......
  .....Agreement, Bond, 0.46, 0.53, 0.71, 31.1
  to condemn defective work or material does
   not imply acceptance.............0.9, 0.28
  to keep premises clean..................0.52
  to maintain rate of progress is breach of
   Contract..........................0.3
Faulty work, replacing.....................0.26
Fences, removal of.......................3.1
  temporary, to be provided for live stock,
   (see also Barriers; Guard rails)..........0.36
  wire.........................68.1, 68.2
Fill, (see Embankment)
Filler, bituminous, for joints in pavement.64.7, 65.6
  grout, for brick pavement...........61.3, 64.7
  screenings for pavement.............61.2, 61.6
Filling, (see Backfilling; Refilling)
Final payment, how computed, when paid....0.21
Finishing embankment slopes..........0.81, 31.2
  excavation slopes.....................0.70
  surfaces of concrete.....0.98, 0.103, 0.104, 42.1
Fire apparatus, contractor must provide.....0.58
First aid to injured......................0.53
First party, defined......................0.1
Flap valves, tide gates...................77.6
Flexible concrete slab revetment........45.1, 45.2
Flood gates, cast iron....................77.2
Flood gates, steel.......................77.3
Floods, control of during construction.......
  ....0.39, 1.1, 1.3
  data on file at Engineer's office...........0.4
  delays caused by......................0.39
  discontinuance of work during...........0.3
  protection of property during.......0.11, 0.104
Floors, concrete in......................43.1
Fly screens, Contractor to provide.........0.50
Footing walls for concrete revetment........44.1
Force, increase or decrease of, may be ordered
  by Engineer....................0.12, 0.19
Foremen and superintendents.........0.29, 0.33
Foremen's wages may be included in cost of
  extra work..........................0.19
Forms for concrete............0.94, 0.104, 71.1
Foundations, draining springs in..0.79, 0.92, 16.1
  drilling and grouting...................54.2
  frozen, how treated...................0.92
  preparation of for concrete..............0.92
  preparation of for embankment..........
  ....0.66, 0.79, 0.80, 5.1
  protection of........................0.64
  removal of..........................5.1
Freezing weather, laying concrete in...0.99, 0.100
  laying stone masonry in................48.1
  no brick lining to be laid in..............52.6
  no macadam road construction permitted
   during...........................61.7
Frost, removing...........0.92, 0.99, 0.104
Frozen foundations, how treated..........0.92
  material not permitted in embankments....0.80
  surfaces, materials must not be placed
   against................0.80, 0.84, 0.92

G

Item

Galvanized wire for fences.................68.2
Garbage, disposal of......................0.2
Gate chambers, concrete in................4.51
  stands, hoists......................76.5
  stems for flood gates..................77.4
Gates and fastenings for fences...........68.2, 82.1
  cast iron flood.......................77.2
  steel flood..........................77.3
  tide..............................77.6
General Specifications....................0.58
Glacial deposits, classification of...........0.65
Grade and line stakes to be preserved.......0.13
Grades and lines to be furnished by Engineer. 0.13
Grades for shallow rock cuts may be readjusted 19.6
Grading borrow pits.....................0.60
  embankment slopes...................34.3
  excavated slopes....................0.70
Grassing and dressing surfaces............
  ....0.62, 34.3, 38.4, 85.1
Gratings, cast iron..................82.1, 82.2
Gravel for concrete......................0.88
Gravel or crushed rock for paving. 60.1, 60.2, 61.1
Gravel pavement........61.1, 61.2, 61.3, 64.6
Grit for road paving................61.2, 63.7
Grout, cement. 0.95, 0.98, 0.104, 51.2, 54.3, 54.4
Grouting foundations or masonry...........
  ....54.1, 54.2, 54.3, 54.4
Grouting macadam pavements.............61.7
Grubbing and clearing..........2.1, 2.2, 2.3
Guard rails, (see also Barriers; Fences).....
  ....67.1, 67.2, 67.3, 67.4
Guarding against accidents.................0.37
Gutter paving..........................59.1

H

Hand rails, iron........................82.1
Harmonizing operations of different Contractors........................0.10, 0.31
Haul, overhaul..............9.2, 12.1, 12.2
Health of employees to be safeguarded.....
  ....0.48, 0.53, 0.54, 0.55
Highway Department specifications, Ohio
  State..................61.1, 61.3, 61.6
Highways, bridges, to be maintained.......
  ....0.36, 0.64, 19.4
  excavation and embankments for.19.1, 19.2, 19.4
  not to be closed, etc., without permit.....0.36
  paving for...................61.1 to 61.7
  traction lines, etc., Contractor not to
   interfere with construction of......0.10, 0.29
Hillside pavers...................61.5, 64.7
Hindrances and delays (see also Delays).....
  ....Agreement, 0.3, 0.4, 0.39
Hoisting devices for gates.................77.5
Hospitals, medical supplies............0.53, 0.54
Hydraulic method of constructing dams......0.73

I

Imperfect work may be retained....0.8, 0.28, 55.1
Imperfect work, replacing................
  ....0.6, 0.7, 0.8, 0.25, 0.101, 0.104
Incinerators for garbage and excreta.........0.52
Incompetent employees to be discharged.....0.35
Increase or decrease in quantities. Bond, 0.16, 0.17
Increase or decrease of plant, equipment, or
  force, may be ordered by Engineer.0.12; 0.19
Indemnify District against damage caused by
  negligence of Contractor.........Bond, 0.25
Infringement of patents.............Bond, 0.42
Initial set, concrete.....................0.93
Injunction affecting part of work does not
  necessarily entitle to extension of time..0.4
Injunctions interfering with work...........0.4
Inspection and right of access..........0.6, 0.40
  of camp buildings, etc..............0.53, 0.56
  of drinking water for employees..........0.56
  of excavated surfaces..................0.68
  of food for employees.................0.56
  of large stones required before placing in
   concrete........................0.90
  of materials not on site of work..0.6, 0.41, 50.3
  tests, etc., of cement.................50.3
  work torn down for, may be paid for as
   extra work.......................0.6
Insurance, Contractor's liability, may be
  included in cost of extra work..........0.19

Item

Interest in Contract debarred to directors or
  employees of District.................. 0.30
  in Contract (persons having Interest).... 0.30
  not figured on amounts properly retained.. 0.23
  on delayed payments.................. 0.23
Interference with collateral work......0.10, 0.29
  with progress of work (see also Delays)...
  .... Agreement, 0.3, 0.4, 0.39
  with roads, railways, telegraph lines, etc..
  ....0.36, 0.64
  with work by failure of District to furnish
    right-of-way...................... 0.5
  with work by floods.................. 0.39
  with work taken over by Board, Contractor
    shall not cause.................... 0.46
  with work, through injunction or orders.0.4, 0.40
Interpretation of drawings or specifications by
  Chief Engineer, final........0.6, 0.11, 0.15
Interruption of work, by failure of District to
  furnish right-of-way................ 0.5
  if ordered, may be compensated for....... 0.39
  to enable Engineer to give lines and grades,
    or make measurements................ 0.11
Intoxicants............................ 0.49
Invoices, bills of lading, etc., to be furnished
  Engineer upon request...0.6, 0.19, 0.26, 50.1
Iron, cast, wrought, etc.................. 82.1
Iron chains for booms...............70.1, 82.1
Items of work defined and explained......,
  ....Specifications, 0.17
Items of work, in schedule...........Agreement

J

Jetting piles.............72.2, 72.4, 73.3, 74.4
Joining new concrete to old, or to rock.....
  ....0.92, 0.95, 0.104
Joining old and new embankments......0.76, 1.3
Joints, face, in masonry............48.1, 52.6
  lead, for cast iron pipe................ 76.1
  mortar, for sewer pipe................ 56.1
  (see Expansion or Contraction Joints)

L

Ladders for inspection.................. 0.6
Ladders, iron.......................... 82.1
Laitance, removal of.................... 0.95
Laws authorize retention of money.....0.20, 0.43
Laws, Contractor's compliance with.......
  ....Bond, 0.34, 0.69
Lead joints for cast iron pipe............ 76.1
Ledge rock, classification of (see also Rock
  excavation; Limestone)............. 0.65
Legal proceedings, cause of delays........ 0.4
Levee culverts....................25.1, 46.1
Levees, embankment for............0.76, 34.1
Levees, finishing slopes of....,
  ....0.81, 0.82, 34.1, 34.2, 34.3, 38.1
Liability insurance may be included in cost
  of extra work. 0.19
Lights for night work, etc.......... 0.37, 0.40
Limestone, classification of.............. 0.65
  (local ledge rock) not generally suitable for
    concrete aggregate.................. 0.87
  not generally suitable for crushed rock..... 60.1
  not generally suitable for rip-rap or paving
    ....57.1, 58.1
  screenings not permitted in concrete..... 0.88
Limits of excavation 0.67, 6.2, 19.7
Lines and grades, stakes to be preserved... 0.13
  to be furnished by Engineer......... 0.13
  to be requested a reasonable time in ad-
    vance.........................0.13
Lining river channels with concrete revet-
  ment.............................. 45.1
Lining sewers with vitrified brick......... 52.1
Liquidated damages . ...Agreement, 0.3
Liquidated damages, payment of, not to
  relieve Contractor or Sureties from their
  obligation ....Agreement
List of exhibits composing Contract . Agreement
Live stock, temporary fences to be built for.. 0.36
Local conditions, Contractor familiar with ..
  ....0.4, 0.11, 0.39
Lodging for Engineer or assistants....... 0.47
Loose rock, removal of 0.59, 0.65, 5.1
Loss from delays, nature of work, weather,
  etc., borne by Contractor........0.39

Item

Loss of life or property, Contractor to protect
  public against..........0.11, 0.12, 0.37, 1.1
Lumber and timber, miscellaneous......... 71.1

M

Macadam, water-bound, paving...61.1, 61.2, 61.6
Maintenance of cut-off trenches............ 6.2
  of camps, not part of extra work cost...... 0.19
  of excavations...................... 0.71
  of highways, streets, etc........0.36, 0.64, 19.4
  of stream control channels....... 6.1, 6.2
  of work in general.............0.8, 0.11, 0.39
Manholes, concrete in.................... 47.1
Manholes, gratings for.................... 82.1
Masonry in concrete....................0.103
Masonry, stone........................ 48.1
Materials and workmanship to be of best
  quality when meaning of drawings or
  specifications is doubtful............... 0.16
Materials, delivered at site of work may be
  paid for in part...................... 0.20
  delivery of......................... 0.41
  excavated, disposal of........0.62, 0.71, 19.2
  for backfilling...................0.84, 19.4
  for embankment.................... 0.80
  for road paving......60.1, 60.2, 61.1, 61.2, 61.3
  frozen, not allowed in embankments...... 0.80
  mortgaging, requires consent of Engineer.. 0.45
  not on site of work, inspection of.0.6, 0.41, 50.3
  not specified or shown in the drawings but
    implied must be furnished............. 0.16
  obtained from construction..........
    ....0.62, 0.71, 2.2, 3.2, 5.2
  selected for dams.................... 0.80
Materials, unsuitable, to be removed from
  site of work..............0.7, 0.62, 50.3
  (see also respective items)
Meals for Engineer and assistants........... 0.47
Meaning of specifications and drawings where
  in doubt shall be interpreted to call for
  best type of construction.............. 0.16
Measurement of areas by planimeter........ 0.14
Measurement of work done does not imply
  acceptance........................ 0.28
Measurements and quantities to be determined
  by Engineer....................... 0.14
Measurements, extra or special, not allowed . 0.14
Measurements for payment (see respective
  items)
Measuring boxes for concrete materials..... 0.95
  proportions of concrete................ 0.93
  water for concrete.........0.93, 0.94, 0.95
Medical service and requirements..0.53, 0.54, 0.55
Metal work, in sheeting left in place....89.1, 89.2
  inspection at factory................. 0.40
  miscellaneous.............68.4, 70.2, 82.1
  placing of in concrete
    ....0.96, 0.97, 0.102, 51.1, 51.2
Methods and equipment, Contractor to design
  and be responsible for......0.4, 0.12, 1.2
  modification of, may be required by En-
    gineer.......................0.12, 0.19
  must be safe, efficient and adequate.0.4, 0.12, 1.2
Mineral aggregates for road paving.......
  ....60.1, 61.1, 61.2, 61.3, 61.6
Miscellaneous cast iron, wrought iron, steel ..
  ....82.1, 82.2
  embankment 0.78, 36.1
  excavation ....r........ 28.1
Misunderstandings as to meaning of drawings
  or specifications............0.10, 0.15, 0.16
Mixed excavation, classification........... 0.65
Mixing and placing grout................. 54.3
Mixing concrete..................0.93, 42.1
Modification of methods and equipment may
  be required by Engineer............. 0.12
Molds for concrete revetment blocks........ 45.1
Monolithic concrete revetment........... 44.1
Mortar..0.87, 0.95, 0.98, 0.104, 0.105, 56.1, 58.1
Mortgaging equipment or materials requires
  consent........................... 0.45

N

Nails, bolts, etc., in lumber and sheeting.71.2, 89.2
Negligence of Contractor, sub-contractors,
  agents, etc., not to become basis of claims
  against District (see also Failure)....
  ....Bond, 0.25, 0.38

Item

Night work............................. 0.40
Noncompletion of work...........Agreement, 0.3
Notices to Contractor, where sent.....0.26, 0.32
  to Contractor, written....0.2, 0.11, 0.12, 0.19,
    0.22, 0.35, 0.46, 0.71, 19.3, 19.7, 31.1, 36.1
    89.1; 89.2
  to Engineer, written
    ....0.4, 0.6, 0.13, 0.26, 0.40, 0.55, 50.3
  to Surety, written.................Bond, 0.46

O

Ohio Laws.....................Bond, 0.34, 0.69
Ohio State Highway Department specifica-
  tions....................61.1, 61.2, 61.3
Oiling concrete forms..................... 0.94
Oiling pavements (penetration method).61.3, 63.7
Operating stands for gates............... 77.5
Order of completion, and precedence......... 0.3
Orders, (see Notices; Written consent)
Ordinances, Contractor's compliance with....
  ....Bond, 0.34, 0.67
Outflow devices, (temporary) at hydraulic fill
  dams................................ 0.73
Outlet conduits, concrete for lining......... 42.1
Outlet works, excavation for.............. 13.3
Outlet works, concrete in.................. 41.1
Overbreak in excavation..............0.67, 6.2
Overfill to allow for settlement...............
  ....0.76, 19.4, 81.3, 34.2, 34.3, 36.2
Overhaul, no payment for................. 0.82
Overhead expense not part of extra work cost. 0.19

P

Painting booms and accessories.........70.2, 70.4
  guard rails......................... 67.4
  metal work......................... 82.2
Parson's manganese bronze......77.2, 77.4, 77.5
Parties to bond........................Bond
Parties to contract and agreement..Agreement, 0.1
Patching concrete.......0.98, 0.101, 0.104, 42.1
Patents, infringement of.............Bond, 0.42
Paving, at dams, concrete in.............. 43.1
  dry rubble......................... 57.1
  for streets and highways..........61.1 to 61.7
  gutter, at dams...................... 59.1
Pavement, bituminous-bound macadam......
  ....61.1, 63.7
  brick, on concrete foundation...........
  ....61.1, 61.2, 64.6 to .6
  brick, on rolled foundation.....61.1, 64.7 to .6
  concrete.............61.2, 61.3, 64.7, 65.6
  concrete curb for.................... 65.7
  gravel.........61.2, 61.3, 61.6, 61.7, 64.6
  materials, specifications for........61.2, 61.
  sandstone curb for................... 61.
  water-bound macadam.........61.1, 61.6, 61.3
Payment, for board and lodging furnished to
  Engineer and assistants............... 0 47
  for cleaning up..................... ab.2
  for defective work or materials does not
    relieve Contractor................... 0.9
  for extra work.........Agreement, 0.19, 90.2
  for labor and material by Contractor.Bond, 0.43
  for removing isolated trees.............. 3.3
  for sheeting left in place............... 89.2
  for stream control.................... 1.3
  for work done (see respective items)
  for work rendered useless by changes...... 0.18
  general clause...................Agreement
  in part, for materials delivered......... 0.20
  in whole or in part, does not waive Con-
    tractor's obligations.............0.9, 0.28
  of expenses, losses or damages, for which
    Contractor is responsible, how made..... 0.25
  of liquidated damages.........Agreement, 0.3
  on progress estimates within 10 days after
    date of estimate..................... 0.20
  only in accordance with Contract......... 0.24
  schedule of prices.................Agreement
Payments, by check..................... 0.22
  delayed draw interest................. 0.23
  progress and final................0.20, 0.21
Penetration method (bituminous macadam)..
  ....61.3, 63.7
Percentage retained on progress payments.... 0.20
Period of preparation............Agreement, 0.3

Item

Permission to complete after time stipulated
  no waiver to collect liquidated damages.. 0.3
Personal attention of Contractor........... 0.31
Persons interested in Contract............. 0.30
Piling, steel sheet...................74.1, 89.1
  timber............................. 72.1
  wood sheet...................73.1, 73.2, 89.1
Pipe and fittings...................54.2, 82.1
Pipe, cast iron......................... 75.1
  castings, special.................... 82.1
  dipped, in tar...................... 82.2
  flanged............................ 82.1
  trenches, excavating and backfilling...... 25.1
  vitrified...............Agreement, 55.1, 56.1
Placing concrete........................ 0.95
Planimeter may be used to measure areas.... 0.14
Plans (see Drawings)
Plant (see Equipment)
Plant drawings to be furnished by Contractor 0.12
Plastering concrete faces 0.100, 0.101, 0.104, 42.1
Plates, steel.......................... 82.1
Plum rock........0.58, 0.85, 0.90, 0.95, 42.1
Pointing up concrete or masonry........... 0.98
Police and sanitary regulations............. 0.48
Possession of equipment, materials, etc.,
  Board may take..................... 0.46
Possession taken by Board of work done does
  not waive Contractor's obligations...... 0.28
Posts for wire fences and guard rails......
  ....67.2, 67.3, 68.2
Powder, dynamite, (see Explosives)
Preparation, period of...........Agreement, 0.3
Prices, unit, schedule of..............Agreement
Progress, delayed......Agreement, 0.3, 0.4, 0.39
  estimates, how computed and paid........ 0.20
  estimates superseded by final estimate..... 0.21
  rate of...............Agreement, 0.3, 0.12, 0.46
  rate of, not waived by failure of Engineer
    to demand increase in force or equip-
    ment............................. 0.12
Protection against claims arising from performance of Contract......Bond, 0.26, 0.38, 0.43
  against fire, Contractor must provide...... 0.58
  against floods, during construction........
  ....0.4, 0.11, 0.104, 1.1, 1.3
  against freezing..........0.99, 0.104, 48.1
Protection against violation of laws, etc.Bond, 0.34
  during emergencies may be assumed by
    District............................ 0.11
  of concrete...................0.6, 0.11, 0.99
  of District may justify increase of hold-
    back.............................. 0.20
  of District may justify omitting progress
    estimates.......................... 0.20
  of existing structures................0.36, 0.64
  of life or property by Contractor..........
  ....0.11, 0.12, 0.37, 0.64, 1.1
  of public water supply or public service by
    Contractor......................... 0.11
  of reinforcing steel..................51.1, 51.2
  of trees, shrubs, etc.................. 0.61
  of work in general.........0.6, 0.11, 0.12, 0.40
Protective booms and accessories.......70.1, 82.1
Providing for emergencies................ 0.11
Public safety and health to be safeguarded...
  ....0.11, 0.48, 0.55
Puddled core in dams................0.73, 0.74
Puddling earth fills (see also Backfilling).....
  ....0.58, 0.82, 19.4, 22.2, 37.1, 37.2
Pumps and drains.........0.59, 0.71, 0.82, 0.104
Pumps, dredge, for hydraulicking........... 0.73

Q

Quality of work..Agreement, 0.9, 0.12, 0.15, 0.16
Quantities and measurements, (see also
  respective items)................... 0.14
  changes in.................Bond, 0.16, 0.17
  estimated, are approximate only......... 0.16
  schedule of estimated..............Agreement
Quarters for employees.....0.50, 0.51, 0.53, 0.56

R

Railroads, roads, bridges, telephone lines,
  etc., to be maintained.............0.36, 0.64
Railroads, traction lines, etc., Contractor
  not to interfere with construction of.0.10, 0.29
Ramming brick pavement................ 64.7

Item

Rate of progress.......Agreement, 0.3, 0.12, 0.46
Rate of progress not waived by failure of
  Engineer to demand increase in force or
  equipment............................. 0.12
Rattler test for brick.............52.2, 52.3
Records, invoices, bills of lading, etc., to be
  furnished Engineer upon request.....
  ....0.8, 0.19, 0.26, 50.2
Reduction of bond....................... 0.2
Refilling (see also Backfilling).........
  ....0.62, 0.71, 0.83, 19.2, 19.4
Regulations, Contractor's compliance with...
  ....Bond, 0.34, 0.69
Regulations for storing, handling explosives.. 0.69
Reinforcing steel..........0.95, 0.102, 51.1, 51.2
Rejected concrete, cement not paid for......0.101
Rejected materials, disposal of.....0.7, 0.62, 50.3
Relations with other contractors........0.11, 0.29
Release of claims against District at time of
  final payment....................... 0.21
Remedies cumulative..................... 0.27
Remedying defective work............... 0.7
Removal of buildings, fences, foundations,
  etc................................ 3.1
  of defective work................0.9, 0.94
  of steel bridges..................... 81.1
  or mortgaging of equipment, requires
    consent.......................... 0.45
Replacing faulty work...............0.7, 0.101
Replacing work disturbed by inspection may
  be paid for as extra work........... 0.6
Responsibility for existing structures....0.36, 0.64
  as to equipment and methods rests with
    Contractor...................0.3, 0.10, 1.2
  for claims arising from performance of
    contract...........Bond, 0.26, 0.37, 0.42
  for failure of temporary or partly completed
    work......................0.11, 0.36
  (of District) for inevitable damages.....
    ....0.11, 0.12, 0.38
  for maintenance of work........0.6, 0.11, 0.39
  for stream control.............0.11, 0.39, 1.3
Retained percentage.................... 0.20
Retaining imperfect work.............0.8, 0.28
  money for defects and damages.......... 0.26
  money for holdback, etc.............. 0.20
  money for protection against claims...0.38, 0.43
  walls, concrete in................... 41.1
  walls, excavation for................ 25.1
Retempering concrete not permitted......... 0.93
Right of access for inspection.............. 0.6
Right-of-way to be furnished by District...... 0.5
Rip-rap.........................58.1, 58.2
Risks and uncertainties assumed by Contractor.........................0.39, 1.3
River and weather conditions as affecting
  rate of progress................... 0.4
River channel excavation, improvements.... 9.1
  excavation, new.................... 6.1
  lining............................ 45.1
River channels, trimming slopes of.........
  ....0.70, 4.1, 4.2, 9.1
River control during construction.........
  ....0.37, 1.1, 1.2, 1.3, 6.1
River, weather and local conditions, data on
  file in Engineer's office.............. 0.4
Riveting steel gates................... 77.3
Road metal.....60.1, 60.2, 61.1, 61.2, 61.3, 61.6
Roads, (see Highways)
Rock, crushed, for paving............60.1, 61.1
  excavation..........0.65, 0.68, 0.70, 19.1, 19.6
  (local limestones) not generally acceptable
    for coarse aggregate............... 0.88
Rods, tie, anchor, truss, etc.............. 82.1
Rollers, for compacting pavement.......... 61.3
Rollers, traction engines, etc., for compacting
  embankments...................... 0.75
Rolling and sprinkling backfilling.......... 0.83
Rolling embankments.................... 0.75
  foundations for concrete.............. 0.92
  macadam pavement.........61.3, 61.6, 61.7
  sand cushion for brick pavement......... 64.7
  sub grade for pavement............... 61.5
Rope, wire.........................84.1, 84.2
Round timber for booms, measurement of... 70.2
Rubble or brick masonry permitted in concrete............................. 0.103

Item

Rubble paving......................... 57.1
Run of bank........................0.88, 60.1
Run of crusher.....................0.88, 60.1
Rust on reinforcing steel............... 51.2
Rusting of castings.................... 82.2

S

Safety of life and property.............
  ....0.11, 0.12, 0.36, 0.37, 0.71, 1.1
Safety of methods and equipment, Contractor
  responsible for.....0.4, 0.11, 0.12, 0.16, 0.39
Safety of temporary roads, bridges, etc..... 0.36
Saloons.............................. 0.49
Samples of materials for concrete.......... 0.91
Samples, (see Tests)
Sand covering for brick pavement......... 64.7
  cushion for brick pavement........61.3, 64.7
  for concrete, mortar, or grout.........
    ....0.87, 0.105, 48.1, 61.3
  for road material.........61.1, 61.2, 61.3, 61.6, 61.7
  washing............................ 0.58
Sandstone curb for pavement............. 61.3
Sanitary and police regulations........... 0.48
Scales, 300-pound, to be provided by Con-
  tractor for weighing material........0.6, 50.2
Schedule of unit prices, estimated quantities
  ....Agreement
Screening, grading and washing aggregates
  ....0.88, 42.1
Screenings for road paving..........61.2, 61.3, 61.6
Screenings, limestone, not allowed in con-
  crete............................. 0.88
Second party, defined.................. 0.1
Seeding (see Surface Dressing and Grassing)
Selected materials for backfilling........... 0.83
Selected materials for dams.............. 0.80
Semi-hydraulic method of constructing dams. 0.74
Settlement, allowance for, in embankments
  ....0.76, 19.4, 34.2, 36.2
Settling ponds for puddling cores in dams
  ....0.73, 0.74
Sewage disposal at camps............... 0.52
Sewer trenches, excavating and backfilling.. 25.1
Sewers, concrete in.................... 47.1
Sewers, lining with vitrified brick......... 52.1
Shale, classification of.................. 0.65
Sheet piling, steel...................74.4, 89.1
Sheet piling, wood...................73.2, 89.1
Sheeting for bracing (see also excavation for
  structures)........................ 71.2
Sheeting left in place.......0.71, 13.2, 22.2, 89.1
Shower baths for employees at camps...... 0.50
Shrinkage, allowance for, in embankments
  ....0.76, 19.4, 34.2, 36.2
Shrubs, trees, protection of.............. 0.61
Signatures to Bond.....................Bond
Signatures to Contract.............Agreement
Signs, lights, etc., to protect public....... 0.37
Slag for road material.............61.1, 61.2, 61.3
Slides, slips, in embankments, correction of. 31.2
Slip joints in concrete.................. 0.97
Slopes, excavated, finishing............. 0.70
  of embankments....0.73, 0.74, 31.2, 31.3, 34.3
  outer, of dams, how made. 0.73, 0.74, 31.1, 31.3
Sluice gates (see Flood gates)
Sluicing, materials in place.......0.73, 0.74, 0.82
Snow, ice, removing................0.92, 0.99
Soil stripping, described................ 5.1
Specifications, defined................. 0.1
  standard, referred to in the Specifications
    are on file at office of Chief Engineer... 0.16
  when of doubtful meaning, shall be inter-
    preted to call for best type of construc-
    tion.............................. 0.16
Specifications and drawings, indicate only
  approximate quantities.............. 0.16
  interpretation of, by Chief Engineer, final
    ....0.9, 0.15, 0.16
  may be corrected by Engineer when not
    clear............................ 0.15
Spillways, concrete in................41.1, 41.2
  excavation for...................... 13.1
Spoil banks, dressing and grassing of.....
  ....0.62, 38.1, 38.4
  grading of......................... 0.62
  right-of-way for.................... 0.5

Item

Springs in foundation............0.79, 0.92, 16.1
Sprinkling and rolling method of constructing
  dams.. 0.75
Sprinkling, backfill........................ 0.83
  concrete...................................0.100
  concrete pavement.................... 65.6
  macadam pavements................61.6, 61.7
  water for................................. 0.58
Stables..............................0.50, 0.52
Stakes and marks to be preserved........... 0.13
Standard specifications referred to in the
  Specifications are on file at office of
  Chief Engineer...................... 0.16
Stands for operating gates.................. 77.5
State Highway Department specifications..
  ....61.2, 61.3
Steam rollers for compacting pavements..... 61.3
Steam rollers, traction engines, etc., for com-
  pacting embankments.................. 0.75
Steel bridges, raising, removing and reerecting. 81.1
  brooms for cleaning rock, etc.........0.92, 0.95
  flood gates.............................. 77.3
  miscellaneous........................... 82.1
  plates, bolts, etc....................... 82.1
  reinforcing........0.95, 0.102, 51.1, 51.2, 51.3
  sheet piling....74.1, 74.2, 74.3, 74.4, 89.1, 89.2
  stems for flood gates.................... 77.4
Stone, broken, for bedding rubble paving..
  ....57.1, 57.2.
  broken, for concrete.................... 0.88
  broken, for paving........61.2, 61.6, 61.7, 63.7
  for rubble paving.....................57.1, 57.2
  large, for imbedding in concrete...
  ....0.58, 0.85, 0.90, 0.95
  masonry................................. 48.1
  size of, permitted in backfilling.......... 0.84
  size of, permitted in levees.............. 0.80
Storage of cement......................... 50.2
Storage of concrete blocks for revetment..... 45.1
Storekeeper, for cement................... 50.2
Straining posts for wire fences............ 68.2
Stream control.......0.3, 0.39, 1.1, 1.2, 1.3, 9.1
Streets, (see Highways)
Stripping, soil............................ 5.1
Structural steel for gates.................. 77.3
Stumps, boulders, etc., in borrow pits........ 0.60
Sub grade for pavements................... 61.5
Subletting requires consent of Board......... 0.44
Substitution of one class of embankment for
  another........................0.71, 31.1
Summer months, relative time value of...... 0.3
Superintendents and foremen.........0.29, 0.33
Sureties' consent.............Bond, 0.1, 0.2
Sureties' consent required when general
  character of work is changed..........Bond
  liability may be reduced................. 0.2
  liability not waived by changes which
  increase contract price less than 25%..
  ....Bond, 0.19
  liability not waived by extension of time..
  ....Bond, 0.2
  may take over plant and equipment. Bond, 0.46
Sureties may sublet contract..........Bond, 0.46
  must exercise option to assume contract
  within two weeks from date of notice
  of suspension.......................... 0.46
  need not be notified in case of changes,
  extension of time, etc............Agreement
  to be notified when work is ordered suspended................................ 0.46
  to be satisfactory to Board......Agreement, 0.2
Surety bond.........................Bond, 0.2
Surface dressing and grassing..............
  ....0.60, 0.62, 31.1, 38.4, 85.1
Suspension of work if Contract is violated.. 0.46
Suspension, when ordered, shall take place
  within two weeks from date of notice.. 0.46

T

Item

Tamping backfilling...................... 0.83
Tamping embankments.................... 0.75
Tar dipped pipe.......................... 82.2
Tarring pavements (penetration method)..
  ....61.3, 63.7
Telephone and telegraph lines, etc., Con-
  tractor not to interfere with construction
  of.................................0.10, 0.29
Telephone and telegraph lines to be main-
  tained...............................0.36, 0.64
Templet for crowning pavements......64.6, 65.6
Temporary roads, bridges, etc., to be safe.... 0.36
Temporary work, payment for (see also re-
  spective items)...Agreement, 0.60, 0.104, 1.3
Terms, to define, and explain plans.... 0.15
Testing foundations, etc...............0.63, 53.1
Testing, steel sheet piling...........74.3, 74.5
Test piles........72.1, 72.6, 73.1, 73.4, 74.1, 74.5
Test pits, Contractor to assist in digging..... 0.63
Tests, cement.................0.86, 50.2, 50.3
  concrete..........................0.85, 0.91
  Contractor to furnish assistance and
  facilities........................0.6, 0.26, 0.63
  (see respective items)
Tide gates, flap valves.................... 77.6
Tie rods, steel........................... 82.1
Timber and lumber, grading............... 71.1
Timber and lumber, miscellaneous.... 71.1
Timber piles............................. 72.1
Timbers, round, for booms, measurement of . 70.1
Time and order of completion....Agreement, 0.3
Time, extension of (see also Extension of
  time)..............................0.3, 0.4
Time of beginning............Agreement, 0.3
Top soil for dressing and grassing....
  ....38.1, 38.2, 38.3, 38.4
Traffic to be kept off green concrete or brick
  pavements..........................64.7, 65.6
Transverse joints in concrete pavement..... 65.6
Trees, isolated, removal of, how paid for..... 2.3
Trees, shrubs, protection of............... 0.61
Trench excavations, concrete pavement over.. 65.6
Trenches, excavating and backfilling..... 25.1
  measurement of........................ 25.3
  pipe, etc., backfilling of................. 0.83
Trestles in embankments, removal of........ 0.80
Trimming excavations.
  ....0.66, 0.70, 4.1, 9.1, 19.1, 19.4
  slopes of levees, embankments..0.81, 0.82, 34.3
  slopes of river channels and excavations..
  ....0.67, 0.70, 4.1, 9.1
Truss rods.............................. 82.1
Tunneling in trench excavation requires permission............................... 25.1

U

Unavoidable delays (see also Delays)......
  ....Agreement, 0.3, 0.4, 0.39
Uncertainties and risks assumed by Con-
  tractor.............................0.39, 1.3
Unit prices, schedule of...............Agreement
Unsuitable materials to be removed from
  work...........................0.7, 0.62, 50.3
Unwatering, pumping, draining excavations..
  ....0.59, 0.71, 0.92, 16.1, 22.2

V

Value of work done...................0.3, 0.21
Valves, (see Gates)
Variation in quantities..........Bond, 0.16, 0.18
Violation of contract may cause work to be
  suspended............................. 0.46
  of laws and ordinances............Bond, 0.34
  of police or sanitary regulations.......... 0.48
Vitrified brick for lining sewers......52.1 to 52.7
Vitrified pipe............Agreement, 55.1, 56.1

W   Item

Waiving obligations of Contractor not within
    province of Engineer..............0.9, 0.28
Walls, concrete retaining.... ....22.1, 41.1
    footing, for concrete revetment..........44.1
    stone masonry........................48.1
Washing sand.... ....0.87
Water-bound macadam paving ....
    .... .61.1, 61.3, 61.6, 61.7
Water, depositing concrete in.............0.95
    for concrete 0.89, 0.94
    for sprinkling concrete.............0.95
    jet for pile driving.............72.4, 73.3, 74.4
    level, classification of materials excavated
        above or below..........6.1, 13.1, 22.1
    supply....................0.50, 0.51, 0.58
Water level, classification of concrete above
    or below 41.1, 43.1, 44.1, 46.1
Wear and tear on plant not part of extra
    work cost........................0.19
Weather and river conditions as affecting
    rate of progress..............0.4, 0.39, 1.1
Weather conditions, etc., responsibility for
    effect of 0.4, 0.11, 0.37, 0.39, 1.3
Weather, river and local conditions, data on
    file in Engineer's office.. ....0.4
Weeper drains, iron...................82.1
Weighing materials, scales for, to be pro-
    vided by Contractor...... .0.6, 50.2, 51.3
Wet excavation. 0.71, 6.1, 13.1, 22.1
Wetting brick lining in sewers.............52.6
    concrete forms.............0.94, 0.104
    concrete pavement. 65.6
    earth covering over brick pavement......64.7
    earth covering over concrete pavement...65.6
    finished concrete.................0.98, 52.6
    foundations for concrete................0.92
    macadam pavements...............61.6, 61.7

                                           Item

Winter months, work during (see Freezing
    Weather, Frost, etc.)
Winter months, relative time value. 0.3
Wire brushes for cleaning foundations, etc...
    ....0.92, 0.95
Wire fence.............................68.1
Wire rope.................................84.1
Wood sheet piling..........73.1, 73.2, 73.3, 89.1
Work done, value of.............0.3, 0.20, 0.21
Work not specified or shown in the drawings
    but implied must be done without
    additional charge...................0.16
    rendered useless by changes or alterations
        ordered, to be paid for by District....0.18
    to be done, stated.................Agreement
Workmanship to be of best quality when
    meaning of drawings or specifications is
    doubtful...........................0.16
Workmen (see Employees)
Written certificates of Engineer to Board,
    when required...............0.21, 0.46
    consent of Board, when required 0.8, 0.44, 0.46
    consent of Engineer, when required .....
        0.36, 0.45, 0.48, 0.67, 0.71, 0.73, 0.99, 1.2,
        13.2, 19.3, 19.4, 19.7, 22.2, 31.1, 37.1,
        42.1, 50.1, 54.2, 54.3, 71.2.
    consent of Surety when required...... . Bond
    notices from Contractor to Engineer, when
        required.0.3, 0.4, 0.13, 0.26, 0.40, 0.55, 50.3
    notices from Engineer to Contractor, when
        required....0.23, 0.11, 0.12, 0.19, 0.32,
        0.33, 0.46, 0.71, 19.3, 19.7, 31.1, 36.1, 89.2
    notices to Contractor, where sent... ... 0.32
    notices to Surety, when required... .Bond, 0.46
Wrought iron chains for booms... ...70.3, 82.1
Wrought iron, steel, etc................. 82.1

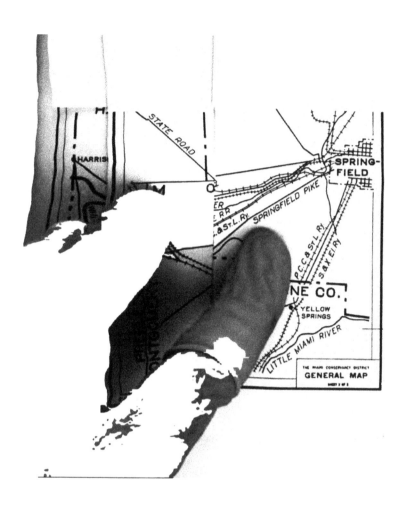

HARRIS

STATE ROAD

SPRING-
FIELD

SPRINGFIELD PIKE

ER
RR

C.&St.L.Ry.

P.C.C.&St.L.Ry.

S.&X.El'Ry.

NE CO.

YELLOW
SPRINGS

LITTLE MIAMI RIVER

THE MIAMI CONSERVANCY DISTRICT

GENERAL MAP

.

STATE OF OHIO

THE MIAMI CONSERVANCY DISTRICT

Hydraulics of the Miami Flood Control Project

BY

SHERMAN M. WOODWARD

M. Am. Soc. C. E; M. Am. Soc. M. E.

Consulting Engineer of the District; Professor of Mechanics and Hydraul'cs.
State University of Iowa

TECHNICAL REPORTS
Part VII

DAYTON, OHIO
1920

THE MIAMI CONSERVANCY DISTRICT

DAYTON, OHIO

———

PREFATORY NOTE

This volume is the seventh of a series of Technical Reports issued in connection with the planning and execution of the notable system of flood protection works now being built by the Miami Conservancy District.

The Miami Valley, which forms a part of the large interior plain of the central United States and comprises about 4000 square miles of gently rolling topography in southwestern Ohio, is one of the leading industrial centers of the country. Out of the great flood of March 1913, which destroyed in this valley alone over 360 lives and probably more than $100,000,000 worth of property, there resulted an energetic movement to prevent a recurrence of such a disaster, by protecting the entire valley by one comprehensive project. The Miami Conservancy District, established in June 1915, under the newly enacted Conservancy Act of Ohio, became the agency for securing this protection. On account of the size and character of the undertaking, the plans of the District have been developed with more than usual care.

A Report of the Chief Engineer, submitting a plan for the protection of the District from flood damage, was printed March 1916 in three volumes of about 200 pages each. After various slight modifications, this Report was adopted by the board of directors as the Official Plan of the District, and was republished in May 1916 under the latter title. This plan for flood protection contemplates the building of five earth dams across the valleys of the Miami River and its tributaries to form retarding basins, and the improvement of several miles of river channel within the towns and cities of the valley. It is estimated that the dams will contain nearly 8,500,000 cubic yards of earth; that their outlet structures will contain nearly 200,000 cubic yards of concrete; that the river channel improvements will involve the excavation of nearly 5,000,000 cubic yards; and that the whole project will cost about $35,000,000.

At the time of publication of this report the flood control works are about two-thirds completed.

In order to plan the project intelligently, many thorough investigations and researches had to be carried out, the results of which have proved of great value to the District and will also, it is believed, be of widespread use to the whole engineering profession. To make the

5

results of these studies available to the residents of the State and to the technical world at large, the District is publishing a series of Technical Reports containing all data of permanent value relating to the history, investigations, design, and construction of the flood prevention works.

The following list shows the titles of the reports published to date and the prices at which they may be purchased.

Part I.—The Miami Valley and the 1913 flood, by A. E. Morgan, 1917; 125 pages, 44 illustrations; 50 cents.

Part II.—History of the Miami flood control project, by C. A. Bock, 1918; 196 pages, 41 illustrations; 50 cents.

Part III.—Theory of the hydraulic jump and backwater curves. by S. M. Woodward. Experimental investigation of the hydraulic jump as a means of dissipating energy, by R. M. Riegel and J. C. Beebe, 1917; 111 pages, 88 illustrations; 50 cents.

Part IV.—Calculation of flow in open channels, by I. E. Houk, 1918; 283 pages, 79 illustrations; 75 cents.

Part V.—Storm rainfall of eastern United States, by the engineering staff of the District, 1917; 310 pages, 114 illustrations; 75 cents.

Part VI.—Contract forms and specifications, by the engineering staff of the District, 1918; 192 pages, 3 folding plates, and index; 50 cents.

Atlas of selected contract and information drawings to accompany Part VI; 139 plates, 11 by 15 inches; $1.50.

Part VII.—Hydraulics of the Miami flood control project, by S. M. Woodward, 1920; 338 pages, 126 illustrations; $1.00.

The following parts of the Technical Reports are in the course of preparation:

Rainfall and runoff in the Miami Valley.

Laws relating to flood control.

Structural design, construction plant and methods.

Methods of appraising benefits to property.

Orders for Technical Reports should be sent to: The Miami Conservancy District, Dayton, Ohio.

ARTHUR E. MORGAN,
Chief Engineer.

Dayton, Ohio,
July 1, 1920.

CONTENTS

PAGE

Officers of the Miami Conservancy District 4
Prefatory Note 5
Contents 7
List of Illustrations 10
List of Tables 14

CHAPTER I. INTRODUCTION

General 17
Magnitude of problem 17
Scope of this Report 19
Acknowledgments 21

CHAPTER II. THE FLOOD PROBLEM

The flood of March 1913 in the Miami Valley 25
Comparison with great floods elsewhere 34
Other floods in the Miami Valley 36
Great rainstorms in the United States 41
Great storms of the past as applied to Miami Valley conditions 42
Maximum possible flood in the Miami Valley 44
Flood discharge 40 per cent greater than that of March 1913 chosen as basis
  for design 45

CHAPTER III. THE RETARDING BASIN PLAN

Various plans investigated 47
Cost of retarding basin plan compared with that of other plans 53
Description of Miami Valley 55
Principal features of plan adopted 58
General description of dams and appurtenances 58
General description of local protection works 67
Selecting the retarding basin sites 84

CHAPTER IV. OPERATION OF RETARDING BASINS

Runoff from maximum storm 86
Size of outlet conduits 89
Capacities of retarding basins 91
Spillways 94
Protection afforded by basins 96
Behavior of basins during a flood like that of March 1913 101
Effect on farm lands in basins 105
Backwater action in retarding basins 109

CHAPTER V. CAPACITIES OF BASINS

Characteristics of curves for area and capacity 112
Exponential curves for area and capacity 113

7

PAGE

CHAPTER VI. FLOW THROUGH OUTLET CONDUITS

General design of outlet conduits 122
Characteristics of individual outlets 123
Superiority of outlet conduits to weirs 126
Effect of tailwater level on conduit outflow 132
Selection of roughness factor n 133
Formulas for outflow, conduits flowing full 135
Formulas for outflow, conduits partly full 147
Formulas for outflow, temporary outlets 152

CHAPTER VII. ROUTING FLOODS THROUGH RETARDING BASINS

Establishing inflow curves at the damsites 156
General principles of routing floods 170
Step methods for routing floods 171
Integration methods for routing floods 185

CHAPTER VIII. FACTORS AFFECTING THE HEIGHT OF DAMS AND SIZE OF OUTLET CONDUITS

Various controlling factors 201
Combining the controlling factors in retarding basin design 204
Study of retarding basin operation 205

CHAPTER IX. DETERMINATION OF CAPACITY OF SPILLWAYS

General principles 221
Types of spillway 222
Step method for solving spillway problems 226
Integration formulas for spillway action 226
Results of studies of spillway action for Englewood and Taylorsville retarding basins. First Method 237
Results of studies of spillway action. Second Method 242
Step method check on approximate spillway computations 242
Action of spillways as finally designed 244
Action of notched spillways 244

CHAPTER X. BEHAVIOR OF RETARDING BASINS DURING LOCALIZED CLOUDBURSTS

Selection of storms 247
Effect on retarding basins 248
Probability of occurrence 255

CHAPTER XI. HYDRAULICS OF CHANNEL IMPROVEMENTS

Channel capacities required with retarding basin control 259
Principles of channel improvement design 261
Backwater curves 280

CHAPTER XII. BALANCING THE FLOOD PROTECTION SYSTEM

Principles and data 286
Graphical balancing of the Miami Conservancy District system 287

PAGE

Lockington cost-outflow curve 287
Combining Lockington basin with Piqua and Troy improvements 289
Combining the Lockington and Taylorsville basins 291
The Lockington-Piqua-Troy-Taylorsville or L. P. T. T. System 292
Combining the improvements above Dayton 295
Combining the channel improvement from Dayton to Franklin 301
Combining all improvements above Twin Creek 301
The relation between the discharges at Dayton, Germantown, and Hamilton.. 302
The effect of the Germantown basin 303
Introducing the Hamilton and Middletown channel improvements 304
Determining the size of the various improvements of the cheapest combination. 306
The elimination of undesirable basins 307

CHAPTER XIII. ALTERNATIVE FLOOD PROTECTION PLANS

City protection by local improvement 309
Protection of entire valley by channel improvement 314
Protection by numerous small retarding basins 319

APPENDIX

Tables 324

INDEX

LIST OF ILLUSTRATIONS

FIGURE   PAGE

1 Rainfall map of March 1913 storm 25

2 Daily rainfall during March 1913 storm, over Miami Valley 28

3 Cumulated rainfall during March 1913 storm, over Miami Valley 29

4 Rainfall and runoff of March 1913 storm for watershed above Dayton, Ohio 30

5 Maximum rates of runoff in Miami Valley during flood of March 1913.. 33

6 Maximum rates of flood discharge for American and foreign streams ... 35

7 Maximum Miami River stages of 11 feet and over at Dayton, Ohio.... 39

8 Frequency curves for Miami River floods 40

9 Location of retarding basins in Miami Valley 54

10 Idealized geological section across Miami Valley 55

11 Condensed profiles of Mad River, Buck Creek, and Twin Creek 56

12 Condensed profiles of Miami River, Stillwater River, Loramie Creek, and Wolf Creek ....Facing 56

13 Cross section through hydraulic fill earth dam forming retarding basin.. 60

14 Longitudinal section on axis of Germantown dam 60

15 General layout of Huffman Dam 61

16 General layout of Germantown Dam 63

17 Map of Piqua, Ohio, showing local flood protection works 70

18 Map of Troy, Ohio, showing local flood protection works 71

19 Map of Tippecanoe, Ohio, showing local protection works 72

20 Map of Dayton, Ohio, showing local flood protection works 74

21 Map of West Carrollton, Ohio, showing local flood protection works 76

22 Map of Miamisburg, Ohio, showing local flood protection works 77

23 Map of Franklin, Ohio, showing local flood protection works 78

24 Map of Middletown, Ohio, showing local flood protection works 80

25 Map of Hamilton, Ohio, showing local flood protection works 82

26 Maximum flood runoff curves adopted for the Official Plan Flood 88

27 Relation between inflow, outflow, and retarding basin capacity when spillway does not overflow 91

28 Relation between inflow, outflow, and retarding basin capacity when spillway overflows 93

29 Intensities and durations of storm runoff which would fill the retarding basins to their spillway levels 96

30 Flood discharge at Dayton for various intensities and durations of storm runoff 98

31 Inflow and outflow hydrographs of Twin Creek at Germantown retarding basin for a flood like that of March 1913 101

32 Inflow and outflow hydrographs of Stillwater River at Englewood retarding basin for a flood like that of March 1913 102

33 Inflow and outflow hydrographs of Loramie Creek at Lockington retarding basin for a flood like that of March 1913 103

34 Inflow and outflow hydrographs of Miami River at Taylorsville retarding basin for a flood like that of March 1913 104

35 Inflow and outflow hydrographs of Mad River at Huffman retarding basin for a flood like that of March 1913 105

FIGURE   PAGE

36 Seasonal occurrence of flooding of retarding basin lands 107
37 Area-elevation curves for the five retarding basins 110
38 Storage-elevation curves for the five retarding basins 111
39 Topographic map of Germantown retarding basin 114
40 Topographic map of Englewood retarding basin 115
41 Topographic map of Lockington retarding basin 116
42 Topographic map of Taylorsville retarding basin 117
43 Topographic map of Huffman retarding basin 118
44 Relation between depths and areas submerged in Englewood basin 120
45 Section through Germantown dam outlet conduits 124
46 Combined outlet and spillway structure at Lockington dam 125
47 Various possible types of retarding basin outlets 127
48 Discharge curves characteristic of various types of retarding basin outlets. 128
49 Comparison of outflow curves for various types of retarding basin outlets. 130
50 Duration of flooding and acreage submerged in retarding basin for
    various types of outlet 131
51 Relation between critical depths of water in Taylorsville outlet conduits
    and elevation of tailwater 132
52 Drop-off curve of water surface at end of flume 137
53 Distribution of internal pressure in drop-off at section Y, figure 52 139
54 Type of cross section of outlet conduit used in preliminary computations. 146
55 Plan and profile of rectangular channel with constriction 147
56 Channel cross section of irregular form 149
57 Typical low-head outflow conditions at outlet conduit 150
58 Typical outflow conditions at Germantown conduit exit with floor de-
    pressed temporarily 153
59 Typical outflow conditions through temporary outlet channel at Tay-
    lorsville dam 154
60 Sliding coefficient diagram for determining variable ratio of runoff from
    drainage area above Englewood dam 161
61 Graphical determination of crest stages in Miami River flood hydro-
    graphs 167
62 Rectilinear and curvilinear hydrographs of February 1909 flood at
    Englewood dam site 168
63 Relation between inflow and outflow at a retarding basin for any time
    interval during filling period 172
64 Relation between mass curves of inflow and outflow, at a retarding basin,
    for any time interval during filling period 173
65 Storage-elevation and outflow-elevation curves for Germantown retard-
    ing basin 174
66 Storage-outflow curve for Germantown retarding basin 176
67 Inflow and outflow hydrographs of Twin Creek at Germantown retarding
    basin for a flood like that of March 1897 177
68 Storage-outflow curve for Lockington retarding basin with temporary
    outlet 180
69 Modified form of storage-outflow curve for Lockington retarding basin,
    with temporary outlet 182
70 Inflow and outflow hydrographs of Loramie Creek at Lockington retard-
    ing basin, assuming a flood like that of March 1913, and the tem-
    porary outlet in operation 183

FIGURE   PAGE
71 Inflow and outflow hydrographs for flood of uniform intensity, illus-
      trating notation used in five-sixths rule 187
72 Example illustrating use of five-sixths rule in replacing a flood of irreg-
      ular hydrograph by an equivalent uniform flood 195
73 Examples of parabolic outflow curves for floods of uniform intensity.... 197
74, 75, and 76 Examples of equivalent uniform floods for three types of
      triangular flood waves 199
77 Inflow and outflow hydrographs of Stillwater River at Englewood retard-
      ing basin, for a flood like that of October 1910 206
78 Inflow and outflow hydrographs of Mad River at Huffman retarding
      basin, for a flood like that of October 1910 207
79 Areas flooded and time of submergence in the five retarding basins for a
      flood like that of March 1913 208
80 Duration and depth of flooding of lands in the Germantown retarding
      basin 209
81 Duration and depth of flooding of lands in the Englewood retarding
      basin 210
82 Duration and depth of flooding of lands in the Lockington retarding
      basin 211
83 Duration and depth of flooding of lands in the Taylorsville retarding
      basin 212
84 Duration and depth of flooding of lands in the Huffman retarding basin. 213
85 Frequency of flooding in Englewood retarding basin 215
86 Frequency of flooding in Huffman retarding basin 217
87 Recovery of capacity of the Englewood and Huffman retarding basins
      after a flood like that of March 1913 219
88 Probability of flooding of lands at different elevations in the Englewood
      and Huffman retarding basins 220
89 Plan and section of spillway at Englewood dam 224
90 Plan and sections of spillway near Germantown dam 225
91 Profile of spillway crest 226
92 Relation between water surface elevations and outflow at Lockington re-
      tarding basin showing effect of spillway overflow 227
93 Inflow and outflow hydrographs at a retarding basin with spillway
      operating 228
94 Hydrograph of retarding basin outflow with spillway operating, and as-
      suming uniform inflow 229
95 Graphical representation of notation used in integration formulas for
      spillway action 230
96 Curves for determining values of M and N in formulas for spillway dis-
      charge 233
97 Depths of spillway discharge at Englewood dam for various intensities
      and durations of storm runoff 238
98 Depths of spillway discharge at Taylorsville dam for various intensities
      and durations of storm runoff 239
99 Maximum rates of outflow at Germantown dam for various total amounts
      and durations of storm inflow 249
100 Maximum rates of outflow at Englewood dam for various total amounts
      and durations of storm inflow 250
101 Maximum rates of outflow at Lockington dam for various total amounts
      and durations of storm inflow 251

FIGURE   PAGE
102 Maximum rates of outflow at Taylorsville dam for various total amounts
        and durations of storm inflow 252
103 Maximum rates of outflow at Huffman dam for various total amounts and
        durations of storm inflow 253
104 Channel cross section illustrating conditions in bend of river 265
105 Profiles of Miami River through Dayton, showing water surface eleva-
        tions along each bank during the floods of February 1, 1916 and
        April 21, 1920 268
106 Cross section of conduit, Umatilla Project, Oregon 270
107 Alignment of Umatilla Project conduit used for determining losses
        through curvature 271
108 Type of channel cross section adopted by Miami Conservancy District in
        bends of rivers 273
109 Plan of bridge pier illustrating standard shape adopted by Miami Con-
        servancy District 276
110 Conditions of flow where bridge piers are built at an angle with the
        current 277
111 Backwater curves for Miami River at Hamilton, Ohio 282
112 Relation between cost and maximum discharge capacity of Lockington
        retarding basin, for a 10-inch storm runoff in 3 days 288
113 Determining the most economical combinations of Lockington retarding
        basin with channel improvements at Piqua and Troy, for various maxi-
        mum outlet capacities at the dam 290
114 Cost of Taylorsville retarding basin for various maximum outlet capaci-
        ties at Lockington dam 291
115 Determining the most economical combination of Lockington and Tay-
        lorsville retarding basins with channel improvements at Piqua and
        Troy, for various maximum outlet capacities at the Lockington and
        Taylorsville dams 293
116 Relation between minimum cost of L. P. T. T. system and maximum outlet
        capacity at Taylorsville dam 294
117 Relation between cost and maximum discharge capacities of the Huffman
        and Englewood retarding basins for a 10-inch storm runoff in 3 days. 295
118 Determining the most economical combination of retarding basins and
        channel improvements above Dayton 297
119 Determining the minimum cost of all improvements, except Germantown
        basin, above Middletown, for various maximum discharges through
        Dayton 300
120 Determining the most economical combination of Germantown retarding
        basin with all other improvements above Middletown 302
121 Determining the most economical combination of the five retarding basins
        and all channel improvements except that at Middletown, for various
        maximum discharges at Hamilton 305
122, 123, and 124 Four early flood protection schemes for Dayton 310
125 Hydrograph of March 1913 flood discharge of Miami River at Hamilton
        and computed hydrograph assuming valley storage eliminated 318
126 Map of Miami River watershed above Dayton showing suggested system
        of 45 small retarding basins 321

LIST OF TABLES

TABLE   PAGE

1 Comparative intensities of 12 great storms for 1-day to 5-day periods of maximum rainfall 41

2 List of existing flood control reservoirs 49

3 Ratios of maximum discharge per square mile for Miami River tributaries, as derived from formulas 159

4 Ratios of maximum discharge per square mile for Miami River tributaries, as derived from flood measurements 160

5 Example of trial-and-error step method of routing floods. March 1897 flood routed through Germantown retarding basin 175

6 Example of direct step method of routing floods, using mass curve data. March 1913 flood routed through Lockington retarding basin, with temporary outlet in operation 181

7 Example of direct step method (condensed form) of routing floods using rating curve data. March 1913 flood routed through Lockington retarding basin with temporary outlet in operation 185

8 Ratio of average outflow to maximum outflow for floods of uniform intensity 190

9 Summary of computations of retarding basin sizes for the Official Plan flood with the aid of the Five-Sixths rule 203

10 Method of calculating spillway capacity required at Englewood dam 240

11 Method of calculating spillway capacity required at Taylorsville dam.... 241

12 Maximum storm rainfall recorded in the northern half of the Mississippi Valley for 1, 2, and 3-day periods averaged over areas comparable in size with those draining into the retarding basins of the Miami Valley. 248

13 Actual rainfall during March 1913 storm over specific drainage areas in the Miami Valley, compared with the maxima observed elsewhere during that storm over areas of like extent 257

14 Computations of backwater curve for 200,000 second foot flood at Hamilton with velocity head changes considered 283

15 Computations of backwater curve for 200,000 second foot flood at Hamilton with velocity head changes neglected 284

16 Summary of cost of local protection for the cities and towns in the Miami Valley from Piqua to Hamilton (without retarding basins) against floods as large as that of March 1913 311

17 Summary of cost of a continuous channel improvement of the Miami River from Piqua to Hamilton (without retarding basins) to take care of floods as large as that of March 1913 316

IN THE APPENDIX

TABLE   PAGE

I Drainage areas of streams in the Miami Valley 324

II Slopes of streams in the Miami Valley 325

III Channel capacities in 1913 in the towns and cities of the Miami Valley compared with the 1913 flood discharge 326

IV   Channel capacities outside of towns and cities in the Miami Valley compared with the 1913 flood discharge 326

V   Valley storage during crest stages of 1913 flood 327

VI   Population and elevation above sea level of towns and cities in the Miami Valley 327

VII   Data pertaining to dams and appurtenances 328

VIII   Conditions at dams for a flood equal to that of March 1898........ 329

IX   Conditions at dams for a flood equal to that of March 1913 329

X   Conditions at dams for an Official Plan Flood 330

XI   Conditions at dams for assumed flood runoff of 14 inches in 3 days.. 330

XII   Flooding conditions in Germantown retarding basin during the period 1893 to 1917 assuming dam to have been in existence 331

XIII   Flooding conditions in Englewood retarding basin during the period 1893 to 1917 assuming dam to have been in existence 332

XIV   Flooding conditions in Lockington retarding basin during the period 1893 to 1917 assuming dam to have been in existence 333

XV   Flooding conditions in Taylorsville retarding basin during the period 1893 to 1917 assuming dam to have been in existence 334

XVI   Flooding conditions in Huffman retarding basin during the period 1893 to 1917 assuming dam to have been in existence 335

XVII   Channel capacities required at Dayton and towns below for an Official Plan flood assuming retarding basins in operation 335

XVIII   Summary of observations on scour in Miami Valley 336

XIX   Convenient conversion factors 337

XX   Special conversion factors for drainage areas in the Miami Valley .. 338

CHAPTER I.—INTRODUCTION

GENERAL

As a result of rainfall almost unprecedented for this region all portions of southern Indiana and Ohio, together with adjacent areas in surrounding states, suffered from phenomenal floods on March 25, 1913. In the Miami Valley this flood submerged much of the residence portion and the greater part of the business and manufacturing area of the cities of Piqua, Troy, Dayton, Franklin, Miamisburg, and Hamilton. The considerable loss of life and the immense damage to property immediately raised the question whether it would be feasible to take steps to prevent any possible recurrence of such a disaster, and the different interested communities gradually drew together to seek relief.

The Miami is a river of steep slope, averaging 3 feet to the mile, draining an area of 4000 square miles of rolling country, having a natural channel with an average capacity of about 10,000 second feet just above Dayton and in its lower portion a capacity of not over 50,000 second feet. Being situated in a region of occasional heavy rainfalls, the river flow is subject to moderate annual floods and to rare extreme floods. The greatest of these during the 125 years since the valley has been settled had a maximum flood flow down the valley of about 10 times the natural capacity of the channel. The flood plain of the river is occupied by numerous vigorous growing, industrial and commercial cities, and several railroads lie in the valley flood plain. With these conditions, the engineering problem presented was to devise a practicable method of protection which should not hamper the activities and growth of the communities being protected, and which should not put an undue financial burden upon the valley.

MAGNITUDE OF PROBLEM

As the result of some three years intensive study of this problem by a large engineering staff, the plan for flood protection explained in this volume was gradually evolved. Protection against floods has been practiced for centuries in various countries of the old world, and during recent years in America; but it is probable that nowhere before has there ever been initiated on as large a scale so complete, elaborate, safe. and intensive a system of flood protection; though with a few

17

exceptions, every particular element of the system has been used for a long time in some other place. The distinguishing features of this problem were its unusual magnitude, the difficulties due to the presence of numerous cities and railroads in the flooded area, the great value of the property affected, and the small capacity of the natural stream channels as compared with the maximum flood runoff to be provided for.

From the first it was recognized that the importance of the problem justified the utmost care and effort in its solution. The novelty of the situation and the magnitude of the project, emphasized the lack of precedents, so it became necessary to undertake many investigations of a somewhat pioneering type. The plan as finally perfected was a gradual evolution. Time and again details of the protective system as at first projected were changed as additional information pointed the way to desirable modifications. Hence many studies and tentative plans had to be given consideration before the best form for various features could be determined.

In the light of the experience gained it is perfectly obvious now that much time and labor could have been saved in preparing the plans, if the best methods of procedure had been known from the beginning. In many cases methods had to be built up starting from elementary principles. Methods that ought to be followed often could not be foreseen until much study had been given to a problem, and sometimes many attempts were made before the best solution was reached. This led, of course, to much loss of time. Such loss cannot be called waste, because no precedent existed as a guide by which the loss could have been avoided. The magnitude of the work and the issue of public security involved justified the large amount of time and engineering effort given to working out orderly methods of approaching and handling the problem. Fortunately, no delay ever resulted in the progress of the project on account of the time consumed in the engineering studies. The plans were always ready for each step in the procedure as soon as the legal and legislative obstacles had been overcome.

Commonly, in engineering projects involving novel features, a large part of the engineers' problem consists in analyzing and determining the economic or financial aspects of the project, in order to show whether or not the proposed expenditure is justifiable. For the Miami Valley this part of the problem was unusually simple. A careful canvass of the situation by obtaining the opinions of many of the men most competent to judge showed conclusively that the depreciation in real estate values in the larger cities of the valley on account of the 1913 flood amounted to at least $70,000,000, which amount would be recoverable if flood protection could be secured. The people were determined to have protection if possible, so it remained for the engineers only to find the best plan and to show that it could be executed at a practicable expenditure.

SCOPE OF THIS REPORT

This volume includes only a small part of the total work done in the preparation of the flood protection plans. The treatment here given is intended to embrace: first, the considerations determining the decision as to the type of protection system to be adopted; and second, the general matters of a hydraulic nature necessary to be used in the design of structures. But no matters referring solely to structural design are included.

In preparing plans the engineers of the Miami Conservancy District sought advice and information from very many sources, including private engineers and those connected with public and private organizations. But for the information received from these many sources the preparation of the plans would have been much more difficult. In view of these contributions from the engineering profession the District is but meeting its moral obligations when it reciprocates by making public any additions it may have made to the knowledge of flood prevention methods, so that in the solution of similar problems the pioneer work of the Conservancy engineers may not have to be repeated. Moreover, as the years pass and the engineers who prepared the plans are no longer at hand to interpret them, it will be important that a clear, orderly statement of the theory on which the work is designed shall be available to those who become responsible for the maintenance and operation of the system. With these objects in mind the material has been arranged in what seems to be the best order for clear and logical presentation; the order has no reference to the actual chronological order of development in the evolution of the Miami Conservancy District plans, and in many cases departs entirely from that order.

A considerable amount of the general theoretical development here given, so far as known, has never before been worked out or published. A considerable portion has been presented previously in more or less similar form in scattered and fugitive publications; frequently, these were not discovered until after their substance had been derived anew and applied on this project. Other parts consist of matters more or less generally known, and included here so as to make this a complete and comprehensive manual for practicing engineers, in an effort to make this volume of the utmost value to engineers engaged upon similar projects. If a treatise of the nature of this and the companion volumes had been available at the beginning of the project, important portions of the work in preparation of plans could probably have been done with one quarter the labor actually expended, and with an aggregate saving equal at least to the lifetime's work of a thoroughly competent engineer.

The order of presentation is arranged with the idea of showing:

First, the general nature of the problem in the Miami Valley. This involves discussion of the physical features of the valley, of the extent of industrial development, and of the magnitude of past and probable future floods.

Second, a general comparison of the various flood protection plans considered, and the reasons for the adoption of the plan determined upon.

Third, a more detailed description of the adopted plan, the way it will function in practice when completed, and its effect upon the activities of the valley.

Fourth, a thorough, detailed, technical discussion of the principles, theories, and methods of making the voluminous hydraulic calculations required in developing the details of the plan. These embrace the capacities of the retarding basins, the flow through outlet conduits, the methods of predicting the effect of the retarding basin upon definite floods, the determination of the size of dams, outlets, and spillways, the principles governing the design of channel improvements, the proper coordination of the different retarding basins and channel improvements, and, finally, the more pertinent details of the various plans that were studied, but rejected, in favor of the combined retarding basin and channel improvement plan.

This volume does not duplicate any other Technical Report but contains numerous references to the other Reports already issued. Occasional quotations are made from the other Reports in order to render this one by itself as complete and useful as possible.

Some portions of this volume are far from easy reading, and perhaps no one will care to read it entirely through. Parts of the text will be useful chiefly as reference for thorough treatment of particular problems. In order to make it as convenient as possible for such use, the whole work is written with the endeavor to have each portion as complete and intelligible by itself as it could be made. To facilitate reference to isolated topics, cross references are freely used in the text, and an analytical and topical index is supplied at the end.

The material in this volume is but a minute fraction of the data and computations accumulated in the files of the District in the course of the preparation of the flood protection plans. In preparing the material for publication considerable effort has been made to condense it to the very minimum consistent with clearness, and to exclude every item whose inclusion could not be justified by the necessity of logical completeness of explanation, or by its applicability to similar situations elsewhere. In this process some details of method have been hit upon and are described which are obviously better than those that were actually used in the preparation of the District's plans. This but illustrates the fact that we never reach perfection in the development of engineering methods.

ACKNOWLEDGMENTS

Probably fifty or more members of the engineering staff of the District have worked upon the material drawn upon for this report. This material has been so rearranged, revised, and condensed, that it is now impossible to assign with any approach to accuracy the relative amounts of credit that ought to be given to the work of different engineers. The engineering work of the District has been distinguished, furthermore, by a most remarkable degree of hearty cooperation among the various members of the staff, leading to an exchange of mutual suggestions and help such as to preclude, oftentimes, any precise determination as to limits of originality of conception.

The following indicates in a general way the chief lines of work of those most closely connected with the preparation of the general plans. O. N. Floyd assisted materially in the determination of the type of plans to be adopted, and had charge largely of considering the effect of the plan upon existing railroads. B. M. Jones made many of the studies relating to spillway calculations. A. B. Mayhew, who later lost his life in the service of the District, made many studies of coordination of the various parts of the flood protection system. H. S. R. McCurdy made investigations of structural materials and methods of construction applicable to the dams. J. H. Kimball had charge of the final development of plans for channel improvement. K. C. Grant made most of the detailed studies of retarding basin operation. Walter M. Smith, as designing engineer, had charge of the structural design. D. W. Mead, consulting engineer of the District, acted for a time as chief engineer during an illness of Mr. Morgan and contributed largely to settling the general outlines of the adopted plan. G. H. Matthes assisted on numerous branches of the hydraulic calculations and was largely instrumental in preparing this volume for publication. E. W. Lane assisted materially in the preparation of the manuscript and illustrations and in insuring numerical accuracy. H. A. Thomas, professor of civil engineering at Rose Polytechnic Institute, gave several months most valuable assistance on the manuscript of this volume. S. M. Woodward outlined the volume, wrote or rewrote a large part of the text, and throughout has been responsible for its final form and content. The work of the District was at all times under the supervision of A. E. Morgan, chief engineer.

CHAPTER II.—THE FLOOD PROBLEM

There could be no assurance either of security or economy in planning protection at great cost against floods, the sizes of which are mere conjecture. One of the essentials, therefore, in the intelligent planning of any system of flood control is to ascertain the magnitude and occurrence of all great floods in the past, and the likelihood of their being exceeded in the future.

Many difficulties usually stand in the way of obtaining full data relating to past floods, particularly the earlier ones, and such knowledge as it may be possible to gather on this subject can rarely be made either concise or complete. The subject is so very important, however, that it is deserving of exhaustive study in any attempt made to arrive at conclusions. The more thorough such a study is made, the more nearly can the protection system based thereon be made to fit the given conditions.

No method is known of predicting with certainty the time or size of any floods that may take place in the future. Their occurrence and magnitude are as uncertain as the meteorological phenomena and contributing factors which cause them. A few aspects of the flood problem are, however, capable of being formulated into fundamental propositions, and these can be made to serve as the basis for further study. For instance, there is abundant proof that small floods occur more frequently than large ones; also, that the probability of floods is greatest at certain seasons of the year, which, in the case of the Miami River, are the winter and early spring. It may be shown that more or less definite relations exist between heavy rainfall, the rate of its run-off from the land, and the flood discharges in the streams. Starting with these elementary considerations, and developing these in the light of all data that may be applicable to the given stream, much can be accomplished towards gaining an idea as to what are likely to be the maximum local flood conditions. In this, it is of great help to make comparison with flood records obtained elsewhere, especially those covering long periods of time and under similar climatic conditions. Records extending over many centuries are available for some foreign rivers, notably the Danube, the Tiber, and the Seine. In the United States, flood records covering 100 years have been compiled for a few streams in the eastern section, including the Ohio, the Delaware, and the Hudson, but for most streams little is known beyond the principal ·happenings of the past 50 years. This is the more unfortunate since great floods are of infrequent occurrence, and records over long periods of time are necessary to establish their frequency.

Since storms are the principal cause of floods, a study of the greatest storms of the past usually is of assistance in estimating the frequency of large floods. In the case of the Miami Valley, this procedure possessed certain distinct advantages, which will doubtless be found to apply to similar investigations elsewhere. These advantages are:

First: Rainfall records in the eastern half of the United States cover longer periods of time than do river stage records of most streams, while many streams have no records at all.

Second: Knowledge of the volume of water which may be precipitated on a watershed materially assists in estimating the amount that the streams of that watershed may be called upon to carry. On streams with drainage areas no larger than that of the Miami River this class of information sometimes is more useful than a mere record of river stages, especially where the latter consists only of readings 24 hours apart, as is usually the case.

Third: Aside from the storms that have caused floods in a drainage area under investigation, many other storms can be taken into account, especially those on neighboring watersheds; in fact any unusual storm rainfall that could logically have taken place over the region in question, whether or not it caused any flood, is of value in estimating the probability of recurrence of such storms. The total number of storms that can be so studied is naturally far in excess of the number of floods for which there are records on any given stream.

In such a study it is well to bear in mind that the greatest storms did not always cause the greatest floods, as intensity of runoff is affected in some measure by the absorptive condition of soil, the character of the topography, and other factors. Only general conclusions can, therefore, be drawn from such a study.

It would be beyond the scope of this Report to review in detail all of the data and sources of information which were utilized by the engineering staff of the Miami Conservancy District in determining the maximum size of flood against which protection should be furnished. Only the more important studies are here described, special stress being laid upon those which contributed most toward shaping the conclusions favoring the plan of protection finally adopted.

THE FLOOD OF MARCH, 1913, IN THE MIAMI VALLEY

The flood of March, 1913, in the Miami Valley was not only the most severe of which there is record in that valley, but as regards dam-

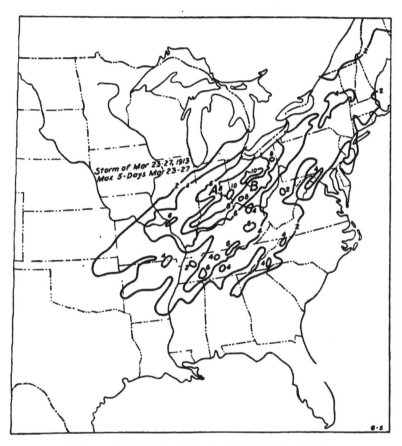

FIG. 1.—RAINFALL MAP OF MARCH 1913 STORM.

The map shows the areas over which 2, 4, 6, 8, and 10 inches of rainfall, respectively, fell during the 5-day period of greatest precipitation, March 23-27, 1913. Maxima of 11.2 inches in 5 days were recorded at the points marked *A* and *B*. The area covered by less than 2 inches of rainfall is not shown.

age was the greatest that has occurred in the eastern half of the United States since the days of first settlement, or when floods first began to attract attention. A description of this flood and the damage which it wrought has been published in a previous report.[*]

It was caused primarily by hard rains which commenced March 23, and continued with scarcely any interruption until the 27th. Contributing factors were a saturated soil, the result of previous rains, and low temperatures which reduced evaporation to insignificant rates. In the popular mind, at that time, there figured another influence, namely, the breaking of the State reservoirs on the headwaters of the Miami and Stillwater Rivers, used for storing water for the operation of the Miami and Erie Canal. Investigation proved this to be false, as neither of the State reservoirs broke. If any thing, these basins reduced in a slight degree the severity of the flood by detaining some of the water.

Rainfall

This storm, while not the most intense, was perhaps the greatest in point of total rainfall that has occurred in the eastern half of the United States in the past quarter of a century. The areas covered by heavy rainfall are shown on the map in figure 1 by means of isohyetal lines, or rainfall contours, drawn at 2-inch intervals. The 2-inch isohyetal extends from Arkansas and Oklahoma to Maine, and includes about 492,000 square miles. The average depth of rainfall over this area was 4.2 inches for the 5-day period. This is equivalent to 32.6 cubic miles of water. The map shows also that the heaviest rainfalls took place in central and western Ohio, including a section of the Miami River valley.

The depths of rainfall that fell day by day over the Miami drainage basin during this storm are shown in figure 2, except for March 27 on which date the rainfall averaged only about a half inch and could not be represented by isohyetal lines. Figure 3 shows the total cumulated rainfall up to 7 p.m. of each day.

It was a fortunate circumstance that in 1913 there were a fair number of well equipped rainfall stations in this part of the United States. Reports from 69 stations were utilized in the preparation of figures 2 and 3. The stations are scattered over an area much larger than shown, some of them being in Indiana and Kentucky.

The amount of the precipitation recorded is shown on the maps by means of figures printed after the name of each station. For the river stations maintained by the Weather Bureau, where the rainfall is measured in the morning, it was necessary to estimate figures to

---

[*] The Miami Valley and the 1913 Flood, by Arthur E. Morgan, Chief Engineer; Technical Reports, Part I, The Miami Conservancy District, Dayton, Ohio, 1917.

correspond with the 7 p.m. readings of the cooperative stations. Wherever this was done such figures are shown enclosed in brackets.

As will be seen in figure 2 the precipitation on March 23 was not very heavy, averaging 1.20 inches in depth on the entire drainage area. It was heaviest in the northern portion. On the 24th heavier rains occurred, averaging 2.20 inches, with the maximum over the headwaters of Twin Creek, further south than on the previous day. The greatest precipitation occurred on the 25th, averaging 4.11 inches, and registering a maximum of 5.61 inches at Bellefontaine, in the northeastern portion. On the morning of this date the river, which had been steadily rising, overtopped the levees in the principal cities, and near the following midnight attained its highest stages. On the 26th an average of 1.62 inches of rain fell, and on the 27th 0.47, the latter being of little consequence as the waters were then everywhere receding rapidly.

Figure 4 shows the rainfall and its resulting runoff for the drainage area above Dayton in terms of inches of depth per day. Since the ordinates of the rainfall curve represent rates of rainfall, and the abscissas represent time, the area under the curve at any point represents the product of the rate of rainfall and its duration, or the total quantity of water which has fallen up to any given moment. The runoff curve portrays in a similar manner the total quantity of water that flowed off.

The rainfall data available is in the form of average rainfall over the watershed for 24-hour periods preceding 7 p.m. of the dates given. As the rate of rainfall throughout the day was not uniform, but became progressively more intense during the first days of the storm and gradually diminished toward the last, the rainfall curve has been constructed in such a way as to represent this variation, keeping the total rainfall up to the end of each 24-hour period the same as the depth actually measured. This method of constructing the rainfall curve represents more nearly the true condition than would the assumption of a uniform rate of rainfall for each day, and portrays the actual conditions as nearly as it is possible to do so with the data available.

Relation of Runoff to Rainfall

To estimate the total runoff from the rainfall of March 23–27, 1913, it is necessary to consider the weather and ground conditions preceding the storm. January, 1913, with a precipitation of over 7 inches was an unusually wet month. February was drier than usual,

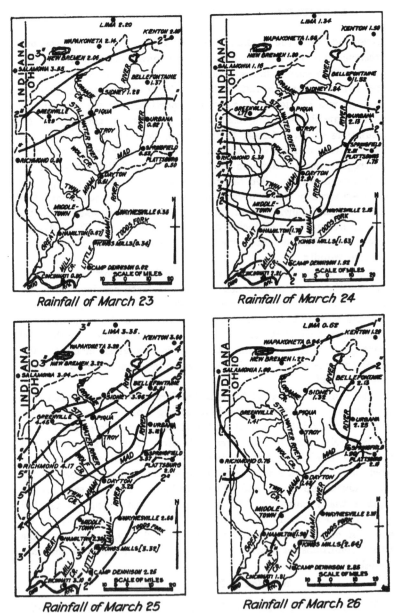

Rainfall of March 23   Rainfall of March 24

Rainfall of March 25   Rainfall of March 26

FIG. 2.—DAILY RAINFALL DURING MARCH 1913 STORM, OVER MIAMI
VALLEY.

Each map shows the rainfall for the 24-hours ending at 7 p.m. of the date
indicated.

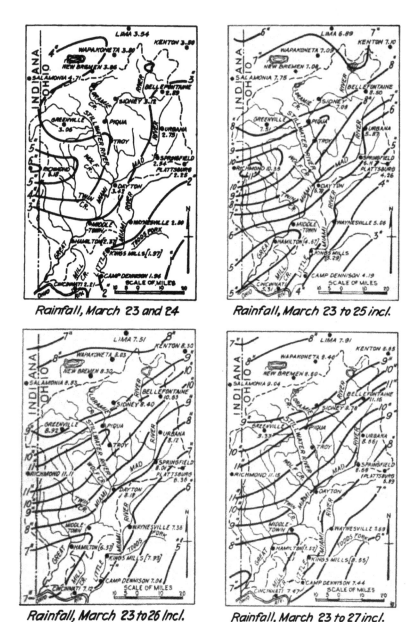

Rainfall, March 23 and 24

Rainfall, March 23 to 25 incl.

Rainfall, March 23 to 26 Incl.

Rainfall, March 23 to 27 incl.

FIG. 3.—CUMU'LATED RAINFALL DURING MARCH 1913 STORM, OVER
MIAMI VALLEY.

Each map shows the total observed rainfall from the beginning of the storm up
to 7 p.m. of the date indicated.

the rainfall totaling an inch less than the normal of 3 inches for that
month, and occurring mostly on the last three days. March was wet
throughout. From the first to the 21st rain was recorded at all of
the observation stations on about 10 days, but these rains were of
moderate intensity. On the 21st it rained nearly half an inch.

It is evident from these conditions that at the beginning of the rain
storm on March 23 the soil was well saturated and the surface wet.

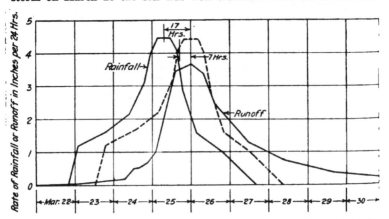

FIG. 4.—RAINFALL AND RUNOFF OF MARCH 1913 STORM FOR WATER-
SHED ABOVE DAYTON, OHIO.

These curves were adapted from the observed data and represent the most probable distribution of rainfall and runoff rates during the storm period.

This would naturally conduce to a high rate and consequently large
total of runoff. The total rainfall of March 23–27 was 9.6 inches.
The 0.4 inch rain which fell on March 21 probably contributed slightly
to the storm runoff which extended over the period March 24–31 and
which totaled 9.3 inches. Assuming that 0.1 inch ran off from that
small rainfall the runoff from the 9.6-inch rainfall of March 23–27
would be 9.2 inches, or a ratio of runoff to rainfall of 95 per cent.

The following table gives the rainfall and runoff above Dayton for
each 24-hour period preceding 7 p.m. of the dates indicated. It is not
extended further because of light rains on March 31 and April 1,
which again increased the runoff rate.

The largest amount of rainfall in 24 hours listed is 4.11 inches.
Ordinarily one would expect the rainfall for the 24-hour period of
greatest precipitation (beginning at some time other than 7 p.m.) to
be much heavier. Measurements of such actual maxima, made from
the diagram in figure 4, for 24-hour, 48-hour, and 72-hour periods

		Rainfall Inches.	Runoff Inches.
March	23	1.20	0.04
"	24	2.20	0.33
"	25	4.11	2.17
	26	1.62	3.27
	27	0.47	1.58
'	28		0.89
'	29		0.49
'	30		0.32
	31		0.22
		9.60	9.31

showed but little difference, however, from the maxima obtained by adding the figures in the foregoing table, exceeding the latter by only about 0.1 inch. Slight differences, also, were found in the case of the runoff maxima. In the following table is given a comparison of rainfall and runoff totals by periods.

	Rainfall, Inches	Runoff, Inches	Percentage of Runoff
24 hours	4.1	3.44	84
48 hours	6.4	5.84	91
72 hours	8.0	7.27	91
96 hours	9.2	8.03	87
120 hours	9.6	8.48	88

It will be noticed that the shape of the runoff curve, figure 4, is quite similar to that of the rainfall, but the runoff lags behind the rainfall. This is due to the interval, between the time when the rain fell, and when its runoff passed through Dayton. The peak rate of runoff came about 17 hours later than the time of maximum rainfall, indicating that the time of collection was about 17 hours.

If we should superimpose the peaks of the rainfall and runoff curves, as shown by the dotted line in figure 4, two important differences in their shape will be noted. Considerable rainfall occurred in the first two days, while in the same period there was comparatively little runoff. Also, when the rainfall ceased rather abruptly, the runoff continued for several days afterwards. The cause of this condition is that the first few inches of rain which fell was retained by the many small depressions and irregularities in the surface of the ground from which it drained gradually, continuing to do so even after the rainfall had ceased.

The curves cross at about 5 p.m. on the 25th, the rate of runoff at that instant having become equal to that of the rainfall. The amount of rainfall which had not run off, or the storage on the surface, in ponds, rivulets, and streams was then a maximum. Since the total rainfall up to that instant is represented by the area inclosed by the rainfall curve up to that point, and the corresponding area under the runoff curve represents the total runoff up to that time, the difference between these areas represents the volume of this storage. This was found to be equal to an average depth of 5.07 inches of rainfall over the drainage basin.

Theoretically, the point of intersection of the curves should be coincident with the instant of maximum rate of runoff. Figure 4 shows the latter to have occurred at 12 midnight of March 25, or 7 hours later, due probably to a certain amount of retardation of the flood crest in reaching Dayton. At the time of maximum rate of runoff the total rainfall which had fallen was 8.02 inches and the amount of the runoff 3.14 inches. The difference, 4.88 inches, must have been spread over the drainage basin. Of this, the volume of water in the main streams above Dayton is estimated at 2.17 inches and the quantity in the small streams and ponds, or on the surface, at 2.71 inches.

Soon after the flood, surveys were made to trace the high water lines, and to determine the maximum rates of flow at various points along the river. The method of determining these maximum rates of discharge is given in detail in another Technical Report.* The results are shown on Figure 5. On this map opposite each point at which a measurement was made, is shown a fraction and quotient. The numerator of the fraction represents the greatest discharge in cubic feet per second that took place at that point. The denominator represents the drainage area in square miles above that point, while the quotient represents the discharge in cubic feet per second per square mile for that drainage area. It will be noticed that these rates of runoff are very high. Especially in the case of the smaller areas these maximum rates continued for but small fractions of a day.

Channel Capacities and Valley Storage

It is needless to say that the channel of the Miami River and those of its principal tributaries were unable to carry the discharge during the high stages of the 1913 flood. The extent to which the channel capacities were exceeded, however, is so extraordinary as to merit more than passing mention here. Besides, channel capacity and excess flood flow are among the determining features which entered into the plan for flood control. Their full significance should be grasped in order that the subsequent development of the plan may be clear.

From actual surveys, and from hydraulic measurements made at

* Calculation of Flow in Open Channels, by Ivan E. Houk, Technical Reports. Part IV, Miami Conservancy District, Dayton, Ohio, 1918.

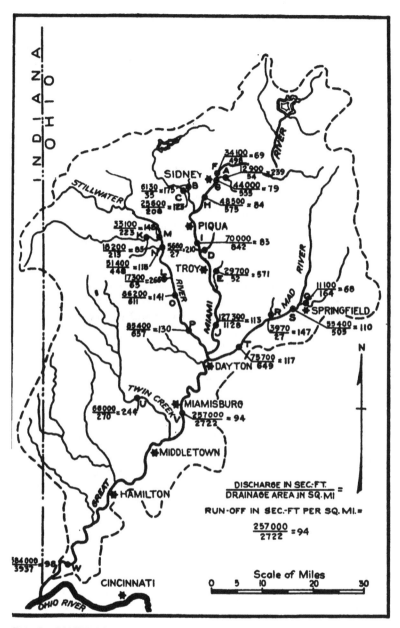

FIG. 5.—MAXIMUM RATES OF RUNOFF IN MIAMI VALLEY DURING
FLOOD OF MARCH 1913.

The figures are written opposite the points at which measurements were
made subsequent to the flood, and give the runoff in second feet per square mile
as indicated in the legend in right-hand lower corner.

3

numerous points along the main river and its tributaries, it was found
that on an average the bank full channel capacity outside the cities
and towns was not quite 10 per cent. of the 1913 crest discharge, while
in the cities, where the channel capacities had been artificially increased by means of levees and other improvements, the channel
capacities average one third of the 1913 crest discharge. In table III
of the Appendix are shown in tabulated form the results obtained at
the various localities.

    Conjointly with this study, computations were made of the volume
of water which the river channels were unable to carry off and which
was temporarily stored on the valley lands during the flood period.
and which for brevity's sake will be referred to here as "valley storage." This was determined from the topographic maps by platting
thereon the highwater contour, taking numerous cross sections of the
valley up to this contour, and computing the volumes of water contained between successive cross sections. In table V of the Appendix
is shown the valley storage for the principal streams, expressed in
terms of acre feet and also in terms of inches of depth over the contributing drainage areas. The latter figure will convey some idea of
the proportion of rainfall represented by the valley storage.

    The figures given involve the assumption that the flood reached
high water mark in all parts of the valley at the same instant. This
was not actually the case, but it was so nearly true that the error introduced by this assumption does not sensibly affect the conclusions
reached.

COMPARISON WITH GREAT FLOODS ELSEWHERE

    Floods in streams other than the Miami also throw light on the
problem of planning a flood control system for the Miami Valley. It
has been the custom to compare great floods on the basis of their
maximum rates of flow in cubic feet per second, or in terms of maximum rate of runoff from the areas drained by them. In these forms
the data is valuable in considering channel capacity, but offers little
information upon which to base retarding basin computations, since
the sufficiency of retarding basins is affected by total volumes of discharge as well as by maximum rates.

    Figure 6 represents graphically the maximum rates of runoff of
great floods in American and foreign rivers. The data used is given
in tabulated form in Technical Report, Part IV, page 66. The diagram is modeled after that of the late Emil Kuichling, and shows the
relation between high rates of runoff and the size of catchment areas.
the ordinates representing the former in terms of cubic feet per sec-

ond per square mile of area, and the abscissas square miles of area drained. American and foreign floods are designated by means of different symbols. The rates for the March, 1913, flood on the Miami

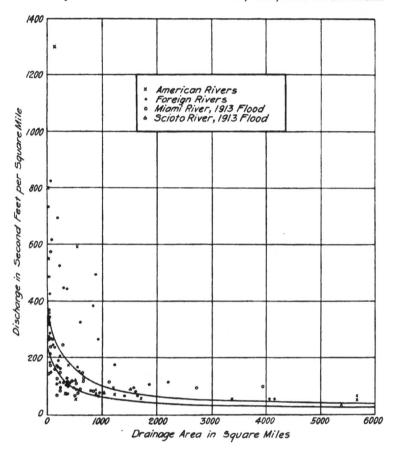

FIG. 6.—MAXIMUM RATES OF FLOOD DISCHARGE FOR AMERICAN
AND FOREIGN STREAMS.

The lower curve represents the size of floods which are likely to occur with more or less frequency, while the upper curve is indicative of those that occur but rarely.

River and tributaries are indicated by means of small circles, and those for the Olentangy and Scioto river floods* of the same date are

* See Report to the Mayor and City Council on Food Protection for the City of Columbus, Ohio, by Alvord and Burdick, Sept. 15, 1913.

indicated by triangles. The diagram shows that the rates for these streams are among the largest on record. For areas of 1000 square miles and more the Miami River rates exceed those of other American streams, but are exceeded occasionally by foreign streams. For areas less than 1000 square miles the Miami River rates present nothing unusual, having been exceeded in numerous instances both at home and abroad. This is as might be expected, since the rainfall during the March, 1913, storm, though very large considering the enormous area which it covered, did not possess the intensities that sometimes occur in case of summer showers and so-called cloudbursts, which cover only restricted areas and cause the greatest flood rates on the smaller streams.

For purposes of comparison the curves used by Kuichling have been placed on the diagram. These curves, it should be remembered. were based on conditions for the most part analogous to those in the Mohawk Valley, New York. The lower curve represents the size of floods which are likely to occur with more or less frequency, while the upper curve is indicative of those that occur but rarely. The run-off conditions in the Miami and Scioto valleys are in some respects comparable with those of the Mohawk, and the position of their flood runoff rates on the diagram therefore emphasizes the magnitude of their floods.

OTHER FLOODS IN THE MIAMI VALLEY

Though the 1913 flood was the greatest recorded in the Miami Valley, a number of other large floods have occurred, each of which showed the inefficiency of the old levee system of protecting the cities. The following has been compiled for the most part from the early histories of Dayton and gives a fairly reliable picture of the more important happenings of that kind.

1805.—The Miami Valley was first settled at Dayton in 1796, and the first flood of record occurred in March, 1805. It was caused by the melting of deep snows and heavy rains on the headwater of the Mad and Miami Rivers. The water rose rapidly, overflowed the banks at the head of Jefferson Street and just west of Wilkinson Street. covering nearly all the town plat, except a small portion bounded by Monument Avenue, Wilkinson, Perry, Third, and Main Streets. No levees had been built at this time, and ordinary freshets covered the low ground east from the present location of the canal. The water was about eight feet deep at the corner of Third and Main Streets. where there was a gully at that time, which has since been filled to a depth of six to seven feet above its original elevation.

The direction and depth of flow has been so changed by the construction of buildings in the main part of Dayton and by the cutting of the forests in the river valley immediately above and below the city that it is not possible to tell whether the flood of 1805 represented a flow as great as or greater than that of 1913. Probably the discharge at Dayton was very much less than in 1913.

1814.—The rivers again reached a dangerous stage in 1814 (no record of the month), and overflowed the low lands to the east of Main Street at the present intersection of First Street and Canal Street. A ferry was maintained for several days, the water being deep enough to swim a horse.

1828.—On January 8, 1828, the rivers were in flood, breaking or passing around all the small levees that had been built up to that time. Considerable damage was caused, small bridges were carried away and a warehouse at the head of Wilkinson Street was destroyed.

1832.—In February 1832 another flood occurred as high as that of 1828. This flood washed out the middle pier of the Bridge Street bridge (now Dayton View bridge).

1847.—The following is quoted from an account of the flood of January 2, 1847, published in Odell's Dayton Directory and Business Advertiser, dated 1850.

"A few minutes after midnight, (Jan. 1, 1847) the insignificant outer levee, that for years had been neglected and weakened by earth being hauled from it to fill up house yards and roads, gave way, near Bridge Street, and the inner levee being insufficient to withstand the torrent suddenly rushing upon it, and rising in a breast two feet above it, soon after fell in. A breach once made, the waters rose rapidly, filling the cellars and covering the ground floors of houses in the vicinity. . . . The levee gave way near the head of Mill Street, about two o'clock (Jan. 2)."

The account states that comparatively little damage was done, and that no lives were lost. It is accompanied by a map of the city, showing the inundated area, from which it appears that nearly the entire section bounded by Water (now Monument), Wilkinson, Fifth, and Jefferson Streets was not covered by water, the principal damage being done west of Perry and east of St. Clair Streets. The account closes with the following significant remark:

"A levee was soon after constructed, which will completely secure the lower parts of town from any such catastrophe for the future."

1866.—Heavy rainstorms starting on September 17, 1866, and lasting for four days, caused a serious flood. The levee gave way east of town and water rushed through the lower parts of the City, backwater from the Miami River through an old ditch aiding in flooding the southern part. At the corner of Third and Jefferson the water was a foot deep on the floor of the Beckel House, and at Third and Main it was four inches deep on the floor of the Phillips House. Railroad communication was cut off and the losses to public and private property were estimated at $250,000.00. After this flood, the river channel was widened by adding a span to the Third Street bridge and to the Bridge St. bridge.

The rainfall causing this flood was recorded at a unmber of points. At Spiceland, Indiana, 8.50 inches fell; at Urbana, Ohio, 7.43; at Westerville, Ohio, 5.98; at Marion, Ohio, 4.54, and at Muncie, Indiana. 4.60 inches. Considerable rain had fallen during the first part of the month and the ground was therefore partly saturated.

The low stage caused by the high rainfall in 1866 as compared with 1913, was doubtless partly due to the fact that in 1866 the swamp lands on the upper watershed were not drained and constituted natural storage reservoirs, and also to the fact that the flood occurred at the end of summer, when the capacity of the soil to hold water naturally would be greatest, and when the runoff would be only half or two-thirds as much as would result from a like storm in March.

1883.—There was a general flood in the Miami on February 3, 4. and 5, the danger being increased by the large amount of ice in the stream. The water did not quite reach the high water mark of 1847. and was two feet below that of 1866. Wolf Creek rose to an unprecedented height. The rivers did not overtop or burst their levees at any point in Dayton, but all of the low lying sections of town were flooded by water which entered through the flood gates of the canal. hydraulics, and sewers, which were frozen to their bearings and could not be shut. The water was twenty-two inches deep at the foot of Ludlow Street. After this flood, the levees were strengthened and extended.

1886.—On May 12, 1886, a heavy rainfall, accompanied by a shower of hail, occurred generally over the drainage area, being especially heavy on Wolf Creek. The southern part of the city was troubled by the collection of storm water. The railway embankment across the Wolf Creek Valley finally gave way, washing a number of houses on the west side from their foundations and flooding a large part of that section. On Fifth Street from Wayne Avenue to Eagle Street, the water covered the streets and the side-walks for nearly the whole distance. Between Wayne and Bainbridge and on Wayne from Fifth south to Burns, the water was about 2½ or 3 feet deep, and the territory bounded by Warren, Buckeye, Chestnut, Wayne, and Park

Streets, was entirely submerged, deep enough in places to swim a horse.

1897 and 1898.—On March 6, 1897, and again on March 23, 1898, there were severe floods. The Miami in both years broke into North Dayton at a sharp bend which has since been cut off, and flooded all the low ground in that section. Riverdale was flooded by backwater through the gates of the old hydraulic race which has since been removed. Storm water collected on the streets in the lower parts of the. city, flooding a number of low spots, to a depth of several feet. The

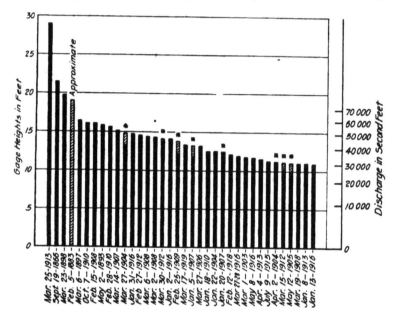

FIG. 7.—MAXIMUM MIAMI RIVER STAGES OF 11 FEET AND OVER AT DAYTON, OHIO.

Based on gage readings at Main Street bridge, Dayton, Ohio. Crest stages marked with stars were estimated.

level of the water in many places reached the top of the then existing levees which were raised about three feet during the next year. The cause of the flood of 1898 was a rainfall of about 4½ inches coming at a time when the ground was thoroughly saturated by previous rains.

Figure 7 shows the flood stages at Dayton of which record has been preserved or which it was possible to compute from reliable high

water marks or other references.   The floods are arranged in order
of magnitude, the earliest flood represented being that of September
1866.   The stars in the diagram indicate those floods for which the
crest stages were determined graphically by the method explained in
chapter VII.

An attempt was made to determine the frequency of Miami River
floods of various magnitudes from the above data.   The results of the
study are shown on figure 8.   From figure 7 were obtained the average number of years intervening between successive recurrences of a
flood which equalled or exceeded a particular size.   Using the average
interval and the given size of flood as coordinates, points were plotted

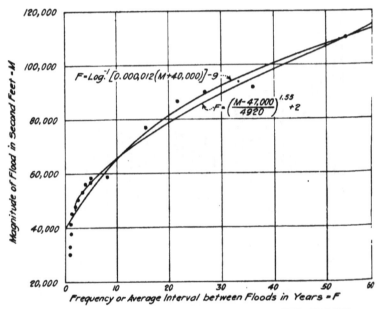

FIG. 8.—FREQUENCY CURVES FOR MIAMI RIVER FLOODS.

Based on floods recorded at Main Street bridge gage, Dayton, Ohio.

as shown on figure 8.   The length of the period covered by accurate
records is so short as to give reliable estimates of the probability of
recurrence of the smaller floods only.   As the size of the flood increases, the reliability of the corresponding frequency decreases, and
for very large floods such as that of 1913 the diagram indicates a wide
range of possibilities.   This is shown by the two curves drawn in the
figure.   These curves are of different mathematical types, but each

was fitted to the plotted points as closely as possible, and its equation was then determined. These equations, using F as the average interval in years between floods, and M as the magnitude of the flood in second feet, are shown on the diagram. According to the equation

$$F = \left\{ \frac{M - 47,000}{4920} \right\}^{1.55} + 2$$

the average interval between floods as large as that of 1913 would be 318 years. According to the other curve

$$F = \text{Log}^{-1}[0.000,012(M + 40,000)] - 9$$

the interval would be about 3000 years. The limitations of this method of determining the probable frequency of recurrence of large floods are evident.

The upward tendencies of the curves should not be construed to mean that there is no limit to the sizes of floods that can occur. Climatic conditions impose limits which, though not mathematically definite, cannot be ignored. To extend the curves beyond these limits would lead to absurdities. A number expressing a probability of occurrence of a flood 40 per cent greater than that of 1913, or even greater, could be ascertained by extending the curves, but this would be no indication that such a flood would be at all possible of occurrence. The studies of storms and floods made by the engineers of the District from all data available to them show that floods exceeding the March 1913 flood by from 15 to 20 per cent are the probable limit of what may be expected in the Miami Valley under present climatic conditions.

GREAT RAINSTORMS IN THE UNITED STATES

The great storm of March 1913 was not a unique occurrence. Other storms comparable in size with it have occurred in the United States, east of the Mississippi River, and a review of such storms made by the engineers of the District gives every indication that a rainfall greater than that of March 1913, over the Miami Valley, is well within the range of possibilities.

All of the large storms that have occurred since 1892, when rainfall stations became sufficiently numerous to give reliable indication of the areas covered by different depths of rain, were studied in detail, as were also a few that occurred earlier than this and for which sufficient data could be obtained.[*] In all, 160 storms were studied, 113

---

[*] Storm Rainfall of Eastern United States, Technical Reports, Part V, Miami Conservancy District, Dayton, Ohio, 1917.

in the southern states and 47 in the northern, all east of the 103d meridian. West of this line rainfall conditions are so different as to throw little light on the problem in the Miami Valley. Of the 160 storms, 17 in the north and 16 in the south were made the subject of special study, a set of isohyetal maps being constructed for each storm, showing the distribution of rainfall, respectively, for the 1, 2. 3, 4, and 5-day periods of maximum precipitation. In addition. curves were platted showing the relation between depth of rainfall and area covered for each of these periods. These studies are the most elaborate of their kind ever undertaken. They are fully described in Part V of the Technical Reports, and the reader is referred to it should he be in search of detailed information as to methods and results. They show conclusively that the March 1913 storm has been equalled or exceeded on areas of 4000 square miles by several of the northern storms, and has been exceeded by practically all of the 16 southern storms.

GREAT STORMS OF THE PAST AS APPLIED TO MIAMI VALLEY CONDITIONS

Of the 33 great storms which were given special study, the 16 in the southern group are not closely applicable to Miami Valley conditions, as the evidence obtained from the studies showed conclusively that these southern storms cannot be duplicated in the Miami Valley. Of the 17 storms in the northern group, 5 were east of the Appalachian Mountains, and could not be duplicated in the Miami Valley on account of the barrier these mountains furnish against the passage of great quantities of water vapor. The remaining 12 storms in the northern group are those which, within the limits of possibility. are considered applicable to the Miami watershed. A list of these giving the date of occurrence and geographical location of center is shown in table 1. The average rainfall for each storm is given for 1-day to 5-day periods of maximum precipitation in inches and also expressed as percentages of that of the March 1913 storm. In each case the area covered is 4000 square miles, this being the approximate size of the Miami River drainage basin.

The table shows that for the maximum rainfall period of one day the storms of August 1903 and June 1905 materially exceeded the storm of March 1913. Both were Iowa summer storms. The first lasted 4 days, but except for the 1-day maximum it was practically the same size as the March 1913 storm over an area of 4000 square miles at the center. The second fell principally in 12 hours, and altogether in 24 hours. For 1-day maximum rainfall those two storms

Table 1.—Comparative Intensities of Twelve Great Storms Applicable to the Miami Valley, for 1-day to 5-day Periods of Maximum Rainfall.

The greatest average depth of rainfall over an area of 4000 square miles at the center of each storm is expressed in inches and also as percentages of the depths for the corresponding rainfall of the storm of March 23-27, 1913.

Date of Storm	Center	Average Maximum Rainfall, by Periods									
		1-Day		2-Day		3-Day		4-Day		5-Day	
		Inches	Per Cent.	Inches	Per Cent.	Inches	Per Cent.	Inches	Per Cent.	Inches	Per Cent.
Dec. 17–20, 1895....	Mo.	5.5	95	8.1	100	9.6	107	9.7	103	....	....
July 14–16, 1900....	Iowa	5.8	100	8.6	108	9.5	106	....	....	....	....
Aug. 25–28, 1903...	Iowa	7.8	133	8.3	104	9.4	105	9.6	102	....	....
June 9–10, 1905....	Iowa	7.8	135	...	...	....	....	...	...	....	....
Sept. 15–19, 1905...	Mo.	5.5	95	7.4	93	9.1	101	9.9	105	10.4	103
July 5–7, 1909......	Mo.	5.0	86	6.1	76	8.6	96	...	...	...	...
July 20–22, 1909....	Wis.	4.9	83	6.2	76	7.2	80	...	...	....	....
Oct. 4–6, 1910......	Ill.	5.9	102	8.7	109	11.7	130	...	...	....	....
July 20–24, 1912....	Wis.	...	...	...	...	...	...	...	...	6.5	64
Jan. 10–12, 1913....	Ark.	4.9	83	6.0	75	7.1	79	...	...	....	....
March 23–27, 1913 .	Ohio	5.8	100	8.0	100	9.0	100	9.4	100	10.1	100
Aug. 17–20, 1915...	Ark.	6.4	110	8.6	108	11.2	125	11.6	123	....	....

exceeded that of 1913 by 33 and ·35 per cent respectively. For a stream of the size of the Miami, a 1-day rainfall is of too short duration to produce a maximum flood condition. A storm of longer duration, although the rainfall on the maximum day might be considerably smaller, could easily cause a more serious flood. In the course of the investigation it was found that the 3-day period of maximum precipitation would place the greatest burden on a system of flood control works in the Miami Valley. A comparison of the 3-day maxima in table 1 is therefore more to the point. This reveals six storms greater than March 1913. The largest are: that of October 1910 which had 30 per cent, and that of August 1915, which had 25 per cent more rainfall on a 4000 square mile area in 3 days than had the storm of March 1913. Neither of these storms occurred in winter. They occurred at seasons of the year when soil absorption and evaporation are greatest and the percentage of runoff usually least. It is doubtful if they would have produced floods as severe as the 1913 flood in the Miami River had they been centered over its drainage basin. But it is not certain that such storms can occur over that basin. Their centers are farther south, and it is probable that in reaching the Miami Valley they would have suffered considerable reduction. It is still less probable that such storms would occur during the winter months when the capacity of the air for carrying moisture is very materially less than it is in summer.

Of the 12 storms listed, all but 3 occurred during the summer or fall months. This in itself is an indication that a winter storm of the magnitude of that of March 1913 is a rare occurrence. One winter storm, that of December 1895, is quite similar to the March 1913 storm, being slightly less for the 1-day and slightly greater for the 3- and 4-day periods. If it had occurred over the Miami Valley, it would have produced a flood very similar and probably no greater than the 1913 flood.

From the foregoing it will be seen that the storm of March 1913 was an unusually great winter storm, considering the region of its occurrence. The storm data collected, which covers more than half a century, does not furnish evidence to indicate that it will ever be greatly exceeded in the Miami Valley as long as climatic conditions remain similar to those of recent centuries.

MAXIMUM POSSIBLE FLOOD IN THE MIAMI VALLEY

The records of European streams are illuminating since they show that floods which occur on an average of once in a century or two, have been exceeded in the course of many centuries. The amounts by which they were so exceeded, however, were usually small. For instance, in the case of the Danube, the flood record of which dates back to 1000 A.D., it appears that the greatest flood occurred in 1501. The second largest, in 1787 was 85 per cent as large in volume, while the difference in stage was trifling. The third largest occurred in 1899, and its discharge was 75 per cent as large.

In the Seine at Paris the highest stage on record in three centuries is 30 feet, in the year 1611. The next highest flood was only one foot less and occurred in 1658, while the third highest was two and a half feet less and occurred in 1910. It is estimated that the maximum rates of discharge in 1658 and 1910 were practically the same. There have been a number of other great floods only three or four feet less.

On the Tiber River the first great flood recorded at Rome was in 413 B.C. The greatest flood occurred 1598 A.D. Another nearly as large took place in 1870. There have been several other floods but little smaller.

Similar flood data is at hand for other streams. Reasoning from analogy we may conclude that the flood of March 1913 in the Miami Valley probably will not be greatly exceeded so long as climatic conditions are similar.

The extensive investigation of great storms, previously described, gives every indication that the flood of March 1913 is one of the greatest floods in centuries in the Miami Valley. Coupled with the evidence obtained from the foreign long term flood records, there is reason to believe that in the course of three or four hundred years, a flood 15 or 20 per cent greater in volume may occur. There is no evidence that would lead to the conclusion that the greatest possible flood discharge could be appreciably more than 20 per cent in excess. These conclusions are based upon the results of investigations probably more thorough than have ever before been made for a similar problem.

If longer records were available for the Miami River or for neighboring streams, a closer estimate could be made and the percentage of excess might be determined within closer limits.

FLOOD DISCHARGE 40 PER CENT GREATER THAN THAT OF MARCH 1913 CHOSEN AS BASIS FOR DESIGN

In the planning of the flood control works for the Miami Valley provision has been made for floods greater than the maximum considered possible. This was done to place the engineering works safe beyond limitations of data or errors in judgment or technique. Such a course is as necessary in the design of flood control works as it is in the design of any structure. It would not be good engineering to provide only for the maximum possible conditions indicated by human judgment.

Accordingly, the Miami Valley flood control works are designed to protect the valley against a flood discharge nearly 20 per cent greater than the maximum flood flow that is considered possible to occur, or nearly 40 per cent larger than the crest discharge during the March 1913 flood.

CHAPTER III.—THE RETARDING BASIN PLAN

This chapter is devoted to a brief description of the plan adopted for controlling floods in the Miami Valley, and the principal reasons favoring its adoption. This plan is usually spoken of as the retarding basin plan, because its principal features are large retarding basins designed to reduce flood flow to the carrying capacity of the river channels.

Each basin is formed by building across the valley an earth dam of the safest and most durable type. Substantial concrete outlets through the base of each dam permit the ordinary river flow to pass through unobstructed. The size of these outlets is such that at times of highest floods only such amounts of water will escape through them as can be safely taken care of in the river channels below the dams. The excess water is held back by the dams, and accumulates temporarily on the valley lands situated above them, to flow off later through the outlets as the flood subsides.

Retarding basins differ from storage reservoirs in that they come into use only at infrequent intervals, and then only for a few days at a time. Water is not held in them in the sense of storage; it is merely delayed or retarded there in its transit downstream. There being no gates, the outlet conduits are wide open at all times to pass the ordinary river flow, and even small rises, without reduction. Except during floods, therefore, the dams do not obstruct the rivers, and thus ordinarily, the valley lands which compose the bottoms of these basins are in their normal condition and can be used for crop growing just as though there were no dams. As floods on the Miami River usually occur in January, February, and March, but little interference with the use of these lands for agricultural purposes is expected.

In the preceding chapter attention has been called to the large amount of water which ordinarily becomes detained temporarily on the valley lands during a high flood, and which was referred to as valley storage. The function of a retarding basin is to confine this water to a comparatively small area, and to relieve from overflow the remainder of the valley below. As in the case of the valley storage, the water in a retarding basin is not brought to a complete standstill, as it would be in a closed reservoir, but continues to flow slowly, its movement merely being greatly retarded. It is this retaining or retarding action that has earned the name of " retarding basins," rather

46

than that of storage reservoirs, for the artificial basins created by the dams.

The retarding basins as designed for the protection of the Miami Valley have a joint capacity equal to 1½ times the valley storage during the peak of the 1913 flood, measured from Hamilton northward. The aggregate area that will be flooded within the boundaries of the retarding basins during a flood like that of March 1913 is less than half the total area that was actually flooded in 1913 above Hamilton, and does not include any highly developed areas such as cities.

VARIOUS PLANS INVESTIGATED

Immediately after the March 1913 flood, when investigations were first undertaken, there was a general belief at Dayton, Hamilton, and other cities in the valley, that the solution of the flood problem should be by means of channel improvements. A number of such plans were considered, descriptions of which will be found in chapter XIII. Their object was to create greater carrying capacity in the river channels by correcting their alignment, by making enlargements, and by increasing the height of the levees. The enormous cost of these projects, in part due to the large amount of valuable property that would have had to be acquired, and more especially the excessive maintenance costs which such disturbances of river regimen would cause, led to the eventual abandonment of all such plans in favor of the retarding basin plan of protection. The advantages of the latter plan, at the outset, had not been fully appreciated. Investigation of the use of dams for detaining excess flood waters had been made in pursuance of a desire to test all feasible plans of protection; its advantages over other plans gradually became apparent.

Different combinations of large basins were investigated, as were also the possibilities of using a large number of small ones scattered over the watershed. The incidental use of the larger dams for the production of water power was considered, but no conditions were found that would make such development either desirable or commercially feasible.

As the studies progressed it became clear that a few large retarding basins located at controlling points on the main river and its principal tributaries would furnish the greatest degree of protection at the least cost. It became evident also, however, that the capacities of the more promising basin sites could not be developed beyond certain physical limits without excessive cost imposed by the presence of towns and railroads, while to a limited extent channel improvement was feasible and economical. Thus the policy was reached of supplementing the protection which the dams afforded by increasing the river channel capacities through the cities. Many different combinations of retarding basins and local channel improvements were worked out, until by comparison and trial a plan was evolved which in terms of value and cost seemed to give approximately the best possible results. It provides for five large retarding basins, so located as to give the greatest feasible degree of protection to both city and country properties. It provides also for certain improvements of the river channels and levees in the principal cities, at a cost representing only a fraction of what the channel projects, contemplated in the earlier studies, would cost. The plan as a whole was made consistently efficient throughout; that is, the respective capacities of retarding basins. their outlets, and the improved channels through the cities were so proportioned as to be a balanced system. As designed it will afford complete protection to the cities against a flood discharge about 40 per cent greater than what occurred in March 1913. The greater part of the lands outside the cities are only partially protected, as in most cases complete protection of low lying farm lands would cost far more than the benefits resulting.

An idea of the magnitude of the system may be gained from the statement that the basins if filled to spillway level would cover an aggregate area of 55½ square miles. A flood similar to that of March 1913 would flood 46½ square miles in the basins.

The use of retarding basins for flood control is not without precedent. They have been used for many years both in this country and in Europe. Two, in France, that are the largest in Europe, have been in use over two hundred years. However, the great majority of retarding basins in Europe are of comparatively recent date, their construction having had to await the enactment of adequate laws for apportioning the costs of such undertakings.

In table 2 is given a list of the better known flood control reservoirs. Of these 45 are in Europe, 1 in India, and 10 in the United States. Included in the latter number are the 5 retarding basins of the Miami system. It will be seen that these are among the largest built and that a combination of five large basins is without precedent.

It will be noted that many of the European reservoirs serve a double purpose, their lower portions being used for storage purposes. and their upper portions being kept empty so as to be available for detaining flood flow. The table, accordingly, was arranged to show. not only total capacity, but also the portion reserved for flood control. The cost per acre foot was figured on the basis of total storage.

Table 2.—List of Existing Flood Control Reservoirs

No.	Name of Reservoir	Stream	Location	Date of Completion	Capacity in Acre Ft.		Cost	Cost per Acre Foot	Height of Dam, Feet	Type of Dam	Drainage Area, Sq. Mi.
					Total	Flood Storage					
1	Furens	Furens	France	1862	1,300	320	$ 318,000	$245	184	Masonry	10
2	Ternay	Ternay	France	1867	2,400	*	204,400	85	128	"	11
3	Chartrain	Tache, Near Roanne	Loire Basin, France	1894	3,700	410	400,000	108	177	"	5
4	Pinay	Loire	France	1711	81,000	81,000	34,000	0.42	69	"	
5	La Roche	"	"	1711	88,000	88,000	8,000	0.09	33	"	
6	Lingese	Lingese	Wupper Basin, W. Germany	1898	2,100	80	256,800	122	80	"	4
7	Brucher	Brucher	Wupper Basin, W. Germany	1913	2,800	*	—	—	82	"	
8	Bever	Bever	Wupper Basin, W. Germany	1898	2,700	410	343,200	127	82	"	9
9	Hemfurt	Eder	Weser Basin, W. Germany	1913	164,000	24,400	4,745,000	29	154	"	552
10	Klingenberg	Weisseritz	Elbe Basin, Central Germany	1913	12,600	*	858,000	68	131	"	35
11	Malter	"	Elbe Basin, Central Germany	1913	7,100	*	742,000	104	119	"	40
12	Königreiche-Walde	Tributary of Elbe	Elbe Basin, Bohemia	1914	7,400	5,900	965,000	130	136	"	200
13	Hlinsko	"	Elbe Basin, Bohemia	1912	1,900	1,400	150,000	79	43	Earth	22
14	Spindelmühle	"	Elbe Basin, Bohemia	1914	2,700	2,400	652,900	242	136	Masonry	81
15	Parizow	"	Elbe Basin, Bohemia	1913	1,400	1,000	300,000	214	101	"	22
16	Kreibitzbach	"	Elbe Basin, Bohemia	1913	810	320	140,000	173	77	Earth	2
17	Horka	"	Elbe Basin, Bohemia	1914	1,100	660	—	—	20	"	17

* Indicates capacity kept empty for flood control purposes not known.

4

Table 2.—List of Existing Flood Control Reservoirs.—Continued

No.	Name of Reservoir	Stream	Location	Date of Completion	Capacity in Acre Ft.		Cost	Cost per Acre Foot	Height of Dam, Feet	Type of Dam	Drainage Area, Sq. Mi.
					Total	Flood Storage					
18	Karlstal	Iser	Tributary of Elbe, Bohemia	1912†	13,000	6,500	$1,990,000	$153	105	Masonry	23
19	Bystrzicka	Bystrziczkabach	Tributary of Beczwa, Bohemia	1911	3,600	*	—		123	"	25
20	Darre	Schwarzen Desse	Tributary of Iser, Bohemia	1909†	5,000	2,300	410,000	82	70	"	6
21	Weissen Desse	Weissen Desse	Tributary of Iser, Bohemia	1909†	210	44	53,500	255	46	"	3
22	Blattney	Blattney	Tributary of Iser, Bohemia	1910†	4,900	2,100	552,500	113	97	"	6
23	Aglsboden	Mareitherbach	Near Sterzing, Austria	1880	410	410	7,150	17	33	"	
24	Marklissa	Queis	Oder Basin, S. E. Germany	1904	12,000	8,100	750,000	62	141	"	118
25	Mauer	Bober	Oder Basin, S. E. Germany	1912	40,500	16,000	1,992,000	49	203	"	467
26	Warmbrunn	Zacken	Oder Basin, S. E. Germany	1908	4,900	4,900	381,000	78	28	Earth	46
27	Herischdorf	Heidewasser	Oder Basin, S. E. Germany	1906	3,200	3,200	219,000	68	28	"	38
28	Buchwald	Bober	Oder Basin, S. E. Germany	1906	1,800	1,800	264,000	147	88	Concrete	23
29	Erdmannsdorf	Gr. Lomnitz	Oder Basin, S. E. Germany	1913	2,400	2,400	274,000	114	36	Earth	19
30	Glatzer Neisse	Glatzer Neisse	Oder Basin, S. E. Germany	1913†	83,000	14,600	3,840,000	46	37	"	908
31	Arnoldsdorf	Goldbach	Oder Basin, S. E. Germany	1913	1,800	1,800	119,000	66	51	"	19
32	Woelfelsgrund	Woelfelsbach	Oder Basin, S. E. Germany	1907	740	740	124,000	168	98	Masonry	10

† Indicates date when construction was started; date of completion not known.

Table 2.—List of Existing Flood Control Reservoirs.—Continued

No.	Name of Reservoir	Stream	Location	Date of Completion	Capacity in Acre Ft.		Cost	Cost per Acre Foot	Height of Dam, Feet	Type of Dam	Drainage Area, Sq. Mi.
					Total	Flood Storage					
33	Seitenberg	Mohre	Oder Basin, S. E. Germany	1908	930	930	$68,000	$73	55	Earth	20
34	Schönau	Steinbach	Oder Basin, S. E. Germany	1913	1,300	1,300	90,000	69	64	"	15
35	Klein-Waltersdorf	Röhrsdorfer Wasser	Oder Basin, S. E. Germany	1913	600	600	58,600	98	45	"	7
36	Grüssau No. 1	Zieder	Oder Basin, S. E. Germany	1906	420	420	83,500	110	16	"	14
37	Grüssau No. 2	"	Oder Basin, S. E. Germany	1906	340	340			21	"	22
38	Friedeberg	Langwasser	Oder Basin, S. E. Germany	1913	2,800	2,800	127,000	45	44	"	24
39	Friedrichswald	Görlitzer Neisse	Oder Basin, N. Bohemia	1906	1,600	810	360,000	225	92	Masonry	2
40	Voigtsbach	"	Oder Basin, N. Bohemia	1906	200	100	94,000	470	52	"	3
41	Mühlscheibe	"	Oder Basin, N. Bohemia	1906	200	100	123,000	615	72	"	3
42	Gorsbach	"	Oder Basin, N. Bohemia	1908	400	200	206,000	515	70	"	5
43	Harzdorf	"	Oder Basin, N. Bohemia	1904	510	190	165,000	324	62	"	6
44	Grünwald	"	Oder Basin, N. Bohemia	1908	2,200	1,100	540,000	246	66	"	10
45	7 Res. at Weidlingau	Wien, Near Vienna	Danube Basin, Austria	1900	1,300	1,300	1,680,000	1290		Concrete	65
46	Gundipet	Musi	Hyderabad, India	1913†	218,000	*	110,000	71	120	Masonry	250
47	Harrisburg	Paxton Creek	Susquehanna Basin, Pa.	1910	1,550	950			12	Earth	23
48	Watervliet	Dry River	Hudson Basin, N. Y.	1914	910	910	165,000	180	{85/25}	Earth-Masonry	3

Table 2.—List of Existing Flood Control Reservoirs—Continued

No.	Name of Reservoir	Stream	Location	Date of Completion	Capacity in Acre Ft. Total	Capacity in Acre Ft. Flood Storage	Cost	Cost per Acre Foot	Height of Dam, Feet	Type of Dam	Drainage Area, Sq. Mi.
49	West Basin	Little River	Mississippi Basin, Mo.	1918	154,000	112,000			30	Earth	546
50	Middle Basin	"	Mississippi Basin, Mo.	1918	43,500	39,000	$695,000		35	"	§1075
51	East Basin	"	Mississippi Basin, Mo.	1918					25	"	§1160
52	Germantown	Twin Creek	Miami Basin, Ohio	1920	106,000	106,000	‡‡‡‡‡‡‡‡‡‡‡	‡‡‡‡‡‡‡‡‡	100	Earth	270
53	Englewood	Stillwater River	"	1921	312,000	312,000			110	"	651
54	Lockington	Loramie Creek	"	1920	70,000	70,000			69	"	255
55	Taylorsville	Miami River	"	1920	186,000	186,000			67	"	§1133
56	Huffman	Mad River	"	1921	167,000	167,000			65	"	671

‡ Indicates cost not known as construction was not completed at time of publication.
§ Includes drainage area of preceding dam.

Retarding basin development has been chiefly on the smaller streams, as will be seen by reference to the size of the drainage areas controlled by the dams, indicated in the last column of table 2. For the most part, however, the dams are units in comprehensive systems of flood control of large rivers. Thus, of the 21 retarding basins on the headwaters of the Oder River, 6 are in the catchment area of one stream in Bohemia known as the Gorlitzer Neisse; of the 9 basins on tributaries of the Elbe River, 5 are in the catchment area of the Iser River.

The Pinay and LaRoche dams built two centuries ago on the Upper Loire, France, are quite crude, yet have justified their existence during the great floods of 1790, 1846, 1856, 1866, and 1907, causing appreciable reductions in the flood stage at the city of Roanne situated 17 miles below the lower of the two dams. Their low construction cost shown in the table is attributable, in part, to the favorable natural dam sites utilized, and, in part, to the small labor costs that prevailed in those early days.

COST OF RETARDING BASIN PLAN COMPARED WITH THAT OF OTHER PLANS

The original estimate for the adopted plan, made before costs had advanced to their present levels, including the river improvements in the cities was $23,750,000. The project for protecting the entire valley by means of a comprehensive channel improvement under the same conditions would have cost $92,370,000. Aside from the great difference in the cost of these two projects, it is important to note that the latter plan was designed to protect against floods not larger than that of 1913, while the adopted plan furnishes protection against floods 40 per cent greater, and that the maintenance of a channel improvement system would have been a very serious burden.

A project for improving the river channels in the cities only, to take care of a flood flow equal to that of 1913, would have cost $45,-750,000, of which $23,630,000 alone would have been required to protect Dayton. This latter item approaches closely the cost of the entire adopted plan.

A comparison of the cost of a system of numerous small retarding basins above Dayton, such as is described in chapter XIII, with the cost of the four large retarding basins above Dayton shows a difference in favor of the latter of $6,500,000, in addition to which the reliability of the protection furnished by the large basins is by far superior to that of the numerous small basins.

All of the figures here given cover first cost of construction, includ-

ing real estate and administrative costs. Annual operation and maintenance costs have, so far, not been mentioned. These would be large
items in the case of the channel improvement schemes, while for the
adopted plan such charges will be light. The retarding dams having
no movable parts, and being constructed in a most substantial manner
of earth and concrete, will be subject to little deterioration. The im-

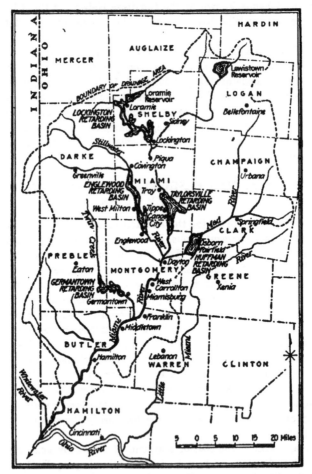

FIG. 9.—LOCATION OF RETARDING BASINS IN MIAMI VALLEY.

The map shows only the portion of the Miami River watershed above the
mouth of Whitewater River. In order to bring out the retarding basins clearly
they are shown like lakes, although in operation they will rarely if ever be filled
with water to the extent shown.

proved channels in the cities, during floods, will be called upon to take
care only of the water discharged by the retarding basins. They will
not be exposed therefore to the full force of the flood flows as in the
case of the other channel projects. Yet even the maintenance of these
moderate channel improvements probably will be more expensive than
the maintenance of all the dams and retarding basins.

DESCRIPTION OF MIAMI VALLEY

To gain a clear conception of the plan of protection as adopted, a
knowledge of the physiographic conditions of the Miami Valley is
essential. Figure 9 shows that portion of the drainage basin of the
Miami River with which we are here concerned. The Whitewater
River, an important tributary entering the Miami near its mouth and
draining a portion of Indiana, is not shown entirely because it is
scarcely affected by the works of the Miami Conservancy District.
The Miami River is about 163 miles in length. Its drainage basin
measures about 120 miles on the longest axis and covers about 4000
square miles, including parts of 15 counties. In outline it resembles
the left half of an oak leaf of which the midrib along the eastern margin and the veins represent the main river and tributaries. The more
important of the latter, following northward up the west side of the
Miami, are: Indian Creek, emptying a few miles below Hamilton;
Four-Mile Creek, a flashy stream entering just above Hamilton; and
Twin Creek with its outlet just below Franklin; four streams, the
Miami, Mad, Stillwater Rivers, and Wolf Creek unite within the city

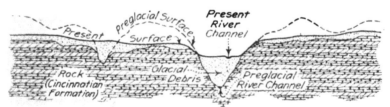

FIG. 10.—IDEALIZED GEOLOGICAL SECTION ACROSS MIAMI VALLEY.

*Illustrating how land surface was changed by glacial action, the hills shown by
dotted outline being worn down and the valleys filled up.*

limits of Dayton; and just above Piqua, 27 miles north of Dayton, the
Miami is joined by Loramie Creek.

The valley is a nearly flat plain fifty to two hundred feet below the
general elevation of the adjacent rolling country, which ranges from
1100 feet above sea level at the headwaters to less than 800 feet near

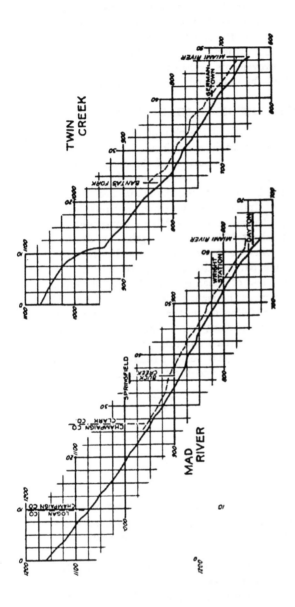

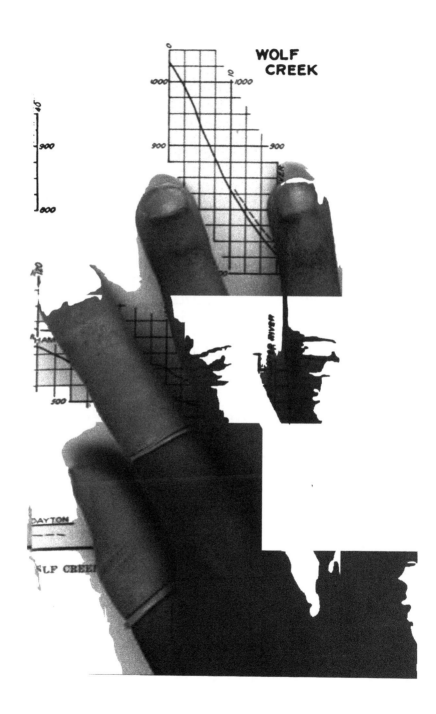

WOLF
CREEK

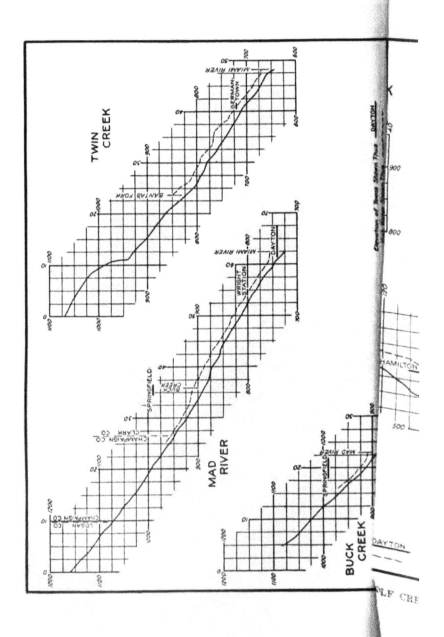

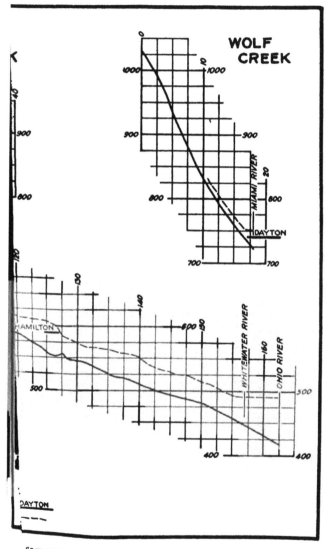

WOLF CREEK.

the mouth of the Miami. The drainage areas of the principal streams are shown in table I of the Appendix.

Except for its southernmost portion the entire basin bears evidence of having been covered by the ice sheet during the glacial period. The preglacial valleys carved into the limestone formation and the crests of the preglacial hills have been almost entirely obliterated by the glacial action, (see figure 10). The Miami River and its principal tributaries flow in the partially filled valleys, in comparatively insignificant channels. The natural tendency of these streams is to meander back and forth over the valley, slowly but persistently shifting and adjusting their channels. This process, so inimical to the interests of man, can be controlled only by carefully planned engineering works.

Table II of the Appendix shows the fall of the various streams of the basin as determined from the United States Geological Survey topographic maps. The Miami has a drop per mile of about 2 feet in the flat uplands, and of 3.3 feet for the major part of its course. Stillwater and Mad Rivers are much steeper, the former sloping more than 4 feet and the latter about 6 feet per mile. Figures 11 and 12 show profiles of beds of the principal streams, and also show by means of dotted lines the high water lines of the March 1913 flood. The rolling topography, together with the fan shaped arrangement of the larger tributaries, presents a singularly favorable combination for quick collection of storm runoff and the formation of high flood crests. Most of the tributaries are flashy, Mad River, Twin Creek, and Four-Mile Creek being noted for the suddenness with which they rise and the short duration of their flood stages. Floods in the main stream rarely last more than a few days. During the great flood of March 1913 the rainfall lasted 5 days, the highwater period 8 days, and the danger stage 2¾ days.* As has been shown in the preceding chapters, the maximum discharge was from two to four times as great as the channels through the cities could carry, and about 10 times as great as the channel capacity outside the cities.

The fact that the principal cities in the Miami basin are located for the most part on the bottom lands along the rivers, accounts for the enormous damage, estimated at over 100 million dollars, that was caused by the great flood. Table VI of the Appendix shows their populations and elevations. Of these cities Dayton is the largest, and, because of its exposed position, was particularly in need of protection. Hamilton, the next largest city, suffered even more heavily in

---

* The danger stage at Dayton has been fixed by the Weather Bureau at 18 feet on the Main Street bridge gage.

proportion by the flood and presented a by no means simple problem. Lesser problems presented themselves at Troy, Miamisburg, Franklin. Piqua, Middletown, and West Carrollton.

On the Upper Miami and on Loramie Creek, are two small reservoirs, the Lewistown and Loramie, built many years ago by the State of Ohio for storing water needed in the operation of the Miami and Erie Canal. They control small drainage areas, and although their effect on flood flow was taken into consideration, it was found to be almost negligible. The canal extends down the valley to Hamilton where it turns southeasterly connecting with the Ohio River at Cincinnati. It is no longer navigated but is still used in a limited way for water supply and power.

PRINCIPAL FEATURES OF PLAN ADOPTED

The plan of flood control which was officially adopted by the Board of Directors* and received the approval of the Conservancy Court on November 24, 1916 is here outlined in brief, the five retarding basins being taken up first and the local channel improvements next.

GENERAL DESCRIPTION OF DAMS AND APPURTENANCES

While the protective works at the various towns and cities differ materially according to the particular conditions existing in each locality, the dams and their appurtenances are more nearly similar. The bodies of all of the dams are of compacted earth, in cross section as indicated in figure 13, which shows the standard design adopted. The diagram shows a theoretically level crest 25 feet wide, but in actual construction this will be given a slightly rounded form to pro-

---

* In accordance with the provisions of the Conservancy Act of Ohio, which required a plan to be prepared to ''include such maps, profiles, plans, and other data and descriptions as may be necessary to set forth properly the location and character of the work, and of the property benefited or taken or damaged, with estimates of cost and specifications for doing the work,'' the Chief Engineer of the Miami Conservancy District, under date of February 29, 1916, submitted such a plan of protection in the form of a report entitled Report of the Chief Engineer. It consisted of 3 volumes of printed matter of about 200 pages each, accompanied by 408 drawings. Volume I contained a synopsis of the data on which the plan is based, a description of its development, and a statement of the plan in detail. Volume II contained a legal description of all lands affected by the plan. Volume III contained the contract forms, specifications, and estimates of quantities and cost. After various slight modifications the Report of the Chief Engineer was adopted by the Board of Directors as the Official Plan of the District, and was republished in May 1916 under that title.

Many of the technical studies mentioned in the Official Plan volumes, particularly those of special interest to engineers, have been published in detail in the Technical Reports series.

vide for a 30-foot roadway. The variable slopes are a feature of the design. These are: 2 to 1 for the upper 20 feet of dam; 2½ to 1 for the next 30 feet; 3 to 1 for the 30 feet below that; 4 to 1 for the remaining portion down to an elevation near the toe; and in the old river bed a slope of 10 to 1. Some of the dams are less than 80 feet in maximum height, and therefore do not possess the 4 to 1 slopes. At each change of slope the faces are provided with level berms 10 feet in width. These berms are intended primarily to break the flow of rain water down the face of the dam, there being gutters on the berm for that purpose, except where the entire surface is paved with rock. They serve also as roadways facilitating inspection and maintenance.

It will be noted that the cross sections of the dams are exceptionally heavy. Because of their being built in a densely populated valley, and in order to relieve the public mind of any apprehension as to their possible failure, the dams were purposely designed so massive as to insure safety beyond any doubt. Watertightness is obtained by making the interior portion of each dam of carefully selected and puddled materials, and extending these down some distance into the underlying gravels by means of a cutoff trench to prevent seepage underneath the fills. The outer portions are made up of coarse sands and gravels which protect the watertight core, and insure proper drainage for the entire structure. The outer surfaces are covered with fertile soil and planted to blue grass which in this region forms a dense sod.

The necessity for building the dams of earth arose from the fact that the solid rock formations which underlie the glacial gravels are at places so deep under the surface as to make their economic use as concrete dam foundations utterly impracticable. Systematically conducted borings, made during the investigation period, indicated those areas at the damsites where solid rock was near the surface, and these were taken advantage of to serve as foundations for the concrete outlet structures. The conditions at the Germantown damsite are illustrated in figure 14.

The openings or outlets for passing the river flow through the dams are of two types. That used at the Lockington, Taylorsville, and Huffman dams consists essentially of a section of dam of the gravity type built of massive concrete, perforated by large tunnel shaped openings at the base, and flanked by heavy retaining walls against which the earthwork of the dam proper abuts. This concrete section has a rounded crest to serve as a spillway, 15 feet lower than the top of the earth dam. The down stream face is of ogee shape, so designed that water spilling over the crest will be discharged into the

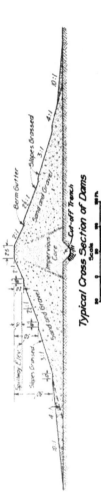

Typical Cross Section of Dams

Scale

Berm Gutter
Slopes Grassed
Sand and Gravel
Cut-off Trench
Impervious Core
Sand and Gravel
Spillway Elev.
Slopes Grassed

FIG. 13.—CROSS SECTION THROUGH HYDRAULIC FILL EARTH DAM FORMING
RETARDING BASIN.

Except for minor modifications, due to local conditions, this cross section is typical of the
dams forming the retarding basins in the Miami Valley.

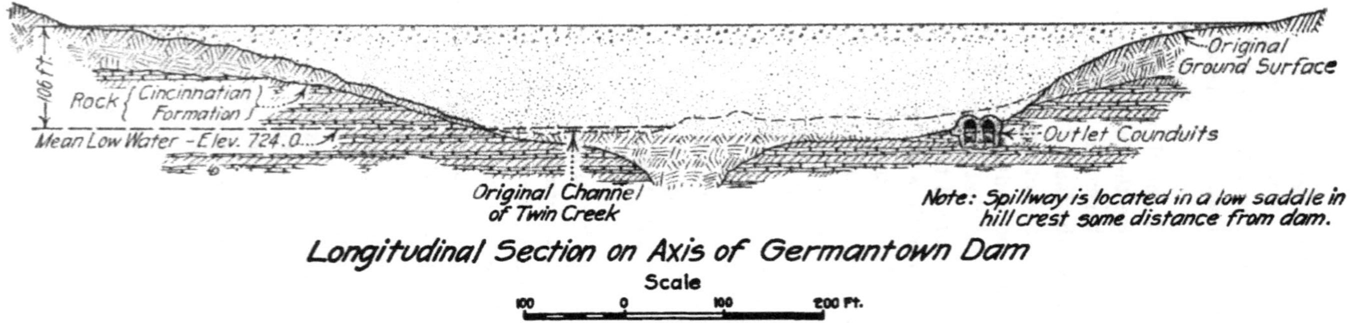

Original
Ground Surface

Outlet Conduits

Note: Spillway is located in a low saddle in
hill crest some distance from dam.

Original Channel
of Twin Creek

Rock { Cincinnati }
{ Formation }

Mean Low Water - Elev. 724.0

106 Ft.

Longitudinal Section on Axis of Germantown Dam

Scale

200 Ft.

FIG. 14.—LONGITUDINAL SECTION ON AXIS OF GERMANTOWN RETARDING BASIN DAM.

The section is across the valley of Twin Creek at the Germantown Dam, and shows the location of the dam and outlet conduits with respect to the position of the shaly limestone rock known as the Cincinnatian formation. The dam is 100 feet high above the original valley bottom, and has a crest length of 1210 feet.

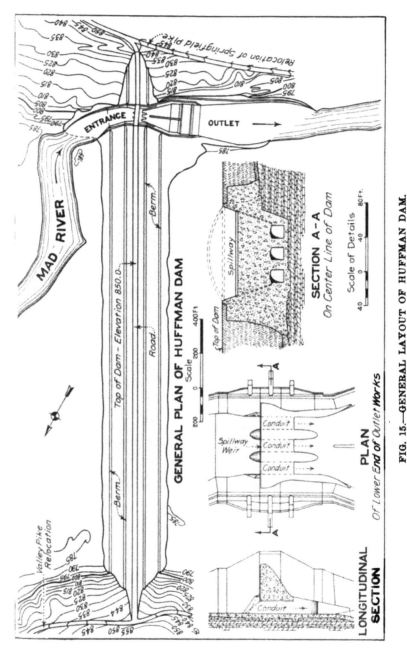

GENERAL PLAN OF HUFFMAN DAM

SECTION A-A
On Center Line of Dam

Scale of Details

PLAN
Of Lower End of Outlet Works

LONGITUDINAL
SECTION

FIG. 15.—GENERAL LAYOUT OF HUFFMAN DAM.

Illustrating type of hydraulic fill dam with combined outlet and spillway structure.

same channel into which the outlet openings discharge. This type of combined spillway and outlets is illustrated in figure 15 showing the general plan of the Huffman dam and its outlet structures. For the two highest dams, those at Germantown and Englewood, this type of construction was not economical, and the outlets were built in the form of twin conduits piercing the base of the earth dam, (see figure 16, showing the Germantown dam and outlet structures) and resting on rock foundation over their entire length. The spillway in such case was made a separate structure. Two conduits rather than one are provided in order to facilitate inspection, maintenance, and repairs.

There are no gates at the outlet openings, the outflow from any basin being controlled solely by the fixed size of the orifices. The discharge through the latter varies, therefore, depending upon the depth of water above the dam. The considerations entering into the design of the outlets represent one of the most interesting features of the engineering work of the District, and are set forth in detail in chapter VI.

Immediately below the outlets is an expanding channel which leads the issuing water into a stilling basin. While ordinarily the velocity of flow through the outlet openings is approximately the same as in the river channel above and below the dams, during floods the outflow, being under head, will attain velocities so high that if unchecked it would erode the river channel below the dam. Erosion of the river bed was not considered permissible, as it might lead eventually to undermining the adjacent toes of the earth structure. It was desirable, therefore, to provide automatic means for destroying the kinetic energy of the outflowing water. In view of the high velocities and large volumes involved this problem was by no means simple of solution. At the Taylorsville dam, for instance, the joint discharge from the four openings, with retarding basin full, will amount to more than 50,000 second feet, moving with an average velocity of nearly 50 feet per second. At the other dams the quantities are less but the velocities are quite as high.

The stilling pool embodies many features that are novel, its basic principle of operation being what is known as the "hydraulic jump." This has been worked out to act with equal efficiency for high as well as for moderate flood discharges, the water in every instance being delivered into the river channel below under velocities approximating those of the natural river flow under like conditions. The utilization of the hydraulic jump for this purpose was previously made the subject of elaborate experiments by the engineers of the District. A full description of these, together with an explanation of the theory of the

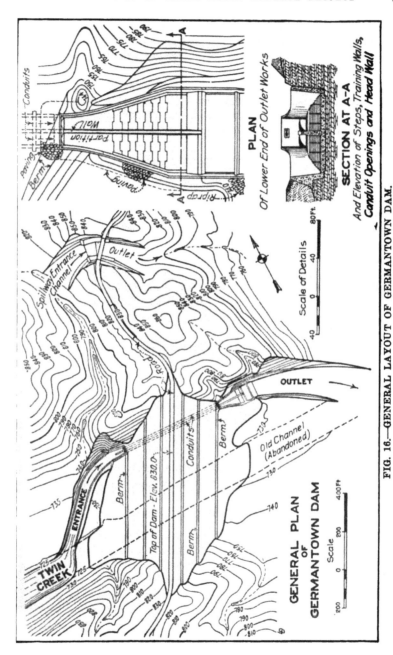

PLAN
Of Lower End of Outlet Works

SECTION AT A-A
And Elevation of Steps, Training Walls,
Conduit Openings and Head Wall

Scale of Details

GENERAL PLAN
OF
GERMANTOWN DAM

Scale

FIG. 16.—GENERAL LAYOUT OF GERMANTOWN DAM.

Illustrating type of hydraulic fill dam with long twin outlet conduits under the earth fill. The spillway is a separate structure.

hydraulic jump, is published in another volume of the Technical
Reports,* and is not, therefore, reviewed in these pages.

In table VII of the Appendix are given the principal dimensions
of the retarding basins, dams, and outlet works. For a detailed description of the manner in which the various parts were proportioned
with respect to each other, the reader is referred to chapters IV
and VIII. Data regarding the conditions in the retarding basins
for floods of various sizes will be found in the Appendix tables.

Germantown Retarding Basin

The Germantown basin, figure 39, located on Twin Creek two miles
west of Germantown, reduces the flood flow of the Miami River at the
cities of Middletown and Hamilton. The reduction so effected greatly
reduces the cost of the necessary local protection work at these cities.
Like the Englewood site on Stillwater River this basin site is capable
of controlling a flood flow vastly in excess of the maximum considered possible. The creek flows through a narrow valley which affords
excellent dam sites. The dam, which has a height of 100 feet above
the valley floor, is notable for its short crest 1210 feet long. It contains 865,000 cubic yards of earth, or less than any other of the Conservancy dams.

In case of a storm like that of March 1913, this basin will reduce
the maximum flow of Twin Creek from 66,000 second feet to 9340
second feet. At its highest stage in such a flood the basin will hold
73,000 acre feet, spread out over 2950 acres of land. Since 2045 acres
within the basin were actually flooded in 1913, when there was no
dam, the additional area needed for the operation of the basin is only
905 acres. In 7½ days after the highest stage the basin will again
be empty.

No railroads or towns are located in the basin, and the only additional works required consist of two miles of highway construction.

Englewood Retarding Basin

The Englewood dam is on Stillwater River about ten miles northwest of Dayton and one mile southeast of the village of Englewood.
With a height of 110.5 feet above the valley floor and a mass of
3,500,000 cubic yards of earth and gravel, this dam is by far the
largest built by the District. In case of a flood like that of March
1913 the Englewood basin will reduce the crest discharge from 85,400

* Hydraulic Jump and Backwater Curves, by S. M. Woodward, R. M. Riegel
and J. C. Beebe. Technical Reports, Part III, The Miami Conservancy District.
Dayton, Ohio, 1917.

second feet to 11,000 second feet, and will hold back about 209,000 acre feet of water spread over 6350 acres of land. Since 3285 acres of this area were actually flooded in March 1913 the additional area required by the building of the dam will be 3065 acres. In 23 days after reaching its maximum stage during such a flood the basin will be empty.

In many ways the Englewood basin, figure 40, is the most favorably situated of the five. There were no railroads or towns in the Stillwater Valley to interfere materially with its construction or operation; the only limiting condition was that the outlet conduits be of sufficient size to empty the basin in a reasonable time after being filled by a flood, so that the agricultural land above the dam would be relieved of water as soon as would be feasible, and also in order to supply capacity for a possible second flood immediately following the first. The configuration of the basin and its damsite are such that a dam could have been built there high enough to control floods much larger than the maximum considered possible.

Highways are required aggregating four miles, to maintain communication when the basin is filled.

Lockington Retarding Basin

To assist in providing adequate channel capacities for the Miami River through the cities of Piqua and Troy, it was desirable to provide a retarding basin above these cities. A number of sites were surveyed and considered before the one on Loramie Creek at Lockington was decided upon, see map in figure 41. Although the valley at this point is wide, the foundation which it afforded for the construction of a dam is perhaps better than at any other damsite; the basin capacity, moreover, is of sufficient size to control the entire flood flow of the creek. The basin will benefit to some extent the entire valley below, but the principal effect will be at Piqua and Troy. It will reduce a flood discharge from Loramie Creek equal to that of 1913, from 33,000 second feet to 8600 cubic feet per second. Even with this reduced discharge, extensive improvements in the channel at Piqua and Troy are necessary in order to insure safety against the maximum possible flood.

The dam is located half a mile northwest of Lockington and four miles north of Piqua. It is a massive earth dam with concrete outlet and spillway structure, and has a maximum height above the valley floor of 69 feet. In case of a storm like that of March 1913 this basin will hold 63,000 acre-feet of water, covering 3600 acres of land. Since 3240 acres of this area were actually flooded in March 1913, the addi-

5

tional area required for the operation of the basin is only 10 per cent more than that covered in 1913. In 7 days after reaching its maximum stage during such a flood the basin will be empty.

The Lockington basin requires the construction of six miles of highways, but does not interfere seriously with railroads. The main line of the Cleveland, Cincinnati, Chicago & St. Louis Railway from Cleveland to St. Louis is 3 feet below spillway elevation for a distance of about half a mile.

Taylorsville Retarding Basin

The Taylorsville dam crosses the Miami River valley about eight miles north of the city of Dayton, and creates a basin with a capacity second only to that at Englewood. It is the largest of all the basins in point of acreage. A map of the basin is shown in figure 42. A flood of the size of that of March 1913 will cover 9650 acres. Of this area 6700 acres were actually flooded in 1913 before any dam was built; the additional area required for the operation of the basin is, therefore, 2950 acres. It will reduce a flood like that of 1913 from 106,400 second feet to about 51,300 second feet. In 4½ days after reaching its maximum stage during such a storm the basin will be empty, the rapid discharge being due to the ample outlet capacity provided, which is larger than at any other of the dams. The Taylorsville dam is of earth, 67 feet high above the valley floor, and contains approximately 1,235,000 cubic yards. The height of the dam was limited by the proximity of the city of Troy to the upper edge of the basin. A higher dam would have increased the opposition in that city due to fear of being flooded by backwater. Under the plan as adopted, it was decided to build a levee along the east side of the town of Tippecanoe to protect against flooding by backwater from the basin during extreme floods, as described on another page.

The construction of this basin made necessary the removal to higher ground of 9.9 miles of the Baltimore & Ohio Railroad tracks, the old location of which was within the basin contour for a distance of 2 miles. About five miles of new highways are required to maintain communication when the basin is filled.

Huffman Retarding Basin

The Huffman dam to control the flood flow of Mad River is located six miles above the mouth of the river, its southern abutment resting against Huffman Hill. The dam is 65 feet high above the valley floor and contains 1,655,000 cubic yards of earth. In case of a storm like that of March 1913, this basin will reduce the maximum flow from 78,300 second feet to 32,600 second feet. At highest stage it will hold 124,000 acre feet of water spread over 7300 acres of land. This is 3880 acres more than were actually flooded within the basin in 1913, before any dam was built. In 5 days after reaching the maximum stage, the basin will be empty.

The Huffman site, see map in figure 43, though physically well adapted for retarding basin purposes, presented a number of disadvantages. The main line of the Erie Railroad, and that of the Cleveland, Cincinnati, Chicago & St. Louis Railroad, had to be removed to higher ground, requiring a change of location in each instance of about 15 miles; and the Ohio Electric Railway tracks likewise required relocating for a distance of about 10 miles. All property in the village of Osborn, a community of about 1200 people, situated approximately in the middle of the basin, had to be purchased in order to make removal possible. Through a special agreement, the property holders in Fairfield will receive compensation for actual damage sustained should damage occur from flooding from the basin. About 4½ miles of new roads are required in order to replace the roads that will be flooded when the basin is filled, and on account of changes made necessary by the relocation of the railroads.

GENERAL DESCRIPTION OF LOCAL PROTECTION WORKS

As before stated the construction of the retarding basins makes possible without undue cost adequate river channel improvement through the cities of the valley for the care of the reduced flood flow. At most points in the valley the retarding basin control has permitted channel changes to be confined to moderate limits except at Hamilton where at some points a channel widening of 300 feet is required involving considerable property damage. Some improvement in the river channel, in order to give the standard degree of protection, is required at Piqua, Troy, Dayton, West Carrollton, Miamisburg, Franklin, Middletown, and Hamilton. As noted in connection with the description of the Taylorsville retarding basin, local protection work is required at Tippecanoe City to protect this village against back water from the basin.

Three types of protection work are required in the cities and villages: First, that in which channel excavation is the essential or most prominent feature, as at Hamilton and Dayton; second, that in which the work is almost entirely confined to levee construction, as at West Carrollton, Miamisburg, Franklin and Middletown; and third, a combination of these two features in about equal proportions. The latter is the case at Troy and Piqua. Where an increase in capacity of the river channel was required, it has been obtained in several ways: by widening and deepening the channel by excavation, by constructing levees, by improving the alignment and cross section of the channel. by developing greater capacity under bridges through their raising and lengthening, and by clearing trees and other obstructions from the channel.

Channel improvement, by nearly all of these methods, leads to increased velocity of the river flow. To prevent subsequent erosion of the bed and banks, protection work has been designed for points along the channel where the velocities are particularly high, or where the direction of flow is toward the bank, and also where, because of the nature of the property involved, erosion would be particularly serious. On levee slopes a concrete facing 6 inches thick, reinforced with wire mesh, is used. The concrete is laid in slabs approximately 8 feet by 12 feet, with expansion joints. To obviate the glaring effect which might be produced by a large expanse of concrete revetment, a small amount of lamp black is added to the cement to darken the color.

As a rule, for protection against erosion, dependence is placed upon the holding power of a well sodded bank. Both river banks and levee surfaces, where not otherwise protected, are covered with a layer of loam and seeded with blue grass. At susceptible points, such as the outside of the bends of curved channels, concrete revetment is used to a vertical height of about 6 feet above the foot of the bank To protect the bed at the foot of the bank a flexible slab revetment is used. This is a mat made of blocks, each 12 inches by 24 inches by 5 inches thick, tied together by galvanized iron wire rope. This mat is anchored to a concrete wall built at the foot of the slope on timber piles. At the outer edge of the mat a larger concrete block is cast in place. The small blocks composing the mat were cast at a central plant.

In cross section the standard river channel has a depressed bottom whose width is about one-fourth of the width of the entire channel and whose elevation is 8 feet lower than the elevation of the foot of the banks. This depressed bottom is located in the center of the channel where the latter is straight, and near the outside of the bends in curved channels. In such a winding channel as that at Dayton. the depressed bottom swings from side to side as the curvature is reversed. Where space is available a berm 30 or more feet wide is left between the top of the channel bank and the base of the levee. At restricted points the excavated bank of the channel continues upward as the slope of the levee. The standard levee is 8 feet wide on top.

For the first 8 feet from the top it has slopes of 2 feet horizontally to one vertically, for the next 10 feet, 2½ to 1 slopes, and for the third 10 feet, 3 to 1 slopes. It has been rarely necessary to construct levees over 15 feet in height.

One of the advantages reaped from using the retarding basin plan is the elimination of wholesale bridge reconstruction which would have been required under any plan of channel improvement without basins. In comparatively few cases it has been necessary to provide a moderate enlargement of the channels by raising or lengthening of the bridges. In other cases, particularly at Dayton, it has been sufficient to utilize the full area under the bridges. At this city several of the bridges were located below curves in the channel or below projecting banks which tended to deflect the flow away from a considerable section of the bridge opening. A prominent feature in the Dayton channel works has been the reduction of these projections. Incident to this work there has been included at several cities a protection against dangerous scour by placing steel sheet piling across the channel immediately below the bridges. The principal features in the local protection at each of the towns are described as follows:

Local Protection at Piqua

The maximum discharge during the flood of 1913 is estimated at 70,000 second feet. The Lockington retarding basin on Loramie Creek will reduce the flood flow so that the quantity to be cared for at Piqua with the margin of safety provided in the Official Plan flood will be 81,000 second feet. After the 1913 flood a certain amount of work was done by the Commissioners of Miami County in constructing and repairing levees, and in removing obstructions, but this work was wholly insufficient as a protection against a repetition of the 1913 flood. The general nature of the protection work at Piqua covered by the Official Plan of the District, is shown in figure 17.

The protection consists of a combination of levee construction and channel enlargement, but the enlargement is confined practically to the amount necessary to supply material for the levee embankments. At the lower end of the city the river bed is of rock. Although affecting the cost of the work to some extent, the amount of rock to be removed is not a serious item. The Miami and Erie Canal where it passes through the city has been practically abandoned for all uses. For this reason the levee at the upper end has been located on the site of the canal, a line of 36-inch conduit being provided to care for the small amount of water flowing in the canal. A portion of the land occupied by the canal will be available for the widening of River

Street which, although having a rather heavy traffic, is a particularly narrow street. The river, at Piqua, makes a long loop around the low eastern part of the city. A levee is provided for around this loop, following the general line of the existing levee, it being impracticable to construct a cutoff channel to avoid the detour of the river. The new levee is from four to eight feet higher than the old one and is in

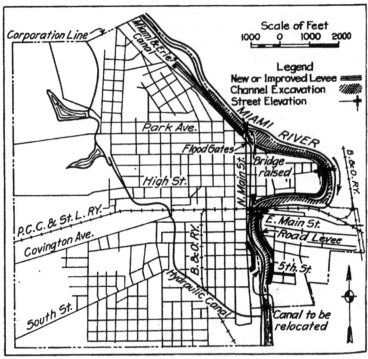

FIG. 17.—MAP OF PIQUA, OHIO, SHOWING LOCAL FLOOD PROTECTION WORKS.

places 19 feet high. Below the loop, the levee south of the Pennsylvania Railroad built by the County to protect East Piqua, is to be raised to give adequate protection. This levee extends far enough south to protect the residential property, but does not include the marble quarries.

The new Pennsylvania Railroad bridge and the new Union Street bridge have sufficient water way for the river at time of flood. It was found necessary for the Main Street bridge to be raised 3 feet to allow for passage of flood flows. The Ash Street bridge provides insufficient channel capacity, but it must be rebuilt in a few years and when rebuilt enlarged waterway will be required.

Local Protection at Troy

The plan of improvement at Troy, see figure 18, provides for a channel capacity of 98,000 second feet which is the reduced river flow for the Official Plan flood. This is obtained by enlarging the river channel and raising the levees. Increased capacity is also provided by means of the construction of a cutoff channel shortening the river

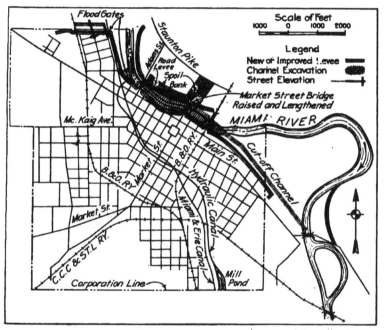

FIG. 18.—MAP OF TROY, OHIO, SHOWING LOCAL FLOOD PROTECTION WORKS.

at the large bend in the lower end of the city. The cutoff channel will not be excavated to full width, but the strong flows through it are depended upon to widen the channel within a short time, and meanwhile ample overflow space is available. The principal excavation in the river is between the B. & O. Railroad bridge and the old timber dam including the stretch of the channel opposite the business section of the city. The waste material from the excavation is utilized to produce attractive building sites on land owned by the District. Between the railroad bridge and the dam the right bank of the river was lined by an irregular series of stone and concrete retaining walls. Because of their light section, it was not possible to raise these and the new bank is made by filling against the old walls, the fill being protected from scour by concrete revetment and carried to the proper height to form the necessary levee embankment. The total excavation in the channel will be about 200,000 cubic yards.

To provide the requisite water way under the Market Street bridge it was necessary to raise it 4 feet and build an additional span of 150

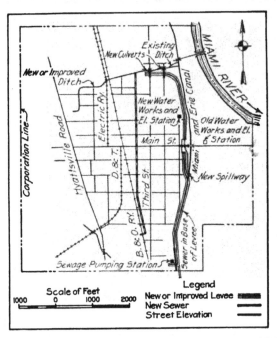

FIG. 19.—MAP OF TIPPECANOE, OHIO, SHOWING LOCAL PROTECTION WORKS.

feet. It was originally planned to permit the Adams Street bridge. a structure of small waterway which was under construction during the 1913 flood, to remain, supplementing its very inadequate waterway by reserving an overflow area on farm land northeast of the channel. This plan, although objectionable in several ways, appeared the cheaper in ultimate cost. It is now contemplated, however, by the County Commissioners to reconstruct this bridge with ample waterway, and the channel improvement plan has been modified to adapt it to the new bridge.

Local Protection at Tippecanoe City

Tippecanoe City is, for the most part, situated on high ground, and was, therefore, but slightly affected by the flood of 1913. The construction of the Taylorsville retarding basin will during large storms back water past the village, making protection necessary. The Miami and Erie Canal separates the higher portion of the town from the low plain on the east. Upon this low land were located a few moderate sized dwellings, and the village electric light and water works station. This land has all been purchased by the District, the dwellings are to be removed, and the electric light and water works station relocated at a point west of the canal.

The protection consists of a levee with its crest elevation 3 feet above that of the spillway at the Taylorsville dam. The route of the levee is along the canal and, for a portion of its length, is on the site of the canal. In order to care for storm water resulting from rains when the water surface in the retarding basin is high, a sewer is provided which intercepts the drainage of the higher part of the village and discharges it above the water surface in the basin. The drainage from the lower section of the village is taken care of by a pumping plant located at the southerly end of the levee. This feature is a provision against emergency conditions which are expected to occur rarely if ever. The general location of the various works is indicated in figure 19.

Local Protection at Dayton

The retarding basins on the Miami, Stillwater, and Mad Rivers, all within a few miles of Dayton, control about 2400 square miles out of the 2500 square miles of the drainage area of the rivers above the city. On account of this complete control much less radical changes are required in the Dayton channel than at some other points. A considerable amount of excavation is necessary, but widening of the channel is required only at isolated points.

Within the city the three rivers and Wolf Creek have a total mileage of 12 miles; there is a total length of over 18 miles of river bank that must be protected against overflow at time of flood. Channel improvement in addition to plain levee work is required in the Miami River from the Island Park Dam to the proposed new bridge at Broadway below the Big Four Railroad Bridge. Between the latter point and Washington Street an improvement had been undertaken by the City of Dayton before the flood, but was never completed. This improvement will be completed upon plans modified to adapt them to the new conditions. Channel excavation will extend over a distance of five miles. Through the center of the city the channel has a bottom width increasing gradually but retaining a section as nearly uniform as can be obtained without undue expense. The bed for a width of 150 feet will be depressed 8 feet below the foot of the river banks. Below Washington Street the channel has a nearly uniform width between toes of bank slopes of about 590 feet. The grade is .065 per

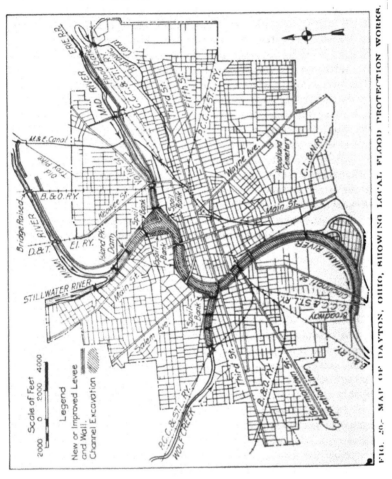

FIG. 20.— MAP OF DAYTON, OHIO, SHOWING LOCAL FLOOD PROTECTION WORKS.

cent above Washington Street and .08 per cent below Washington Street. Throughout its length the depressed bottom because of continuous curvature of the channel is to one side of the center of the channel, being within about 75 feet of the outer bank. In the section of the improvement below Washington Street, which is for the most part in an area but little improved, the material excavated will be used to provide roadways 50 feet in width on each side of the river.

Throughout the center of the city the levees are built to a height 2 or 3 feet above their old elevations. Above the Island Park Dam the levee along the left bank of the river required raising from 4 to 6 feet. The improvement allows for a depth of flow in the improved channel of about 26 feet, leaving 3 feet of free board below the tops of the levees. Practically no channel excavation is required along Stillwater or Mad Rivers, but a material increase in the capacity of Wolf Creek, a stream of 70 square miles drainage area, is obtained through deepening the bed.

As already stated, concrete revetment is provided along the Miami River on the outside of the curves to a height about 6 feet vertically above the bottoms of the bank. Above this height the banks and levee slopes are to be grassed. A flexible block revetment to a width of about 20 feet is provided at the foot of the plain concrete revetment.

The elevation of the levees at Dayton with respect to that of the land back of the levees has been such as to require the installation of gates on the outlets to the city's sewers, and pumping stations to care for the surface water and sanitary drainage during the high stages of the river. The improvement made by the District does not materially change this condition, and the pumping system will be operated by the City as before.

Dayton did not suffer from the loss of bridges during the flood as much as other cities in the valley. Since the flood four new bridges have been built according to plans approved by the District, three of which were to replace bridges destroyed in 1913. In addition to these, the B. & O. Railroad bridge above the city has been raised to provide enlarged waterway, as well as to obtain the elevation necessary for the road to clear the Taylorsville retarding basin. The B. & O. Railroad bridge over Mad River will be raised when the general elevation of the railroad tracks through the city is made.

The improved channel through Dayton will have a capacity sufficient to carry the discharge from the retarding basins during an Official Plan flood in addition to the unregulated flow entering the rivers between the dams and the city. The total discharge provided for is 75,000 second feet in the Miami River above the mouth of Mad River, 110,000 second feet between Mad River and Wolf Creek, and 120,000 second feet below the mouth of Wolf Creek. The tops of the levees will be 3 feet above the computed maximum flood level in such a storm. The greatest discharge during the 1913 flood was about

250,000 second feet. A sketch of the Dayton improvement is shown in figure 20.

Local Protection at West Carrollton and Alexandersville

These two villages, located on the left bank of the Miami about 6 miles below Dayton, are protected from future flooding by a levee on the west surrounding the low areas, as shown in figure 21.

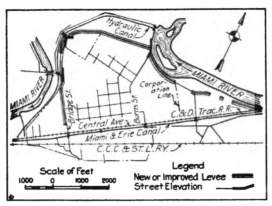

FIG. 21.—MAP OF WEST CARROLLTON, OHIO, SHOWING LOCAL FLOOD PROTECTION WORKS.

Local Protection at Miamisburg

This town is situated principally on the left bank of the Miami River about 11 miles south of Dayton at a point where the river valley is narrow and where the river channel fortunately is very wide. The maximum flow against which protection is provided is 130,000 second feet, this quantity being the estimated flow from the Official Plan flood with the retarding basin control. The flood flow of 1913 was about double this amount.

Protection is secured by the construction of a levee starting near the corporation line on the east bank of the river and extending to about 2000 feet below the corporation line at the south end. The retarding basin control reduces the flood flow to such an extent that the levee will not be required for the whole length of the river bank. Not far from the center of the village, Sycamore Creek enters the river from the east. This is not a large stream, but its runoff is rapid at times of storm. In order to give it a free outlet the levees are brought from the river on each bank of the creek as far as the Big

Four Railroad embankment and protection is not needed beyond this point. On the west bank of the river a large degree of protection is already afforded by the embankment of the B. & O. Railroad. In order to get the full value of this it proved necessary to raise an existing highway at the southerly end of the village from the railroad embankment to the higher ground on the west. This protection will be practically sufficient for a storm as large as that of 1913, but has not

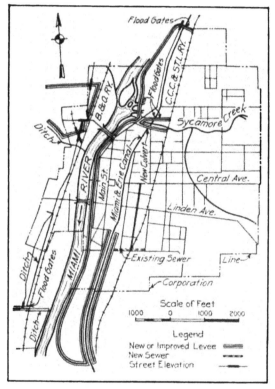

FIG. 22.—MAP OF MIAMISBURG, OHIO, SHOWING LOCAL FLOOD PROTECTION WORKS.

the standard margin of safety. At the northerly portion of the village west of the river, full protection is afforded by constructing a levee adjacent to the railroad and several feet higher with lines extending north and south of the protected district from the railroad to the high ground.

The Miami and Erie Canal runs through Miamisburg at an elevation, with respect to ground level, lower than at the other towns south of Dayton. The canal crosses Sycamore Creek on an aqueduct, which also acts as a spillway. Flood gates are required at three points where the levee and the canal intersect. A small amount of drainage work is required at the village. A general view of the levee system is shown in figure 22.

Local Protection at Franklin

The village of Franklin is situated in part on each side of the Miami River about three miles above the mouth of Twin Creek. Al-

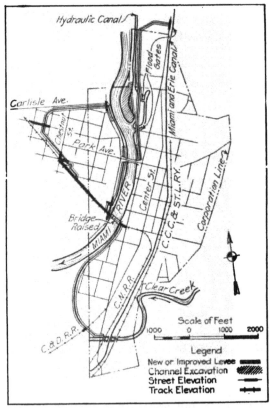

FIG. 23.—MAP OF FRANKLIN, OHIO, SHOWING LOCAL FLOOD PRO-
TECTION WORKS.

though the river channel at the village is narrow the overflow plain is very wide. The section of the village west of the river lies wholly within this plain, but because of its extent it is impracticable to provide full protection beyond the portion at present largely occupied by dwellings. It is also necessary, because of the narrow channel, to make use of this low plain for overflows during large floods. Standard protection is given the village within the leveed area against an estimated flow of 135,000 second feet. In addition to the levees, which are located as shown in figure 23, the protection work requires the enlargement of the channel at a point above the suspension bridge, where the channel is restricted and where the past flows have been concentrated against the west bank thereby creating considerable damage.

The Miami and Erie Canal conveys but a small quantity of water through the village, but gates are required at the entrance of the canal into the village to shut off what might be a dangerous flow at flood times. Flood gates are also provided at the intersections of the levee and the head and tail-race of the Franklin hydraulic, a canal used locally for water power. There are but two bridges at Franklin. One is a suspension bridge well above the water surface, and the other, the Cincinnati Northern Railroad Bridge, must be raised about 2 feet.

In addition to the protection given the village proper, a short levee is to be built from the east end of the dam, about a mile above the village, to the high land on the east. This is to prevent overflow of low ground, which has in the past been a source of damage to farm lands and to the banks of the hydraulic.

Local Protection at Middletown

This city is situated below the mouth of Twin Creek, and, therefore, is benefited by the effect of all five of the retarding basins. The greatest flow during the 1913 flood at Middletown was about 300,000 second feet. The flow to be provided against is 150,000 second feet. The river channel, including available overflow area, is very wide at Middletown, and the retarding basins give such effective control that the local protection needed is small compared with that at the other cities of the valley. The river bank between the northerly and southerly limits of the levee has a length of about 4 miles. For about 3 blocks at the center of the city, the land is so high that no levee is required. The maximum height of the levee is 17 feet. The only bridge at Middletown is a concrete arch structure about 1800 feet in length. In order to lead the river to near the center of this bridge a narrow cut-off channel is built. This channel was constructed 40 feet wide, but it has been widened rapidly by the water since its construc-

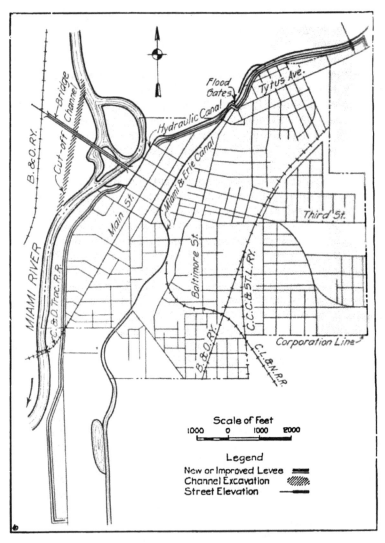

FIG. 24.—MAP OF MIDDLETOWN, OHIO, SHOWING LOCAL FLOOD PRO-
TECTION WORKS.

tion. Flood gates are required at the intersection of the levee and
the Miami and Erie Canal. The location of the levee at Middletown
is shown in figure 24.

Local Protection at Hamilton

The maximum flood flow at Hamilton in 1913 was about 350,000 second feet. The Official Plan flood, assuming all five retarding basins in operation, would deliver an estimated maximum flow at Hamilton of about 200,000 second feet. This flow is provided for in the plans, with a free-board on the levees varying from 3 feet at the south end of the city to 10 feet at the north end, at which point the most damaging flood flow entered the city in 1913. The situation at Hamilton presented the most difficult problem of any of the cities in the valley. Above this point the river has a drainage area of 3500 square miles, about 50 per cent greater than the drainage area above Dayton. Notwithstanding this fact, the channel at Hamilton at its narrowest point was only 390 feet wide, while the river at Dayton at no point was less than 550 feet in width. However, the channel at Hamilton was several feet deeper than that at Dayton. A large section of Hamilton is situated on each side of the river, and along both banks buildings stood close to the edge of the channel. In some instances these buildings were industrial works of considerable size and importance. The removal of many of these buildings was involved in the improvement, which consists principally in enlarging the channel to a minimum width, between toes of river banks, of 520 feet. At the lower end of the city the new bank is moved about 300 feet back from the old one. The entire improvement calls for the excavation of about 2,000,000 cubic yards of material. · Some of this has been utilized in building levees, but the larger part of it has been placed on land purchased by the District at the north and south ends of the city, totaling 110 acres. The depth from the bottom of the channel to the top of the new levees is about 34 feet. The general location of the improvements is shown in figure 25.

The great channel depth and steep surface slope will cause high velocities at time of high water. For this reason, concrete protection is provided along the east bank of the river above Main Street up to the estimated extreme water level. At the foot of the bank, flexible block revetment will be built over a width of 30 feet on the channel bed. On the west bank the slopes are to be revetted for about 6 feet of vertical height and extending down to the lower end of the bend in the river just below the city proper.

The Miami and Erie canal passes through the city at an elevation above maximum flood level. A local canal for water power, which entered the city from the north and passed through an important industrial quarter discharging its tail water into the river near Main

6

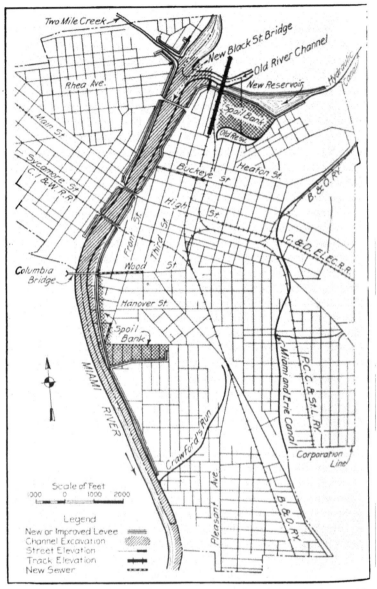

FIG. 25.—MAP OF HAMILTON, OHIO, SHOWING LOCAL FLOOD PRO-
TECTION WORKS.

Street, has been removed from the protected area to a point north of and outside of the city, and a new hydro-electric plant has been built for the utilization of the entire canal flow.

Flood Protection at Other Localities

The reduction in flood stages due to retarding basin control is the only protection furnished the farming properties along the river. In most cases this reduction is such that during ordinary floods little trouble from overflow need be looked for. During very high floods, such as that of March 1913, the higher land will be completely protected from overflow and the land lying very low will be benefited by an appreciably less depth of water and a markedly shorter duration of overflow. To have built special protective works for farm land in most cases would have cost more than the resulting benefits.

The same remarks apply to a few small villages where the area or the aggregate value of property subject to inundation is small, and where the reduction by retarding basin control is sufficient for almost complete protection. The village of Cleves situated in Hamilton County, about 5 miles above the mouth of the river, is affected more by back water from Ohio River flood stages than by high water in the Miami. The retarding basin control does benefit Cleves and other parts of Hamilton County, however, by removing in large part the danger of damage by swift currents in the Miami, which have been a source of considerable injury in the past to lands, roads, and bridges.

Modification of Existing Structures at Owner's Expense

Under Section 20 of the Conservancy Act, structures such as bridges, sewers, and pipe lines encroaching upon waterways must be changed or modified so as to conform to the District's plan of protection, at the expense of the owners.

Changes of this nature were required at nearly all of the cities in the District. In some instances bridges had to be raised in order to secure sufficient clearance for high water stages; in other cases water mains and gas pipes crossing river channels had to be lowered in order to be below the established grades for the channel bed. In some cases, as the result of contracts made with municipalities, these changes are being made by the District.

Railroad Changes

Aside from the relocation of railroad tracks made necessary by the construction of the retarding basins, as described in previous

pages, railroad tracks at a number of points in the valley are required
to be raised in order to be above future high water levels. In most
cases the readjustments so required are of minor proportions, but they
are important, nevertheless, to preserve the integrity and consistency
of the plan as a whole.

SELECTING THE RETARDING BASIN SITES

Before a final decision could be reached as to the number and
location of basins, it was necessary to study many sites. Aside from
the general scheme of distribution of the basins so as to control the
largest possible amounts of drainage area above the principal cities,
other considerations entered into these studies, chief among which
were the securing of the largest possible amount of storage capacity
with the least expense for building dams, and with as little interference as possible with existing public service utilities, cities, farms,
and other valuable property. The former consideration required the
best possible utilization of the valley topography. The United States
Geological Survey maps, on a scale of approximately one mile to the
inch, were found to be of great assistance in locating retarding basin
sites, but in order to arrive at dependable figures for making estimates it was necessary to make topographic surveys of the sites, and
to plat the results on maps to a scale of 1 inch equal 500 feet. In
order to determine the exact location for each dam it was necessary to
make larger maps, using a scale of 1 inch to 200 feet. Finally, still
larger maps were used for working out the details of these structures.

In the earlier stages of the investigation the advisability of providing a retarding basin on the upper Miami River above Port Jefferson, and another on upper Mad River above the mouth of Buck Creek
near Springfield, was considered. The capacity of the first was limited by the existence of large areas of flat lands near the upper end of
the reservoir, which would have become covered with water if the dam
were raised beyond a certain height. Below this elevation the capacity
of the basin would have been too limited to reduce the flood flow by
more than 10,000 second feet at Troy and Piqua. It was found that a
saving of about half a million dollars could be effected by abandon-
ing this basin, enlarging the channel capacities at Troy and Piqua,
and increasing the height of the Taylorsville Dam to give the equivalent protection to Dayton and the cities below. A number of complications, some of them having no relation to engineering difficulties
resulted in eliminating the basin at Springfield. Equivalent results
were secured by increasing the height of the Huffman Dam. The
studies which led to the system as finally adopted are given in detail
in chapter XII.

In general, it has not been found advantageous to build dams above each other on the same stream, if it was possible to provide a single retarding basin large enough to give the desired protection. An illustration of this is the case of the Stillwater River, where the possibility of using a smaller dam at Englewood and another dam above at Pleasant Hill or at Clayton was studied and rejected, because the two basins together were not as economical as a single large one at Englewood.

The final results proved that a great deal of preliminary investigation in picking out dam sites was justifiable. For example, at Huffman seven combinations were studied involving various heights of dam and types of spillway. At Taylorsville, the cost was computed for several locations, for different heights and locations of dam, and various types of spillway. The present location was chosen because at that point more economically than elsewhere the combined spillway and conduit structure could be built on a rock foundation. On Loramie Creek a site above Lockington was investigated, but the capacity of the basin was found inadequate. A dam also was considered on the Miami just below the mouth of Loramie Creek, to back water up both this creek and the Miami River. It was rejected because of objections from the people of Sidney who would fear backwater, though from an engineering standpoint it perhaps was the best location. Its construction would have eliminated the need for local protection works at Piqua, and would have reduced the local work required at Dayton. Many combinations were studied on the Stillwater River. In addition to the one in the vicinity of Englewood, possible sites at West Milton and Little York were examined. In the vicinity of Englewood six dam locations and thirteen spillway arrangements were investigated. At Germantown, six locations for the dam and three types of spillway were considered.

The methods used in finally deciding on the heights of dams, capacities of basins, and sizes of outlets, so as to proportion these various elements with respect to the channel capacities at the principal towns, are described in detail in chapters VII and VIII.

CHAPTER IV.—OPERATION OF RETARDING BASINS

This chapter outlines in general terms the principles upon which the design and operation of the retarding basins is based, and discusses briefly the effects to follow from their construction. The succeeding chapters deal in much greater detail with the various steps taken in the design.

RUNOFF FROM MAXIMUM STORM

In chapter II the reasons were given for taking as the basis of design a flood discharge 40 per cent greater than that of March 1913. It will now be shown by what method a flood wave, or flood discharge. of this magnitude was proportioned so as to be characteristic of Miami River conditions and also be capable of expression in flow and time units convenient for use in computation.

The method adopted consists in assuming rates of runoff, in terms of inches of depth on the drainage area, over given periods of time. Two distinct advantages result from this. In the first place, the data in this form is applicable with but little or no modification to any section of the Miami drainage area, and from it the discharge in cubic feet per second can be computed readily for any point on the main stream or its tributaries for which the drainage area is known. Second, it is a comparatively simple matter to decide on the shape of the runoff curve because it can be patterned after that of the 1913 storm. The latter had been determined with great care from the actual flood and rainfall data, as described in chapter II, and offered therefore a convenient and reliable starting point. The use of runoff in preference to rainfall, aside from the standpoint of convenience, is justified in view of the fact that in the 1913 storm practically all of the rainfall reached the streams in the form of runoff, and it is reasonable to expect that the same will hold true for a storm 40 per cent larger.

The first step is to decide on the duration and the amount of storm runoff. The studies of great storms, referred to in the second chapter. brought out the fact that in this section of the United States the periods of intense storm rainfall are not likely to exceed three consecutive days. This was the case in March 1913, and most of the runoff occurred within a similar period. It was therefore decided to make the period of maximum runoff three days.

The maximum 3-day runoff in March 1913 averaged 7.27 inches over the catchment area above Dayton. As a result of the experience and judgment of the engineers of the Conservancy District in flood control, and of their studies of rainfall in the eastern United States, it was finally concluded that a runoff of 10 inches in three days for the smaller drainage areas and of 9½ inches for the larger areas should be assumed as a standard, which the retarding basins should be designed to handle without overflowing the spillways. This standard is referred to hereinafter as the Official Plan flood. With such a design the retarding basins system will give the full protection intended in case of a storm about 40 per cent greater than that of 1913.

The next step is to determine the distribution of the assumed runoff during the 3-day period. To do this it is necessary to assume a distribution in a manner analogous to that which has been observed to take place during large storms, making the expressions of these variations so simple, however, as to prevent their mathematical use from becoming cumbersome. The importance of the latter requirement will not fail to impress one, when in reading chapter VII he becomes aware of the many times that this data had to be used in working out the proportions of the retarding basins.

Figure 26 shows in diagrammatic form the distribution for the 10-inch and 9½-inch standard runoffs as finally adopted in the Official Plan flood, and affords comparison with the actually observed runoff from the 1913 storm. As a flood does not start from an initial flow of zero but from some intermediate stage of the stream, it was decided to start in each case with a rate of runoff of ½ inch per 24 hours over the drainage area. This rate was then increased uniformly over a period of 16 hours, at the end of which it reached a maximum of 4½ inches per 24 hours over the drainage area. This is indicated by the heavy dotted line in figure 26 which shows the distribution for the 10-inch runoff. The solid line similarly indicates a maximum rate of 4 inches per 24 hours for the 9½-inch runoff. These maximum rates were assumed to remain constant for 24 hours, at the end of which they were made to taper off uniformly through a period of 32 hours to a final rate of 1¼ inches per 24 hours over the drainage area, in the case of the 10-inch runoff, used in retarding basins with small outlet, and of 2 inches for the 9½-inch runoff used in retarding basins with large outlets. There is no necessity in this connection for carrying this discharge further, as at the lower limits the outflow will equal or exceed the inflow.

It will be noted that both diagrams show a quick rise to the maximum rate, and allow twice as much time for falling as for rising. In each case the final rate is much larger than the initial rate. The reason for this is that the water in a retarding basin reaches its maximum stage toward the latter part of a flood, when the inflow into the

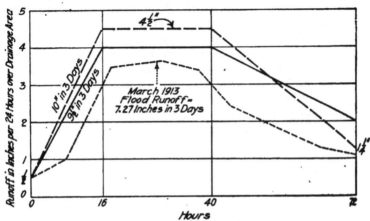

FIG. 26.—MAXIMUM FLOOD RUNOFF CURVES ADOPTED FOR THE OFFICIAL PLAN FLOOD.

The assumed runoff rates shown give approximately 40 per cent more total discharge in 3 days than did the March 1913 flood, and were used as a standard, known as the Official Plan flood, for the purpose of designing the retarding basins. The runoff rates assumed are no implication that floods of such magnitudes are considered as at all possible of occurrence.

basin has decreased to the same rate as the outflow through the conduits.

If the 9½-inch runoff curve be extended, at its final rate it yields 10 inches in 3⅓ days. In other words, under the assumptions as made, it would take the larger drainage areas a third of a day longer to contribute the same amount of water as the smaller areas. In the light of the data collected such an assumption appears justifiable for maximum flood conditions.

As actually applied, the 10-inch runoff was used in the design of the Germantown, Lockington, and Englewood basins, and the 9½-inch runoff in the design of the Taylorsville and Huffman basins. As the Englewood retarding basin has nearly as large a drainage area as the Huffman, it might appear that the 9½-inch runoff would have been sufficient in its case. The outlet of the Englewood dam is, however, relatively small, so that while for a 3-day storm the 9½-inch runoff would leave it in as favorable condition as the Huffman retarding basin, in case of a storm of longer duration than three days its smaller outlet capacity would give the basin less capacity for handling the inflow than the Huffman basin with its much larger outlets. On account of this it became necessary to give the Englewood basin a capacity greater than required for a 3-day storm, in order to provide the desired margin of safety in case of a storm of longer duration. The 10-inch runoff in three days was therefore adopted for Englewood. It is believed that the rates of runoff, as applied, provide approximately the same margin of safety for both large and small retarding basins.

It is not here contended that the exact form of runoff curve shown in figure 26 possesses any special merit, aside from simplicity. Nevertheless it fulfills admirably the requirements of the case, as in actual flood hydrographs for the Miami River the runoff increases rapidly during the first part of a flood, remains nearly constant during the high stages, and then decreases gradually. The drainage from small areas being always more rapid than from large ones, it follows that the maximum rates of runoff from the former should be higher than for the latter. It also follows that at the end of the flood period, the drainage having been more thorough from the small areas, their final runoff rates should be less than from the larger areas. All of these conditions are embodied in the Official Plan flood.

In the course of the studies a number of runoff curves were tried out, some of them with much higher peaks, others of quite elaborate form, as described in chapter VII. It was found that so long as the runoff in each case totaled the same for a 72-hour period, and in a general way conformed to the shape of hydrographs characteristic of Miami River floods, there was but little difference in the ultimate result. This is natural, as the action of any large retarding basin during a flood is in essence an equalizing process in which fluctuations of inflow tend to be obliterated. The time element, on the other hand, was found to be of considerable importance. Shortening or lengthening the period covered by heavy runoff affects materially the proportions of outlet conduits and basin capacity required.

SIZE OF OUTLET CONDUITS

As the purpose of a retarding basin is to hold back temporarily all flow in excess of the capacity of the river channel below the dam, the outlets must be designed to give a maximum discharge not exceeding this channel capacity, taking into account inflow from uncontrolled tributaries below the dam. This maximum discharge occurs when the basin is filled to its highest level, and the hydrostatic head is greatest.

Theoretically, an ideal plan would be to regulate the outflow by means of gates, so as to maintain a uniform rate throughout a flood period, equal to the maximum determined upon. Such a plan would effect the largest aggregate outflow consistent with safety, and naturally would require a minimum of basin capacity. However, such mechanical regulating devices would require manipulations during flood, and constant care during the long periods between floods.

The tendency to keep the gates closed during small floods, so as to avoid the minor inconvenience and damage of small freshets, might result in their being kept closed too long in case of what should prove to be a great flood.

Such regulating devices, if properly used, probably would come into operation only at intervals of several years. Because of the seriousness of the result if they should fail to operate properly, it was decided to omit all gates or other devices for manipulating the openings, leaving the outlets fully open at all times, so that there could be no failure to respond promptly at the critical moment. The system should be to the greatest possible degree dependable, automatic, and fool proof.

Such automatic control implied the adoption of outlet openings of fixed dimensions, somewhat smaller than would have been possible if provision had been made for opening and closing them. This, in turn, made necessary larger retarding basin capacity and consequently higher dams. These objections were considered to be more than offset by the elimination of operating mechanisms, and the consequent attendance and upkeep costs for all time of equipment of this kind. Independence from human control and absence of movable parts, in the operation of these retarding basins, are significant features characterizing the Miami Valley flood control works. The total maximum discharge of the outlets is only $\frac{1}{6}$ less than that of the theoretically best possible. What difference this represents in the total amount of water to be held in a retarding basin during a big flood depends upon local conditions. In the case of each of the Conservancy basins it proved to be a small item when measured in terms of added height of dam and additional lands to be flooded. This is especially true of the Englewood and Germantown basins, where the conduits are so small that all but a small fraction of the total inflow during a large flood will be stored, the net difference in storage as between the two types of outlets being almost negligible. A more detailed discussion of the design of these outlet works is given in chapter VI.

Consideration was given to the possibility of the clogging of the outlets by drift or ice. At high stages considerable drift may come down the streams, unless it is regularly removed by the forces of the District. Provisions will be made to prevent large pieces of drift from lodging at the entrances of the conduits. These provisions will vary at the different dams, depending on local conditions.

CAPACITIES OF RETARDING BASINS

Having determined the carrying capacity of the improved channel below the dam, and therefore the allowable capacity of the outlet, the next step is to determine the required basin capacity.

The relations of inflow, outflow, and basin capacity are readily visualized by referring to diagram 27, in which $A\ B\ C\ D\ E\ F$ represents the total inflow, equal to the 72-hour runoff from the drainage area above the basin. The area inclosed by any curve in the diagram shows the total volume of water up to any instant, being the product of rate of flow multiplied by time. The curve $A\ I\ E$ shows the variation in rate of flow through the outlet conduits, the rate growing rapidly at first, as the retarding basin fills, then increasing quite

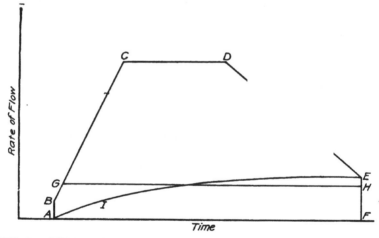

FIG. 27.—RELATION BETWEEN INFLOW, OUTFLOW, AND RETARDING BASIN CAPACITY WHEN SPILLWAY DOES NOT OVERFLOW.

slowly until the maximum is reached. This maximum is represented by the ordinate $F\ E$, and corresponds to the instant when the rate of inflow has decreased to an amount equal to the rate of outflow. Obviously, at that moment water has ceased being added to that held in the retarding basin, the water level in the latter has reached its maximum elevation, and the head on the outlets is a maximum. The ordinate $E\ F$ must equal the maximum outflow capacity of the outlet conduits.

It follows from the above that the area $A\ I\ E\ F$ represents the total amount of water discharged during the filling period. The storage capacity of the basin, being naturally the difference between total inflow and total outflow, is represented by the area $A\ B\ C\ D\ E\ I$.

If, by means of gates or other controlling devices, the same total outflow were to be secured by means of a uniform flow throughout the flood period, such constant rate would appear as a horizontal line $G\ H$ and ordinate $H\ F$, and the area $A\ B\ G\ H\ F$, representing the total volume discharged, would equal the area $A\ I\ E\ F$.

In designing the retarding basins of the Miami Conservancy District, $H\ F$, in all cases, was found to be close to ⅚ of $E\ F$. It may, therefore, be stated as a convenient rule, that, under assumptions similar to those here made, *the approximate capacity of a retarding basin may be arrived at quickly by computing the total outflow during the filling period on the basis of a uniform rate equaling ⅚ the maximum outlet capacity, and deducting this total outflow from the total inflow during that period.*

Throughout the preceding discussions it has been held that the limit of capacity of a retarding basin is at the crest of its spillway. It is true that when water flows over the spillway the depth on the crest represents a vast amount of additional storage in the basin. The fact remains, however, that maximum efficiency in retarding basin design is secured when the spillway is located at such an elevation that it will be reached, but not overtopped, by the water above the dam during the greatest flood the basin ever will be called upon to control. Flood damage is determined, especially in case of urban property, not by duration of flooding, but by depth of flooding. Levees will be broken down and buildings and personal property damaged about as much by a flood lasting for a few hours, as by one lasting several days at the same stage. Therefore, the estimated capacity of the channel below the dam must never be exceeded, or the work will fail of its purpose. To make the outlets of such capacity as only to partially fill the channel below the dam, depending upon flow over the spillway to add to the flow through the outlets, is a wasteful method.

The channel below the dam can carry the maximum flow throughout the whole flood period about as safely as it can for a few hours at the flood crest. If the outlets are made large enough to fill the channel below the dam to the safe limits, this capacity will be serving to relieve the basin throughout the whole flood period, which in effect adds to the capacity of the basin. If the outlets are made smaller, with the expectation that the channel below will be further filled by water flowing over the spillway, a larger amount will be retained in the basin, which in effect reduces its capacity. The margin of safety in that case actually is reduced, because the basin fills more quickly, and rises to a higher level over the spillway; and the total amount that must be carried in the channel below is increased. Overflow of the spillway means a shorter but a higher river stage below the dam, where maximum elevation and not duration is the limiting condition.

It should, therefore, be a fundamental principle in retarding basin design to place the spillway at such elevation as will just clear the highest water level in the basin during the assumed maximum flood, which in the case of the Miami Valley system is the Official Plan flood. To place it higher would add a margin of safety that is not needed, since the maximum assumed flood is selected so as to provide the nec

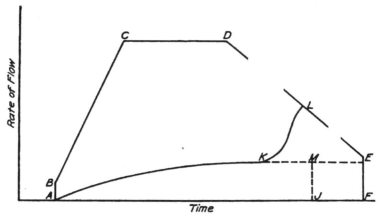

FIG. 28.—RELATION BETWEEN INFLOW, OUTFLOW, AND RETARDING
BASIN CAPACITY WHEN SPILLWAY OVERFLOWS.

essary margin of safety. To place it lower would result in spillway overflow into the river channel when the latter already should be carrying the full amount for which it is designed.

Figure 28 shows the condition which would result from providing too small an outlet without additional basin capacity to offset it. The total inflow has been assumed the same as in figure 27. The maximum discharge through the outlets is represented by the ordinate MJ

which is less than *EF*. The outflow curve *AK* shows the rates of discharge through the outlets up to the time when the basin is filled to spillway level and begins to spill. *KL* indicates the combined discharge of outlet and spillway, with a maximum at *L* exceeding the safe carrying capacity of the channel, *EF*. It is scarcely necessary to point out that such a condition would jeopardize the safety of the levees and would be likely to result in considerable damage from overflow.

SPILLWAYS

Since the spillway of a retarding basin should be so located that its crest will be just above the water level in the basin during the assumed maximum flood, theoretically, the spillways will never come into use. They have no function in the usual operation of the retarding basins, and the capacities of the improved channels below do not provide for large flows over the spillways. Their only function is to further insure the safety of the dam structures. Their insertion in the design of the dams should be looked upon as a factor of safety against imperfection of design or the most remote contingencies.

They are so proportioned as to make the earth dams safe against overtopping in a flood twice as great as that of March 1913, or equal to 14 inches of runoff in 72 hours. Thus, if a storm could occur great enough to fill the retarding basins above the spillway crests, the dams themselves would be safe against damage. A slight flow over the spillways would not necessarily result in material damage below, because the levees through the cities have an elevation 3 feet above the flow line of the assumed maximum flood. Should flow over the spillways result in overtopping the levees, the damage would be very serious. but still far less than though the basins were not in operation.

In calculating the discharge over the spillway crests, it was decided that the maximum level of water reached in the retarding basins in case of a 14-inch runoff from the drainage area in 72 hours, should not exceed 10 feet above the level of the spillway crest, leaving as a still further factor of safety a freeboard of 5 feet to the top of the dam. The sum of the storage capacity and the total amount discharged through the outlet conduits was subtracted from the total 14-inch runoff to obtain the amount that must be discharged over the spillways. The detailed method of making these determinations is given in chapter IX. The lengths and elevations of spillway crests for the various dams are given in table VII of the Appendix.

In designing the spillways of the Miami Conservancy dams many practical considerations had to be taken into account, which operated to modify the dimensions dictated by theoretical requirements. Thus.

at the Lockington, Taylorsville, and Huffman dams the length of crest had to be increased in order to fit in with the design for the combined outlet and spillway structures needed at these points. Lengthening the spillways did not materially affect the cost of these structures, the principal dimensions of which were governed by the size and number of outlet conduits to be provided in them, and of course added to the safety of the dams.

In the case of Lockington and Taylorsville dams it developed that a great deal of local opposition to the construction of these structures due to backwater could be removed by lowering the spillway elevations by two feet. Investigation showed that the Official Plan flood would then fill the basins to an elevation nearly two feet above the spillways, but that the discharge from the spillways under so small a head would be very small and could safely be taken care of by the channels below. The design was accordingly changed.

At the Englewood basin the spillway crest was placed several feet above the stage that would be reached by the Official Plan flood. This allows the storage in the basin to be increased to such an extent that a 14-inch runoff in 72 hours can be taken care of by a 100-foot spillway, without exceeding the safe head of 10 feet. It was found to be more economical to pay the cost of the greater land damage and higher dam required by such a scheme than to build a longer spillway. At the Germantown basin a similar solution was worked out. As the spillway for this dam consists of a channel across a saddle in the hills bounding the basin, greater economy was secured by increasing the height of the dam and paying the resulting land damage, than by excavating a large spillway channel through the rock to a lower elevation.

In both cases increasing the height of the spillway had another beneficial effect on the operation of the basins. Figure 29 shows the size of floods of various durations that the basins are capable of handling without raising the stages above the maximum flow line or spillway level. For runoffs of short duration, the Huffman and Taylorsville basins will be seen to have the least capacity. Since they control large drainage areas, the most rapid runoffs are not likely to occur. Because of their large outlets, the size of storm which can be handled in them increases rapidly for longer periods of runoff, and it is evident that they are amply safe for storms of long duration. The Englewood basin, however, while its capacity for short duration floods is large, has but little more capacity for storms of great duration. By raising its spillway, the capacity is so increased that its ability to take care of large floods of great duration is considerably extended. The same is true for the Germantown basin, but

because its outlet conduits are larger in proportion to its drainage
area than is the case at Englewood, the raising of its crest was relatively less important.

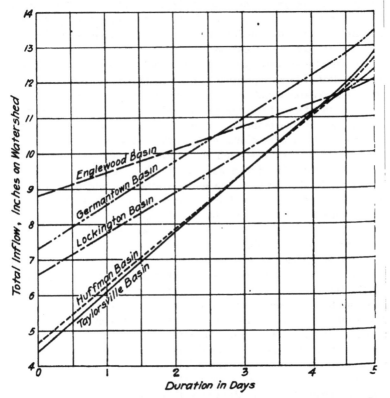

FIG. 29.—INTENSITIES AND DURATIONS OF STORM RUNOFF
WHICH WOULD FILL THE RETARDING BASINS TO THEIR SPILL-
WAY LEVELS.

PROTECTION AFFORDED BY BASINS

In the preceding pages it has been explained that the retarding
basins were proportioned individually as well as in combination to
take care of the Official Plan flood, which assumes a runoff 40 per
cent greater than that of 1913. It should not be inferred from this
that they are designed to protect against large floods only. Any
flood which under former river conditions would have exceeded appreciably the river channel capacities, will have its crest discharge
diminished by the retarding action of the basins.

The reduction in peak flood flow brought about by the basins is not proportional, however, to the magnitude of the floods, the effect being to reduce the larger floods much more in ratio to their size than the smaller ones, since the outlet conduits discharge essentially as orifices, and their outflow therefore varies as the square root of the hydrostatic head on them. Hence, for small depths of water in a retarding basin, the rates of outflow show a rapid increase with increase of depth, but as these depths become greater the increments in outflow decrease until at the larger depths the outflow rates become nearly constant. Thus, the retarding basins when half full discharge nearly as much as when completely filled. Another factor that enters into the matter is the shape of the retarding basins themselves. These roughly are V-shaped valleys, the bottoms having a slope of several feet to the mile toward the dam. Therefore, the capacity per foot of depth in the wide top portion is much greater than at the bottom. Minor storms will fill the basins to depths greater in proportion to their rainfall than will large storms.

The combined effect of these two factors is to give relatively large discharges for small floods and relatively small discharges for large floods. This phase of retarding basin action is of especial advantage, operating as it does to relieve the overflowed bottom lands in the basins during small floods in a short space of time, thereby causing the least possible interference with agricultural operations. This is the more important since small floods have the greater frequency. In the case of larger floods the rates of outflow from any basin are very nearly uniform, approaching the maximum for which the outlets are designed.

The effect, as described above, of the four retarding basins above Dayton in regulating the maximum flow through the city for a great variety of floods of different magnitudes and durations is shown by the diagram, figure 30. For the purpose of making this diagram each individual flood considered is assumed to have the same duration and to produce the same total depth of flood runoff over the catchment areas above each of the four retarding basins. The effect of certain specific selected floods of high runoff over restricted areas, when centered over the different catchment areas, will be discussed subsequently in chapter X. It seems advantageous first to consider here by means of figure 30 floods of such wide-spread extent as to be practically uniform over all the drainage area above Dayton.

The lowest curved line in the diagram, figure 30, is for flood runoffs lasting only 1 day. The curve is drawn through successive points, each determined by using the depth in inches of total flood runoff as an ordinate and the corresponding maximum value of the regulated

7

flood flow through Dayton as an abscissa. Thus, a 6-inch flood run-off in 1 day would produce a flood flow through Dayton of 110,000 second feet, the safe channel capacity. The tops of the levees will be 3 feet above the flow line for such a discharge, and the ultimate channel capacity, before overflow would begin, is about 135,000 second feet.

While it is not supposed that greater flood runoffs can ever occur in this region, it is a necessary part of the procedure in the engineering design of the protection system, to determine the margin of safety

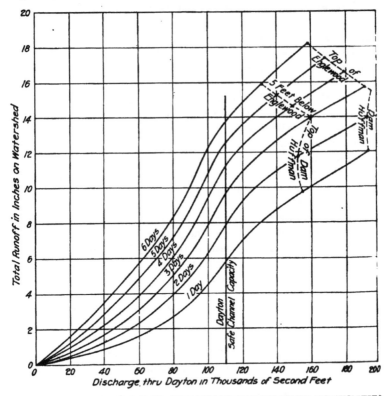

FIG. 80.—FLOOD DISCHARGE AT DAYTON FOR VARIOUS INTENSITIES AND DURATIONS OF STORM RUNOFF.

The discharges shown are peak rates, and all four retarding basins above Dayton are assumed to be in operation.

of the plans by considering the effects of hypothetical flood runoffs much greater than any that may possibly occur. Thus, the curve for 1-day floods is continued, and the diagram shows that it would require a runoff of 8 inches in 1 day to bring the spillways at the dams into operation sufficiently to produce a discharge at Dayton of 135,000 second feet. Similarly, a flood runoff of 10 inches in 1 day would produce a discharge of 160,000 second feet. The point at the intersection of the 1-day curve with the broken line marked ''5 feet below top of dam,'' indicates that, if such an event were possible, the Huffman basin would be filled thereby to a point 5 feet below the top of the dam, but no other basin would be filled to an equal extent. In order to have a conservative factor of safety, the dams as engineering structures are built to be perfectly safe against such a depth of water, and the outlets and spillways are strong enough to handle the resulting discharge.

The curve is continued, showing that a runoff of 12 inches would be required before the top of any of the dams would be reached, the first one for this case being Huffman.

Similar curves are drawn for flood runoffs spread uniformly over longer periods, up to 6 days. A heavy vertical line is drawn at a discharge of 110,000 second feet, the safe channel capacity in Dayton. The diagram shows that to produce this discharge will require a total flood runoff of about 8 inches in 2 days, or 10 inches in 3 days, or 14 inches in 6 days. This indicates that Dayton is protected against any storm that is within the range of probability for this section of the United States.

Figure 30 indicates by the inclinations of the lines in the lower portion, that the maximum rates of flood flow at Dayton, with the retarding basins in operation, at first increase rapidly as the rates of runoff increase. As the lines grow steeper the increases in flood discharge are seen to become less rapid so that for large depths of runoff the rate of increase tends to become constant. This is due to the orifice-like behavior of the conduits at the dams as previously explained. It will be noted that the lines in the diagram take a flatter slope where they cross the vertical line which represents the limit of safe channel capacity in Dayton, showing that the limit of greatest efficiency of the retarding basins is reached at that stage. Beyond this line there is a more rapid increase in flood flow for given increases in runoff, due to spillway overflow being added to the discharge from the outlets.

The diagram shows in a graphical manner that the flood protection system is so designed that in the Official Plan flood the reduced discharge through Dayton will just equal the channel capacity, leaving a margin of 3 feet from the water level to the tops of the levees. The diagram is of interest also in showing the various combinations of storm intensity and duration which the system is capable of handling.

For very high rates of runoff a point is reached where the head on the spillway is the maximum for which the dams and spillways are designed. This has been fixed at 5 feet below the tops of the dams. The depth of runoff which will cause this stage to be reached is shown on the curves by a dotted line accompanied by the name of the retarding basin at which this point would first be reached. It has already been explained, see figure 29, that the ability of the various basins to control floods of various durations, without overtopping the spillways, differs somewhat. Similar differences obtain at maximum stage. Thus a 10-inch runoff in 1 day would fill the Huffman basin slightly above the maximum stage, and would nearly reach that stage in the Taylorsville basin. These two basins, though large, are less able to absorb short floods of great intensity than some of the other basins, because their capacities are small relative to the drainage areas controlled by them. Because of their large conduits, however, they are especially adapted to retarding great floods of long duration. Twelve inches of runoff in 2 days time would be required to fill the Huffman Basin to maximum stage, i. e. 5 feet below top of dam.

As has been previously stated a 14-inch runoff in 3 days was used as the criterion for proportioning the spillways, and such a runoff would produce the maximum stage in the Englewood and Huffman basins. At Lockington and Taylorsville the resulting stage would be slightly below the maximum because of the 2-foot reduction in spillway height.

The Englewood basin has the largest capacity of the five basins. Owing to its small conduit discharge capacity, large storms, even when distributed over long periods, will fill this basin, whereas they would not do so in the case of the other basins equipped with relatively larger conduits. This explains why in the diagram the 4, 5, and 6-day runoffs are shown to fill this basin to maximum stage sooner than the other three.

As regards filling the retarding basins level with tops of dams, the diagram shows the relation between the basins to be the same as for the maximum stage conditions just described, the Huffman basin giving less protection against short and heavy runoffs, and the Englewood basin against long protracted ones.

Throughout the foregoing discussions it should be borne in mind that all assumptions of runoff necessary to fill any of the retarding basins to tops of dams or within 5 feet of the tops are purely theoretical considerations, such floods being considered wholly impossible of occurrence in the Miami Valley. As the years pass, the public knowledge that the works are safe far beyond any burden that may be put upon them will be the basis of confidence on the part of the

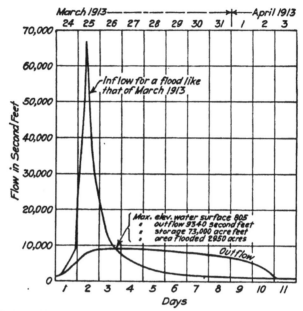

FIG. 31.—INFLOW AND OUTFLOW HYDROGRAPHS OF TWIN CREEK AT GERMANTOWN RETARDING BASIN FOR A FLOOD LIKE THAT OF MARCH 1913.

people of the valley, and this complete confidence in a very large factor of safety will justify the unusual precautions which have been taken, and the expense which these precautions have incurred.

BEHAVIOR OF RETARDING BASINS DURING A FLOOD LIKE THAT OF MARCH 1913

The behavior of the retarding basins is best visualized in figures 31 to 35, which show in hydrograph form the inflow and outflow rates for the five retarding basins, assuming a repetition of the March 1913 flood to occur. The irregular peaked line, in each instance, represents the unmodified river discharge or inflow into the basin, and the curved line the resulting discharge through the outlets in the dams. The contrast between the flashy nature of the high-crested flood wave entering the basin, and the smoothly appearing outflow curve, devoid of any sharp crest whatsoever, is striking.

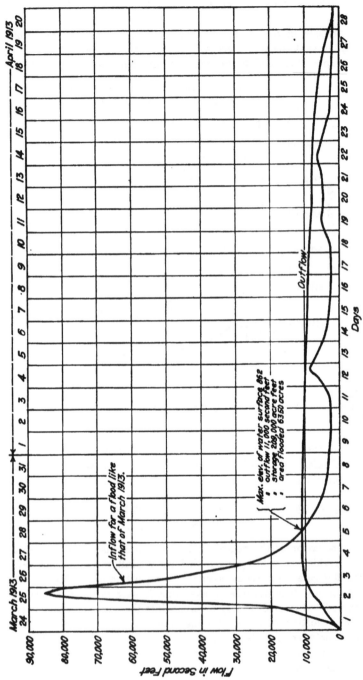

FIG. 32.— INFLOW AND OUTFLOW HYDROGRAPHS OF STILLWATER RIVER AT ENGLEWOOD RETARDING BASIN FOR A FLOOD LIKE THAT OF MARCH 1913.

Figure 32 shows the crest discharge of the Stillwater River in 1913 to have been 85,400 second feet on March 25, the entire flood period lasting 5 days. It also shows that the effect of the retarding basin would have been to reduce the crest discharge to ⅛ this amount, and to extend the flood wave over a period of 27 days. Figure 34 shows that the Taylorsville basin would have reduced the flood crest in the Miami River from 127,000 second feet to 51,300 second feet, or to about 40 per cent of the maximum, while the period of flood flow would have been lengthened but little if any. These diagrams illustrate the two extremes in retarding basin behavior, the Englewood basin with its huge capacity and small outlets effecting a great reduction in flood crest and extending the outflow over nearly a month,

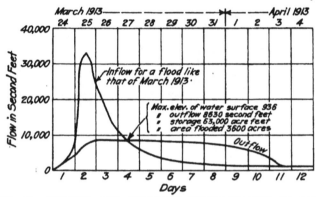

FIG. 33.—INFLOW AND OUTFLOW HYDROGRAPHS OF LORAMIE CREEK AT LOCKINGTON RETARDING BASIN FOR A FLOOD LIKE THAT OF MARCH 1913.

while the Taylorsville basin with its large outlet capacity effects less of a reduction in the flood crest and empties in a few days time. The behavior of the other retarding basins during a flood like that of March 1913 is between these extremes; in a general way the Germantown and Lockington basins, shown in figures 31 and 33, are in behavior similar to the Englewood, while the Huffman basin curves, see figure 35, bear close analogy to those of the Taylorsville. In chapter VIII, figures 77 and 78 show similar hydrographs for the large flood of October 1910.

Since the peaked line represents inflow and the smooth curve outflow, the difference in length of the ordinates of the two lines at any point represents the rate of storage. The point of intersection of the two lines marks the instant where inflow and outflow are equal,

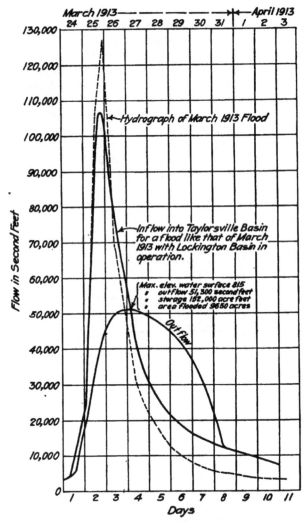

FIG. 34.—INFLOW AND OUTFLOW HYDROGRAPHS OF MIAMI RIVER AT TAYLORSVILLE RETARDING BASIN FOR A FLOOD LIKE THAT OF MARCH 1913.

and water ceases to accumulate in the basin. Beyond this point the position of the lines is reversed, the rate of outflow exceeds the rate of inflow, and the vertical intercept between them at any point gives the rate at which the stored water is being diminished.

The area included between the inflow and outflow curves, from the point where they first separate to where they intersect represents the amount of water temporarily stored in the retarding basin. Similarly the area to the right of this point of intersection, included be-

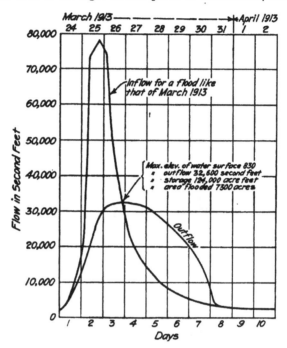

FIG. 35.—INFLOW AND OUTFLOW HYDROGRAPHS OF MAD RIVER AT HUFFMAN RETARDING BASIN FOR A FLOOD LIKE THAT OF MARCH 1913.

tween the curves, represents the release of this stored water, and is, of course, equal to the area first mentioned.

In figure 34 there is shown by means of a dotted line the flood flow as it passed the Taylorsville damsite in March 1913. The Lockington retarding basin would modify this flood flow in the manner shown by the full line representing the inflow into the Taylorsville basin.

EFFECT ON FARM LANDS IN BASINS

There is every reason to believe that the farm lands composing the bottoms of the retarding basins, except for small areas immediately adjoining some of the dams, will be benefited rather than injured by

FIG. 36.—SEASONAL OCCURRENCE OF FLOODING OF RETARDING BASIN LANDS.

Had the dams been completed in 1893 the acreage overflowed within the retarding basins during the 25-year period 1893–1917 would have been as here shown. There is a notable scarcity and even absence of floods during the spring, summer, and fall.

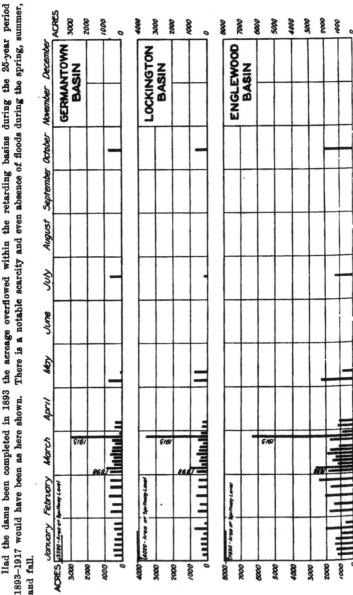

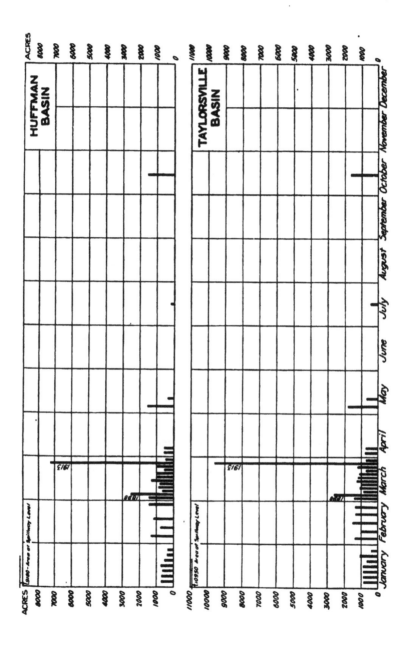

the occasional submergence. As pointed out, most recorded floods have occurred in January, February, and March. Had the dams been built these floods would not have affected any crops, except possibly winter wheat and alfalfa. During the 25-year period, 1893 to 1917, there were 30 floods or freshets of sufficient size to be affected by the dams; 24 occurred between January 1 and March 31, half of them in the month of March.

The accompanying diagram, figure 36, brings out strikingly the fact that but few floods happen during the growing season. It shows, grouped by months, all floods that would have overflowed land in the retarding basins, each flood being represented by a vertical line, the ordinates showing the acreage submerged. Thus, in the upper tier, representing the flooding in the Germantown basin, the 7 January floods that took place during the 25-year period referred to above are arranged alongside each other; likewise, the 5 February floods and the 12 March floods. Only 2 April floods would have submerged any land in the basin, and both would have been attributable to preceding March floods which had not yet subsided and which aggravated what otherwise would have been small rises. Two floods occurred in May, none in June, one in July, none in August and September, one in October, and none in November and December. Of the summer floods, one in May and the one in July would have been too small to have affected crops to an appreciable extent. If the two remaining floods had occurred the same year a corn crop could have been grown and harvested between them.

With the exception of extraordinary floods of rare occurrence, such as those of March 1913 and October 1910, the area overflowed within the basin would have been small. Thus, the 7 January floods would not have covered to exceed 500 acres each, the February floods would have covered a little more, and among the March floods, except that of 1913, no flood would have covered as much as 1000 acres. In the other basins the conditions, as shown in the diagram, are very similar, larger areas, of course, being submerged in the larger basins. In chapter VIII the frequency of flooding in the basins is enlarged upon.

The dams will not be responsible for all the flooding within the basins. In 1913, within the Germantown basin, Twin Creek overflowed 2045 acres, nearly 70 per cent of the acreage that would be submerged in the basin during a flood similar to that of 1913 with the dam in place. In the Lockington basin this proportion is about 90 per cent, and in the other basins about 50 per cent.

A discussion of the value of silt deposits to the soil would be out of place here, except to bring out the fact that the manner in which water is held back in a retarding basin promotes settlement of nearly all materials held in suspension. Such deposits will be greater than in the case of overflow by rapidly flowing flood water as in the past. Also, the comparatively still water will make impossible repetition of the destruction of farm land by strong currents.

Chapter VIII contains a description of the methods used in estimating the extent of submergence for such floods as have happened in the past.

BACKWATER ACTION IN RETARDING BASINS

A retarding basin, when filled to high levels by a great flood, presents practically a level water surface, much as in the case of a lake fed and drained by a river. The lowest portion of the city of Troy being but a few feet above the spillway level of the Taylorsville dam, it was of interest to ascertain the exact nature of the transition curve from the slope of the stream to the water in the basin, that would be developed at the upper end of this basin during maximum flood conditions; and to determine what effect, if any, the construction of the Taylorsville Dam would have upon this low land in Troy.

Computations showed that the greatest increase in stage at the upper end of the basin, due to this transition curve, amounts to less than two feet when the water surface at the lower end is at the crest of the spillway. At this stage the inflow from the assumed maximum flood is no longer the maximum flood flow of the stream, but has dropped to a rate equal to the outflow capacity of the outlet conduits. Larger backwater effects result during the earlier periods of the storm when the inflow is very large but when the basin is filled only in part, and such transition effects are later covered by the water in the basin. For maximum conditions it was found that the surface slope in the lower portion of the basin, for about 7 miles above the dam, would total but a fraction of an inch. Only in its upper portions, where the cross sections are much reduced by the shallower depths, would the water surface show any rise.

Inasmuch as the flow entering and passing through the Taylorsville and Huffman retarding basins will be larger in proportion to the normal flow than in any of the other basins, the backwater effect in these basins will be greater than in the others. It should be especially small in the Englewood basin where the movement of the water through the basin is greatly reduced by the small discharging capacity of the outlets.

CHAPTER V.—CAPACITIES OF BASINS

The alternation of broad and narrow sections in the valleys of the Miami watershed, affords favorable opportunities for the storage of large volumes of flood water by the construction of dams at reason-

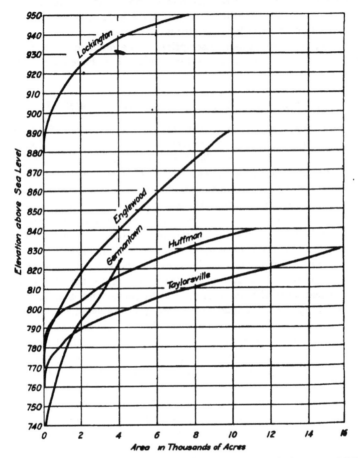

FIG. 37.—AREA-ELEVATION CURVES FOR THE FIVE RETARDING BASINS.

These curves indicate the acreage submerged in any basin for different water surface elevations.

able cost. The five basins of the Miami Conservancy District develop only a portion of the storage possibilities available on the Miami River and its tributaries. These natural topographic advantages for reservoir location were, however, somewhat lessened by the necessity for

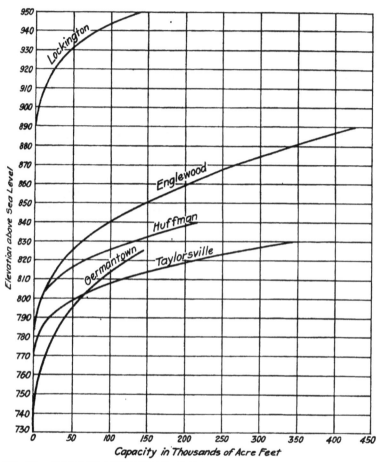

FIG. 38.—STORAGE–ELEVATION CURVES FOR THE FIVE RETARDING BASINS.

These curves indicate the capacity in acre feet in any basin for different water surface elevations.

voiding excessive interference with cities, villages, and railroads. Questions regarding the effects of backwater on such existing developments required careful study and economic balancing before a final decision could be made as to the most desirable storage capacity to be developed at each of the various sites.

CHARACTERISTICS OF CURVES FOR AREA AND CAPACITY

The area and capacity curves of figures 37 and 38 show for each basin the area submerged in acres and the capacity in acre feet for all depths of water. The data used in plotting them was obtained from planimeter measurements on detailed topographic maps of the basin locations. As explained in a previous Technical Report,[*] these maps were based on complete topographic surveys made by stadia methods. Small scale topographic maps, showing the general features of each basin are given in figures 39 to 43 inclusive.

The principal characteristics of the area and capacity curves of the different basins are as follows.

The area below spillway elevation is greatest in the Taylorsville basin, the other basins in order of area being Huffman, Englewood. Lockington, and Germantown. The area below spillway elevation varies from about 17 square miles at Taylorsville to 5½ square miles at Germantown. The area curves of the five basins are of the same general type, the surface area varying roughly as the three-halves power of the depth. The Taylorsville and Huffman basins bear close resemblance in the rapidity with which their surface broadens out as the depth increases, while the curves for the Englewood and Germantown basins show a more gradual rate of expansion. The area curve for the Lockington basin is intermediate between these. For every basin, except that at Germantown, the curves of area are remarkably smooth and regular. The depression in the Germantown curve near elevation 800 is due to the presence of a considerable amount of flat bottom land in the valley near this elevation.

The storage capacity at spillway level is greatest at Englewood. the other basins in order being Taylorsville, Huffman, Germantown. and Lockington. The figures of the following table aid in visualizing the storage capacities of the different basins at spillway elevation.

Basin	Surface Area at Spillway Level, Sq. Mi.	Average Depth of Water, Feet	Storage, Feet Depth Over One Sq. Mi.
Englewood..........	12½	39½	487
Taylorsville.........	17	17	291
Huffman...........	14½	18	261
Germantown........	5½	30	166
Lockington.........	6½	17½	109

[*] History of the Miami Flood Control Project, by C. A. Bock; Technical Reports, Part II, The Miami Conservancy District, Dayton, Ohio. 1918.

The rapid broadening of the Taylorsville basin surface causes its storage volume to increase at a high rate after spillway level is passed. The Huffman storage also increases rapidly, while the lowest rate of increase is at Germantown. The capacity-depth curves of figure 38 do not show the minor depressions which appear in the corresponding area-depth curves. Without exception they are remarkably smooth and regular.

EXPONENTIAL CURVES FOR AREA AND CAPACITY

Inspection of the area and capacity curves for the different basins, figures 37 and 38, suggests the possibility of approximating these curves by lines having definite mathematical equations of the exponential type. This operation permits calculus methods to be used in solving problems regarding inflow and outflow relations for the basins, and makes possible a great saving of time and labor in certain kinds of computations concerning retarding basin action.

For any depth of water in a retarding basin, let

A = area of water surface, in acres,

S = storage volume of retarding basin at same depth, in acre feet,

H = corresponding depth of water in basin at dam, in feet,

O = corresponding outflow rate, in acre feet per hour,

C = constant, depending on form and dimensions of retarding basin,

B = constant, depending on size and construction of outlet tunnel.

In the basins of the Miami Conservancy District, it was found practicable to approximate the area curves by semi-cubic parabolas having the type equation:

$$A = CH^{\frac{3}{2}} \tag{1}$$

When the approximate area curves are established by this formula, the corresponding storage curves are automatically fixed as five-halves power parabolas, since $dS = AdH$, giving

$$S = \int AdH = \tfrac{2}{5}CH^{\frac{5}{2}} \tag{2}$$

In order to have these equations available for use on any retarding basin it is necessary to determine two quantities; first the constant C, which will be different for each basin; and second, the elevation which will be considered as the zero depth in the basin. In fixing these quantities there are two conflicting sets of considerations which must be kept in mind, each being applicable to its own class of problems. These may be classified under: **Case 1,** where the zero depth may be

8

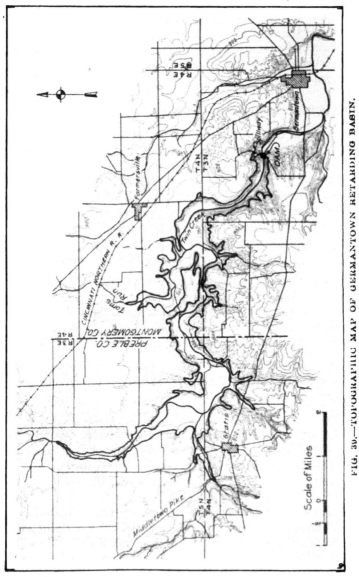

FIG. 30.—TOPOGRAPHIC MAP OF GERMANTOWN RETARDING BASIN.

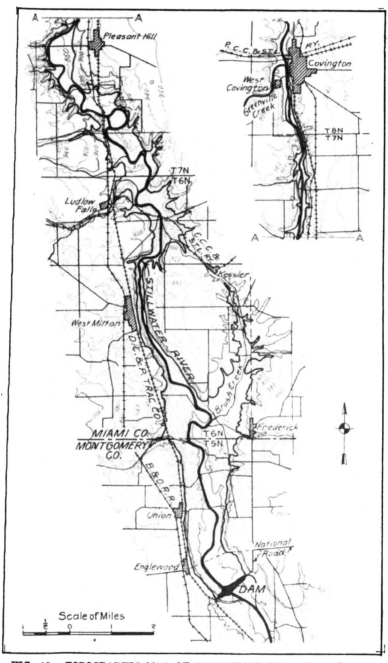

FIG. 40.—TOPOGRAPHIC MAP OF ENGLEWOOD RETARDING BASIN.

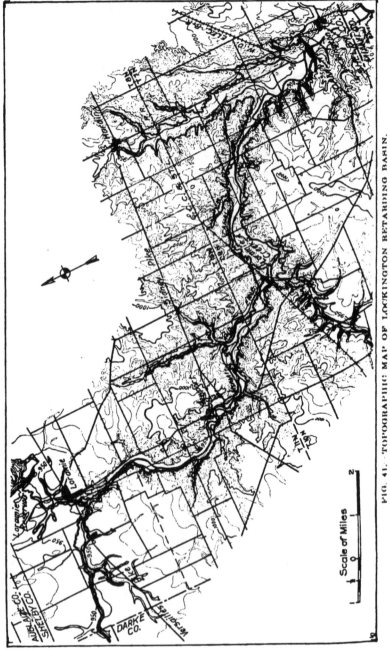

FIG. 41. TOPOGRAPHIC MAP OF LOCKINGTON RETARDING BASIN.

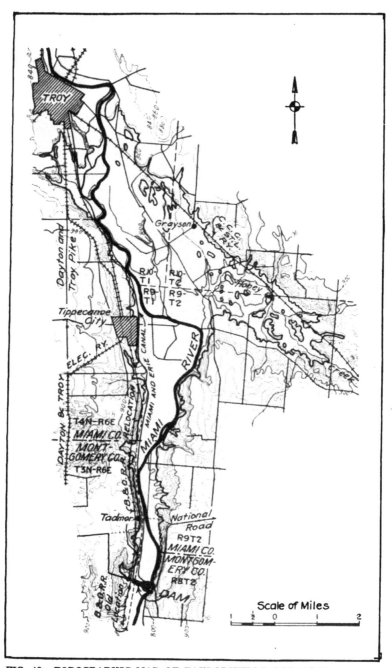

FIG. 42.—TOPOGRAPHIC MAP OF TAYLORSVILLE RETARDING BASIN.

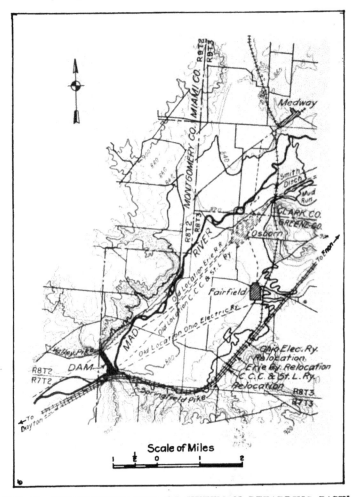

FIG. 43.—TOPOGRAPHIC MAP OF HUFFMAN RETARDING BASIN.

varied to adjust the curve; and **Case 2**, where the zero depth is fixed
by outlet conditions.

Adjusting Curves: Case 1, Zero Depth Variable

Closest possible correspondence between the assumed semi-parabolic
curves and the actual curves for the areas and capacities of the basins
can be secured more satisfactorily by adjusting both the constant C
and the elevation of zero depth, than by holding one of these fixed.

In general, the zero depth will be known approximately from the ground elevation in the neighborhood of the damsite, but two things must be taken into consideration in fixing the value to be used in the equation. First, for very shallow depths the area and volume in the retarding basin are small, uncertain, and relatively unimportant. Second, the capacity of the basin at the upper levels, or when nearly full, is much the more important, and therefore in fitting the above equations to the given data the values for the upper levels should be given the most weight in determining the constants in the equations.

The procedure used in fixing the constants for the basins of the Miami Conservancy District was as follows: The areas for a given basin were platted as abscissas and the corresponding elevations as ordinates on ordinary coordinate paper, and from a curve sketched through the points a tentative elevation of the zero depth was determined. Using this zero point, the depths and corresponding areas were platted on logarithmic cross section paper. The points thus obtained were found to lie in general on a flat curve concave either upward or downward. If the curve was concave upward it showed that the elevation used for the zero depth was too low, or that a higher zero should be used in reducing the values of the depths previously platted. Hence the values were replatted on the logarithmic paper reducing all depths by some constant amount such as five feet or whatever seemed most suitable. If the first curve was concave downward, then all elevations were increased by a constant amount in going through the above process. The trials were repeated until the value of the zero elevation to the nearest foot was determined such that the resulting points came nearest possible to falling on a straight line. The straight line averaging these points was drawn.

Starting with an equation of the form $A = CH^n$ and taking the logarithm of both sides we obtain $\log A = \log C + n \log H$. Now if $\log A$ and $\log H$ are platted as variables on ordinary coordinate paper, a straight line is obtained. This is exactly the same straight line that is obtained when logarithmic cross section paper is used. The intercept on the area axis is $\log C$ and the tangent of the angle which the line makes with the H axis is n. The values of n obtained by the methods of the preceding paragraph were in general not exactly $\frac{3}{2}$, although they came surprisingly close to this value. However, a slight change in the value of n could be partly compensated by a suitable change in C, hence it was next necessary to determine the best equation when n was arbitrarily fixed at $\frac{3}{2}$.

Besides having the equation fit all the points on the curve as well as possible, there were two other criteria of particular importance.

First, the equation for area should be particularly accurate just below the elevation corresponding to the maximum flow line in the retarding basin. Second, the total capacity of the basin at the maximum flow line should closely agree with the value given by the equation. The first criterion was satisfied by making the straight line go through a point representing the area corresponding to an elevation five feet below the maximum flow line. The second test was met by drawing the straight line so that the sum of the plus discrepancies,

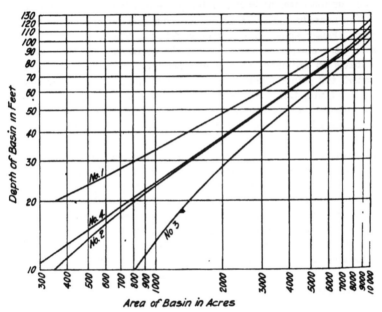

FIG. 44.—RELATION BETWEEN DEPTHS AND AREAS SUBMERGED IN ENGLEWOOD BASIN.

The curves were platted on logarithmic paper, and represent various assumptions as to point of zero depth.

with the measured areas at different depths, just equaled the sum of the minus discrepancies. The equations of both these lines were determined, or if possible, one equation that would satisfy both conditions, and from these the ones to use in subsequent calculations were selected.

Figure 44 illustrates the application of the above method in determining an exponential equation to fit the area-depth relation for the Englewood retarding basin. Lines 1, 2, and 3 of this figure represent the curves obtained on the logarithmic coordinate paper by assuming the elevation of zero depth to be 770, 780, and 790 feet. Of these, line 2 is seen to correspond most nearly to a straight line. Line 4 is drawn to fit the data of line 2 as nearly as possible and to have the slope corresponding to the exponent $n = \frac{3}{2}$. The equation of this line is:

$$A = 8.6 (\text{Elev.} - 780)^{\frac{3}{2}}$$

This line passes within a foot of the point of zero head for the outlet of this basin as actually designed. The line through this zero point, which agrees best with the head-area data was found to have a slope of 1.47, showing the close agreement with the assumed value $\frac{3}{2}$. This same study was not made for all the other basins, and probably some of them would show the most nearly straight line to have a slope diverging more from $\frac{3}{2}$ than was found to be the case with the Englewood data.[*]

Adjusting Curves: Case 2, Zero Depth Fixed

In inflow and outflow computations by calculus methods, where the outflow is expressed in terms of the head on the outlet conduit by equations of the type $O = BH^{\frac{1}{2}}$, it is very desirable, for the sake of simplicity, to have this value of H equal to that in the area and capacity formulas, equations (1) and (2). This condition fixes the position of the point of zero depth with reference to the floor of the proposed conduit, as explained on page 119.

In this case it is not, in general, possible to adjust the exponential curves to correspond with the areas and depths at a variety of points, but the nature of the problem to be solved determines the particular points which should be given most weight in adjusting the curves. In computations for the design of outlet conduits, it is important to have the total storage volume of the basin accurate at the greatest depth which will be attained under the conditions for which the computations are made, usually the depth at spillway. It is found by experience that when this condition is fulfilled, the particular shape of the storage curve makes relatively little difference in the results of the outflow computations. The method of adjusting the curves in Case 2 requires no detailed explanation. The use of logarithmic cross-section paper is convenient for this purpose.

[*] For a much more complete discussion of the process of using logarithmic paper to derive empirical equations to fit experimental data, see " System makes Easy Determination of Empirical Formulas," by E. W. Lane in Engineering News Record of September 20, 1917, Vol. 79, page 554, and Cornell Civil Engineer of February 1919.

CHAPTER VI.—FLOW THROUGH OUTLET CONDUITS

GENERAL DESIGN OF OUTLET CONDUITS

The outlet conduits of the five retarding basins of the Miami Conservancy District are of two types, as shown in figures 15 and 16, pages 61, 63. Those at the Germantown and Englewood dams are culverts or tunnels several hundred feet long, passing under the earthwork of their respective dams as shown in figure 16; while those at the Lockington, Taylorsville, and Huffman dams are relatively short, extending only through the concrete spillway structures as illustrated in figure 15. The former design avoids the difficulty and cost of constructing open spillway and outlet channels through the high dams at Englewood and Germantown.

In all cases the outlet conduits are constructed so as to minimize resistance to flow. The entrance edges are rounded off in easy curves in order to avoid loss of head at this section. An especially rich mixture of concrete is used in the walls of the conduits, and unusual precautions were taken during construction to secure smooth, hard, and uniform inside surfaces.

In order to take full advantage of the desirable features of the orifice type of retarding basin outlet, it was necessary to set the outlet conduits as low in the dams as could be done without danger of obstruction by silting or freezing. This was accomplished by locating the permanent tunnel floors at or slightly below ordinary low water mark. The floors are level in the three short outlet structures, and slightly sloping in the two longer ones.

At the Englewood and Germantown outlets a single tunnel would have formed a less expensive construction than the twin conduits actually adopted. The latter were chosen, however, in order to permit the flow through one of them being diverted to the other during the dry season, in case it should become desirable at some future period to make examinations of, and repairs to, the lining.

In the design of all outlet structures of the system, the water approaches the conduit entrances by way of a gradually narrowing channel with concrete walls and floors. The transition from channel to conduit occurs without offset or abrupt change in direction of bottom or sides.

At their downstream ends the conduits in every case discharge into a gradually broadening concrete channel. Beginning at a short distance below the conduit exits, the floor of this channel curves downward and becomes a stepped incline leading into a deep pool confined between the side-walls. As described in detail in Part III of the Technical Reports,* the purpose of this pool is to cause the formation of a hydraulic jump or standing wave which will dissipate the energy of the water issuing from the conduits. Below the pool the outlet channel is crossed by two concrete weirs whose purpose is to regulate the transverse distribution of the outflowing water.

CHARACTERISTICS OF INDIVIDUAL OUTLETS

Germantown Outlet Conduits

The permanent outlet from the Germantown retarding basin consists of two parallel conduits each 546 feet long, as shown in figure 16. The floors of the conduits are flat, and the roofs are formed by parabolic arches which curve down nearly to the floor on either side, the side walls being only about four feet high. Each culvert is 13 feet wide by 9 feet 1 inch high inside, and has a cross-sectional area of 91 square feet. The conduit floors are inclined so that the downstream ends are one foot lower than the upstream ends, the slope being 1 in 560.

As indicated in figure 45, an important feature of the design of the outlet conduits for this basin consists in the provision for temporary conduits much larger than the final permanent outlets, to protect the earth dam from floods during the period of construction. The temporary conduits are 22 feet 10 inches high from floor of invert to crown of arch, and have a cross-sectional area of 252 square feet each. At the close of the construction period the lower portions of the temporary conduits are to be filled with unscreened gravel or crushed rock, and a permanent concrete floor built, resting on the 8-inch shoulders specially provided for that purpose. The transition from the temporary to the permanent floor at the lower end of the conduits is indicated in figure 58. A similar arrangement was adopted at the intake end.

Englewood Outlet Conduits

The outlet from the Englewood retarding basin consists of two conduits, similar in construction to those at the Germantown dam. The principal dimensions of each permanent conduit are: length 709 feet, inside width 13 feet, inside height 10½ feet, cross-sectional area

* Hydraulic Jump and Backwater Curves, by S. M. Woodward, R. M. Riegel, and J. C. Beebe; Technical Reports, Part III, The Miami Conservancy District, Dayton, Ohio, 1917.

108½ square feet, slope 1 in 709. The temporary outlets are 22 feet 6 inches high inside, and have each 252 square feet cross-sectional area. This area will be reduced by filling the lower portion with

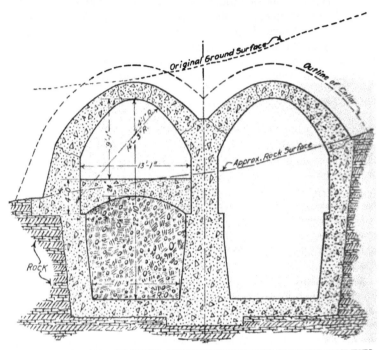

FIG. 45.—SECTION THROUGH GERMANTOWN DAM OUTLET CONDUITS.

The conduit on the right shows the large capacity opening built to safeguard the dam during the construction period. On the left is shown the final form, with restricted opening, required for the proper operation of the basin.

gravel or crushed rock covered with a concrete slab, as in the case of the Germantown outlets.

Lockington Outlet Structure

The outlet from the Lockington retarding basin, shown in figure 46, consists of two parallel conduits each 46 feet long, 9 feet wide, and 9 feet 2 inches high. The floors are level and the side walls vertical, while the roofs are slightly arched. Each opening has a cross-sectional area of 79 square feet. The partition or pier between the conduits is 9 feet wide and 80 feet long, extending somewhat both upstream and

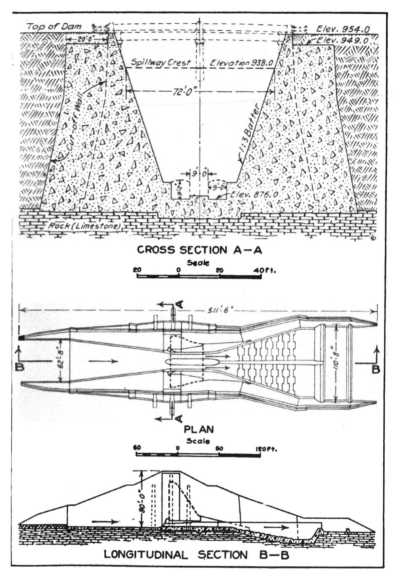

FIG. 46.—COMBINED OUTLET AND SPILLWAY STRUCTURE AT LOCK-
INGTON DAM.

Cross section *A–A* shows the form of outlet provided for temporary use during the construction of the dam. The dotted lines show the final spillway crest,
bridge over spillway, and location of the two outlet tunnels. The relative positions of spillway and tunnels are further shown in the plan and longitudinal
section.

downstream from the covered portion of the conduits in order to give proper support to the spillway above. These piers are rounded off in streamline curves at their upper and lower ends.

The combined spillway and outlet structure of this basin is similar to those built at the Taylorsville and Huffman dams, the one at Huffman being shown in figure 15. The structure is built across a channel formed by two massive retaining walls which cut through the earth dam. This channel is left open for flood protection during the construction period, as shown in figure 46, and is not obstructed by the spillway structure until the earthwork is nearly completed. The bottom width of the temporary outlet channel, at its narrowest portion, is 27 feet, the walls have a side slope of 1 horizontal to 3 vertical, and their maximum height is 78 feet.

Taylorsville Outlet Structure

The Taylorsville retarding basin-outlet is similar in all respects to that at Lockington, except that it has four parallel outlet conduits instead of two. The principal dimensions of each of these conduits are: length 40 feet, width 15 feet, height 19 feet 2 inches, area 279 square feet. The piers separating the conduits are each 11 feet thick and 86 feet long over all.

The temporary outlet channel or sluiceway provided at the Taylorsville dam for the construction period is of the same type as at Lockington. Its base width is 93½ feet, slope of side walls 1 horizontal to 3 vertical, and extreme depth 77 feet.

Huffman Outlet Structure

The outlet conduit structure at the Huffman retarding basin, see figure 15, is of the same type as those at Lockington and Taylorsville. except that three conduits are required. The chief dimensions of each conduit are: length 40 feet, width 15 feet, height 16 feet 4 inches, area 235 square feet. The piers between the conduits are 11 feet wide and 82 feet long.

The concrete sluiceway in which the spillway and outlet structure is set, and which is to be left open for flood protection during the construction period, has a base width of 67 feet, side slopes of 1 horizontal to 3 vertical, and extreme depth of 73 feet.

SUPERIORITY OF OUTLET CONDUITS TO WEIRS

Various kinds of retarding basin outlets were considered in the early studies made in connection with the Miami Valley flood protection project. Much attention was given to different types of notched or stepped spillway weirs. As the studies progressed it became evident that, for purposes of pure flood control, the orifice type of retarding basin outlet possesses two important advantages over all other types available. These may be stated as follows:

(1) For a given maximum flood and retarding basin capacity, the orifice type of outlet secures the smallest peak outflow rate.

(2) For floods less than the maximum the orifice type of outlet minimizes the extent and time of land flooding in the retarding basin area.

In deciding upon the final plans for the five basins of the Miami Conservancy District, these advantages led to the choice of conduit outlets.

Comparison of Types of Outlets

Figures 47 to 50 are given to illustrate the reasons for the superiority of orifices to other types of retarding basin outlets. Figure 47 shows the general nature of several possible types of outlets, and figure 48 gives the characteristic discharge curves for these outlets. The important features of each type are summarized as follows:

(1) **Orifice outlet: tunnel, conduit, or gate.** Equation: $Q = C_1 H^{1/2}$. The discharge curve for outlets of this type is convex downward, indi-

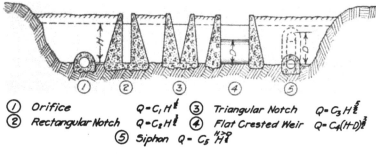

① Orifice $\quad\quad Q = C_1 H^{\frac{1}{2}}$ ③ Triangular Notch $\quad Q = C_3 H^{\frac{5}{2}}$
② Rectangular Notch $\quad Q = C_2 H^{\frac{3}{2}}$ ④ Flat Crested Weir $\quad Q = C_4 (H+D)^{\frac{3}{2}}$
⑤ Siphon $\quad Q = C_5 \frac{H \cdot D}{H^{\frac{1}{2}}}$

FIG. 47.—VARIOUS POSSIBLE TYPES OF RETARDING BASIN OUTLETS.

The various types of outlet studied are here shown in a group; also the discharge formula for each.

cating a large discharge at low heads, and small variation of discharge with head at high heads. The outlet thus acts vigorously during the early stages of a flood, getting rid of much water before the arrival of the flood peak, and yet holds down the outflow rate to a relatively uniform value while the flood peak enters the retarding basin and

raises the water level to its greatest height. The strong action of the
orifice at low heads prevents the basin from filling deeply or long
during a minor flood.

(2) **Vertical slit: sluiceway.** Equation: $Q = C_z H^{\frac{1}{2}}$. In this type
of rectangular outlet, both the velocity and the area of flow are reduced at low heads, making the discharge curve concave downward
and therefore unfavorable for flood control for reasons opposite to
those given under (1).

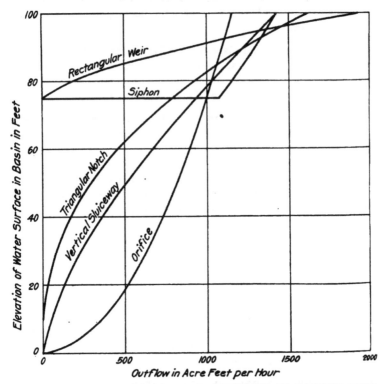

FIG. 48.—DISCHARGE CURVES CHARACTERISTIC OF VARIOUS TYPES
OF RETARDING BASIN OUTLETS.

The diagram shows the relation between head and discharge for each of the
types of outlet shown in figure 47. The orifice type is seen to give the greatest
discharge for low heads and the least for maximum heads.

(3) **Triangular notch.** Equation: $Q = C_z H^{\frac{1}{2}}$. This type of outlet
represents in a general way the class of notched or stepped spillways.
or series of several spillways at different levels. The discharge curve
is concave downward, and is even more unfavorable for flood control
than the curve of the preceding case.

(4) **Level-crested weir: ordinary spillway.** Equation: $Q = C_4(H-D)^{3/2}$. This outlet is not suitable for pure flood control, but is used where the water below spillway level is stored for power or other purposes. The discharge curve is very unsatisfactory from the standpoint of retarding basin operation, having a low outflow rate during the initial stages of a flood and high rate of increase of outflow with head at maximum discharge.

(5) **Siphon outlet.** Equation: $Q = C_5 H_{H>D}^{1/2}$. This kind of outlet is suitable for flood protection only in cases where the storage capacity of the basin below the siphon inlet is reserved for power development or other purposes. The discharge curve is very favorable for flood control under these conditions, having all the desirable characteristics of case 1. When the siphon once starts working it acts vigorously and uniformly.

Examples Showing Action of Different Outlets

It is difficult to visualize the complete action of a retarding basin outlet during a flood, on account of the close interrelation between the effects due to time and storage and those due to the peculiarities of the outlet. To aid in this, the following arbitrarily chosen example is worked out, showing the effect of an outlet of each of the above types, on a uniform flood passing through a basin of given capacity.

Data assumed: Maximum flood, 2000 acre feet per hour, lasting 20 hours. Maximum allowable head in basin, 100 feet. Storage capacity of basin at this head, 20,000 acre feet. Storage in basin varies as 2.5 power of head.

Found by trial: The size of each type of outlet (neglecting friction) necessary to develop the full storage of the basin during the maximum flood. These values are as follows:

(1) Orifice, circular opening 11.9 feet in diameter; $C_1 = 116$.
(2) Vertical sluiceway, opening 3.24 feet wide, the full height of the dam; $C_2 = 1.43$.
(3) Triangular notch, angle of opening 5 degrees, 10 minutes; $C_3 = 0.016$.
(4) Rectangular weir: crest elevation assumed at 75 feet, length 48.7 feet; $C_4 = 15.35$.
(5) Siphon: top of inlet assumed at 75 feet elevation, diameter of circular opening 18.8 feet; $C_5 = 185$.

Data desired: (*A*) Outflow-time curves showing the degrees of protection afforded by the basin during the maximum flood, using outlets of the different types.

9

(*B*) Curves showing the extent and duration of land flooding in the basin itself during a flood less than the maximum, using outlets of the different types.

Method used: The step method used in construction of the outflow-time curves is explained in the following chapter.

Results: (*A*) Figure 49 shows the outflow curves for the different outlets during the maximum flood. The ratios of maximum outflow rate to maximum inflow rate for the different outlets in this case are

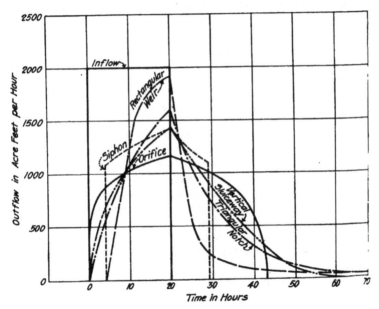

FIG. 49.—COMPARISON OF OUTFLOW CURVES FOR VARIOUS TYPES OF
RETARDING BASIN OUTLETS.

Each outflow curve is drawn for a flood wave of 2000 acre feet per hour lasting 20 hours, as shown by the solid black lines, and assuming a retarding basin capacity of 20,000 acre feet. The outlets are of the several types shown in figure 47, with dimensions as given on page 129. The orifice type is seen to give the least maximum discharge and require the least time for emptying the basin.

as follows: orifice, 0.58; vertical slit, 0.71; triangular notch, 0.80; rectangular weir, 0.96; siphon, 0.68. The protection afforded by the orifice outlet is seen to be much greater than that of its nearest competitor. The curves bring out clearly the superiority of the orifice outlet in beginning to act on the flood promptly, discharging the crest uniformly, and emptying the basin rapidly after the peak passes.

(*B*) Figure 50 shows flooding curves in the retarding basin using outlets of the different types, during a flood one-half as intense and lasting one-half as long as the assumed maximum flood. These curves illustrate the advantage of the orifice outlet in minimizing land flooding in the retarding basin during minor floods. The curves for the

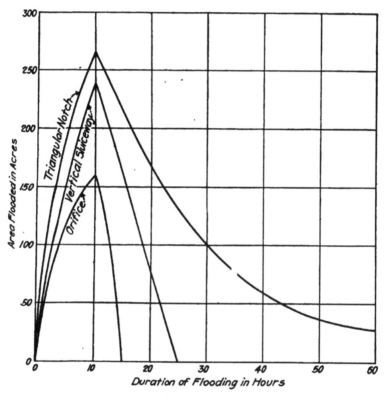

FIG. 50.—DURATION OF FLOODING AND ACREAGE SUBMERGED IN RE-
TARDING BASIN FOR VARIOUS TYPES OF OUTLET.

The time of submergence is shown for different areas flooded, assuming a flood inflow of 1000 acre feet per hour lasting 10 hours, and a retarding basin capacity of 20,000 acre feet. The orifice type is seen to submerge not only the least amount of land but to empty the basin in the shortest time.

spillway and the siphon are omitted from this diagram as the conditions under which they act in this case are not strictly comparable to those of the other three types.

EFFECT OF TAILWATER LEVEL ON CONDUIT OUTFLOW

Since the floors of the outlets of the basins of the Miami Conservancy District are located at or below mean low-water mark, these outlets during large floods will be partially or wholly below tailwater level. From this fact it might be incorrectly inferred that the usual formulas for submerged orifices are applicable to these cases, and that the tailwater elevations have an important influence on the discharge from the outlets. In the early studies for the project careful investigations were made to determine the probable tailwater rating curve at each of the damsites. As the plans of the system progressed to more advanced stages, however, it became evident that on account of

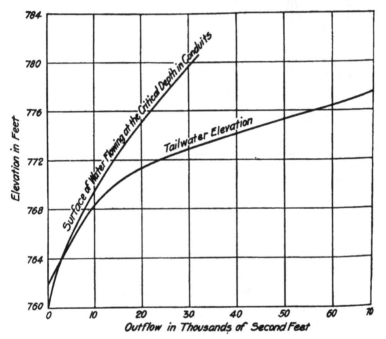

FIG. 51.—RELATION BETWEEN CRITICAL DEPTHS OF WATER IN TAYLORSVILLE OUTLET CONDUITS AND ELEVATION OF TAIL-WATER.

The curves are drawn for low discharges and show that except at very low heads the tailwater level is always below the outflowing water level.

the high conduit velocities and the special design of the outflow channels to secure the hydraulic jump, the discharge rates from the conduits, at high heads, are entirely independent of tailwater depths.

The theoretical analysis of the reasons for this independence of conduit outflow and tailwater depth is fully given in Part III of the Technical Reports, previously referred to. In brief, the water issuing from the conduit exits spreads out on the downward-sloping incline below the conduits with a velocity greater than the critical velocity corresponding to its depth, and rises suddenly to tailwater level as soon as the tailwater becomes deep enough to produce the hydraulic jump. Variations in tailwater level within certain limits, therefore, only move the position of the jump up and down the incline and do not affect the discharge from the conduits.

For the case where the conduits are flowing only partly full, an additional reason for the independence of conduit discharge and tailwater level is found in the location of the actual tailwater surface with respect to the surface of limiting discharge from the conduits. When a constriction occurs in the cross section of an open channel, and the water surface above is held at constant elevation, the amount of water passing the constriction increases as the surface level of the water below is lowered, until a critical or limiting discharge is reached which cannot be increased by further lowering of the downstream surface. The surface elevation corresponding to this limiting discharge at the exit of a typical outlet of the Miami Conservancy District, is plotted in figure 51. As this curve falls entirely above the tailwater rating curve for the river at this section, except at very low heads, it is evident that in this case the tailwater elevation cannot limit the conduit discharge. The methods of computing the quantities shown in figure 51 are further discussed later in this chapter in connection with the derivation of formulas for outflow in conduits partly full.

SELECTION OF ROUGHNESS FACTOR n

In connection with the design of the outlet tunnels a study was made of all the available data on the flow of water through large concrete conduits. In the case of these tunnels it was as important that the discharge should not exceed the estimated amount, as that it should not be less than this amount. An error either way might seriously affect the efficiency of the retarding basin system. If the actual outflow should prove greater than the calculated values, the estimated protection to cities below would be partially destroyed; while if it should prove smaller, the water might rise to a dangerous height behind the dams. It appeared probable that if the concrete surfaces were to be made as smooth as practicable the value of n in Kutter's formula would be below .013, and quite likely less than .012, but there was considerable probability that the smoothness of the surface would deteriorate with time. In the actual design a value of $n = .013$ was adopted as a conservative mean between present and future conditions.

Although the collection of data on which this choice of n was based is too extensive for reproduction in this report, a few of the representative results are given for purposes of general comparison. For a more comprehensive study of this subject the reader is referred to U. S. Department of Agriculture Bulletin No. 194,[*] from which a large part of the following data is taken.

Kern River Conduit: Rectangular, 7 by 8 feet, mostly concrete lined tunnel. Mean measured velocity at 6.5 feet depth was 2.3 per cent greater than computed with $n = .012$.

San Gabriel Power Tunnels: 4.5 feet wide and 4 feet deep, mortar-lined. Tunnel No. 15, $n = .0127$. Tunnel No. 23, $n = .0115$.

Los Angeles Supply Conduit: Mortar-lined; $n = .0112$ to .0109.

Umatilla Project Reinforced Concrete Pipe: Diameter 46 inches; $n = .0112$.

Jersey City Aqueduct: Concrete lined; average of tests, $n = .0127$.

Sulphur Creek Wasteway, Sunnyside Project, U. S. Reclamation Service: Concrete; circular section, radius 4 feet; velocity 20.6 feet per second; $n = .0108$.

Dry Creek Flume, Handy Canal Loveland, Colorado: Rectangular concrete channel lined with cement mortar, trowel finish; $n = .0115$.

Ridenbaugh Canal, Nampa—Meridian Irrigation District, Idaho: Trapezoidal concrete channel; very smooth, hand troweled, cement wash on base of concrete. Tangent, $n = .0110$; tangent and curve, $n = .0121$.

Long Pond Chute, Ft. Collins, Colorado: Reinforced concrete rectangular chute; velocity 12.9 and 19.7 feet per second; $n = .0125$ and .0123 respectively.

While the data summarized above, indicates in a general way the proper value of n for use in the design of the outlet tunnels, it was found difficult in the more detailed study to obtain consistent results regarding the exact effects of variations in the shape or surface finish of the conduits. This difficulty may be due to the presence of errors in the published experiments, or to the fact that Kutter's formula does not give precise results when applied to large covered conduits.

[*] Flow of Water in Irrigation Channels, by Fred C. Scobey, Bulletin No. 194, U. S. Department of Agriculture, 1915.

FORMULAS FOR OUTFLOW, CONDUITS FLOWING FULL

Notation

Q = outflow discharge rate, in second-feet,

A = combined cross-sectional area of conduit openings, in square feet,

H = effective head, in feet (see page 127),

L = length of conduit, in feet,

R = hydraulic radius of conduit section, in feet,

C = Chezy coefficient for conduits (see page 145),

c = coefficient of discharge for conduits (see page 145),

D = depth of flow at governing section of outflow channel, in feet,

B = width of flow at governing section of outflow channel, in feet,

w = weight of a cubic foot of water,

n = roughness term in Kutter's formula.

Fundamental Principles

When the depth of water in the retarding basins of the Miami Conservancy District becomes great enough to cause the conduits to flow full, the rate of discharge may be computed by a formula similar to that used for the discharge from ordinary orifices:

$$Q = cA\sqrt{2gH} \tag{4}$$

The derivation of the coefficient c to make proper allowance for friction in the conduits is taken up later. We will first consider the correct head H to be used in such a formula.

In the case of an ordinary orifice discharging freely into the air, the head to be used is measured from the surface of the water in the reservoir to the center of gravity of the orifice. If such an orifice, however, discharges into an open trough or flume, having a cross section of the same shape and size as the orifice, and having a longitudinal profile level or with only sufficient slope to maintain the initial velocity, the discharging stream has no free fall after efflux, but, on the contrary, its weight is fully supported by the flume, and the proper head to use in this case is measured from the surface in the reservoir to the surface of the water in the flume at the point of efflux.

If such a flume should discharge at its lower end freely into the air, a drop-off curve would be formed on the surface of the water near that end. If the flume were short enough, this drop-off curve would extend back to the orifice, and might lower the water surface there, or reduce the pressure in the stream, so that the effective head would be greater than that measured to the water surface.

An analogous phenomenon is visible when a stream discharging from the end of a horizontal pipe has a free fall in the air. If the velocity is low, the pipe does not remain full to the end, but the greater the velocity of efflux, the more nearly full the pipe remains at the end.

The conduits through the dams empty into open conduits having a floor practically level for a short distance, 24 feet, with a cross section at first just the same as the closed conduits, but immediately beginning to widen laterally. The question to be determined, then, is whether the lateral expansion below the outlet ends of the conduits and the steepened floor some distance downstream could operate to increase the effective head at the conduit portal. For the light that it may throw upon the general relation of the quantities involved, the simplest possible ideal case will first be considered, namely, a uniform. frictionless, rectangular, horizontal, open conduit, carrying water at a high velocity, and discharging freely into the air.

Drop-off Curve at the End of a Horizontal Flume

In accordance with the conditions that will prevail in the Miami Conservancy District conduits when carrying floods, let it be assumed that the initial velocity in the flume is considerably higher than the critical velocity.

Then, as shown in Technical Reports, Part III, previously referred to in this chapter, if friction be ignored, and if the channel be of uniform width, there will be according to the usual backwater theory three cases of surface slope, as follows:

(*a*) If the bottom of the channel be level, the surface will also be a level parallel line.

(*b*) If the channel slope slightly downward, the surface curve will be almost straight but sloping downward a little more than the bottom, Case I, figure 9, page 35 of Part III.

(*c*) If the bottom of the channel slope. slightly upward, the surface curve will rise more steeply than the bottom, Case VI, figure 9. page 35.

If friction be considered, then the surface backwater curve will be either:

(*d*) Case *C*, figure 25, page 51, when the bottom slopes downward.

(*e*) Case *J*, figure 32, page 55, when the bottom is level,

(*f*) Case *L*, figure 34, page 57, when the bottom slopes upward.

In the last three cases the depth increases in the direction of flow.

If our horizontal flume has a free drop-off at its end, there must be some sort of a drop-off curve, which the ordinary backwater theory fails to give. The reason is plain,—the ordinary backwater theory ignores vertical components of velocity, and it is these vertical components which give the drop-off curve under consideration. The following is an attempt to establish a rational theory for computing such drop-off curves.

Since the drop-off curve to be studied is relatively very short, the importance of friction is correspondingly unimportant, and for the present will be ignored. Let figure 52 represent the horizontal flume carrying water at a uniform velocity V_1 until it approaches the free drop at the end of the flume.

Let $y_1 =$ the uniform depth,

$y_2 =$ the depth at the end of the flume,

$y =$ the depth at any intermediate point of the drop-off curve,

$K =$ velocity head corresponding to the velocity $V_1 = V_1^2/2g$.

At the brink the velocity of the water will be speeded up by the additional head available. The upper filament has an additional head amounting to $y_1 - y_2$ to be converted into velocity; while the lower filament has an additional available head amounting to y_1. It is to be noticed that the direction of the velocity of the upper filaments at the

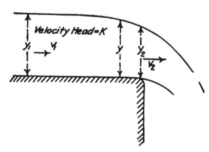

FIG. 52.—DROP-OFF CURVE OF WATER SURFACE AT END OF FLUME.

brink will be inclined in an amount corresponding to the suddenness of the drop-off curve. The fact that the velocity is inclined raises an obstacle to the use of Bernoulli's theorem in its ordinary simple form. It will be shown later, however, that by the use of Newton's Second Law of Motion we may calculate y_2 and v_2 at the brink, where

$y_2 =$ the vertical depth (not normal to the actual velocity), and

$v_2 =$ average horizontal component of velocity at the brink (not actual inclined velocity).

Distribution of Internal Pressure

It seems desirable first to study conditions prevailing generally throughout the drop-off curve, in order to derive some basic relations necessary to establish the length and amount of the drop-off curve. Let us consider in detail the conditions prevailing at a section whose depth is y, figure 52. Let a equal the vertical acceleration of the upper filament. The bottom filament has no vertical acceleration. Intermediate filaments are assumed to have a vertical acceleration proportional to their intermediate position. Let z be the depth from the surface to any such intermediate filament, and a_0 its acceleration.

Then, at any depth, z,

$$a_0 = a \left(1 - \frac{z}{y} \right). \tag{5}$$

When any matter has a downward acceleration its apparent weight is reduced by an amount proportional to the acceleration. If the acceleration becomes equal to g, the weight reduces to zero. Since the upper filaments have a vertical acceleration, their weights are reduced and they press less heavily on the lower filaments. This reduction of pressure permits the lower filaments to be accelerated as well as the upper filaments.

The average acceleration from the surface down to the depth z, is $a - a(z/2y)$. The unit pressure, therefore, at any depth z is

$$\left[1 - \frac{a}{g} \left(1 - \frac{z}{2y} \right) \right] wz = \left[1 - \frac{a}{g} + \frac{a}{g} \frac{z}{2y} \right] wz. \tag{6}$$

At the top, where $z = 0$, the pressure is zero, while at the bottom, where $z = y$, the pressure is $(1 - \frac{1}{2}[a/g])wy$. At any point the horizontal pressure is the same as the vertical pressure. Hence the total horizontal pressure acting on the section y may be found by integrating the above expression with respect to z between the limits 0 and y. Hence,

$$\text{Total horizontal pressure} = \int_0^y \left(1 - \frac{a}{g} + \frac{a}{g} \frac{z}{2y} \right) wz\,dz$$

$$= w \left[\frac{y^2}{2} \left(1 - \frac{a}{g} \right) + \frac{a}{g} \frac{y^2}{6} \right] = w \left(\frac{1}{2} - \frac{a}{3g} \right) y^2. \tag{7}$$

Equation (6) is illustrated by the curve AE in figure 53; it is the equation of a parabola. Using for convenience w equal 1, CB is drawn equal to AB. The point D is so taken that $CD/AB = a/g$.

Then the line AD represents $(1 - [a/g])z$. The point E is midway between C and D, and the curve from A to E is so drawn as to represent the parabolic distribution of pressure along the vertical y. The parabola is tangent to AD at A, and at E is tangent to a line parallel to AC.

Figure 53 looks reasonable enough, but equation (7) shows that, if the total horizontal pressure is assumed to be zero at the brink, a/g must equal ⅔. It then follows that near the upper filaments the pres-

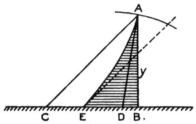

FIG. 53.—DISTRIBUTION OF INTERNAL PRESSURE IN DROP-OFF AT SECTION Y, FIGURE 52.

sure will actually become less than zero. It is probably more reasonable to assume that at the brink, on account of the convergence of the upper and lower filaments, the total pressure is not quite zero. If it may be assumed that at the brink $a = g$, the total horizontal pressure there, by equation (7), would be $(w/6)y_2^2$.

Shape of the Surface Curve

We may now apply Newton's Second Law to a mass of the stream beginning upstream where the depth is y_1 and ending at any point in the drop-off curve where the depth is y and the corresponding horizontal velocity is V. During a brief time dt the change in momentum of such a mass is $(wy_1V_1dt/g)(V - V_1)$. The unbalanced force producing the change of momentum is

$$\frac{w}{2}y_1^2 - w\left(\frac{1}{2} - \frac{a}{3g}\right)y^2;$$

$$\therefore \frac{w}{g}y_1V_1(V - V_1)dt = w\left(\frac{y_1^2}{2} - \frac{y^2}{2} + \frac{a}{3g}y^2\right)dt. \qquad (8)$$

Since $y_1V_1 = yV$, $\qquad V - V_1 = V_1(y_1 - y)/y$.

Substituting in (8) and simplifying

$$6\frac{V_1^2}{g}\frac{y_1}{y}(y_1 - y) = 3y_1^2 - 3y^2 + 2\frac{a}{g}y^2. \qquad (9)$$

Substituting K for $V_1^2/2g$ equation (9) becomes

$$12Ky_1^2 - 12Ky_1y = 3y_1^2y - 3y^2 + 2\frac{a}{g}y^3, \tag{10}$$

which may be written

$$y = \frac{4Ky_1}{4K + y_1} + \frac{3 - \frac{2a}{g}}{3y_1(4K + y_1)}y^3. \tag{11}$$

For numerical values of a size such as are important in these conduits, the last term in the second member of equation (11) is quite small compared with the first term. Hence the equation may easily be solved by trial for any given value of a by first obtaining an approximate solution by ignoring the last term of equation (11) and then correcting the value first obtained by successive trials. When $a = g$, the value obtained for y is the depth y_2 at the brink.

The approximate shape of the drop-off curve may now be obtained easily as follows:

Since, for the cases in which we are interested, y never differs much from y_1, it is in some respects advantageous to let $z = (y_1 - y)/y_1$, in which z will vary from zero to a small value never exceeding a few per cent. Then $y = y_1(1 - z)$. Substituting this in equation (11), there is easily obtained

$$z = \frac{y_1}{4K + y_1} - \frac{1 - \frac{2a}{3g}}{4K + y_1}y_1(1 - z)^3. \tag{12}$$

Solving this equation for a,

$$a = \frac{3g}{2y_1}\left(\frac{4Kz + y_1z - y_1}{(1 - z)^3} + y_1\right). \tag{13}$$

When z is a small quantity, $1/(1 - z)^3$ equals approximately $1 + 3z$, whence, neglecting all powers of z higher than the first,

$$a = 3g\left(2\frac{K}{y_1} - 1\right)z. \tag{14}$$

If $v =$ vertical component of motion, then using one of the fundamental differential equations of motion

$$vdv = -ady = ay_1dz = 3g(2K - y_1)zdz \tag{15}$$

$$v^2 = 3g(2K - y_1)z^2. \tag{16}$$

But $v = -(dy/dt) = y_1dz/dt$.

Hence,

and

$$dt = \frac{y_1 dz}{v} = -\frac{y_1 dz}{\sqrt{3g(2K - y_1)}z} \tag{17}$$

$$t_2 - t_1 = \frac{y_1}{\sqrt{3g(2K - y_1)}} \log_e \frac{z_2}{z_1} \tag{18}$$

Since the time becomes infinite if z_1 is zero, the surface drop-off curve must be of infinite length approaching the depth y_1 as an asymptote. The shape of the curve will be fairly well shown by locating the three points: where $z = z_2$ at the brink, where $z = 0.5z_2$, and where $z = 0.1z_2$.

Since the horizontal velocity remains nearly constant, the horizontal distance of the various points from the brink may be obtained after the time is known, by using the initial horizontal velocity.

For the Englewood conduits the maximum velocity will be about 50 feet per second, and the depth at the lower portal is about 8 feet. The following table shows the calculated values for terminal drop-off curves having an initial depth of 8 feet and various initial velocities from 20 to 50 feet per second.

1. V_1 (feet per second).....	20	25	30	35	40	45	50
2. K (feet)...............	6.22	9.70	13.99	19.02	24.85	31.45	38.82
3. $\dfrac{4Ky_1}{4K + y_1}$	6.26	6.63	7.00	7.24	7.40	7.52	7.61
4. y_1 (feet)...............	6.61	6.93	7.25	7.45	7.57	7.66	7.73
5. z_2.....................	.174	.134	.094	.069	.053	.042	.034
6. x for point where $z = \frac{1}{2}z_2$.	5.35	4.18	3.98	3.61	3.50	3.42	3.38
7. x for point where $z = 0.1z_2$	17.78	13.88	12.59	11.97	11.62	11.37	11.20
8. a at point where $z = 0.1z_2$	1.00	1.96	2.28	2.54	2.73	2.86	2.93
9. y at point where $z = 0.1z_2$	7.86	7.89	7.92	7.94	7.96	7.97	7.97
10. $\sqrt{2\left(\dfrac{1}{2} - \dfrac{a}{3g}\right)}y^2$ (feet)...	7.75	7.73	7.72	7.70	7.69	7.69	7.68

In this table line 1 shows the initial velocity in feet per second; line 2 is the initial velocity head in feet; line 3 is the first term in the second number of equation (11), and is the chief term in making up the value of y_2; line 4 shows the complete value of y_2 at the brink; line 5 is the corresponding value of z_2, which may be conveniently thought of as the per cent by which the original depth has been reduced at the brink; line 6 is the horizontal distance measured upstream from the brink to the point where z equals one-half of z_2; similarly, line 7 is the horizontal distance to the point where z equals one-tenth of z_2; line 8 is the surface vertical acceleration at the same point, calculated by means of equation (13); so long as z remains less than one per cent practically the same value will be obtained from equation (14).

It is of interest to note particularly the conditions prevailing at the point where $z = 0.1z_2$. Line 9, in the table, gives the depth of the water at this point, which is to be compared with the original depth of 8 feet. By using equation (7) the total horizontal pressure on a vertical cross section at this point may be calculated. This pressure, of course, is less than the pressure in still water of the same depth would be. For convenient comparison we have next calculated the depth of still water which would give the same pressure as actually exists by equation (7). This depth is given in line 10 of the table, showing values always somewhat less than those in line 9.

The significance of the figures in the table may be explained a little more at length using the values in the last column as follows:

As water flowing 8 feet deep with a velocity of 50 feet per second in a horizontal flume approaches the brink or free end of the flume, the surface will drop until at the brink it is 7.73 feet deep or 3.4 per cent less than its original depth. One-half of this drop will occur in the last 3.38 feet, and nine-tenths of the drop occurs in the last 11.20 feet. At the point 11.20 feet back from the brink, the surface will be dropping with a vertical acceleration of 2.93 feet per second which is nearly one-tenth the acceleration of a freely falling body. On account of this drop the water exerts less weight and pressure than would still water of the same depth, and the total horizontal pressure on a vertical cross section of the stream would be the same as would be exerted by quiet water 7.68 feet deep. This means that for the purpose of computing the average velocity of efflux the total effective head from a reservoir source to this cross section should be measured from the water surface in the reservoir to an elevation 7.68 feet above the bottom of the flume.

The calculations in the preceding table will all fail if the initial velocity is assumed less than the critical velocity corresponding to the assumed initial depth, that is, if the initial velocity head is less than half the initial depth. In the table the critical velocity corresponding to the initial depth of 8 feet is 16 feet per second. Hence, if an initial velocity of 15 feet per second is tried, impossible results will be secured. According to equation (14) the acceleration will be negative, and equation (17) will contain the square root of a negative term, or an imaginary. This corresponds to the fact that near the end of an open flume the velocity cannot be less than the critical velocity corresponding to the depth. It is analogous to the fact that at the controlling section over a smooth broad-crested weir, where friction and velocity of approach may be neglected, the depth must be two-thirds the head on the weir and the velocity head half the depth.

one of our conduits, for which purpose the Englewood conduits will be chosen.

Application to Outlet Conduits

Each of the Englewood twin conduits is 13 feet wide throughout its entire length, and has an area of 107 square feet. The interior cross section of the conduits is of constant shape throughout their whole length, except for a gradual transition in the lower 20 feet of each. Throughout the main length of the conduits the arch is of horse shoe shape with low vertical side walls, and the height at the center of the cross section is 10.45 feet. The experiments conducted on the model for producing the hydraulic jump, described in Technical Reports, Part III, previously referred to in this chapter, showed that there was great difficulty in expanding the issuing jet laterally without serious impact against the side walls, when the jet as it emerged from the conduit had such a curved form that in the center it stood much higher than at the sides. To obviate this difficulty the lower 20 feet of each conduit is used as a transitional section in which, while maintaining the same constant cross section, the shape is changed by dropping the crown of the arch 1.38 feet and raising the springing line of the arch 3.04 feet, so that at the portal the height at the center is only 9.07 feet, and the top of the issuing jet is a flat circular curve having only 2.17 feet rise at the center.

The average depth of the issuing jet is 107 divided by 13, or 8.23 feet. The floor of the conduit has a slight uniform grade throughout and the floor of the channel below the portal continues at the same grade for a distance of 24 feet, and then is given a vertical curve approximating the parabola which a freely falling body would follow if it had an initial horizontal velocity of about 60 feet per second. Hence, if the water were moving fast enough, this vertical curve would tend to produce conditions approximating the brink considered in the previous theoretical discussion.

Immediately below the portal the side walls begin to expand laterally by the use of very gradual curves. Ten feet from the portal the width is 13.36 feet; at 24 feet from the portal, where the vertical curve begins, the width is 16 feet.

A natural way to study conditions in this outlet channel will be to consider the effect of one element at a time. If the velocity of the emerging water should not change during the first 10 feet, the area of the cross section would remain constant and the depth at the 10-foot point would be 107 divided by 13.36 or 8.01 feet. The surface would then have dropped $8.23 - 8.01 = .22$ feet. The average hydraulic radius of this channel is 107 divided by 29.42 or 3.64 feet. Using $n = .013$ in Kutter's formula, $C = 140$ approximately, and for a velocity of 50 feet per second the friction head in 10 feet is .35 feet. Hence, the friction would require more than the available drop, and the result would be that kinetic energy would be taken from the moving water to overcome friction. The velocity would then be slowed down very slightly in this section. Making the calculation in a slightly different way it is found that if the initial velocity is 40 feet per second the surface drop of .22 is just sufficient to overcome friction in this 10-foot distance.

Considering the next stretch of 14 feet to the beginning of the vertical curve, if the 50-foot per second velocity were maintained unchanged in this section, the depth at the end of the section would be 107 divided by 16 or 6.69 feet, with a corresponding surface drop of 1.32 feet. In this section the average hydraulic radius would be 3.64 as before, and the corresponding friction head would be .49 feet. This leaves available .83 feet which would increase the original 50-foot velocity to about 50.5 feet per second. With a uniform velocity the 14-foot distance would be traversed in .28 second. If the surface particles dropped vertically with an acceleration equal to g while traversing this stretch, they could drop only 1.25 feet; hence, it is apparent that so far as the surface particles are concerned, there would exist somewhere in this stretch a condition somewhat approaching a brink such as was discussed on the previous pages. This would give an additional increment of velocity to the moving water in addition to that calculated above. This brink, of course, does not exist so far as the bottom filaments are concerned. From the previous theoretical discussion it seems evident that no appreciable effect of this apparent brink could be exerted back through the 10-foot section to the portal of the conduit. It follows, therefore, that the portal of the tunnel is substantially the determining or controlling section which fixes the flow, and that the head to be used in calculating the flow is but slightly greater than the head measured to the crown of the portal. Although this head cannot be determined precisely, it seems probable that it should in all cases be measured to an elevation between the crown and the average top of the curved portal.

Another element, not so far discussed, which would also affect somewhat the above results is the effect of the transition section just inside the conduit portal, but this effect would only strengthen the conclusion stated in the preceding sentence.

Losses of Head

Four sources of loss of head may be recognized in the flow through the outlet tunnels: (1) friction losses in the approach channel; (2) losses due to eddies formed at the conduit entrances; (3) friction losses in the conduits; (4) friction and eddy losses occurring between the conduit exits and the critical section of the outlet channel.

Every effort was made in the design of the outlets to minimize entrance and exit losses by the avoidance of projections or angles which would interfere with the stream-line motion of the water entering or leaving the conduits. For this reason, the friction in the conduits themselves was the only item considered in the ordinary computations concerning tunnel outflow. As previously mentioned, the Chezy coefficient C for the outlet tunnels was computed by Kutter's formula, using $n = .013$.

Derivation of Outflow Formulas

The formula for outflow from the conduits when flowing full is derived by Bernoulli's theorem, taking section 1 across the supply reservoir and section 2 across the outflow channel at its critical section. The reference plane of zero elevation may be taken at the surface of the water at the critical section of the outflow channel.

Data: Section 1: pressure head at point in reservoir surface $= 0$, elevation head $= H$, velocity head $= 0$.

Section 2: pressure head at point in surface of water at critical section $= 0$, elevation head $= 0$, velocity head $= Q^2/2gA^2$. Loss of head between sections 1 and $2 = (L/C^2R)(Q^2/A^2)$ by Chezy's formula. Substituting in Bernoulli's theorem,

$$H = \frac{Q^2}{2gA^2} + \frac{LQ^2}{C^2RA^2}. \tag{19}$$

Solving,

$$Q = \frac{A\sqrt{H}}{\sqrt{\dfrac{1}{2g} + \dfrac{L}{C^2R}}}. \tag{20}$$

This equation may also be written

$$Q = cA\sqrt{2gH} \quad \text{where} \quad c = \sqrt{\frac{1}{1 + \dfrac{2gL}{C^2R}}}. \tag{21}$$

In making computations for the outflow from the tunnels of the Miami Conservancy District, tables were prepared for the convenient solution of this formula for a variety of conditions.

10

Areas of Conduits

In the preliminary stages of the work, before the detailed design of the conduits had commenced, it was necessary to make computations to obtain the approximate size of the outlet opening required at each of the various dams. For these computations the assumption was made that the cross section of the conduit would consist of a semi-

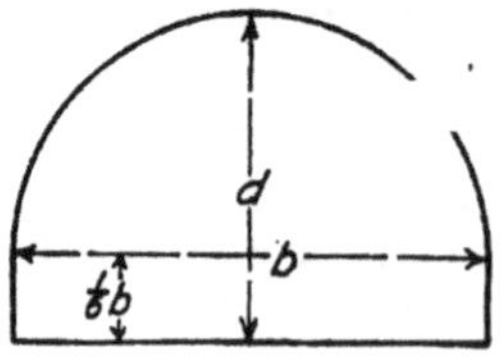

$$Perimeter = \left(\tfrac{\pi}{2}+\tfrac{4}{3}\right)b = \tfrac{3}{2}\left(\tfrac{\pi}{2}+\tfrac{4}{3}\right)d$$

$$Area = \left(\tfrac{\pi}{8}+\tfrac{1}{6}\right)b^2 = \tfrac{9}{4}\left(\tfrac{\pi}{8}+\tfrac{1}{6}\right)d^2 = 1.259 d^2$$

$$\substack{Hydraulic \\ radius} = \left(\tfrac{3\pi+4}{3\pi+8}\right)\tfrac{b}{4} = \tfrac{3}{8}\left(\tfrac{3\pi+4}{3\pi+8}\right)d = 0.289 d$$

FIG. 54.—TYPE OF CROSS SECTION OF OUTLET CONDUIT USED IN PRELIMINARY COMPUTATIONS.

This type was selected for convenience in computation. The form of cross section as finally built differs only in the shape of the arch.

circle and rectangle combined, as shown in figure 54. On this assumption an expression for the size of the conduit in terms of the given head and discharge, was determined as follows:

Let d = height of conduit in feet,

$a = A \div d^2$ = ratio of area to square of height,

$r = R \div d$ = ratio of hydraulic radius to height.

Take other notation as on page 135.

Using this notation, equation (21) may be written

$$Q = \frac{A\sqrt{H}}{\sqrt{\dfrac{1}{2g}+\dfrac{L}{C^2 R}}} = \frac{a d^2 \sqrt{H}}{\sqrt{\dfrac{1}{2g}+\dfrac{L}{C^2 r d}}}. \tag{22}$$

Transposing,

$$d = \sqrt[5]{\dfrac{1}{a^2}\dfrac{Q^2}{H}\left(\dfrac{d}{2g}+\dfrac{L}{C^2 r}\right)}. \tag{23}$$

For the particular type of cross section shown in figure 54 the values of a and r are found to be 1.259 and 0.289 respectively, and equation (23) becomes,

$$d = \sqrt[5]{.00983\dfrac{Q^2}{H}\left(d+223\dfrac{L}{C^2}\right)}. \tag{24}$$

This formula may be solved for d by trial, if the other quantities are given.

In using the above formulas in the early computations for Miami Valley flood protection studies, the head H was taken as the difference in elevation between headwater and tailwater, as previously explained. When the design developed along lines which rendered the conduit discharge independent of tailwater conditions, it became necessary to determine the value of H by successive approximations. This could be done by first making a rough estimate of the height required for the proposed conduit and from this obtaining a trial value of H. Using the latter, the required conduit size could be determined with a reasonably close degree of approximation, and from this the desired location for the point of zero head could be found by the methods given in the preceding paragraphs.

FORMULAS FOR OUTFLOW—CONDUITS PARTLY FULL

When the depth of water in a retarding basin is not sufficient to cause the outlet conduits to run full, the preceding formulas are not applicable, on account of the variation in the area of cross section of flow. In order to make clear the factors which govern the relation between head and outflow under these conditions, a brief summary is given of the theory of critical flow at a constricted section of a channel. A more detailed discussion of the general subject of critical velocity is given in Part III of the Technical Reports.

Principle of Critical Flow

Consider a rectangular channel, having a level floor and vertical side walls which converge and diverge so that the channel is narrower at some sections than at others, as in figure 55. Neglect the effects of friction and consider the changes of cross section to be so gradual that

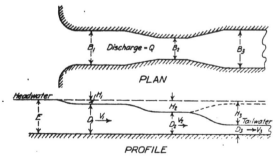

FIG. 55.—PLAN AND PROFILE OF RECTANGULAR CHANNEL WITH CONSTRICTION.

the effects of curvature of the stream lines may be neglected. Use the notation given in figure 55.

From the law of continuity of flow,

$$Q = V_1 B_1 D_1 = V_2 B_2 D_2 = V_3 B_3 D_3. \tag{25}$$

By Bernoulli's theorem, disregarding friction,

$$E = H_1 + D_1 = H_2 + D_2 = H_3 + D_3, \tag{26}$$

$$E' = \frac{Q^2}{2g B_1^2 D_1^2} + D_1 = \frac{Q^2}{2g B_2^2 D_2^2} + D_2 = \frac{Q^2}{2g B_3^2 D_3^2} + D_3. \tag{27}$$

Transposing,

$$\frac{Q^2}{2g} = EB_1^2 D_1^2 - B_1^2 D_1^3 = EB_2^2 D_2^2 - B_2^2 D_2^3 = EB_3^2 D_3^2 - B_3^2 D_3^3. \tag{28}$$

If the headwater depth E is maintained constant, the maximum discharge which can pass the narrowest section of the channel (Section 2 in figure 55) is found by differentiation of equation (28)

$$\frac{d}{dD_2} \left(\frac{Q^2}{2g} \right) = 0 = 2EB_2^2 D_2 - 3B_2^2 D_2^2. \tag{29}$$

Solving:

$$D_2 = \tfrac{2}{3}E, \tag{30}$$

$$H_2 = E - D_2 = \tfrac{1}{3}E, \tag{31}$$

$$V_2 = \sqrt{2gH_2} = \frac{1}{\sqrt{3}} \sqrt{2gE} = 0.577 \sqrt{2gE}, \tag{32}$$

$$Q = B_2 V_2 D_2 = \frac{2B_2}{3\sqrt{3}} E \sqrt{2gE}. \tag{33}$$

This result may be summarized in the following statement. The maximum discharge which can pass the constricted section of a frictionless rectangular channel with constant headwater level, is that due to a critical velocity-head of $\frac{1}{3}$, and a critical depth of $\frac{2}{3}$ of the depth of the headwater above the floor of the constricted section.

The discharge obtained by applying this rule cannot be increased by any amount of lowering of the tailwater level. The rule may also be stated: The velocity-head at any section of a rectangular channel in which the water is flowing at critical velocity and depth, is one-half the depth of the water at that section.

Where the floor of a channel is not level, or there are other peculiarities in the variation of the cross section, it may not be possible to decide by inspection which cross section is the critical or controlling one. In this case it is necessary to determine the maximum discharge which could pass each section regarded as a possible critical section, and to take the least of these maximum values as the actual critical discharge, the section giving this least maximum discharge being the actual critical or controlling section.

If, as in figure 56, the cross section of a channel is not rectangular, the critical area and velocity are determined as follows:

FIG. 56.—CHANNEL CROSS SECTION OF IRREGULAR FORM.

Let $A =$ area of cross section at any stage D measured above some arbitrary datum,

$W =$ surface width at this stage, and

$H =$ velocity head measured from the water surface to the fixed headwater level.

The discharge at any stage is given by

$$Q = A\sqrt{2gH}. \qquad (34)$$

For a maximum,

$$\frac{dQ}{dH} = 0 = \sqrt{2gH}\frac{dA}{dH} + \frac{1}{2}\frac{\sqrt{2g}A}{\sqrt{H}} . \qquad (35)$$

Solving,

$$H = -\frac{A}{2}\frac{dH}{dA} = \frac{A}{2W} . \qquad (36)$$

This result may be stated: The critical velocity for maximum discharge at any cross section of a channel is that due to a head equal to half the *average* depth of water at the cross section.

After the water passes the constricted section it is possible for it to flow in the last section at either one of two alternative stages, shown by the solid and dotted tailwater lines in figure 55, the transition from the critical section being accomplished by smooth backwater curves. The relation between the depths and velocities for these stages in a level-bottomed rectangular channel is given by the cubic equation, formula (28) above. If the water surface in the last section is held by other means at some level between these two stages, the flow will proceed down the lower backwater curve for a portion of the distance in the transitional section, and then form a hydraulic jump to reach the required tailwater stage. If the tailwater level in the last section is held artificially at some higher level than the upper line shown, then the level of flow throughout the whole system will be raised, with a corresponding reduction in the quantity flowing.

Selection of Critical or Controlling Section of Outlet Conduits

In applying the principle of critical flow to the case of low-head outflow from the retarding basins of the Miami Conservancy District, the first difficulty which arises is that of selecting the critical cross section of the outlet tunnels. This question may be best discussed by reversing the order of procedure used in deriving the formulas of the preceding paragraphs. Figure 57 shows the conditions considered.

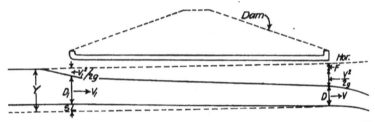

FIG. 57.—TYPICAL LOW–HEAD OUTFLOW CONDITIONS AT OUTLET CONDUIT.

Assume a value for the depth D of the flow at the critical section of a rectangular conduit. The velocity head will be $\frac{1}{2}D$, and the total effective head E at the section will be given by:

$$E = \tfrac{3}{2}D. \tag{37}$$

The discharge will be given by:

$$Q = BD\sqrt{g\bar{D}}. \tag{38}$$

Let F be the head lost in friction above the critical section, Y_0 the elevation of the floor of the critical section, and Y the elevation of the headwater surface. We may then write

$$Y = Y_0 + \tfrac{3}{2}D + F. \tag{39}$$

If this expression is evaluated for all possible cross sections of the conduit, that cross section which gives the largest value of Y will be the actual critical section which limits the discharge to the given value Q. For a conduit with level floor the critical section is evidently at the downstream exit, where F reaches its largest value.

If the slope S of the floor of the conduit happens to be such that the depth of flow is the same at all sections, the friction loss F can be expressed by the equation,

$$F = \frac{gLA}{WC^2R}.\tag{40}$$

This is obtained from the relations

$$V^2 = C^2R\frac{F}{L} = 2g\frac{A}{2W}.\tag{41}$$

For a wide rectangular channel equation (40) becomes approximately

$$F = \frac{Lg}{C^2}.\tag{42}$$

With the aid of these formulas the following rules may be deduced for locating the critical section corresponding to a given depth of flow in an outlet conduit with uniform cross section and sloping floor.

(1) The critical section is at the exit end when

$$S] < \frac{gA}{C^2WR}.\tag{43}$$

(2) The critical section is at the entrance end when

$$S > \frac{gA}{C^2WR}.\tag{44}$$

(3) The critical section is anywhere in conduit when

$$S = \frac{gA}{C^2WR}.\tag{45}$$

For a wide rectangular cross section the last term in each of these rules reduces to g/C^2.

In the outlet tunnels of the Miami Conservancy District the critical section as determined by these formulas occurs in every case at the exit end of the tunnel.

Evaluation of Friction Term

Where the depth and discharge are determined at one section of a channel of known cross section and roughness, as in the above case, the depth at any other section can be computed by the use of backwater formulas. If this is done for all portions of the outlet tunnel, after the critical section has been found as described, the friction term F can be evaluated with any necessary degree of accuracy, and the desired relation established between corresponding values of the outflow Q and the head Y.

If the curvature of the backwater surface within the conduit is not large, we may make the usual assumption of considering the average friction loss to be equal to the mean of the friction losses corresponding to the depths of water at the entrance and exit. That is, using the notation of figure 57:

$$F = \frac{L}{2C^2}\left(\frac{V^2}{D} + \frac{V_1^2}{D_1}\right) = D_1 - D + SL + \frac{1}{2g}(V_1^2 - V_2). \quad (46)$$

Substituting the values $V_1 D_1 = VD = D\sqrt{gD}$ in this equation, we obtain, for a rectangular conduit

$$F = \frac{Lg}{2C^2}\left(1 + \frac{D^3}{D_1^3}\right) = D_1 + SL + \frac{D}{2}\left(\frac{D^2}{D_1^2} - 3\right). \quad (47)$$

The numerical value of F may be determined in any given case by finding by trial a value of D_1 which will satisfy this equation. The corresponding expression for a conduit in which the cross section of flow is not rectangular is somewhat more complex, but can be readily derived from the same principles as the above.

Summary for Flow at Low Heads

The successive steps of the method which has been given for establishing a point on the rating curve for a conduit flowing partly full may be summarized as follows:

(1) Assume a depth of flow D.

(2) Determine the critical section by equation (43), (44), or (45).

(3) Compute the outflow rate Q by equation (38).

(4) Compute the friction loss by equation (47).

(5) Compute the headwater elevation by equation (39).

It was found by experience that moderate errors in computing the outflow at low heads have a negligibly small effect on the final result in routing a large flood through a retarding basin. For this reason, in many of the computations made for the Miami Conservancy District it was not considered worth while to go to great refinement in computing the outflow from the conduits at low heads. In such cases the depth of flow and velocity head in the tunnels were assumed to be ⅔ and ⅓ respectively of the depth from the floor to the headwater surface, and the friction correction was disregarded.

FORMULAS FOR OUTFLOW: TEMPORARY OUTLETS

It has already been noted that temporary outlets were provided at each of the dams to protect the earthwork from being overtopped by floods which might occur during the period of construction. These

:emporary outlets were of two kinds: the "tunnel" type, used at Englewood and Germantown, and the "sluiceway" type used at the other basins.

Temporary Tunnel Outlets

The temporary outlets of the Englewood and Germantown basins consist of large tunnels whose cross sections are to be partly filled in at the close of construction to form the permanent outlets. Hydraulic conditions at the exits of the temporary tunnels are indicated diagrammatically in figure 58. More detailed views are given in figure

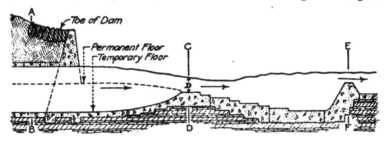

FIG. 58.—TYPICAL OUTFLOW CONDITIONS AT GERMANTOWN CONDUIT EXIT WITH FLOOR DEPRESSED TEMPORARILY.

The concrete conduit floor shown is that used temporarily during the construction of the dam. The final position of the floor is shown by the dotted line.

45. The water leaving the temporary tunnels is compelled to rise and pass over a concrete bar or very flat weir, at section CD in figure 58.

The effective head on the temporary outlets when they are flowing full may be approximately computed by the methods explained on pages 135 to 147. If the change of velocity between tunnel exit and the bar CD is disregarded, the depth D of the water over the bar may be computed by

$$D = A/B, \qquad (48)$$

where A is the cross-sectional area of the temporary conduits, and B the width of the outflow channel at the bar.

For extreme low-head conditions the critical cross section for determining the outflow from the temporary conduits is at the crest of the main regulating weir, EF in figure 58, below the large pool. The stage at which the critical section recedes from this weir to the tunnel exits could be computed, if desired, by the methods previously given. In the actual cases the nature of the computations made concerning the temporary outlets did not justify this refinement, the approximate outflow at low heads being estimated by the $\frac{1}{3}$ and $\frac{2}{3}$ rule as in the case of the permanent outlets.

Temporary Sluiceway Outlets

At the Taylorsville, Huffman, and Lockington basins, temporary outlets were secured by postponing the placing of the combined outlet and spillway weirs until near the completion of the earthwork, thus leaving an immense concrete open channel, or sluiceway, through each of these dams during the construction period. Hydraulic conditions at these temporary sluiceways are indicated diagrammatically in

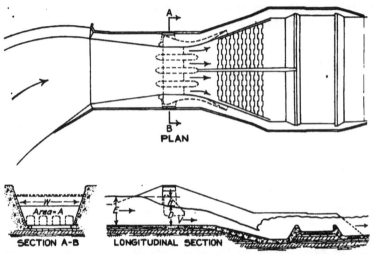

FIG. 59.—TYPICAL OUTFLOW CONDITIONS THROUGH TEMPORARY OUTLET CHANNEL AT TAYLORSVILLE DAM.

The plan shows the wide throat provided during the construction of the dam, capable of passing large volumes of water. The dotted outlines show the position of the concrete spillway weir and the four outlet conduits in the base of the latter.

figure 59. The concrete sidewalls of the sluiceways are inclined at 1 to 3 from the vertical, and, as shown in the diagram, converge and form a contracted section at the place where the final spillway and outlet weir is to be constructed.

From the discussion given on pages 147 to 150 regarding the theory of critical flow, it is evident that the critical section in these temporary sluiceways is at their contracted portion.

Let Q, V, D, W, and A be respectively the discharge, velocity. depth, surface width, and area, of the flow at the critical section and let E be the elevation of the reservoir surface above the floor at this section. Let $H_v = Q^2/2gA^2$ be the velocity head at the critical section. and assume that the head lost in friction by the water before reaching this section can be expressed in the form $H_f = KH_v$, where K is a coefficient depending on the form and roughness of the channel. Let

$$H = H_v + H_f = (1 + K)\frac{Q^2}{2gA^2}. \tag{49}$$

We may then write for any depth of flow

$$E = D + H = D + \frac{Q^2(1 + K)}{2gA^2}. \tag{50}$$

For maximum discharge at constant headwater depth,

$$\frac{dE}{dD} = 0 = 1 - \frac{2Q^2(1 + K)}{2gA^3}\frac{dA}{dD} = 1 - \frac{2WH}{A}. \tag{51}$$

Solving,

$$H = A/2W,$$

$$Q = A\sqrt{\frac{2g}{1 + K}\frac{A}{2W}}, \tag{52}$$

$$E = D + \frac{A}{2W}. \tag{53}$$

These equations permit the construction of a rating curve for the sluiceway, giving the relation between Q and E for all desired values.

The friction term K can be approximately evaluated on the assumption that practically all the friction loss occurs in the contracted section, where the velocity is high. If L is the length of the contracted section this assumption leads to the following formula (derived as on page 150).

$$K = \frac{2gL}{C^2D}. \tag{54}$$

The friction becomes negligibly small when the flow in the temporary sluiceways reaches any considerable depth. With the walls and floors of smooth concrete, if the Chezy coefficient C is assumed at 160, the friction loss when $D = \frac{1}{2}L$ amounts to about half of one per cent of the effective head. In planning the temporary sluiceways it was at one time intended to construct at an early stage of the work a few feet of the lower portion of the piers between the conduit openings, thus greatly increasing the effective roughness of the contracted portion of the channel. Even in this case, taking $C = 58$, the friction term K was found to be only about two per cent of the effective head at the maximum discharge of the 1913 flood.

CHAPTER VII.—ROUTING FLOODS THROUGH RETARDING BASINS

ESTABLISHING INFLOW CURVES AT THE DAMSITES

In predicting the action of the various basins suggested, one of the first steps was to estimate the intensity and duration of floods which might be expected to occur above each of the damsites. Since the unprecedented flood of March 1913 was the immediate cause of the undertaking, it was considered important to estimate the action of the proposed retarding basins in such a flood. A knowledge of their probable effects upon lesser floods of the past also was desired. However, no direct measurements of the flow of the 1913 flood or previous floods were available at or near any of the damsites under consideration, the only direct information of this kind which could be secured being the daily record of gage readings on the Miami River at Dayton. The problem thus presented was that of dividing up, or pro-rating, the Dayton runoff of the 1913 flood, and other floods, among the various watersheds on which the retarding basins were to be located.

In the following discussion it will be necessary to use many times several related quantities expressing total flows and rates of flow. In order to lessen confusion of these quantities the following nomenclature is consistently used throughout this chapter. The word runoff is used to refer to the total amount of water flowing off during a day, or during a storm, and is usually measured in inches of depth over the drainage area. In considering the action of retarding basins, the terms total inflow and total outflow are used to represent the total quantity of water which has flowed into or out of a basin. Discharge and inflow or outflow will be the words used to represent rates of flow when measured in cubic feet per second or acre feet per hour. Rate of discharge, rate of inflow or outflow, and rate of runoff will be used exclusively to represent discharge per unit area, as second feet per square mile or acre feet per hour per square mile.

The symbols used are as follows:

A = area of watershed of main river, in square miles,

a = area of watershed of tributary, in square miles,

Q_A = discharge from main river during flood, in second feet or acre feet per hour,

Q_a = discharge from tributary during flood,

156

q = maximum rate of discharge from main river, in second feet per square mile or acre feet per hour per square mile, .

q_a = maximum rate of discharge from tributary, in second feet per square mile or acre feet per hour per square mile,

q = average rate of discharge from main river during any step,

t = duration of above step for main river hydrograph, in hours,

q' = average rate of discharge from tributary during step corresponding to the above,

t' = duration of above step for tributary hydrograph,

$r_0 = q_a/q_A$ = ratio of maximum discharge rates,

$r = q'/q$ = ratio of average discharge rates during any step.

In determining the most probable hydrograph of a tributary from the discharge record on the main stream, three relations which may reasonably be supposed to exist between them should be kept in mind. They are as follows:

1. The total runoff from the flood, in inches of depth on the drainage area of the tributary, should be the same as from the main stream.

There is no reason for supposing that during a widespread storm the total rainfall will be materially greater on one tributary basin than another, and the topography of the region does not warrant the assumption of· different total runoff coefficients in pro-rating the flood-flow among them. This proportionality between total runoffs is expressed by the formula:

$$\Sigma q't' = \Sigma qt \qquad (55)$$

2. The duration of the tributary flood should be equal to the main flood. Not to do this is equivalent to assuming zero flow from the tributary during part of the period of the main flood, which condition is extremely improbable. This equality of duration is expressed by

$$\Sigma t' = \Sigma t \qquad (56)$$

3. The peak discharge rate on the tributary differs from that on the main stream, being usually larger since small areas, in general, give rise to higher rates of runoff than large ones. This relation may be expressed algebraically by

$$q_a = r_0 q_A \qquad (57)$$

where r is a coefficient which depends upon the size and other characteristics of the drainage areas in question.

During the progress of the studies three different methods were used in subdividing the Dayton flood hydrographs among the tributaries. These were known as, (1) the "Direct Area-Ratio Method," (2) the "Sliding Coefficient Method" and (3) the "Rainfall Method."

Inflow Computed from Dayton Discharge by Direct Ratio of Drainage Areas

This method is based on the assumption that all portions of the Miami River drainage area contribute to a flood in the direct proportion which their respective areas bear to the entire drainage area, and that for any given time interval the rate of runoff is the same over the entire drainage area.

The basin inflow curves computed by this method agree with the first and second of the criteria previously mentioned, since the total depth of runoff and the total duration of the flood on the tributary is the same as the runoff and duration of the main flood. It fulfills the third condition, however, only if r_0 is assumed to be unity, or that the peak rate of runoff on the tributary is the same as that on the main stream. Such an assumption is notoriously wrong for normal rivers. the headwaters of which drain regions of considerable declivity. In the case of the Miami River, however, it may give a fair approximation to what actually takes place, for its headwaters drain a rather level country which on the whole does not favor rapid collection; besides, the Lewistown and Loramie Reservoirs are important retarding factors. These facts would lead one to expect a degree of uniformity in runoff rates in the upper watershed wholly unlike the conditions normally found to exist in river basins of similar size. Our study of the February 1916 and July 1915 floods on the Miami and its tributaries, in which comparison is made between the inflow as computed from actual gage records and as derived from the Dayton record by different methods involving ratios of drainage areas, proves that the direct ratio method gives fairly close results for the large tributaries, notably the Stillwater at Englewood and the Miami at Taylorsville. On the other hand, it gives poor results for small tributaries which have a rapid runoff, as in the case of Twin Creek and Upper Mad River.

Generally speaking, the direct ratio method, though commending itself on account of its simplicity, in the end must give peak rates of inflow less than the actual, for the reason that no allowance whatever is made for increased rate of runoff from such portions of the watersheds as do favor rapid collection. The effect of the latter is to increase by a small amount the rate of runoff during the peak of the flood. This feature has an important bearing on the rapidity of filling of retarding basins, and should therefore receive full recognition.

Inflow Computed from Dayton Discharge by the Variable or Sliding Coefficient Method

This method is similar to the direct ratio method, but takes care in an ingenious manner of the increased rate in runoff from small areas, above referred to. In order to establish this method studies were first made to compare the results obtained by numerous published formulas which express the relation between peak rates of flood discharge from different watersheds having similar hydrologic characteristics but unequal areas. Some of these formulas, together with the results

Table 3.—Ratios of the Maximum Discharge per Square Mile of Drainage Area for certain Miami River Tributaries, to the Maximum Discharge of the Miami River at Dayton, as derived from various formulas.

Stream	Drainage Area, in Square Miles	Ratio: $a = \dfrac{a}{A}$ Drainage Area at Dayton / Drainage Area	Ratio: $\dfrac{\text{Maximum Discharge Rate on Tributary}}{\text{Maximum Discharge Rate at Dayton}} = \dfrac{q_a}{q_A}$									Adopted Values
			March, 1913 Flood	March, 1914 High Water	Kutohling's Curve No. 1 $q = \dfrac{44,000}{M+170}+20$	Kutohling's Curve No. 2 $q = \dfrac{127,000}{M+370}+7.4$	Burkli-Ziegler, $q = \dfrac{C}{M^{1/4}}$	McMath, $q = \dfrac{C}{M^{1/5}}$	McMath, $q = \dfrac{CS^{1/5}}{M^{1/5}}$	Fanning, $q = \dfrac{C}{M^{1/6}}$	Discharge Rate Ratio $\dfrac{q_a}{q_A}$	Discharge Ratio $\dfrac{Q_a}{Q_A} = \dfrac{q_a \, a}{q_A \, A}$
	1	2	3	4	5	6	7	8	9	10	11	12
Miami River at Dayton	2525	1.00	1.00	1.00	1.00	1.00	1.00	1.00	1.00	1.00	1.00	1.00
Mad River	671	.266	1.28	....	1.12	1.14	1.38	1.29	1.54	1.24	1.3	.335
Loramie Creek	255	.089	1.78	....	2.08	2.00	1.83	1.63	1.77	1.51	1.8	.160
Miami River below Upper Dams	778	.308	....	....	1.03	1.03	1.33	1.26	1.42	1.21	1.26	.390
Stillwater River	651	.258	1.31	1.20	1.17	1.19	1.41	1.31	1.31	1.26	1.3	.335
Twin Creek	270	.107	2.46	....	1.90	1.86	1.75	1.56	1.76	1.45	2.0	.214

obtained by applying them to the watersheds tributary to the retarding basins of the Miami Conservancy District, are given in table 3. In applying these formulas it was necessary to take into consideration the peculiar fanlike shape of the Miami watershed and its effect on the runoff rates at Dayton. This table also gives the ratios of the maximum discharge rates for the different watersheds in the 1913 flood, as computed from high water mark data at the various damsites, proper adjustment being made for the differences of rainfall on the various watersheds, and also a few ratios obtained by direct measurements in 1914. For purposes of further comparison, additional data is given in table 4, showing the results of actual gagings obtained subsequent to the construction of table 3. These gagings have the usual irregularity of data of this nature, but agree in a general way with the calculated values of table 3, except in the case of the Mad River

Table 4.—Ratios of the Maximum Discharge, per Square Mile of Drainage Area, for certain Miami River Tributaries to the Maximum Discharge of the Miami River at Dayton, as derived from Actual Flood Measurements.

Stream and Gaging Station	Drainage Area	Flood Dates					Average
		July, 1915	Jan. 2, 1916	Jan. 13, 1916	Feb., 1916	Mar., 1916	
Miami River at Dayton	2525	1.00	1.00	1.00	1.00	1.00	1.00
Mad River at Wright Station	652	0.76	0.60	0.97	0.46	0.97	0.75
Loramie Creek at Lockington	255	....	2.24	1.06	1.94	1.62	1.72
Miami River at Tadmor	1128	1.13	1.24	0.88	1.18	1.06	1.10
Stillwater River at West Milton	600	1.69	1.43	0.88	1.98	0.83	1.36
Twin Creek at Germantown	272	3.96	1.56	2.21	1.64	0.97	2.07

at the Huffman damsite. The observed runoff rates for this watershed are uniformly lower than the computed ones. It is probable that these low flood runoff rates of the Mad River watershed are caused by extensive deposits of gravel and porous material. The effect of such underground storage would be proportionately greater in moderate floods, such as those given in table 4, than in extreme floods such as that of 1913.

Columns 11 and 12 of table 3, give the official runoff ratios adopted for use in routing floods through the basins of the District. The values in column 11 represent the ratios of the peak runoff rates of the various tributary watersheds, to the peak runoff rate at Dayton during any given flood, while the values in column 12 represent the corresponding ratios of the peak discharges on the tributaries to the total peak discharge at Dayton. It will be noticed from inspection of the figures given in table 3, that there are considerable discrepancies in the ratios obtained by the various formulas and measurements. Such discrepancies are not uncommon in runoff studies and are to be expected on account of the irregularity of rainfall, soil absorption, and other factors which affect maximum flood discharges. The officially adopted values of column 11 were chosen by judgment in the light of all evidence which was available at the time the early studies for the design of the basins were in progress.

The peak inflow for a given retarding basin in any particular flood, as determined by the methods just given, is of no more importance in the study of basin operation than the amount of the total inflow during the flood and its distribution with respect to time. The ratio r or q'/q evidently varies between its maximum value r_0, determined as previously explained, and a value of unity at low stages, the reason for the latter being that during ordinary weather conditions there is no reason for assuming that the flow comes more from one tributary than another. In a considerable number of the computations at the Miami Conservancy District, the assumption was made that the ratio r varies uniformly from r_0, at the maximum discharge rate q_A, to unity at an arbitrarily assumed low discharge rate q_B, as shown by line (1) in figure 60. This low discharge was taken equal to the re-

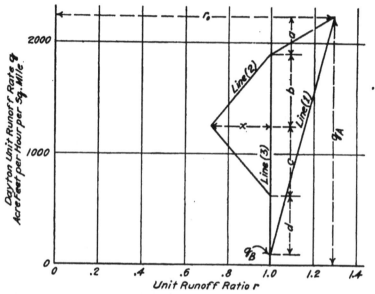

FIG. 60.—SLIDING COEFFICIENT DIAGRAM FOR DETERMINING VARIABLE RATIO OF RUNOFF FROM DRAINAGE AREA ABOVE ENGLEWOOD DAM.

corded flow at the beginning of the 1913 flood. The equation for the ratio r as located by line (1) is,

$$r = \frac{q'}{q} = \frac{t}{t'} = (r_0 - 1)\frac{(q - q_B)}{(q_A - q_B)} + 1. \tag{58}$$

With r assumed in this way, the values of q' and t' for each step of the tributary hydrograph are easily computed, giving all the data needed for platting this hydrograph.

The above straight-line assumption for the variation of r did not

11

in all cases give results which were considered satisfactory. One of the reasons for this was that this assumption violates the condition expressed by equation (56), namely: that the durations of the main and tributary floods should be equal. Since on this straight line assumption the value of r is greater than unity for all steps considered during the flood, it follows that the new duration t' obtained for each step will be less than the original duration t of the corresponding step, and the entire duration $\Sigma t'$ obtained for the tributary flood will be less than the duration Σt of the main river flood. The straight line variation of r thus tacitly leads to the assumption of zero flow from the tributary during certain days of the flood. This throws an excessive amount of the total runoff into the higher stages of the flood, and thereby imposes an unnecessarily heavy burden on the proposed retarding basin. While such a distribution of runoff is not mathematically impossible, it seems highly improbable that it should occur during a widespread flood of considerable duration.

These difficulties of the straight-line assumption for r may be avoided by letting r vary according to some curve or broken line, such as line (2) in figure 60, which will satisfy the condition $\Sigma t' = \Sigma t$. Such a curve must have at moderate stages of the flood a sufficiently large range of values of r less than unity, to balance the values of r greater than unity at the high stages. If the general shape of the desired curve is assumed, its dimensions may be computed by trial and error methods. For instance: if the curve is assumed to consist of four broken lines, as in figure 60, having $b = 2a$, the value of x, which will permit equation (56) to be satisfied, can be found by trial.

There is room for much difference of opinion as to the best method for determining the curve of r, the final criterion being that of experienced judgment regarding the probable shape of the flood hydrograph on a tributary of the size considered. Results fully as reliable as those obtained by the use of the above arbitrary rules for determining the curve of r, might be secured by sketching in by eye a hydrograph for the tributary, which would conform to the conditions of equations (55), (56), and (57). The arbitrary rules, however, have the advantage that they give consistent results when applied by different computers.

Attention should be called to the fact that in constructing inflow hydrographs by the sliding coefficient method, it is possible to introduce a considerable error into the computations by taking from the curves equivalent to figure 60 a value of r for each step corresponding to the initial discharge rate instead of the average rate during the step. While this error would practically compensate in a flood symmetrical about the peak, it does not do so in floods which are of longer duration on the falling than the rising stage, as in the majority of actual cases.

From the results of the studies of the Miami Conservancy District it was found, in general, that if a straight line variation for this coefficient is assumed between some arbitrary low discharge and the maximum discharge, the coefficient is on the average somewhat excessive and gives consequently too high total discharge. In fact, it exaggerates all of the flood peak except the very crest, and consequently it introduces too much time contraction. This will be clear from the following considerations.

High rates of runoff from small areas are caused by two concomitant factors; great intensity of precipitation, and conditions favoring rapid collection. Their joint effect is to produce flood peaks which rise abruptly and subside nearly but not quite as suddenly. During severe storms, one or more heavy downpours may occur, but their duration is nearly always short, rarely lasting over half an hour. The high rates of discharge which are characteristic of small streams and to which the McMath ratio and similar formulas apply, are the immediate result of such downpours, and are therefore also of short duration. Flood peaks of this kind, when platted on hydrographs, present quite needle-like appearances. Good illustrations of this may be found on the charts of automatic gage records obtained on small streams.

During the March 1913 storm probably several hard downpours took place, and the severest of these was responsible for the excessive rates of runoff obtained on Twin Creek and some of the lesser tributaries of the Miami and Stillwater rivers. Had these downpours lasted for a protracted time, the total precipitation in the localities affected would have exceeded by far the average total rainfall which fell on the watershed. This, while not impossible, would be an exceptional circumstance not supported by the records. During the 1913 storm, especially, the rainfall appears to have been fairly evenly distributed, and no unusual maxima were observed at any point. In all of the studies of routing floods through reservoirs heretofore made, such exceptional conditions have been assumed not to exist.

As may be inferred from the above reasoning, the variable coefficient method should give the best results when used to obtain the flow for storms of short duration and from small areas subject to high rates of runoff. Referring again to our study of the February 1916 and July 1915 floods, it is seen that this method fitted closely the conditions on Twin Creek and Upper Mad River, but gave excessive values for the larger tributaries of the Miami River. This is corroborated by other investigations, which proved that the sliding coefficient method gave excessive values for the March 1913 flood on the Stillwater River and on the Miami above Dayton.

Inflow Computed from Rainfall

This is a time honored method full of uncertainties and pitfalls. It requires a series of assumptions as to the percentages of rainfall retained by soil and vegetation, and evaporated by the air, for different time intervals throughout the duration of a storm and also after the latter has ceased. It involves assumptions as to the time required for the remaining water to drain into the watercourses, and varying rates of arrival at the particular locality under consideration. So many and so variable are the factors which affect such considerations that no one stereotyped method can be laid down to fit all cases, but new assumptions must be made for each storm and each drainage area.

Analysis of our figures of inflow, worked up from rainfall data of the 1913 storm, show that the selection of percentages of runoff was influenced by forehand knowledge of the maximum inflow rates which had been computed from measurements of the channels and highwater marks.* Such a plan of procedure may be entirely justifiable and good engineering practice, but furnishes no independent check on inflow computations based on Dayton discharge. For the purposes of this report the rainfall method, therefore, will not be considered in detail.

The difficulties and uncertainties of the operations discussed in the preceding paragraphs express the common sense fact that a single hydrograph of a main river during a flood does not furnish sufficient data to determine the hydrographs of individual tributaries which contributed to that flood. Fortunately this matter had no direct bearing on the design of the flood protection system for the Miami Conservancy District. The actual design of the dams, outlets, and channel improvements was based not upon the 1913 flood but upon an assumed Official Plan flood having a simple and definite hydrograph at each of the damsites. However, at many times during the planning of the system, especially in connection with estimates of possible damage to property in and near the basins, it was desired to answer questions as to how high the 1913 flood, or some other flood, would have risen at various points if the dams had been built. The assumptions described in the preceding appeared to furnish the most reasonable basis

* Calculation of Flow in Open Channels, by Ivan E. Houk, Technical Reports, Part IV, The Miami Conservancy District, Dayton, Ohio, 1918.

which could be secured for giving an approximate answer to questions of this kind.

Effect of Loramie and Lewistown Reservoirs

An additional complication was introduced into the determination of probable inflow hydrographs at the damsites of the Miami Conservancy District by the presence of the old Loramie and Lewistown reservoirs in the upper watershed of the Miami River. Of the total of 2525 square miles of watershed above Dayton, an area of 181 square miles, or 7.16 per cent of the whole, is tributary to these reservoirs. Unquestionably, during the 1913 flood, and other floods considered, the runoff from these 181 square miles was retarded to a greater or less extent by the reservoirs, and did not contribute its full intensity to the flood peak at Dayton.

This fact was allowed for in different ways by different members of the engineering staff of the District. In some of the computations made for routing floods through the retarding basins, the presence of the Lockington and Loramie reservoirs was disregarded, and the entire drainage area of 2525 square miles was used in the denominator of the fraction for pro-rating the Dayton flood among the tributary watersheds. Other computations went to the opposite extreme of subtracting out the entire 181 square miles and using 2344 as the denominator of the pro-rating fraction. It is probable that the correct course lies somewhere between the two. Investigations made during the progress of the work indicated that the peak outflow from the Lewistown dam in the 1913 flood was probably about 77 per cent of the peak flow which would have occurred at the same point if the dam were not in existence. The corresponding figure at the Loramie dam was estimated at 70 per cent. As these outflows were not only reduced but retarded with respect to the Dayton flood peak, it would appear that not more than about half of the 181 square miles tributary to the Lewistown and Loramie reservoirs should be included in the effective drainage area above Dayton in estimating the peak inflow rates at the various damsites. However, in estimating the total inflow during each flood considered at each of these damsites, it appears correct to include in the Dayton drainage area the entire amount of the above 181 square miles, since the Lewistown and Lockington reservoirs do not hold back the water permanently but only retard it.

Error Introduced through the Rectilinear Method of Platting River Stages

The gage records at Dayton are made up of morning readings, usually taken at 7 a.m. Of recent years the Weather Bureau has made it a practice to obtain special readings during flood stages, including the 'determination of the maximum or crest stage. The earlier records, however, rarely make mention of the crest stage, even though it may have been observed at the time. How much higher the water rose during the interval between observations cannot be gleaned by inspecting the published records. There is danger, therefore, of introducing material errors in computing river discharge by indiscriminately taking the morning readings as representing average daily conditions during flood stages. In this respect the Dayton record presents the same dangers as do the many early river stage records published by the Signal Service, Weather Bureau, and Geological Survey.

An approximation to the maximum or crest stage may be obtained by platting the gage readings in hydrograph form and connecting them with a curved line. This method is illustrated in figure 61, showing hydrographs of Miami River floods at Dayton. The small circles indicate the gage readings, and the lines drawn through them enable the probable stage to be found for any given time. The dotted portions represent the interpolated flood stages. It will be seen that the probable crest stage is in most cases defined within narrow limits, and is not likely to be in error by more than a few tenths. Thus, the flood crest of January 18, 1913, in the figure, can be obtained in a number of different ways according to individual judgment, the two curves in the diagram representing probable extremes, yet the crest stages do not vary by more than 0.2 of a foot.

An inspection of figure 61 shows that there is much similarity among the hydrographs of the Miami River floods. The parallelism in the ascending and descending curves is very striking, particularly in the latter. In the former it will be found that the steepest lines are characteristic of the highest floods. There is also similarity in the shape of the crests, even between such widely different floods as those of March 1913, and July 1915.

The usual practice in constructing flood hydrographs is to connect observed points by straight lines. Platted on cross section paper this gives rise to a rectilinear diagram as shown in a number of cases in figure 61. The hydrographs of this figure bring out clearly the fact that, if the observed points are connected by continuous curves instead of straight lines, the crest stage in a number of cases will be increased by a considerable amount. The effect of these increments is to add materially to the peak discharge as ordinarily computed.

In order to determine how much error is introduced in the routing of floods through retarding basins by using the rectilinear method of computing inflow, the February 1909 flood, which represents an' extreme case, was routed through the Englewood basin by both methods. The solid line in figure 62 shows the inflow into the basin computed

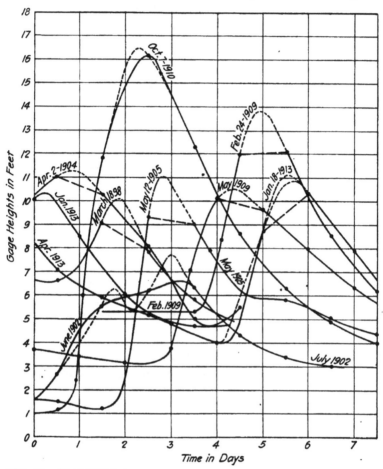

FIG. 61.—GRAPHICAL DETERMINATION OF CREST STAGES IN MIAMI RIVER FLOOD HYDROGRAPHS.

Observed river stages are indicated by black dots. The maximum stages were obtained by sketching in the curves and extending these by means of the dotted lines.

by the rectilinear method, assuming direct ratio of drainage areas. The dotted line represents a curvilinear hydrograph based on the same observed points, also computed by the direct ratio of drainage areas. The outflow curves obtained by routing each of these flood waves through the Englewood Basin are shown in the figure. The principal differences between the two cases are summarized as follows:

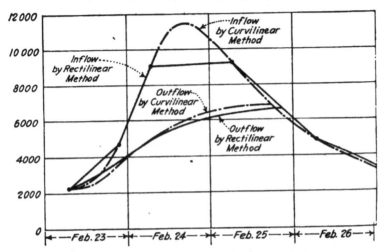

FIG. 62.—RECTILINEAR AND CURVILINEAR HYDROGRAPHS OF FEBRUARY 1909 FLOOD AT ENGLEWOOD DAM SITE.

The resulting differences in storage and outflow from the Englewood retarding basin obtained by using the two inflow hydrographs are shown by the outflow curves in lower portion of diagram.

Assumptions: Inflow into Englewood reservoir $= 651/2525$ Q (Dayton); conduit capacity $= 108.8 \sqrt{H}$ acre feet per hour; tailwater elevation assumed as fixed at elevation 778.

	Rectilinear Inflow Curve.	Curvilinear Inflow Curve.
Maximum surface elevation, feet	803.6	805.3
Maximum head H, feet	25.6	27.3
Maximum inflow rate, second feet	9,250	11,500
Maximum outflow rate, second feet	6,660	6,870
Maximum storage, acre feet	11,140	12,980
Time of filling, hours	62.5	59

The difference in maximum storage by the two methods, 1840 acre feet, represents a deficiency for the rectilinear method of 14 per cent. It is only fair to state that this flood wave represents a much more unfavorable condition for the rectilinear method than obtains on the average. At least one instance was found in which the rectilinear method gave a total discharge in excess of the curvilinear. In general, however, the tendency is for the rectilinear method to give results that are too small.

Conclusions Regarding Different Methods of Platting Inflow Curves

The following conclusions summarize the previous discussion of the various methods of drawing inflow curves.

1. The method of computing inflow from Dayton discharge by direct ratio of drainage areas, tends to give results less than the actual for such of the tributaries as possess small drainage areas subject to rapid runoff, as, for example, the Upper Mad River or Twin Creek. The method appears to be well adapted to the conditions which obtain on the Stillwater River, the Miami River, the Miami River above Dayton, and the Lower Mad River. Its use, though involving but small error, provides no margin of safety in the final result, and, when applied to inflow figures computed from rectilinear discharge diagrams, will in the long run give results appreciably smaller than the actual.

2. The method of computing inflow from Dayton discharge by applying a uniformly 'varying sliding coefficient to the ratio of the drainage areas, appears to be well suited to the small drainage areas subject to rapid runoff, but gives results from 15 to 18 per cent too high for tributaries draining large areas.

3. The method of computing inflow from rectilinear discharge diagrams, i. e. by assuming the flow to vary uniformly between observed stages, gives as a rule too small a peak discharge and most often too small a total discharge.

4. The method of computing inflow from rainfall is of secondary value, and it does not appear wise to resort to it except where stream flow data are not available.

In connection with these conclusions it may be remarked that the objections urged against the sliding coefficient method apply only to the particular form of this method in which a straight line variation for r is assumed. In this form it is doubtful if the method is satisfactory even for small drainage areas with uniform runoff. In its more general sense the sliding coefficient method is applicable to all conditions, the direct area-ratio method being merely a special case with r equal to unity for all rates of flow.

Retarding Basins in Series

Where two retarding basins occur one above the other on the same stream, the inflow curve for the lower one may be found by adding together the ordinates of the discharge curve for the drainage area tributary to the stream between the basins, and the corresponding ordinates of the outflow curve for the upper basin, making the necessary allowance for the time required for the outflow from the upper basin to reach the lower one. In the final system of retarding basins adopted by the Miami Conservancy District, this condition occurs in the case of the main branch of the Miami River, where the Taylorsville retarding basin receives the outflow from the Lockington basin and from the old Lewistown reservoir, and also in the case of the Lockington basin which receives the outflow from the old Loramie reservoir.

Studies made to determine the rate of travel of the 1913 flood crest between dam sites indicated that there was no steady progression of the flood crest down the river from source to mouth, but rather that the main stream over its entire length was filled simultaneously with flood water from its numerous tributaries, due, no doubt, to the fan-like arrangement of the drainage as a whole, and the relatively short lengths and steep slopes of most of the tributaries.

Examples of Inflow Curves

Typical inflow curves used in routing floods through the different retarding basins are shown in figures 31 to 35. A very large number of curves of this nature were constructed in connection with studies made to determine the operating characteristics of the basins during past floods. Additional examples are shown in figures 77 and 78.

GENERAL PRINCIPLES OF ROUTING FLOODS

The expression "routing a flood through a retarding basin" refers to the operation of computing the probable effect which the retarding basin will have on the hydrograph of the flood. For any given flood the data required for this operation and the result desired may be stated as follows:

Given: 1. The inflow-time curve, or inflow hydrograph, showing the flow into the retarding basin at each instant of the flood.

2. The storage-elevation curve of the retarding basin, showing the total amount of water stored at each foot of elevation of the surface.

3. The outflow-elevation curve, or rating curve, for the outlet. showing the outflow corresponding to each foot of elevation of the water surface.

Required: to construct the outflow-time curve, or outflow hydrograph, and to obtain the maximum water surface elevation reached in the basin.

Two different types of procedure may be used for routing floods; (1) step methods, and (2) calculus methods. In the former the flood is divided up into a number of finite steps or time intervals, while in the latter these steps are infinitesimal in size. Step methods readily adapt themselves to irregularities in the given data and are used where precise results are required. The calculus method is very rapid and convenient, and is suitable for use where only approximate re-sults are wanted, as in preliminary studies.

STEP METHODS FOR ROUTING FLOODS

The first operation in the step method for routing a flood consists in dividing the flood up into steps so short that the inflow and outflow curves will not vary sensibly from straight lines during the time in-terval represented by any step. The durations of the steps chosen may be equal or unequal according to the preference of the computer. In routing floods it is only necessary to consider one step at a time. The problem to be solved thus reduces itself to the prediction of the outflow at the end of any step if the outflow at the beginning of that step is given. The relation between the .variables met with in this operation is clearly stated in the following quotation from an article describing the work, published in Engineering News:[*]

"The outflow from the basin in a flood depends upon the height to which the basin is flooded. This height in turn depends on the volume stored, which is the difference between the total inflow and the total outflow. Thus the quantities involved are inter-related. The problem of computing the outflow had to be solved by trial and error."

Trial and Error Methods

There are many possible variations in the details of the process by which this operation may be carried out. It is an interesting fact that nearly every member of the engineering corps of the Miami Con-servancy District, who worked extensively on flood routing problems, developed a method of his own for obtaining the desired results. As each of these methods appeared to have its own special advantages and disadvantages, it seems desirable to discuss the subject in some detail for the benefit of those who may have to make such computa-tions in the future. As an aid in this discussion the following nota-tion is used:

Notation. The most satisfactory units for use in routing floods

[*] Study of Retarding Basin Operation, the fourth of a series of 5 staff articles on The Miami Valley Flood-Protection Works, Engineering News of February 1, 1917, Vol. 77, pages 186-8.

are the acre foot for expressing quantity of water, and the acre foot per hour for expressing rate of flow. This is due partly to the fact that the area flooded in a retarding basin is ordinarily expressed in acres, and the drainage area often so, and partly to the fact that the computations are less bulky and less laborious than if the cubic foot and second are used. In this connection it is convenient to remember that one acre foot per hour equals 12.1 second feet.

In the following notation absence of subscript indicates average value of quantity during step, or value of quantity at middle of step.

Subscript 1 denotes value of quantity at beginning of step.

Subscript 2 denotes value of quantity at end of step.

Let E = surface elevation of water in basin, in feet,

A = surface area of water in basin, in acres,

T = total time since beginning of flood, in hours,

I = accumulated inflow since beginning of flood, in acre feet,

O = accumulated outflow since beginning of flood, in acre feet,

S = accumulated storage since beginning of flood, in acre feet,

t = duration of steps, in hours,

i = inflow, in acre feet per hour,

o = outflow, in acre feet per hour,

s = rate of storage, in acre feet per hour.

Figures 63 and 64 show the geometric relations between the inflow and outflow curves during any step, the latter figure being the integral curve or mass curve of the former.

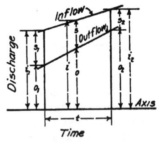

FIG. 63.—RELATION BETWEEN INFLOW AND OUTFLOW AT A RETARDING BASIN FOR ANY TIME INTERVAL DURING FILLING PERIOD.

Figure 65 shows the nature of the storage-elevation and outflow-elevation curves required in flood routing computations.

Example: Step method with _t_ assumed. The following example illustrates that class of step methods in which the time interval _t_ for each step is assumed at the beginning of the work. The example gives the details of routing the flood of March 1897 through the outlet conduits of the Germantown retarding basin. Figure 65 gives the

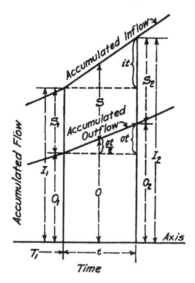

FIG. 64.—RELATION BETWEEN MASS CURVES OF INFLOW AND OUTFLOW, AT A RETARDING BASIN, FOR ANY TIME INTERVAL DURING FILLING PERIOD.

preliminary curves required for this operation. Methods for obtaining these curves have been discussed in preceding chapters of this report. It is convenient to combine the storage-elevation and outflow-elevation curves to form a single storage-outflow curve as shown in figure 66. The inflow curve estimated for this flood at the damsite is given in figure 67. Table 5 gives a summary of the computations for the various steps of this example, and figure 67 shows the outflow curve finally obtained. The trial-and-error method used in obtaining the outflow for a given step may be explained, as follows.

The routing will be started at 7 a.m., March 5, when the inflow is 146 acre feet per hour. It will be assumed that at this point the outflow is equal to the inflow. This involves the assumption that the total inflow prior to that time has exceeded the total outflow sufficiently to raise the stage in the reservoir to 734.7, corresponding to a storage of

70 acre feet. Any error introduced by such an assumption has been proved by experience to make no difference in the results.

Assume a time increment $t = 6$ hours, giving inflow $i_2 = 603$ acre feet per hour. Make a guess at the outflow six hours later than the

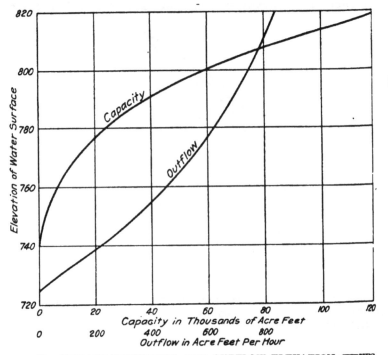

FIG. 65.—STORAGE–ELEVATION AND OUTFLOW–ELEVATION CURVES FOR GERMANTOWN RETARDING BASIN.

The storage-elevation curve indicates the capacity in acre feet, and the outflow-elevation curve the outflow in acre feet per hour, for different elevations of the water surface.

starting point, basing this guess on the probable position of the outflow curve.

Trial outflow $o_2 = 265$ acre feet per hour.

Corresponding storage rate $s_2 = i_2 - o_2 = 603 - 265 = 338$ acre feet per hour.

Mean storage rate during step $= s = (s_2 + s_1)/2 = (338 + 0)/2 = 169$ acre feet per hour.

Storage increment during step $= st = (169)(6) = 1014$ acre feet.

Accumulated storage $S_2 = S_1 + st = 70 + 1014 = 1084$ acre feet.

Table 5.—Example of Trial-and-Error Step Method of Routing Floods. The March 1897 Flood routed through Germantown Retarding Basin.

1	2	3	4	5	6	7	8	9	10	11
Step	Time at End of Step	Duration of Step	Final Inflow	Final Outflow	Final Rate of Storage	Mean Rate of Storage	Increment of Storage	Final Storage (Computed)	Final Surface Elevation	Final Storage (from Curve)
	T_2	$t = T_2 - T_1$	i_2	o_2	$s_2 = i_2 - o_2$	$s = \frac{1}{2}(s_1 + s_2)$	st	$S_2 = S_1 + st$	E_2	S_2'
	Hrs.	Hrs.	Ac. Ft. per Hr.	Ac. Ft. per Hr.	Ac. Ft. per Hr.	Ac. Ft. per Hr.	Ac. Ft.	Ac. Ft.	Ft.	Ac. Ft.
0	0	—	146	146	0*	—	—	70	734.7	70
1	6	6	603	290	313	157	942	1012	745.5	1000
2	12	6	1050	413	637	475	2850	3862	755.8	3700
3	17	5	850	463	387	512	2560	6422	760.5	6400
4	22	5	669	483	186	286	1430	7852	762.8	7800
5	27	5	489	489	0	93	465	8317	763.5	8300
6	29	2	413	487	− 74	− 37	− 74	8243	763.4	8100
7	40	11	281	470	−189	−132	−1452	6797	761.0	6900
8	51	11	149	421	−272	−230	−2530	4261	756.6	4050
9	61	10	114	347	−233	−252	−2520	1741	749.4	1800
10	68	7	99	240	−141	−187	−1309	432	740.8	440
11	74.2	6.2	—	—	0	− 70	434	−2	732.2	—

* Found by trial.

Entering the storage outflow curve of figure 66 with 1084 acre feet gives an outflow of 302 acre feet per hour.

Hence the guess of 265 was not very close.

It is generally thus with the first point after the start, as the direction of the outflow curve has not yet been indicated.

Assume a second trial outflow, which should be somewhere between 265 and 302, and nearer the latter.

Trial outflow $o_2 = 290$.

Corresponding rate of storage $s_2 = i_2 - o_2 = 603 - 290 = 313$ acre feet per hour.

Mean storage rate during step $= s = (s_2 + s_1)/2 = (313 + 0)/2 = 157$ acre feet per hour.

Storage increment during step $= st = 157 \times 6 = 942$ acre feet.

Accumulated storage $s_2 = s_1 + st = 70 + 942 = 1012$ acre feet.

Entering the storage-outflow curve, figure 66, with an outflow of 290 acre feet per hour gives a storage of 1000 acre feet, which checks reasonably close. The corresponding elevation of the reservoir surface is found by entering the storage-elevation curve of figure 65 with 1012 acre feet, or the outflow-elevation curve with 290 acre feet per hour.

It will be readily seen that the order of working in the method illustrated by the above example may be varied almost indefinitely.

Of the eight quantities o, s, o_2, s_2, O, S', O_2, S_2, and one might be chosen in the first guess of the trial-and-error method, and the final check made on any other, or the solution might be further changed

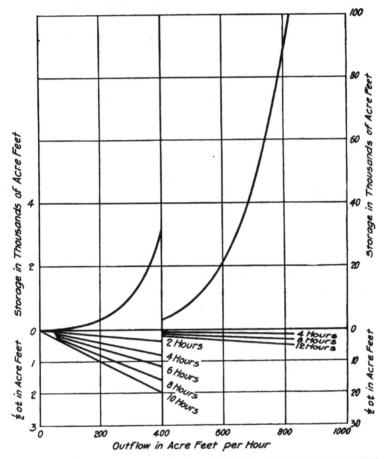

FIG. 66.—STORAGE–OUTFLOW CURVE FOR GERMANTOWN RETARDING
BASIN.

The heavy curved lines are obtained by combining the data shown in figure 65, thus giving a direct relation between storage in the basin and outflow from the latter. The lines below the horizontal axis of coordinates give the outflow corresponding to average outflow rates and given periods of time in hours.

by bringing into consideration the elevations and areas of the water surface.

In Table 5, the last column has been included for the purpose of furnishing a general check on the precision of the work. It will be seen that during the particular computation given in this table, the trial-and-error process was not carried to the point of securing an absolute check in every step. Where residual errors are left in the computations, as in this case, it is very desirable to include a checking column in the tabulation of results in order to afford a measure of the degree of accuracy secured in the work.

An important property of the routing curves is that the inflow and outflow hydrographs intersect at the maximum ordinate of the latter. In other words, the quantity of water stored and the outflow reach their maximum values at the instant when the outflow equals the inflow. This fact should be remembered in drawing the outflow

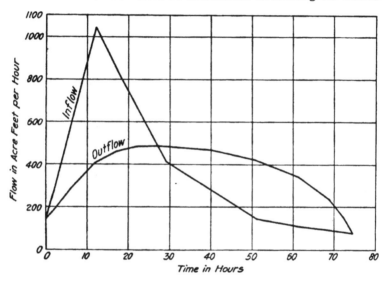

FIG. 67.—INFLOW AND OUTFLOW HYDROGRAPHS OF TWIN CREEK AT GERMANTOWN RETARDING BASIN FOR A FLOOD LIKE THAT OF MARCH 1897.

The outflow curve is obtained with the aid of the diagram in figure 66.

hydrograph, in case the step method used does not happen to furnish a point for platting the exact maximum outflow. It should also be noted that the two areas intercepted between the inflow and outflow curves, to the left and right respectively of their point of intersection, are each equal to the maximum storage developed in the basin.

12

Example: Step method with *t* found by trial. In the work done for the Miami Conservancy District, a number of computers used a step method in which the time *t* was determined by trial. Table 5 may be used to illustrate this method of working, by considering the various columns to be computed in the following order: Column 10: assume the elevation E_2 of the water surface at the end of the step considered. Column 11: find the corresponding storage volume from the storage-elevation curve, figure 65. Column 3: assume an increment of time *t*. Column 2: compute the total time T_2 to the end of the step. Column 4: find the corresponding inflow i_2 from the inflow hydrograph, figure 67. Column 5: find the outflow o_2 corresponding to the surface elevation E_2 by the outflow-elevation curve, figure 65. Column 6: find the storage rate at the end of the step by $s_2 = i_2 - o_2$. Column 7: find the mean rate of storage during the step by $s = \frac{1}{2}(s_1 + s_2)$. Column 8: find the increment of storage during the step by multiplying the average rate *s* by the assumed time increment *t*. Column 9: find the total accumulated storage by $S_2 = S_1 + st$. If this value does not check that previously obtained in column 11, assume a new value of *t* and repeat the computation. It is evident that the details of this method could be varied almost indefinitely, as is the cases where a constant *t* was assumed.

Comparison of trial-and-error methods. There is little choice between the various possible trial-and-error methods as far as expenditure of time and labor for obtaining the required results is concerned. Those methods in which a definite time increment *t* is assumed at the beginning, however, have the advantage of enabling the computer to pick the breaks in the inflow hydrograph, especially the peak, thus utilizing the points on that hydrograph which are most accurately determined.

A method of routing floods* which was used by the Morgan Engineering Company on previous work, consisted in preparing for the given outlet a series of curve sheets giving the complete relation between various uniform rates of inflow, outflow, and storage at different elevations of the reservoir surface. Using these sheets it was possible to solve flood routing problems by direct interpolation, without the use of trial-and-error methods. While this process is satisfactory for a reservoir in which the outlet is definitely designed or actually constructed, it is not suitable for studies in which the conduit design is unsettled and changes frequently as the work progresses.

Summary of experience with trial-and-error methods. The expe-

---

* See A Graphical Solution of the Problem of Storm Flow Through a Reservoir, by Albert S. Fry, Engineering and Contracting, of April 19, 1916, Vol. 45, pages 370–372.

rience of the engineers of the Miami Conservancy District in routing floods by trial-and-error methods may be summarized as follows: Those who are first beginning such calculations require considerable time to acquire a sufficient grasp of the subject to be thoroughly at home in the process. It is important, therefore, that the simplest and most direct method of procedure be followed.

In investigations involving a large number of computations of this kind it is desirable that a uniform method be observed by all computers, so as to have a standard office practice.

Under some circumstances the calculations should be based on assumed increments of gage height in the retarding basin; under other circumstances the time increments should be assumed; and in some instances both methods might be advantageously used in the same table.

The columns should be arranged in the logical order in which they naturally occur in making the calculations; and second, they should be in the most convenient order for the numerical work. These two requirements are often conflicting and must in each case be carefully weighed in deciding upon the most satisfactory order.

The carrying of additional columns for the purpose of comparing assumed and derived values is desirable, and enables the discrepancies to be balanced, so that a cumulative error may be avoided and the total final discrepancy kept as small as desired.

Direct Graphical Methods

The use of trial-and-error methods in extending the outflow hydrograph through any time interval may be avoided by the following method of working:[*] Construct the storage-outflow curve, figure 66, for the retarding basin and outlet under consideration, and on the same sheet construct another curve, (over the first and not shown in figure), whose ordinate corresponding to any abscissa o is $S + \frac{1}{2}ot$. If the time interval t varies for different steps, one such curve will have to be constructed for each value of t used. The quantity $S + \frac{1}{2}ot$ is seen from figure 64 to be equal to the accumulated inflow at the middle of the step minus the accumulated outflow at the beginning of the step, that is: $S + \frac{1}{2}ot = I - O_1$. The desired average outflow o for any step may thus be determined without the aid of trial-and-error method by entering such a curve as described above with a value of $S + \frac{1}{2}ot$ equal to the value of $I - O_1$ for the step.

Instead of using the curve last described, it is found convenient in practice to vary this method by drawing a straight line passing

[*] See Flood-Retarding Reservoir Problem Directly Solved, by Harold A. Thomas, Engineering News-Record, Vol. 79, August 2, 1917, page 226.

through the origin of coordinates and sloping below the outflow axis at such a rate that its ordinates represent values of $-\frac{1}{2}ot$. A set of such lines for different values of t is indicated in the lower portion of figure 66. The value of o corresponding to a given value of $S + \frac{1}{2}ot$ or $I - O_1$ is then readily found by setting off the latter quantity to scale with a pair of dividers, or strip of paper, and fitting it to the

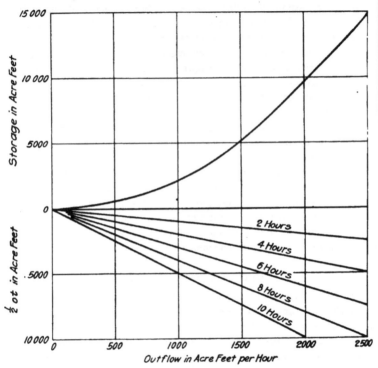

FIG. 68.—STORAGE–OUTFLOW CURVE FOR LOCKINGTON RETARDING BASIN WITH TEMPORARY OUTLET.

The significance of the various curves is the same as for figure 66.

space between the upper curve and the appropriate sloping line of figure 66.

Example using mass curve data. Table 6 and figures 68 and 70 give the details of a computation made by this method to determine the probable action of the temporary outlet of the Lockington retarding basin during a flood similar to that of March 1913. The temporary outlet in this case was an open channel or sluiceway as described on page 154, the rating curve being computed by the formulas on page 155.

Table 6.—Example of Direct Step Method of Routing Floods, using Mass Curve Data. The March 1913 Flood routed through Lockington Retarding Basin with Temporary Outlet in Operation.

1	2	3	4	5	6	7	8	9	10	11	12	13	14	15
Step	Time at End of Step T_1	Duration of Step $t = T_1 - T_1$	Final Inflow i_1	Average Inflow $i = \frac{i_1 + i_1}{2}$	Increment of Inflow u	Final Accumulated Inflow I_1	Average Accumulated Inflow $I = \frac{I_1 + I_1}{2}$	Initial Accumulated Outflow O_1	Working Ordinate $S + \frac{1}{2}ot = I - O_1$	Average Outflow o	Increment of Outflow ot	Final Accumulated Outflow $O_1 = O_1 + ot$	Final Storage $S_1 = I_1 - O_1$	Final Surface Elevation E_1
	Hrs.	Hrs.	Ac. Ft. per Hr.	Ac. Ft. per Hr.	Ac. Ft.	Ac. Ft.	Ac. Ft.	Ac. Ft.	Ac. Ft.	Ac. Ft. per Hr.	Ac. Ft.	Ac. Ft.	Ac. Ft.	Ft.
0	0	0	72	72	0	50	50	0	50	72	0	0	50	883
1	7	7	216	144	1,008	1,058	554	0	554	140	980	980	78	885
2	13	6	346	281	1,686	2,744	1,901	980	921	240	1,440	2,420	324	890
3	17	4	521	434	1,736	4,480	3,612	2,420	1,192	400	1,600	4,020	460	891
4	21	4	1,248	884	3,536	8,016	6,248	4,020	2,228	680	2,720	6,740	1,276	895
5	25	4	2,600	1,924	7,696	15,712	11,864	6,740	5,124	1,140	4,560	11,300	4,412	903
6	31	6	3,020	2,810	16,860	32,572	24,142	11,300	12,842	1,770	10,620	21,920	10,652	911
7	35	4	2,850	2,935	11,740	44,312	38,442	21,920	16,522	2,230	8,920	30,840	13,472	913
8	41	6	2,060	2,455	14,730	59,042	51,677	30,840	20,837	2,380	14,280	45,120	13,922	914
9	46	5	1,800	1,930	9,650	68,692	63,867	45,120	18,747	2,310	11,550	56,670	12,022	912
10	53	7	1,404	1,602	11,214	79,906	74,299	56,670	17,629	2,070	14,490	71,160	8,746	909
11	60	7	1,082	1,243	8,701	88,607	84,256	71,160	13,096	1,720	12,040	83,200	5,407	905
12	70	10	842	962	9,620	98,227	93,417	83,200	10,217	1,280	12,900	96,100	2,127	898
13	80	10	653	748	7,480	105,707	101,967	96,100	5,867	870	8,700	104,800	907	893
14	90	10	496	574	5,740	111,447	108,577	104,800	3,777	600	6,000	110,800	647	92
15	100	10	381	438	4,380	115,827	113,637	110,800	2,837	470	4,700	115,500	327	890

In every method of routing floods it is necessary to adopt some special or arbitrary assumptions and expedients at the beginning of the computations in order to get the regular procedure under way. Various devices may be used, and since they are applied to relatively small numerical quantities at the beginning of the flood, any variation in the exact procedure followed in initiating the calculations produces no appreciable variation in the subsequent results. The following process is a convenient one. Assume that at the beginning of the flood period the rates of inflow and outflow are equal; they are given, then, by the inflow hydrograph at the time decided upon to be considered as the beginning of the flood period. This assumption necessitates a certain corresponding amount of initial storage as shown by figure 68. Imagine a very brief initial step of zero duration which

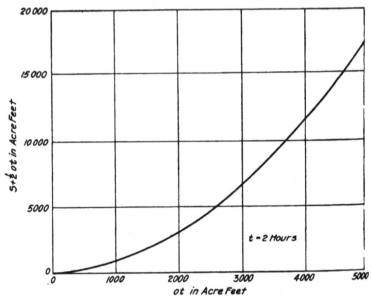

FIG. 69.—MODIFIED FORM OF STORAGE–OUTFLOW CURVE FOR LOCK-INGTON RETARDING BASIN, WITH TEMPORARY OUTLET.

The curve is derived from figure 68 by adding to the ordinates of the storage curve the ordinates of the 2-hour curve for values of ½ ot; and the horizontal scale is changed to read the total outflow in acre feet for 2-hour time intervals. This diagram can be used, therefore, only for values of t equal 2 hours. Similar diagrams can readily be prepared for use with other time intervals.

may be called step zero in the table. At the end of every step $S_2 = I_2 - O_2$. Therefore, we start with $O_2 = 0$ at the end of step zero, and $S_2 = I_2$.

In table 6 at the end of step zero, i_2 and o_2 as obtained from figure 70 are 72 acre feet per hour. The corresponding storage S_2 from figure 68 is 50 acre feet. Accordingly I_2 is also 50 acre feet, and O_2 is zero. This value of $S_2 = I_2$ is usually so small that it is difficult to read from the curve, and for practical purposes it may be assumed to be any small number, the error thus introduced being inappreciable in the final results.

The various columns in table 6 are then found as follows: Column 2, from the original inflow hydrograph; column 3, differences between successive values in column 2; column 4, from the original inflow hydrograph; column 5, average between successive values in column 4; column 6 = column 3 times column 5; column 7, summation of column 6; column 8, average of successive values in column 7; column 9, final outflow from preceding step as shown in column 13;

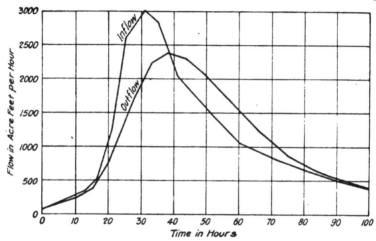

FIG. 70.—INFLOW AND OUTFLOW HYDROGRAPHS OF LORAMIE CREEK AT LOCKINGTON RETARDING BASIN, ASSUMING A FLOOD LIKE THAT OF MARCH 1913, AND THE TEMPORARY OUTLET IN OPERATION.

The diagram shows the results of routing the flood through the basin with the aid of the storage outflow curve in figure 68 and as worked out in detail in table 6.

column 10 = column 8 minus column 9; column 11, found with aid of diagram, figure 68; column 12 = column 3 times column 11; column 13 = column 9 plus column 12; column 14 = column 7 minus column 13; column 15, from storage-elevation curve for basin.

The degree of accuracy which can be obtained by the method just described, depends upon the amount of care used in constructing the storage-outflow curve, figure 68. If it is desired to obtain close results at very low heads it may be necessary to redraw to a larger scale that portion of this curve which is nearest the origin.

Almost as many variations of details are possible as in the trial-and-error methods. For example, column 5 may be read directly from the inflow hydrograph rendering column 4 unnecessary. If equal time increments are used throughout the computations, a procedure that would have some other advantages, then by preparing the proper scale, column 6 could be read directly from the inflow hydrograph, dispensing with both columns 4 and 5. Columns 9 and 13 are identical so that one may easily be omitted. Again, figure 68 may be modified by using as the horizontal coordinates values of ot. and as vertical coordinates values of $S + \frac{1}{2}ot$ as shown in figure 69. instead of values of o and S respectively as in figure 68. With this modification column 12 may be read directly from the diagram dispensing with column 11. Another useful variation is explained in the next paragraph.

Example using inflow curve data. The computations for the direct method as given in table 6 may be kept in somewhat less bulky form if it is not desired to use the columns of accumulated inflow and outflow. From figure 64 it is seen that $S_1 + \frac{1}{2}it = S + \frac{1}{2}ot$. Since in routing floods the first member of this equation can be evaluated at the beginning of the computations for any step, this value may be substituted for the working ordinates $S + \frac{1}{2}ot$ in figure 68, and the desired value o of the average outflow for the step thus determined without the use of mass curve data. Table 7 gives the details of a computation by this method, the problem being the same as that of table 6. The same variations as explained in the preceding paragraphs could also be applied in the case.

In none of the step methods explained in the preceding, either by trial-and-error or direct methods, has it been thought worth while to carry an extra column for the final outflow o_2 of any step. Although this could be obtained approximately by averaging the average outflow o of the step with that of the following step, the desired outflow hydrograph is platted as conveniently and accurately by using the mid-step values of the outflow as by using final values.

The processes given above for constructing the hydrograph of the outflow from a retarding basin without the use of trial-and-error methods, were not brought to the attention of the engineers of the Miami Conservancy District until the greater part of the routing computations for the work had been completed.

Table 7.—Example of Direct Step Method (Condensed Form) of Routing Floods, using Rating Curve Data. The March 1913 Flood routed through Lockington Retarding Basin with Temporary Outlet in Operation.

1	2	3	4	5	6	7	8	9	10	11	12
Step	Time at End of Step	Duration of Step	Final Inflow	Average Inflow	Half Increment of Inflow	Initial Storage	Working Ordinate	Average Outflow	Increment of Outflow	Final Storage	Final Surface Elevation
	T_1	$t = T_2 - T_1$	i_2	$i = \frac{i_2+i_1}{2}$	$\frac{1}{2}ti$	S_1	$S + \frac{1}{2}ot = S_1 + \frac{1}{2}ti$	o	ot	$S_2 = S_1 + \frac{1}{2}ti - ot$	E_2
	Hrs.	Hrs.	Ac. Ft. per Hr.	Ac. Ft. per Hr.	Ac. Ft.	Ac. Ft.	Ac. Ft.	Ac. Ft. per Hr.	Ac. Ft.	Ac. Ft.	Ft.
0	0	0	72	72	—	50	50	72	0	50	883
1	7	7	216	144	504	50	554	140	980	78	885
2	13	6	346	281	843	78	921	240	1,440	324	890
3	17	4	521	434	868	324	1,192	400	1,600	460	891
4	21	4	1248	884	1768	460	2,228	680	2,720	1,276	895
5	25	4	2600	1924	3848	1,276	5,124	1140	4,560	4,412	903
6	31	6	3020	2810	8430	4,412	12,842	1770	10,620	10,652	911
7	35	4	2850	2935	5870	10,652	16,522	2230	8,920	13,472	913
8	41	6	2060	2455	7365	13,472	20,837	2380	14,280	13,922	914
9	46	5	1800	1930	4825	13,922	18,747	2310	11,550	12,022	912
10	53	7	1404	1602	5607	12,022	17,629	2070	14,490	8,746	909
11	60	7	1082	1243	4350	8,746	13,096	1720	12,040	5,406	905
12	70	10	842	962	4810	5,406	10,216	1290	12,900	2,126	898
13	80	10	653	748	3740	2,126	5,866	870	8,700	906	893
14	90	10	496	574	2870	906	3,776	600	6,000	646	892
15	100	10	381	438	2190	646	2,836	470	4,700	326	890

INTEGRATION METHODS FOR ROUTING FLOODS

The operation of routing a flood through a retarding basin consists essentially in combining the data furnished by three curves:

(a) the area-depth curve of the basin,

(b) the outflow-depth curve for the outlet, and

(c) the inflow hydrograph of the flood,

to form a fourth curve: the outflow hydrograph.

If each of the three given curves could be defined by a mathematical equation, the fourth could be determined by exact or integration methods. While it is, of course, out of the question precisely to duplicate such curves as area-depth curves or inflow hydrographs by lines having simple mathematical equations, it was nevertheless found possible in the case of the basins of the Miami Conservancy District to do this with a sufficient degree of accuracy to furnish results of great value in preliminary studies and computations.

In making comparisons among the multitude of possible combinations of reservoir locations and capacities which presented themselves during the different stages of the work, it was necessary to make a very large number of approximate estimatès of retarding basin action under various conditions. A vast amount of time and labor which would have been required to make these computations by step methods. was saved by basing many of the estimates on the following simple assumptions:

(1) The area depth curve for a retarding basin is a semi-cubic parabola: $A = CH^{\frac{3}{2}}$.

(2) The outflow-depth curve for the outlet is an ordinary parabola: $o = BH^{\frac{1}{2}}$.

(3) The inflow hydrograph of the flood may be replaced by a rectangle representing a uniform inflow rate I, lasting for a ·length of time T.

These assumptions lead directly to a simple working rule or theorem, whose proof is given in the following paragraphs, namely: *The average rate of outflow during a uniform flood is approximately five-sixths of the maximum rate of outflow during this flood.*

Proof of Five-Sixths Rule for Routing Floods:
Notation: (See figure 71)

$h =$ Depth of water in retarding basin, or effective head on the outlet orifice,

$H =$ Maximum value of h during flood,

$a =$ Surface area of basin at depth h,

$A =$ Maximum value of a during flood,

$t =$ Time elapsed,

$t_0 =$ Time of beginning of flood,

$T =$ Total time from beginning to end of uniform flood, or time required for outflow to obtain its maximum value,

$o =$ Outflow at depth h or time t,

$O =$ Maximum outflow at depth H or time T,

$i =$ Inflow in general at time t,

$I =$ Inflow assumed in uniform flood,

$s =$ Storage in retarding basin at depth h,

$S =$ Storage in retarding basin at final depth H,

$C =$ Constant in area-depth equation, $a = Ch^{\frac{3}{2}}$ and in $s = \frac{2}{5} Ch^{\frac{5}{2}}$.

$B =$ Constant in outflow-depth equation, $o = Bh^{\frac{1}{2}}$,

$z = h^{\frac{1}{2}}$,

$x = O/I = BH^{\frac{1}{2}}/I$,

$r =$ Ratio of average outflow rate from retarding basin during filling period to maximum outflow rate,

$r_1 =$ Ratio of average outflow rate during emptying period to maximum outflow rate.

Since in all retarding basin studies the inflow at any instant is equal to the outflow plus the storage rate, the following differential

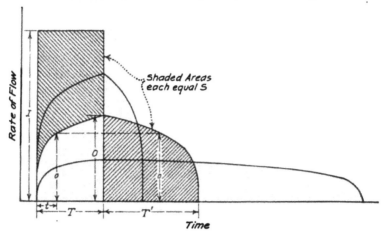

FIG. 71.—INFLOW AND OUTFLOW HYDROGRAPHS FOR FLOOD OF UNIFORM INTENSITY, ILLUSTRATING NOTATION USED IN FIVE-SIXTHS RULE.

The rectangle represents the flood of uniform intensity I and duration T. For an outlet with maximum capacity equal to O the amount of storage in the basin is indicated by the shaded areas S, the curved lines bounding same showing the varying rates of outflow from the basin. The period during which the retarding basin fills is T and the emptying period T'. Two other sets of outflow curves are shown corresponding to an outlet capacity in the one case 50 per cent greater than O and in the other equal to one half O.

equation is always true, regardless of the shape of the curves for area, outflow, or inflow.

$$idt = odt + adh \qquad (59)$$

Substituting the approximate expressions given above for area and outflow in terms of depth, the differential equation becomes,

$$idt = Bh^{1/2}dt + Ch^{1/2}dh \qquad (60)$$

This equation will take different forms according to the rate of variation of the inflow. In the first part of a flood, if the rate of inflow is increasing uniformly, the equation becomes

$$i_1(t - t_1)dt = Bh^{1/2}dt + Ch^{1/2}dh \qquad (61)$$

where i_1 and t_1 are chosen to give the required rate of increase. During the middle of a flood, when the inflow has a constant value i_2, the equation is

$$i_2 dt = Bh^{1/2}dt + Ch^{1/2}dh \tag{62}$$

During the last part of a flood, if the inflow is decreasing at a uniform rate, the equation is

$$[i_2 - i_3(t - t_2)]dt = Bh^{1/2}dt + Ch^{1/2}dh \tag{63}$$

where i_3 and t_2 are chosen to secure the required rate of decrease.

When the inflow has a constant value I during the entire flood, the differential equation may be integrated as follows:

$$Idt = Bh^{1/2}dt + Ch^{1/2}dh \tag{64}$$

Let $h^{1/2} = z$, $h = z^2$, $dh = 2zdz$.
Then,

$$Idt = Bzdt + 2Cz^4dz \tag{65}$$

Giving

$$dt = 2C\frac{z^4}{I - Bz}dz = \frac{2C}{B}\frac{z^4}{\dfrac{I}{B} - z}\,dz. \tag{66}$$

Dividing the denominator into the numerator, we obtain,

$$dt = \frac{2C}{B}\left[\left(\frac{I}{B}\right)^4\frac{1}{\dfrac{I}{B} - z} - \left(\frac{I}{B}\right)^3 - \left(\frac{I}{B}\right)^2 z - \left(\frac{I}{B}\right)z^2 - z^3\right]dz. \tag{67}$$

Integrating, we obtain

$$[t]_{t_0}^t = \frac{2C}{B}\Bigg[-\left(\frac{I}{B}\right)^4\log_e\left(1 - \frac{B}{I}z\right) - \left(\frac{I}{B}\right)^3 z - \frac{1}{2}\left(\frac{I}{B}\right)^2 z^2 \\ - \frac{1}{3}\left(\frac{I}{B}\right)z^3 - \frac{1}{4}z^4\Bigg]_{z_0}^z. \tag{68}$$

If t_0 and z_0 are both zero, this becomes

$$t = \frac{2C}{B}\Bigg[-\left(\frac{I}{B}\right)^4\log_e\left(1 - \frac{B}{I}z\right) - \left(\frac{I}{B}\right)^3 z - \frac{1}{z}\left(\frac{I}{B}\right)^2 z^2 \\ - \frac{1}{3}\left(\frac{I}{B}\right)z^3 - \frac{1}{4}z^4\Bigg]. \tag{69}$$

The logarithmic term in this expression may be expanded by McLaurin's series as follows:

$$\log(1 - x) = -\left(x + \frac{x^2}{2} + \frac{x^3}{3} + \frac{x^4}{4} + \frac{x^5}{5} + \cdots\right) \tag{70}$$

$$-\left(\frac{I}{B}\right)^4\log\left(1 - \frac{B}{I}z\right) = \frac{I^3}{B^3}z + \frac{1}{2}\left(\frac{I}{B}\right)^2 z^2 + \frac{1}{3}\left(\frac{I}{B}\right)z^3 \\ + \frac{1}{4}z^4 + \frac{1}{5}\left(\frac{B}{I}\right)z^5 + \frac{1}{6}\left(\frac{B}{I}\right)^2 z^6 + \cdots. \tag{71}$$

Substituting this value in equation (69), we obtain

$$t = \frac{2C}{B}\left[\frac{1}{5}\frac{B}{I}z^5 + \frac{1}{6}\left(\frac{B}{I}\right)^2 z^6 + \frac{1}{7}\left(\frac{B}{I}\right)^3 z^7 + \frac{1}{8}\left(\frac{B}{I}\right)^4 z^8 \right.$$
$$\left. + \frac{1}{9}\left(\frac{B}{I}\right)^5 z^9 + \cdots \right] \tag{72}$$

$$= \frac{2}{5}\frac{Cz^5}{I}\left[1 + \frac{5}{6}\frac{Bz}{I} + \frac{5}{7}\left(\frac{Bz}{I}\right)^2 + \frac{5}{8}\left(\frac{Bz}{I}\right)^3 + \frac{5}{9}\left(\frac{Bz}{I}\right)^4 + \cdots \right]. \tag{73}$$

Substituting the values $s = \frac{2}{5}Cz^5$ and $o = Bz$, this equation becomes

$$t = \frac{s}{I}\left[1 + \frac{5}{6}\frac{o}{I} + \frac{5}{7}\left(\frac{o}{I}\right)^2 + \frac{5}{8}\left(\frac{o}{I}\right)^3 + \frac{5}{9}\left(\frac{o}{I}\right)^4 + \cdots \right]. \tag{74}$$

The desired outflow hydrograph of the flood issuing from the retarding basin may be plotted by equation (74), different values of the outflow being assumed and the corresponding times t calculated by the equation. For relatively large values of o this series will converge very slowly, and the computation can then be made more conveniently by using the form in equation (69); the latter is convenient for numerical computation only when o is relatively large. It should be noted that if equation (69) were not desired, equation (74) could be obtained more directly from equation (66) by dividing so as to change the fraction into an infinite series and integrating term by term.

At the end of the uniform flood, when the elapsed time has reached the value T, the storage of water in the basin will have attained its maximum value S, and the outflow its maximum value O. At this instant equation (74) takes the form,

$$T = \frac{S}{I}\left[1 + \frac{5}{6}\frac{O}{I} + \frac{5}{7}\left(\frac{O}{I}\right)^2 + \frac{5}{8}\left(\frac{O}{I}\right)^3 + \frac{5}{9}\left(\frac{O}{I}\right)^4 + \cdots \right]. \tag{75}$$

This equation may be made less cumbersome by substituting the value $x = O/I =$ ratio of the maximum outflow to the average inflow. Using this abbreviation, the total inflow during the flood is given by,

$$IT = S[1 + \tfrac{5}{6}x + \tfrac{5}{7}x^2 + \tfrac{5}{8}x^3 + \tfrac{5}{9}x^4 + \cdots] \tag{76}$$

The total outflow which occurs during the reservoir filling period T, is equal to the total inflow minus the quantity of water stored in the basin, that is:

$$\int odt = IT - S = S[\tfrac{5}{6}x + \tfrac{5}{7}x^2 + \tfrac{5}{8}x^3 + \tfrac{5}{9}x^4 + \cdots] \tag{77}$$

The average outflow during the period T is

$$\frac{\int odt}{T} = \frac{IS[\tfrac{5}{6}x + \tfrac{5}{7}x^2 + \tfrac{5}{8}x^3 + \tfrac{5}{9}x^4 + \cdots]}{S[1 + \tfrac{5}{6}x + \tfrac{5}{7}x^2 + \tfrac{5}{8}x^3 + \cdots]}. \tag{78}$$

The ratio r of average outflow to the maximum outflow is given by

$$r = \frac{\int o\,dt}{OT} = \frac{\frac{5}{6} + \frac{5}{6}x + \frac{5}{6}x^2 + \frac{5}{6}x^3 + \cdots}{1 + \frac{5}{6}x + \frac{5}{7}x^2 + \frac{5}{8}x^3 + \cdots}. \tag{79}$$

By actual division this fraction is found to be equivalent to the following series:

$$r = \frac{5}{6} + \frac{5}{6\cdot6\cdot7}x + \frac{5\cdot32}{6\cdot6\cdot6\cdot7\cdot8}x^2 + \frac{5\cdot8712}{6\cdot6\cdot6\cdot6\cdot7\cdot7\cdot8\cdot9}x^3 + \cdots \tag{80}$$

$$= \frac{5}{6}\left[1 + \frac{1}{42}x + \frac{1}{63}x^2 + \frac{121}{10584}x^3 + \cdots \right]. \tag{81}$$

$$= \frac{5}{6}[1 + 0.02381x + 0.01587x^2 + 0.01143x^3 + \cdots]. \tag{82}$$

For small values of x the right hand member of this equation may be considered as practically equal to five-sixths, thus proving the five-sixths rule stated on page 186. Table 8 gives values of the ration r corresponding to different values of x, as computed by equation (82). The value for $x = \frac{3}{4}$ is not exactly as calculated by equation (82) because more terms of the infinite series would be necessary to give an accurate value; the value shown is what is estimated to be correct.

Table 8.—Ratio of Average Outflow to Maximum Outflow for Floods of Uniform Intensity.

Ratio of Maximum Outflow to Average Inflow. x	Ratio of Average Outflow to Maximum Outflow.
0	$\frac{5}{6}$ (1.000) = .833
$\frac{1}{8}$	$\frac{5}{6}$ (1.003) = .836
$\frac{1}{4}$	$\frac{5}{6}$ (1.007) = .840
$\frac{3}{8}$	$\frac{5}{6}$ (1.012) = .844
$\frac{1}{2}$	$\frac{5}{6}$ (1.018) = .848
$\frac{5}{8}$	$\frac{5}{6}$ (1.027) = .856
$[\frac{3}{4}]$	$\frac{5}{6}$ (1.037) = .865
1	$\frac{5}{6}$ (1.200) = 1.000

It is seen from this table that the five-sixths rule for routing floods is most accurate in cases where the outlet conduits are relatively small in comparison with the size of the retarding basin, so that the maximum outflow is small in proportion to the average inflow. The rule becomes less accurate as the relative size of the openings becomes larger, and fails altogether when they are so large that their retarding effect on the flood is nil, that is, when $x = 1$.

Time of Emptying Retarding Basin

The time required to empty a retarding basin after the cessation of the uniform inflow assumed in the derivation of the preceding formulas is found as follows:

Letting i equal zero in equation (60), the differential equation becomes for this case,

$$o = Bh^{1/4}dt + Ch^{1/4}dh \qquad (83)$$

Transposing,

$$dt = -\frac{C}{B}hdh. \qquad (84)$$

Integrating,

$$[t]_T^t = -\frac{C}{B}\int_H^h hdh = -\frac{C}{B}\left[\frac{h^2}{2}\right]_H^h. \qquad (85)$$

$$t - T = \frac{C}{2B}(H^2 - h^2). \qquad (86)$$

Substituting the values $o = Bh^{1/4}$, $O = BH^{1/4}$, and $S = \frac{2}{5}CH^{1/4}$ in this equation we obtain

$$t - T = \frac{5}{4}\frac{S}{O}\left[1 - \left(\frac{o}{O}\right)^4\right]. \qquad (87)$$

This equation furnishes the values needed for plotting the outflow curve during the emptying of the basin. The total time required to empty the basin after the uniform inflow ceases is found by making o equal zero in the above equation giving

$$T' = \frac{5}{4}\frac{S}{O}. \qquad (88)$$

It is readily seen from equation (88) that during the emptying period the average outflow is $\frac{4}{5}$ of the maximum outflow, that is

$$r' = \frac{S}{T'O} = \frac{4}{5}. \qquad (89)$$

In figure 71 are shown three examples of outflow curves corresponding to three different sizes of outlets, the uniform flood in each instance being assumed the same. The latter is indicated by the rectangle of height I and base T. For the medium sized outlet the figure shows by means of shading the amount of flood water stored, at the left as a portion of the uniform flood flow, and at the right in the form of outflow after the cessation of the uniform flood, the two quantities being, of course, equal in amount. O in this case represents the outlet capacity, and o the average rate of outflow during the emptying period, which will be found to be 4/5 of O. The time of emptying T' will be seen to vary materially, depending upon the size of outlet capacity assumed.

It is well to bear in mind, in this connection, that a uniform flood assumption, such as represented in figure 71, takes no account of any inflow that may enter the basin after the end of the filling period T, and that therefore the outflow curve for the emptying period and its time T', as here represented, are purely theoretical quantities subject, in practice, to appreciable modification by the inflow into the basin after cessation of uniform flood. The discrepancy caused by neglecting this inflow is illustrated in figure 72, which shows two outflow curves for the emptying period, one with inflow during the latter neglected and the other with inflow taken into consideration.

Examples of Use of Five-Sixths and Four-Fifths Rules

In order to illustrate the method of using the five-sixths and four-fifths rules in making estimates of retarding basin action during the filling and emptying periods, the two following numerical examples are given.

Example 1. Given: (*a*) The duration and average intensity of the flood to be controlled: 20 hours and 2000 acre feet per hour respectively; and (*b*) the maximum permissible outflow: 600 acre feet per hour.

Required: The storage capacity needed in the proposed basin.

Solution: Average outflow during the filling period = 5/6 of 600 = 500 acre feet per hour; total outflow during this period = average outflow times duration = 500 × 20 = 10,000 acre feet; storage capacity required = total inflow minus total outflow = 20 × 2000 — 10,000 = 30,000 acre feet.

Example 2. Given: (*a*) The duration and average intensity of flood to be controlled, 20 hours and 2000 acre feet, as before, and (*b*) the maximum available storage capacity of the given basin, 30,000 acre feet.

Required: (*a*) The maximum outflow from the basin, and (*b*) the time needed to empty the basin after the flood is over.

Solution: (*a*) Total outflow during filling period = total inflow minus total storage = 20 × 2000 — 30,000 = 10,000 acre feet; average outflow during filling period = total outflow divided by time = 10,000 ÷ 20 = 500 acre feet per hour; maximum outflow = 6/5 of 500 = 600 acre feet per hour; (*b*) average outflow during emptying period = 4/5 of 600 = 480 acre feet per hour; time to empty basin = total storage divided by average outflow = 30,000 ÷ 480 = 62.5 hours.

Determination of Equivalent Uniform Flood

The "equivalent uniform flood" corresponding to a given flood in a given retarding basin may be defined as an imaginary flood possessing a uniform rate of flow, having the same total inflow as the given flood, and capable of producing the same total amount of storage with the same maximum rate of outflow from the basin as the given flood. The duration of an equivalent uniform flood is usually shorter than that of its original flood, as indicated in figure 72. Since the derivation of the five-sixths rule is based on the assumption of a uniform inflow during the time the basin is filling, it follows that this rule will give accurate results for floods with irregular or peaked hydrographs only if correct methods are used in replacing these floods by their equivalent uniform floods. The methods used for this purpose may be classified as: (1) arbitrary methods, (2) exact methods, and (3) approximate methods.

(1) **Arbitrary selection of equivalent uniform flood.** In designing a retarding basin system for a hypothetical flood which is larger than any which has ever been recorded on the river or rivers considered, there is no reason for assuming that the hydrograph of this flood will take one shape rather than another, within a wide range of limits. In such a case the most that can be done with any degree of certainty is to determine from the rainfall and runoff records the duration and average intensity of the hypothetical flood to be provided for. For example, the final designs of the basins of the Miami Conservancy District are based on the assumption of a runoff of 10 inches in 3 days for the Englewood, Germantown, and Lockington Basins, and 9½ inches in 3 days for the Taylorsville and Huffman Basins.

Under such conditions, where the length and average intensity of the maximum flood are arbitrarily assumed, the five-sixths rule works to the greatest advantage. All ordinary problems regarding the action of retarding basins in a flood of this kind can be solved by this rule with great speed and convenience. This is illustrated in numerical examples given in the following chapter.

(2) **Exact determination of equivalent uniform flood.** Exact methods for the determination of the correct duration and intensity of an equivalent uniform flood involve the routing of the given original flood through the retarding basin by step methods. If the final design of a basin could be settled by a single routing computation, there would be nothing gained by determining the uniform flood which would duplicate the results of this computation. In actual cases, however, it is often desirable to replace a given flood by equiva-

13

lent uniform floods once for all, in order to avoid the inconvenience and labor of revising the computations by step methods whenever a change is made in the plans for the height of the dam or hydraulic properties of the outlet structure. Many such changes are necessary and unavoidable in balancing the several retarding basins of a flood system to obtain maximum protection at minimum cost.

The method of replacing a flood of irregular hydrograph by an equivalent inflow, is illustrated in figure 72. It is necessary to replace the actual hydrograph *b p e* by an equivalent rectangular hydrograph *a c d f* which will produce the same maximum outflow and develop the same total storage. Routing the flood through the basin by step methods will establish the end point *f* of the filling period and the maximum outflow *e f*. It will also establish the duration of the filling period *g f*, and the total inflow, total outflow, and the total storage, occurring during this period, if the flood is assumed to begin at some arbitrary instant *g*. This assumption is referred to as assumption *A* in the following. (As previously explained in discussing the step method, the point of beginning is ordinarily assumed at a low stage during the early part of the flood when the inflow and outflow can be regarded as practically equal.) The duration *a f* of the equivalent uniform flood is found by dividing the total actual outflow during the filling period by five-sixths of the maximum outflow *e f*. If considerable refinement is desired in the computation, the values of *r* taken from table 8 may be used instead of the constant five-sixths, in this division. The intensity *a c* of the equivalent uniform flood is found by dividing the total actual inflow by the time *a f* just computed.

Slightly different values for the duration and magnitude of the equivalent uniform flood will be obtained, if, instead of starting from an arbitrary low stage, as in assumption *A*, the duration of the original flood is assumed to be equal to that of the equivalent flood, that is, if the total outflow and inflow of the original flood are considered to begin at *a* in figure 72. The latter assumption may be designated as assumption *B*. It has the advantage of avoiding the use of the arbitrarily located point *g* on the inflow hydrograph. If this assumption is used the exact duration of the corresponding equivalent uniform flood can be determined only by trial-and-error methods, first assuming a value for *a f* and then revising it if the total storage as computed by the five-sixths rule does not check the step method value. With certain flood hydrographs it is impossible to find any equivalent uniform flood which will satisfy assumption *B*.

The following example illustrates the use of the exact method for

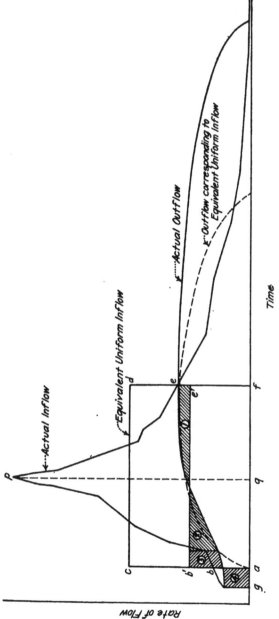

FIG. 72.—EXAMPLE ILLUSTRATING USE OF ⅚TH RULE IN REPLACING A FLOOD OF IRREGULAR HYDROGRAPH BY AN EQUIVALENT UNIFORM FLOOD.

The equivalent uniform inflow shown by the rectangle $a c d f$, replaces the original flood hydrograph only up to the instant that the basin reaches its maximum water level, which is at e. The ordinate $e f$ represents the maximum outflow through the conduits. The average outflow during the filling period is shown by $e' f$ which is five-sixths of $e f$.

determining an equivalent uniform flood as defined by assumption *A*. The hydrographs for this example are given in figure 34, page 104.

Taylorsville retarding basin during March 1913 flood.

Step method results:

End of filling period at 66 hours.
- Maximum outflow, 4240 acre feet per hour.
Total inflow during filling period, 334,000 acre feet.
Maximum storage 152,000 acre feet.
Total outflow during filling period, 182,000 acre feet.

Data for equivalent uniform flood:

$\frac{5}{6}$ of maximum outflow $= 3530$ acre feet per hour.
Duration of equivalent uniform flood $= 182,000/3540$
$= 51.4$ Hours.
Intensity of equivalent uniform flood $= 334,000/51.4$
$= 6500$ acre feet per hour.

(3) **Approximate methods for establishing equivalent uniform flood.** In retarding basin studies occasions often arise on which it is desired to obtain an approximate estimate of the action of a particular flood in a particular retarding basin without going to the labor of making an exact step-method solution for the problem. In such cases an experienced computer can estimate by judgment the duration of a uniform flood which will be practically equivalent to the given flood. Under favorable circumstances the results of estimates made in this way will check the step method values with remarkable closeness. This method of working is in general most successful in the case of a sudden flood occurring in a basin where the outlets are relatively small and the ratio of maximum outflow rate to maximum inflow rate correspondingly low. The difficulty of estimating an equivalent uniform flood by judgment is much increased if the flood rises gradually or the retarding effect of the basin is slight.

The first step in the process of estimating an equivalent uniform flood is to locate the point *e*, figure 72, on the inflow hydrograph. This point determines the end of the filling period, or instant at which the outflow attains its maximum value *e f*. In many retarding basin problems the maximum permissible outflow is determined beforehand, from considerations regarding the degree of protection required and the cost of channel improvement in the cities to be protected. In such cases the point *e* can be definitely fixed at the point on the falling side of the inflow hydrograph where the inflow equals the permissible maximum outflow. If the nature of the problem to be solved is such that the maximum outflow *e f* is not known at the beginning. the proper location of *e* can be established only by trial computations, first assuming a location for *e* and then seeing if the value obtained for the final outflow rate checks the ordinate *e f*. It should be noticed that some variation in the position of the point *f* makes no appreciable error in the computed value of the maximum outflow or the maximum stage reached in the retarding basin.

The principal difficulty in determining an equivalent uniform flood by judgment, occurs in the location of the initial ordinate, *a c*, figure 72. No simple rule applicable to all cases can be given for establishing the position of this ordinate. Where a flood rises suddenly, it is true in general that the start of the equivalent uniform flood is at or very near the beginning of the sudden rise. The outflow hydrograph corresponding to a uniform inflow in a typical retarding basin with conduit outlets is approximately a fifth-power

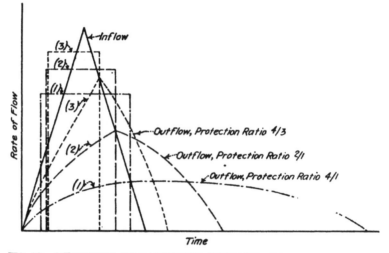

FIG. 73.—EXAMPLES OF PARABOLIC OUTFLOW CURVES FOR FLOODS
OF UNIFORM INTENSITY.

Each rectangle represents a uniform flood and is equivalent in area to the triangular flood less the portion to the right of the ordinate representing maximum outflow. The three cases selected give protection ratios as indicated, these ratios being the proportion which the maximum flood inflow bears to the maximum outflow through the conduits.

parabola. The exact shape of this curve for three different cases is shown in figure 73. The flood hydrograph is represented by the isosceles triangle indicated by the full lines. Three alternative uniform floods, each exactly equivalent to the inflow from the triangular flood during the filling period, are shown by rectangles numbered

1, 2, and 3. The outflow curves corresponding to these uniform floods are similarly numbered, and to facilitate further identification each is drawn with the same symbol used in drawing the rectangular flood of which it is the outflow. When the initial ordinate of the equivalent uniform flood is correctly located, the area under its semi-parabolic outflow curve will equal the area under the actual outflow curve for the filling period. If the computer can form a mental picture of probable shapes of and relation of these two curves without actually drawing them, he can from this often locate the initial point of the former with sufficient precision for many practical purposes.

Where the total inflow is considered as starting from some arbitrary point of low stage, as g in figure 72 (assumption A), the following procedure is helpful in locating the starting point of the equivalent uniform flood. In figure 72 sketch in by eye the probable position of the actual outflow curve; make e' f equal ⅚ of e f and draw in the horizontal line e' b'; estimate area (2) equal to area (1), and locate the line a c so that area (4) will equal area (3). The proof of this rule follows readily from the fact that the area under the actual outflow curve, during the filling period, is practically equal to five-sixths of the maximum outflow times the duration of the equivalent uniform flood.

An approximate rule which may be used to locate the position of the initial ordinate of the equivalent uniform flood, is to make a b, figure 72, a definite fraction of f e. It was found by experience in routing the March 1913 flood through the basins of the Miami Conservancy District, that the results of the step method could be closely checked by the five-sixths rule if a b was taken as five-sixths of e f, and the total inflow computed by assumption B, see page 194. This rule, however, does not hold good for all types of inflow hydrographs and should not be used unless it can be checked by step methods or by inspection for the particular type of inflow curve under consideration.

When the location of the initial ordinate a c, figure 72, has been decided upon by one method or another, the length of this ordinate, or intensity of the equivalent uniform flood, is found by dividing the total inflow by the duration of the filling period.

Figures 74, 75, and 76 illustrate the selection of equivalent uniform floods for three different types of triangular flood waves. The dotted lines in these figures show the equivalent uniform floods as computed under assumption A. The use of assumption B would give somewhat shorter and higher rectangles. In figure 76, however, it is impossible to determine any equivalent uniform flood if assumption B is used.

Two extreme cases are shown: figure 74, in which the final ordinate of the uniform flood must coincide with the vertical side of the triangular flood wave, thereby fixing the position of the initial ordinate; and figure 76, in which it becomes necessary for the initial ordinate to precede by a brief time interval the beginning of the flood

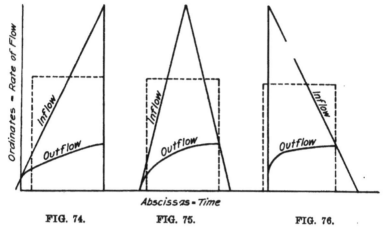

FIG. 74.   FIG. 75.   FIG. 76.

EXAMPLES OF EQUIVALENT UNIFORM FLOODS FOR THREE TYPES OF TRIANGULAR FLOOD WAVES.

wave. These two cases illustrate the difficulty of finding any rule of universal application to supplement the judgment of the computer in establishing the beginning of an equivalent uniform flood.

Appreciable differences are apparent also in the outflow curves of the three triangular flood types shown, calling ·for values of r differing from the ⅝th value previously discussed.

Possibility of using equivalent triangular floods. The flood hydrographs of some rivers are more nearly triangular than rectangular in shape. The curves of figures 74, 75, and 76, suggest the possibility of representing such hydrographs by equivalent triangular floods, whose outflow could be computed by the aid of a table giving values of r for triangular flood peaks with different shapes and protection ratios. For example: the value of r for an isosceles inflow triangle in a basin with a protection ratio of 4 to 1, as in figure 75, would be ¾. If a sufficiently extensive table of such values of r were constructed for various triangular hydrographs, a considerable degree of accuracy might be attained in approximate flood routing computations by replacing the given inflow curves by their equivalent triangular floods. However, since the element of judgment could not in any case be eliminated in selecting the equivalent flood, it appears on the whole simpler in routing floods by approximate methods to depend only on rectangular equivalent floods, having the practically constant outflow ratio $r = \frac{5}{6}$.

CHAPTER VIII.—FACTORS AFFECTING THE HEIGHT OF DAMS AND SIZE OF OUTLET CONDUITS

Many of the considerations which affect the height of dam and size of outlet required for a given retarding basin have been discussed in detail in this and preceding volumes of the Technical Reports. In the present chapter the principal results of these discussions are reviewed and summarized, and the method used in combining them in the design of any individual basin is illustrated. In this connection a brief description is also given of extensive studies which were made to determine the effect which the design of the basins would have upon their operation during floods of different magnitudes. The actual final selection of the heights of dams and size of outlet conduits for the basins is explained in chapter XII, on the balancing of the system as a whole.

VARIOUS CONTROLLING FACTORS

Four principal factors control the height of dam and size of outlet conduit required for an individual retarding basin. These are:

(1) the nature of the maximum flood to be provided for;
(2) the maximum outflow rate permitted;
(3) the location of the dam, as affecting the storage obtainable at different depths; and
(4) the type of outlet to be used.

In designing the basins of the Miami Conservancy District, each of these factors was made the subject of extensive studies and investigations. The final conclusions reached as a result of these researches are given in the following paragraphs.

(1) The maximum inflow, or Official Plan flood, to be provided for in designing the outlet conduits and fixing the spillway elevations, was taken at 10 inches runoff in 3 days for the Englewood, Germantown, and Lockington basins, and 9½ inches in 3 days for the Taylorsville and Huffman basins. This runoff is about 40 per cent in excess of that of the 1913 flood. These figures represent the outcome of the very extensive researches on rainfall and runoff, which are described in Parts IV and V of the Technical Reports.*

* Calculation of Flow in Open Channels, by Ivan E. Houk, Technical Reports, Part IV; and Storm Rainfall of Eastern United States, Technical Reports, Part V; The Miami Conservancy District, 1917 and 1918.

(2) The maximum permissible outflow rates from the different basins were fixed by various considerations. The combined outflow rate permitted from the basins of the system as a whole, was determined by balancing the cost of channel improvements in the cities to be protected against the cost of additional basin development; but the outflow from individual basins was in some cases controlled by other factors than cost. Certain basins in the system are situated so favorably with regard to topography and existing improvements that it was economical to develop them to the fullest extent possible, using the less favorably situated basins only to make up the total of the required protection. The former type is approached by the basins at Englewood, Germantown, and Lockington, and the latter type by those at Taylorsville and Huffman. In the first three, the chief factors which limited the degree of development and fixed the minimum size of the conduits were: (a) the necessity of providing openings sufficiently large to avoid danger from floating trash; and (b) the desirability of having the basins empty themselves with reasonable rapidity after a flood, both to provide protection in case of a second flood and to avoid prolonged flooding of farm lands. In the last two of the basins mentioned the maximum permissible outflows were established by careful estimates of cost and studies of the action of the system as a whole.

The fifth line of table 9 gives the maximum outflow rates assumed for the various basins in making the final design. The degree of development to which the flood retarding possibilities are carried in each basin is indicated in the thirteenth line of this table, which gives the protection ratio, or ratio of maximum inflow rate to maximum outflow rate, as used in the final plan.

(3) The locations of the damsites, as affecting the storage obtainable at different depths, were decided upon after exhaustive surveys had been made of surface and subsurface conditions. The area and capacity curves at the various basins have been given in figures 37 and 38. These curves can be approximately reproduced by using in the equations $a = Ch^{\frac{1}{2}}$ and $s = \frac{2}{3}Ch^{\frac{3}{2}}$.

(4) The types of outlets used at the various basins were chosen as the result of extensive studies made by the designing engineers of the District. For each outlet as designed it was found possible to express the approximate outflow rate at high heads by an equation of the form $o = Bh^{\frac{1}{2}}$, or for outlets of similar design and shape but with the linear dimensions of the cross-section slightly increased or decreased in the ratio d/d_1, by an equation of the form $o = B_1(d/d_1)^2 h^{\frac{1}{2}} = Bh^{\frac{1}{2}}$. Values of B used in the approximate final design of the basins are given in the twelfth line of table 9.

Table 9.—Summary of Computations of Retarding Basin Sizes for the Official Plan Flood with the Aid of the Five-Sixths Rule.

	Symbols	Retarding Basins				
		German-town	Englewood	Lockington	Taylors-ville	Huffman
1. Area of watershed, square miles........		270	651	255	878	671
2. Total flood runoff, inches on watershed ...		10	10	10	$9\frac{1}{2}$	$9\frac{1}{4}$
3. Duration of flood runoff, hours......	T	72	72	72	72	72
4. Total inflow, acre feet.....	IT	144,000	347,200	121,200*	474,000†	340,000
5. Maximum permissible outflow rate, acre feet per hour.	O	826	990	744	4,550	2,890
6. Maximum permissible outflow, rate second feet......		10,000	12,000	9,000	55,000	35,000
7. Approximate total outflow, acre feet.....	$5/6OT$	50,160	59,400	44,640	273,000	173,400
8. Maximum storage, acre feet.........	$S = IT -5/6OT$	93,840	287,800	76,560	201,000	166,600
9. Maximum surface elevation, feet above sea level........	E	812	873	940	819	834
10. Elevation of zero head, feet above sea level........	E_o	732	781.5	885	779	793
11. Maximum head, feet....		80	91.5	55	40	41
12. Constant B in outflow depth equation, to give discharge in acre feet per hour.....	B	93	103	100	720	451
13. Ratio: $\dfrac{\text{Maximum inflow rate}}{\text{Maximum outflow rate}}$		3.3	6.6	3.4	1.9	2.1

* Allows for storage in Loramie State Reservoir.
† Includes outflow from Lockington Basin and allows for storage in Lewistown State Reservoir.

COMBINING THE CONTROLLING FACTORS IN RETARDING BASIN DESIGN

When the four chief factors which control the design of a retarding basin have been determined, as in the above summary, there is no room for farther choice regarding the size of the dam or outlets, for the data assumed leads to single definite values for both the maximum depth of water stored in the basin, and the size of the conduits required. To determine these values approximately by the five-sixths rule, if the necessary preliminary data is at hand, requires but the work of a few moments. The process consists of five steps.

(1) Estimate from the assumed maximum flood hydrograph the duration of the equivalent uniform flood (this was assumed as 72 hours), and compute the total inflow during this period.

(2) Estimate the total outflow during the period just computed as five-sixths of the maximum outflow rate times the duration of the period.

(3) Find the maximum storage required by subtracting the total outflow from the total inflow.

(4) Find the corresponding maximum head of water in the basin by the storage elevation curve or formula. This fixes the maximum flood water elevation in the basin, and this, in turn, is the chief factor in determining the height of the dam.

(5) From the outflow depth formulas for the type of conduit used, determine the dimensions of the outlet necessary to discharge the assumed outflow at the head computed in (4).

Table 9 gives the details of this process as applied to the five basins of the Miami Conservancy District. The part of the process which deals with the determination of the dimensions of the outlet conduits is not shown in table 9, having been explained in detail in chapter VI.

It should not be assumed that the actual computations for the design of any one of the retarding basins of the Miami Conservancy District were as simple as those indicated in table 9. Although this table summarizes the essential steps in the process of design, it makes no record of the many tentative plans and calculations which were necessary before satisfactory values could be obtained for the final rates of maximum outflow and the outflow depth curve coefficients.

In general, it is desirable to check and revise by step methods any important preliminary computations made by the five-sixths rule. In the case of the computations of table 9, however, there is found to be very little difference between the values obtained by the approximate and exact methods. This is true on account of the nature of the arbitrarily assumed maximum flood on which these computations are based.

STUDY OF RETARDING BASIN OPERATION

Knowledge of the probable behavior of a retarding basin in floods of different magnitudes is in ordinary cases prerequisite for the economic design of such a basin. It has already been pointed out that in a flood control system by dry reservoirs it is usually economical to develop the flood retarding properties of certain sites to the fullest possible extent. However, as it is out of the question to develop a basin to the extent of shutting off the entire flow from its tributary watershed, it becomes necessary in the course of the design to determine by studies of operative conditions how far it is safe or practicable to carry such development.

The factor which ordinarily governs the minimum size permissible for the outlet conduits of a retarding basin, is the length of the period required for the basin to become empty after a flood. It is desirable for two reasons that this period should not be excessively long: first, that protection may be available in case of another flood coming soon after the first; and second, that interference with farming operations in the basin area may be avoided as far as possible. In order to investigate this subject quantitatively during the design of the basins of the Miami Conservancy District, comprehensive studies were made to determine the probable action of these basins during all floods for which records could be obtained. These studies were also intended to furnish a basis for securing land easements, or flooding rights, and the appraisal of damages within the basin areas.

Sources of Data

At the beginning of the studies of retarding basin operation, practically no gagings were available on the tributaries of the Miami River. Daily readings of the river gage at Dayton had been recorded since 1893. In this record 28 principal floods were noted as having occurred during the 24 years from 1893 to 1916 inclusive. In the absence of other data an estimate of the probable occurrence of floods, at the various damsites during this 24-year period, was made by prorating the hydrographs of the 28 Dayton floods among the tributary watersheds according to the sliding-coefficient method explained in chapter VII.

It must be recognized that this method of estimating tributary floods, even at its best, gives results which are comparative only, and not literally true in any particular case. Assuming that the sliding coefficient can be adjusted to give the proper proportions of different runoff rates during a flood, there is no assurance that it gives the correct times at which these rates occurred. For example: the estimated hydrograph for the 1913 flood at the Englewood damsite, figure 32, indicates that at 6 p.m. on March 25, 1913, the Stillwater river at this location reached its peak discharge rate of 85,000 second feet. As a matter of fact, there is no evidence to show that this tributary actu- ally attained this particular discharge at this particular time. At this instant in actual history it may have been true that other tribu- taries were making up a large part of the Dayton discharge, and the flow from the Stillwater river was below its peak stage. Besides being misleading in matters of detail, this method of estimating past floods has a tendency to give less than the true values for the larger floods. In spite of these limitations, however, the method probably represents the most practicable system available for using the existing data in the study of retarding basin operation, and gives results which are valuable when properly interpreted.

Construction of Outflow Hydrographs

After constructing inflow curves in the above manner for the 28 floods at each of the five damsites, and after determining the neces- sary area-depth and outflow-depth curves for the basins as then de- signed, the next step in the detailed study of retarding basin opera- tion was to route each estimated flood through its corresponding basin, obtaining as a result 140 different outflow hydrographs. Fig-

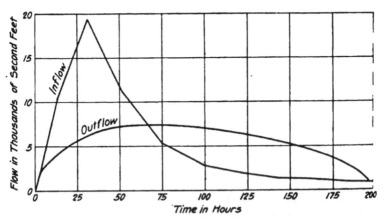

FIG. 77.—INFLOW AND OUTFLOW HYDROGRAPHS OF STILLWATER RIVER AT ENGLEWOOD RETARDING BASIN, FOR A FLOOD LIKE THAT OF OCTOBER 1910.

ures 31 to 35 inclusive, on pages 101 to 105, chapter IV, show the inflow and outflow curves obtained for the five basins during the flood of March 1913. Figures 77 and 78 give the hydrographs for the flood of October 1910 in the Englewood and Huffman basins. The latter represents a typical dry-weather flood, such as might occur during the farming season. The retarding effect of the basins is seen to be much less pronounced in the case of the smaller flood than in that of the 1913 flood.

The computations for this study were made by step methods, using the trial-and-error process explained in the preceding chapter. A summary of the principal data obtained from the outflow hydrographs constructed for the five retarding basins, is given in tables XII, XIII,

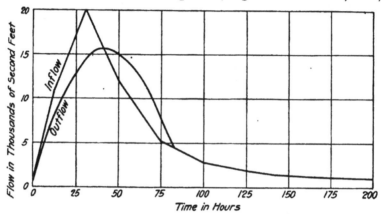

FIG. 78.—INFLOW AND OUTFLOW HYDROGRAPHS OF MAD RIVER AT HUFFMAN RETARDING BASIN FOR A FLOOD LIKE THAT OF OCTOBER 1910.

XIV, XV, and XVI of the Appendix. Only 27 of the 28 floods mentioned above appear in table XIII, the reason being that on account of the long emptying period of the Englewood basin, the floods of March 2 and March 6, 1908, combined in it to form a single outflow peak.

Construction of Duration-Depth Curves

Using the data furnished by the 140 outflow hydrographs, duration-depth curves for the flooding caused by the 28 floods, in each of the retarding basins, were next platted. Figure 80 shows these curves for the Germantown basin. The method used in plotting each curve was as follows: The interval of time during which the outflow rate of a given flood exceeded any chosen value, was found by scaling the

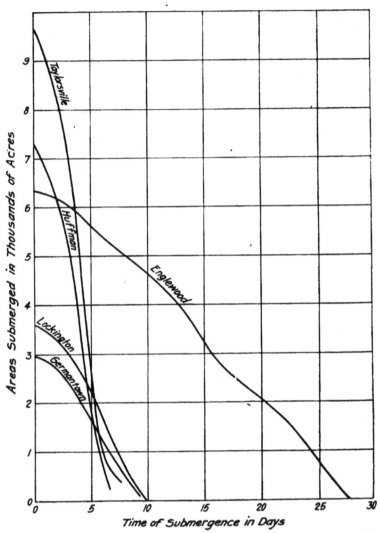

FIG. 79.—AREAS FLOODED AND TIME OF SUBMERGENCE IN THE FIVE
RETARDING BASINS FOR A FLOOD LIKE THAT OF MARCH 1913.

horizontal intercept corresponding to that outflow rate on the outflow
hydrograph. The corresponding elevation of water surface in the
basin was read from the outflow-elevation or rating curve for the outlet. Platting the interval of time as abscissa and the corresponding

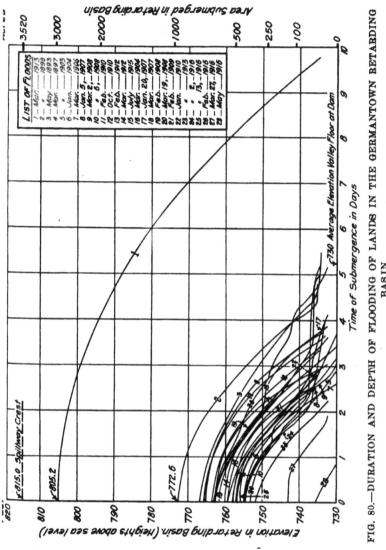

FIG. 80.—DURATION AND DEPTH OF FLOODING OF LANDS IN THE GERMANTOWN RETARDING BASIN.

To be interpreted as explained under figure 83.

surface elevation as ordinate, established a point on the desired duration depth-curve for the given flood in the given basin.

It is seen from figure 80 that the duration-depth curves of the various floods in the Germantown retarding basin show a tendency to assume a parabolic shape and to parallel one another in a general way. A number of the floods shown in the diagram, however, do not

14

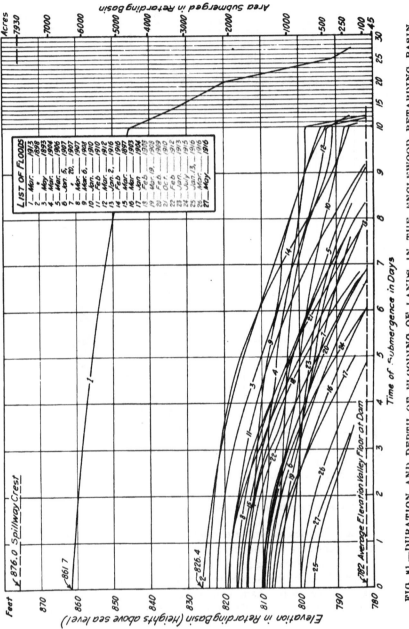

FIG. 81.—DURATION AND DEPTH OF FLOODING OF LANDS IN THE ENGLEWOOD RETARDING BASIN.

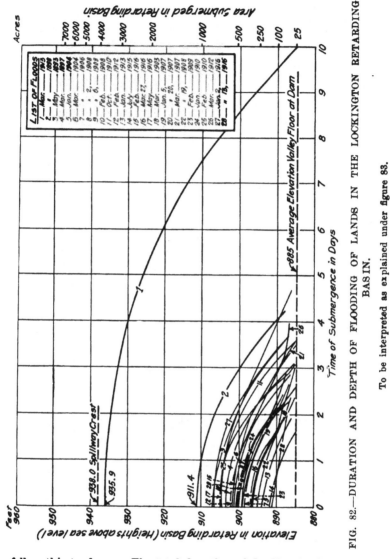

FIG. 82.—DURATION AND DEPTH OF FLOODING OF LANDS IN THE LOCKINGTON RETARDING BASIN.

To be interpreted as explained under figure 83.

follow this tendency. The total duration of flooding in the basin for the typical floods appears to be roughly proportional to the maximum depth attained, but the relation is not sufficiently definite to be used as the basis for a mathematical rule. The curves in figures 81, 82, 83, and 84 for the other retarding basins of the system show

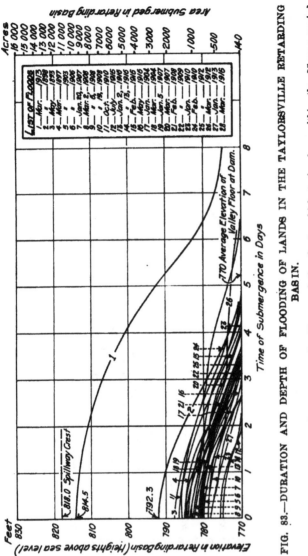

FIG. 83.—DURATION AND DEPTH OF FLOODING OF LANDS IN THE TAYLORSVILLE RETARDING BASIN.

Each curve is numbered to correspond with one of the floods which took place within the 25-year period 1893-1917, listed in the diagram, and shows the flooding conditions within the basin that would be caused by a flood of that size. The diagram also shows the acreages submerged up to any particular elevation, their depths and durations of submergence, and the relative frequencies of submergence at different levels. Thus, lands in the Taylorsville basin situated below elevation 790 appear to be subject to a frequent flooding, while those above that elevation appear to be but rarely flooded.

characteristics similar to those noted in the case of Germantown. The durations of flooding are considerably longer in the Englewood basin on account of the higher protection ratio adopted for that basin.

In order to visualize the duration of flooding in the various basins with respect to the amount of land flooded within them, the scale of

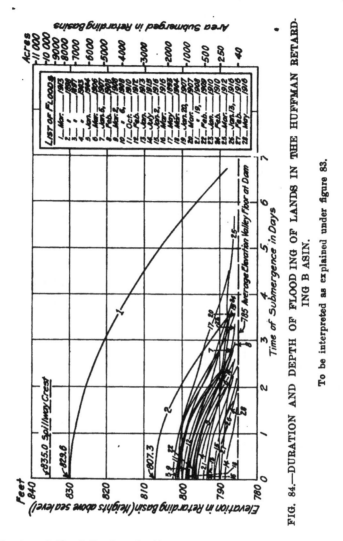

FIG. 84.—DURATION AND DEPTH OF FLOODING OF LANDS IN THE HUFFMAN RETARDING BASIN.

To be interpreted as explained under figure 83.

ordinates at the left of each diagram was translated into terms of areas submerged by consulting the area-depth curves. The resulting figures, expressed in acres, appear on the right hand side of the diagrams, and enable one to determine for any particular flood what extent of land would be submerged below a given level and the length of time of such submergence.

The combination of data presented in figures 80 to 84, renders

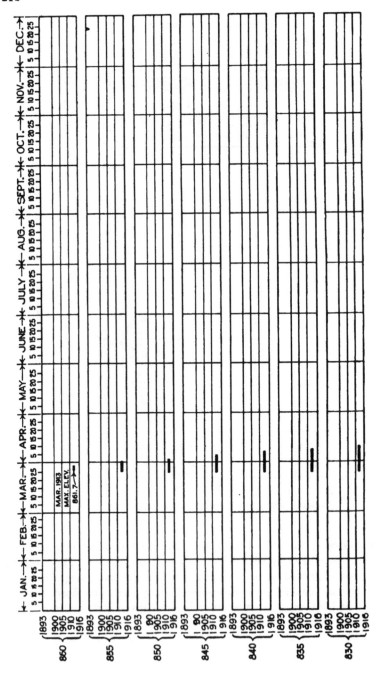

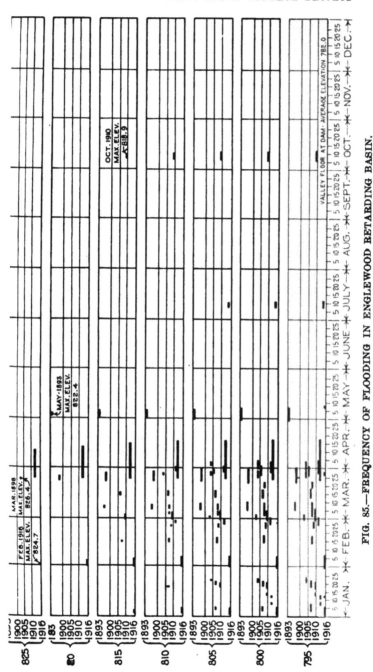

FIG. 85.—FREQUENCY OF FLOODING IN ENGLEWOOD RETARDING BASIN.

Each flood occurring during the 24-year period 1893–1916 is represented by a vertical series of heavy black lines, whose lengths indicate the date and duration of flooding at successive 5-foot levels in the basin.

them of special value in bringing out the differences in the operating characteristics of the various retarding basins of the Miami Conservancy District. Thus, it is seen that the two basins which flood the largest areas, Taylorsville and Huffman, empty themselves in the shortest time during great floods. The Englewood basin, with its small outlet conduits and large storage capacity, is seen to require by far the longest time to empty. In the construction of its diagram it became necessary to change the scale of abscissas beyond the 10-day point in order to keep the entire diagram within a reasonable compass. As a result, the parabolic shape of the curves is not so apparent there.

The Pianola Diagrams

From the duration-depth curves for the 140 floods included in this study, frequency-of-flooding charts were constructed for lands at different elevations in the basin areas. These charts cover the 24-year period during which continuous river stage readings at Dayton were available. Figures 85 and 86 reproduce the charts as constructed for lands lying within the Englewood and Huffman basins. Each chart is made up of bands or strips, one to each five feet of elevation in each retarding basin, requiring 57 bands for the entire system. From their similarity in appearance to player piano rolls these charts were known as the "pianola diagrams."

In figures 85 and 86, the lowest bands show the flooding conditions which would have prevailed at farm lands on the valley floor not far above the damsite, if the dam had existed during the 24-year period covered by the studies. The upper bands show corresponding conditions for lands situated in the upper levels of the basins, but considerably below spillway level, while the other bands show conditions at intermediate elevations. The frequency-of-flooding diagrams for the other retarding basins of the system, though differing in matters of detail, are similar in general appearance to those in figures 85 and 86, and lead to the same general conclusions regarding damage to farm lands within the basin areas.

By charting the frequency of flooding in this manner a remarkably complete picture of probable conditions of land submergence in the retarding basins was obtained. The diagrams bring out with great clearness the relative infrequency of floods during the growing season for crops. In figure 86 only two dry-season floods are shown as occurring during the entire 24 years of record, one of them, shown in the lowest line of the figure, being barely sufficient to submerge the lowest lands in the basin. Most of the floods during the period covered are seen to have occurred during early spring and late winter.

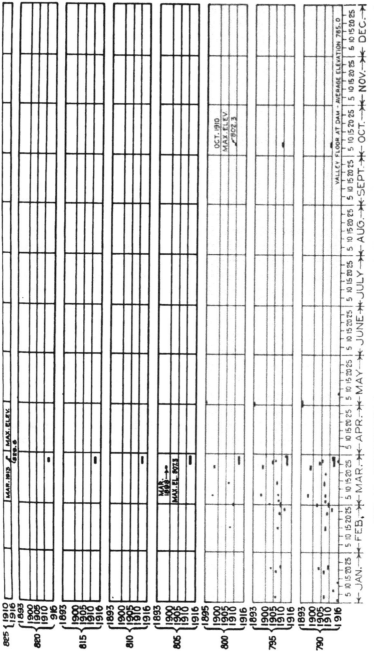

FIG. 86.—FREQUENCY OF FLOODING IN HUFFMAN RETARDING BASIN.

Such floods are considered to be a benefit rather than a detriment to the land, on account of the increase in fertility which they produce.

It is interesting to note that in carrying out the final plans of the Miami Conservancy District, it was not possible to persuade the owners of the accuracy of forecasts of floods, based upon information furnished by the pianola diagrams. Practically all lands in the basins lying below the level which would have been attained by the 1913 flood were purchased outright by the District, in order to avoid litigation and dispute over the value of easements. In selling the lands subject to flooding, however, these diagrams were generally relied upon by the purchaser as being representative, and the prices secured were materially affected by these estimates. The purchase of these lands by the District itself does not, of course, obviate the necessity for designing the dams and outlets in a way to avoid unnecessary interference with farming operations in the basins.

A careful study of figures 85 and 86, and the corresponding charts for the other basins, however, shows that it is not within the power of the designer to control interference with farming operations in the basins to any appreciable extent. As noted before, only two floods would have occurred in the Huffman basin in the farming season during the 24 years of record. To have lowered the crests of these floods enough to reduce to any considerable extent the area of land submerged, would have required an increase in the size of the outlet conduits out of all proportion to the benefit produced. Even at the Englewood basin, where the conduits are much smaller than at Huffman and the retarding effect on small floods proportionately greater. it is difficult to determine from the pianola diagrams that any appreciable reduction in interference with farm operation would be secured by a considerable enlargement of the conduits. From this standpoint the results of the detailed study of retarding basin operation were largely negative.

Basin-Recovery Curves

An important question influencing the design of a retarding basin is how it would act in the case of two large storms occurring in close succession. In order to study this question, the 140 outflow-hydrographs mentioned above were used to construct basin-recovery curves similar to those shown in figure 87. These curves show not only the total time required to empty the basins after a flood, but the times at which different percentages of the total storage capacity of the basin become available after the passing of the maximum stage. The large safety factor provided in the design of the basins of the District is

well illustrated in these recovery diagrams. For example, as shown
in figure 87, about 25 per cent of the total capacity of the Huffman
basin below spillway level would remain empty even at the peak stage
of a flood similar to that of March 1913. Within three days 60 per
cent of the basin capacity would be available to receive a second flood,

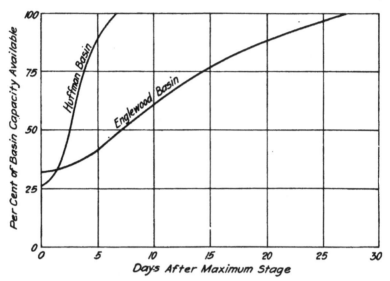

FIG. 87.—RECOVERY OF CAPACITY OF THE ENGLEWOOD AND
HUFFMAN RETARDING BASINS AFTER A FLOOD LIKE THAT OF
MARCH 1913.

and in five days after peak stage practically the whole protecting
power of the basin would be recovered. While the recovery of the
Englewood basin is much slower than that of the Huffman, it is seen
from figure 87 that within a week after the passing of the peak stage
of the 1913 flood, 50 per cent of the protecting capacity of the basin
would be restored.

Probability-of-flooding Diagrams

The detailed study of retarding basin operation, during past
floods, also furnished data for constructing probability diagrams,
figure 88, giving the probable intervals of time at which lands lying
at different levels in the basins might be submerged. The lower part
of these curves was constructed by using frequencies computed from
the 28 floods occurring during the 24 years covered by the studies,

while the upper portions of the curves were based on previous studies
of the probable frequency of extreme floods. On account of the seasonal distribution of the floods in the Miami valley—only a small
number of those shown on the pianola diagrams occurring in the

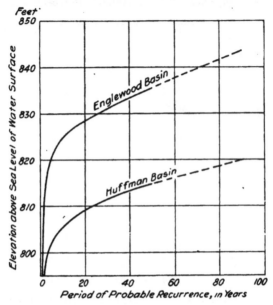

FIG. 88.—PROBABILITY OF FLOODING OF LANDS AT DIFFERENT
ELEVATIONS IN THE ENGLEWOOD AND HUFFMAN RETARDING
BASINS.

growing season—it is difficult to make use of probability-of-flooding
diagrams as shown in figure 88 for answering questions concerning
damage to farm lands within the basin areas.

CHAPTER IX.—DETERMINATION OF
CAPACITY OF SPILLWAYS

GENERAL PRINCIPLES

Each of the dams of the Miami Conservancy District is provided with a spillway near its top. It is extremely improbable that any of these spillways will ever come into use. Their function is to stand ready to protect the dams from breaking should a calamitous flood occur far greater than any known in the experience of the present and past generations.

The selection of the size of a great imaginary flood to be used in designing the protection works for the safety of the retarding basin system was a problem which could not be solved on a purely mathematical basis. It involved elements of judgment and of imagination. This was true for the reason that in storing enormous volumes of flood water above the cities and towns of a populous valley, it is justifiable and right to provide a degree of safety far in excess of the demands of probability.

It has been repeatedly mentioned in the Technical Reports that the flood of March 1913 was unprecedented in the Miami Valley. Not only was this true, but in the monumental research into the rainfall of the past made by the engineers of the Miami Conservancy District, no storm was found in the records of the entire upper Mississippi valley which would have produced a flood greater than that of 1913 if centered over the Miami watershed. It is true that heavier rainfalls have occurred over equal areas on several occasions, but always during the summer or fall, when the proportion of runoff would have been much lower than in the early spring. The runoff of the 1913 flood from the watershed above Dayton amounted to 7.27 inches in 3 days. This value represents the extreme limit of actual knowledge regarding great floods in this region. However, after a careful study of storms occurring at other places and seasons, the engineers of the District assumed a flood having a runoff of from 9.5 to 10 inches in 3 days as one which in all human probability would never be exceeded in all future time at this location. This runoff was used in the design of the outlet conduits and in determining the heights of the spillway crests, and has been referred to in preceding chapters as the Official Plan flood.

Although a considerable element of imagination was required in the choice of this imaginary Official Plan flood, on account of its exceeding all past records, far more imagination was required in selecting a yet greater flood to serve as a basis for spillway design. Its magnitude was arbitrarily taken at 14 inches in 3 days, or nearly twice the 3-day runoff of the 1913 flood.

The occurrence of such a flood would not exhaust the reserve capacity of the design, for the spillways provide in such event for a depth of water of from 10 to 14 feet over their crests. The actual tops of the earth dams are about five feet above this level. It is well known that the discharge of a spillway weir is nearly doubled when the head upon it is increased one-half. Therefore, if the basins should already be filled to the spillways, before the first trickle of water could be forced across the paved tops of the dams, a second deluge far greater than even this assumed flood would have to occur. Results of computations concerning such a cataclysmic flood for the various basins are given in the latter part of this chapter.

Attention should be called to the fact that the primary purpose of the spillway of a retarding basin is to prevent overtopping of the dam, and not to reduce flood heights in the cities below. In chapter VI it was shown that the characteristics of a spillway are much less favorable for purposes of flood control below the dams than those of an orifice outlet, on account of the fact that the outflow rate from the spillway grows very rapidly as the head increases. If the Official Plan flood should ever occur in the Miami Valley, its flow would be so reduced in passing through the great retarding basins, that it would traverse the improved channels in the cities below with a liberal margin against overtopping the levees. For imaginary floods greater than this, the computations show that the outflow from the basins would increase rapidly as the spillways would come into action. If imagination pictures a sufficiently great flood, the corresponding outflow could be made to exceed the capacity of the leveed channels below. They would, however, still be protected by the spillways from the failure of the dams, and would also be in much better position than if no dams existed. The amount of outflow at the different dams for various extremely high rates of flood runoff is further discussed in the next chapter.

TYPES OF SPILLWAY

Two types of spillway structures are used for the retarding basins of the Miami Conservancy District. At the Taylorsville, Huffman. and Lockington dams the combined spillway and outlet structure is of solid concrete built across the concrete outlet channel which cuts through the body of the dams. This construction is shown in figures

15 and 46. At the Englewood and Germantown basins the spillway weirs are constructed on the hillsides at or near the ends of the dams and are entirely separate from the outlet conduits which pass under the body of the earthwork, see figure 16. In all cases the spillways are founded upon solid rock, the possibility of doing this being one of the primary considerations which governed the choice of the damsites. In the early stages of the designs made for the District it was planned to have the spillways separate from the outlets at all the basins, but it was later found more economical to adopt the combined structure for all dams except the two highest of the system.

The imaginary flow over the spillways at Taylorsville, Huffman, and Lockington would combine with the outflow from the conduits, and its energy would be dissipated by the hydraulic jump in the deep pool of the outflow channel. At Englewood the spillway overflow would pass into a channel 100 feet wide and about 1500 feet long, figure 89, which empties into a small natural valley joining the main valley below the damsite. This outflow channel is excavated in rock, has concrete side walls for its entire length, and is floored with concrete for its upper 300 feet. The Germantown spillway, figure 90, is located in a saddle about 500 feet north of the dam. Its discharge would pass into a short steep outflow channel and return to the river by way of a natural ravine.

The principal dimensions for the spillways of the system, as finally adopted, are given in table VII of the Appendix.

Four of the five spillways of the District are concrete weirs of the ogee type. The fifth, that at Germantown, consists of a short channel crossing a saddle between two hills and has no abrupt dropoff, as will be seen in figure 90. The profiles of the ogee weirs were designed to conform approximately to the profile of the lower nappe of the overflow from a sharp-crested weir as determined by Bazin's experiments. The profiles as designed agree approximately in their upper portions with the formula $x^2 = 1.8\ hy$, where x and y are horizontal and vertical coordinates measured from the crest of the weir, located as shown in figure 91, and h is the maximum effective head on the weir including that due to velocity of approach. The discharge over the spillway weirs per foot of length was computed by the formula: $q = 3.8\ h^{3/2}$. The discharge from an open sluiceway, such as the spillway at Germantown, may be computed by formula 52, page 155.

The coefficient 3.8 used in computing the discharge over the spillways, while somewhat higher than the value usually used for such weirs, is believed to be conservative and amply justified by the existing conditions.

Since the profile of the ogee section is made to fit the shape which would be taken by the underside of the nappe from a sharp-crested weir, aside from the effect of friction, which would probably be small. the flow will take place in the same manner as from a sharp-crested

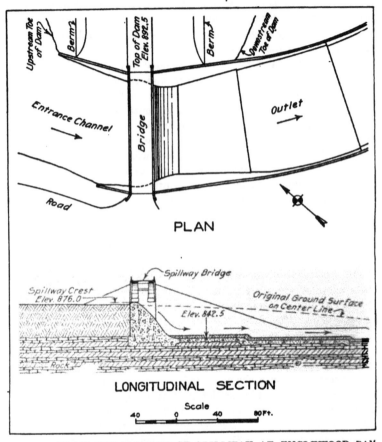

PLAN

LONGITUDINAL SECTION

Scale

FIG. 89.—PLAN AND SECTION OF SPILLWAY AT ENGLEWOOD DAM.

weir and the discharge will be 3.33 $H^{3/2}$, the head H in this case being measured up from the point corresponding to the sharp crest of the weir. It will be seen from figure 91 that H will be equal to 1.124 h. where h is the head measured from the crest of the ogee section. The discharge in terms of h will therefore be 3.33 (1.124 h)$^{3/2}$ or 3.97 $h^{3/2}$ This coefficient will apply only for the maximum head on the spillway or the head at which the concrete section just fits the undersurface

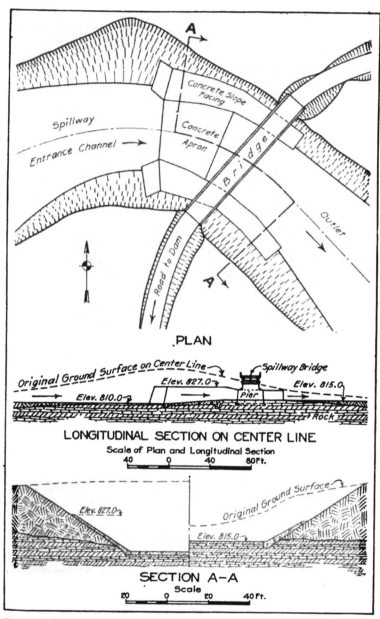

PLAN

LONGITUDINAL SECTION ON CENTER LINE

Scale of Plan and Longitudinal Section

SECTION A-A

Scale

FIG. 90.—PLAN AND SECTIONS OF SPILLWAY NEAR GERMANTOWN DAM.

15

of the nappe as it would flow from a sharp-crested weir. For lower heads it will be somewhat less but since the greater part of the total discharge from the spillways will take place when the head is near

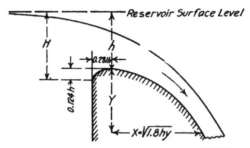

FIG. 91.—PROFILE OF SPILLWAY CREST.

the maximum, the value 3.8 for the average is conservative. The correctness of the 3.97 coefficient for high heads on ogee weirs of this type has been indicated by the results of numerous published experiments.

STEP METHOD FOR SOLVING SPILLWAY PROBLEMS

The step method solution for spillway problems, whether on the assumption that the outlet conduits are open or blocked, does not differ in principle from the step methods described in detail in chapter VII. The only modification necessary occurs in constructing the elevation-outflow or storage-outflow curve, which must have the proper characteristics for spillway outflow or for combined spillway and conduit outflow. Figure 92 shows the elevation-outflow curve for the Lockington retarding basin, as modified by the presence of the spillway. On account of the uncertain nature of the flood assumed in designing the spillways, the labor of using the step methods for tentative computations concerning the spillways of the Miami Conservancy District did not appear justifiable except in rare cases.

INTEGRATION FORMULAS FOR SPILLWAY ACTION

Since the flood assumed for spillway design so far exceeds all floods actually experienced that its magnitude can only be fixed by judgment in the most general way, it follows that no great refinement is needed in computing the detailed action of such a flood in passing through the retarding basins. It is sufficient for practical purposes if the method of computation is correct in its broader fea-

Outflow in Acre Feet per Hour

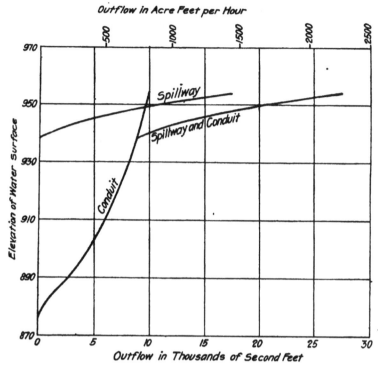

FIG. 92.—RELATION BETWEEN WATER SURFACE ELEVATIONS AND OUTFLOW AT LOCKINGTON RETARDING BASIN SHOWING EFFECT OF SPILLWAY OVERFLOW.

The steep curve at the left shows the outflow through the conduits only, its upper portion above elevation 938 (spillway crest) being drawn assuming no overflow to take place over the spillway. The curve marked ''Spillway'' likewise shows only spillway discharges. The rapid increase in the latter is in striking contrast with the more uniform outflow through the conduits. The curve marked ''Spillway and Conduit'' shows the sum of spillway and conduit discharges.

tures. Many of the tentative calculations of spillway capacity required for the basins of the Miami Conservancy District were based on the four simple assumptions given below. In certain cases, where the results of such calculations were checked by the step method, it was found that the values for the time of filling obtained by the two methods agreed so closely that the scale used in plotting the curves was not sufficiently large to show the difference.

A typical hydrograph showing the combined action of conduits and a spillway is given in figure 93. This figure furnishes a good illustration of the unsuitability of a spillway for purposes of pure flood protection.

Assumptions

The assumptions used in deriving the integration formulas for spillway action are: (1) that the rate of inflow is constant; (2) that the discharge over the spillway varies as the three-halves power of the head; (3) that the discharge of the tunnels is constant while the surface of the water is above spillway level; and (4) that the area of the water surface is constant at all elevations above the spillway crest. The first of these assumptions is purely arbitrary, for any other

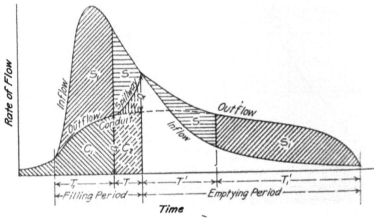

FIG. 93.—INFLOW AND OUTFLOW HYDROGRAPHS AT A RETARDING BASIN WITH SPILLWAY OPERATING.

The usual parabolic outflow curve is seen to possess a sharp-pointed protuberance representing spillway overflow. For the short time that the spillway operates its discharge causes the maximum outflow from the basin to become greatly increased, hence lessening the protection afforded by the retarding basin.

shape of inflow hydrograph might be chosen for the assumed maximum flood. The second agrees closely with actual facts, except where the spillway is relatively short compared with the maximum head and the end walls have a considerable batter. The third assumption is not so far from the truth as might be inferred at first thought. One of the principal reasons for the choice of orifice outlets for the retarding basins, was that at spillway elevation the outflow rate from such an outlet increases very slightly with the head. The fourth assumption violates actual conditions more seriously than any of the others.

and might lead to appreciable errors in computing the time required to fill a fraction of the total storage above spillway level. It can, however, introduce no serious error in computing the total time required for the water to rise to its maximum depth. In approximate computations the constant area referred to in assumption (4) was usually taken as the mean area flooded between spillway elevation and the maximum depth attained.

Notation

H = maximum rise or head above spillway level, in feet,

T_1 = time required for basin to fill to spillway level, in hours,

T = time required for the water surface to rise from the spillway level to its maximum head, H, in hours,

T' = time required for basin to empty down to spillway level, in hours,

T'_1 = time required for basin to empty from spillway level down to normal flow, in hours,

I_1 = gross uniform inflow rate, assumed constant, in second feet,

O_1 = outflow rate from tunnels, assumed constant at all heads above spillway depth, in second feet,

$I = I_1 - O_1$ = net uniform inflow rate used in spillway design, in second feet,

Q = maximum overflow rate from spillway, in second feet,

A = average area of water surface in reservoir above spillway level, in acres,

B = spillway coefficient in formula $Q = BH^{3/2}$,

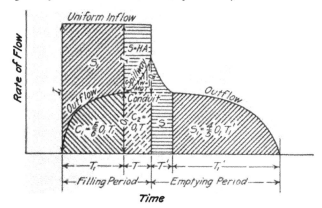

FIG. 94.—HYDROGRAPH OF RETARDING BASIN OUTFLOW WITH SPILLWAY OPERATING, AND ASSUMING UNIFORM INFLOW.

The diagram illustrates the notation used in spillway studies.

$IT =$ total net inflow into reservoir, while the surface is rising from spillway level to its maximum stage. This is the amount of water which must be taken care of by the storage above spillway level, plus the discharge over the spillway,

$S_1 =$ total storage volume below spillway level, in acre feet,

$S = HA =$ total storage volume above spillway level, in acre feet,

$W =$ total outflow over spillway weir while surface is rising from spillway level to maximum stage, in acre feet,

$N = IT/HA =$ ratio of total net inflow to total storage above spillway,

$R = Q/I =$ ratio of maximum overflow rate to maximum inflow rate,

$M = W/QT =$ ratio of average overflow rate to maximum overflow rate,

$t =$ any interval of time after water reaches spillway level, in hours,

$h =$ corresponding head on spillway, in feet,

$q = Bh^{3/2} =$ corresponding outflow rate, in second feet.

$r =$ corresponding value of q/I,

$C_1 =$ total discharge of conduits while basin is filling to spillway level,

$C_2 =$ total discharge of conduits while basin is filling above spillway level, in acre feet.

The relations existing between many of the above quantities are shown graphically on the hydrographs given in figure 94.

Analogy with Rectangular Weir Box

It is convenient in studying and interpreting the formulas derived from the above assumptions to have in mind some definite

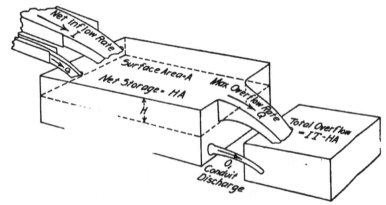

FIG. 95.—GRAPHICAL REPRESENTATION OF NOTATION USED IN INTEGRATION FORMULAS FOR SPILLWAY ACTION.

physical conception of the conditions which they represent. These formulas would lead to exact results for a reservoir of the type shown in figure 95. This figure represents a basin with vertical sides and rectangular outflow weir, supplied by a uniform net inflow rate I. The fact that the tunnel discharge rate does not enter directly into the spillway analysis, is indicated in this figure by showing that the net inflow rate I may be considered as being obtained by subtracting from the gross inflow rate an amount sufficient to just balance the tunnel outflow O_1. The problem to be solved is to determine the time T required to fill the basin from spillway level to the top.

A full discussion of this problem, and the various formulas proposed for solving it, is given by R. E. Horton in Water-Supply and Irrigation Paper No. 200 of the United States Geological Survey.

Series Formulas for Spillway Action

Equations expressing the relation between the various factors involved in spillway action may be written in the series form or in transcendental form. The series formulas will be developed first.

·The fundamental differential equation for spillway action states that, at any instant, the net inflow rate into the reservoir is equal to the net outflow rate plus the rate of storage.

Under the conditions assumed above this may be written:

$$I dt = B h^{1/2} dt + A dh \qquad (90)$$

Transposing,

$$dt = \frac{A dh}{I - B h^{3/2}}. \qquad (91)$$

Let $h^{1/2} = z$, $h = z^2$, and $dh = 2z\, dz$. Then,

$$dt = \frac{2Az\, dz}{I - Bz^3} = \frac{2A}{I} \frac{z\, dz}{1 - \frac{B}{I} z^3}. \qquad (92)$$

Expanding the last fraction by division,

$$dt = \frac{2A}{I} \left[z + \frac{B}{I} z^4 + \left(\frac{B}{I}\right)^2 z^7 + \left(\frac{B}{I}\right)^3 z^{10} + \cdots \right] dz. \qquad (93)$$

Integrating between the limits 0 and t, and 0 and z,

$$t = \frac{2A}{I} \left[\frac{z^2}{2} + \frac{B}{I} \frac{z^5}{5} + \left(\frac{B}{I}\right)^2 \frac{z^8}{8} + \left(\frac{B}{I}\right)^3 \frac{z^{11}}{11} + \cdots \right]. \qquad (94)$$

Substituting the values $z = h^{\frac{1}{2}}$ and $q = Bh^{\frac{1}{2}}$

$$= \frac{Ah}{I}\left[1 + \frac{2}{5}\left(\frac{B}{I}\right)h^{3/2} + \frac{2}{8}\left(\frac{B}{I}\right)^2 h^{6/2} + \frac{2}{11}\left(\frac{B}{I}\right)^3 h^{9/2} + \cdots\right] \quad (95)$$

$$= \frac{Ah}{I}\left[1 + \frac{2}{5}\left(\frac{q}{I}\right) + \frac{2}{8}\left(\frac{q}{I}\right)^2 + \frac{2}{11}\left(\frac{q}{I}\right)^3 + \cdots\right]. \quad (96)$$

At the completion of the filling period this equation becomes,

$$T = \frac{AH}{I}\left[1 + \frac{2}{5}\left(\frac{Q}{I}\right) + \frac{2}{8}\left(\frac{Q}{I}\right)^2 + \frac{2}{11}\left(\frac{Q}{I}\right)^3 + \cdots\right] \quad (97)$$

$$= \frac{AH}{I}[1 + \tfrac{2}{5}R + \tfrac{2}{8}R^2 + \tfrac{2}{11}R^3 + \cdots]. \quad (98)$$

This equation determines the desired duration of the filling period. The ratio N of the total net inflow during this period to the storage capacity of the basin above spillway level is found by transposing,

$$N = \frac{IT}{AH} = 1 + \tfrac{2}{5}R + \tfrac{2}{8}R^2 + \tfrac{2}{11}R^3 + \cdots. \quad (99)$$

The series in the right hand member of this equation converges very slowly for values of R which approach unity, and is inconvenient for purposes of computation in such cases. A curve showing the relation between R and N is given in figure 96. The points necessary for plotting the extreme right hand portion of this curve, were computed by the transcendental formulas given in the following pages. By using this curve, equation (99) can be readily solved for any of its variables.

For purposes of computation another series, which converges somewhat more rapidly than equation (99), may be obtained as below. The ratio M of the average outflow rate to the maximum outflow rate is determined by

$$M = \frac{IT - AH}{QT} = \frac{AH(1 + \tfrac{2}{5}R + \tfrac{2}{8}R^2 + \tfrac{2}{11}R^3 + \cdots) - AH}{\dfrac{QAH}{I}(1 + \tfrac{2}{5}R + \tfrac{2}{8}R^2 + \tfrac{2}{11}R^3 + \cdots)} \quad (100)$$

$$= \frac{\tfrac{2}{5} + \tfrac{2}{8}R + \tfrac{2}{11}R^2 + \cdots}{1 + \tfrac{2}{5}R + \tfrac{2}{8}R^2 + \tfrac{2}{11}R^3 + \cdots} \quad (101)$$

$$= 0.400 + 0.090R + 0.0458R^2 + 0.0293R^3 \cdots. \quad (102)$$

The last equation is obtained from the preceding one by actual division. Although this series converges more rapidly than that of equation (99), it is not convenient for use when R approaches unity. The formula indicates that during a sharp, sudden flood of much greater intensity than the maximum spillway outflow rate, the average spillway discharge is about two-fifths of the maximum.

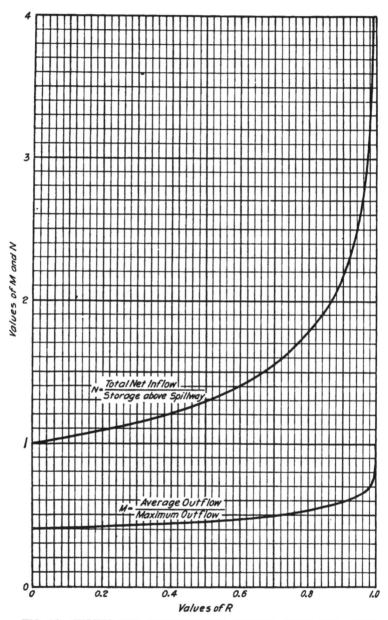

FIG. 96.—CURVES FOR DETERMINING VALUES OF M AND N IN
FORMULAS FOR SPILLWAY DISCHARGE.

After M has been determined, the value of N may be found from the following:

$$M = \frac{IT - AH}{QT} = \frac{I}{Q}\left(1 - \frac{AH}{IT}\right) = \frac{1}{R}\left(1 - \frac{1}{N}\right). \quad (103)$$

The relation between M and R as expressed by equation (102) is plotted graphically in figure 96.

Transcendental Formulas for Spillway Action

By the use of the following methods, it is possible to integrate the differential equation for spillway action without resorting to the expansion into a series. Rewriting equation (92):

$$dt = \frac{2Az\,dz}{I - Bz^3} = \frac{z}{I - Bz^3}2A\,dz. \quad (104)$$

Let $I = a^3$, and $B = b^3$, then $\dfrac{z}{I - Bz^3} = \dfrac{z}{a^3 - b^3z^3}.$ (105)

Let $\dfrac{z}{a^3 - b^3z^3} = \dfrac{E}{a - bz} + \dfrac{F}{a^2 + abz + b^2z^2} + \dfrac{Gz}{a^2 + abz + b^2z^2}.$ (106)

By the method of indeterminate coefficients find

$$E = \frac{1}{3ab}, \qquad F = -\frac{1}{3b}, \qquad G = \frac{1}{3a}. \quad (107)$$

Substituting in equation (104),

$$dt = 2A\left[\frac{1}{3ab(a - bz)} - \frac{1}{3b(a^2 + abz + b^2z^2)} + \frac{2}{3a(a^2 + abz + b^2z^2)}\right]dz, \quad (108)$$

$$t = \frac{2A}{3ab}\left[\int \frac{dz}{a - bz} - a\int \frac{dz}{a^2 + abz + b^2z^2} + b\int \frac{z\,dz}{a^2 + abz + b^2z^2}\right]. \quad (109)$$

The three integrals in this equation are evaluated as follows:

$$\int \frac{dz}{a - bz} = -\frac{1}{b}\log_e(a - bz), \quad (110)$$

$$-a\int \frac{dz}{a^2 + abz + b^2z^2} = -\frac{2}{\sqrt{3}b}\tan^{-1}\frac{2b^2z + ab}{\sqrt{3}ab}, \quad (111)$$

$$b\int \frac{z\,dz}{a^2 + abz + b^2z^2} = \frac{1}{2b}\log_e(a^2 + abz + b^2z^2) - \frac{a}{2}\int \frac{dz}{a^2 + abz + b^2z^2} \quad (112)$$

$$= \frac{1}{2b}\log_e(a^2 + abz + b^2z^2) - \frac{1}{\sqrt{3}b}\tan^{-1}\frac{2b^2z + ab}{\sqrt{3}ab}. \quad (113)$$

Substituting these values in equation (109),

$$t = \frac{2A}{3ab^2}\left[-\log_e(a - bz) + \tfrac{1}{2}\log_e(a^2 + abz + b^2z^2) \right.$$
$$\left. - \sqrt{3}\tan^{-1}\frac{2bz + a}{\sqrt{3}a}\right]_0^z. \tag{114}$$

$$= \frac{2A}{3ab^2}\left[-\log_e\left(1 - \frac{bz}{a}\right) + \log_e\sqrt{1 + \frac{bz}{a} + \frac{b^2z^2}{a^2}} \right.$$
$$\left. - \sqrt{3}\left(\tan^{-1}\frac{\dfrac{2bz}{a} + 1}{\sqrt{3}} - \tan^{-1}\frac{1}{\sqrt{3}}\right)\right]. \tag{115}$$

The last term is the difference of two angles. The tangent of the first angle is $\dfrac{\dfrac{2bz}{a} - 1}{\sqrt{3}}$. The tangent of the second is $1/\sqrt{3}$. The tangent of the difference is,

$$\frac{\dfrac{2bz}{a}}{\sqrt{3}} - \dfrac{1}{\sqrt{3}} \bigg/ \left(1 + \dfrac{\dfrac{2bz}{a} + 1}{\sqrt{3}} \cdot \dfrac{1}{\sqrt{3}}\right) = \frac{\sqrt{3}\dfrac{bz}{a}}{z + \dfrac{bz}{a}}. \tag{116}$$

Substituting this value in equation (115),

$$t = \frac{2A}{3ab^2}\left[-\log_e\left(1 - \frac{bz}{a}\right) + \log_e\sqrt{1 + \frac{bz}{a} + \frac{b^2z^2}{a^2}} \right.$$
$$\left. - \sqrt{3}\tan^{-1}\frac{\sqrt{3}\dfrac{bz}{a}}{2 + \dfrac{bz}{a}}\right]. \tag{117}$$

Substituting in this equation the values

$$\frac{bz}{a} = \left(\frac{Bh^{3/2}}{I}\right)^{1/3} = \left(\frac{q}{I}\right)^{1/3} = r^{1/3},$$

and

$$\frac{1}{ab^2} = \frac{1}{I^{1/3}B^{2/3}} = \frac{h}{I^{1/3}q^{2/3}} = \frac{h}{I}\frac{I^{2/3}}{q^{2/3}} = \frac{h}{Ir^{2/3}},$$

$$t = \frac{Ah}{I}\frac{2}{3r^{2/3}}\left[-\log_e(1 - r^{1/3}) + \log_e\sqrt{1 + r^{1/3} + r^{2/3}} \right.$$
$$\left. - \sqrt{3}\tan^{-1}\frac{\sqrt{3}\,r^{1/3}}{2 + r^{1/3}}\right]. \tag{118}$$

At the completion of the filling period this equation becomes:

$$T = \frac{AH}{I}\frac{2}{3R^{2/3}}\left[-\log_e (1 - R^{1/3}) + \log_e \sqrt{1 + R^{1/3} + R^{2/3}} \\ - \sqrt{3}\tan^{-1}\frac{\sqrt{3}\,R^{1/3}}{2 + R^{1/3}} \right]. \qquad (119)$$

The ratio N of the total net inflow to the storage above spillway level is given by:

$$N = \frac{TI}{AH} = \frac{2}{3R^{2/3}}\left[-\log_e (1 - R^{1/3}) + \log_e \sqrt{1 + R^{1/3} + R^{2/3}} \\ - \sqrt{3}\tan^{-1}\frac{\sqrt{3}\,R^{1/3}}{2 + R^{1/3}} \right]. \qquad (120)$$

Equations (119) and (120) give the exact relation which would be obtained by summing the series of equations (98) and (99). By the aid of figure 96 any of these equations can be quickly solved numerically for any of the variables involved.[*]

Examples Illustrating Investigation of Spillway Action

As stated above, the labor of substituting in the preceding formulas is much reduced by the use of the curves given in figure 96. With this diagram two methods of investigating the action of spillways are open, both of which are illustrated in the following examples. The first is to assume a maximum capacity of spillway and a given intensity of net inflow, and compute the time it would take under these conditions to fill the retarding basin to the maximum head on the spillway. This is equivalent to assuming R (since $R =$ the ratio of the maximum spillway discharge rate to the net rate of the inflow) and obtaining N. The time of filling is then computed by $T = AHN/I$, I and $A H$ being known.

For example, how long would it take an inflow of 4 inches per day to fill the Englewood retarding basin to elevation 886, or 10 feet above the spillway crest, if the spillway is 100 feet long.

Assumed: Inflow rate = 4 inches per day from tributary drainage area. Spillway discharge capacity at 10 feet depth = 0.688 inches per day.

Known: Capacity of the retarding basin to spillway elevation = 9.01 inches. Capacity of the basin to 10 feet above spillway elevation = 11.22 inches. Outflow of tunnels when basin is filled to spillway elevation = 0.69 inches per day.

[*] *Note:* Since the above was written, an extensive discussion has appeared under the title Determining the Regulating Effect of a Storage Reservoir, by Robert E. Horton, in the Engineering News-Record of September 5, 1918, Vol. 81, pages 455–458.

Required: Time to fill basin to 10 feet above spillway elevation. Solution: Average tunnel outflow rate while filling to spillway elevation $= \frac{5}{6} \times 0.69 = 0.58$ inches per day. (See chapter VII for five-sixths rule.) Net inflow rate while filling to spillway elevation $= 4.00 - 0.58 = 3.42$ inches per day. Time of filling to spillway level $= 9.01/3.42 = 2.64$ days. Tunnel outflow while spillway is discharging $= 0.69$ inches per day, assumed the same as the discharge of the tunnels with the basin full to the spillway elevation. Net inflow rate while spillway is discharging $= 4.00 - 0.69 = 3.31$ inches per day. $R = $(maximum spillway discharge capacity)$\div$(Net inflow rate) $= 0.688 \div 3.31 = 0.208$. Corresponding value of N from curve (figure 96)$= 1.09$. Time for filling from spillway to 10 feet above $= T = AHN/I = (11.22 - 9.01)/3.31(1.09) = 0.72$ days. Total time required to fill the basin to 10 feet above spillway crest $= 2.64 + 0.72 = 3.36$ days.

The second method used in this investigation was to assume a storm of a given intensity and duration, and to determine the size of spillway necessary in order that the water surface should not rise more than 10 feet above the spillway crest. This is equivalent to assuming $N = IT/AH$ and determining the required spillway capacity from the corresponding value of r.

This method is illustrated by the following example.

Assumed: Inflow rate of 4 inches per day, lasting 3½ days.

Known: Capacity of retarding basin to spillway level and to 10 feet above spillway level $= 9.01$ and 11.22 inches respectively, as before. Outflow from tunnels when basin is filled to spillway elevation $= 0.69$ inches, as before.

Required: Minimum spillway capacity to discharge the assumed flood without exceeding 10 foot head.

Solution: Time to fill basin to spillway level $= 2.64$ days, as before; time to fill basin from spillway level to 10 feet above spillway level $= 3.50 - 2.64 = 0.86$ days. $N = IT \div AH = 4(0.86) \div (11.22 - 9.01) = 1.56$. Corresponding value of $R = 0.705$. Required discharge of spillway at maximum head $= Q = RI = (0.705 \times 4) = 2.82$ inches per day.

RESULTS OF STUDIES OF SPILLWAY ACTION FOR ENGLEWOOD AND TAYLORSVILLE RETARDING BASINS—FIRST METHOD

In studying the action of the spillways of the basins of the Miami Conservancy District with different assumptions of inflow and spillway capacity, computations were made by each of the above methods.

The results of the computations by the first method are shown graphically in figures 97 and 98. These curves give the time required to fill the retarding basins up to spillway level and to various depths above spillway level (10 feet in the curve for the Taylorsville basin and 10 feet and 16.5 feet in the curves for the Englewood basin), by any uniform inflow up to 7 inches per day. The solid lines show the time

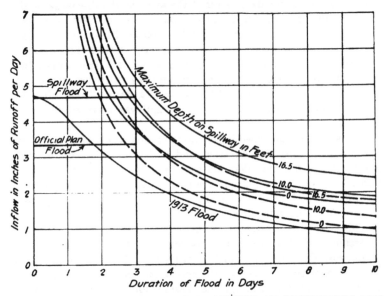

FIG. 97.—DEPTHS OF SPILLWAY DISCHARGE AT ENGLEWOOD DAM FOR VARIOUS INTENSITIES AND DURATIONS OF STORM RUNOFF.

The full lines (above the 1913 flood curve) indicate the time required to fill to the given maximum depths on spillway assuming the conduit outlets open, and the dotted lines give this information assuming all conduits closed. The rectangle marked "Spillway Flood" illustrates the storm runoff of 4⅔ inches per day for 3 consecutive days, which was assumed as the criterion for computing spillway capacities at all of the retarding basin dams.

required with the conduits open, and the dotted lines show the corresponding times with the conduits closed. These curves, and the computations which accompany them, are based on the size of spillways adopted, 100 and 132 feet for the Englewood and Taylorsville basins respectively. In platting the curves of figure 98 for the time of filling of the Taylorsville retarding basin, it was necessary to make allowance for the effect of the Lockington basin, which is located above it on the same stream. This was done by the assumption that the Lockington basin fills to the spillway elevation, and from the spillway to an elevation 10 feet above, simultaneously with the Taylorsville basin, and that the outflow from the upper basin during each of these periods is constant. (The details of a computation made under this assumption are given in table 11.)

In making use of curves for spillway action, similar to those of figures 97 and 98, the time required to fill the retarding basin from spillway to maximum stage for any given rate of inflow, may be readily found by taking the difference between the length of time required to fill to the maximum elevation and that required to fill to the spillway elevation. On the diagrams this is the horizontal intercept, for the given rate, between the zero curve and the curve of maximum stage.

Referring to figure 97, it is seen that the dotted lines, representing conditions with the conduits closed, are relatively close to the corre-

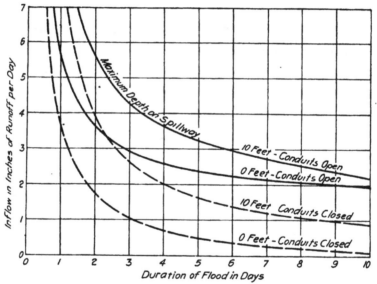

FIG. 98.—DEPTHS OF SPILLWAY DISCHARGE AT TAYLORSVILLE DAM FOR VARIOUS INTENSITIES AND DURATIONS OF STORM RUNOFF.

The significance of the full lines and dotted lines is the same as for figure 97.

sponding solid lines representing conditions with the conduits open. This relation is typical for basins with relatively small outlets, such as those at Englewood, Germantown, and Lockington. Opposite relations occur at the basins with relatively large outlet conduits, such as

Table 10.—Method of Calculating Spillway Capacity required at Englewood Dam.
DATA.

Drainage area above dam .... 615 square miles
Elevation of spillway above sea level .... 876 feet
Storage in retarding basin below spillway elev. ,........ 312,000 acre feet
Maximum discharge through conduits .... 12,000 second feet
Maximum elevation of water surface .... 887.4 feet
Storage in retarding basin below elev. 887.4 .... 413,000 acre feet

| Description | Symbol | Conduits | | H w Obtained |
		Closed	Open	
Duration of storm, hours..	$T + T_1$	72	72	Given
Inflow in 3 days, inches of runoff................		14	14	Given
Total inflow, acre feet.....	$I_1(T + T_1)$	486,000	486,000	Given
Average inflow, acre feet per hour...............	I_1	6,740	6,740	$\dfrac{I_1(T + T_1)}{(T + T_1)}$
Conduit discharge with water surface above spillway elev., acre feet per hour...................	O_1	0	992	Given
Average conduit discharge for basin filling to spillway elev., acre feet per hour.................	O_2	0	827	$5/6 O_1$
Storage below spillway elev., acre feet........	S_1	312,000	312,000	Given
Time of filling to spillway, hours................	T_1	46.3	52.7	$\dfrac{S_1}{I_1 - O_2}$
Time of filling above spillway, hours............	T	25.7	19.3	$(T + T_1) - T_1$
Amount discharged over spillway plus storage above spillway, acre feet	IT	174,000	110,800	$(I_1 - O_1)T = IT$
Storage above spillway, acre feet.............	S	101,000	101,000	Given $= HA$
Ratio N...............	N	1.72	1.10	$\dfrac{(I_1-O_1)T}{S} = \dfrac{IT}{HA}$
Ratio R...............	R	0.785	0.207	From fig. 96
Maximum spillway discharge, acre feet per hour	Q	5,290	1,190	$R(I_1-O_1) = RI$
Length of spillway required, feet...........	L	440	99	

those at Taylorsville and Huffman. As shown in figure 98, the dotted
lines are widely separated from the corresponding solid lines, indicating that in these cases it would be extremely expensive and difficult
to provide spillways sufficient to care for large floods, were the conduits to remain closed.

The curves of figures 97 and 98 bring out clearly the immensity of
the storage afforded by the basins of the Miami Conservancy District
in their upper levels. From the first of these figures it is seen that

Table 11.—Method of Calculating Spillway Capacity required at Taylorsville Dam.

DATA.

Drainage Area above Taylorsville Dam .... 1,133 square miles
Elevation of Taylorsville spillway above sea level .... 818 feet
Storage in Taylorsville Basin below spillway elevation .... 186,000 acre feet
Maximum discharge through Taylorsville conduits .... 53,600 second feet
Maximum elevation of water, above sea level .... 832 feet
Storage in Taylorsville Basin below elevation 832 .... 386,000 acre feet
Elevation of Lockington spillway above sea level .... 938 feet
Storage in Lockington Basin below spillway elevation .... 70,000 acre feet
Maximum elevation of water surface, above sea level .... 948.4 feet
Storage in Lockington Basin below elevation 948.4 ...... 126,000 acre feet

| Description | Symbol | Conduits | | How Obtained |
		Closed	Open	
Duration of storm, hours .	$T + T_1$	72	72	Given
Inflow in 3 days, inches of runoff..................		14	14	Given
Total inflow, acre feet	$I_1(T+T_1)$	845,000	845,000	Given
Average inflow, acre feet per hour..............	I_1	11,740	11,740	$I_1 \dfrac{(T + T_1)}{(T + T_1)}$
Conduit discharge with water surface above spillway elev., acre feet per hour.................	O_1	0	4,430	Given
Average conduit discharge with water surface below spillway elev., acre feet per hour..............	O_2	0	3,690	$5/6 O_1$
Storage in both basins below spillway elev., acre feet..................	S_1	256,000	256,000	Given
Time of filling to spillway, hours................	T_1	21.8	31.8	$\dfrac{S_1}{I_1 - O_2}$
Time of filling above spillway, hours............	T	50.2	40.2	$(T + T_1) - T_1$
Amount discharged over spillway, plus storage in both basins above spillways, acre feet........	IT	589,000	294,000	$(I_1 - O_1)T = IT$
Storage in both basins above spillways, acre feet	S	256,000	256,000	Given
Ratio N	N	2.300	1.148	$\dfrac{IT}{S}$
Ratio R	R	0.920	0.310	From fig. 96
Maximum spillway discharge at Taylorsville, acre feet per hour......	Q	10,800	2,265	$R(I_1 - O_1) = RI$
Length of spillway required, feet...........	L	628	132	

16

even with conduits closed, the Englewood retarding basin would have absorbed the entire runoff of the 1913 flood without having the water reach the spillway level. The greatest 3-day runoff at Dayton during this flood has been previously given as 7.27 inches total or 2.42 inches per day. This would only partly fill the basin to spillway level even with conduits closed. The entire runoff of the 1913 flood for the period of about 9 days while the discharge was greater than the ordinary April flow amounted to about 9.4 inches. If this is considered equivalent to a uniform flood of any particular duration, for instance 1.57 inches for 6 days, and this uniform flood is plotted as a rectangle on figure 97, this rectangle will be found to lie below the zero curve in the figure, indicating that the basin would not be filled above spillway level by a flood of this magnitude, even assuming the conduits to be closed.

RESULTS OF STUDIES OF SPILLWAY ACTION—SECOND METHOD

The results of computations made by the second method for studying spillway action, mentioned on page 237, are given for ehe Englewood and Taylorsville basins in tables 10 and 11. In computing the latter table, the effect of the Lockington retarding basin was allowed for in accordance with the same assumption that was used in constructing the curves of figure 98, namely: that the water surfaces in both upper and lower basins simultaneously reach spillway elevation and also simultaneously reach their maximum elevation above spillway level; and that while the retarding basin is filling to the spillway the outflow is constant, and while it is filling above the spillway it is also constant but at a different rate. The investigations given in tables 10 and 11 were made for the assumed flood of 14 inches runoff in 3 days used in designing the spillways, and for a maximum depth of 11.4 feet over the spillway at Englewood and of 14 feet over the spillway at Taylorsville. The latter values were decided upon as a result of numerous trial calculations by various methods. It is seen that the required spillway capacities obtained by the calculations of tables 10 and 11 agree closely with the values actually adopted in the final design.

STEP METHOD CHECK ON APPROXIMATE SPILLWAY COMPUTATIONS

After the final dimensions for the spillways of the five basins had been tentatively decided upon, on the basis of approximate computations similar to the preceding, additional studies were made by step

methods to determine more accurately the action of the spillways during floods of various magnitudes. The following general assumptions
were used in these studies:

1. Head on conduits measured from top of conduits at their downstream end.
2. Conduit discharge, (a) below spillway $Q = BH^{1/2}$; (b) above spillway, constant.
3. Spillway discharge $Q = 3.8 \, LH^{1/2}$, ends considered vertical.
4. Runoff for assumed flood of 14 inches in 3 days proportioned as
follows, the rates given being in inches per day.

	0 Hrs.	16 Hrs.	40 Hrs.	72 Hrs.
Germantown................	0.75	6.125	6.125	2.25
Englewood................	0.75	6.375	6.375	1.5
Lockington................	0.5	6.0	6.0	2.75
Taylorsville*................	0.75	6.125	6.125	2.25
Huffman................	0.5	6.0	6.0	2.75

5. No allowance was made for the regulative effect which the Lewistown and Loramie State reservoirs might have under certain
conditions.

The final results of these step-method computations are given in
table XI of the Appendix.

For convenience of reference, the description of the various real
and imaginary floods considered in this table may be reviewed as
follows:

1898 Flood: Second largest measured flood in history of Miami Valley; 2.47 inches runoff at Dayton in 3 days, or about one-third
the 1913 flood.
1913 Flood: Largest measured flood in history of Miami Valley; 7.27
inches runoff at Dayton in 3 days.
Official Plan Flood, used in fixing conduit sizes and spillway elevations of basins; 9.5 to 10 inches in 3 days; about 40 per cent
greater than 1913 flood.
Flood used in designing spillways: 14 inches in 3 days, or about twice
as great as the 1913 flood.
Flood which would cause water in basins to reach tops of dams; duration 3 days.

* To obtain the Taylorsville inflow, the above rates were applied to the 878
square miles below the Lockington dam, and to this runoff the outflow of the
Lockington retarding basin was added. No allowance was made for the time lost
in the flow reaching Taylorsville.

ACTION OF SPILLWAYS AS FINALLY DESIGNED

The dimensions of the spillways decided upon are given in the Appendix, see table VII. The following statements will serve to bring out the large degree of safety provided by the system as finally designed. The storage furnished by the retarding basins is so large, that in every case they could have absorbed the entire maximum 3-day runoff of the 1898 flood with the conduits closed. The 1913 flood would pass the basins without raising the water to approach spillway level if the conduits were open, and would not fill the Germantown and Englewood basins to the spillway even if the conduits were closed. In the other basins, only a small fraction of the available spillway capacity would be required to pass this flood in case the conduits were closed, except at Taylorsville where a little more than half of the spillway capacity would come into use under these conditions. The Official Plan flood, which fills the basin to spillway level with conduits open, would in no case be sufficient to overtop the dams if the conduits were closed. Under these circumstances the freeboard remaining for additional protection varies from 3 feet at Huffman to 12½ feet at Englewood. The assumed flood of 14 inches in 3 days, used in the design of the spillways, would leave a freeboard of over 5 feet at all the basins except Huffman, where the freeboard is 4.2 feet. This imaginary deluge would not overtop the Lockington or Englewood dams even if the conduits were closed.

An imaginary flood that would just exhaust the capacity of the basins with conduits open would call for an inflow of 19.2 inches in 3 days at Lockington to 15.9 inches at Englewood, giving safety factors of between 2 and 3 over the corresponding runoff of the 1913 flood. The figures for such an imaginary flood are somewhat lower on the assumption that the conduits are closed. It should be remembered, however, that this assumption is speculative and improbable in the highest degree, especially in the case of the large outlets at Taylorsville and Huffman. The studies of spillway action with conduits closed were made as a matter of interest in investigating the general safety of the system, rather than to determine conditions which might actually occur. The computations show that, even at Taylorsville and Huffman, the basins with conduits entirely closed would be safe during an inflow of more than 1.5 times the 3-day discharge of the 1913 flood.

ACTION OF NOTCHED SPILLWAYS

During the early stages of the preparation of plans for the retarding basins of the Miami Conservancy District, much attention was given to the possible use of notched or stepped spillways. In a typical design of this character the middle third of each spillway was depressed 4 feet below the remainder of the crest, forming a rectangular notch. The principal object of using notches of this kind was to decrease the amount of land which it might be necessary to purchase, it being assumed that no damage would have to be paid on land lying above the bottom of the notch. There was also a possibility that this device might secure large reductions in the cost of relocating roads and railroads, and protecting other property in or on the borders of the basins.

The chances of a flood equal to the Official Plan flood actually occurring are so extremely remote, that the present value of possible damage in the far distant future to lands and structures in the basins near spillway level, is negligible. On the other hand the particular elevations chosen for the spillway crests would have an overwhelming weight in the popular mind in deciding matters connected with land damages or relocation of improvements. For these reasons, if the spillway crests could be lowered several feet by means of notches, without appreciably affecting the flood control characteristics of the system, much needless expense might be saved for the taxpayers of the District.

The advisability of using a spillway notch and its most economical size may be easily determined with the aid of cost-outflow curves, similar to those used in the balancing of the retarding basin system, as described in chapter XII. A cost-outflow curve is constructed for a basin having no notch. The cost and maximum discharge for the same basin with various sized notches is then computed. If the points representing these values of cost and maximum discharge, when plotted, fall below the cost-outflow curve for the basin with no notch, the use of a notch is desirable. The size of notch which gives the lowest point, other things being equal, is the most economical. In no case, however, should the notch be made so deep that the owners of land, which will receive appreciable damage from flooding, do not receive proper compensation. It should be noticed that while two basins, one with, and one without a notch, may be designed to give the same protection against the Official Plan flood, for somewhat greater inflows, other conditions being equal, the basin having the notch will give less protection. An apparent saving resulting from the use of a notch may, therefore, be largely offset by the decreased protection which the basin gives against larger floods. This is particularly true if the size of the capacity flood is not very conservative.

In some cases a basin with a spillway notch may be used to effect favorably the synchronizing of peaks. Because of certain conditions, this phase of retarding basin control did not enter into the design of the Miami Valley system, although, had conditions been slightly different, the most economical system could not have been secured without its consideration. As will be shown later, two peaks pass through Hamilton, the first being larger, while only one large peak occurs at Middletown. If the peaks at Middletown were similar to those at Hamilton, or if no protection was required at Middletown, the size of the second peak could be increased until it equalled the first, by means of a notch in the spillway at Germantown. A saving in the cost of the Germantown basin would thus be made, while no greater expense would be incurred for the channel improvements.

CHAPTER X.—BEHAVIOR OF RETARDING BASINS DURING LOCALIZED CLOUDBURSTS

It was shown in chapter II that the flood of March 1913 was the greatest recorded in the Miami Valley since its first settlement over 100 years ago; that it was apparently also one of the greatest general floods on record in the whole eastern part of the country; that the storm producing it was in the class of the largest general winter storms occurring in this section of the country; and that the Miami Valley flood protection works are designed to furnish complete protection against a general uniform flood runoff about 40 per cent greater than the flood runoff in 1913. It remains to be shown how the retarding basin system will be affected by any possible cloudburst runoff resulting from the intense local summer thunderstorm rainfall more or less common throughout the whole Mississippi watershed.

SELECTION OF STORMS

The basis for such a study is readily obtained from Technical Reports, Part V. During the 25-year period, 1892–1916, there occurred over the northern half of the Mississippi Valley 11 large and intense storms, whose records are studied in detail in chapters VI, VII, and VIII of Part V. For each of these storms the rainfall data was plotted on maps, isohyetals were drawn and measured, and so-called Time-Area-Depth curves were constructed. The locations of these storms and some of their isohyetals are shown on the small scale maps reproduced in Part V. Table 12 shows for each of these storms the maximum rainfall recorded at any station, and the greatest average rainfall at the storm centers on areas of 250, 650, 1100, 1800, and 2500 square miles during 1, 2, and 3-day periods. These values are read from the Time-Area-Depth curves for storms over northern states in figures 94, 95, and 96, pages 182–187, Part V. The various areas used were chosen to correspond closely with the drainage areas above the different dams. Table 12 also shows for each storm its key number used throughout Part V, the date of its occurrence, and the location of the center of the storm.

The method of selecting greatest storms, as described in chapter VI, Part V, was devised with the intention of securing the storms having greatest average intensity over areas of approximately 2500 square

Table 12.—Maximum storm rainfall recorded in the northern half of the Mississippi Valley for 1, 2, and 3-day periods averaged over areas comparable in size with those draining into retarding basins of the Miami Valley.

Area in Square Miles	Winter Storms			Summer and Fall Storms							
	No. 15 1895 Dec. 17-20 Mo.	No. 130 1913 Jan. 10-12 Ark.	No. 132 1913 Mar. 23-27 Ohio	No. 51 1900 July 14-16 Iowa	No. 72 1903 Aug. 25-28 Iowa	No. 83 1905 June 9-10 Iowa	No. 86 1905 Sept. 15-19 Mo.	No. 109 1909 July 5-7 Mo.	No. 110 1909 July 20-22 Wis.	No. 114 1910 Oct. 4-6 Ill.	No 151 1915 Aug. 17-20 Ark.
1-day rainfall	In.	In.	In.	In.	In.	In.	In.	In.	In.	In.	In.
1	5.8	5.8	7.0	7.0	14.6	12.1	8.1	9.7	10.7	8.5	7.0
250	5.8	5.7	6.9	6.9	10.9	11.5	7.7	8.6	9.8	7.9	6.7
650	5.8	5.6	6.7	6.8	10.4	10.6	7.2	7.8	8.6	7.3	6.6
1100	5.7	5.4	6.5	6.6	9.9	9.8	6.7	7.1	7.7	7.0	6.5
1800	5.7	5.2	6.2	6.4	9.2	9.1	6.3	6.3	6.7	6.6	6.5
2500	5.6	5.1	6.0	6.2	8.7	8.6	5.9	5.7	6.0	6.3	6.5
2-day rainfall											
1	9.6	7.4	9.5	13.0	14.7	—	8.2	9.8	11.0	12.9	10.9
250	9.4	7.0	9.1	12.7	12.3	—	8.1	9.0	10.4	11.8	10.5
650	9.2	6.8	8.7	11.4	11.4	—	8.1	8.2	9.4	10.7	10.0
1100	9.0	6.6	8.4	10.6	10.7	—	8.0	7.6	8.4	10.1	9.6
1800	8.7	6.4	8.2	9.8	9.9	—	7.8	6.8	7.6	9.6	9.2
2500	8.4	6.3	8.1	9.3	9.3	—	7.7	6.4	7.0	9.2	9.0
3-day rainfall											
1	11.9	7.5	10.2	13.7	15.5	—	10.5	11.2	12.8	15.2	14.0
250	11.1	7.5	9.9	12.8	13.0	—	10.4	10.5	10.7	14.4	13.5
650	10.6	7.4	9.7	12.0	12.0	—	10.2	10.0	9.4	13.4	13.0
1100	10.4	7.4	9.5	11.3	11.3	—	10.1	9.6	8.6	12.7	12.5
1800	10.2	7.3	9.3	10.7	10.7	—	9.8	9.2	8.1	12.2	12.1
2500	9.9	7.2	9.2	10.2	10.2	—	9.6	9.0	7.8	12.0	11.8

miles. This method would not, therefore, necessarily include storms limited to much smaller areas, even though the rainfall should be phenomenally intense. For the present discussion, however, the method that was used seems a reasonably satisfactory basis of selection. This point will be taken up later for further discussion.

It is necessary, first, to distinguish between winter and summer storms in table 12. There are but 3 winter storms: No. 15 in December 1895, No. 130 in January 1913, and No. 132 in March 1913. These are grouped in the first part of the table. As indicated by the figures printed in heavy type, for a 1-day period the last of these had the greatest rainfall, but for 2 and 3-day periods the first had the greatest rainfall. As shown in chapter II, the maximum actual rates of runoff during the 1913 flood were about 90 per cent of the corresponding rates of rainfall. Therefore, we may safely use a 90 per cent factor to obtain rates of runoff from rainfall for winter storms in this region.

EFFECT ON RETARDING BASINS

In order to be able to test each retarding basin as it might be affected by these various maximum storms, a diagram has been prepared for each basin, figures 99 to 103, showing the discharge from the basin that would be produced by an inflow of any amount from zero up to 20 inches, occurring in any number of days from 1 up to 6. These diagrams are similar to figure 30 in chapter IV, except that the latter showed for various runoffs the total flow entering the city of Dayton as made up of the combined discharges from the various dams

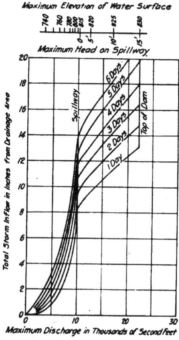

FIG. 99.—MAXIMUM RATES OF OUTFLOW AT GERMANTOWN DAM FOR VARIOUS TOTAL AMOUNTS AND DURATIONS OF STORM INFLOW.

The discharge is seen to increase rapidly after the spillway begins to overflow. The depths of overflow are indicated at top of diagram together with the corresponding height of the water surface above sea level.

and from the unreserved area below the dams. The results obtained from these diagrams are indicated below.

The Germantown and Lockington basins have each a drainage area of little more than 250 square miles. Applying the 90 per cent coefficient to the 1913 rainfall on this area for 1 day gives 6.2 inches as the possible runoff. Figures 99 and 101 show that this amount would not fill the Germantown basin to spillway level, but would just reach the spillway at Lockington.

Englewood and Huffman drainage areas are a little over 650 square miles. Figures 100 and 103 show that the possible 1-day runoff indicated by table 12 would not fill either of these basins to spillway level.

Taylorsville basin has a drainage area of over 1100 square miles,

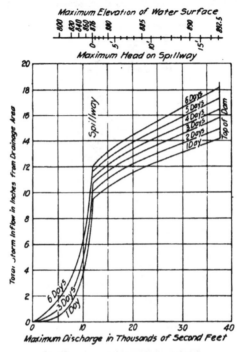

FIG. 100.—MAXIMUM RATES OF OUTFLOW AT ENGLEWOOD DAM FOR VARIOUS TOTAL AMOUNTS AND DURATIONS OF STORM INFLOW.

The descriptive matter under figure 99 applies also to this diagram.

and as shown by figure 102 would be filled a little above spillway level by the corresponding runoff.

A similar comparison on the maximum 2-day runoff, for the storm of December 1895, using again the 90 per cent coefficient, shows that such a flood would not exceed spillway level at the Germantown and Englewood dams, but that it would rise to a depth of about 3 feet on the Lockington spillway, about 2 feet on the Huffman spillway, and about 3 feet on the Taylorsville spillway. These figures are obtained in each case by assuming the heaviest rainfall to be centered in succession over the drainage area above each dam, and the shape of the area receiving the heaviest rainfall to conform exactly to the shape of each of the various drainage areas above the different dams. If the heaviest rainfall had been divided between the Taylorsville and Huffman drainage areas, the depth on each crest would have been less. In any event, the amount discharging over the spillways under

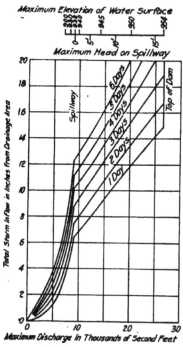

FIG. 101.—MAXIMUM RATES OF OUTFLOW AT LOCKINGTON DAM FOR VARIOUS TOTAL AMOUNTS AND DURATIONS OF STORM INFLOW.

The descriptive matter under figure 99 applies also to this diagram.

such circumstances would have increased the flow entering Dayton by only three or four thousand second feet.

A similar study of the maximum 3-day runoff for the December 1895 storm shows that there would be no flow over any spillway except at Lockington, where it would not be of great importance.

Next, the summer storms may be considered. These are arranged in chronological order in the latter part of table 12. Again the heaviest precipitations are in heavy type. For a period of 1 day these figures show that over an area of 250 square miles the greatest rain-

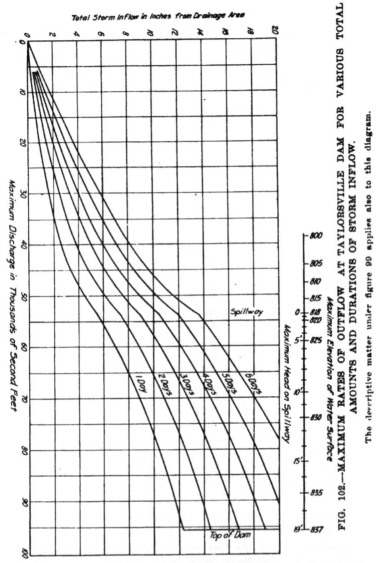

fall occurred during the storm of June, 1905, centered over Iowa, and amounted to 11.5 inches over the drainage area.

During the summer time it seems to be well established that the runoff from a given amount of rainfall is less than for the same rainfall during the winter season. In lack of definite information it is assumed for the purpose of these calculations that for every summer storm period, whether 1, 2, or 3 days, the runoff is 2 inches less than the rainfall. It is certain that the total amount of rainfall that disappears by absorption into the ground and evaporation into the air, and that is retained by surface storage to produce delayed runoff under ordinary summer conditions, is greater in amount than 2 inches

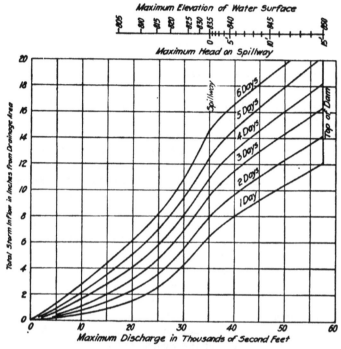

FIG. 103.—MAXIMUM RATES OF OUTFLOW AT HUFFMAN DAM FOR VARIOUS TOTAL AMOUNTS AND DURATIONS OF STORM INFLOW.

The descriptive matter under figure 99 applies also to this diagram.

for such heavy storms; so using this amount seems to err, if at all, in putting too severe a test upon the protective capacity of the retarding basin system.

Deducting 2 inches from the 11.5 inches leaves 9.5 inches as the maximum 1-day runoff to be taken care of by the Germantown and Lockington retarding basins. Figure 99 shows that the Germantown basin would then be filled to about 4 feet above the spillway crest, and that the combined outflow from conduits and spillway would be about 12,000 second feet, or 2000 feet in excess of the conduit discharge. Similarly, figure 101 shows that Lockington basin would be filled to about 8 feet depth on the spillway, with a combined discharge of about 15,000 second feet, or 6000 second feet more than the conduit discharge.

By the same process, for a drainage area of 650 square miles the storm of June 1905, would have a maximum rainfall of 10.6 inches. The corresponding runoff of 8.6 inches would not fill the Englewood basin to spillway crest level according to figure 100, but according to figure 103 would fill the Huffman basin to a depth of about 7 feet above the spillway crest, giving a combined discharge at the dam of about 42,000 second feet, or 7000 second feet in excess of the conduit discharge.

Similarly, for 1100 square miles the storm of August 1903, had the greatest rainfall, and the runoff would be 9.9 minus 2.0, or 7.9 inches, which would fill the Taylorsville basin to about 8 feet above the spillway crest, with a total discharge of about 64,000 second feet, or about 10,000 second feet in excess of the conduit discharge alone.

If the storm center were distributed equally over the 1800 square miles including both Taylorsville and Huffman basins, the runoff would become 7.2 inches, giving a depth on Huffman spillway of about 3 feet, and on Taylorsville spillway of about 6 feet, producing at the two places together a discharge of 9000 second feet in excess of the discharge through the conduits.

For a 2-day heaviest storm rainfall the storm of July 1900 was greatest for an area of 250 square miles, the storms of July 1900 and August 1903 were equal for 650 square miles, and the storm of August was greatest for larger areas. Deducting, as before, 2 inches from each total rainfall to obtain the corresponding runoff, the following results are obtained at the various basins. The runoff of 10.7 inches in 2 days would fill Germantown basin to 4 feet above the spillway with a maximum discharge of 12,000 second feet, or Lockington basin to a depth of 8 feet over the spillway with a discharge 6000 second feet in excess of the conduit discharge. The runoff of 9.4 inches would not fill Englewood to spillway level, but would fill Huffman to 5 feet over the spillway, producing an excess discharge of 4000 second feet. A runoff of 8.7 inches would fill Taylorsville to 5 feet over the spillway with an excess discharge of 5000 second feet. If spread over both Huffman and Taylorsville, Huffman would be filled barely to spillway level, and Taylorsville would reach only 3 feet over the spillway.

For a 3-day runoff the storm of October 1910 was the greatest for all the areas under consideration. Deducting a 2-inch retention as before, with a runoff of 12.4 inches, Germantown basin would be filled 6 feet over the spillway, giving an excess discharge of 3000 second feet, and Lockington 8 feet over the spillway, giving an excess discharge of 6000 second feet. With a runoff of 11.4 inches, Englewood basin would be filled 3 feet above the spillway, giving an excess discharge of 2000 second feet, and Huffman to 5 feet above the spillway with an excess discharge of 5000 second feet.

With a runoff of 10.7 inches Taylorsville basin would be filled to 6 feet above the spillway, giving an excess discharge of 7000 second feet.

With a runoff of 10.2 inches over Englewood and Taylorsville drainage areas together, the former basin would not fill to spillway level, and the latter 5 feet above spillway, giving an excess discharge of about 5000 second feet. The same runoff over Huffman and Taylorsville together would fill the Huffman basin to about 2 feet above the spillway with a combined excess discharge of 6000 second feet.

With a runoff of 10.0 inches in 3 days over the whole drainage area above Englewood, Taylorsville, and Huffman dams, Englewood basin would not be filled to spillway level, and the combined excess discharge from the other two would be only about 5000 second feet.

Before discussing the above figures further it will be useful to consider somewhat more in detail the possibility of the occurrence of storms small in area but of greater intensity than those shown in table 12. For this purpose we may properly confine our attention to the records of the states of Kentucky, Ohio, Indiana, Illinois, Iowa, and Missouri, or the area included between latitude 37 and 43 degrees north and longitude 81 and 97 degrees west. Throughout this region, for many years, there has been a Weather Bureau rainfall gaging station for about every five or six hundred square miles of territory. Hence, the chance is remote that storms covering two or three hundred square miles could have occurred materially greater in intensity than those that have been recorded.

PROBABILITY OF OCCURRENCE

In Technical Reports, Part V, figures 38, 39, and 40 show the most intense rainfall ever recorded (up to December 31, 1915) at any station during 1, 2, or 3 days in each quadrangle 2 degrees square. From these diagrams it is seen that the greatest single day rainfall in the area under consideration occurred in Iowa and was 14.6 inches. This occurred during the storm of August 1903, and is shown in table 12. The next greatest was 12.1, also in Iowa, during the storm of

June, 1905, and is also shown in table 12. Similarly, the diagrams show that the largest two 2-day records and the largest two 3-day records ever obtained at any station in this region also occurred during the storms listed in table 12, and hence appear in that table. This proves conclusively that table 12 includes the most intense rainfall ever recorded in this region.

Of all the figures for summer storms, as discussed above, the largest calculated excess discharge which would affect the-flow through Dayton is 10,000 second feet, which is much less than the available margin of safety of the Dayton channel. So, assuming that these storms could occur over the Miami Valley, the resulting runoff would not be a serious menace. But the chance that any such summer runoff can occur in the Miami Valley seems almost negligible, because of the extreme assumptions involved in the above calculations. Let us review the mitigating conditions seriatim.

First, it will be noticed that all the summer storms enumerated in table 12 were centered over the states of Iowa, Wisconsin, Illinois, Missouri, and Arkansas. From this it might be argued that no such storms can occur over the Miami Valley. On the other hand, one large summer flood has occurred over the Miami Valley, that of September 1866. So there seems a chance, albeit a very minute one, that such storms may occur in Ohio.

Second, in the above calculations nothing has been deducted for the reduced discharge from basins not filled to spillway level. This would not usually be an important factor.

Third, in the above calculations the tacit assumption has been made that for each case being considered the storm center would so coincide with the drainage area that the isohyetals would have exactly the same shape as the boundary of the drainage area. It is practically impossible for this to be true. The maps of these storms in Technical Reports, Part V, show that invariably the isohyetals are more or less elongated elliptical curves with the longer axis in a general east and west direction, in agreement with the usual eastern movement of thunderstorms. On the contrary, the drainage areas of the Miami Valley have their longer axes in a general north and south direction. This renders the chance almost infinitesimal that the isohyetals and drainage boundaries could exactly coincide. To evaluate numerically the effect of this discrepancy is very difficult on account of lack of data, but some light may be thrown on the matter by a study of the detailed records of the 1913 flood rainfall in this valley.

Table 13 shows the actual maximum 1, 2, and 3-day precipitations over the various drainage areas of the Miami Valley as obtained from the daily rainfall maps, figures 2 and 3, pages 28 and 29, in comparison with the maximum precipitation for similar areas in the same storm as shown in table 12.

Table 13.—Actual rainfall during March 1913 storm on specific drainage areas in the Miami Valley, compared with the maxima observed elsewhere during that storm over areas of like extent.

Name of Basin	Drainage Area	For 1 Day			For 2 Days			For 3 Days		
		Maximum Precipitation from Table 12	Actual Greatest Precipitation from Figure 2	Ratio	Maximum Precipitation from Table 12	Actual Greatest Precipitation from Figure 3	Ratio	Maximum Precipitation from Table 12	Actual Greatest Precipitation from Figure 3	Ratio
	Sq. miles	In.	In.	%	In.	In.	%	In.	In.	%
Germantown............	270	6.9	4.4	64	9.1	8.0	88	9.9	9.0	91
Englewood..............	651	6.7	4.2	61	8.7	6.2	71	9.7	7.7	79
Huffman...............	671	6.7	3.7	55	8.7	6.0	69	9.7	7.8	80
Taylorsville............	1131	6.5	4.2	65	8.4	6.1	73	9.5	7.6	80
Englewood & Taylorsville.	1784	6.2	4.2	68	8.2	6.1	74	9.3	7.6	82
Huffman & Taylorsville ..	1804	6.2	4.0	65	8.2	6.1	74	9.3	7.7	83
Englewood, Huffman, and Taylorsville..........	2455	6.0	4.0	67	8.1	6.1	75	9.2	7.7	84

The belt of most intense rainfall in this storm crossed the Germantown, Englewood, and Taylorsville drainage areas, and must have given nearly the greatest average precipitation over these areas possible for such a storm. The table shows, nevertheless, that the actual was for 1-day precipitation only about 65 per cent the amount from the time-area-depth diagram; for 2-day precipitation about 75 per cent; and for 3-day precipitation about 85 per cent. It is evident, then, that the maximum amounts in table 12 could never be reached on our drainage areas.

Fourth, we have above deducted only 2 inches from the total summer rainfall to obtain the corresponding runoff. In 1913, at the time of maximum runoff, about 5 inches in depth over the drainage areas was stored on the ground surface and in stream channels. The drainage basins will occupy only a little of this storage capacity, the remainder being available as before. Hence, only the most rare combination of unfortunate circumstances, including heavy previous rains sufficient to saturate the ground and fill up the stream channels, followed by rains of the most extreme intensity given in table 12, could produce a condition corresponding to that assumed in the calculations.

Taking all the above qualifications into consideration, it would

17

seem as though there were no chance that the system could ever be submitted to as severe a test as shown in the cloud burst calculations above.

Assuming, for the moment, that the calculations are not too severe, as made above, for the possible effect on the Miami Valley protection system of the occurrence in this region of a summer storm as heavy as the greatest on record anywhere in the Mississippi Valley, it is of interest to consider the chance of the occurrence of such a storm in this locality. Table 12 indicates that heavy summer rains may occur at points throughout the states of Kentucky, Ohio, Indiana, Illinois, Iowa, and Missouri. The combined area of these six states is close to 300,000 square miles. The drainage area above Dayton is approximately 2500 square miles or one-one hundred twentieth part of the total area of the six states. If the 25-year record studied in Part V is representative of all similar periods; if it be assumed that the greatest summer storm on record would put an equal test on the flood protection system if it occurred anywhere over the drainage area above Dayton, (which, in reality, is much too severe an assumption); and, lastly, if the chance of such a storm occurring over the Miami Valley is as great as of its occurrence over any other part of the Mississippi Valley (which does not seem to be true according to the records); then we may predict that once in 120 times 25 or 3000 years the protection works will be subjected to such a test as discussed above by the occurrence of a summer storm comparable in magnitude with the greatest shown in table 12. The flood protection system has been so designed as to handle such a storm if it ever should come.

CHAPTER XI.—HYDRAULICS OF CHANNEL IMPROVEMENTS

The design of the improvements at the various cities is so intimately connected with local conditions, such as the position of bridges, buildings, railroads, and canals, that it is possible in this volume to discuss them but briefly. For this reason the scope of this chapter is limited to describing the method of determining the channel capacities required with retarding basin control, and a discussion of the general principles of channel improvement.

CHANNEL CAPACITIES REQUIRED WITH RETARDING BASIN CONTROL

Because of the uncertainty of the rates of runoff from the unreservoired areas, a mathematically rigid determination of the channel capacity required at each of the various cities to take care of a flood forty per cent larger than that of March 1913 with retarding basins built, was not possible. In the 1913 flood the topography of the watershed exerted a material influence on the rates of flow at the different points. This influence would be the same in the case of another flood. The distribution of rainfall was another principal factor. The precipitation in the vicinity of Troy was very heavy and caused intense local runoff rates. This was shown not only by the results of the measurements but also by the damage done by the floods in the small streams in that section. In another flood the rainfall distribution would, of course, differ and a design for channel improvement to protect against future floods should take this into account. The flood measurements showed at Piqua a runoff of 70,000 second feet from 842 square miles, and at Tadmor 127,300 from 1128 square miles. The maximum inflow between those points was, then, 57,300 second feet from 286 square miles or about 200 second feet per square mile. If this rate is applied to the 66 square miles of drainage area between Troy and Piqua, it gives 13,000 second feet. To make the 1913 flood discharges represent more nearly what they would be likely to be from another such storm, we would probably increase the measured flow at Piqua a little and possibly decrease it somewhat at Tadmor. An increase of 12,000 instead of 13,000 was therefore adopted for the area between Piqua and Troy, or a rate from the 66 square miles of 182 cubic feet per second per square mile. This same unit rate was

also assumed for 12 square miles above Piqua. The computation of
the required capacity at Piqua and Troy would be as follows:

For Piqua

Miami River 1913 flow above mouth of Loramie Creek, from 575 square miles	48,500	second feet
From 12 additional square miles, at 182 second feet per square mile	2,200	" "
	50,700	" "
Add 40 per cent	20,300	" "
From Lockington Basin, including discharge over spillway two feet deep	9,800	" "
Discharge at Piqua	80,800	" "

For Troy

Miami River 1913 flow above mouth of Loramie Creek, 575 square miles	48,500	second feet
From 78 additional square miles at 182 second feet per square mile	14,200	" "
	62,700	" "
40 per cent additional	25,100	" "
From Lockington Basin, including discharge over spillway two feet deep	9,800	" '
Discharge at Troy	97,600	" "

Flood protection by retarding basins is most effective at points
immediately below the basins. As we consider points further and
further down the valley from the dams, the discharge from the unreservoired area becomes an increasingly larger factor until finally it
may become the dominant one.

At Dayton, after the retarding basins are finished, the crests of
floods will be caused primarily by the maximum retarding basin- discharges, which will occur somewhat later than the heaviest rain, at a
time when the runoff from the unreservoired area will have decreased
materially.

At Hamilton conditions will, in general, be quite different. There
will be two separate flood peaks, the first caused primarily by the
flow from the 947 square miles below the dams, and the second caused
principally by the sum of the maximum outflows through the dams.
As shown below, the first of these peaks will be the larger.

For points between Dayton and Hamilton it appears that the flood
crests will be caused principally by the maximum retarding basin
outflows, and that the discharge from the unreservoired area will not
become the leading factor until the mouth of Four Mile Creek is
reached a short distance above Hamilton.

Table XVII of the Appendix shows the results of a method of estimating the needed channel capacity at Dayton, and points below, after the retarding basins are completed, based upon the principles stated above. The rates of runoff from the unreservoired areas are based upon the measurements of the 1913 flood. For a storm like that of 1913, at the time the peak caused by the maximum discharge from the dams reaches the various cities, it is assumed that the simultaneous unit rate of runoff from the corresponding unreservoired area is, at Dayton, 120 second feet per square mile; at Middletown, 75 second feet per square mile, varying at intermediate points in direct proportion to the increase in the unreservoired area; at Hamilton, 60 second feet per square mile. For the Official Plan flood these rates are increased by 40 per cent.

The crests at these cities will occur shortly before the discharges from the retarding basins reach their maximum values. This is allowed for by the slight reduction shown in column 7 below the values in column 6. The values in column 8 are the sum of those in columns 5 and 7. The larger peak at Hamilton is obtained by using the maximum rate of runoff, 120 second feet per square mile, from the 947 square miles of unreservoired area, combined with 45 per cent of the total maximum discharge from the retarding basins.

In planning the actual channel improvements at the various cities it was found in some cases that to provide a capacity sufficient for a flood 40 per cent greater than that of 1913, with 3 feet of freeboard on the levees, would entail a cost very great in proportion to the value of the property protected. This was true at Piqua and West Carrollton. At Franklin and West Carrollton the depth of flooding would be slight, the menace being one of inconvenience rather than of great injury. At these places the margin of protection was reduced, though in all cases a flood greater than 1913 is protected against. At Troy the margin of 40 per cent over 1913 is provided against the entrance of water at the upper limits of the city, thus even in case of such a flood reducing overflow solely to quiet backwater.

•

PRINCIPLES OF CHANNEL IMPROVEMENT DESIGN

To have the maximum flood carrying capacity, an ideal open channel should have a cross section constant in area and shape, a straight alignment, smooth bed and banks, a uniform slope to the bed, should be free from obstructions, and should be secure against damage from scour or deposit. Since in actual practice it is possible to approach only approximately to the ideal conditions, it is desirable to discuss the effects of departing from them in various respects, to show the best ways of minimizing the deleterious effects of such departure as may be unavoidable, and to determine so far as possible their relative importance.

Variation in Area of Cross Section

A uniform area of cross section in all portions of the channel is highly desirable; some variation, however, is permissible, in fact, is often unavoidable. A change in the area of cross section entails an inverse change in the average velocity and in the velocity head. When the cross section is diminished the velocity of the moving water must increase, and this increase can be obtained only by the consumption of pressure or head. Hence, when the water moving in an open channel increases in velocity, the surface of the water must have an increase in slope by an amount exactly equal to the increase in the amount of the velocity head of the moving water. In such a change some of the potential energy of the water has been converted into kinetic energy and added to the previous kinetic energy. In such a change there is not necessarily any impact or any additional consumption of energy by friction more than would exist in a uniform channel. Such a change, therefore, takes place easily and naturally under appropriate circumstances without any unusual loss or dissipation of energy in the form of heat.

It is quite otherwise, however, with a change in the opposite direction. If the cross section of a channel is increasing, the average velocity is being reduced. To accomplish this reduction requires the action of a force opposing the moving water. Theoretically, such a force may be an increasing head or height in the water surface through a flattening of the slope down stream, the opposite of the condition described in the preceding paragraph. But such a condition of rising surface is always relatively unstable and uncertain in action. The water, initially moving faster, strikes the slower moving water ahead of it producing the state generally spoken of as internal impact, leading to dissipation of kinetic energy into heat. This unstable condition may be partly explained as follows. The water has unequal velocities in different portions of the cross section. The head to stop entirely the velocity of one part will only somewhat reduce the velocity of another portion. This renders it difficult to maintain proper equilibrium between the different portions.

It is found that in closed pipes under pressure an expansion of cross section, if sufficiently smooth and gradual, can be used to convert velocity into pressure, that is, kinetic energy into pressure energy, with only a negligible loss of energy in the form of heat. This fact is made use of in the expanding tube of a Venturi meter. But it is also found that unless the expansion is quite gradual the internal impact is considerable.

In an open channel, the very fact that the water surface is open to the air and thus relatively unrestrained on half its boundary, makes a gradual expansion more difficult to secure. If the expansion is too rapid the moving water seems, in a sense, to break loose from restraint, to leave some portion of the enlarged cross section unoccupied, and to move on with velocity undiminished except as it is retarded by surface friction. That portion of the cross section unoccupied by the moving stream is filled with stagnant water or water moving in an eddy whose motion is maintained by friction with the moving current. To sum up, although in both gradual contraction and gradual expansion the changes of velocity and surface height may take place in accordance with Bernouilli's Law for the mutual interconvertibility of pressure and velocity, we may say that such changes are also subject to another law, namely, that the change from pressure to velocity takes place easily and readily, while the change in the opposite direction is secured only by care and effort to overcome the tendency of the water to avoid the change.

In gradual expansion there is another agency which may often be utilized to advantage, that is bank friction. It is a force always tending to retard velocity. If the expansion is made so gradual as to use friction as the sole means of retarding velocity, undoubtedly it will operate satisfactorily. Instead of the usual surface gradient as found in a uniform channel, the gradient of the water surface would then be flatter or might even be level throughout the expanding section, and the tendency to produce eddies or internal impact would be a minimum.

Where changes in the area of cross section are unavoidable, reduction of area accompanied by increase in velocity may be relatively abrupt, but expansion of cross section should be relatively much more gradual and should be limited to such a rate as will maintain the surface gradient only level rather than rising throughout the expanding section.

Enlarging the area of cross section of a channel, in general, increases its maximum carrying capacity. Enlargement may be secured by widening the channel, by deepening it, or by building levees on the banks. All of these methods are limited in their application. In addition to the expense of handling the excavated materials, widening a channel in some locations may not be feasible on account of damage to valuable adjacent property. Deepening a channel is of limited application, because there are objections to giving any portion of the bottom of a channel a lower elevation than the bottom in adjacent portions of the channel. Any such sag or pocket in the bottom is liable to filling, and hence such construction lacks permanence. Building levees on the banks may be objectionable because of the room they occupy, because they are unsightly, and because they are obstacles to passage across the river. Obstruction to view caused by high levees in cities tends to depreciate the value of adjacent property. Any project for channel enlargement becomes a complicated problem of carefully balancing the relative advantages and disadvantages of the different methods. · ·

Although it may be much easier and cheaper to enlarge a channel in one portion of its length than in another, it does not follow that the amount of enlargement should be greater in the portion where it is cheaper. Variation of cross section should be kept as small as possible. It follows, therefore, that a given amount of excavation to enlarge the cross sections of smallest area is much more effective than the same amount of excavation in places where the area of cross section is larger. The first endeavor in enlarging a channel, then, should be to locate the constrictions existing in the channel and to enlarge them as much as practicable. Removing the pinched places is one of the most effective means of increasing a channel's carrying capacity. If a channel has a considerable amount of rather abrupt variation in cross section it is doubtful whether any excavation in the larger cross sections is of value so far as the increased area goes. On the contrary, it may be wise to fill a portion of an abnormally large cross section, especially in cases where such filling will furnish a convenient and economical means of disposal of material excavated in other portions of the channel.

Curves in Alignment

One of the troublesome but unavoidable characteristics of open channels is the presence of curves in alignment. When water flows in a straight smooth channel the transverse surface profile is probably a straight horizontal line. Various observers have claimed that the water in the middle of the stream is higher than at the banks. If waves are present, they are much larger in the middle than near the banks, and their crests are therefore correspondingly higher; but it has not yet been satisfactorily proved that the average surface elevation is any greater at the center.

On a curve or bend the transverse surface profile cannot be level. Water, like all matter, when in motion moves in accordance with Newton's First Law of Motion; that is, it moves in a straight line unless deflected by the action of some unbalanced force. If water moves in a curve, there must be an unbalanced force acting against the water and directed towards the center of curvature. Let figure 104 represent the transverse cross section of a stream at a bend, A being at the inner bank, and B at the outer bank. Consider that portion of the stream whose cross section is represented by the area $C D H G$. As this portion flows around the bend it is deflected toward the center of curvature on the side toward A. This deflection is caused by the excess of pressure on the face $D H$ over the pressure on the face $C G$. The excess of pressure can exist only when the water

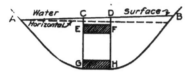

FIG. 104.—CHANNEL CROSS SECTION ILLUSTRATING CONDITIONS IN BEND OF RIVER.

surface at D is slightly higher than at C. The amount of the excess pressure can be exactly calculated by the formula for centrifugal force when the velocity of the moving water and the radius of curvature of its path are known. For an actual stream such calculations are subject to some degree of uncertainty because the velocity varies throughout the cross section, and because the radius of curvature cannot be precisely determined. Subject to these uncertainties, however, the calculated differences in surface elevation between A and B, on the inner and outer banks at a bend, agree with the observed difference within the limit of possible error in the observation. The surface at the outer bank B is rarely higher than the surface at A by more than a few inches. The complete mathematical discussion of the calculation of the transverse surface slope at a bend is given on page 271.

Another detail regarding the curved motion around a bend should be noted at this place. Referring again to figure 104, the excess of pressure required at F over that at E varies as the square of the velocity of the moving water, other things being equal. But the layer near the bottom $G H$ has a much slower velocity than a central layer like $E F$. It follows then that the same surface elevation at D cannot furnish precisely the required excess pressure for both layers $E F$ and $G H$. What really happens is that the two layers do not bend around the same center of curvature and do not have the same radius of curvature. The excess pressure at D takes an intermediate value, and the motion of each layer adjusts itself to this pressure difference. The layer $G H$ has a shorter and the layer $E F$ a greater radius of curvature than the average. As a result the fastest moving water gradually shifts toward the outer bank as it moves around the bend, and there is a compensating creep of the slower moving water near the bed of the channel towards the inner bank. This effect cumulates with the length of the bend, so that in a long bend, such as a semi-circular curve, finally all the water with highest velocity is near the outer bank while in the inner half of the cross section the velocity of the water is much slower. This phenomenon was observed and studied in a small experimental channel by Professor James Thompson about 1870.

The next point for consideration is the condition at the junction points of straight channels and bends. In actual rivers there is always some sort of gradual transition curve at both the beginning and ending of a curve, but it will be simpler to discuss a theoretical channel in which two tangents are connected by a circular curve of constant radius. Since the transverse profile is level on the approaching tangent but must be inclined around the bend, the question arises as to how the change takes place from one state to the other. Perhaps, theoretically, the level cross section could be transformed to the slanting shape most simply by raising the water surface at the outer bank and lowering it at the inner bank by equal amounts. Such a rising at the outer bank would be open to the difficulties and complications previously discussed for a rising water surface. As a matter of fact, during careful observations on numerous sharp bends in actual river channels, no case has been found in which there was plainly a rise in the water surface at the outer bank at the beginning of a curve. But there is always a noticeable tendency for the longitudinal slope to be flattened at the outer bank at this point and it frequently becomes practically level for a short distance. The remainder of the required transverse slope results from a rather sudden depression in the water surface at the inner bank at the beginning of the curve.

An important secondary effect of the changes in surface elevation at the beginning of a curve is the disturbance of the velocity distribution in the cross section. The relative surface elevation at the outer bank is accompanied by a reduction in velocity in that region, while the surface depression at the inner bank produces a local acceleration of velocity. This condition may usually be observed at the beginning of a sharp bend in a river channel, and produces an appearance as though the rapidly moving water had moved over from the center of the channel to a course near the inner bank. It should be remembered, however, that this appearance is partly deceptive. The rapidly moving water near the inner bank is in large part the same water which was near the bank farther up stream, moving slowly, and which has been accelerated at the approach of the curve.

At the lower end of the curve another adjustment of the transverse surface profile must be effected in joining to the straight channel. As before, it seems to be easier for the higher surface, now at the outer bank, to be depressed, than for the lower surface, now at the inner bank, to be raised, in order that the two may be adjusted to a common level. But, as before, observation of actual channels shows that the longitudinal slope often becomes nearly level at the inner bank for a short length, even though the major part of the necessary surface adjustment is obtained by a rather sudden drop of the water surface along the outer bank. During the high waters of January 2 and February 1, 1916, and April 21, 1920, profile levels were taken along both banks of the Miami River through Dayton. The resulting profiles for the latter two dates are shown in figure 105. These show that the water surface on the inside of a curve drops suddenly at the beginning of the curve, and then assumes a flatter slope, while the surface at the outside of the curve has a flat slope in the first part of the curve, with a rather sudden drop at the lower end.

The profiles for the two different floods show considerable differences in details. During the flood of 1920 a much greater proportion of the total discharge came from Mad River than during the flood of 1916. This implied that Mad River was at a relatively high stage as compared with the upper Miami. Hence, there was a decided local flattening in the surface slope above the junction with Mad River and extending as far as Island Park dam.* At the time of the 1920 flood a large amount of channel improvement had been completed along the stretch of channel covered in these profiles. At various places, however, uncompleted portions made very perceptible local distortions in the profile. Thus, just below Herman Avenue bridge, on the left bank, was a construction and storage basin for barges and a large pile of gravel projecting into the channel. This acted as an obstacle to the current and formed a large eddy on its downstream side; these show a decided sudden wave in the profile. Again, just above the junction with Wolf Creek, on the right bank, there were some large piles of gravel and two dragline excavators in the current producing a similar distortion in the surface profile along that bank. It is to be

* At the time of the February 1916 flood the Island Park dam had not been built, but the old Steele dam, located 600 feet farther upstream, was still in existence. This accounts for the difference in the locations of the drop-off over the dam at upper ends of profiles in figure 105.

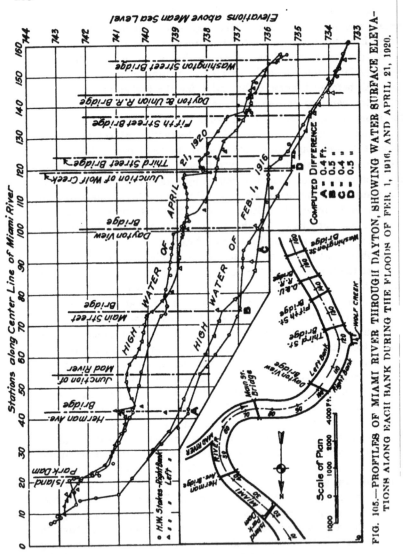

FIG. 105.—PROFILES OF MIAMI RIVER THROUGH DAYTON, SHOWING WATER SURFACE ELEVATIONS ALONG EACH BANK DURING THE FLOODS OF FEB. 1, 1916, AND APRIL 21, 1920.

expected that when the channel improvements are completed, the surface profile will show much less irregularity.

Just as was true at the beginning of the curve, the sudden drop of the water surface is necessarily accompanied by a corresponding acceleration of velocity, but with this important difference. At the upper end of the curve, the water receiving the acceleration was in that part of the stream near the bank which had previously possessed a relatively low velocity; therefore, even after being accelerated its velocity was not conspicuously high. At the lower end of the curve, however—particularly on a long curve of 90 degrees or more—the water receiving the sudden acceleration near the outer bank is the water already possessing the highest velocity of any in the cross section, having been thrown outward towards the outer bank by centrifugal action during its progress around the curve; therefore, after being accelerated, its velocity is conspicuously high, often greater than at any other point within a considerable distance either up or down stream. This high velocity close to the outer bank often produces heavy erosion. While erosion often takes place near the inner bank at the beginning of a curve, it is generally much less extensive than that to be seen near the outer bank at the lower end of a curve.

The superelevation of the water surface at the outer bank of a bend is approximately proportional to the width of the stream and inversely proportional to the radius of the bend. To reduce as much as possible the disturbing effect of the bend upon the normal velocity distribution throughout the stream, it is desirable, then, to keep the radius of curvature as large as possible. The inner bank should have a radius not less than twice the width of the stream and should be as much larger as practicable.

The effect of the variable velocity distribution throughout a cross section is, probably, to increase slightly the actual radius of curvature followed by the moving water in going around a bend, especially for a short curve. Thus, for a curve of 90 degrees the effective radius for the whole stream is probably as great as the radius of the outer bank; but for a curve of 180 degrees the effective or actual radius cannot much exceed the average of the radii of the inner and outer banks.

Where compound curves are possible they may be of some use. There would not seem to be much advantage in having a transition curve of large radius at the beginning of a sharp curve, because the erosion at the beginning of a curve is not generally a serious matter; but a curve of gradually increasing radius, as a transition from a sharp bend to a tangent, apparently is of considerable value in reducing the tendency to set up injurious erosion.

Curves in open channels are objectionable, first, because by disturbing the normal velocity distribution they tend to increase frictional losses; and second, because they increase the danger of serious local erosion. Where bank protection becomes necessary, it is most needed on the outer bank in the lower part of the curve, and to a less degree on the inner bank at the beginning of a curve.

The loss of capacity of a channel due to curvature is not as great, however, as has ordinarily been assumed. The roughness coefficient for the channel through Dayton, at the time the profiles of figure 105 were taken, was found to differ little, if any, from that which would be expected for a similar straight channel. Probably the best data on the loss of head due to curvature is that obtained under the direction of E. G. Hopson, on a concrete lined irrigation canal near Echo, Oregon, and described by him in the Engineering Record of October 21,

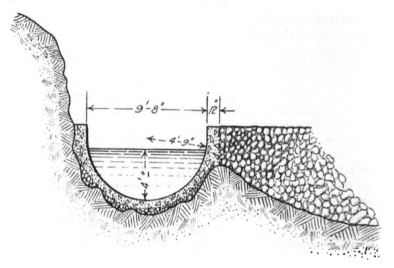

FIG. 106.—CROSS SECTION OF CONDUIT, UMATILLA PROJECT, OREGON.

1911, Vol. 64, page 480. Figure 106 shows a cross section and figure 107 a plan of the portion of the canal upon which the experiments were made. The curves were made up of 10-foot tangents, which, no doubt, increased the friction loss somewhat over that which would have occurred with smooth curves. Assuming that, if there were no curves present, the loss per unit length of the entire canal would have been the same as that measured in the straight portion of the canal. the results show that the total loss due to curvature and irregularities at bends was only 0.36 feet out of a total friction loss of 3.65 feet, or less than 10 per cent. It seems probable, therefore, that, except in such extreme cases as of channels with very flat slopes, the error due to neglecting the loss of capacity of a channel due to curvature is less than that which is likely to be made in selecting a roughness factor.

The mathematical formulas for the transverse slope of the water surface of a stream flowing around a curve are derived as follows: Consider the forces acting on an element of the stream represented by $C\ D\ H\ G$, figure 104. For the purposes of this discussion let it be assumed that all parts of this element have the same velocity and that all parts of the stream are moving in circular paths around the same center of curvature.

Let $b =$ breadth of the stream,

$r =$ radius of curvature,

$r_1 =$ radius of curvature of inner bank,

$r_2 =$ radius of curvature of outer bank,

$dr =$ distance $C\ D$,

$y =$ depth of element $C\ G$,

$dy =$ height of D above C,

$V =$ velocity of water,

$w =$ weight of unit volume of water,

$R =$ radius of curvature at center of stream.

The centrifugal force acting on the element $C\ D\ H\ G$ is equal to the excess pressure on the face $D\ H$ over the pressure on the face $C\ G$, due to the height of the surface at D above the surface at C. Considering the elementary volume of water to have a length of unity up and down stream, the volume of the element is $y\ dr$, its weight is $w\ y\ dr$, and its mass is $(w/g)y\ dr$. The excess pressure on the face $D\ H$ is $w\ y\ dy$. From mechanics it is known that centrifugal force $= mV^2/r$. Therefore,

$$\frac{wyV^2dr}{gr} = wydy$$

or

$$dy = \frac{V^2dr}{gr}. \qquad (121)$$

To integrate this equation the value of the velocity at all points across the river must be stated in terms of r. Formulas may be obtained under a variety of assumptions.

A fairly good approximation is obtained by assuming V constant at the average velocity, and assuming r to be constant at the value for the center of the stream. Then the

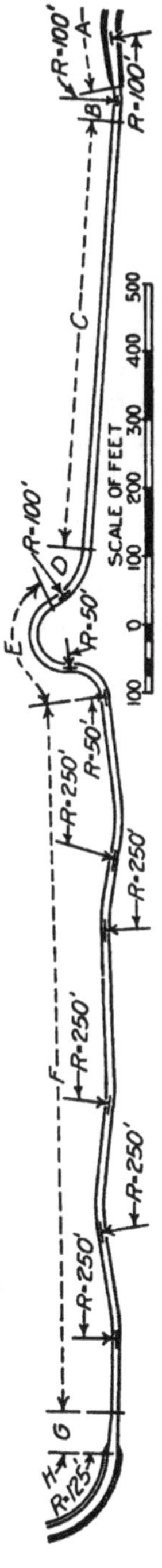

FIG. 107.—ALIGNMENT OF UMATILLA PROJECT CONDUIT USED FOR DETERMINING LOSSES THROUGH CURVATURE.

$$\text{Difference in elevation of the two banks} = \frac{V^2b}{gR}. \qquad (122)$$

This will always give too small a value due to the fact that the effect of the filaments with the higher velocities more than offsets the effect of the slower filaments, since the velocity enters to the second power in the formula.

If the actual velocity distribution across the stream is known, the width may be divided into several sections, the difference in surface elevation computed for each section using its appropriate velocity and radius, and the total difference found as the sum of the differences for the separate sections.

If V is assumed constant while r is variable, the differential equation integrates into:

$$\text{Total difference in surface elevation} = \frac{V^2}{g} \log_e \frac{r_2}{r_1}. \qquad (123)$$

If the angular velocity of the water is constant, as in a closed whirling vessel containing water,

$$V = k\,r \text{ and}$$

$$\text{Total difference in surface elevation} = \frac{k^2}{2g}(r_2^2 - r_1^2). \qquad (124)$$

If the velocity of the water varies inversely as r, a distribution of velocities which has received among mathematicians the name of vortex motion, $V = P/r$, then:

$$\text{Total difference in surface elevation} = \frac{P^2}{2g}\left(\frac{1}{r_1^2} - \frac{1}{r_2^2}\right). \qquad (125)$$

If the velocity is zero at each bank and has a maximum value Vm at the center, varying in between according to a parabolic curve: Total difference in surface elevation =

$$\frac{V_m^2}{g}\left[\frac{20}{3}\frac{R_2}{b^4} - 16\frac{R^2}{b^3} + \left(\frac{4R^2}{b^2} - 1\right)^2 \log_e \frac{2R+b}{2R-b}\right]. \qquad (126)$$

The numerical value of the transverse slope may be illustrated by taking the most extreme case that will be approached in the improved channel in the upper part of Dayton during the occurrence of an extreme flood.

Using as data, $b = 600$ feet, $R = 1300$ feet, area of cross section $= 12,500$ square feet, maximum discharge $= 110,000$ second feet; the average velocity will be 8.8 feet per second; and on the assumption of a parabolic distribution, $V_m = 13.2$ feet per second.

Formula (122) then gives as the difference of the elevation at the two banks 1.12 feet.

If the total width is then divided into six sections, each 100 feet wide, and there are used the appropriate radius for the center of the section and the corresponding velocity according to the parabolic law, the following differences will be obtained for the six sections beginning at the one on the inner bank of the curve .05, .27, .41, .38, .21, and .03, or a total difference of 1.35 feet. This result is 20 per cent greater than the approximate result given above.

If formula (126) is used the result obtained is 1.31 feet, about 17 per cent greater than the first approximate result above.

On figure 105 are shown the differences of elevation of the two sides of the river as computed for the conditions existing at the time the measurements were taken in February 1916. Considering the difficulties of measurement of the elevation of the water surface, and the determination of the radii of curvature, the observed and computed results are in reasonable agreement.

Irregularities in Banks

The banks of an improved channel should not only be smooth in the ordinary sense but, so far as feasible, should be free from angles or irregularities of a larger size. Such irregularities are in effect roughness on a larger scale. Angles or sudden changes in shape or curvature, in surfaces otherwise smooth, are open to the objections previously urged to changes of cross section and bends, even though the irregularities are of so limited an extent as to affect only a portion of a stream.

Shape of Cross Section

A modified trapezoidal shape as shown in figure 108 is probably best for the cross section of an improved channel several hundred feet wide. It should have a central level bottom portion $C\,D$ relatively narrow and following a uniform longitudinal gradient. Such

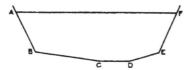

FIG. 108.—TYPE OF CHANNEL CROSS SECTION ADOPTED BY MIAMI
CONSERVANCY DISTRICT IN BENDS OF RIVERS.

The low portion $C\,D$ is located in the exact center of the section where the river is straight, and nearer the outer bank where the channel is curved, conforming approximately to the natural location of the deepest part of the channel.

18

a shape provides satisfactorily for both low water and flood flow. The portion C D should not always be in the exact center of the whole section, but around bends should be located where the deepest part of a channel naturally occurs. This will conduce to satisfactory maintenance and permanence.

If at any point the width A F must necessarily be reduced slightly, the constant area of cross section may be preserved by widening the portion C D. Changes of shape, if made gradually, and if they do not change the area of cross section, are not highly objectionable.

Bridges

Under the best of circumstances bridges with channel piers must be somewhat of an obstruction to the maximum flow in a channel; but by careful arrangement it is possible to reduce by a considerable extent the amount of the obstructive effect. The retarding effect of the piers is due: (a) to their reducing the cross section of the water way; (b) to the increased friction on the additional area in contact with the moving water along the ends and sides of the piers; (c) to the loss of energy in impact of the stream against relatively quiet water at the ends of the piers and, in arch bridges, against the lower portion of the arch ring and spandrel walls; (d) to the disturbances of flow produced by the obliquity of the piers when their length is not parallel to the direction of flow of the stream.

During heavy floods a river channel is generally scoured around bridge piers by the high velocities produced in the reduced cross section. This suggests that, in an improved channel, the net area under bridges should be made to approach the area of the cross section of the channel at other points as nearly as possible. In the case of an improved channel having the shape of cross section shown in figure 108, the area at the bridge can be increased by widening the central level portion CD even to nearly the whole width of the bottom BE, if necessary. Some sort of transition then becomes necessary from the standard cross section of the channel to the cross section at the bridge.

When the net cross section at a bridge is less than in the channel at other points, the increased velocity of flow through the diminished cross section can be obtained only at the expense of head. The head required equals the increase in velocity head of the moving stream corresponding to the increase of velocity through the bridge. This head manifests itself as a sudden drop in the elevation of the water surface as the water enters the bridge. The head is obtained by backing up the water above the bridge until it is high enough to furnish the additional velocity needed.

Observations of the water flowing at the time of floods, and of the scour produced by bridges at time of floods, seems to show that the increase of velocity takes place in a relatively short space just above the bridge; while below the bridge the water loses its increased velocity only slowly, and hence the scour often shows for some distance down stream.

This indicates that a transitional channel, from the standard cross section to the bridge cross section, should not extend more than 50 feet upstream from the bridge at the point where the change is greatest, nor more than 200 feet downstream from the bridge. These distances should be tapered gradually to zero at the points where the bottom elevation is not changed.

Sometimes it is possible to have the width between abutments at maximum high water level greater than the width of the standard channel. But it would certainly not be profitable to try to compensate for the area of numerous bridge piers by lengthening a bridge and putting abrupt curves into the banks, to widen them out near the bridge so as to preserve the same clear width of water way through the bridge as at other points in the channel. The most that it would ever be wise to do in this line would be to curve each bank, as it approaches an abutment, away from the channel by a distance not greater than the width of one pier, and to set the abutments back from the standard channel line by a distance not greater than the thickness of one pier. In computing the area through a bridge, piers of cellular or columnar construction should be treated as though they were solid within the straight tangent lines bounding and enclosing them.

To reduce the surface friction, pier surfaces should be as smooth as possible, with no sharp or re-entrant angles, and with easy entrance and exit. As flowing water approaches the upstream nose of a bridge pier, a roll or local rise of the water surface always occurs. This shows that there exists a local area in which there is slowing up of the velocity of the water. Such a surface rise is necessary in order to produce the pressure to deflect the water away from the piers and into the openings. This lateral motion tends to produce a contraction loss if there is a sharp corner at the upstream ends of the parallel sides of the piers. Hence the upstream noses of the piers should be semi-circular or elliptical rather than square or triangular.

In taking up the impact losses due to piers, it is necessary to consider separately the conditions at the upstream and at the downstream ends of the piers. The water which approaches the upstream end of a pier, and which would strike the pier unless it were deflected from

its straight course, suffers a slight loss of energy through impact in
having its velocity reduced as it enters the area of local surface elevation just above the nose of the pier. This loss is relatively unimportant both because the quantity of water concerned is only a small
fraction of the whole stream, and because the water affected has its
velocity reduced by only a small amount. The main body of the
stream suffers no loss of energy in having its velocity accelerated as it
enters the span openings because such a change, as already explained
in discussing variation in area of cross section, is not one which leads
ordinarily to any loss by conversion of energy into heat through friction or impact.

At the downstream end of each pier, or, as it might be called, in
the lee of each pier, there is always a considerable body of water
which is relatively quiet and stationary. This water includes irregularly distributed, turbulent boils and moves with uncertain and halt-
ing motion upstream towards the pier. At the end of the pier the
water moving rapidly through the bridge along the sides of the pier
comes into surface contact with this relatively quiet water, and immediately commences to mix with it. The mixing causes eddies and

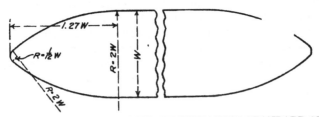

FIG. 109.—PLAN OF BRIDGE PIER ILLUSTRATING STANDARD SHAPE
ADOPTED BY MIAMI CONSERVANCY DISTRICT.

boils of a turbulent type. It may be compared to the mixing of the
rapidly issuing smoke from the top of a tall chimney or from a loco-
motive smoke stack with the surrounding air; the detached whirling
masses of smoke are plainly visible, and are analogous to the eddies
below a pier. In such mixing there is inevitably a large loss of energy through impact. The mass of stagnant water has a roughly
triangular shape gradually becoming narrower as the distance from
the pier increases. The impact loss continues downstream as far as
the stagnant water reaches, and until the water throughout the whole
cross section of the river is again moving with a uniform, regularly
distributed velocity. The amount of the impact loss apparently depends upon the size of the stagnant water segments, and any reduc-

tion of this size would reduce the impact loss. If a smooth solid tail could be attached to each pier reaching down stream far enough to prevent the formation of any stagnant water mass, the impact loss would become negligible. We should then have secured an ideal gradual enlargement of the cross section. It is wise, therefore, to approximate this shape in pier construction so far as practicable. In most cases, a rather flat elliptical section may be obtained. Figure 109 shows the standard shape of pier adopted for use by the Miami Conservancy District.

Where sharp corners at the upstream ends of the piers cause any perceptible contraction effect, then all the above described phenomena of dead water space, eddies, boiling, and whirlpools, with the corresponding impact losses will exist in the lee of each such sharp corner.

When a bridge crosses a stream on a skew, and the piers are set at such an angle that their length is not parallel to the natural direction of flow, as illustrated in figure 110, the piers act as a greater obstruction than they would if set parallel to the natural direction of flow of the stream. No experiments are available to furnish data as to the proper method of computing the obstructive effect of such oblique

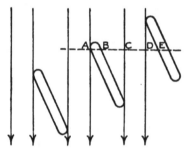

FIG. 110.—CONDITIONS OF FLOW WHERE BRIDGE PIERS ARE BUILT AT AN ANGLE WITH THE CURRENT.

piers. Undoubtedly, the piers deflect the moving water to some extent, so that the effective span opening is greater than CD, figure 110, but it is equally certain that the effective opening is less than the width BE. Probably, for purposes of calculation, as reasonable a procedure as any is to use the average of CD and BE as the effective span width. This is equivalent to using as the thickness of a pier the average of AB and AC.

The ideal entrance to a bridge opening would be funnel-shaped. This, however, it would be wholly impracticable to provide in the case of an arch bridge. In a truss or girder bridge, the obstruction to flow consists only of the piers and abutments which may be so shaped as to give the funnel-shaped opening. This is also true of an arch bridge in which the springing line is above high water level. In an arch bridge, of which the springing line is low, a large surface of the bridge structure is exposed to the current, and if the cross section through the bridge is enlarged by widening the standard narrow channel floor, the cross section above the face of the bridge is necessarily enlarged above the normal. The effect is therefore, first, an enlargement of the channel as it approaches the bridge; second, a sudden contraction at the upstream face of the bridge; third, a sudden enlargement at the downstream face of the bridge; and finally, a gradual contraction below the bridge to the normal channel cross section.

If it were possible to make the cross section of the approaching channel equal in area to that of the bridge opening, and the channel and bridge so shaped that the river flow could be led to the bridge opening without meeting the resistance of piers, arch ring, and spandrel walls, no head would be lost at the face of the bridge. This, however, is not the case, for not only does the bridge masonry present a large surface against the current and a different area of cross section, but the area of the bridge opening is very differently distributed from that of the standard channel. In the case of a channel of straight alignment, the area of cross section per unit of width increases from the sides to the center, while with a curved channel the maximum area per unit of width is at some point near the outside of the curve. The area of the openings of the different spans is usually about the same for all the spans. It is not probable that the flow in the deeper portion of the approaching channel can be readily forced over to a side span of the bridge. For this reason, if the area of bridge opening is distributed very differently from that of the channel at the bridge, the transition in the approach channel should not be made in too short a length.

It is not altogether clear what the effect has been in the attempt of the river to scour out a large waterway under the bridges at the time of the flood. In some cases, a considerable scour took place, while at others there is not a definite record left of the scour. As might be expected, the evidence of erosion is more pronounced below the bridges than above, but some indication is given of the distance above the bridges that the channel was affected. At the Big Four bridge below Miamisburg, where the entire river flow was confined to the bridge opening, the bed was scoured to a depth at least 7 feet below the previous elevation of the bed downstream from the bridge, and this scour extended to a distance of more than 500 feet above the bridge. At the High-Main Street bridge at Hamilton, the depression extended for at least 100 feet above the bridge. At the Dayton View bridge in Dayton, the scour is indicated for a distance of from 100 to 200 feet above the bridge. Apparently a depression was scoured in the bed above the Troy bridges for one hundred or more feet. The bed above the Pennsylvania Railroad bridge at Piqua was scoured out for several hundred feet above the bridge. This case is complicated, however, by the effect of the sharp bend in the river at that point.

The records left by the flood are not sufficiently definite, or uniform, or free from complications to permit of a free determination of the approach channel made by the river itself. As a guide to the shape of the cross section of the approach channel, it may be said that any section of the channel directly above any span of an arch bridge should have an area at least equal to the cross sectional area of such span opening. If this is obtained there will be required no deflection of the stream necessary beyond the limits of any one span.

Slope

The effective slope of a channel in carrying flood flow is not the slope of the bottom of the channel, but the slope of the water surface. If frequent changes are made in the bottom grade they will not be followed closely by the slope of the water surface, and hence cannot be relied upon in calculating maximum discharging capacity. In the interest of simplicity, and for the purpose of securing the highest degree of natural maintenance of the channel during ordinary low water conditions, the slope of the bottom of the channel should be kept as uniform as possible.

Allowable Velocities

An important factor entering into the design of the channel improvements was the maximum velocity which could be safely allowed in the channels. Because of the high cost of acquiring land for right of way along the banks of the rivers through the cities, it was desirable to limit the cross section by using high velocities. Moreover, to secure larger openings at existing bridges required that the bottom be excavated between the piers. If considerable enlargement at the bridges should be required, underpinning of the piers might be necessary, a very expensive operation.

To determine the maximum safe velocities for the improved channels, a number of investigations were made. These may be divided into three classes: (a) Results of observations on the 1913 flood; (b) Results of current meter measurements; (c) Miami River cut-off channel observations. The results of these observations are given in table XVIII of the Appendix.

The methods of determining the discharges during the 1913 flood, from which velocities were computed, are given in detail in another of the Technical Reports,* and will therefore not be repeated here. While evidence of scour was apparent at but a few of these bridges, it is possible that during the period of maximum velocities some scour of the bed may have occurred, which was filled again by the lower velocities during the latter part of the flood. The measurements of the second class were made with a current meter, their primary purpose being the determination of discharges at high stages of the river. The observations on the Miami River cut-off are described fully in Part IV, of the Technical Reports. There, also, velocities were determined by means of a current meter.

While the evidence is somewhat conflicting, because of the many variable factors which enter into scouring, it appears that in case of coarse gravel, like that forming the bed of the Miami River, if the bridge piers are of good design, and have reasonably deep foundations, temporary maximum velocities of 10 feet per second are safe. Where velocities higher than this were found necessary in the flood control plans, the foundations are protected by a line of steel sheet piling across the river just below the piers. The estimated maximum velocity in any of the bridge openings, for a runoff forty per cent greater than that of March 1913, is 13.2 feet per second at the High-Main Street bridge in Hamilton.

The greatest velocity which was assumed to be safe in a straight channel was also 10 feet per second. This is considerably above the velocity which caused movement in the Miami River cut-off. The conditions at that point were extremely favorable to erosion, however, and the results are hardly applicable to the conditions existing in an improved channel. Where the mean velocity will exceed 10 feet per second, the sides of the channel are paved with concrete slabs, and the bottom for some distance out from the bank is covered with a flexible concrete block revetment. Where the mean velocity exceeds 6 feet per second at bends, this protection also is placed on the outside bank.

BACKWATER CURVES

In the design of the channel improvement through the cities a large number of backwater curves were worked out to determine the

* Calculation of Flow in Open Channels, by Ivan E. Houk. Technical Reports. Part IV; The Miami Conservancy District, Dayton, Ohio. 1918.

flow line for various discharges and channel conditions. The situation at some of the cities of the District, however, particularly at Hamilton, was unusual, and necessitated the use of a slightly different form of computation from the usual one for backwater curves. In this computation the velocity head at the various sections of the stream was considered, as well as the effect of friction.

Under ordinary conditions of flow in channels, when the velocity is low and the cross section is nearly uniform, the changes of velocity head are so slight that they may be safely neglected. With high velocities, however, a slight change in cross section causes relatively a much greater change of velocity head. At Hamilton the velocities of flow in the improved channel are high and the changes of cross section considerable. To neglect the velocity head in this case would give rise to considerable error. This has been pointed out by I. P. Church in the Engineering Record of August 9, 1913, Vol. 68, page 168.

Figure 111 shows the profiles of the water surface for a flow of 200,000 second feet in the improved channel: (a) with velocity head considered; and (b) with velocity head changes neglected. In the lower part of the channel the cross section is uniform and the profiles are identical. Farther up, however, changes of cross section occur and the profiles diverge, the greatest difference (1.35 feet) being at the upper end, where water enters the improved channel. Because of the large cross section above this point, the water flows at low velocity and to increase this to the high velocity of the improved channel requires the expenditure of considerable head.

In tables 14 and 15 the computations for the two profiles are given. While the method of computing backwater curves is familiar to most engineers, in order to furnish a guide to those who are not acquainted with the process, as well as to illustrate the difference between the two forms, the computations for both assumptions are worked out in detail.

'The channel is divided into a number of sections, and the elevations of the water surface at the end of each of the sections is determined with the use of Kutter's formula, by a cut and try process, beginning with an assumed surface elevation at the lower end of the section farthest downstream. The first step is to estimate the elevation of the upper end of this section or the slope of the water surface in it. The discharge which would take place in the section is then computed, different elevations or slopes being tried until the discharge obtained is equal to the flow for which the backwater curve is desired. The computations may conveniently be arranged as in the

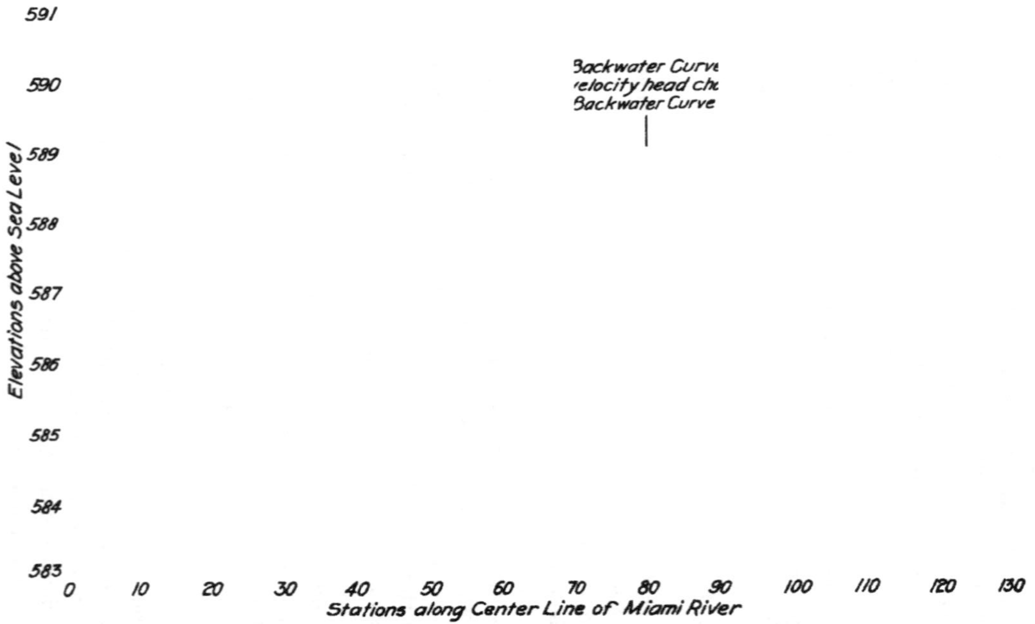

FIG. 111.—BACKWATER CURVES FOR MIAMI RIVER AT HAMILTON, OHIO.

The full line shows the computed profile taking velocity head into consideration. The dotted line shows the profile with velocity head changes neglected. Both profiles are for the improved channel assuming a discharge of 200,000 second feet. The depth of water, over 25 feet, makes it impossible to show river bottom elevations; the profile, therefore, shows water surface elevations only

Table 14.—Computations of Backwater Curve for 200,000 Second foot Flood at Hamilton with Velocity Head Changes Considered

Sta.	Length of Section Ft.	Surface Elevation Ft.	Area, Sq. Ft.	Velocity, Ft. Per Sec.	Vel. Head, Ft.	Change Vel. Head, Ft.	Elev. Head, Ft.	Friction Head, Ft.	Friction Slope, Ft.	Perimeter, Ft.	Hyd. Radius, Ft.	Ave. Rad., Ft.	C.	Ve-locity, Ft.	Average Area, Sq. Ft.	Discharge, Sec. Ft.
125		583.50	17,750	11.27	1.97	0.00	0.50	0.50	.00050	670	26.5	26.5	98	11.28	17,750	200,000
115	1000	584.00	17,750	11.27	1.97	0.00	0.50	0.50	.00050	670	26.5	26.5	98	11.28	17,750	200,000
105	1000	584.50	17,750	11.27	1.97	0.00	0.50	0.50	.00050	670	26.5	26.5	98	11.28	17,750	200,000
95	1000	585.00	17,750	11.27	1.97	0.00	0.50	0.50	.00050	670	26.5	26.5	98	11.28	17,750	200,000
85	1000	585.50	17,750	11.27	1.97	0.00	0.50	0.46	.00046	670	26.5	26.5	98	11.28	18,200	200,000
75	1000	586.15	18,650	10.72	1.78	-0.19	0.65	0.46	.00046	700	26.8	26.7	99	10.98	18,320	200,000
65	1000	586.47	17,995	11.11	1.92	-0.14	0.32	0.49	.00098	680	26.5	26.8	99	10.99	17,030	201,000
60	500	586.47	16,070	12.45	2.41	+0.49	0.00	0.38	.00076	610	26.3	26.4	98	15.76	15,625	209,000
*55	500	586.67	15,180	13.17	2.69	+0.18	0.20	0.31	.00062	590	25.7	26.0	98	13.77	15,645	215,000
*55	500	586.60	15,220	13.15	2.69	+0.18	0.13	0.32	.00064	590	25.8	26.1	98	12.47	15,650	195,000
55	500	586.02	15,230	13.13	2.68	+0.17	0.15	0.28	.00056	575	25.8	26.1	98	12.65	15,700	198,000
*50	500	587.20	1,660	12.38	2.38	-0.30	0.58	0.34	.00068	575	28.1	27.0	98	12.05	15,710	189,200
50	500	587.25	15,085	12.35	2.37	-0.29	0.63			575	28.2	27.0	98	12.78		201,000
Bridge																
*50		587.25	15,210	13.15	2.69	+0.37	0.00	0.40	.00080	575	28.5	27.4	97	14.36	15,770	226,000
50	500	587.57	6,350	12.23	2.32	+0.38	0.67	0.70	.00070	575	26.4	26.3	97	13.30	15,100	201,000
45	500	587.63	1,450	12.21	2.31	+0.40	0.50	0.79	.00079	545	26.2	25.1	97	13.65	14,690	201,000
35	500	588.30	15,060	13.20	2.71	-0.03	1.50	0.50	.00050	660	24.1	25.3	98	11.03	15,920	175,500
25	500	588.30	4,320	13.27	2.74	+0.29	1.67	0.64	.00064	660	26.5	25.4	98	12.50	15,970	199,500
*15	500	590.30	17,510	13.97	3.03	+1.00	0.58	0.53	.00076	670	26.7	26.7	97	13.81	17,750	245,000
15	700	590.47	17,620	11.42	2.03	+1.03	0.35	0.33	.00047	670	26.4	26.6	98	10.95	17,670	193,500
*8		591.05	17,870	11.35	2.00	-0.05	0.38	0.35	.00050	670	26.5	266	98	11.30	17,680	200,000
*8		590.82	17,720	11.20	1.95	-0.02	1.35	0.42								
8	300	590.85	17,740	11.29	1.98	-0.03										
*5		592.25	24,450	11.27	1.97	-0.93	1.05	0.14	.00047	910	26.6	28.6	99	11.09	20,960	232,000
5		591.90	24,170	8.18	1.04	-0.91	1.02	0.11	.00037	910	26.5	26.5	99	9.80	20,940	205,000

* Trial computation.

Table 15.—Computations of Backwater Curve for 200,000 Second foot Flood at Hamilton with Velocity head changes neglected

Sta.	Length of Section, Feet	Water Surface Slope	Water Surface Elevation, Ft.	Area, Sq. Ft.	Wetted Perimeter, Ft.	Hydr. Radius, Ft.	Av. Hydr. Radius, Ft.	C.	Velocity Ft. per Sec.	Ave. Area, Sq. Ft.	Discharge, Sec. Ft.
125	1000	.00050	583.50	17,750	670	26.5				17,750	200,000
115	1000	.00050	584.00	17,750	670	26.5	26.5	98	11.28	17,750	200,000
105	1000	.00050	584.50	17,750	670	26.5	26.5	98	11.28	17,750	200,000
95	1000	.00050	585.00	17,750	670	26.5	26.5	98	11.28	17,750	200,000
85	1000	.00050	585.50	17,750	670	26.5	26.5	98	11.28	18,140	200,000
75	1000	.00048	585.98	18,530	680	26.5	26.5	98	11.05	18,260	201,000
65	1000	.00048	586.46	17,990	610	26.5	26.6	98	11.05	17,110	202,000
60	500	.00064	586.73	16,230	590	26.6	26.5	98	11.72	15,900	201,000
*55	500	.00058	587.02	15,580	590	26.4	26.5	98	12.15	15,900	193,000
55	500	.00062	587.04	15,590	590	26.4	26.5	98	12.56	15,910	200,000
†50	500	.00060	587.34	16,230	575	28.2	27.3	98	12.55	15,910	200,000
Under bridge				15,190	575	28.6					
Above bridge				16,430	575	26.7					
45	500	.00060	587.73	15,360	575	26.5	27.7	98	13.17	15,900	200,000
*35	1000	.00062	588.03	15,250	595	26.6	26.6	98	12.64	15,300	201,000
35	1000	.00066	588.65	15,270	660	24.6	26.6	98	12.60	15,310	193,000
25	1000	.00071	588.69	14,650	660	26.2	25.6	98	12.98	14,960	199,000
*15	1000	.00064	589.40	17,320	595	26.3	25.4	98	13.21	15,980	198,000
15	1000	.00052	590.04	17,340	660	26.0	25.4	98	12.10	15,990	193,000
8	700	.00036	590.40	17,440	670	26.0	26.1	98	12.50	17,390	200,000
*5	300	.00039	590.51	22,980	910	25.2	25.6	98	11.42	20,210	198,000
5	300		590.52	22,990	910	25.2	25.6	99	9.48	20,210	191,000
									9.88		200,000

* Trial computation.
† Immediately below High-Main St. bridge.

examples given, the sequence of the steps being the same as the arrangement of columns from left to right. Some error probably will be made in the estimate of the surface elevation at the lower end of the channel, but as the computations are carried further upstream the effect of this error becomes less and less. To secure an accurate profile, it is therefore necessary to begin some distance below that portion of the channel for which accuracy is desired. For the profiles at Hamilton, the elevation assumed at station 125 was determined by previous calculations.

In the computations given, the hydraulic radius of the section was assumed to be the mean of the hydraulic radii of the two end cross sections. Another assumption which would give possibly as good results is that the hydraulic radius for the section is equal to the mean of the two end areas divided by the mean of the wetted perimeters of the end sections. The number of steps in the computation is the same for either of these assumptions.

In considering the changes of velocity head it was assumed that in no case could the decrease in velocity head cause the surface to slope upward in the direction of flow. In sections 65–60 and 50–45, therefore, where the change in velocity head is more than sufficient to overcome friction, the surface slope was assumed to be zero.

The method of computing the drop due to the effect of the bridge is also illustrated. This is based on the assumption that the surface, as the water enters the bridge openings, drops to the elevation of that just below the bridge, and the heading up which occurs is just sufficient to increase the velocity of the water approaching the bridge to the higher velocity which exists under the bridge, and to overcome the friction through the bridge. None of this head is assumed to be recovered as the velocity of the water is again reduced in the section below the bridge. It is believed that this method of computation gives results which are amply conservative.

CHAPTER XII.—BALANCING THE FLOOD PROTECTION SYSTEM

PRINCIPLES AND DATA

The term balancing as applied to the flood-control system of the Miami Conservancy District refers to the simultaneous adjustment of the heights of the various dams and the amounts of channel improvements at different cities for the purpose of securing the desired protection at minimum cost. In the preliminary studies for the system this operation was carried out on several occasions, each time with increasing accuracy, as the data from surveys and the details of the design of the dams and outlets became more complete. The earlier balances sufficed to eliminate from consideration a number of the numerous basin sites which were investigated at the beginning of the studies, and established in a general way the principal dimensions required for the structures of the remaining basins. Certain of the proposed reservoirs, however, could not be definitely eliminated until a relatively advanced stage in the design had been reached, while a considerable degree of uncertainty as to the economic height of dam to be developed at the Taylorsville and Huffman sites was not removed until the final balance for the entire system was completed.

The method used was largely a cut-and-try process, in which a large number of different possible combinations were tried out, their cost computed, and the most economical combination chosen. Later a graphical method was developed by which it was possible to proceed directly to the most economical system of basins and channel improvements. By this method it is possible to determine which dams to use. the height and size of each, and the capacity of each of the channels at the cities, so as to give the desired degree of protection at all points with a minimum total cost.

This method of balancing is based on the assumption that all the factors entering into the design of the system can be evaluated on a dollars and cents basis. For the flood prevention works themselves this can be done with comparative accuracy since it is largely a matter of quantities and unit costs. For some of the other factors, as for example, the purchase of right of way or of town sites, the element of judgment enters to a large extent; while the importance of opposition to the flood prevention plan and the extent to which it may be advisable to go to mitigate this opposition are even less capable of

286

exact determination. In any method of balancing, however, these factors in reality are considered, and, consciously or unconsciously, they are balanced on a dollars and cents basis. The balancing method does not eliminate the necessity for engineering judgment; it is essentially, as its name implies, a method of balancing the various factors after their relative weights have been determined, dollars and cents being the unit in which all these weights are expressed. This method enables the engineer to proceed in a logical way to a result which he would otherwise attempt to reach in a more or less haphazard fashion.

One of the principal virtues of this graphical method of balancing is the aid it gives one in grasping the complex interrelation of the various parts of a retarding basin system. These relationships are so varied and involved that to grasp them without the aid of a method such as this, is an exceedingly difficult undertaking. By means of these diagrams the important relations are emphasized, while the less important ones are kept in the background, and the likelihood of placing false emphasis upon a particular relation is thus greatly lessened.

GRAPHICAL BALANCING OF THE MIAMI CONSERVANCY DISTRICT SYSTEM

As the purpose of this chapter is to explain the graphical method of balancing a retarding basin and channel improvement system, rather than to give in detail its application to the Miami Conservancy District system, only those retarding basins which were finally selected will be considered, and the method by which others were eliminated will be briefly indicated. The assumptions of degree of protection required and time of flood peaks will be the same as those used in the design of the Miami system. For instance, the retarding basins will be assumed to have inflows of 9½ or 10 inches as previously discussed, while the runoff from the areas below the basins will be the same as used in the preceding chapter.

LOCKINGTON COST-OUTFLOW CURVE

The method of attack is to begin with the improvement furthest upstream, in this case the Lockington basin. This basin affects the flow at all the cities below it, and to determine its size, therefore, requires that its effect at all these points be considered. For a given inflow, as, for example, the 10 inches in 3 days adopted for this basin, the smaller the conduit used, the smaller will be the outflow rate, and the greater will be the volume of water retained in the basin. To retain a larger volume of water would increase the cost of the basin, since it would not only flood more land but would also require a higher and more expensive dam. The smaller the outflow from the basin, therefore, the greater will be its cost. Assuming that no water is allowed to escape over the spillway, we may construct a curve, figure 112, showing the relation between the maximum rate of outflow

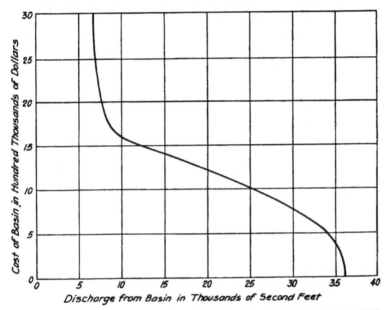

FIG. 112.—RELATION BETWEEN COST AND MAXIMUM DISCHARGE CAPACITY OF LOCKINGTON RETARDING BASIN, FOR A 10-INCH STORM RUNOFF IN 3 DAYS.

from the Lockington Basin for this 10-inch inflow in 3 days and the total cost of the basin, including the cost of the dam, outlets, spillway, land damages, etc. For the Official Plan flood, if no dam were built, the discharge at the dam site would be 140 per cent of the 1913 discharge, or 36,000 second feet. The curve, therefore, begins where the coordinates represent the cost as zero and the discharge as 36,000 second feet.

When the discharge becomes small, several factors enter to make the cost increase very rapidly. One of these is the slow emptying of a basin with small conduits. With very small outflows the length of time necessary for a basin to empty is very great, and the possibility of another storm coming before the basin is empty requires that additional storage capacity be provided to cover this contingency. This larger capacity would also be required in the case of floods of great duration. It has already been shown in chapter IV that, other things being equal, basins with small outflows have less protection against floods of great duration than do those from which the discharge is large. The smaller conduits also cause greater damage to the land in the basin from flooding. With small discharges, not only is the area flooded by a given storm greater, but both the frequency and duration of flooding, especially in the lower part of the basin, are greatly augmented. The cost of measures for protecting small conduits from clogging also increases the cost of basins having small outflows. For these reasons the cost of a basin with small outflows increases very rapidly as the capacity of the outlets is decreased.

COMBINING LOCKINGTON BASIN WITH PIQUA AND TROY IMPROVEMENTS

The second step in the balancing process is to combine the cost of the Lockington dam and the channel improvements at Piqua and Troy. As was shown in the preceding chapter, the maximum runoff rate from the total drainage area of the Miami River above Piqua, except that portion tributary to the Lockington basin, for the Official Plan flood, would be 80,800 second feet. At Troy it would be 98,000 second feet. To protect these cities against such a flood, it is necessary to provide them with channels of sufficient capacity to take care of these discharges and, in addition, the discharge from the Lockington basin. For example, if the Lockington basin were built to discharge, at maximum stage, 20,000 second feet, a channel at Piqua of 81,000 + 20,000 = 101,000 second feet capacity would need to be provided, and that required at Troy would be 98,000 + 20,000 = 118,000 second feet capacity. Corresponding to each discharge from the Lockington basin there is, therefore, a definite channel capacity required at Piqua and Troy, entailing a certain cost for channel improvement at these places.

A curve ($A\ B$ in figure 113) can therefore be constructed giving, for each value of the discharge from the Lockington basin, the combined cost of the improvements at Piqua and Troy which are necessary to provide channels of the required capacity. As stated, when the Lockington discharge is 20,000 second feet, the channel capacity required is 101,000 second feet at Piqua and 118,000 at Troy. The cost of these two improvements was estimated to be $830,000. The curve $A\ B$ is then combined with the Lockington cost-outflow curve $C\ D$, by adding the ordinates of the two curves for each value of the abscissa. The resulting curve $E\ F$ represents the total cost of the

19

Lockington basin and the channel improvements at Piqua and Troy, for various discharges from the Lockington basin. Using the previous example, if the Lockington discharge is 20,000 second feet, the cost at Piqua and Troy would be $830,000 and at Lockington $1,-210,000, giving a total combined cost of $2,040,000.

It will be noticed that this combined cost decreases as the discharge from the Lockington basin increases, and becomes least when the discharge from this basin is 36,000 second feet. This discharge

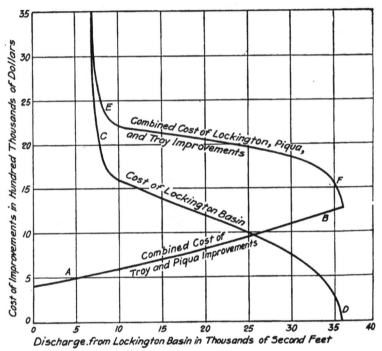

FIG. 113.—DETERMINING THE MOST ECONOMICAL COMBINATION OF LOCKINGTON RETARDING BASIN WITH CHANNEL IMPROVEMENTS AT PIQUA AND TROY, FOR VARIOUS MAXIMUM OUTLET CAPACITIES AT THE DAM.

would occur if no dam at all were built at Lockington, and it is, therefore, evident that, if only Troy and Piqua were to be protected, this could most cheaply be done by channel improvement without the aid of the Lockington basin. It will be shown later, however, that because this basin, while assisting in the protection of Piqua and Troy, also reduces the flow through the cities below, its use as a part of the whole system is economical.

COMBINING THE LOCKINGTON AND TAYLORSVILLE BASINS

The next step is to combine the effect of the Lockington and Taylorsville basins. With a given maximum discharge through the Taylorsville dam, for example 60,000 second feet in the Official Plan flood,

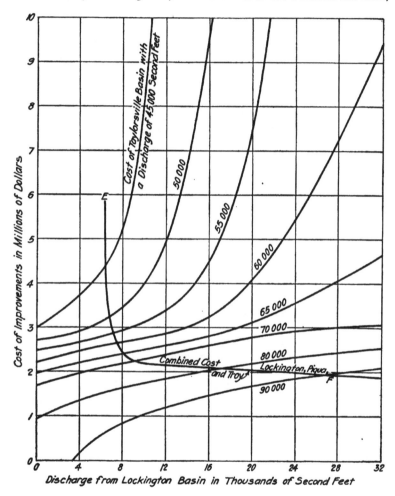

FIG. 114.—COST OF TAYLORSVILLE RETARDING BASIN FOR VARIOUS MAXIMUM OUTLET CAPACITIES AT LOCKINGTON DAM.

the water detained in the retarding basins may be divided between the Lockington and Taylorsville basins in varying proportions. If the Lockington outflow is large, with a relatively small storage in the Lockington basin, the capacity of the Taylorsville basin will have to be correspondingly increased, requiring a correspondingly high and expensive dam. If, on the contrary, a larger amount be retained at Lockington, the dam at Taylorsville may be lower, preserving the same maximum discharge through its outlet conduits by enlarging their dimensions. The latter case would not require so expensive a dam at Taylorsville, but would require a greater expenditure at Lockington. By the process described in previous chapters, the heights of the two dams required in each of these cases may be determined, and the cost of each of the basins may then be estimated. When the cost of the basin at Taylorsville is thus determined for a given discharge at maximum stage of 60,000 second feet, for a number of different quantities stored at Lockington, a curve may be plotted representing the relation between the cost of the Taylorsville basin and the corresponding storage which is necessary at Lockington. For each quantity retained with the 10-inch inflow at Lockington, however, there is a definite maximum outflow rate for which the conduits must be designed in order that the basin may retain this quantity of water. A curve may, therefore, be constructed showing the cost of the Taylorsville basin, with a maximum outflow of 60,000 second feet, for various maximum discharges from the Lockington basin.

Figure 114 shows a series of such "cost of Taylorsville-Lockington discharge" curves, for maximum discharges at Taylorsville ranging from 45,000 to 90,000 second feet. In all cases the cost of the Taylorsville basin increases as the discharge from the Lockington basin is increased. This is to be expected, since the greater the discharge at Lockington, the smaller will be the quantity of water retained there, and the greater will be the storage and cost required at Taylorsville. For the smaller discharges from Taylorsville, the cost increases rapidly as the discharge from Lockington is increased. This is caused by the rapid increase in the cost of the Taylorsville basin for height of water surface above 818 on account of the objection which would be met in Troy to flooding above this elevation.

THE LOCKINGTON-PIQUA-TROY-TAYLORSVILLE OR L. P. T. T. SYSTEM

On figure 114, in addition to the curves showing the relation between the cost of the Taylorsville basin and the discharge of the Lockington basin, is curve *E F*, developed in figure 113, showing the relation of the combined cost of the Lockington basin and the channel improvements at Piqua and Troy to the discharge of the Lockington basin. By combining this curve E F with the curves for the various discharges of the Taylorsville basin, by the same process of adding the ordinates as used in figure 113, a series of curves shown in figure 115 is obtained. These show the relation of the total cost of the Lockington and Taylorsville basins, and the Piqua and Troy channel improvements, to various discharges from the Taylorsville and Locking-

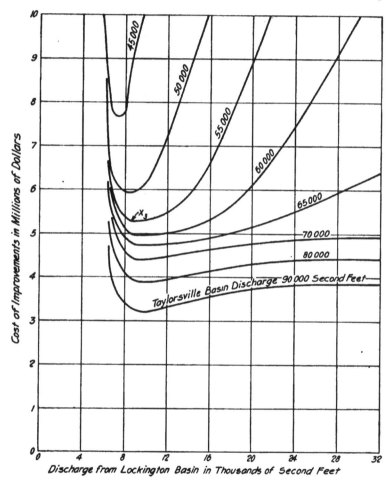

FIG. 115.—DETERMINING THE MOST ECONOMICAL COMBINATION OF LOCKINGTON AND TAYLORSVILLE RETARDING BASINS WITH CHANNEL IMPROVEMENTS AT PIQUA AND TROY, FOR VARIOUS MAXIMUM OUTLET CAPACITIES AT THE LOCKINGTON AND TAYLORSVILLE DAMS.

ton basins. For example, figure 114 shows that if the maximum discharge from the Taylorsville basin is 60,000 second feet, and that from the Lockington basin is 20,000 second feet, the cost of the Taylorsville basin would be $4,100,000; and curve *E F* shows that for the 20,000 second foot discharge from Lockington, the cost of the Lockington, Piqua, and Troy improvements is $2,040,000. The combined cost of the two basins and two channel improvements for these discharges is, therefore, $4,100,000 + $2,040,000 = $6,140,000 as shown in figure 115. For brevity, in the following explanation, the combination of the improvements at Lockington, Piqua, Troy, and Taylorsville will be called the L. P. T. T. system. Figure 115 shows that for each discharge from the Taylorsville basin there is a discharge from the Lockington basin for which the total cost of the four improvements is a minimum. These discharges at Lockington all are between 7,000 and 10,000 second feet, which shows that for all discharges at Taylorsville it is economical to have a high dam and small discharge at Lockington.

From the minimum points on the curves of figure 115 it is possible to construct a curve shown in figure 116 giving the minimum cost of

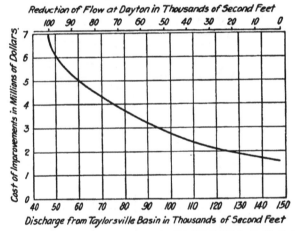

FIG. 116.—RELATION BETWEEN MINIMUM COST OF L. P. T. T. SYSTEM AND MAXIMUM OUTLET CAPACITY AT TAYLORSVILLE DAM.

the L. P. T. T. system for the various discharges from the Taylorsville basin. Thus, for a discharge at Taylorsville of 60,000 second feet, figure 115 shows that the minimum combined cost of the L. P. T. T. system would be about $5,000,000, and in figure 116 the value of $5,000,000 is platted at the abscissa 60,000 second feet. The curve, in figure 116, is drawn to the point where the Taylorsville discharge is 147,000 second feet (140 per cent of the 1913 discharge), which would occur when the height of both Taylorsville and Lockington dams is zero. The total cost of the improvement would then be about $1,550,000, the cost for the protection of Piqua and Troy from a storm 140 per cent as large as that of 1913, by channel improvement alone.

COMBINING THE IMPROVEMENTS ABOVE DAYTON

The next step in the balancing process is to combine the Huffman and Englewood basins with the L. P. T. T. system, to obtain the combination which will give the desired reduction of flood flow at Dayton and below, with a minimum cost. The discharge at Dayton for a flood 140 per cent as large as that of March 1913 would be 350,000 second feet. Allowing 12,000 second feet for the runoff from the drainage area between the city and the dams, and dividing the remaining 338,000 second feet between the basins in proportion to their actual 1913 discharges, gives the discharges effective at Dayton for the various damsites in a flood 40 per cent larger than that of 1913 as

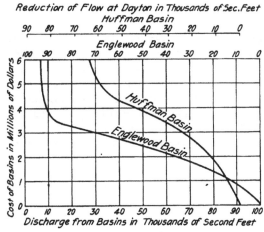

FIG. 117.—RELATION BETWEEN COST AND MAXIMUM DISCHARGE CAPACITIES OF THE HUFFMAN AND ENGLEWOOD RETARDING BASINS FOR A 10-INCH STORM RUNOFF IN 3 DAYS.

This diagram is essentially similar to that for the Lockington basin in figure 112, but in addition indicates the amount by which the maximum river discharge at Dayton, in the assumed Official Plan flood, will be reduced by each basin for different outlet capacities.

follows: Taylorsville 147,000 second feet, Huffman 91,000 second feet, and Englewood 100,000 second feet. For the Englewood and Huffman basins cost-discharge curves (figure 117) are plotted similar to that for the Lockington basin (figure 112).

While these cost-discharge curves show the relation between the cost of the basins and the discharge which they would permit to flow through Dayton, they also show the relation between the cost of the basins and the reduction which they would cause in the flow at Dayton. For example: if the Englewood basin were not constructed, the flow at Dayton from its drainage area would be 100,000 second feet. If it were built so that its discharge was 40,000 second feet, the reduction which it would cause would be 100,000 — 40,000 = 60,000 second feet, indicated at top of figure 117. These cost-outflow curves may also, therefore, be interpreted as cost-reduction curves, by considering the origin to be at the discharge corresponding to dams of zero height or basins of zero cost, and the magnitude of the reduction to be represented by the distance from this point measured to the left, as shown by the two upper coordinate scales.

Figure 118 shows the effect of the operation of these basins on the Dayton discharge. If no basins were built, the discharge through the city would be 350,000 second feet, and the cost of the improvements above would be the $1,550,000 required for the protection of Piqua and Troy by channel improvement. Suppose it is desired to reduce the discharge at Dayton 50,000 second feet, or to 300,000 second feet, at the same time protecting Piqua and Troy. This might be done by building the channel improvements at Piqua and Troy and reducing the flow at Dayton 50,000 second feet by building the Huffman dam. The cost of the Piqua and Troy improvements would be $1,550,000 and to build the Huffman basin with sufficient capacity to reduce the flow at Dayton 50,000 second feet, is shown by the cost-reduction curve of this basin (figure 117) to be $4,250,000. The total cost is therefore $5,800,000. The total cost of this combination for any desired reduction at Dayton can easily be determined by drawing the cost-reduction curve for the Huffman basin as shown in figure 118 with its origin at the point 0, the ordinate of which is $1,550,000, or the cost of protecting Piqua and Troy. The ordinate of any point on the Huffman curve in this position represents the combined cost of these improvements for the reduction of flow at Dayton represented by the abscissa of that point. For a reduction of flow at Dayton of 50,000 second feet, corresponding to an actual flow of 300,000 second feet, the combined cost is shown by this curve to be $5,800,000 as previously computed.

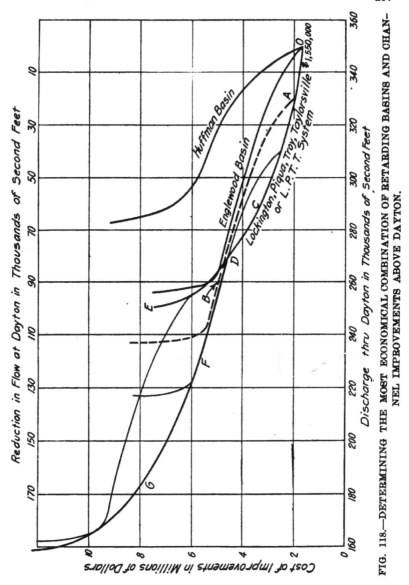

FIG. 118.—DETERMINING THE MOST ECONOMICAL COMBINATION OF RETARDING BASINS AND CHANNEL IMPROVEMENTS ABOVE DAYTON.

The Englewood basin might be used instead of the Huffman. Applying its cost-reduction curve in the same manner as that of the Huffman basin, the combined cost is shown in figure 118 to be \$3,-930,000, or \$1,870,000 less than with the Huffman basin. A still cheaper method is to use the L. P. T. T. system, as represented by the curve $O\,A\,C\,D\,E$. This curve is made to start from the same point O, since the Piqua and Troy improvements are included in the cost of the system at this point. The protection of Piqua and Troy and a reduction at Dayton of 50,000 second feet by this system is shown by the curve to be $2,850,000.

If it were desired to reduce the flow at Dayton 90,000 second feet, or to a discharge of 260,000 second feet, it is seen from figure 118 that the cost of the L. P. T. T. system would be $5,300,000; while to build the Piqua and Troy channel improvements and reduce the flow at Dayton to 260,000 second feet by means of the Englewood basin would cost the same amount. The cost of using the Huffman dam instead of the Englewood is shown to be prohibitive. For a flow of 262,000 second feet at Dayton the channel improvements at Piqua and Troy, combined with the Englewood basin, would be slightly cheaper than the L. P. T. T. system, while for a discharge of 258,000 second feet the reverse would be the case.

Neither the L. P. T. T. system alone nor the Englewood basin with the Piqua and Troy channel improvements is the cheapest method, however, of securing the protection of Piqua and Troy and a reduction of the flow at Dayton to 260,000 second feet. For example, if the L. P. T. T. system were built so that it would give protection to Piqua and Troy and reduce the flow at Dayton 20,000 second feet, figure 116 shows that the cost would be $1,900,000. Figure 117 shows that the cost of reducing the flow at Dayton 70,000 second feet by the Englewood basin would be $3,000,000. If both these improvements were built, a reduction of 90,000 second feet in the flow at Dayton would be secured at a cost of $1,900,000 + $3,000,000 = $4,900,000, or about $400,000 less than the cost by either of the two methods separately, as previously discussed. This combination is illustrated graphically by placing the origin of the Englewood cost-reduction curve at point A, figure 118, on the line representing the L. P. T. T. system, which point indicates a reduction by that system of the flow at Dayton of 20,000 second feet, the Englewood curve taking the position of the dashed line $A\,B$. The additional reduction of flow at Dayton of 70,000 second feet caused by the Englewood basin, and represented by the horizontal distance from A to B, is thus graphically added to the 20,000 second foot reduction caused by the L. P. T. T. system, while the cost of the Englewood basin to give this reduction ($3,000,000), represented by the vertical distance between A and B, is also graphically added to the cost of the L. P. T. T. system. The total cost of this combination of improvements is, therefore, represented by the ordinate of the point B, while the abscissa of this point represents the discharge through Dayton, or, measuring to the left from O, represents the total reduction which this combination would cause in the flow at Dayton. From the position of point B, the cost for this reduction of 90,000 second feet is shown to be $4,900,000 as determined by the previous method.

The minimum cost, however, for a 90,000 second feet reduction in flow at Dayton by means of the L. P. T. T. system and the Englewood basin is not necessarily secured by combining them in just these proportions. Possibly a reduction of 40,000 second feet by the L. P. T. T. system, and a 50,000 second feet from the Englewood basin would give a smaller total cost. This may be easily determined by moving the origin of the Englewood cost-reduction curve to the point on the L. P. T. T. curve representing a reduction of 40,000 second feet. If the ordinate at B is less than in the previous case, then this combination is cheaper. The combination which will give the 90,000 second foot reduction at a minimum cost can easily be found by moving the origin of the Englewood curve along the curve of the L. P. T. T. system until the ordinate at B reaches a minimum value. The minimum points may be found for other reductions in the same manner and a minimum cost-reduction curve drawn for the combination of the L. P. T. T. system and the Englewood basin. This curve $O A C D F G$ will be the envelope of the system of curves generated by moving the Englewood curve along that of the L. P. T. T. system. The proportion of the reduction which each basin produces in this minimum cost combination may be easily determined by obtaining the position of the two curves which generated the point on the envelope the abscissa of which is the total reduction. Thus for a reduction of 90,000 second feet at Dayton, the point on the envelope is obtained by placing the origin of the Englewood curve at the point on the L. P. T. T. line corresponding to 40,000 second feet reduction. This shows that the minimum cost for a 90,000 second foot reduction is secured by building the L. P. T. T. system to reduce the flow at Dayton 40,000 second·feet and the Englewood basin in such a manner that it will reduce the flow the remaining 50,000 second feet. It will be noticed that for reductions of 80,000 second feet, or less, the envelope is the same as the L. P. T. T. curve, indicating that the cheapest method of securing these reductions is to build only the L. P. T. T. system.

The combination of the Englewood basin and the L. P. T. T. system may now be considered as a unit and the envelope $O A C D F G$ may be combined with the Huffman curve in the same manner as the Englewood curve was combined with that of the L. P. T. T. system,

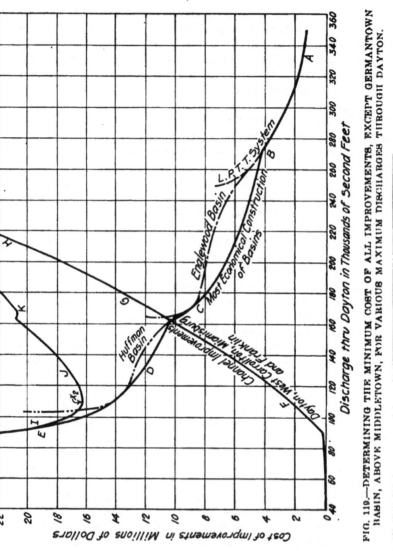

FIG. 119.—DETERMINING THE MINIMUM COST OF ALL IMPROVEMENTS, EXCEPT GERMANTOWN BASIN, ABOVE MIDDLETOWN, FOR VARIOUS MAXIMUM DISCHARGES THROUGH DAYTON.

another envelope, *A B C D E*, figure 119, being obtained which represents the minimum cost of securing the various reductions of flow at Dayton by the combination of all these improvements. When the curves of several basins are combined in this manner, the resulting envelope or curve will be the same, regardless of the order in which they are combined.

COMBINING THE CHANNEL IMPROVEMENTS FROM DAYTON TO FRANKLIN

The next step in the balancing process is to combine the cost of channel improvement at Dayton, West Carrollton, Miamisburg, and Franklin. The method is the same as that used in the case of the improvement at Piqua and Troy. The costs at these four cities are combined into one curve as follows: Table XVII of the Appendix shows that for a discharge at Dayton, above Wolf Creek, of 110,800 second feet, the discharge at West Carrollton would be about 125,000 second feet, at Miamisburg 134,800, and at Franklin 139,600. The cost of the improvement at each of these points for these discharges is totaled, giving in this case $2,475,000, and the total is platted on curve $F G H$, figure 119, with an abscissa of 110,800 second feet, the discharge through Dayton. Similarly, the combined cost of these improvements is computed for other discharges through Dayton, the discharge at the other cities differing from the Dayton discharge by the amount of the runoff from the unreservoired area between, which is a constant for each city. A curve $F G H$ is thus constructed representing the combined cost of these improvements for various discharges at Dayton.

COMBINING ALL IMPROVEMENTS ABOVE TWIN CREEK

Having now a curve representing the minimum cost of the improvements above Dayton for various discharges at that place, and another curve showing the cost of the channel improvements for the four cities above the mouth of Twin Creek for these same discharges at Dayton, it is a simple matter, by the addition of the ordinates of these two curves for the various discharges, to obtain the curve $I J K$, figure 119, representing the cost of the most economical method of protecting all the cities above the mouth of Twin Creek for the different discharges at Dayton. This curve reaches a minimum point at X_2 representing a discharge at Dayton of 111,000 second feet. If no improvements below Franklin were contemplated, the desired protection to the cities could be secured by constructing the retarding basin system so that the discharge through Dayton will be 111,000 second feet. The discharge at West Carrollton would then be about 125,200 second feet, at Miamisburg 135,000, and at Franklin 139,800, as previously shown. Since improvements below Franklin are intended, it does not necessarily follow that the system thus proportioned will give the minimum total cost for the entire system, and it is necessary to investigate further before this can be definitely determined.

THE RELATION BETWEEN THE DISCHARGES AT DAYTON, GERMANTOWN, AND HAMILTON

The method of balancing the system below Franklin is somewhat complicated by the location of the Germantown basin, and the effect of the peak flows from Four and Seven Mile Creeks which empty into the Miami River just above Hamilton. As explained in the previous chapter, there are two flood peaks at Hamilton. The first of these comes early in the flood and is due to the runoff from the area below the retarding basins, principally from Four and Seven Mile Creeks. When the discharge from this area is greatest, the flow through Hamilton from the retarding basins is 45 per cent of their maximum discharge. The second peak is due to the maximum outflow from the basins, and comes near the end of the flood. At the time of this peak the runoff from the unreservoired area is at the rate of 84 second feet per square mile. For the plan adopted it was shown in the previous chapter that at Hamilton the first of these peaks was the largest. We

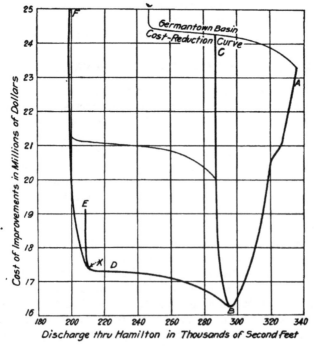

FIG. 120.—DETERMINING THE MOST ECONOMICAL COMBINATION OF GERMANTOWN RETARDING BASIN WITH ALL OTHER IMPROVEMENTS ABOVE MIDDLETOWN.

will therefore start on the assumption that this will be the case in the system determined by the graphical method.

Table XVII of the Appendix shows that for the Official Plan flood the maximum discharge at Hamilton from the area below the four retarding basins would be 159,100 second feet. From the area above the Germantown damsite, if no dam were built, we would expect 140 per cent of 66,000 or 92,000 second feet, giving a total from the area below the Englewood, Taylorsville, and Huffman dams of 251,000 second feet. In the previous chapter it was shown that for this flood the runoff from the unreservoired area above Dayton, at the time the basins were discharging their maximum, would be about 12,000 second feet. With a discharge through Dayton of 111,000 second feet, therefore, the discharge from the basins would be 99,000 second feet. When the unreservoired area above Hamilton, including the whole Twin Creek drainage area, is discharging 251,000 second feet, however, the discharge at Hamilton due to the basins is only 45 per cent of their maximum discharge, or $0.45 \times 99,000 = 44,600$ second feet. A combination of the basins above Dayton which would cause a flow through that city of 111,000 second feet, for a storm 40 per cent larger than that of 1913, would reduce the peak flow at Hamilton to $251,000 + 44,600 = 295,600$ second feet. Similarly, the discharge at Hamilton corresponding to 100,000 second feet at Dayton would be $251,000 + 0.45(100,000 - 12,000) = 290,600$ second feet. A curve $A \ B \ C$ in figure 120 can therefore be constructed showing the relation between the least cost of the improvements above Twin Creek, as determined from curve $I \ J \ K$ in figure 119, and the corresponding discharge which would result at Hamilton.

THE EFFECT OF THE GERMANTOWN BASIN

To investigate the effect of the Germantown basin it is necessary to construct a curve for this basin showing the relation between its cost and the reduction in flow which it would cause at Hamilton. The shape of this curve is also influenced by the effect of the two peaks. As has already been shown, the discharge at Germantown without the dam would be 92,000 second feet. If we build the dam to such a height that its maximum discharge is 10,000 second feet, the flow at the damsite would be reduced $92,000 - 10,000 = 82,000$ second feet. While the flow in the river at Hamilton is largest, however, only 45 per cent of this 10,000 or 4,500 second feet from Germantown is flowing through Hamilton. The reduction in the flow at Hamilton would therefore be $92,000 - 4,500 = 87,500$ second feet. Similarly, if the discharge at the damsite were 20,000 second feet, the reduction at

Hamilton would be $92,000 - 20,000 \times 0.45 = 83,000$ second feet. A curve, $B\ D\ E$ in figure 120, can therefore be constructed on which the horizontal distance, measured to the left from the origin B, is the reduction in the flow at Hamilton caused by the Germantown basin, and the vertical distance above the origin B is the cost of the basin which will give this reduction.

The cost-reduction curve of the Germantown basin is then combined with curve $A\ B\ C$ of figure 120. This may be done by moving the origin of the cost-reduction curve along the curve $A\ B\ C$, and finding the envelope $A\ B\ D\ F$ of the system of curves thus generated. The reasoning in this case is similar to that used in combining the retarding basins above Dayton. For example, if the basins above Dayton were so constructed that the flow at Hamilton is reduced to 295,600, the minimum cost of the improvements above Twin Creek is shown by curve $A\ B\ C$ to be $16,250,000. If in addition the Germantown basin were built so that it would reduce the flow at Hamilton 75,000 second feet, or to 220,600 second feet, the additional cost is represented by the vertical distance of point D above B, while the 75,000-second foot reduction is represented by the horizontal distance of D to the left of B. The minimum cost for which it is possible to secure a given flow at Hamilton is therefore shown by the envelope $A\ B\ D\ F$ of the system of curves generated by moving the cost-reduction curve for the Germantown basin along the curve $A\ B\ C$.

INTRODUCING THE HAMILTON AND MIDDLETOWN CHANNEL IMPROVEMENTS

The envelope $A\ B\ D\ F$ is then combined in figure 121 with the cost curve for the Hamilton channel improvement by a direct addition of ordinates as in figures 113 and 119, giving curve $M\ N\ O$. This curve shows at point X a minimum total cost of $20,000,000 with a discharge through Hamilton of 208,000 second feet. The final step is to introduce the cost of the Middletown improvement. Table XVII of the Appendix shows that the peak at this point, unlike that at Hamilton, results from the maximum discharge of the retarding basins. As the cost of the improvement at Middletown is small in comparison with that at Hamilton, it is not likely materially to change the discharge at Hamilton, which will give the minimum total cost. We may, therefore, compute, as was done in deriving table XVII, the maximum flow at Middletown which will result from the systems of basins which give discharges close to 208,000 second feet at Hamilton. The costs for the Middletown and Hamilton improvements may then be added, at their respective discharges at Hamilton, to the curve $M\ N\ O$,

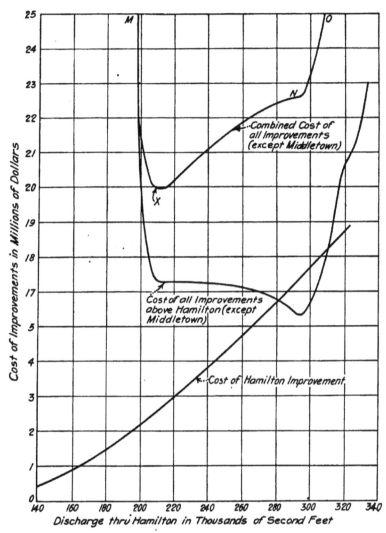

FIG. 121.—DETERMINING THE MOST ECONOMICAL COMBINATION
OF THE FIVE RETARDING BASINS AND ALL CHANNEL IMPROVE-
MENTS EXCEPT THAT AT MIDDLETOWN, FOR VARIOUS DISCHARGES
AT HAMILTON.

and another curve drawn through the points thus obtained, which
will give a point differing only slightly from X, representing the
minimum total cost of the entire system of basins and channel improvements.

20

It will be remembered that this minimum cost was based upon the assumption that the peak caused by the runoff from the unreservoired area was the larger of the two peaks at Hamilton. Having determined, by the method shown in the following paragraphs, the size of the basins which will give this minimum cost, it is necessary to check up to see if the relative size of the two peaks agrees with the original assumption. A rigid proof would require that the balancing be carried out for the other assumption also, namely, that the maximum discharge from the basins might cause the larger peak at Hamilton. For the Miami system it would be found that with the combination of improvements which would give the minimum cost with this latter assumption, the runoff from the unreservoired area would still cause the greatest discharge at Hamilton. In some cases the combination determined by the second assumption might be such that this assumption would also be justified and the most economical combination would be the one of the two systems determined with these assumptions which gave the lowest minimum cost.

DETERMINING THE SIZE OF THE VARIOUS IMPROVEMENTS OF THE CHEAPEST COMBINATION

Having found the discharge at Hamilton which would produce the minimum total cost of the system, it now remains to follow back through the development and determine the size of the various dams and channel improvements in the system which reduce the flow at Hamilton to this amount. Figure 120 shows that the most economical method of securing 208,000 second feet discharge at Hamilton is to build the dams above Dayton so that they will reduce the flow at Hamilton to 295,600 second feet, and to secure the balance of the reduction i. e. 87,600 second feet, by building the Germantown dam. From the discussion on page 303, it will be seen that this corresponds to a maximum discharge of $(92,000 - 87,600) \div 0.45 = 9.800$ second feet at the Germantown dam, or practically that given by the Official Plan.

From the relation between the Dayton and Hamilton discharges previously discussed, it will be seen that the flow at Dayton from the system of basins which will give a discharge of 295,600 second feet at Hamilton, assuming Germantown dam not built, would be 111,000 second feet. The Dayton channel, above Wolf Creek, should therefore have a capacity of 111,000 second feet, and the capacities for Miamisburg, Franklin, and Middletown channels can be computed as in table XVII of the Appendix.

The next step is to determine what sizes of the retarding basins above Dayton gave the least combined cost for a flow of 111,000 second feet at Dayton. To do this it is necessary to determine what combination of basin curves produced point X_2, at 110,000 second feet on the envelope $A\ B\ C\ D\ E$ of figure 119. This is shown by the dashed lines representing the position of the three basin curves which produced this point. These dashed lines show that the L. P. T. T. system should be built to effect a reduction of $350,000 - 258,000 = 92,000$ second feet, the Englewood dam a reduction of $258,000 - 170,000 = 88,000$ second feet, and the Huffman dam a reduction of $170,000 - 111,000 = 59,000$ second feet. Since, as shown on page 295, the discharges, with no reduction at these points, are at Taylorsville 147,000, at Englewood 100,000, and at Huffman 91,000 second feet, the above reductions correspond to outflows of 55,000 second feet at Taylorsville, 12,000 second feet at Englewood, and 32,000 at Huffman. These values, with the exception of that at Huffman, were those adopted in the Official Plan. The difference of 3000 second feet, at Huffman, between the discharge obtained by the graphical method and that from the plan adopted, is introduced by the assumption made in the computations for the plan, as shown in table XVII of the Appendix, that only 99,000 of the 103,400 second feet discharge from the basins would be effective at Dayton. In order to avoid complications, this assumption was not considered in the graphical balancing.

The discharge at Taylorsville, for the cheapest system, was found to be 55,000 second feet. Figure 115 shows that, with this discharge, the minimum cost of the L. P. T. T. system occurs at X_3 when the discharge at Lockington is 9000 second feet, which is the figure adopted in the Official Plan. The size of channels required at Piqua and Troy are then computed as shown on page 289.

THE ELIMINATION OF UNDESIRABLE BASINS

The method of eliminating the undesirable basins is to determine the cost of the improvement on the assumption that they will be used. If the cost of the entire system, including a given basin, is more than that of some other combination without that basin, then that particular basin is clearly uneconomic. For example, to determine the advisability of building the Port Jefferson basin on the Miami above Piqua, it would be necessary to combine the effect of this basin with the improvements at Lockington, Piqua, Troy, and Taylorsville, forming a system similar to the L. P. T. T. group described. If, by using this in place of the L. P. T. T. system, a greater minimum cost of the entire system was obtained, the use of the Port Jefferson basin would be undesirable. In the same way, two basins on the Mad or

Stillwater rivers might be combined, as were the Lockington and Taylorsville basins. By using these combinations in place of the single basin on these streams, the desirability of their use could be determined. Usually, it is not necessary to carry the balancing entirely through to the end to eliminate a basin, as it is often evident at some intermediate step that the basin is undesirable.

CHAPTER XIII.—ALTERNATIVE FLOOD PROTECTION PLANS

In the foregoing chapters the design of the protection works for the Miami Valley has been discussed in detail, with only a brief mention of the various alternative plans which were investigated (see chapter III). While the problem was approached with ideas of what seemed to be the most likely solution, there was also a determination to find the best method, whatever that might be. The acceptance of this principle led to a complete change of view, and the plan which at first seemed to be most practical was discarded in favor of one which proved much better.

The investigation of any flood prevention system for which retarding basins are adopted, if undertaken in the same spirit as was that for the Miami Valley, would probably take much the same course, and a discussion of retarding basin design would therefore hardly be complete here without showing, at least briefly, some of the alternative methods that were considered, and the reasons why, under certain conditions, they were less desirable.

CITY PROTECTION BY LOCAL IMPROVEMENT

The plan to protect the whole Miami Valley was developed coincidently with studies for flood prevention at Dayton alone. A number of different methods were investigated for this city, but it was soon evident that the whole valley could be protected by retarding basins as cheaply and more effectively than Dayton alone could be by any other method. The later work of the engineers was therefore concerned with the development of the retarding basin plan. In order, however, that the relative cost of the various projects could be presented to the Conservancy Court, in obtaining their approval for the construction of the retarding basins, it was necessary to work out the channel improvement plans in some detail. The cost of all such schemes proved greater than for retarding basins, but their real merits, as compared with the adopted plan, are not shown simply by a direct comparison of the figures, for it should be noticed that while the unit costs used were the same, the channel improvement schemes provide for flood discharges equal only to those which occurred in March 1913, while the adopted plan gives ample protection from a flood 40 per cent larger. As will be shown later, if the plan using

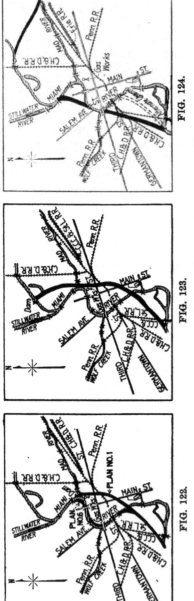

FIG. 122.   FIG. 123.   FIG. 124.

FOUR EARLY FLOOD PROTECTION SCHEMES FOR DAYTON.

Plan 1, figure 122, contemplated the diversion of Mad River through a channel along the Miami and Erie Canal in the eastern part of Dayton, in conjunction with two retarding basins on Stillwater river. Plan 2, figure 123, would have utilized the same basins, but provided for a joint diversion of both Miami and Mad Rivers, partly along the route of Plan 1. Plan 3, figure 124, shows a diversion of Mad River into the Miami north of Dayton, and a cut-off channel to carry the three rivers through the western part of Dayton. Plan 6, shown in figure 122, provided for channel improvement through the main part of Dayton.

channel improvements alone were to be constructed, the discharge at Dayton would be increased, and a repetition of the 1913 flood would give rise to larger discharges than those for which these channels were designed.

Because of the sinuous course of the Miami River through Dayton, the use of cut-off channels immediately suggested itself. A number of possible schemes were investigated. Figures 122, 123, and 124 show four, but a number of other combinations of various parts of these were looked into. Unfortunately, these schemes, to be successful, would necessitate wide channels through thickly built up portions of the city, and would entail excessive damage to property. Another disadvantage is the way in which such cut-offs would divide up the city and become a hindrance to future development. To make the sections thus created accessible to each other would call for a large number of new bridges, the present layout of sewers, water works, and railroads would be seriously interfered with, and other extensive changes necessitated. The disadvantages of this method of improvement were so obvious, and the cost so great, that the investigation was not carried beyond the preliminary stage. A preliminary estimate made in 1913 indicated the more obvious elements of cost to aggregate $14,000,000. In most such estimates, especially where many interferences occurred with existing improvements, the final estimates, based on detailed plans, aggregated at least twice as much as these preliminary estimates based on only the more obvious elements of cost.

The protection of all the cities of the valley by improvement of the existing channels was worked out in some detail. A great objection to these plans is their great first cost, but a more serious objection would be the excessive and continuing cost of maintenance. Table 16 shows the estimates for the various cities to protect against dis-

Table 16.—Summary of cost of local protection, without retarding basins, for the cities and towns in the Miami Valley, against floods as large as that of March 1913.

City or Town.	Total Cost.
Piqua	$361,800
Troy	853,900
Dayton	23,634,300
West Carrollton	346,400
Miamisburg	1,071,400
Franklin	1,500,000
Middletown	532,200
Hamilton	17,444,000
Total	$45,744,000

charges equal to those which occurred in the 1913 flood. Compared with those for the retarding basin plan which gives 40 per cent more protection, the conclusion is obvious, but it may be worth while to show some of the factors which enter into this cost. Because of the many details which would have to be included, it will be impossible to consider the cities separately, but a general discussion will be sufficient to indicate the sources of expense in such a system.

In increasing the carrying capacity of a river through a city, either one of two methods or a combination of them may be used. (*a*) Levees along the sides may be built, or (*b*) the channel may be enlarged by excavating earth from the bottom and sides. There are a number of disadvantages to the first of these methods. The carrying of the river between levees through the city, at an elevation above the surrounding land, is very difficult and expensive on account of the high cost of right of way, the obstruction to traffic due to having all streets approach the levees with steep grades, the damage to adjoining property by the elevation of streets approaching the bridges, the cost of rebuilding bridges at a higher elevation, and the cost of gates at all sewer outlets. Very high levees are also very objectionable in appearance.

The cost of a levee increases rapidly with increase in height, not only because of the greater quantity of material, but also because of the greater area of land occupied, the cost of which, particularly in a city, becomes an important factor. Another factor which operates to swell the cost of raising a levee through a city is the extent to which it has to be extended both above and below the city to connect with high ground. This may be explained as follows: To raise the water surface through the city, since the river must always have a downward slope, it is necessary to begin the levees some distance upstream from the town, at a point where the elevation of the water surface is somewhat higher than that which it is desired to maintain through the city. The higher the water level in the town is raised, the farther upstream must these levees be extended. Not only is it necessary that they be extended in this way on the main stream, but also up the tributaries, until an elevation is reached higher than that of the water surface in the main stream. Downstream from a city the end of the levees must be tied in to high ground, in order to prevent back flooding. In addition to these considerations is that of raising highway and railroad bridges, entailing costly grade elevations. The latter items alone, in the case of the cities in the Miami Valley, made it undesirable to raise the levees above the elevations of the principal existing bridges.

On the other hand, increasing the carrying capacity of a channel by removing earth has many disadvantages. If the channel is enlarged by taking the earth from the sides, the cost of the land taken may be great. A narrow deep channel can be constructed, but in order that this may not be filled up by earth washed down from above, the bottom at the upper end must slope gradually to meet the natural bed of the river, on a grade so flat that the earth will not be washed down from above. This involves continuing the excavation a long distance above the city. An alternative plan is to build a weir at the upper end of the excavation, over which the water would drop without carrying the earth down into the deeper channel below. One of these two plans must also be used wherever tributaries enter the deepened channel. Below a city it is necessary to continue the excavation to a sufficient distance down stream so that the bottom of the improved channel, which must still have a flat downward slope, may intersect the natural bed of the river. If this is not done, the water will form a pond in the deepened channel, which will allow silt to deposit and gradually fill up. The sewage which is discharged into the river might collect in such a pool in periods of low flow and create a nuisance.

Disposing of the earth excavated from the channels is a big item of cost. While it may sometimes be placed in low, waste areas or used in strengthening levees, it is usually necessary to purchase land for spoil banks. In cities this may be so expensive that it is often more economical to transport the materials considerable distances to cheaper land. The river channels leading in and leading out of the cities are usually through agricultural regions. The spoil areas along these must be purchased because the deposits of the excavated material on the land destroy its productiveness.

A combination of the two types of improvement is usually the most economical, and its application at most of the cities in the valley was investigated. It was found that at West Carrollton and Miamisburg levees alone would be required, while at Piqua, Troy, Franklin, and Middletown levees would form the principal part, with only a limited amount of channel excavation. At Dayton and Hamilton large quantities of excavations would be necessary in addition to levees of considerable height. It was found advisable to use approach channels rather than weirs. At Dayton these improvements would extend 3 miles above the city on the Miami, and 2½ miles on both Mad and Stillwater Rivers. The "get away" channel would extend 4 miles below Stewart Street. In Hamilton they would extend 5 miles above the city and 2 miles below. In many cases in addition to building new levees, the removal of the old ones would be required.

At nearly all the cities it would be necessary to raise and lengthen the bridges, raise their approaches, and repave portions of the streets. In the case of railroads, in order to raise the bridges it would be necessary to raise the roadbed for some distance on each side. In the cities this also would involve raising all the street crossings over this portion of the track.

At most of the cities, besides the Miami and Erie Canal, there are power canals leading to mills or manufacturing plants. The improvements would seriously interfere with several of these power developments and in some cases it would be necessary to buy them outright. Where the canals pass through the levee systems flood gates would be required, and gates are also necessary in the tailraces to prevent the water backing up through them into the city. Extensive alterations in the water works and sewerage systems in some cases would have to be made. At one of the towns it would involve an entirely new water works pumping station. A great many gates on the various sewers which discharge below the high water line would be required.

When the cost of all these changes is considered, it is evident that this plan of improvement would be exceedingly costly. Not only would the first cost of this type of improvement be high but the maintenance charges also would be excessive. Willows and trees would grow up in the channel and have to be removed, while the water flowing at high velocity would form pools and bars during every flood, making the waterway irregular and reducing its capacity.

PROTECTION OF ENTIRE VALLEY BY CHANNEL IMPROVEMENT

In order to dispel any doubt in the minds of the Conservancy Court as to the impracticability of the protection of the entire valley by channel improvement, a fairly complete plan for this method was worked out and the cost estimated. The channel begins at the upper end of Piqua and extends through Hamilton. In the cities it is very similar to the plan proposed for the protection of the cities alone. Since it is continuous, the approach and get-away channels are eliminated except the approach channel at Piqua and the get-away channel at Hamilton. Step-off weirs are provided at each tributary, and on most of them, levees along both sides of the stream leading to high ground were necessary. All existing dams in this stretch of the river would have to be removed and practically all the bridges would have to be lengthened, or raised, or both. A good idea of the magnitude of the work involved can be gained from the following summary.

Channel Length.

Extent of improvement, from north end of Piqua to south end of Hamilton, total distance, 75 miles.

Excavation.

Total excavation,* 167,000,000 cubic yards.

Maximum cut, 35 feet.

Right-of-way.

Through cities	1,500 acres
Outside of cities	12,500 "
Total	14,000 acres

This is divided as follows: 8000 acres for channel, berms, and levees; 6000 acres for spoil banks and borrow pits.

Water Powers.

Existing water power plants would be affected at Troy, Miamisburg, Franklin, Middletown, Excello, Woodsdale, and Hamilton. Seven dams would have to be removed.

Tributaries.

Fifty tributaries would require improvement in the shape of step-off weirs and levees.

Railroads and Highways.

Miles of railroads affected, 10.

Miles of highways to be relocated, 9.

Additional miles of highways and city streets affected, 7.

Bridges.

Affected, on Miami River	43
Affected, on tributaries	28
Total	71

Of this total, 17 are railroad bridges. Nine highway bridges, also included in this total, would have to be removed.

Buildings.

Requiring removal	550
Disturbed by levees or spoil banks	110
Total affected	660

Table 17 gives the estimated cost of such an improvement, by sections, based on prewar prices; it covers only the more obvious elements of cost, such as construction cost, rights-of-way and property damage, and items of 10 per cent for engineering, legal, and administrative expense, and 17.28 per cent for contingencies, the latter figure being taken for purposes of comparison from the Official Plan cost estimates.

It should be noted that in table 17 the improvements at cities include not merely that part within corporation limits but the entire section of the plan so designated. These figures show conclusively that the complete improvement of the Miami River, without retarding basins, is not warranted from a cost standpoint.

* The total excavation in the Panama Canal was 195,000,000 cubic yards.

Table 17.—Summary of cost of a continuous channel improvement of the Miami
    River from Piqua to Hamilton (without retarding basins), to take
    care of floods as large as that of March 1913.

Section.	Total Cost.
Piqua	$1,066,000
Piqua and Troy	2,010,000
Troy	2,043,000
Troy to Dayton	10,456,000
Dayton	14,604,000
Dayton to Hamilton	50,950,000
Hamilton	11,240,000
	$92,369,000

It should be noted that the figures in table 16 include the same
allowances for engineering and administrative expense and contingencies as in table 17, and are based also on prewar prices.

It is noteworthy, further, that both the estimates in tables 16 and
17 exceed by far the Official Plan estimates, although neither provides
protection against floods any larger than that of 1913, while the
Official Plan provides protection against floods 40 per cent greater.

Increased Flow Resulting from Channel Improvement

While the channel improvement plans were projected to take care
of discharges equal to those which occurred in the 1913 storm, their
construction would so change conditions that a repetition of this storm
would cause materially larger discharges, and it is doubtful if any of
the channel improvement plans investigated would be safe in case of
such a flood.

During the 1913 storm the rivers of the Miami Valley overflowed
their banks for practically their entire length, flooding large areas of
land to considerable depths, temporarily storing or retarding the
flow of great quantities of water in this way. The valley therefore
acted as an immense retarding basin, storing that part of the water
which the stream channel could not accommodate, and discharging it
gradually and over a long period of time. An idea of the quantity
of water in the valley may be gained from the statement that throughout the distance of 75 miles along the river, from Piqua to Hamilton,
the flood plain on the average is about a mile wide, and was covered
to an average depth of 9 feet. The total quantity of water stored
above Hamilton, at the peak of the flood, was equivalent to a depth of
2.9 inches on the catchment area, or three-tenths of the total rainfall
on the watershed.

If a channel improvement scheme were constructed, a part of this
storage would be eliminated, for although the capacity of the new
channel would be larger than that of the old river channel proper,

it would be less than the joint capacities of river channel and flood plain. The real reduction of storage would be the difference between the capacity of river and flood plain and that of the improved channel. Not only would a reduction in storage capacity be caused on the main stream but also on the tributaries, since for some distances above the improved channel the water surface heights would be reduced by the lowering of the water level in the improved section.

In the computation of the increased flow caused by the channel improvement for the entire valley, it was assumed that the valley storage affected would be that part which lay below the Lockington, Englewood, Huffman, and Germantown damsites. In this distance the storage, when the river reached its maximum stage in 1913, was 451,000 acre feet, while the capacity of the improved channel would have been only 194,000 acre feet. The difference, 257,000 acre feet, or 57 per cent of the total storage, would be eliminated by the improvement. This is equivalent to a depth of 1.32 inches of water over the entire drainage area.

This storage began at about the time the river overtopped its banks, which at Hamilton is equivalent to a discharge of 40,000 second feet. The Hamilton hydrograph, figure 125, shows this to have occurred about 5 p.m. on March 24. The maximum stage was reached at 3 a.m. on the 26th or 34 hours later. During this interval 257,000 acre feet of runoff was stored in the valley. If the channel had been improved this could not have been stored, and therefore would have been discharged through Hamilton during that interval. The flow of the improved channel would, therefore, have been 257,000 acre feet greater between 5 p.m. on March 24 and 3 a.m. on the 26th than the actual 1913 discharge. During the actual flood this 257,000 acre feet passed through Hamilton after the peak discharge occurred and before the river again fell to a bank full stage. Between 3 a.m. Monday the 26th and midnight on March 30, when the flow again reached 40,000 second feet, the discharge with the improved channel in operation would, therefore, have been 257,000 acre feet less than the actual 1913 discharge.

The extent to which this increase in flow during the first part, and decrease in flow during the latter part, of the flood would have affected the discharge rates can best be determined graphically. Figure 125 shows the 1913 Hamilton hydrograph and also the computed hydrograph with channel improvement constructed. The criteria which determine the hydrograph for the improved channel are four: (1) It must diverge from the 1913 hydrograph at A, where the discharge is 40,000 second feet and valley storage begins; (2) it must

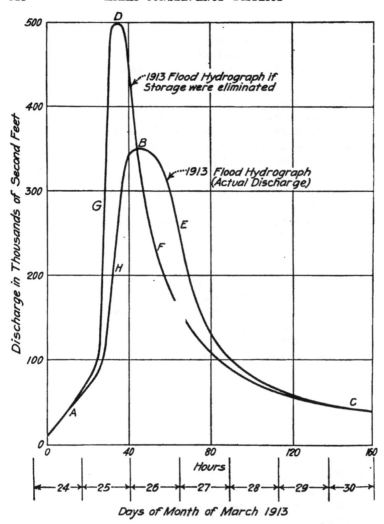

FIG. 125.—HYDROGRAPH OF MARCH 1913 FLOOD DISCHARGE OF
MIAMI RIVER AT HAMILTON AND COMPUTED HYDROGRAPH ASSUM-
ING VALLEY STORAGE ELIMINATED.

intersect again at *B*, the peak of the 1913 curve; (3) it joins the actual
hydrograph again at *C* where the discharge again reaches 40,000 second feet; and (4) the areas *A D B* and *B E C F* must each be equivalent to 257,000 acre feet. The basis for these criteria may perhaps
best be understood by considering the different rates of flow as the

action of a retarding basin. Since the improved channel prevents the storage of water in the valley, its discharge rate must be the same as that of the inflow into it. Its discharge curve may, therefore, be considered as the curve of inflow into the valley, while the actual 1913 hydrograph represents the outflow curve, the valley itself being the retarding basin. The problem is, therefore, the reverse of that with the retarding basins, for here we are to determine the inflow curve, having given the outflow curve. Storage in the valley starts at the 40,000-second foot stage, see point A, figure 125, and here the outflow should become less than the inflow. The maximum outflow should occur when the storage, represented by the area between the curves $A G D B$ and $A H B$, is a maximum, or where the inflow and outflow are equal, see point B. When the discharge again becomes 40,000 second feet or at C, the basin is empty and the inflow and outflow are therefore the same. Since the storage in the valley is the difference between the inflow and outflow, its total is represented by the area $A G B H$, which must have an area, at the scale of the diagram, equivalent to 257,000 acre feet. Since this storage begins to decrease as soon as the maximum is reached, and when the valley is empty the total storage has been discharged, the area $B E C F$ must be equal to $A G D B H$ or also be 257,000 acre feet.

While a number of curves fulfilling these four conditions could be drawn, with the shape of retarding basin inflow and outflow curve in mind, it is possible to draw a curve which will represent very closely the true rates of flow. The 1913 flood discharge, with storage eliminated, is shown by figure 125 to be 500,000 second feet, as compared with the actual 1913 discharge of 350,000 second feet. In the same way it may be shown that if the channel improvement had been built above Dayton, the discharge of the 1913 flood there would have been 340,000 instead of 250,000 second feet, and at the other cities it would have increased proportionately.

This same effect would be noticed in the construction of channel improvement schemes for the cities alone. The elimination of only that storage which occurred in the city of Dayton would have increased the discharge at Dayton from 250,000 to 285,000 second feet.

PROTECTION BY NUMEROUS SMALL RETARDING BASINS

In following out the intention to investigate all possible methods of flood control for the Miami Valley, a study was made of the possibilities of the use of numerous small retarding basins. It was thought that in addition to the possibility that they might be cheaper, small basins might possess several other advantages. Thus, they might appear to the general public as less of a menace to life and property in the valleys below than would large dams. The land on which they would be situated might be cheaper than that required by larger basins. Since many of the small basins investigated on the Miami River were above Piqua and Troy, they would more effectively reduce the flood flow at these points.

In order that a comparison of the merits of the small basin systems could be made with those of the plan adopted, it was necessary that the ultimate degree of protection furnished be the same in either case. Because of the many factors which entered into the problem, it was found to be impossible to secure exactly the same degree of protection, and the following conditions were therefore adopted: (1) That, if possible, the total storage in the small basins should equal that of the large ones, and (2) that the inflow assumed for the small basins should be the same as that used in the adopted plan. For this reason, the basins on the Stillwater River and its tributaries were assumed to receive an inflow of 10 inches in 3 days while those in the Miami and Mad River basins received $9\frac{1}{2}$ inches in 3 days.

The desirability of a system of small basins may be considered from two standpoints, (a) cost, and (b) reliability of the flood protection obtained. In the following discussion the comparison will be first made on the basis of cost. It should be kept in mind that the cost given in the case of the small basins is a preliminary estimate, and an assumption of very favorable conditions was made in regard to the foundations for structures, so that the figures are no doubt in many cases considerably below the true cost. On the other hand the costs given for the large basins are estimates made after very thorough investigation of the dam sites and of the probable cost of construction, and are much higher in proportion.

Figure 126 shows the location of the 45 basins considered. The sites of 14 small basins were found on the Stillwater tributaries. Their combined capacity was considerably less than that of the Englewood basin. The use of 13 of these with a large dam at Pleasant Hill or West Milton was investigated. The cost of either of these systems was over $3,000,000 more than that for the Englewood basin. The use of 9 of the best of these small basins with a basin at the site of, but smaller than, the Englewood was found to cost $2,500,000 more than the Englewood alone.

Twenty small reservoir sites were found on the drainage basin of the Miami River above the mouth of the Stillwater. These had a combined capacity about equal to that of the Lockington and Taylorsville basins, but their cost was found to be about $3,000,000 more than that of the two dams adopted. In this plan there were 4 small basins on Loramie Creek, one of them at the site of the present Lockington basin. While the combined storage capacity of the 4 was about equal to that of the Lockington as adopted, their cost would have been 18 per cent greater. A system of 18 small basins with one large one just

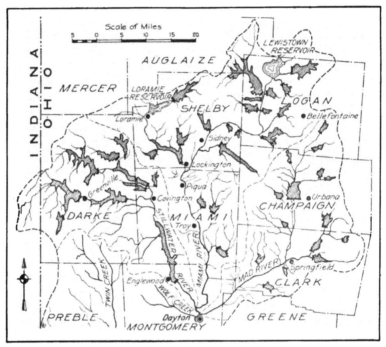

FIG. 126.—MAP OF MIAMI RIVER WATERSHED ABOVE DAYTON SHOW-
ING SUGGESTED SYSTEM OF 45 SMALL RETARDING BASINS.

above the mouth of Loramie Creek, and another with 11 small ones and a large one at the present site of the Taylorsville dam was tried, but the cost was 50 per cent greater in the first case and 24 per cent in the second case, than the cost of the basins in the plan adopted. In all the cases, if we assume that the entire cost of protecting Troy and Piqua is removed by using small basins, the balance is still heavily in favor of the large basins.

On the Mad River, 14 small basins were located on the tributaries. Two systems were worked out: the first included the 14 on the tributaries and a small one on the main stream just above Urbana; the second contained 9 of the small basins and a large one at the same

21

point. The estimate of the first of these plans showed a cost slightly greater than that of the Huffman basin. The estimate for the second showed a very slightly less cost, but as has already been pointed out, the estimate for the small basins is probably much lower, relatively, than that of the plan adopted. The principal reason for not adopting this plan for the Mad River is that the protection given by it is not so reliable, as will be presently shown.

It will be seen from the foregoing that, except on the Mad River, it is impossible to obtain a small basin system without costs greatly exceeding those of large basins. On the Mad River the cost is approximately the same, but the cheapest of the systems contains a large dam at Urbana, which deprives this plan of one of the advantages peculiar to small basins, namely, that of apparently less menace. This plan also has the disadvantage that it leaves about 250 square miles more of unreservoired area than does the Huffman basin.

The cost of the land in the small basins would not be less per acre than that above the large dams. In the small basins it is usually good farming land, and in many cases more valuable than that required for the large ones. Since the average depth above the small dams is less, to have storage equal to that above the large ones, much more land is required. This results in increased cost and is responsible for a large part of the unfavorable comparisons shown in the preceding paragraphs.

While small basins have some advantages, they are open to so many serious objections that it is doubtful if such a system should be adopted, even if the cost of properly located large basins is considerably greater. The advantages of less menace claimed for the small basins is more fictitious than real. No system of basins of any size or extent should be constructed above cities if there is a shadow of doubt as to their safety.

It is a well known fact that intense rainfalls are more likely to occur over small areas than over large ones; also, that the probability of occurrence of such rainfalls is greater than for general storms covering large areas. An intense local rainfall occurring above one of the small basins might tax its entire capacity, while it would cover only a part of the drainage area of a large basin and, therefore, not fill the latter. The small basins would have to be made sufficiently large to take care of this, and this would very much increase their cost. It is evident that the assumption of 10 or 9½ inches of runoff in 3 days would not give these small basins the factor of safety that it gives the larger retarding basins. The discussion in chapter X of the effect of localized cloud bursts on the five large retarding basins emphasizes this point.

A series of small basins cannot be as reliable in its action as a few large ones, for the rainfall from even the largest storms is not equally distributed, and those basins over whose drainage area the heaviest precipitation falls would be filled completely, and possibly discharge over the spillways, while the basins in the region of lighter rainfall would only be partly filled. Since the discharge over the spillway increases rapidly as the depth on it increases, the combined discharge from the basins in this case would be more than if they were all filled just to their spillways. The same thing is true, but to a much less extent, of a series of large basins, for in the latter the variations in rainfall and runoff which would be felt at the small basins in the same watershed, are equalized. Other things being equal, the larger the drainage area controlled by a retarding basin, therefore, the more reliable is the protection afforded by that basin.

Another disadvantage of a system of small basins is the large proportion of drainage area from which the runoff remains uncontrolled by dams. Thus, for the small basin system, the uncontrolled catchment area above Dayton is about 650 square miles, as compared with only 70 square miles in the adopted plan. Figure 5, page 33, shows that in one case the maximum rate of discharge from a 650 square mile area during the 1913 flood was 85,400 second feet. The discharge from the maximum local storm would no doubt be greater. This means that the flow which it would be necessary to provide for in Dayton and the cities below, if the small basin system were used, would be much more than with the adopted plan, and, therefore, the cost of the channel improvements at these places would be correspondingly increased.

It is evident, therefore, that small basins should not be used on the Stillwater or Miami rivers, on the score of both excessive cost and unreliability. While in the case of Mad River the cost is approximately the same as in the proposed plan, the area remaining uncontrolled would be larger. The cost figures given do not take into consideration the extra channel improvement which the additional 250 square miles of unreservoired area, in such a plan, would necessitate at Dayton and the cities below. If this is taken into consideration, together with the fact that the estimates for the small basin system are low, for the reasons stated previously, it appears that for the Miami Valley as a whole the cost of the small basins in all cases would be much greater, and the protection furnished by them less reliable, than for the five large basins.

APPENDIX

Table I.—Drainage Areas of Streams in the Miami Valley

MIAMI RIVER

	Square miles
Above Lewistown Reservoir	100
Above Tawawa Creek	499
At Sidney	555
Above Mill Branch	575
At Piqua	842
At Troy	908
At Tadmor	1128
At Taylorsville Damsite	1133
Above Stillwater River	1162
At Dayton, above Wolf Creek	2525
At Dayton including Wolf Creek	2598
Above Bridge at West Carrollton	2640
At Big Four Crossing below Miamisburg	2722
At Franklin	2785
Above Twin Creek	2797
At South Line of Middletown	3162
Above Four Mile Creek	3330
At South Line of Hamilton	3672
At Venice	3800
About 2 Miles below Miamitown	3937
At Mouth	5433

STILLWATER RIVER

Above Greenville Creek at Covington	223
At Sugar Grove	448
At West Milton	611
At Englewood Damsite	651
At Little York	657
At Mouth	674

MAD RIVER

Above Buck Creek	324
Above Mill Creek	488
Above Big Four R. R. Bridge No. 122	505
About 2 Miles below Osborn	649
At Huffman Damsite	671
At Mouth	689

MIAMI RIVER TRIBUTARIES

Tawawa Creek at Mouth	56
Loramie Creek at State dam	81
Loramie Creek above Mouth of Turtle Creek	208
Loramie Creek at Damsite	255
Loramie Creek at Mouth	262
Turtle Creek at Mouth	36
Shawnee Creek at Mouth	15
Spring Creek at Mouth	27
Lost Creek at Mouth	58

	Square miles
Honey Creek at Mouth	89
Wolf Creek at Mouth	73
Bear Creek at Mouth	53
Clear Creek at Mouth	51
Twin Creek at Damsite	270
Twin Creek at Mouth	313
Dick's Run at Mouth	49
Elk Creek at Mouth	59
Gregory Creek at Mouth	30
Four Mile Creek above Seven Mile Creek	180
Four Mile Creek at Mouth	322
Pleasant Run at Mouth	15
Indian Creek at Mouth	80
Banklick Creek at Mouth	6
Bluerock Creek at Mouth	8
Paddy's Run at Mouth	16
Taylor Creek at Mouth	27

STILLWATER RIVER TRIBUTARIES

Greenville Creek at Mouth	219
Ludlow Creek at Mouth	66

MAD RIVER TRIBUTARIES

Buck Creek at Mouth	164
Mill Creek at Mouth	17
Donnells Creek at Mouth	29

Table II.—Slopes of Streams in the Miami Valley

Stream	From	To	Distance	Difference in Elevation	Drop per Mile
			Miles	Feet	Feet
Miami River	Lewistown Reservoir	Sidney	29.2	62.0	2.1
Miami River	Sidney	Piqua	14.0	78.0	5.6
Miami River	Piqua	Troy	8.8	30.0	3.4
Miami River	Troy	Dayton	27.2	90.0	3.3
Miami River	Dayton	Miamisburg	14.5	50.0	3.4
Miami River	Miamisburg	Middletown	15.0	50.0	3.3
Miami River	Middletown	Hamilton	16.5	64.0	3.9
Miami River	Hamilton	11 miles above the Ohio River	26.0	85.0	3.3
Loramie Creek	Loramie Reservoir	Miami River	22.7	72.0	3.2
Spring Creek	Source	Miami River	12.1	226.0	18.7
Lost Creek		Miami River	16.7	267.0	16.0
Honey Creek		Miami River	10.1	113.0	11.2
Stillwater River		Englewood	52.0	218.0	4.2
Stillwater River	Englewood	Miami River	10.2	45.0	4.4
Mad River	North line of Clark County	Buck Creek	8.8	56.0	6.4
Mad River	Buck Creek	Dayton	26.7	156.0	5.8
Buck Creek		Mad River	12.2	126.0	10.3
Wolf Creek	Source	Miami River	17.2	301.0	17.5
Twin Creek	Euphemia	Miami River	35.2	282.0	8.0

Table III.—Channel Capacities in 1913 in the Towns and Cities of the Miami Valley Compared with the 1913 Flood Discharge

Locality	Description of Channel	Channel Capacity, Sec. Ft.	Flood Discharge, Sec. Ft.	Ratio, Per Cent.	Drainage Area, Sq. Mi.
Sidney........	High banks, small levees in places	10,000	44,000	22.7	555
Piqua........	Above Penn. R. R. small levees	25,000	70,000	35.7	842
Piqua........	Below Penn. R. R., small levees in places	15,000	70,000	21.4	842
Troy.........	Floods low areas on left of main channel	60,000	90,000	66.7	908
Troy.........	To top of levee above railroad	20,000	90,000	22.2	908
Troy.........	To top of bank below railroad	12,000	90,000	13.3	908
Dayton.......	Above Wolf Creek levees	90,000	250,000	36.0	2525
Dayton.......	Below Wolf Creek levees	100,000	252,000	39.7	2598
Miamisburg...	B. & O. R. R. fill acts as levee	65,000	257,000	25.3	2722
Franklin......	Practically no levees	65,000	267,000	24.3	2785
Middletown...	No levees, wide flood channel	115,000	304,000	37.8	3162
Hamilton.....	Levees above the city	100,000	352,000	28.4	3672

Table IV.—Channel Capacities Outside of Towns and Cities in the Miami Valley Compared with the 1913 Flood Discharge

Locality	Description	Channel Capacity, Sec. Ft.	Flood Discharge, Sec. Ft.	Ratio, Per Cent.	Drainage Area, Sq. Mi.
Miami River above Sidney.....	Practically no levees	5,000	34,100	14.7	498
Miami River below Sidney.....	No levees	5,000	48,500	10.3	575
Miami River below Piqua.......	No levees	10,000	70,000	14.3	842
Miami River at Tadmor.......	To top of bank	8,000	127,300	6.3	1128
Miami River at Tadmor.......	To top of levee	12,000	127,300	9.4	1128
Miami River below Dayton....	No levees	25,000	252,000	9.9	2598
Miami River below Miamisburg	No levees	35,000	257,000	13.6	2722
Miami River below Hamilton ..	No levees	25,000	352,000	7.1	3672
Miami River below Miamitown	No levees	20,000	384,000	5.2	3937
Loramie Creek northwest of Lockington.................	Practically no levees	1,600	25,600	6.3	208
Stillwater River above Covington......................	Practically no levees	1,200	33,100	3.6	223
Stillwater River below Covington......................	Practically no levees	6,000	51,400	11.7	448
Stillwater River above West Milton....................	Practically no levees	7,000	86,200	8.1	600
Mad River west of Springfield..	No levees	5,000	55,400	9.0	488
Mad River below Osborn......	To top of bank	6,500	75,700	8.6	649
Mad River below Osborn......	To top of levee	13,500	75,700	17.8	649
Twin Creek west of Germantown......................	No levees	3,000	66,000	4.5	272

Table V.—Valley Storage During Crest Stages of 1913 Flood

	Acre Feet	Inches
Twin Creek Valley	34,810	2.09
Stillwater River Valley below Covington	57,010	1.59
Miami River Valley above Dayton	119,730	1.93
Mad River Valley	40,250	1.10
In City of Dayton	83,700	...
Total above lower end of Dayton	300,690	2.17
Miami Valley—Dayton to Hamilton	232,440	...
Total above Hamilton	567,940	2.90

Table VI.—Population and Elevation in Feet above Sea Level of Cities in the Miami Valley

Name of City	Population	Elevation
Cleves	1,500	500
Dayton	170,000	740
Englewood	300	930
Fairfield	300	835
Franklin	3,000	680
Germantown	2,000	720
Hamilton	65,000	590
Lockington	200	930
Miamisburg	5,000	705
Middletown	20,000	650
Osborn	1,200	830
Piqua	15,000	870
Sidney	10,000	950
Springfield	60,000	970
Tippecanoe City	2,400	825
Troy	7,300	880
West Carrollton	1,300	720
West Milton	1,200	910

Table VII.—Data pertaining to Dams and Appurtenances

	Germantown Dam	Englewood Dam	Lockington Dam	Taylorville Dam	Huffman Dam
Drainage area above dam, square miles	270	651	255	1,133	671
Average elevation bed of river, feet above sea level	723	768	876	759	777
Average elevation valley floor, feet above sea level	730	782	885	770	785
Elevation of spillway crest, feet above sea level	815	876	938	818	835
Elevation of top of dam, feet above sea level	830	892.5	954	837	850
Maximum height of dam above valley floor, feet	100	110.5	69	67	65
Length of crest of dam, feet	1,210	4,716	6,400	2,980	3,340
Thickness of dam at average elevation of valley floor, feet	655	739	409	397	385
Number of conduits	2	2	2	4	3
Length of each outlet conduit, feet	546	709	46	40	40
Vertical diameter of permanent outlet conduits, feet	9.1	10.5	9.2	19.2	16.3
Maximum width of outlet conduits, feet	13	13	9	15	15
Combined sectional area of permanent outlet conduits, square feet	182	217	158	1,118	705
Elevation of conduit invert at entrance	724	772	876	760	777
Length of spillway crest, feet	*70	100	72	132	100
Volume of earth in dams, cubic yards	865,000	3,500,000	1,135,000	1,235,000	1,655,000
Volume of concrete in dams and appurtenances, cubic yards	16,300	36,800	38,000	48,000	40,000
Storage capacity to spillway crest, acre feet	106,000	312,000	70,000	186,000	167,000
Area in basin below elevation of spillway crest, acres	3,520	7,930	4,020	11,000	9,180
Outflow through conduits with water level at spillway crest, second feet	10,000	12,000	8,800	53,600	35,000
Velocity of outflow with water level at spillway crest, feet per second	55	55	56	48	50

* The spillway at Germantown dam is a concrete flume 70 feet wide built in a low saddle on the hills, and has a capacity equivalent to a 50-foot free fall spillway crest.

Table VIII.—Conditions at Dams for a Flood equal to that of March 1898

	Germantown Dam	Englewood Dam	Lockington Dam	Taylorsville Dam	Huffman Dam
Runoff from drainage area in 3 days, inches of depth.	2.47	2.47	2.47	2.47	2.47
Maximum rate of inflow into retarding basin, second feet....................	20,700	30,200	10,600	35,100	31,200
Maximum rate of outflow through outlet conduits, second feet..............	6,820	8,100	6,160	30,500	19,600
Maximum elevation attained by water surface, feet above sea level.........	773	826	911	792	807
Vertical distance of water surface below spillway crest, feet..............	42	50	27	26	28
Vertical distance of water surface below top of dam, feet....................	57	66	43	45	43
Maximum area flooded, acres..................	940	2,620	1,040	2,600	2,490
Maximum storage, acre feet	15,900	51,500	11,380	22,900	20,000

Table IX.—Conditions at Dams for a Flood equal to that of March 1913

	Germantown Dam	Englewood Dam	Lockington Dam	Taylorsville Dam	Huffman Dam
Runoff from drainage area in 3 days, inches of depth.............	6.94	6.94	6.94	6.94	6.94
Maximum rate of inflow into retarding basin, second feet...........	66,000	85,400	33,000	106,400	78,300
Maximum rate of outflow through outlet conduits, second feet ..	9,340	11,000	8,630	51,300	32,600
Maximum elevation attained by water surface, feet above sea level ...	805	862	936	815	830
Vertical distance of water surface below spillway crest, feet............	10	14	2	3	5
Vertical distance of water surface below top of dam, feet...........	25	30.5	18	22	20
Maximum area flooded, acres................	2,950	6,350	3,600	9,650	7,300
Maximum storage, acre feet................	73,000	209,000	63,000	152,000	124,000

22

Table X.—Conditions at Dams for Official Plan Flood

	German- town Dam	Englewood Dam	Locking- ton Dam	Taylorsville Dam	Huffman Dam
Runoff from drainage area in 3 days, inches of depth..........	10	10	10	9.5	9.5
Maximum daily rate of inflow into retarding basin, inches of depth on drainage area.......	4.5	4.5	4.5	4.0	4.0
Maximum rate of outflow through outlet conduits, second feet...................	9,800	12,000	9,000	55,000	35,000
Maximum elevation attained by water surface, feet above sea level......................	812	873	940	820	835
Vertical distance of water surface below top of dam, feet...	18	19.5	14	17	15
Maximum area flooded, acres...	3,400	7,600	4,400	12,000	9,180
Maximum storage, acre feet....	95,000	289,000	79,000	217,000	167,000

Table XI.—Conditions at Dams for Assumed Flood Runoff of
14 Inches in 3 Days

	Germantown Dam	Englewood Dam	Lockington Dam	Taylorsville Dam	Huffman Dam
Runoff from drainage area in 3 days, inches of depth............	14	14	14	14	14
Maximum daily rate of inflow into retarding basin, inches of depth on drainage area......	6⅛	6⅜	6	6⅛	6
Maximum rate of combined outflow through outlet conduits and over spillway, second feet.................	16,100	26,500	18,100	79,400	48,400
Maximum discharge over spillway, second feet ..	6,100	14,500	9,300	25,800	13,400
Maximum elevation attained by water surface, feet above sea level................	824.5	887.4	948.4	832.0	845.8
Vertical distance of water surface below top of dam, feet............	5.5	5.1	5.6	5.0	4.2
Maximum head on spillway, feet............	9.5	11.4	10.4	14.0	• 10.8
Maximum area flooded, acres...............	4,100	9,400	7,150	16,900	13,500
Maximum storage, acre feet.................	142,000	413,000	126,000	386,000	297,000

Table XII.—Flooding Conditions in Germantown Retarding Basin during the Period 1893 to 1917 Assuming Dam to have been in Existence. The various Floods are Arranged in Order of Size

Date of Flood Crest	Maximum Elevation of Water Surface	Maximum Storage, Acre Ft.	Maximum Outflow, Sec. Ft.	Area Flooded, Acres	Days Required to Empty
March 25, 1913......	805.2	73,400	9,340	2,950	7.4
March 23, 1898......	772.6	15,900	6,820	940	2.9
May 2, 1893.........	766.8	10,720	6,260	750	2.3
October 7, 1910......	765.3	9,640	6,120	730	2.3
February 28, 1910....	764.9	9,380	6,080	710	2.3
March 6, 1897........	763.5	8,330	5,920	680	2.0
February 1, 1916.....	763.0	8,040	5,880	670	1.6
July 9, 1915.........	762.8	7,940	5,850	660	2.6
March 14, 1907.......	762.7	7,860	5,850	660	2.0
February 27, 1912....	761.5	7,060	5,710	630	1.7
February 15, 1908....	761.4	6,995	5,700	630	2.1
March 30, 1912.......	759.7	5,940	5,500	580	2.0
March 27, 1904.......	759.3	5,690	5,450	560	1.6
March 2, 1908........	759.1	5,575	5,440	560	1.5
February 25, 1909....	758.1	4,975	5,300	530	1.7
March 6, 1908........	757.5	4,680	5,250	510	1.7
January 5, 1907......	756.8	4,310	5,150	490	1.5
January 18, 1910.....	756.6	4,240	5,140	480	1.7
March 27, 1906.......	756.3	4,080	5,090	470	1.6
January 20, 1907.....	755.7	3,830	5,020	450	1.5
January 22, 1904.....	755.1	3,555	4,950	440	1.6
January 2, 1916......	755.1	3,555	4,950	440	1.2
January 13, 1916.....	754.9	3,510	4,950	430	1.0
March 1, 1903........	754.6	3,370	4,880	425	1.3
March 19, 1908.......	753.0	2,795	4,670	380	1.1
January 8, 1913......	750.7	2,110	4,370	290	1.2
March 27, 1916.......	742.5	620	3,100	110	.9
May 7, 1916..........	735.5	92	2,050	40	.6

Table XIII.—Flooding Conditions in Englewood Retarding Basin during the
Period 1893 to 1917 Assuming Dam to have been in Existence.
The various Floods are Arranged in Order of Size

Date of Flood Crest	Maximum Elevation of Water Surface	Maximum Storage, Acre Ft.	Maximum Outflow, Sec. Ft.	Area Flooded, Acres	Days Required to Empty
March 25, 1913......	861.7	209,000	11,000	6,350	23.2
March 23, 1898......	826.4	51,500	8,100	2,620	9.1
February 1, 1916.....	824.7	47,050	7,940	2,460	6.7
May 2, 1893.........	822.4	41,800	7,720	2,280	6.3
January 2, 1916......	819.3	35,000	7,400	2,040	4.8
February 28, 1910....	819.1	34,560	7,400	2,040	6.5
October 7, 1910......	818.9	34,200	7,380	2,010	5.5
March 14, 1907.......	817.2	30,600	7,200	1,900	5.6
March 6, 1897........	816.8	29,900	7,150	1,850	6.3
February 15, 1908....	815.8	28,000	7,030	1,790	4.8
March 6, 1908.......	814.7	26,000	6,900	1,720	6.1
March 30, 1912.......	814.4	25,400	6,870	1,700	8.3
February 27, 1912....	814.2	25,250	6,860	1,690	4.6
March 27, 1904......	812.4	22,200	6,650	1,560	10.2
February 25, 1909....	810.5	19,450	6,420	1,450	4.6
January 20, 1907.....	810.2	19,100	6,400	1,430	4.2
January 18, 1910.....	809.7	18,320	6,320	1,390	4.7
July 9, 1915.........	809.4	17,960	6,300	1,370	3.9
January 5, 1907......	809.0	17,450	6,230	1,350	6.6
March 27, 1906......	808.6	16,960	6,200	1,330	7.5
January 22, 1904.....	807.2	15,200	6,000	1,240	3.5
March 1, 1903.......	806.4	14,180	5,870	1,190	3.8
March 19, 1908......	804.8	12,330	5,640	1,100	4.6
January 8, 1913......	803.2	10,730	5,420	1,000	6.9
March 27, 1916.......	800.0	7,850	4,950	830	3.0
May 7, 1916.........	798.9	6,910	4,750	780	2.2
January 13, 1916.....	796.4	5,140	4,300	650	2.9

Table XIV.—Flooding Conditions in Lockington Retarding Basin during the
Period 1893 to 1917 Assuming Dam to have been in Existence.
The various Floods are Arranged in Order of Size

Date of Flood	Maximum Elevation of Water Surface	Maximum Storage, Acre Ft.	Maximum Outflow, Sec. Ft.	Area Flooded, Acres	Days Required to Empty
March 25, 1913.......	935.9	62,500	8,630	3,600	7.0
March 23, 1898.......	911.4	11,380	6,160	1,040	2.7
May 7, 1916.........	907.3	7,345	5,620	800	2.3
January 2, 1916......	906.8	7,000	5,550	770	2.4
May 2, 1893.........	906.5	6,720	5,500	760	2.2
February 28, 1910....	905.2	5,785	5,340	690	2.2
October 7, 1910.......	905.2	5,785	5,340	690	1.8
March 6, 1897........	903.8	4,955	5,150	610	1.9
March 14, 1907.......	903.7	4,860	5,150	610	2.0
February 15, 1908....	902.4	4,040	4,900	550	1.7
February 27, 1912....	901.9	3,855	4,870	530	1.5
March 30, 1912.......	901.8	3,760	4,860	520	1.8
February 1, 1916.....	901.4	3,535	4,780	510	1.8
March 2, 1908........	900.7	3,190	4,700	480	1.5
March 27, 1904........	900.3	3,005	4,640	470	1.6
March 6, 1908........	900.0	2,820	4,550	450	1.6
February 25, 1909....	899.5	2,615	4,500	420	1.5
January 18, 1910.....	898.6	2,275	4,350	390	1.5
March 27, 1906.......	898.5	2,205	4,300	380	4.1
January 20, 1907.....	898.5	2,205	4,300	380	1.4
January 5, 1907......	898.1	2,065	4,250	370	1.5
March 27, 1916......	897.2	1,770	4,100	330	1.4
March 1, 1903........	897.0	1,700	4,050	320	1.3
January 22, 1904.....	896.9	1,650	4,050	310	1.4
March 19, 1908.......	895.4	1,240	3,750	260	1.0
January 8, 1913......	894.0	918	3,500	230	1.2
July 9, 1915.........	892.9	689	3,250	190	1.2
January 13, 1916.....	891.2	459	2,850	150	2.6

Table XV.—Flooding Conditions in Taylorsville Retarding Basin during the
Period 1893 to 1917 Assuming Dam to have been in Existence.
The various Floods are Arranged in Order of Size

Date of Flood Crest	Maximum Elevation of Water Surface	Maximum Storage, Acre Ft.	Maximum Outflow, Sec. Ft.	Area Flooded, A Acres	Days Required to Empty
March 25, 1913	814.5	150,000	51,100	9,550	4.4
March 23, 1898	792.3	22,900	30,500	2,600	2.7
May 2, 1893	788.0	14,385	25,400	1,780	2.4
October 7, 1910	787.0	12,620	23,700	1,620	3.2
February 28, 1910	786.2	11,660	23,300	1,500	3.0
March 6, 1897	785.5	10,700	22,600	1,400	2.5
February 15, 1908	785.2	10,320	22,200	1,350	2.4
February 1, 1916	785.0	9,725	21,700	1,340	2.2
February 27, 1912	784.9	9,450	21,500	1,300	2.1
March 14, 1907	784.5	9,150	21,100	1,270	2.5
March 30, 1912	784.0	8,900	20,900	1,210	2.4
March 2, 1908	783.6	8,200	20,300	1,160	2.0
March 6, 1908	783.3	7,900	20,000	1,100	2.4
March 27, 1904	783.1	7,750	19,700	1,090	2.0
January 2, 1916	783.1	7,750	19,700	1,090	3.1
February 25, 1909	781.9	6,450	18,300	960	1.8
March 27, 1906	781.7	6,290	18,000	950	2.0
January 18, 1910	781.4	6,010	18,000	900	2.3
January 20, 1907	781.1	5,740	17,500	890	2.4
January 5, 1907	781.0	5,620	17,200	870	2.1
March 1, 1903	780.5	5,145	16,500	830	2.2
January 22, 1904	780.5	5,145	16,500	830	1.8
March 27, 1916	779.1	4,310	15,100	700	2.7
March 19, 1908	779.0	4,110	15,000	690	1.2
May 7, 1916	778.6	3,720	14,000	660	1.5
January 8, 1913	778.5	3,670	13,950	650	1.6
July 9, 1915	776.0	2,340	11,250	430	1.5
January 13, 1916	775.2	2,205	10,550	390	1.8

Table XVI.—Flooding Conditions in Huffman Retarding Basin during the Period 1893 to 1917 Assuming Dam to have been in Existence. The various Floods are Arranged in Order of Size

Date of Flood Crest	Maximum Elevation of Water Surface	Maximum Storage, Acre Ft.	Maximum Outflow, Sec. Ft.	Area Flooded, Acres	Days Required to Empty
March 25, 1913.......	829.6	123,000	32,600	7,250	4.3
March 23, 1898.......	807.3	20,000	19,600	2,490	2.1
October 7, 1910......	802.3	9,850	15,700	1,550	1.8
May 2, 1893..........	802.2	9,750	15,600	1,525	1.8
March 6, 1897........	801.8	9,000	15,250	1,420	1.7
February 28, 1910....	801.8	9,000	15,250	1,420	2.2
February 15, 1908....	801.1	8,150	14,900	1,310	1.5
March 14, 1907.......	801.0	7,825	14,700	1,300	1.7
February 27, 1912....	800.0	6,700	14,000	1,140	1.5
March 2, 1908........	799.1	5,830	13,300	1,000	1.6
March 6, 1908........	798.9	5,540	13,100	990	1.4
March 30, 1912.......	798.9	5,540	13,100	990	2.1
March 27, 1904.......	798.2	5,025	12,600	890	1.4
March 27, 1906.......	797.2	4,175	11,750	775	1.2
February 25, 1909....	797.2	4,175	11,750	775	1.3
January 18, 1910.....	797.0	4,015	11,600	750	1.3
January 22, 1904.....	796.5	3,580	11,050	700	1.7
January 5, 1907......	796.3	3,515	11,000	690	1.2
January 20, 1907.....	796.0	3,280	10,800	650	1.7
March 1, 1903........	795.7	3,055	10,400	625	1.0
March 19, 1908.......	794.5	2,340	9,800	530	1.1
January 8, 1913......	794.0	2,065	9,300	500	1.4
March 27, 1916.......	793.3	1,770	8,700	460	1.1
January 2, 1916......	791.4	1,125	7,450	325	1.2
January 13, 1916.....	791.4	1,125	7,450	325	.8
May 7, 1916..........	791.4	1,125	7,450	325	1.2
February 1, 1916.....	789.9	711	6,100	250	1.5
July 9, 1915.........	788.7	528	5,350	180	1.7

Table XVII.—Channel Capacities Required at Dayton and Towns below, for the Official Plan Flood Assuming Retarding Basins in Operation

1	2	3	4	5	6	7	8
		Unreservoired Area			Retarding Basins		
Location	Drainage Area in Sq. Mi.	1913 Flood Runoff Rate, Sec. Ft. per Sq. Mi.	Official Plan Flood Runoff Rate, Sec. Ft. per Sq. Mi.	Total Runoff, Sec. Ft.	Maximum Outflow, Sec. Ft.	Outflow Contributing to Peak, Sec. Ft.	Channel Capacity Required, Sec. Ft.
Dayton, above Wolf Creek...........	70	120	168	11,800	103,400	99,000	110,800
Dayton, below Wolf Creek...........	143	111	155	22,200	103,400	99,000	121,200
Miamisburg........	267	96	134	35,800	103,400	99,000	134,800
Franklin...........	330	88	123	40,600	103,400	99,000	139,600
Middletown........	437	75	105	45,800	113,400	107,600	153,400
Hamilton (smaller peak)...........	947	60	84	79,600	113,400	113,400	193,000
Hamilton (larger peak)...........	947	120	168	159,100	113,400	51,000	210,100

Table XVIII.—Summary of Observations on Scour in Miami Valley Streams

RESULTS OF OBSERVATIONS ON THE 1913 FLOOD

Stream and Location	Average Velocity at Max. Stage, Ft. per Sec.	Mean Depth, Feet.	Stream-bed Material	Amount of Scour
Miami River at Big Four R. R. Bridge at Sidney.	12.4	15.7	Earth and gravel with probably rock in river bed.	Around e. end of bridge and s. side of embankment. None in river bed.
Miami River above the B. & O. R. R. bridge at Troy	6.11	18.6	Clay, sand, and gravel with some broken rock.	None.
Miami River at the B. & O. R. R. bridge at Troy	11.1	16.3	Clay, sand, and gravel with some broken rock.	Considerable around piers.
Miami River overflow area above Tadmor	5.87	13.3	Alluvial soil.	None.
Miami River channel above Tadmor	8.17	23.4	Gravel.	None.
Miami River below Tadmor at Miami & Erie Canal opening	16.6	30.6	Gravel and rock.	Extensive.
Miami River at Main Street bridge, Dayton.	11.0	*24.6	Gravel.	Considerable around piers.
Miami River channel below Miamisburg	9.23	31.1	Clay and gravel.	None.
Miami River overflow areas below Miamisburg	6.67	16.7	Alluvial soil.	None.
Miami River at Big Four R. R. bridge below Miamisburg	15.2	27.7	Clay, gravel, and rock.	Extensive.
Miami River overflow area at Miamitown	11.5	24.6	Alluvial soil.	None.
Tawawa Creek at Big Four R. R. concrete arch opening	23.1	8.2	Clay and gravel.	Some.
Turtle Creek at Miami & Erie Canal stone arch opening	13.7	10.1	Clay and gravel.	Some.
Stillwater River above highway bridge at West Milton	12.0	27.7	Clay and gravel.	None.
Stillwater River at highway bridge at West Milton	14.6	26.3	Clay and gravel with broken limestone around abutments.	Extensive.
Mad River at 1st Big Four R. R. bridge west of Springfield	12.2	17.2	Clay and gravel with some bed rock in middle of channel.	None.
Mad River at 2nd Big Four R. R. bridge west of Springfield	15.0	18.0	Clay and gravel with some bed rock in middle of channel.	Extensive.
Buck Creek at Big Four R. R. bridge in Springfield	13.7	6.7	Clay, sand, and gravel with some loose rock around abutments.	None.

* Average depth under bridge.

Table XVIII.—Summary of Observations on Scour in Miami Valley Streams

Stream and Location	Average Velocity at Max. Stage Ft. per Sec.	Mean Depth, Feet	Stream-bed Material	Amount of Scour
Buck Creek at Limestone Street bridge in Springfield..........	10.1	11.5	Clay, sand, and gravel with some loose rock around abutments.	None.
RESULTS OF CURRENT METER MEASUREMENTS				
Miami River at Sidney..	3.90	5.06	Probably gravel.	None.
Miami River at Tadmor	4.63	9.85	Gravel.	None.
Miami River at Main St., Dayton..........	4.54	10.1	Gravel.	None.
Miami River just above Wolf Creek, Dayton ..	5.0	5.2	Gravel.	Slight shifting of material.
Stillwater River at West Milton..............	6.36	8.70	Gravel.	None.
Seven Mile Creek at Seven Mile...........	5.44	2.58	Limestone.	None.
MIAMI RIVER CUT-OFF CHANNEL OBSERVATIONS				
Miami River cut-off channel below Stewart St., Dayton..........	5.5	1.8	Gravel and clay.	None.
Miami River cut-off channel below Stewart St., Dayton..........	6.1	3.2	Gravel and clay.	Extensive.

Table XIX.—Convenient Conversion Factors

(Computed to Four Significant Figures)

Discharge			Volume		
Second Feet	Acre-feet per Hour	Inches in Depth per Day per Sq. Mile	Cubic Feet	Acre-feet	Inches in Depth on Sq. Mile
1,000,000	82,640	37,190	1,000,000	22.96	0.4304
12.10	1	0.450	43,560	1	0.018,75
26.89	2.222	1	2,323,000	53.33	1

Table XX.—Special Conversion Factors for Drainage Areas in the Miami Valley

	Drainage Area; Sq. Mi.	Discharge Equivalent to One Inch in Depth per Day		Volume Equivalent to One Inch in Depth	
		Sec. Ft.	Ac. Ft. per Hr.	Ac. Ft.	Millions, Cu. Ft.
One square mile....................	1	26.89	2.222	53.33	2.323
Germantown Dam...................	270	7,260	600	14,400	627.3
Englewood Dam....................	651	17,500	1,447	34,720	1,512
Lockington Dam					
Excluding Loramie Res............	174	4,679	386.6	9,279	404.2
Total drainage area..............	255	6,857	566.7	13,600	592.4
Taylorsville Dam					
Excluding Lockington and Lewistown basins....................	778	20,920	1,729	41,491	1,807
Excluding Lockington basin........	878	23,610	1,951	46,830	2,040
Total drainage area..............	1133	30,470	2,518	60,430	2,632
Huffman Dam.....................	671	18,040	1,491	35,790	1,559
Dayton..........................	2525	67,900	5,611	134,700	5,866
Hamilton........................	3672	98,740	8,160	195,800	8,531

INDEX

A

PAGE

Action, of notched spillways................ 244
  of spillways........................... 244
Adjusting curves by logarithmic methods.... 119
Advantages of retarding basin plan......47, 53
Alexandersville, local protection work at..... 76
Alternative flood protection plans.......... 309
Alvord and Burdick...................... 35
Area covered by different floods in retarding
  basins.............................. 107
Area-duration of flooding curves in retarding
  basins.............................. 207
Area elevation curves for retarding basins.... 110

B

Backwater, action in retarding basins....... 109
  curves............................... 280
Curves and Hydraulic Jump..........64, 123
Balancing the flood protection system....... 286
Baltimore and Ohio Railroad.............. 66
Bank protection, revetment for............ 68
Basis for design of flood protection......... 45
Basin's weir experiments................. 223
Beebe, J. C........................64, 123
Behavior of retarding basins during a 1913
  flood............................... 101
  during localised cloudbursts............. 247
Bock, C. A............................. 112
Bridge pier, plan of standard.............. 276
Bridges, effect of, on channel improvement... 274

C

Calculation of flow in open channels..32, 201, 280
Capacities of channels................... 32
  of channels in the Miami Valley......... 326
  of retarding basins..................91, 110
Capacity curves for retarding basins....... 110
Capacity of spillways.................... 221
  relation between inflow, outflow and retarding basin............................ 91
Channel capacities...................... 32
  required with retarding basin control..... 259
  erosion, protection against............. 68
  improvement, effect of bridges.......... 274
  improvement, allowable velocities in...... 279
  improvement, cost of protection of entire
   valley by........................... 316
  improvement, effect of curves........... 264
  improvement, general description........ 68
  improvement, increased flow resulting from. 316
  improvement, principles of design........ 261
  improvement, protection of entire valley by. 314
  improvement, scour in................. 280
  improvement, shape of cross section...... 273
  improvements, hydraulics of............ 259
  in the Miami Valley................... 326
  required for Official Plan flood.......... 335
  shape of cross section................. 273
  scour in............................. 280
Church, I. P........................... 281
City protection by local improvement....... 309
  cost of.............................. 311
Cleveland, Cincinnati, Chicago, & St. Louis
  Railroad............................ 67
Cleves................................ 83
Clogging of outlet conduits............... 90
Cloudbursts, effect of, on retarding basins... 247
  probability of occurrence............... 255
Columbus, Ohio........................ 35
Comparative intensities of twelve great storms 43
Concrete revetment for bank protection..... 68
Conduit outflow, effect of tailwater on...... 132
  formulas for.......................... 135

PAGE

Conservancy Court approves O. o al Plan.... 58
Controlling section in outlet conduits....... 150
Conversion factors, tables of.............. 337
Cornell Civil Engineer................... 121
Cost-discharge curve at Hamilton.......... 305
Cost of city protection by local improvements. 311
  of combined improvement above Dayton... 296
  of combined improvements above Midde-
   town.........................300, 302
  of protection of entire valley by channel
   improvement......................... 316
  of retarding basin plan................. 53
  of various flood protection plans......... 53
Cost-outflow curve at Englewood.......... 295
  at Huffman.......................... 295
  at Lockington.....................288, 290
  at Taylorsville....................... 293
Crest stages, graphical determination of..... 167
Critical flow........................... 147
  section in outlet conduits.............. 150
Cross section, shape of, in channel improvement............................... 273
Curves, in channels, effect of.............. 264
  transverse water surface, slope in......... 270

D

Damage, in 1913 flood.................19, 57
Dam, hydraulic fill earth, cross section...... 60
Dams and appurtenances, general description. 58
  table of data pertaining to............. 328
Danube, floods in....................... 44
Dayton, affected by 1913 flood............ 17
  channel capacity required.............. 261
  cost of combined improvement above..... 296
  high water profiles.................... 268
  local protection work at............... 73
  map of.............................. 74
  protection by channel improvement alone.. 310
  protection by retarding basins.......... 97
Degree of protection by retarding basin plan. 48
Depth-duration of flooding curves in retarding
  basins.............................. 207
Description of Miami Valley............... 55
Design based on flood discharge........... 45
  of outlet conduits.................... 122
  of retarding basins................... 204
Diagram of Official Plan flood............. 88
Diagrammatic representation of spillway
  action.............................. 230
Distribution, seasonal, of floods in Miami
  Valley.............................. 106
Drainage areas in the Miami Valley......... 324
Drop-off curve at the end of horizontal flume. 136
Duration-depth of flood curves in retarding
  basins.............................. 207

E

Earth dam, hydraulic fill, cross section...... 60
Empirical formulas...................... 121
Engineering and Contracting.............. 178
  News............................... 171
News Record......................121, 179
  Record..........................270, 281
Englewood cost-outflow curve............. 295
  dam, calculation of spillway capacity..... 240
  dam, curves showing depths of spillway discharge for various storm runoffs......... 238
  dam, general description............... 64
  dam, outlet conduits, dimensions of....... 123
  dam, plan and section of spillway........ 224
  retarding basin, area curve............. 110
  retarding basin, duration-depth of flooding
   curves.............................. 210

339

PAGE

retarding basin, frequency of flooding and
    season of flooding diagram..........214, 215
retarding basin, general description........   64
retarding basin hydrographs for a 1913 flood 102
retarding basin, inflow and outflow hydro-
    graphs for flood of October 1910........ 206
retarding basin, map of.................... 115
retarding basin, maximum outflow for vari-
    ous storm inflows...................... 250
retarding basin storage curve............. 111
retarding basin, table showing conditions in,
    for various floods.................... 332
Erie Railroad............................. 67
Erosion of channel, protection against....... 68
Examples illustrating spillway action........ 236

F

Factor of roughness n..................... 133
    of safety of earth dams................ 94
Fairfield................................. 67
Farm lands in retarding basins............ 105
Five-sixths rule, illustrated by inflow and out-
    flow hydrographs of a uniform flood.... 187
    for computing retarding basin capacity or
        outlet conduit discharge.......... 92
    for routing floods................... 186
    table showing computations by use of.. 203
Flexible revetment for bank protection....... 68
Flood control reservoirs, list of.......... 49
    damage in 1913....................19, 57
    discharge as basis for design......... 45
    discharges, various, used in design... 244
    frequency............................. 23
    of March 1913........................ 25
Flooding, frequency diagrams..........214, 215
    probability of, diagrams.............. 219
    season of, diagrams...............214, 215
Flood protection at Dayton by retarding
    basins................................ 97
    by small retarding basins............ 319
    plans, cost of various................ 53
    plans investigated................... 47
Flood-retarding reservoir problem directly
    solved................................ 179
Flood-stages at Dayton, plat of........... 39
Floods, frequency of in Miami River........ 40
    in the Miami Valley, history of....... 36
    in the Miami Valley, seasonal distribution.. 106
    relation of, to rainfall.............. 24
Flow, calculation of, in open channels.32, 201, 280
    increased, resulting from channel improve-
        ment.............................. 316
Floyd, O. N.............................. 21
Four-Mile Creek.......................55, 57
Franklin affected by 1913 flood........... 17
    local protection work at.............. 78
    map of............................... 78
Freeboard on earth dams.................. 94
    on levees............................ 99
Frequency of flooding diagram in Englewood
    retarding basin....................... 215
    in Huffman retarding basin........... 217
Frequency of floods...................... 23
    of Miami River floods................ 40
Fry, Albert S............................ 178

G

Gates for outlet conduits................. 89
Geology of Miami Valley...............55, 57
Germantown dam and outlet structures, plan
    of.................................... 64
    general description................... 64
    longitudinal section.................. 60
    plan and sections of spillway......... 225
    outlet conduits, dimensions of........ 123
    outlet conduits, section through...... 124
    retarding basin area curve............ 110
    retarding basin, duration-depth of flooding
        curves............................ 209
    retarding basin, general description... 64
    retarding basin hydrographs for a 1913 flood 101
    retarding basin inflow and outflow hydro-
        graphs for flood of March 1897..... 175
    retarding basin, map of.............. 114

PAGE

retarding basin, maximum outflow for vari-
    ous storm inflows..................... 243
retarding basin storage outflow curve...... 170
retarding basin storage curve............. 111
retarding basin storage elevation and out-
    flow elevation curves................. 174
retarding basin, table showing conditions in,
    for various floods.................... 331
temporary outlet......................... 153
Gorlitzer Neisse River.................... 53
Grant, K. C.............................. 21
Graphical determination of flood crest stages. 167
    solution of storm flow through a reservoir. 178
Great rainstorms in the United States...... 41
    storms of the past.................... 42

H

Hamilton affected by 1913 flood........... 17
    backwater curves..................... 282
    channel capacity required............ 260
    cost-discharge curves................ 305
    increased flow through, produced by channel
        improvements..................... 318
    local protection work at.............. 81
    map of............................... 82
History of floods in the Miami Valley...... 36
    of the Miami Flood Control Project.... 112
Hopson, E. G............................. 270
Houk, Ivan E......................32, 201, 280
Huffman cost-outflow curve............... 295
    dam and outlet structure, plan of..... 61
    dam, general description.............. 66
    outlet structure, dimensions of...... 126
    retarding basin area curve........... 110
    retarding basin, duration-depth of flooding
        curves............................ 213
    retarding basin, frequency of flooding and
        season of flooding diagram........ 217
    retarding basin, general description.. 66
    retarding basin, hydrographs for a 1913 flood 105
    retarding basin, inflow and outflow hydro-
        graphs for flood of October, 1910. 207
    retarding basin, map of.............. 118
    retarding basin, maximum outflow for vari-
        ous storm inflows................. 253
    retarding basin storage curve........ 111
    retarding basin, table showing conditions in,
        for various floods................ 335
Hydraulic fill earth dam, cross section.... 60
Jump and Backwater curves...........64, 123
Hydraulics of channel improvements........ 259
Hydrographs at Englewood retarding basin for
    flood of October 1910................. 206
    for a 1913 flood..................... 102
    at Germantown retarding basin for a 1913
        flood.............................. 101
    at Huffman retarding basin for flood of
        October 1910...................... 207
    at Huffman retarding basin for a 1913 flood. 105
    at Lockington retarding basin for a 1913
        flood.............................. 103
    at retarding basin with spillway operating,
                                     228, 229
    at Taylorsville retarding basin for a 1913
        flood.............................. 104
    inflow and outflow, for flood of March 1897
        at Germantown retarding basin..... 175
    inflow and outflow, for flood of March 1913 at
        Lockington retarding basin........ 183
    inflow and outflow, of a uniform flood to
        illustrate five-sixths rule....... 187

I

Increased flow resulting from channel improve-
    ment.................................. 316
Indian Creek............................. 55
Inflow and outflow at Englewood retarding
    basin for a 1913 flood................ 102
    at Germantown retarding basin for a 1913
        flood.............................. 101
    at Huffman retarding basin for a 1913 flood. 105
    at Lockington retarding basin for a 1913
        flood.............................. 103

	PAGE
at retarding basins, diagram for	172
at retarding basins, mass curves	173
at Taylorsville retarding basin for a 1913 flood	104
hydrographs at Englewood retarding basin for flood of October 1910	206
hydrographs at Huffman retarding basin for flood of October 1910	207
hydrographs at retarding basin with spillway operating	228, 229
hydrographs for flood of March 1897 at Germantown retarding basin	175
hydrographs for flood of March 1913 at Lockington retarding basin	183
hydrographs of a uniform flood to illustrate five-sixths rule	187
Inflow curves at damsites	156
Inflow, outflow, and retarding basin capacity, relation between	91
Inflows, storm, and corresponding maximum outflows at Englewood retarding basin	250
at Germantown retarding basin	249
at Huffman retarding basin	253
at Lockington retarding basin	251
at Taylorsville retarding basin	252
Integration formulas for spillway action	226
Interval pressure in drop-off curve	138
Iser River	53

J

Jones, B. M.	21

K

Kimball, J. H.	21
Kuichling, Emil	34

L

Lane, E. W.	21, 121
LaRoche dam, France	53
Levees for flood protection	68
Lewistown Reservoir	26, 58, 165
Little York	85
Local protection work at Alexandersville	76
at Dayton	73
at Franklin	78
at Hamilton	81
at Miamisburg	76
at Middletown	79
at Piqua	69
at Tippecanoe City	73
at Troy	71
at West Carrollton	76
general description	67
Lockington cost-outflow curve	288, 290
dam, general description	65
outlet and spillway, plan and sections of	125
outlet structure, dimensions of	124
retarding basin area curve	110
retarding basin, duration-depth of flooding curves	211
retarding basin, elevation-outflow curves	227
retarding basin, general description	65
retarding basin hydrographs for a 1913 flood	103
retarding basin inflow and outflow hydrographs for flood of March 1913	183
retarding basin, map of	116
retarding basin, maximum outflow for various storm inflows	251
retarding basin storage curve	111
retarding basin storage-outflow curve	180, 182
retarding basin, table showing conditions in various floods	333
Logarithmic method of adjusting curves	119
Loire River	53
Loramie Creek	55
Reservoir	26, 58, 165

Mc

McCurdy, H. S. R.	21

M

Mad River	57
Map of Dayton	74
of Englewood retarding basin	115
of Franklin	78

	PAGE
of Germantown retarding basin	114
of Hamilton	82
of Huffman retarding basin	118
of Lockington retarding basin	116
of Miamisburg	77
of Miami Valley	54
of Middletown	80
of Piqua	70
of Taylorsville retarding basin	117
of Tippecanoe City	72
of Troy	71
of West Carrollton	76
showing 45 proposed small retarding basins	321
showing maximum rates of runoff	33
Maps of 1913 storm	25, 28, 29
Matthes, G. H.	21
Maximum outflow for various storm runoffs at Englewood retarding basin	250
at Germantown retarding basin	249
at Huffman retarding basin	253
at Lockington retarding basin	251
at Taylorsville retarding basin	252
possible flood in Miami Valley	44
rates of runoff	33, 35
storm run off	86
Mayhew, A. B.	21
Mead, D. W.	21
Miami and Erie Canal	58
Miamisburg affected by 1913 flood	17
local protection work at	76
map of	77
Miami Valley and the 1913 flood	26
description of	55
geology of	55, 57
map of	54
Middletown, cost of combined improvement above	300, 302
map of	80
local protection work at	79
Mohawk River	36
Morgan, A. E	22, 26

N

Notched spillways, action of	244

O

Oder River	53
Official Plan Flood	45, 86
diagram of	88
table showing conditions in retarding basins for	330
Ohio Electric Railway	67
Olentangy River	35
Open channels, calculation of flow in	32, 201, 280
Operation of retarding basins	86, 205
Osborn	67
Outflow and inflow at Englewood retarding basin for a 1913 flood	102
at Germantown retarding basin for a 1913 flood	101
at Huffman retarding basin for a 1913 flood	103
at Lockington retarding basin for a 1913 flood	103
at retarding basins, diagram for	172
at retarding basins, mass curves	173
at Taylorsville retarding basin for a 1913 flood	104
hydrographs at Englewood retarding basin for flood of October 1910	206
hydrographs at Huffman retarding basin for flood of October 1910	207
hydrographs at retarding basin with spillway operating	228, 229
hydrographs for flood of March 1897 at Germantown retarding basin	175
hydrographs for flood of March 1913 at Lockington retarding basin	183
hydrographs of a uniform flood to illustrate five-sixths rule	187
Outflow, inflow, and retarding basin capacity, relation between	91
Outflow, maximum, for various storm inflows at Englewood retarding basin	250
at Germantown retarding basin	249
at Huffman retarding basin	253

PAGE

at Lockington retarding basin............ 251
at Taylorsville retarding basin........... 252
through conduits, effect of tailwater on.... 132
formulas for........................... 135
Outlet conduit discharge, five-sixths rule for
computing........................... 92
Outlets, comparison of various types....... 127
discharge curves for various types....... 128
operation of various types.............. 130
temporary............................ 153
Outlet conduits, clogging of............... 90
critical section in...................... 150
design of............................. 122
determination of size.................. 201
dimensions of......................... 123
flow through.......................... 122
gates for............................. 89
size................................. 89
superior to weirs...................... 126

P

Pianola diagrams......................... 216
Pinay dam, France....................... 53
Piqua affected by 1913 flood.............. 17
channel capacity required............... 260
local protection work at................ 69
map of............................... 70
needs retarding basin protection......... 65
Population of cities in the Miami Valley.... 327
Port Jefferson........................... 84
Precedent for retarding basin plan......... 48
Probability of flooding diagrams........... 219
of occurrence of cloudbursts........... 255
Profiles of streams in Miami Valley........ 56
Protection afforded by retarding basins..... 96
against channel erosion................. 68
by retarding basin plan, degree of....... 48
of cities by local improvement.......... 309
of cities, cost of, by local improvements.. 311
of entire valley by channel improvement.. 314
of entire valley by channel improvement,
cost of............................ 316

R

Railroad changes...................66, 67, 83
Rainfall of March 1913 storm............. 26
relation of, to floods................... 24
relation of, to runoff.................. 27
table of maximum storm................ 248
Rejected flood protection plans............ 309
Relation between inflow, outflow, and retarding basin capacity.................... 91
of rainfall to floods................... 24
of runoff to rainfall................... 27
Reservoirs for flood control, list of........ 49
Retarding basin capacity, five-sixths rule for
computing........................... 92
relation between inflow, outflow, and.... 91
dams, determination of height.......... 201
inflow and outflow hydrographs with spillway operating.................228, 229
plan, advantages of.................47, 53
plan, cost of.......................... 53
plan, general description................ 46
plan, degree of protection.............. 48
plan, precedent for.................... 48
recovery curves........................ 218
sites, selecting the.................... 84
Retarding basins, area covered by different
floods in........................... 107
area-duration of flooding curves in....... 207
area-elevation curves for............... 110
backwater in......................... 109
behavior of, during a 1913 flood........ 101
behavior of, during localized cloudbursts.. 247
capacities of.....................91, 110
channel capacities required with......... 259
design of............................. 204
diagram for inflow and outflow at........ 172
duration-depth of flood curves in........ 207
farm lands in......................... 105
maps of............114, 115, 116, 117, 118
mass curves of inflow and outflow at. 173
operation of......................86, 205
protection afforded by................. 96

routing floods through.................. 156
runoff to fill to spillway level............ 95
small, flood protection by............... 319
small, map showing 45 proposed......... 321
storage elevation curves for............. 111
table showing conditions in, for flood of
March 1898......................... 329
table showing conditions in, for flood of
March 1913......................... 329
table showing conditions in, for Official Plan
flood.............................. 330
table showing computation of sizes by five-
sixths rule......................... 203
time of emptying...................... 191
Revetment for bank protection............ 68
Riegel, R. M.......................64, 123
Roanne, France......................... 53
Roughness factor n..................... 133
Routing floods, definition of.............. 170
direct graphical methods............... 179
five-sixths rule....................... 186
integration methods................... 185
step methods......................... 171
through retarding basins................ 156
trial and error methods................ 171
Runoff from maximum storm............. 86
of March 1913 storm.................. 31
maximum rates of..................33, 35
relation of, to rainfall................. 27
to fill retarding basins to spillway level.... 95

S

Safety of earth dams..................94, 100
Scioto River............................ 35
Scour in channel improvement............ 280
in Miami Valley streams, table of........ 336
Seasonal distribution of floods in Miami Valley 106
Season of flooding diagram in Englewood re-
tarding basin....................... 214
in Huffman retarding basin............. 217
Seine, floods in......................... 44
Selecting the retarding basin sites......... 84
Series formulas for spillway action......... 231
Size of outlet conduits.................. 89
Sliding coefficient method for inflow curves .. 159
Slopes of streams in Miami Valley......56, 325
Small retarding basins, flood protection by... 319
map showing 45 proposed............... 321
Smith, Walter M........................ 21
Spillway action, diagrammatic representation
of................................ 230
during cloudbursts.................... 248
examples illustrating.................. 236
integration formulas for................ 226
series formulas for..................... 231
transcendental formulas for............. 234
capacity, calculation of, for Englewood dam 240
capacity, calculation of, for Taylorsville dam 241
crest, profile of....................... 226
discharge, effect on flood protection...... 92
discharge, formulas for................. 223
discharge for various storm runoffs at Englewood dam.......................... 238
discharge for various storm runoffs at Taylorsville dam....................... 239
level, runoff to fill retarding basins to..... 95
problems, curves for solving............. 233
problems, step method for.............. 226
transcendental formulas for............. 234
Spillways, action of..................... 244
capacity of.......................94, 221
function of........................... 221
general function of.................... 94
notched, action of.................... 244
types of............................. 222
Springfield............................. 84
Step method for spillway problems......... 226
Storage elevation curves for retarding basins. 111
in valley............................. 33
Storm inflows and corresponding maximum
outflows at Englewood retarding basin.. 250
at Germantown retarding basin.......... 249
at Huffman retarding basin............. 253
at Lockington retarding basin........... 251
at Taylorsville retarding basin........... 252

PAGE

Storm rainfall of eastern United States....41, 201
  table of maximum...................... 248
Streams, slopes of, in the Miami Valley....56, 325

T

Taylorsville cost-outflow curve.............. 293
  dam, calculation of spillway capacity...... 241
  dam, curves showing depths of spillway discharge for various storm runoffs......... 239
  dam, general description................. 66
  outlet structure, dimensions of........... 126
  retarding basin, area curve............. 110
  retarding basin, duration-depth of flooding curves............................... 212
  retarding basin, general description...... 66
  retarding basin, hydrographs for a 1913 flood.............................. 104
  retarding basin, map of.............. 117
  retarding basin, maximum outflow for various storm inflows..................... 252
  retarding basin, storage curve........... 111
  retarding basin, table showing conditions in, for various floods..................... 334
  temporary outlet..................... 154
Temporary outlets....................... 153
Thomas, H. A..........................22, 179
Tiber, floods in........................ 44
Tippecanoe City....................... 66
  ocal protection work at................. 73
  map of............................. 72

PAGE

Transcendental formulas for spillway action.. 234
Transverse water surface slope in curves..... 270
Troy affected by 1913 flood............... 17
  backwater effect on................... 109
  channel capacity required................ 260
  local protection work at................. 71
  map of.............................. 71
  needs retarding basin protection.......... 65
Twin Creek...........................55, 57

U

U. S. Geological Survey.................... 84
  Weather Bureau....................... 27

V

Valley storage......................... 32
  during 1913 Flood..................... 327
Various flood-protection plans investigated... 47
Velocities, allowable, in channel improvement 279

W

Weather Bureau, U. S..................... 27
Weirs inferior to outlet conduits........... 126
West Carrollton, local protection work at.... 76
  map of.............................. 76
West Milton........................... 85
Wolf Creek............................ 73
Woodward, S. M...................22, 64, 123

Lightning Source UK Ltd.
Milton Keynes UK
UKHW042316030119

334852UK00022B/949/P

9 781528 310581